ANIMAL BEHAVIOUR IN EGYPTIAN ART

The Australian Centre for Egyptology

A Division of the Macquarie University Ancient Cultures Research Centre

Studies 9

ANIMAL BEHAVIOUR IN EGYPTIAN ART

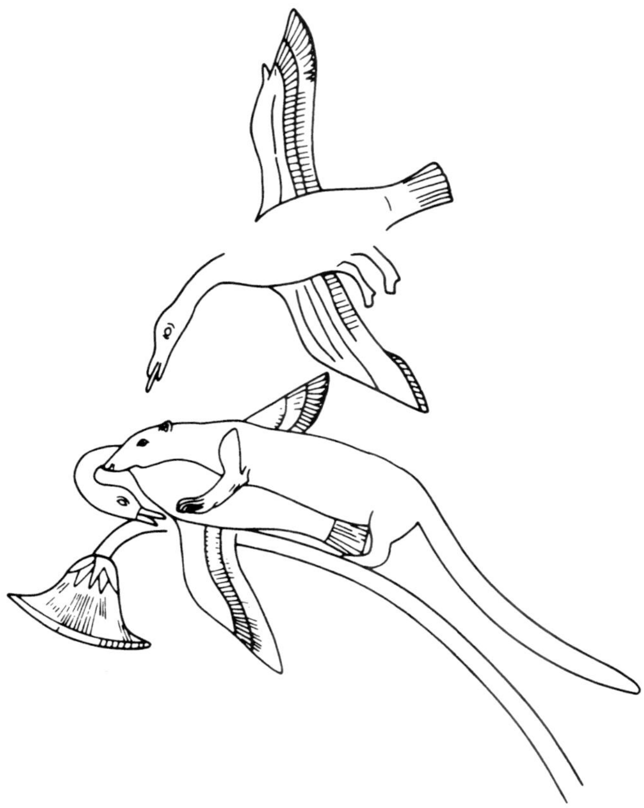

Representations of the Natural World in Memphite Tomb Scenes

Linda Evans

For
Christopher Stuart Evans

Cover illustration: Kanawati-Abder-Raziq, *Nikauisesi*, pl. 50.

ISBN: 978-0-85668-826

Published in England by Aris and Phillips Ltd.
Park End Place, Oxford OX1 1HN

I would like to express my deep appreciation to Naguib Kanawati, Director of the Australian Centre for Egyptology (ACE; now a division of the Macquarie University Ancient Cultures Research Centre) for giving me the opportunity to publish my work as part of the ACE study series. This volume is an edited version of my PhD thesis, submitted in June 2006, for which Naguib was also my supervisor. I thank him most sincerely for his support and careful guidance throughout my time as a graduate student and, more recently, as a postdoctoral fellow in the Department of Ancient History, Macquarie University. In addition, I would also like to thank both Naguib and Alexandra Woods for granting me permission to reproduce images from their past and forthcoming ACE excavation reports.

I am very grateful to have received a Macquarie University Research Fellowship (MQRF), which has enabled me to see this work to fruition and to pursue further study of the animals in Egyptian art. The Department of Ancient History has been an ideal place in which to carry out my research; it is a privilege to be surrounded by fellow scholars who are passionate about the study of the past and eager to embrace new ideas. I am especially indebted to our department head, Alanna Nobbs, for providing both financial and academic support.

The project would have been impossible without the prompt and friendly help of the Macquarie University Library. Special thanks go to the staff in Document Supply who managed to find many obscure papers and volumes for me and to Bill Stinson for his interest and useful advice. I am especially thankful to have had limitless access to the Brunner collection of books and papers, an Egyptological treasure-trove.

Many people provided invaluable assistance throughout the study. In particular, I would like to thank Bob Kilgour (Department of Agriculture, NSW) and Joseph Stookey (Department of Large Animal Clinical Sciences, University of Saskatchewan) for information regarding the behaviour of cattle, Cathy Alexander (Taronga Zoo, Sydney) for advice about birds, and Cynthia Mazer-Druckenbrod (Cleveland Botanical Gardens) for her expertise on praying mantids. I also wish to thank the many web-owners from around the world who gave me permission to reproduce their marvellous photographs of animals in the following pages: the Babcock Institute for International Dairy Research and Development, Dawn Dalton, Lynne Gelston, Giuliano Gerra, Albert Gremy, Andreas Hennig, Brian Jones, Yves Lanceau, Demetris Mina, Howard Norfolk, Silvio Sommazzi, and the Western Watersheds Project.

My heartfelt thanks go to Susanne Binder, Leonie Donovan, and Joyce Swinton for assisting with the final preparation of the volume. I greatly appreciate the time and patient care that Leonie devoted to transforming my clumsy manuscript into a professional publication and the thorough proofreading that Joyce cheerfully provided. I have also valued enormously the continuing support of Jaye McKenzie-Clark, Jana Jones, Lesley Kinney, and Elizabeth Thompson, as well as that of my family.

My partner throughout the editing process has been archaeological artist, Mary Hartley. Mary's keen eye and meticulous attention to detail have produced the exquisite drawings that now enrich this book. I cannot thank her enough for her constant encouragement and friendship.

Many of the animals that are described in this volume no longer exist in Egypt. Hunting, climate change, and loss of habitat have all taken their toll, wiping out species that once roamed the deserts and graced the Nile Valley. Neglect has also contributed to their disappearance, for like elsewhere in the world it has been taken for granted that they would always be there, no matter what they were forced to endure. Many more animals are now under threat in Egypt, including the Nubian ibex, barbary sheep, and dorcas gazelle. To preserve her fauna will require constant vigilance and a commitment to protecting her natural environment. I very much hope that this book will help to increase appreciation of Egypt's extraordinary creatures, which are undoubtedly equal to the beauty and mystery of her ancient wonders.

Finally, I wish to thank my husband, Christopher Evans, Director of the Centre for the Integrative Study of Animal Behaviour, Macquarie University, for opening *my* eyes to the wonders of the animal world, a transforming experience for which I am profoundly grateful. He laid the seed that brought forth this volume, and so - with love and admiration - I dedicate it to him.

AA: *Artibus Asiae*
AEB: *Annual Egyptological Bibliography*
AeUl: *Ägypten und Levante: Zeitschrift für ägyptische Archäologie und deren Nachbargebiete*
Agric. Hist.: *Agricultural History*
AH: *Aus der Heimat*
AJA: *American Journal of Archaeology*
AJE: *African Journal of Ecology*
Albright-Arave, *Behaviour of Cattle*: Albright, J.L. and Arave, C.W. (1997). *The Behaviour of Cattle.* Wallingford, CAB International.
Alexander, *Exploring Biomechanics*: Alexander, R.M. (1992). *Exploring Biomechanics: Animals in Motion.* New York, Scientific American Library.
Alexander, *Locomotion*: Alexander, R.M. (1982). *Locomotion of Animals.* Glasgow, Blackie and Sons.
Allaby, *Dictionary of Zoology*: Allaby, M. (1991). *The Concise Oxford Dictionary of Zoology.* Oxford, Oxford University Press.
Altenmüller, *Jagd im Alten Ägypten*: Altenmüller, H. (1967). *Darstellungen der Jagd im Alten Ägypten.* Hamburg, Paul Parey.
Altenmüller, *Mehu*: Altenmüller, H. (1998). *Die Wanddarstellungen im Grab des Mehu in Saqqara.* Mainz am Rhein, P. von Zabern.
Am. Antiq.: *American Antiquity*
Amph. Rep.: *Amphibia Reptilia*
Anderson, *Zoology of Egypt*: Anderson, J. (1902). *The Zoölogy of Egypt: Mammalia.* London, Hugh Rees.
Anim. Behav.: *Animal Behaviour*
Ann. Entom. Soc. Am.: *Annales of the Entomological Society of America*
Ann. Fac. Lett. Univ. Ibrahim: *Annales de la Faculté des Lettres Université Ibrahim*
Anthropos: *Anthropos*
Anthropozoologica: *Anthropozoologica*
Anthrozoös: *Anthrozoös*
Antike Welt: *Antike Welt*
Antiquity: *Antiquity*
AOF: *Altorientalische Forschungen*
App. Anim. Beh. Sci.: *Applied Animal Behaviour Science*
App. Anim. Ethol.: *Applied Animal Ethology*
Apps, *Wild Ways*: Apps, P. (1992). *Wild Ways: Field Guide to the Behaviour of Southern African Mammals.* Cape Town, Struik Publishers.
Arch. Mus. d'Hist. Nat. Lyon: *Archives Museum d'Histoire Naturelle de Lyon*
Art Hist.: *Art History*
ASAE: *Annales du Service Antiquités de l'Égypte*
Assmann, *Search for God*: Assmann, J. (2001). *The Search for God in Ancient Egypt.* Translated by D. Lorton. Ithaca, Cornell University Press.
Aufrère, *Encyclopédie Religieuse*: Aufrère, S., ed. (1999). *Encyclopédie Religieuse de l'Univers Végétal: Croyances Phytoreligieuses de l'Égypte Ancienne.* Montpellier, Université Paul Valéry.
Auk: *Auk*

BA: *The Biblical Archaeologist*
BACE: *Bulletin of the Australian Centre for Egyptology*
Badawy, *Tombs of Iteti*: Badawy, A. (1976). *The Tombs of Iteti, Sekhemankh-Ptah, and Kaemnofert at Giza.* Berkeley, University of California Press.
Baer, *Rank and Title:* Baer, K. (1960). *Rank and Title in the Old Kingdom: The Structure of the Egyptian Administration in the Fifth and Sixth Dynasties.* Chicago, University of Chicago Press.
Bailey et al., *Human Ecodynamics*: Bailey, G., Charles, R. and Winder, N., eds. (2000). *Human Ecodynamics: Proceedings of the American Association for Environmental Archeology.* Oxford, Oxbow Press.
Baldassarre-Bolen, *Waterfowl Ecology*: Baldassarre, G.A. and Bolen, E.G. (1994). *Waterfowl Ecology and Management.* New York, John Wiley and Sons.
Bandi et al., *La Contribution*: Bandi, H.-G., Huber, W., Sauter, M.-R. and Sitter, B., eds. (1984). *La Contribution de la Zoologie et de l'Éthologie à l'Interprétation de l'Art des Peuples Chasseurs Préhistoriques*, Fribourg, Editions Universitaires.
Barber, *Vampires, Burial, and Death*: Barber, P. (1988). *Vampires, Burial, and Death: Folklore and Reality.* New Haven, Yale University Press.
Baud, *Le Caractère du Dessin*: Baud, M. (1978). *Le Caractère du Dessin en Égypte Ancienne.* Paris, A. Maisonneuve.
BCH: *Bulletin de Correspondance Hellénique*
Beaux, *Le Cabinet de Curiosities*: Beaux, N. (1990). *Le Cabinet de Curiosities de Thoutmosis III: Plantes et Animaux du "Jardin Botanique" de Karnak.* Leuven, Peeters.
Beh. Proc.: *Behavioural Processes*
Behaviour: *Behaviour*

Behrmann, *Das Nilpferd*: Behrmann, A. (1996). *Das Nilpferd in der Vorstellungswelt der Alten Ägypter.* 2 vols. Frankfurt am Main, Peter Lang.

Bekoff, *Enc. Anim. Behav.*: Bekoff, M., ed. (2004). *Encyclopedia of Animal Behavior.* 3 vols. Westport, Greenwood Press.

BES: *Bulletin of the Egyptological Seminar*

BIE: *Bulletin de l'Institut d'Égypte*

Biewener, *Animal Locomotion*: Biewener, A.A. (2003). *Animal Locomotion.* Oxford, Oxford University Press.

BIFAO: *Bulletin de l'Institut Français d'Archéologie Orientale*

Biol. Cons.: *Biological Conservation*

Bird Behav.: *Bird Behaviour*

Blackman, *Rock Tombs of Meir*: Blackman, A.M. (1914-1953). *The Rock Tombs of Meir.* 6 vols. London, Egypt Exploration Fund.

Blanco-Tolbert Roberts, *Herodotus The Histories*: Blanco, W. and Tolbert Roberts, J. (1992). *Herodotus The Histories: New Translation, Selections, Backgrounds, Commentaries.* New York, Norton.

BMFA: *Bulletin of the Museum of Fine Arts (Boston)*

BMMA: *Bulletin of the Metropolitan Museum of Art (New York)*

Bodson, *L'Histoire*: Bodson, L., ed. (1993). *L'Histoire de la Connaissance du Comportement Animal.* Colloques d'Histoire des Connaissance Zoologiques. Liège, Université de Liège.

Boessneck, *Die Haustiere in Altägypten*: Boessneck, J. (1953). *Die Haustiere in Altägypten.* Munich, J. Pfeiffer.

Boessneck, *Tierwelt*: Boessneck, J. (1988). *Die Tierwelt des Alten Ägypten.* München, Verlag C.H. Beck.

Bolshakov, *Man and his Double*: Bolshakov, A.O. (1997). *Man and his Double in Egyptian Ideology of the Old Kingdom.* Wiesbaden, Otto Harrassowitz.

Bonn. Zool. Beitr.: *Bonner Zoologische Beiträge*

Borchardt, *Aegyptische Pflanzensäule*: Borchardt, L. (1897). *Die Aegyptische Pflanzensäule: Ein Kapitel zür Geschichte des Pflanzenornaments.* Berlin, Ernst Wasmuth.

Borchardt, *Denkmäler*: Borchardt, L. (1964). *Denkmäler des Alten Reiches Im Museum von Kairo (ausser den Statuen).* 2 vols. Berlin, Reichsdrucherei.

Borchardt, *Sahure*: Borchardt, L. (1910-1913). *Das Grabdenkmal des Königs Sahure.* 2 vols. Leipzig, J. C. Hinrichs.

Bost. Mus. Bull.: *Boston Museum Bulletin*

Bothma, *Larger Carnivores*: Bothma, J.D.P. (1999). *Larger Carnivores of the African Savannas.* Berlin, Springer.

Boudet, *Man and Beast*: Boudet, J. (1964). *Man and Beast: A Visual History.* New York, Golden Press.

Boulenger, *Fishes of the Nile*: Boulenger, G.A. (1907). *Zoology of Egypt: The Fishes of the Nile.* London, Hugh Rees.

Brewer et al., *Dogs in Antiquity*: Brewer, D., Clark, T. and Phillips, A. (2002). *Dogs in Antiquity. Anubis to Cerberus: The Origins of the Domestic Dog.* Warminster, Aris and Phillips.

Brewer et al., *Domestic Plants and Animals*: Brewer, D.J., Redford, D.B. and Redford, S. (1994). *Domestic Plants and Animals: The Egyptian Origins.* Warminster, Aris and Phillips.

Brewer-Friedman, *Fish and Fishing*: Brewer, D.J. and Friedman, R.F. (1989). *Fish and Fishing in Ancient Egypt.* Warminster, Aris and Philips.

Brion, *Animals in Art*: Brion, M. (1959). *Animals in Art.* New York, Franklin Watts.

Brit. J. Ophthalmol.: *British Journal of Ophthalmology*

Brodrick, *Animals in Archaeology*: Brodrick, A.H., ed. (1972). *Animals in Archaeology.* London, Barrie and Jenkins.

Brooke-Birkhead, *Ornithology*: Brooke, M. and Birkhead, T., eds. (1991). *The Cambridge Encyclopedia of Ornithology.* Cambridge, Cambridge University Press.

Brovarski, *Senedjemib Complex Part I*: Brovarski, E. (1999). *The Senedjemib Complex Part I: The Mastabas of Senedjemib Inti (G 2370), Khnumenti (G2374), and Senedjemib Mehi (G2378).* Boston, Museum of Fine Arts.

Brunner-Traut, *Seschemnofers III*: Brunner-Traut, E. (1977). *Die Altägyptische Grabkammer Seschemnofers III. aus Gîsa.* Mainz am Rhein, P. von Zabern.

Burton, *Bird Behaviour*: Burton, R. (1985). *Bird Behaviour.* London, Granada Books.

Burton-Burton, *International Wildlife*: Burton, M. and R. (1990). *Marshall Cavendish International Wildlife Encyclopedia.* New York, Marshall Cavendish.

Buxton, *Persuasion in Greek Tragedy*: Buxton, R.G.A. (1982). *Persuasion in Greek Tragedy: A Study of Peitho.* Cambridge, Cambridge University Press.

BV: *Beiträge zür Vogelkunde*

CAJ: *Cambridge Archaeological Journal*

Capart, *Rue de Tombeaux*: Capart, J. (1907) *Une Rue de Tombeaux à Saqqarah.* 2 vols. Brussels, Vromant.

Carr, *Handbook of Turtles*: Carr, A. (1952). *Handbook of Turtles: The Turtles of the United States, Canada, and Baja California.* Ithaca, Comstock Publishing.

CdÉ: *Chronique d'Égypte*

CdK: *Cahiers de Karnak*

Champollion, *Monuments de l'Égypte*: Champollion, J.F. (1970-1971). *Monuments de l'Égypte et de la Nubie.* 4 vols. Geneve, Belles-Lettres.

Clark, *Animals and Men*: Clark, K. (1977). *Animals and Men: Their Relationship as Reflected in Western Art from Prehistory to the Present Day*. New York, William Morrow and Company.

Clarysse et al., *Egyptian Religion*: Clarysse, W., Schoors, A. and Willems, H., eds. (1998). *Egyptian Religion: The Last Thousand Years. Studies Dedicated to the Memory of Jan Quaegebeur*. 2 vols. Leuven, Peeters.

Cogger-Zweifel, *Reptiles and Amphibians*: Cogger, H.G. and Zweifel, R.G. (1998). *Encyclopedia of Reptiles and Amphibians*. Sydney, UNSW Press.

Collins, *A History of the Animal World*: Collins, B.J., ed. (2002). *A History of the Animal World in the Ancient Near East*. Leiden, Brill.

Condor: *Condor*

Copeia: *Copeia*

Cosmos: *Cosmos*

Cramp-Simmons, *Handbook*: Cramp, S. and Simmons, K.E.L., eds. (1977-1994). *Handbook of the Birds of Europe, the Middle East and North Africa: The Birds of the Western Palearctic*. 9 vols. Oxford, Oxford University Press.

Cuneo, et. al., *Injuries and Diseases*: Cuneo, S.P., Card, C.S. and Bicknell, E.J. (1993). *Injuries and Diseases of Beef Cattle Associated with Calving*. Animal Care and Health Maintenance, University of Wisconsin.

Dalton, *Borne on the Wind*: Dalton, S. (1975). *Borne on the Wind: The Extraordinary World of Insects in Flight*. Stanmore, Cassell Australia.

D'Auria et al., *Mummies and Magic*: D'Auria, S., Lacovara, P. and Roehrig, C.H. (1988). *Mummies and Magic: The Funerary Arts of Ancient Egypt*. Boston, Museum of Fine Arts.

Davies, *Ancient Egyptian Paintings*: Davies, N.d.G. (1936). *Ancient Egyptian Paintings*. 3 vols. Chicago, Chicago University Press.

Davies, *Antefoker*: Davies, N.d.G. (1920). *The Tomb of Antefoḳer, Vizier of Sesostris I, and of his Wife, Senet (No. 60)*. London, Allen and Unwin.

Davies, *Five Theban*: Davies, N.d.G. (1913). *Five Theban Tombs (being those of Mentuherkhepeshef, User, Daga, Nehemawäy, and Tati)*. London, Egypt Exploration Fund.

Davies, *Ptahhetep i*: Davies, N.d.G. (1900). *The Mastaba of Ptahhetep and Akhethetep at Saqqareh. Part I: The Chapel of Ptahhetep and the Hieroglyphs*. London, Egypt Exploration Fund.

Davies, *Ptahhetep ii*: Davies, N.d.G. (1901). *The Mastaba of Ptahhetep and Akhethetep at Saqqareh. Part II: The Mastaba. The Sculptures of Akhethetep*. London, Egypt Exploration Fund.

Davies et al., *Saqqara Tombs i*: Davies, W.V., El-Khouli, A., Lloyd, A.B. and Spencer, A.J. (1984). *Saqqara Tombs I: The Mastabas of Mereri and Wernu*. London, Egypt Exploration Society.

Davies-Gardiner, *Tutankhamun's Painted Box*: Davies, N.M. and Gardiner, A.H. (1962). *Tutankhamun's Painted Box*. Oxford, Griffith Institute.

Davis, *Canonical Tradition*: Davis, W. (1989). *The Canonical Tradition in Ancient Egyptian Art*. Cambridge, Cambridge University Press.

Davis, *Masking the Blow*: Davis, W. (1992). *Masking the Blow: The Scene of Representation in Late Prehistoric Art*. Berkeley, University of California Press.

DE: *Discussions in Egyptology*

de Silva et al., *Man and Animal*: de Silva, A., von Simson, O. and Troutman, P., eds. (1965). *Man through his Art. Vol. 3: Man and Animal*. London, E.P. Publishing.

de Vore, *Primate Behavior*: de Vore, I. (1965). *Primate Behavior: Field Studies of Monkeys and Apes*. New York, Holt, Reinhart and Winston.

DEA Research Discussion Paper: *Directorate of Environmental Affairs Research Discussion Paper*

Decker-Herb, *Bildatlas*: Decker, W. and Herb, M. (1994). *Bildatlas zum Sport im Alten Ägypten: Corpus der Bildlichen Quellen zu Leibesubungen, Spiel, Jagd, Tanz und verwandten Themen*. Leiden, Brill.

Der Manuelian-Freed, *William Kelly Simpson*: Der Manuelian, P. and Freed, R.E., eds. (1996). *Studies in Honor of William Kelly Simpson*. 2 vols. Boston, Museum of Fine Arts.

Dominicus, *Gesten und Gebärden*: Dominicus, B. (1994). *Gesten und Gebärden in Darstellungen des Alten und Mittleren Reiches*. Heidelberg, Heidelberger Orientverlag.

Donovan-McCorquodale, *Egyptian Art*: Donovan, L. and McCorquodale, K., eds. (2000). *Egyptian Art: Principles and Themes in Wall Scenes*. Giza, Foreign Cultural Affairs Department.

Dorst, *Life of Birds*: Dorst, J. (1974). *The Life of Birds*. 2 vols. New York, Columbia University Press.

Duell, *Mereruka*: Duell, P. (1938). *The Mastaba of Mereruka*. 2 vols. Chicago, University of Chicago Press.

Dunham-Simpson, *Mersyankh III*: Dunham, D. and Simpson, W.K. (1974). *The Mastaba of Queen Mersyankh III*. Boston, Museum of Fine Arts.

EA: *Egyptian Archaeology*

EAWJ: *East African Wildlife Journal*

Ecology: *Ecology*

Econ. Bot.: *Economic Botany*

Edel-Wenig, *Jahreszeitenreliefs Ne-user-re*: Edel, E. and Wenig, S. (1974). *Die Jahreszeitenreliefs aus dem Sonnenheiligtum des Konigs Ne-user-re*. Berlin, Akademie-Verlag.

Eg. Art in the Age of the Pyramids: Arnold, D., Grzymski, K. and Ziegler, C., eds. (1999). *Egyptian Art in the Age of the Pyramids*. New York, Metropolitan Museum of Art.

Eggebrecht, *Schlachtungsbräuche*: Eggebrecht, A. (1973). *Schlachtungsbräuche im Alten Ägypten und ihre Wiedergabe im Flachbild bis zum Ende des Mittleren Reiches*. München, Ludwig-Maximilians-Universität zu München.

Eliade, *Encyclopedia of Religion*: Eliade, M., ed. (1987). *The Encyclopedia of Religion*. 16 vols. New York, Macmillan.

Eltringham, *The Hippos*: Eltringham, S.K. (1999). *The Hippos: Natural History and Conservation*. London, Academic Press.

Emmison-Stephens, *Tribute*: Emmison, F. and Stephens, R., eds. (1976). *Tribute to an Antiquary: Essays Presented to Marc Fitch By Some of his Friends*. London, Leopard's Head Press.

Env. Biol. Fishes: *Environmental Biology of Fishes*

Epron et al., *Ti i*: Epron, L., Daumas, F. and Goyon, G. (1939). *Le Tombeau de Ti. Vol. i*. Cairo, Institut Français d'Archéologie Orientale.

Erman, *Ägypten*: Erman, A. (1885). *Ägypten und Ägyptisches Leben im Altertum*. 2 vols. Tübingen, Mohr.

Erman, *Reden*: Erman, A. (1919). *Reden, Rufe und Lieder auf Gräberbildern des Alten Reiches*. Berlin, Akademie der Wissenschaften in Kommission bei G. Reimer.

Estes, *Behavior Guide*: Estes, R.D. (1991). *The Behavior Guide to African Mammals: Including Hoofed Mammals, Carnivores, Primates*. Berkeley, University of California Press.

Ethol. Ecol. Evol.: *Ethology, Ecology and Evolution*

Ethology: *Ethology*

Ethos: *Ethos*

Études: *Études d'Égyptologie*

Evans-Pritchard, *The Nuer*: Evans-Pritchard, E.E. (1963). *The Nuer: A Description of the Modes of Livelihood and Political Institutions of a Nilotic People*. Oxford, Clarendon Press.

Ewer, *The Carnivores*: Ewer, R.F. (1998). *The Carnivores*. Ithaca, Cornell University Press.

Fazzini et al., *Art for Eternity*: Fazzini, R.A., Romano, J.F. and Cody, M.E. (1999). *Art for Eternity: Masterworks from Ancient Egypt*. New York, Brooklyn Museum of Art.

Fieldiana: *Fieldiana: Zoology*

Firth-Gunn, *Teti Pyr.*: Firth, C.M. and Gunn, B. (1926). *Excavations at Saqqara: Teti Pyramid Cemeteries*. 2 vols. Cairo, Service des Antiquités de l'Égypte.

Fischer, *Turtles*: Fischer, H.G. (1968). *Ancient Egyptian Representations of Turtles*. New York, Metropolitan Museum of Art.

Fox, *Abnormal Behavior*: Fox, M.W. (1968). *Abnormal Behavior in Animals*. Philadelphia, Saunders.

Fox, *Wild Canids*: Fox, M.W. (1975). *The Wild Canids: Their Systematics, Behavioral Ecology and Evolution*. New York, Van Nostrand Reinhold.

Francesconi, *Striped Mullet*: Francesconi, J. (1997). *Striped Mullet* (*Mugil cephalus*). Raleigh, NC Wildlife Resources Commission and NC Division of Marine Fisheries.

Frankfort, *Ancient Egyptian Religion*: Frankfort, H. (1961). *Ancient Egyptian Religion: An Interpretation*. New York, Harper & Row.

Frankfort, *Kingship and the Gods*: Frankfort, H. (1948). *Kingship and the Gods: A Study of Ancient near Eastern Religion as the Integration of Society and Nature*. Chicago, University of Chicago Press.

Fraser, *Reproductive Behaviour*: Fraser, A.F. (1968). *Reproductive Behaviour in Ungulates*. London, Academic Press.

Fraser-Broom, *Farm Animal Behaviour*: Fraser, A.F. and Broom, D.M. (1990). *Farm Animal Behaviour and Welfare*. Wallingford, CAB International.

FS: *Field and Stream*

Gaillard et al., *Recherches sur les Poissons*: Gaillard, C., Loret, V. and Kuentz, C. (1923). *Recherches sur les Poissons Représentés dans Quelques Tombeaux Égyptiens de l'Ancien Empire*, *MIFAO* 51, 1-136.

Gaillard-Daressy, *La Faune Momifiée*: Gaillard, C. and Daressy, G. (1905). *Catalogue General des Antiquités Égyptiennes du Musée du Caire: La Faune Momifiée de l'Antique Égypte*. Cairo, Imprimerie de l'Institut Français d'Archéologie Orientale.

Gamer-Wallert, *Fische und Fischkulte*: Gamer-Wallert, I. (1970). *Fische und Fischkulte im Alten Ägypten*. Wiesbaden, Otto Harrassowitz.

Gardiner, *Egyptian Grammar*: Gardiner, A. (1927). *Egyptian Grammar*. Oxford, Griffith Institute.

Geist-Walther, *Behaviour of Ungulates*: Geist, V. and Walther, F., eds. (1974). *The Behaviour of Ungulates and Its Relation to Management*. 2 vols. Morges, Switzerland, International Union for Conservation and Natural Resources.

Geo Aust.: *Geo Australasia*

Geoarchaeology: *Geoarchaeology: An International Journal*

Gerfaut: *Gerfaut*

Germond, *Egyptian Bestiary*: Germond, P. (2001). *An Egyptian Bestiary*. London, Thames and Hudson.

Gill, *Applied Sheep Behaviour*: Gill, W. (2005). *Applied Sheep Behaviour*. Agricultural Extension Service, University of Tennessee.
Gill, *Ornithology*: Gill, F.B. (1995). *Ornithology*. New York, W.H. Freeman.
Giza Archives Project: *Giza Archives Project* (www.gizapyramids.org/code/emuseum.asp)
GM: *Göttinger Miszellen*
Goodman-Meininger, *Birds of Egypt*: Goodman, S.M. and Meininger, P.L. (1989). *The Birds of Egypt*. Oxford, Oxford University Press.
Gordon-Schwabe, *Quick and the Dead*: Gordon, A. and Schwabe, C.W. (2004). *The Quick and the Dead: Biomedical Theory in Ancient Egypt*. Leiden, Brill.
Gore, *Mires*: Gore, A.J.P. (1983). *Mires: Swamp, Bog, Fen and Moor*. 2 vols (4A & 4B) Amsterdam, Elsevier Scientific.
Grandin, *Genetics*: Grandin, T. (1998). *Genetics and the Behavior of Domestic Animals*. San Diego, Academic Press.
Gray, *Animal Locomotion*: Gray, J. (1968). *Animal Locomotion*. London, Weidenfeld and Nicolson.
Grier, *Biology*: Grier, J.W. (1984). *Biology of Animal Behavior*. St Louis, Times Mirror/Mosby College Publishing.
Griffith, *Hieratic Papyri*: Griffith, F.L. (1898). *Hieratic Papyri from Kahun and Gurob*. London, Bernard Quaritch.
Groenewegen-Frankfort, *Arrest and Movement*: Groenewegen-Frankfort, H.A. (1972). *Arrest and Movement. An Essay on Space and Time in the Representational Art of the Ancient Near East*. New York, Hacker Art Books.
Grzimek's Animal Life Encyclopedia: Grzimek, B. (2004). *Grzimek's Animal Life Encyclopedia*. 17 vols. 2nd edition. Detroit, Gale.
Grzimek's Encyclopedia: Grzimek, B., ed. (1990). *Grzimek's Encyclopedia of Mammals*. 5 vols. New York, McGraw-Hill.
Guggisberg, *Crocodiles*: Guggisberg, C.A.W. (1972). *Crocodiles: Their Natural History, Folklore and Conservation*. London, Wren.
Guthrie, *Nature of Paleolithic Art*: Guthrie, R.D. (2005). *The Nature of Paleolithic Art*. Chicago, University of Chicago Press.

Hafez, *Behaviour of Domestic Animals*: Hafez, E.S.E. (1969). *The Behaviour of Domestic Animals*. London, Baillière, Tindall & Cassell.
Hafez-Hafez, *Reproduction*: Hafez, B. and Hafez, E.S.E., eds. (2000). *Reproduction in Farm Animals*. 7th ed. Philadelphia, Lippincott Williams & Wilkins.
Haltenorth-Diller, *Field Guide*: Haltenorth, T. and Diller, H. (1980). *A Field Guide to the Mammals of Africa Including Madagascar*. London, Collins.
Hancock et al., *Storks, Ibises and Spoonbills*: Hancock, J.A., Kushlan, J.A. and Kahl, M.P. (1992). *Storks, Ibises and Spoonbills of the World*. London, Academic Press.
Hancock-Kushlan, *Herons*: Hancock, J. and Kushlan, J. (1984). *The Herons Handbook*. London, Croom Helm.
Harpur, *Decoration*: Harpur, Y. (1987). *Decoration in Egyptian Tombs of the Old Kingdom: Studies in Orientation and Context*. London, KPI.
Harpur, *Nefermaat and Rahotep*: Harpur, Y. (2001). *The Tombs of Nefermaat and Rahotep at Maidum: Discovery, Destruction and Reconstruction*. Cheltenham, Oxford Expedition to Egypt.
Harpur-Scremin, *Kagemni*: Harpur, Y. and Scremin, P. (2006). *The Chapel of Kagemni: Scene Details*. Reading, Oxford Expedition to Egypt.
Hart, *Behavior of Domestic Animals*: Hart, B.L. (1985). *The Behavior of Domestic Animals*. New York, W.H. Freeman.
Hassan, *Giza ii*: Hassan, S. (1936). *Excavations at Giza, Vol. ii: 1930-1931*. Cairo, Egyptian University.
Hassan, Giza *iii*: Hassan, S. (1941). *Excavations at Giza, Vol. iii: 1931-1932*. Cairo, Egyptian University.
Hassan, *Giza iv*: Hassan, S. (1932-1933). *Excavations at Giza, Vol. iv:* Cairo, Egyptian University.
Hassan, *Giza v*: Hassan, S. (1944). *Excavations at Giza, Vol. v: 1933-1934*. Cairo, Egyptian University.
Hassan, *Giza vi*: Hassan, S. (1950). *Excavations at Giza, Vol. vi: 1934-1935*. Cairo, Egyptian University.
Hassan, *Giza vii*: Hassan, S. (1953). *Excavations at Giza, Vol. vii: 1935-1936. The Mastabas of the Seventh Season and their Description*. Cairo, Egyptian University.
Hassan, *Hemet-Re*: Hassan, S. (1975). *Excavations at Saqqara, 1937-1938: Vol 3. Mastabas of Princess Hemet-Re and others*. Cairo, Arab Republic of Egypt, Antiquities Dept.
Hassan, *Neb-Kaw-Her*: Hassan, S. (1975). *Excavations at Saqqara, 1937-1938: Vol 1. The Mastaba of Neb-Kaw-Her*. Cairo, Arab Republic of Egypt, Antiquities Dept.
Hassan, *Ny-Ankh-Pepy*: Hassan, S. (1975). *Excavations at Saqqara, 1937-1938: Vol 2. Mastabas of Ny-Ankh-Pepy and Others*. Cairo, Arab Republic of Egypt, Antiquities Dept.
Hayes, *Scepter*: Hayes, W.C. (1953). *The Scepter of Egypt: A Background for the Study of the Egyptian Antiquities in the Metropolitan Museum of Art*. 2 vols. New York, The Metropolitan Museum of Art.
Heath, *The Talking Greeks*: Heath, J. (2005). *The Talking Greeks: Speech, Animals, and the Other in Homer, Aeschylus, and Plato*. Cambridge, Cambridge University Press.
Hendrickx et al., *Egypt at Its Origins*: Hendrickx, S., Friedman, R.F., Cialowicz, K.M. and Chlodnicki, M., eds. (2004). *Egypt at Its Origins: Studies in Memory of Barbara Adams*. Leuven, Peeters.
Hiero. Texts: Dept. of Egyptian and Assyrian Antiquities (1911-1970). *Hieroglyphic Texts from Egyptian Stelae, etc., in the British Museum*. 9 parts. London, British Museum.

Hinton-Dunn, *Mongooses*: Hinton, H.E. and Dunn, A.M.S. (1967). *Mongooses: Their Natural History and Behaviour.* Edinburgh, Oliver and Boyd.

Hoath, *Field Guide*: Hoath, R. (2003). *A Field Guide to the Mammals of Egypt.* Cairo, The American University in Cairo Press.

Hornung, *Conceptions of God*: Hornung, E. (1996). *Conceptions of God: The One and the Many.* Ithaca, Cornell University Press.

Houlihan, *Animal World*: Houlihan, P.F. (1996). *The Animal World of the Pharaohs.* London, Thames and Hudson.

Houlihan, *Birds*: Houlihan, P.F. (1986). *The Birds of Ancient Egypt.* Warminster, Aris and Phillips.

Ibis: *The Ibis*

Ikram, *Choice Cuts*: Ikram, S. (1995). *Choice Cuts: Meat Production in Ancient Egypt.* Leuven, Peeters.

Immelmann-Beer, *Dictionary of Ethology*: Immelmann, K. and C. Beer (1989). *A Dictionary of Ethology.* Cambridge, MA, Harvard University Press.

Ingold, *What Is an Animal?*: Ingold, T., ed. (1988). *What Is an Animal?* London, Unwin Hyman.

J. Anim. Ecol.: *Journal of Animal Ecology*

J. Anim. Sci.: *Journal of Animal Science*

J. App. Ecol.: *Journal of Applied Ecology*

J. Arch. Sci.: *Journal of Archaeological Science*

J. Comp. Physiol. A: *Journal of Comparative Physiology A: Sensory, Neural, and Behavioral Physiology*

J. Comp. Physiol. Psychol.: *Journal of Comparative and Physiological Psychology*

J. Comp. Psychol.: *Journal of Comparative Psychology*

J. Dairy Sci.: *Journal of Dairy Science*

J. East Afr. Nat. Hist.: *Journal of East African Natural History*

J. Herp.: *Journal of Herpetology*

J. Linn. Soc.: *Journal of the Linnean Society*

J. Mamm.: *Journal of Mammalogy*

James, *Khentika*: James, T.G.H. (1953). *The Mastaba of Khentika called Ikhekhi.* London, Egypt Exploration Society.

Janssen-Janssen, *Egyptian Household Animals*: Janssen, R. and J. (1989). *Egyptian Household Animals.* Aylesbury, Shire Publications.

JAOS: *Journal of the American Oriental Society*

JARCE: *Journal of the American Research Center in Egypt*

JEA: *Journal of Egyptian Archaeology*

Jensen, *Ethology*: Jensen, P. (2002). *The Ethology of Domestic Animals: An Introductory Text.* Wallingford, CAB International.

JEOL: *Jaarbericht van het Vooraziatisch-egyptisch Genootschap Ex Oriente Lux (Leiden)*

Jequier, *Monument Funéraire*: Jequier, G. (1936-1940). *Fouilles à Saqqarah: Le Monument Funéraire de Pepi II.* 3 vols. Cairo, L'insititut Français d'Archeologie Orientale.

JNES: *Journal of Near Eastern Studies*

Johnsgard, *Waterfowl*: Johnsgard, P.A. (1968). *Waterfowl: Their Biology and Natural History.* Lincoln, University of Nebraska Press.

Junker, *Giza i*: Junker, H. (1929). *Giza i: Die Mastabas der IV. Dynastie auf dem Westfriedhof.* Wien, Hölder-PichlerTempsky.

Junker, *Giza ii*: Junker, H. (1934). *Giza ii: Die Mastabas der Beginnenden V. Dynastie auf dem Westfriedhof.* Wien, Hölder-Pichler-Tempsky.

Junker, *Giza iii*: Junker, H. (1938). *Giza iii: Die Mastabas der Fortgeschrittenen V. Dynastie auf dem Westfriedhof.* Wien, Hölder-Pichler-Tempsky.

Junker, *Giza iv*: Junker, H. (1940). *Giza iv: Die Mastaba des **kȝjmʿnḫ** (Kai-em-anch).* Wien, Hölder-Pichler-Tempsky.

Junker, *Giza v*: Junker, H. (1941). *Giza v: Die Mastabas des* snb *(Seneb) und darumliegende Gräber.* Wien, Hölder-Pichler-Tempsky.

Junker, Giza *vi*: Junker, H. (1943). *Giza vi: Die Mastabas des* nfr *(Nefer),* ḳdf.jj *(Kedfi),* kȝḥjf *(Kahjef) und die Westlich Anschließenden Grabanlagen.* Wien, Hölder-Pichler-Tempsky.

Junker, *Giza viii*: Junker, H. (1947). *Giza viii: Der Ostabschnitt des Westfriedhofes. Zweiter Teil.* Wien, Hölder-Pichler-Tempsky.

Junker, *Giza xi*: Junker, H. (1953). *Giza xi: Der Friedhof Südlich der Cheopspyramide. Ostteil.* Wien, Hölder-Pichler-Tempsky.

Kamstra et al., *Funerary Symbols and Religion*: Kamstra, J.H., Milde, H. and Wagtendonk, K., eds. (1988). *Funerary Symbols and Religion*, Kampen, J.H. Kok.

Kanawati, *Inumin*: Kanawati, N. (2006). *The Teti Cemetery at Saqqara: Vol. vii. The Tomb of Inumin.* Oxford, Aris and Phillips.

Kanawati, *Kaiemankh*: Kanawati, N. (2001). *Tombs at Giza: Vol. i. Kaiemankh (G4561) and Seshemnefer I (G4940).* Warminster, Aris and Phillips.

Kanawati, *Tomb and Beyond*: Kanawati, N. (2001). *The Tomb and Beyond: Burial Customs of Egyptian Officials.* Warminster, Aris and Phillips.

Kanawati-Abder-Raziq, *Hesi*: Kanawati, N. and Abder-Raziq, M. (1999). *The Teti Cemetery at Saqqara: Vol. v. The Tomb of Hesi.* Warminster, Aris and Phillips.

Kanawati-Abder-Raziq, *Iynefert*: Kanawati, N. and Abder-Raziq, M. (2003). *The Unis Cemetery at Saqqara: Vol. ii. The Tombs of Iynefert and Ihy (reused by Idut).* Warminster, Aris and Phillips.

Kanawati-Abder-Raziq, *Meryteti*: Kanawati, N. and Abder-Raziq, M. (2004). *Mereruka and His Family: Part I. The Tomb of Meryteti.* Warminster, Aris and Phillips.

Kanawati-Abder-Raziq, *Neferseshemre*: Kanawati, N. and Abder-Raziq, M. (1998). *The Teti Cemetery at Saqqara: Vol. iii. The Tombs of Neferseshemre and Seankhuiptah.* Warminster, Aris and Phillips.

Kanawati-Abder-Raziq, *Nikauisesi*: Kanawati, N. and Abder-Raziq, M. (2000). *The Teti Cemetery at Saqqara: Vol. vi. The Tomb of Nikauisesi.* Warminster, Aris and Phillips.

Kanawati-Hassan, *Ankhmahor*: Kanawati, N. and Hassan, A. (1997). *The Teti Cemetery at Saqqara: Vol. ii. The Tomb of Ankhmahor.* Warminster, Aris and Phillips.

Kaplony, *Studien Methethi*: Kaplony, P. (1976). *Studien zum Grab des Methethi.* Riggisberg, Abegg-Stiftung Bern.

Kastenbaum, *Encyclopedia of Death*: Kastenbaum, R., ed. (2002). *Macmillan Encyclopedia of Death and Dying.* 2 vols. New York, Macmillan Reference.

Keeling-Gonyou, *Social Behaviour*: Keeling, L.J. and Gonyou, H.W., eds. (2001). *Social Behaviour in Farm Animals.* Wallingford, CAB International.

Kemi: *Kemi: Revue de Philologie et d'Archeologie Egyptiennes et Coptes*

Kilgour-Dalton, *Livestock Behaviour*: Kilgour, R. and Dalton, C. (1984). *Livestock Behaviour: A Practical Guide.* London, Granada.

King, *Weasels and Stoats*: King, C. (1989). *The Natural History of Weasels and Stoats.* Ithaca, Comstock Publishing.

Kingdon, *East African Mammals*: Kingdon, J. (1977). *East African Mammals: An Atlas of Evolution in Africa.* 3 vols. New York, Academic Press.

Klebs, *Reliefs*: Klebs, L. (1915). *Die Reliefs des Alten Reiches (2980-2475 V. Chr.): Material zur Ägyptischen Kulturgeschichte.* Heidelberg, C. Winter.

Klingender, *Animals in Art*: Klingender, F. (1971). *Animals in Art and Thought to the End of the Middle Ages.* London, Routledge and Kegan Paul.

Kozloff, *Animals in Ancient Art*: Kozloff, A.P., ed. (1981). *Animals in Ancient Art from the Leo Mildenberg Collection.* Mainz am Rhein, P. von Zabern.

KronoScope: *KronoScope*

Krzyszkowska, *Klados*: Krzyszkowska, O., ed. (1995). *Klados: Essays in Honour of J.N. Coldstream.* London, University of London, Institute of Classical Studies.

Kuschel, *Die Haustiere*: Kuschel, P. (1911). *Die Haustiere Ägyptens im Altertum.* Görlitz, Görlitzer Nachrichten und Anzeiger.

LÄ: Helck, W. and Otto, E. (1986). *Lexikon der Ägyptologie.* 7 vols. Wiesbaden, Otto Harrassowitz.

Labrousse-Lauer, *Complexe Funéraires d'Ouserkaf*: Labrousse, A. and Lauer, J.-P. (2000). *Les Complexes Funéraires d'Ouserkaf et de Néferhétepès.* Cairo, Institut Français d'Archéologie Orientale.

Labrousse-Moussa, *Complexe Funéraire du Roi Ounas*: Labrousse, A. and Moussa, A. (2002). *La Chaussée du Complexe Funéraire du Roi Ounas.* Cairo, Institut Français d'Archéologie Orientale.

Langton-Langton, *The Cat in Ancient Egypt*: Langton, N. and B. (1940). *The Cat in Acient Egypt.* Cambridge, Cambridge University Press.

L'Anthropologie: *L'Anthropologie*

Larsen, *Butterflies*: Larsen, T. B. (1990). *Butterflies of Egypt.* Svendborg, Apollo Books.

LD ii: Lepsius, R. (1897-1913). *Denkmaler aus Aegypten und Aethiopien. Vol. ii [pts. 3-4].* Leipzig, J. C. Hinrichs.

LD Erganz.: Lepsius, R. (1897-1913). *Denkmäler aus Aegypten und Aethiopien: Ergänzungsband.* Leipzig, J. C. Hinrichs.

Le Naturaliste: *Le Naturaliste*

Le Nature: *Le Nature*

Leahy, *Companion*:Leahy, C. (1982). *The Birdwatcher's Companion.* New York, Bonanza.

Lee-De Vore, *Kalahari Hunter-Gatherers*: Lee, R.B. and De Vore, I., eds. (1976). *Kalahari Hunter-Gatherers: Studies of the !Kung San and their Neighbors.* Cambridge, MA, Harvard University Press.

Leeds-Vayda, *Man, Culture, and Animals*: Leeds, A. and Vayda, A.P. (1965). *Man, Culture, and Animals: The Role of Animals in Human Ecological Adjustments.* Washington, DC, American Association for the Advancement of Science.

Leuthold, *African Ungulates*: Leuthold, W. (1977). *African Ungulates: A Comparative Review of their Ethology and Behavioral Ecology.* Berlin, Springer-Verlag.

Livest. Prod. Sci.: *Livestock Production Science*

Lloyd et al., *Saqqara Tombs ii*: Lloyd, A.B., Spencer, A.J. and El-Khouli, A. (1984). *Saqqara Tombs II: The Mastabas of Meru, Semdenti, Khui and Others.* London, Egypt Exploration Society.

Loret, *Deux Études*: Loret, V. (1896). *Deux Études Égyptologiques.* Paris, E. Bouillon.

LRRD: *Livestock Research for Rural Development* (www.cipav.org.co/lrrd/lrrd12/2/agan122.htm)

Luft, *Festschrift Kákosy*: Luft, U. (1992). *The Intellectual Heritage of Egypt: Festschrift Lászlό Kákosy*. Budapest, Studia Aegyptiaca.

MacDonald, *Encyclopaedia of Mammals*: Macdonald, D., ed. (1984). *Encyclopaedia of Mammals*. London, Allen & Unwin.
Macramallah, *Idout*: Macramallah, R. (1935). *Le Mastaba d'Idout*. Cairo, Service des Antiquités de l'Égypte.
Malek, *The Cat*: Malek, J. (1993). *The Cat in Ancient Egypt*. London, British Museum Press.
Mam. Rev.: *Mammal Review*
Mam. Sp.: *Mammalian Species*
Mammalia: *Mammalia*
Manning-Serpell, *Animals and Human Society*: Manning, A. and Serpell, J., eds. (1994). *Animals and Human Society: Changing Perspectives*. London, Routledge.
Mariette, *Voyage*: Mariette, A. (1893). *Voyage dans la Haute Égypte*. 2. vols. Paris, Werner.
Martin, *Hetepka*: Martin, G.T. (1979). *The Tomb of Hetepka and other Reliefs and Inscriptions from the Sacred Animal Necropolis, North Saqqara, 1964-1973*. London, Egypt Exploration Society.
McFarlane, *Irukaptah*: McFarlane, A. (2000). *The Unis Cemetery at Saqqara Vol. i. The Tomb of Irukaptah*. Oxford, Aris and Phillips.
McFarlane, *Kaiemheset*: McFarlane, A. (2003). *Mastabas at Saqqara. Kaiemheset, Kaipuneset, Kaiemsenu, Sehetepu and Others*. Oxford, Aris and Phillips.
McKilligan, *Herons*: McKilligan, N. (2005). *Herons, Egrets and Bitterns: Their Biology and Conservation in Australia*. Collingswood, CSIRO Publishing.
MDAIK: *Mitteilungen des Deutschen Archäologischen Instituts Abteilung Kairo*
MDOG: *Mitteilungen der Deutschen Orient-Gesellschaft zu Berlin*
Meinertzhagen, *Nicoll's Birds*: Meinertzhagen, C.R. (1930). *Nicoll's Birds of Egypt*. 2 vols. London.
MIFAO: *Mémoires publiés par les members de l'Institut Français d'Archéology Orientale du Cairo.*
MIOC, Rio de Janeiro: *Memórias do Instituto Oswaldo Cruz, Rio de Janeiro*
MMJ: *Metropolitan Museum Journal*
Mogensen, *Ny Carlsberg*: Mogensen, M. (1930). *La Glyptothèque Ny Carlsberg: La Collection Égyptienne*. Copenhagen, Levin & Munksgaard.
Mohr, *Hetep-her-akhti*: Mohr, H.T. (1943). *The Mastaba of Hetep-her-akhti: Study on an Egyptian Tomb Chapel in the Museum of Antiquities Leiden*. Leiden, Brill.
Montet, *Scènes*: Montet, P. (1925). *Scènes de la Vie Privée dans les Tombeaux Égyptiens de l'Ancien Empire*. Paris, Strasbourg University.
Morgenland: *Morgenland: Darstellungen aus Geschichte und Kultur des Ostens*
Morphy, *Animals into Art*: Morphy, H., ed. (1989). *Animals into Art*. London, Unwin Hyman.
Morris, *Primate Ethology*: Morris, D. (1967). *Primate Ethology*. London, Weidenfeld and Nicolson.
Mourant-Zeuner, *Man and Cattle*: Mourant, A.E. and Zeuner, F.E., eds. (1963). *Man and Cattle: Proceedings of a Symposium on Domestication at the Royal Anthropological Institute, 24-26 May 1960*. London, Royal Anthropological Institute of Great Britain & Ireland.
Moussa-Altenmüller, *Nefer and Ka-hay*: Moussa, A.M. and Altenmüller, H. (1971). *The Tomb of Nefer and Ka-hay*. Mainz am Rhein, P. von Zabern.
Moussa-Altenmüller, *Nianchchnum*: Moussa, A.M. and Altenmüller, H. (1977). *Das Grab des Nianchchnum und Chnumhotep*. Mainz am Rhein, P. von Zabern.
Moussa-Junge, *Two Tombs*: Moussa, A.M. and Junge, F. (1975). *Two Tombs of Craftsmen*. Mainz am Rhein, P. von Zabern.
Müller, *Ägyptische Kunstwerke*: Müller, H.W. (1964). *Ägyptische Kunstwerke, Kleinfunde und Glas in der Sammlung E. und M. Kofler-Truniger, Luzern*. Berlin, B. Hessling.
Müller, *Sheep*: Müller, H.A. (1989). *Sheep: A Complete Owner's Manual*. Hauppauge, Barrons.
Munich. Äg. Sammlung: *Ägyptische Sammlung des Bayerischen Staates: Ausstellung in den Ausstellungsräumen der Staatlichen Graphischen Sammlung* (1966). Munich, Bayrischer Rundfunk.
Munro, *Unas-Friedhof*: Munro, P. (1993). *Das Unas-Friedhof Nord-West: Topographisch-Historische Einleitung*. Mainz am Rhein, P. von Zabern.
Murray, *Saqqara Mastabas*: Murray, M.A. (1904). *Saqqara Mastabas Vol. i*. London, British School of Archaeology in Egypt.

NABB: *North American Bird Bander*
Nature: *Nature*
NAWG: *Nachrichten der Akademie der Wissenschaften zu Göttingen*
Nibbi, *Lapwings and Libyans*: Nibbi, A. (1986). *Lapwings and Libyans in Ancient Egypt*. Oxford, DE Publications.
Nickel et al., *Locomotor System*: Nickel, R., Schummer, A., Seferle, E., Frewein, J., Wilkens, H. and Wille, K.-H. (1986). *The Locomotor System of the Domestic Mammals*. Berlin, Paul Parey.
NJAS: *Netherlands Journal of Agricultural Science*
NouvClio: *La Nouvelle Clio*

Nowak, *Walker's Mammals*: Nowak, R.M. (1999). *Walker's Mammals of the World.* 2 vols. Baltimore, Johns Hopkins University Press.
NZ J. Ag. Res.: *New Zealand Journal of Agricultural Research*
NZ J. Exp. Ag.: *New Zealand Journal of Experimental Agriculture*
NZ J. Zool.: *New Zealand Journal of Zoology*

O'Casey, *Making of Papyrus*: O'Casey, I. (1973). *The Nature and Making of Papyrus.* Barkston Ash, The Elmete Press.
Oikos: *Oikos*
OJA: *Oxford Journal of Archaeology*
Or.: *Orientalia*
Ornis Scand.: *Ornis Scandinavica*
Osborn-Helmy, *Contemporary Land Mammals*: Osborn, D.J. and Helmy, I. (1980). *The Contemporary Land Mammals of Egypt (including Sinai)*, Field Museum of Natural History, *Fieldiana Zoology* no. 5, 1-579.
Osborn-Osbornová, *Mammals*: Osborn, D.J. and Osbornová, J. (1998). *The Mammals of Ancient Egypt.* Warminster, Aris and Phillips.
Ostrich: *The Ostrich*

Panofsky, *Meaning in the Visual Arts*: Panofsky, E. (1955). *Meaning in the Visual Arts.* Chicago, University of Chicago Press.
Parkinson-Quirke, *Papyrus*: Parkinson, R. and Quirke, S. (1995). *Papyrus.* Austin, University of Texas Press.
Paton, *Animals of Ancient Egypt*: Paton, D. (1925). *Animals of Ancient Egypt: Materials for a 'Sign-List' of Egyptian Hieroglyphs.* Princeton, Princeton University Press.
Petrie, *Medum*: Petrie, W.M.F. (1892). *Medum.* London, D. Nutt.
Petrie-Murray, *Seven Memphite*: Petrie, H.F. and Murray, M.A. (1952). *Seven Memphite Tomb Chapels.* London, British School of Egyptian Archaeology.
Phillips, *Cattle Behaviour*: Phillips, C.J.C. (1993). *Cattle Behaviour.* Ipswich, Farming Press.
Phys. Beh.: *Physiology & Behavior*
Phys. Biochem. Zool.: *Physiology and Biochemical Zoology*
Pinch, *Magic in Ancient Egypt*: Pinch, G. (1994). *Magic in Ancient Egypt.* London, British Museum Press.
Plant Ecol.: *Plant Ecology*
P-M III: Porter, B. and Moss, R.L.B. (1970-1981). *Topographical Bibliography of Ancient Egyptian Hieroglyphic Texts, Reliefs and Paintings. Vol. iii. Memphis.* Oxford, Clarendon Press.
Pollard, *Birds in Greek Life*: Pollard, J. (1977). *Birds in Greek Life and Myth.* London, Thames and Hudson
Proc. 37th Ann. Pac. Northwest Anim. Nutrit. Conf.: *Proceedings of the 37th Annual Pacific Northwest Animal Nutrition Conference.* Vancouver, BC, Pacific Northwest Animal Nutrition Conference, October 9-10, 2002.
Proc. AAE: *Proceedings of the American Association of Equine Practitioners*
Proc. Zool. Soc. Lond.: *Proceedings of the Zoological Society of London*
PS: *Pollen Spores*
Pullin-Lowe-McConnell, *Tilapias*: Pullin, R.S.V. and Lowe-McConnell, R.H., eds. (1982). *The Biology and Culture of Tilapias.* Manila, ICLARM.

Quart. Rev. Biol.: *Quarterly Review of Biology*
Quibell, *Ptah-hetep*: Quibell, J.E. (1896). *The Ramesseum. And, The Tomb of Ptah-hetep* (copied by Paget, R.F E. and Pirie, A.A. with comments by Griffith, F.Ll.). London, Histories & Mysteries of Man Ltd.
Quibell, *Saqqara (1907-1908)*: Quibell, J.E. (1909). *Excavations at Saqqara (1907-1908).* Cairo, l'Institut Française d'Archéologie Orientale.
Quibell, *Saqqara (1911-1912)*: Quibell, J.E. (1913). *Excavations at Saqqara (1911-1912). The Tomb of Hesy.* Cairo, l'Institut Française d'Archéologie Orientale.
Quirke, *Temple in Ancient Egypt*: Quirke, S. (1997). *The Temple in Ancient Egypt: New Discoveries and Recent Research.* London, British Museum.

RAI News: *Royal Anthropological Institute News*
Ranke, *Art of Ancient Egypt*: Ranke, H. (1936). *The Art of Ancient Egypt: Architecture, Sculpture, Painting, Applied Art.* London, Phaidon Press.
Rawson, *Animals in Art*: Rawson, J., ed. (1977). *Animals in Art.* London, British Museum.
RdÉ: *Revue d'Égyptologie*
REA: *Revue de l'Égypte Ancienne*
Reese, *Pleistocene and Holocene Fauna*: Reese, D.S., ed. (1996). *Pleistocene and Holocene Fauna of Crete and Its First Settlers.* Madison, Prehistory Press.
Reeve, *Hedgehogs*: Reeve, N. (1994). *Hedgehogs.* London, T & A D Poyser.
Reisner, *Giza Necropolis*: Reisner, G.A. (1942). *A History of the Giza Necropolis.* 2 vols. Cambridge, MA, Harvard University Press.

RES: *Revue d'Ethnographie et de Sociologie*

Resh-Cardé, *Insects*: Resh, V.H. and Cardé, R.T., eds. (2003). *Encyclopedia of Insects*. Amsterdam, Academic Press.

Rice, *Swifter than the Arrow*: Rice, M. (2006). *Swifter than the Arrow*. London, I.B. Tauris.

Ridley, *Unification of Egypt*: Ridley, R.T. (1973). *The Unification of Egypt*. Deception Bay, Shield Press.

Robins, *Proportion and Style*: Robins, G. (1994). *Proportion and Style in Ancient Egyptian Art*. Austin, University of Texas Press.

Rock Art Res.: *Rock Art Research*

Ross, *Crocodiles*: Ross, C.A., ed. (1989). *Crocodiles and Alligators*. Silverwater, Golden Press.

Roth, *Cem. Pal. Att.*: Roth, A.M. (1995). *A Cemetery of Palace Attendants: Including G2084-2099, G2230+2231 and G2240*. Boston, Museum of Fine Arts.

Roy, *The Calf*: Roy, J.H.B. (1970). *The Calf*. 2 vols. London, Iliffe Books.

RT: *Recueil de Travaux Relatifs à la Philologie et à l'Archéologie Égyptiennes et Assyriennes*

Rudnai, *Social Life of the Lion*: Rudnai, J.A. (1973). *The Social Life of the Lion*. Lancaster, Medical and Technical Pub. Co.

Rundle-Clark, *Myth and Symbol*: Rundle-Clark, R.T. (1959). *Myth and Symbol in Ancient Egypt*. London, Thames and Hudson.

SAJZ: *South African Journal of Zoology*

SAK: *Studien zur Altägyptischen Kultur*

Saleh, *Amphibians and Reptiles*: Saleh, M.A. (1997). *Amphibians and Reptiles of Egypt*. Cairo, Al Azhar University.

Säugetier. Mitteil.: *Säugetierkundliche Mitteilungen*

Säve-Soderbergh, *Egyptian Representations of Hippopotamus Hunting*: Säve-Soderbergh, T. (1953). *On Egyptian Representations of Hippopotamus Hunting as a Religious Motif*. Uppsala, Lund.

SCC: *Small Carnivore Conservation*

Schäfer, *Principles*: Schäfer, H. (1986). *Principles of Egyptian Art*. Translated and edited by J. Baines. Oxford, Griffith Institute.

Schäfer-Andrae, *Kunst*: Schäfer, H. and Andrae, W. (1942). *Die Kunst des Alten Orients*. Berlin, Propylaen.

Schaller, *Serengeti Lion*: Schaller, G.B. (1972). *The Serengeti Lion: A Study of Predator-Prey Relations*. Chicago, University of Chicago Press.

Schürmann, *Die Reliefs*: Schürmann, W. (1983). *Die Reliefs aus dem Grab des Pyramidenvorstehers Ii-nefret*. Karlsruhe, C.F. Muller.

Schwedler et al., *Raising Catfish*: Schwedler, T.E., Berry, M.L. and King, D.R. (1989). *4-H Aquatic Science Project: Guide to Raising Catfish in a Cage*. SRAC Publication No. 340, Southern Region Aquaculture Center.

Sci. Am.: *Scientific American*

Science: *Science*

Sculpt. Rev.: Sculpture Review

Sebeok, *How Animals Communicate*: Sebeok, T.A., ed. (1977). *How Animals Communicate*. Bloomington, Indiana University Press.

Shepard, *The Others*: Shepard, P. (1996). *The Others: How Animals Made Us Human*. Washington, DC, Island Press.

Shortridge, *Mammals of South West Africa*: Shortridge, G.C. (1934). *The Mammals of South West Africa*. 2 vols. London, Heinemann.

Sillero-Zubiri et. al., *Canids*: Sillero-Zubiri, C., Hoffmann, M. and Macdonald, D.W., eds. (1990). *Canids: Foxes, Wolves, Jackals and Dogs. Status and Conservation Action Plan*. Cambridge, ICUN.

Simpson, *Kawab*: Simpson, W.K. (1978). *The Mastabas of Kawab, Khafkhufu I and II (G7110-20, 7130-40 and 7150 and Subsidary Mastabas of Street G 7100)*. Boston, Museum of Fine Arts.

Simpson, *Kayemnofret*: Simpson, W.K. (1992). *The Offering Chapel of Kayemnofret in the Museum of Fine Arts, Boston*. Boston, Museum of Fine Arts.

Simpson, *Mast. West. Cem. Part I*: Simpson, W.K. (1980). *Mastabas of the Western Cemetery: Part I*. Boston, Museum of Fine Arts.

Simpson, *Qar and Idu*: Simpson, W.K. (1976). *The Mastabas of Qar and Idu (G7101 and 7102)*. Boston, Museum of Fine Arts.

Simpson, *Sekhem-ankh-ptah*: Simpson, W.K. (1976). *The Offering Chapel of Sekhem-Ankh-Ptah in the Museum of Fine Arts, Boston*. Boston, Museum of Fine Arts.

Smith, *Art and Architecture*: Smith, W.S. (1981). *The Art and Architecture of Ancient Egypt*. London, Penguin Books.

Smith, *HESPOK*: Smith, W.S. (1978). *A History of Egyptian Sculpture and Painting in the Old Kingdom*. London, Oxford University Press.

Smith, *Interconnections*: Smith, W.S. (1965). *Interconnections in the Ancient Near East: A Study of the Relationships Between the Arts of Egypt, the Aegean, and Western Asia*. New Haven, Yale University Press.

Smithers, *Mammals of Botswana*: Smithers, R.H.N. (1971). *The Mammals of Botswana*. Salisbury, The Trustees of the National Museum of Rhodesia.

Sphinx: *Sphinx*

Spinage, *Antelopes*: Spinage, C.A. (1986). *The Natural History of Antelopes*. London, Croom Helm.

Stevenson-Hamilton, *Wild Life*: Stevenson-Hamilton, J. (1947). *Wild Life in South Africa*. London, Cassell & Co.

Stoof, *Hasendarstellungen*: Stoof, M. (2005). *Hasendarstellungen im alten Ägypten* (Archäologische Forschungsergebnisse, Bd. 33). Hamburg, Verlag Dr Kovac.
Stoof, *Skorpion und Skorpiongöttin*: Stoof, M. (2002). *Skorpion und Skorpiongöttin im alten Ägypten.* Hamburg, Verlag Dr Kovac.
Strandberg, *Gazelle*: Standberg, A. (2009). *The Gazelle in Ancient Egyptian Art: Image and Meaning.* Uppsala, Uppsala Universitet.

Thear-Fraser, *Complete Book*: Thear, K. and Fraser, A. (1980). *The Australia and New Zealand Complete Book of Raising Livestock and Poultry.* Sydney, Australian and New Zealand Book Company.
Thomson, *Dictionary of Birds*: Thomson, S.A.L. (1964). *A New Dictionary of Birds.* New York, McGraw-Hill.
Toynbee, *Animals in Roman Life*: Toynbee, J.M.C. (1973). *Animals in Roman Life and Art.* Baltimore, Johns Hopkins University Press.
Trans. Zool. Soc. Lond.: *Transactions of the Zoological Society of London*
Trop. Anim. Health Prod.: *Tropical Animal Health and Production*
Trop. Biol.: *Tropical Biology*
Turner-Bateson, *Domestic Cat*: Turner, D.C. and Bateson, P., eds. (1988). *The Domestic Cat: The Biology of its Behaviour.* Cambridge, Cambridge University Press.

van de Walle, *Neferirtenef*: van de Walle, B. (1978). *La Chapelle Funéraire de Neferirtenef.* Brussels, Musées Royaux d'Art et d'Histoire.
van Elsbergen, *Fischerei im Alten Ägypten*: Elsbergen, M.J. (1997). *Fischerei Im Alten Agypten: Untersuchungen zu den Fischfangdarstellungen in den Grabern der 4. bis 6. Dynastie.* Berlin, Achet Verlag.
van Lawick-Goodall, *Innocent Killers*: van-Lawick-Goodall, H. and J. (1971). *Innocent Killers.* Boston, Houghton Mifflin.
van Voss et al., *Studies in Egyptian Religion*: van Voss, M.H., Hoens, D.J., Mussies, G., van der Plas, D. and te Velde, H., eds. (1982). *Studies in Egyptian Religion: Dedicated to Professor Jan Zandee.* Leiden, Brill.
van Walsem, *Iconography*: van Walsem, R. (2005). *Iconography of Old Kingdom Elite Tombs: Analysis & Interpretation, Theoretical and Methodological Aspects.* Leiden, Peeters.
Vandier, *Manuel*: Vandier, J. (1952-1978). *Manuel d'Archéologie Égyptienne: Bas-reliefs et Peintures Scènes de la Vie Quotidienne.* 6 vols. Paris, Picard.
Verh. Internat. Verein. Limnol.: *Verhandlungen Internationale Vereinigung Limnologie*
Verheyen, *Hippopotame*: Verheyen, R. (1954). *Monographie Éthologique de l'Hippopotame (Hippopotamus amphibius* Linné*).* Brussels, Institut des Parcs Nationaux du Congo Belge.
Vet. Clin. North Am. Equine Pract.: *Veterinary Clinics of North America: Equine Practice*
Vis. Res.: *Vision Research*
VNG Zürich: *Vierteljahrsschruft Naturforschende Gessellschaft Zürich*
von Bissing, *Gem-ni-kai*: von Bissing, F.W. (1905-1911). *Die Mastaba des Gem-ni-kai.* 2 vols. Berlin, A. Duncker.
von Droste zu Hülshoff, *Der Igel*: von Droste zu Hülshoff, V. (1980). *Der Igel im alten Ägypten.* Hildesheim, Gerstenberg Verlag.
Vrba-Schaller, *Antelopes, Deer, and Relatives*: Vrba, E.S. and Schaller, G.B., eds. (2000). *Antelopes, Deer, and Relatives: Fossil Record, Behavioral Ecology, Systematics, and Conservation.* New Haven, Yale University Press.

Wassell, *Ancient Egyptian Fauna*: Wassell, B.A. (1991). *Ancient Egyptian Fauna: A Lexicographical Study.* 2 vols. Unpublished PhD thesis, School of Oriental Studies, University of Durham.
Wb: Erman, A. and Grapow, H. (1971). *Wörterbuch der Ägyptischen Sprache.* 7 vols. Berlin, Akademie-Verlag.
Weeks, *Cemetery G6000*: Weeks, K.R. (1994). *Mastabas of Cemetery G6000: Including G6010 (Neferbauptah), G6020 (Iymery), G6030 (Ity), G6040 (Shepseskafankh).* Boston, Museum of Fine Arts.
Weeks, *Egyptology and the Social Sciences*: Weeks, K.R. ed. (1979). *Egyptology and the Social Sciences: Five Studies.* Cairo, The American University in Cairo Press.
Wemmer, *Comparative Ethology*: Wemmer, C.M. (1977). *Comparative Ethology of the Large-Spotted Genet, Genetta tigrina and Some Related Viverrids.* Smithsonian Contributions to Zoology 239, 1-93.
Wenig, *Jahreszeitenreliefs*: Wenig, S.A.M.B. (1966). *Die Jahreszeitenreliefs aus dem Sonnenheiligtum des Konigs Ne-User-Re.* Berlin, Staatliche Museen zu Berlin.
Whittier-Thorne, *Assisting the Beef Cow*: Whittier, J.C. and Thorne, J.G. (1995). *Assisting the Beef Cow at Calving Time.* MU Agricultural Guide, University of Missouri-Columbia.
Wild, *Ti ii*: Wild, H. (1953). *Le Tombeau de Ti. Vol. ii: La Chapelle* [part 1]. Cairo, Institut Français d'Archéologie Orientale.
Wild, *Ti iii*: Wild, H. (1966). *Le Tombeau de Ti. Vol. iii: La Chapelle* [part 2]. Cairo, Institut Français d'Archéologie Orientale.
Wilkinson, *Complete Gods*: Wilkinson, R.H. (2003). *The Complete Gods and Goddesses of Ancient Egypt.* London, Thames and Hudson.
Wilkinson, *Manners and Customs*: Wilkinson, J.G. (1837). *Manners and Customs of the Ancient Egyptians.* 3 vols. London, J. Murray.
Wilkinson, *Symbol and Magic*: Wilkinson, R.H. (1994). *Symbol and Magic in Egyptian Art.* London, Thames and Hudson.

Williams, *Decoration of the Tomb of Per-Neb*: Williams, C.R. (1932). *The Decoration of the Tomb of Per-Neb: The Technique and the Color Conventions.* New York, Metropolitan Museum of Art.
Wilson, *Ecophysiology*: Wilson, R.T. (1989). *Ecophysiology of the Camelidae and Desert Ruminants.* Berlin, Springer-Verlag.
Woods et al., *Egyptian Culture and Society*: Woods, A., McFarlane, A. and Binder, S., eds. (2010). *Egyptian Culture and Society: Studies in Honour of Naguib Kanawati.* Cairo, Supreme Council of Antiquities.
World Arch.: *World Archaeology*

YF&ESB: *Yale F* & *ES Bulletin*

Z. Säugetier.: *Zeitschrift für Säugetierkunde*
Z. Tierpsychol.: *Zeitschrift für Tierpsychologie*
ZÄS: *Zeitschrift für Ägyptische Sprache und Altertumskunde*
Ziegler, *Akhethetep*: Ziegler, C. (1993). *Le Mastaba d'Akhethetep: Une Chapelle Funeraire de l'Ancien Empire.* Paris, Réunion des Musées Nationaux.
Ziegler, *Catalogue*: Ziegler, C. (1990). *Catalogue des Steles, Peintures et Reliefs Égyptiens de l'Ancien Empire et de la Premiere Periode Intermediaire vers 2686-2040 avant J.-C.* Paris, Réunion des Musees Nationaux.
Zoo Biol.: *Zoo Biology*

CHAPTER 1

ANIMAL IMAGES IN EGYPTIAN WALL SCENES

1.1 ANIMALS IN ART

Depictions of animals dominate the visual records of ancient and modern cultures worldwide, reflecting the fundamental role they have played in the lives of people for thousands of years.[1] Indeed, animals have been at the very heart of human development and cultural evolution. Throughout history, their myriad forms have generated endless fascination and have provoked a range of extreme reactions - from fear to awe, pleasure to disgust, veneration to abuse. We associate them with the most raw and uncivilized aspects of life, but animals have also inspired profound ideas about the nature of the divine, for although we rely upon them for many mundane necessities, such as food and transportation, their pervasive presence has equally fed our minds and carried our imaginations in new directions. Part of our world yet separate, they have helped us, through comparison with their inherent 'otherness', to define ourselves as human beings.[2]

Since the dawn of human creativity, this complex relationship has been communicated via the medium of art. Animal imagery has consequently been the subject of detailed examination by archaeologists and art historians because it has the potential to reveal unique information about both the practical and metaphysical lives of ancient peoples.[3] Such analyses are usually driven by one of two interdependent motivations: a desire to either understand the meaning of animal images (i.e., determine what they represent) or to assess their aesthetic value (i.e., evaluate how they are depicted). The animal-rich art of ancient Egypt has thus been investigated from both of these perspectives.

Animals influenced every aspect of Egyptian life. Beyond their importance as a primary source of food, materials, and strength, their images and qualities also contributed to the Egyptians' idiosyncratic form of writing,[4] their religious concepts,[5] symbolism,[6] architectural features, and decorative elements. Such a pronounced response was perhaps inevitable from a people who lived in the narrow yet lush confines of the Nile Valley, a geographic environment that brought them into close and constant contact with the profusion of animals that also dwelled there. Egyptian art reveals that every class of animal was acknowledged: mammals and birds, reptiles and amphibians, fish and invertebrates. All were celebrated in a variety of art forms: painted and inscribed upon walls and ceramics, modelled in sculpture, wrought in jewellery, embossed upon metal, and carved into furniture and seals.[7] These images reveal that the Egyptian people not only watched animals very closely, but that they endeavoured to reproduce as accurately as possible their physical appearance and defining characteristics. This is particularly apparent in representations of animals that are found in tombs of the Old Kingdom period.

Wall scenes, rendered in paint and carved relief, provide a detailed pictorial record of many significant events in the lives of the Egyptian people. During the Old Kingdom, these images were

---

1 von Simson, O. (1965). "Introduction" in de Silva et al., *Man and Animal*, 9.

2 We are what they are not. See Shepard, *The Others*.

3 For general overviews, see Boudet, *Man and Beast*; Brion, *Animals in Art*; de Silva et al., *Man and Animal*; Klingender, *Animals in Art*; Brodrick, *Animals in Archaeology*; Clark, *Animals and Men*; Rawson, *Animals in Art*; Kozloff, *Animals in Ancient Art*; Morphy, *Animals into Art*.

4 Houlihan, P.F. (2002). "Animals in Egyptian art and hieroglyphs" in Collins, *A History of the Animal World*, 97-143.

5 te Velde, H. (1980). "A few remarks on the religious significance of animals in ancient Egypt", *Numen* 27, 76-82; Teeter, E. (2002). "Animals in Egyptian religion" in Collins, *A History of the Animal World*, 335-360.

6 Peck, W.H. (2003). "Animal symbolism in ancient Egyptian art", *Sculpt. Rev.* 52(2), 8-15.

7 Arnold, D. (1995). "An Egyptian bestiary", *BMMA* 52(4), 3-65.

devoted to eight basic themes: namely, marsh pursuits, agriculture, pastoral activities, horticulture, workshop pursuits, banqueting, the funerary meal, and funeral ceremonies for the deceased.[8] Despite their static media and the frozen gestures of their participants, these scenes reliably impart a sense of agitated movement. They therefore offer the modern viewer a rare opportunity to effectively glimpse ancient life 'in action'.[9]

Animals take part in almost all scene types; in fact, they are frequently the focus of the human activity we see represented. The way in which they are exhibited - what they do and how human figures use and respond to them - can consequently enlighten our understanding of the practical role of animals in Egyptian society. An abundance of ecologically appropriate fauna populates each of the environments (marsh, riverine, agrarian, desert, etc.) in which the various scenes are set. Collectively, then, wall reliefs incorporate an extremely diverse range of species,[10] which are frequently portrayed in a variety of poses and precise morphological detail.[11] The depiction of such a heterogeneous array of body-shapes in a clear and unambiguous fashion undoubtedly presented artists with different challenges from those associated with the human form; the treatment of animal figures is hence also potentially informative about aspects of the Egyptians' method of two-dimensional representation.

Scholars have acknowledged the value of examining the animal images in wall scenes since the earliest days of Egyptological enquiry. The nature of their analyses is the subject of the following overviews in which both the *information content* and the *form* of animal iconography are considered.[12]

1.2 ANIMAL IMAGES IN EGYPTIAN WALL SCENES: CONTENT

One of the most common forms of investigation undertaken by scholars in the study of Egyptian animal images has been the zoological identification of the vast number of species that appear in tomb scenes and hieroglyphs. This can be a challenging exercise because the painted details of wall reliefs have often long since disappeared; identification must therefore be accomplished on the basis of morphological features alone, without the diagnostic aid of colour. Furthermore, many of the represented species no longer exist in Egypt, having become extinct in the country either in ancient times or more recently, so living exemplars must be sought much further south, in north-eastern and central Africa.

During the 19th century, this task was virtually impossible, as the African continent was still largely unexplored and many of the animals that are represented in Egyptian art were either completely unknown in Europe or at least not well understood. Nevertheless, some of the earliest work of this kind was carried out in the 1860s by the German explorer and anthropologist, Robert Hartmann. Hartmann travelled with a scientific expedition from Egypt to the Sudan between 1859 and 1860. Upon his return to Berlin, he endeavoured to compare the animals that he had seen on the walls of Egyptian tombs and temples, and those that he could examine in the published works of early Egyptologists, such as Rosellini, Lepsius, and Wilkinson, with the skeletons and preserved specimens of different species he had collected while on his travels. In 1864, he produced a paper that listed in precise taxonomic order the mammals he had been able to identify.[13] For each of

---

8 Harpur, *Decoration*, 175.

9 However, the degree to which wall scenes can be viewed as representations of reality is debatable. See Robins, G. (1990). "Problems in interpreting Egyptian art", *DE* 17, 45-58. It is also unclear whether our interpretation of wall scenes generates accurate information about Egyptian daily life. For example, see Samuel, D. (1993). "Ancient Egyptian cereal processing: Beyond the artistic record", *CAJ* 3, 276–283.

10 Some species do not occur until after the Old Kingdom (e.g. domestic cats and horses).

11 Nevertheless, mistakes did occur. For numerous examples, see Osborn-Osbornová, *Mammals*. See also Keimer, L. (1953). "Les limites de l'observation naturaliste dans quelques représentations animales de l'Égypte antique", *Ann. Fac. Lett. Univ. Ibrahim* 2(May), 121-134; Reed, C.A. and Osborn, D.J. (1978). "Taxonomic transgressions in Tutankhamun's treasures", *AJA* 82(3), 273-283; Edwards, I.E.S. (1979). "Zoomorphic anomalies in Tutankhamun's treasures", *AJA* 83(2), 205-206; Osborn, D.J. (1985). "A rebuttal of criticisms of identifications of animals portrayed in the tomb of Tutankhamun", *AJA* 89(1), 157-161.

12 The following is offered as a general summary of the way in which Egyptian animal images have been assessed. The literature cited is for example only and is not intended as a full review of all available sources.

13 Hartmann, R. (1864). "Versuch einer systematischen Aufzählung der von den alten Aegyptern bildlich dargestellten Thiere (mit Rücksicht auf die heutige Fauna des Nilgebietes)", *ZÄS* 2(1), 7-12; Hartmann, R.

these, he offered both the Latin and hieroglyphic name of the species, a brief description of its representation upon Egyptian monuments, details of its preferred habitat in the present-day, and its distribution on the African continent.

Hartmann's blend of zoology and Egyptology was something that had never before been attempted, but as subsequent expeditions slowly revealed more of the fauna of the African continent, European scholars in the late 19th and early 20th centuries were able to compare Egyptian images directly with mounted specimens displayed in museums and university collections, as well as drawings and engraved plates of newly discovered species that were published in scientific monographs.[14] One invaluable reference was John Anderson's series, *The Zoology of Egypt*, which, although intended primarily to provide a comprehensive description of the extant vertebrates of the country, also sought to "throw light on the crude representations of the ancient Egyptians, determining the species where possible".[15]

The French scholar, P. Hippolyte Boussac, referred to preserved specimens in an early series of papers devoted to the identification of birds in Egyptian art.[16] His investigative approach, which was adopted by many subsequent researchers, was to focus upon a distinctive attribute displayed by a depicted animal, such as a crest, unusual feather patterning or beak-shape, to determine which species it represented. A brief description of the physical appearance of the living animal was always included for readers, most of whom would be unfamiliar with the African species Boussac discussed, but in some of his papers, photographs of taxidermically-prepared birds were also included alongside line drawings of the Egyptian images to confirm his identification visually.[17]

Despite this systematic procedure, the lack of firsthand accounts of the living animals they studied forced Boussac and others of the period to rely almost exclusively upon comments and anecdotes by classical authors, such as Herodotus, Pliny, Aristotle, and Aelian, for information about natural history. As scientific studies and published accounts of African species gradually increased, however, researchers began to adopt a more rigorous method in their interpretation of the depicted fauna. The burgeoning interest in mummified animals at this time[18] may also have contributed to a more empirical and quantitative treatment of two-dimensional animal images. Following the scientific works they consulted, many studies began to include detailed descriptions of the morphology of each identified species, including measurements of their physical features (e.g., height, length of tail, horn-shape, etc.), general appearance, geographic distribution, and habits. In particular, plates reproduced from natural history volumes were often included to facilitate comparison with Egyptian depictions. Throughout the first half of the 20th century, scholars such as Dawson, Hilzheimer, Gaillard, but especially Keimer,[19] examined, identified, and described a diverse range of animals found in both art and hieroglyphs.[20] These included: scimitar oryx,[21] gerenuks,[22] deer,[23] pigs,[24] lions,[25] leopards,[26] caracals,[27] monkeys,[28] porcupines,[29] various bird

(1864). "Versuch einer systematischen Aufzählung der von den alten Aegyptern bildlich dargestellten Thiere (mit Rücksicht auf die heutige Fauna des Nilgebietes)", *ZÄS* 2(2), 19-28.

14 See Bogaert-Damin, A.-M. (1993). "La représentation animale dans les livres illustré de zoologie (XVIe - XXe siècles)" in Bodson, *L'Histoire*, 233-260.

15 de Winton, W.E. (1902). "Introduction" in Anderson, *Zoology of Egypt*, viii-ix.

16 Boussac produced some 20 papers on the subject of depicted birds, as well as a few studies on bats, crocodiles, and the Seth animal. For example Boussac, P.H. (1904). "Les ansérídés dans l'Égypte pharaonique", *Le Naturaliste* 26(42), 209-211; Boussac, P.H. (1905). "L'hirondelle dans les mythes Égyptiens", *Cosmos* (1089), 661-665; and Boussac, P.H. (1909). "Le pigeon dans l'ancienne Égypte", *La Nature* 37, 264-267.

17 For example, see Boussac, P.H. (1912). "L'oie de Gambie", *RT* 34, 166-167.

18 See Gaillard-Daressy, *La Faune Momifiée.*

19 Louis (or Ludwig) Keimer produced over 100 papers devoted to the subject of animals in Egyptian culture, which were published from the 1920s until his death in 1957.

20 Paton, *Animals of Ancient Egypt.*

21 Bonnet, A. (1909). "L'oryx dans l'ancienne Égypt", *Arch. Mus. d'Hist. Nat. Lyon* 10, 159-173.

22 Keimer, L. (1942). "Sur deux représentations Égyptiennes du gerenuk (*Lithocranius walleri*)", *ASAE* 41, 161-181.

23 Dawson, W.R. (1934). "Deer in ancient Egypt", *J. Linn. Soc.* 39(264), 137-145; Houlihan, P.F. (1987). "Some remarks on deer (Cervidae) in ancient Egypt", *JEA* 73, 238-243.

species[30] (e.g., geese,[31] ibis,[32] hoopoes,[33] pied kingfishers,[34] and guineafowl[35]), chameleons,[36] fish,[37] beetles,[38] and grasshoppers.[39]

Species-specific studies of this kind have declined in frequency as the number of unidentified animals in Egyptian art has diminished; nevertheless, representations of unusual species, and those for which the identification remains undecided or contentious, have continued to attract scrutiny, such as: otters,[40] foxes,[41] zorillas,[42] Arabian oryx,[43] duikers,[44] lizards,[45]and praying mantids.[46] In recent years, the publication of general overviews of ancient Egyptian zoology[47] and monographs summarising current knowledge about individual species (e.g., dorcas gazelle,[48] hedgehogs,[49] hares,[50] turtles,[51] scorpions,[52] etc.), as well as entire taxonomic classes of depicted species, such as birds,[53] fish,[54] and mammals,[55] have indicated continued interest among Egyptologists in this line of investigation.

---

[24] Keimer, L. (1935). "Le sanglier Égyptien", *Cd'É* 10(19), 26-33; Keimer, L. (1937). "Remarques sur le porc et le sanglier dans l'Égypte ancienne", *BIE* 29, 147-156.

[25] Garnot, J.S.F. (1937). "Le lion dans l'art Égyptien", *BIFAO* 37, 75-91.

[26] Jèquier, G. (1913). "La panthère dans l'ancienne Égypte", *RES* 4, 353-372.

[27] Keimer, L. (1948). "*Inb*, le caracal (*Caracal caracal* Schmitzi Matsch.) dans l'Égypte ancienne", *ASAE* 48, 373-390.

[28] Vandier d'Abbadie, J. (1964). "Les singes familiers dans l'ancienne Égypte (peintures et bas-reliefs). I. L'ancien empire", *Rd'É* 16, 147-177; Vandier d'Abbadie, J. (1965). "Les singes familiers dans l'ancienne Égypte (peintures et bas-reliefs). II. Le moyen empire", *Rd'É* 17, 177-188; Vandier d'Abbadie, J. (1966). "Les singes familiers dans l'ancienne Égypte (peintures et bas-reliefs). III. Le nouvel empire", *Rd'É* 18, 143-201; Houlihan, P.F. (1997). "Harvesters or monkey business?" *GM* 157, 31-47.

[29] Keimer, L. (1949). "Le porc-épic dans l'Égypte ancienne", *ASAE* 49, 393-415.

[30] Moreau, R.E. (1930). "The birds of ancient Egypt" in Meinertzhagen, *Nicoll's Birds*, 57-77.

[31] Gaillard, C. (1907). "Les oies de Meidum", *RdÉ* 12, 212-215.

[32] Keimer, L. (1954). "Interprétation de plusieurs représentations anciennes d'ibis", *Cd'É* 29(58), 237-250.

[33] Keimer, L. (1931). "Quelques remarques sur la huppe (*Upupa epops*) dans l'Égypte ancienne", *BIFAO* 30, 305-331.

[34] Gaillard, C. (1931). "Quelques représentations du martin-pêcheur pie (*Ceryle rudis*), sur les monuments de l'Égypte ancienne", *BIFAO* 30(1), 249-271.

[35] Davies, N.M. (1940). "Some notes on the NH-bird", *JEA* 26, 79-81.

[36] Keimer, L. (1936). "Sur quelques représentations de caméléon de l'ancienne Égypte", *BIFAO* 36, 85-95.

[37] Montet, P. (1914). "Les poissons employés dans l'ecriture hiéroglyphique", *BIFAO* 11, 39-48; Gaillard et al., *Recherches sur les Poissons.*

[38] Keimer, L. (1931). "Pendeloques en forme d'insectes faisant partie de colliers Égyptiens A. Pendeloques en forme de coléoptères", *ASAE* 31, 145[1]-186[42].

[39] Keimer, L. (1932). "Pendeloques en forme d'insectes faisant partie de colliers Égyptiens B. Pendeloques et pieces de colliers en forme de sauterelles", *ASAE* 32, 129[43]-150[64].

[40] Houlihan, P.F. (1996). "On Herodotus (II, 72) and the question of the existence of the otter (*Lutra* sp.) in ancient Egypt", *GM* 153, 45-53; Bouvier-Closse, K. (2005). "Une nouvelle représentation de loutre", *GM* 204, 21-36; Evans, L. (2010). "Otter or mongoose? Chewing over the evidence" in Woods et al., *Egyptian Culture and Society*, 119-129.

[41] Janssen, J.J. (1990). "On the scent of a fox", *DE* 16, 43-51.

[42] Aufrère, S. (1985). "Études de lexicologie et d'histoire naturelle VII: De nouvelles representations de zorilles de Libye, *Poecilictis libyca* Lataste", *BIFAO* 85, 23-32.

[43] Manlius, N. (2000). "Did the Arabian oryx live in Egypt during pharaonic times?" *Mam. Rev.* 30(1), 65-72.

[44] Manlius, N. (2001). "Were there duikers in ancient Egypt?" *Science* 291(5509), 1701.

[45] Beaux, N. and Goodman, S.M. (1993). "Remarks on the reptile signs depicted in the White Chapel of Sesostris I at Karnak", *CdK* 9, 109-120; Evans, L. (2002). "Poisonous geckos: The validity of an ancient Egyptian belief", *BACE* 13, 47-55.

[46] Evans, L. (2004). "The praying mantis in ancient Egypt", *BACE* 15, 7-18.

[47] For example, Boessneck, *Tierwelt*; Houlihan, *Animal World*; Germond, *Egyptian Bestiary*.

[48] Standberg, *Gazelle*.

[49] von Droste zu Hülshoff, *Der Igel*.

[50] Stoof, *Hasendarstellungen*.

[51] Fischer, *Turtles*.

[52] Stoof, *Skorpion und Skorpiongöttin*.

[53] Houlihan, *Birds*.

[54] Brewer-Friedman, *Fish and Fishing*; Gamer-Wallert, *Fische und Fischkulte*.

[55] Osborn-Osbornová, *Mammals*.

Many of the animals displayed in wall scenes are not exotic, however, but more familiar domestic species (e.g., cattle, donkeys, sheep, and goats). These are immediately recognizable and so do not require scientific identification. Scholars tracing the history of domestication have nevertheless examined the physical appearance of these animals in art for confirmation of morphological changes noted in zooarchaeological remains.[56] For example, differences in the horn-shapes of depicted cattle[57] and sheep[58] have been attributed to the presence of sub-species in Egypt, as well as the effects of controlled breeding.[59] Illustrations of cats[60] in wall scenes and the physical diversity of represented dogs[61] have similarly revealed the Egyptians' apparent relationship with companion animals. Images that appear to show the domestication of uncommon species, such as desert ungulates and striped hyenas, have also been the subject of much speculation.[62]

Concurrent with studies of individual species, scholars have also turned to wall reliefs to understand the function of animals in Egyptian society. In investigations of this kind, the context in which they are depicted is often of more importance than the animal images themselves. The degree to which animals were a part of Egyptian life has meant that they figure prominently in everyday scenes reproduced on tomb walls; the manner in which they are depicted has consequently provided practical information that would otherwise be unattainable. Beginning with the early expositions of Wilkinson,[63] and Erman,[64] and followed later by the influential studies of Montet[65] and Vandier,[66] Egyptologists have consistently drawn upon representations of animals to understand and clarify how many routine tasks were carried out, particularly those associated with food production. These include agricultural tasks (e.g., ploughing, sowing, threshing), pastoral activities (e.g., herding, breeding, calving, milking, force-feeding),[67] methods of slaughter and butchering,[68] hunting,[69] trapping,[70] fishing,[71] and bee-keeping.[72] The physical appearance of some animals, such as cattle with distorted horns[73] or brand-marks, have also hinted at idiosyncratic husbandry procedures. Animals represented in the domestic sphere,[74] and activities associated with leisure, such as bull-

56 Boessneck, *Die Haustiere in Altägypten.*

57 Dürst, *Die Rinder*; Andersson, E. (1912). "La dénomination égyptienne des boeufs sans cornes", *Sphinx* 16, 145-164.

58 Thilenius, G. (1900). "Das Ägyptische Hausschaf", *RT* 22, 199-212; Dürst, J.U. and Gaillard, C. (1902). "Studien über die Geschichte des Ägyptischen Hausschafes", *RT* 24, 44-76.

59 Brewer, et al., *Domestic Plants and Animals*, 82-90.

60 de Morant, H. (1937). "Le chat dans l'art Égyptien", *Cd'É* 12, 29-40; Langton-Langton, *The Cat in Ancient Egypt*; Arkell, A.J. (1962). "An early pet cat", *JEA* 48, 158; Baldwin, J.A. (1975). "Notes and speculations on the domestication of the cat in Egypt", *Anthropos* 70, 428-448; Malek, J. (1993). *The Cat*; Aufrère, S.H. (1999). "Notes et remarques au sujet du chat", *DE* 44, 5-18; Linseele, V., van Neer, W. and Hendrickx, S. (2007). "Evidence for early cat taming in Egypt", *J. Arch. Sci.* 34, 2081-2090.

61 Houtart, A. (1933-1934). "Les chiens dans l'ancienne Égypte", *Cd'E* 9, 28-34; Handoussa, T. (1986). "Le chien d'agrément en ancienne Égypte", *GM* 89, 23-41; Brewer et al., *Dogs in Antiquity*; Rice, *Swifter than the Arrow*.

62 Gaillard, C. (1912). "Les tâtonnements des Égyptiens de l'ancien empire à la recherche des animaux à domestiquer", *RES* 3(11-12), 329-348.

63 Wilkinson, *Manners and Customs.*

64 Erman, *Ägypten.*

65 Montet, *Scènes.*

66 Vandier, *Manuel.*

67 Brewer, et al., *Domestic Plants and Animals.*

68 Montet, P. (1910). "Les scènes de boucherie dans les tombes de l'ancien empire", *BIFAO* 7, 41-65; Eggebrecht, *Schlachtungsbräuche.*

69 Wreszinski, W. (1932). "Löwenjagd im Alten Ägypten", *Morgenland* 23, 3-27; Brunner-Traut, E. (1959). "Jagd im Alten Ägypten auf das Wild der Wüste und die Grosstiere im Nil", *AH* 67(6), 210-215.

70 Dunham, D. (1937). "Two parallels to ancient Egyptian scenes: The bird trap", *BMFA* 35, 52-54; Grdseloff, B. (1938). "Zum Vogelgang im alten Reich", *ZÄS* 74, 136-139; Scott, N.E. (1940). "An Egyptian bird trap", *BMMA* 35(8), 163-164; Altenmüller, *Jagd im Alten Ägypten.*

71 Daumas, F. (1964). "Quelques remarques sur les représentations de pêche à la ligne sous l'ancien empire", *BIFAO* 62, 67-85; Brewer-Friedman, *Fish and Fishing*; van Elsbergen, *Fischerei im Alten Ägypten.*

72 Kuény, G. (1950). "Scènes apicoles dans l'ancienne Égypte", *JNES* 9(2), 84-93.

73 Schwabe, C.W. (1984). "A unique surgical operation on the horns of African bulls in ancient and modern times", *Agric. Hist.* 58(2), 138-156.

74 Evans, L. (2000). "Animals in the domestic environment" in Donovan-McCorquodale, *Egyptian Art*, 73-82.

fighting[75] and other forms of entertainment, have provided valuable insights about the personal lives of the Egyptians, as well as their treatment of, and attitude towards, particular species.

The animals in wall reliefs have been considered a potential source of information about the religious and symbolic beliefs of the Egyptian people. For example, scholars have sought an erotic meaning in the placement of particular species, such as monkeys, cats, and geese, in relation to the human figures with which they share some scenes.[76] The inclusion of symbolically significant animals, such as the tilapia fish, in tomb decoration may have served as visual allusions to ideas about rebirth and regeneration,[77] while recurrent images of conflict between animals and humans, such as the harpooning of hippopotamuses, fowling in the marshes, and the desert hunt, have been interpreted as iconographic references to the deceased's efforts in overcoming chaotic forces.[78] The abundance and procreative activities of animals represented upon the walls of royal mortuary temples, such as that of Niuserre, have also provided insights about the nature of sun worship and the cult of Re in the Old Kingdom.[79]

Animal motifs have also been examined for the purpose of tracing the development of scene content and stylistic changes. As the available wall space slowly increased in Old Kingdom tombs, so the repertoire of scenes was expanded by the addition of small motifs illustrating various events and activities. Many of these depict animals that adopt distinctive poses or are engaged in unique behaviours - small cameos that enriched the major scene types, such as those depicting agricultural, pastoral, and marsh pursuits. These images provide useful data-points with which to date and compare tombs. For example, it has been noted that the motif of a dorcas gazelle nursing her calf was first introduced into desert scenes during the late 4th dynasty,[80] while that of a hippo attacking a Nile crocodile was not included in marsh scenes until the late 5th.[81] Animal motifs have thus been described and catalogued by many scholars as part of their iconographic analyses of tomb scenes.[82]

1.3 ANIMAL IMAGES IN EGYPTIAN WALL SCENES: FORM

Depictions of animals in Egyptian art have been widely acclaimed for their aesthetic qualities.[83] The diversity of species combined with their various and lively poses offer general visual appeal, but scholars and laymen alike have marvelled at the Egyptians' particularly well-observed and sympathetic handling of the animal form. Indeed, animal images are frequently quite beautiful. The careful yet confident treatment of their physical attributes - body-shape, proportions, musculature, species-specific attributes and markings, and distinctive outer coverings of feathers, fur, and scales - indicate close and thoughtful study by the artists of their natural models, as well as an exceptional technical ability to reproduce these features effectively, both in relief and paint. As William Stevenson Smith noted, ". . . the Egyptian was interested in the outward look of living and inanimate things, and consequently endeavoured to perfect his skill in indicating their surface details".[84] Artists were clearly adept at

75 Kanawati, N. (1991). "Bullfighting in ancient Egypt", *BACE* 2, 51-58; Galan, J. (1994). "Bullfight scenes in ancient Egyptian tombs", *JEA* 80, 81-96.

76 Derchain, P. (1975). "La perruque et le cristal", *SAK* 2, 55-74; Derchain, P. (1976). "Symbols and metaphors in literature and representations of private life", *RAI News* (15), 7-10; te Velde, H. (1982). "The cat as sacred animal of the goddess Mut" in van Voss et al., *Studies in Egyptian Religion*, 127-137. For a critical appraisal of the apparent symbolism in marsh scenes, see Feucht, E. (1992). "Fishing and fowling with the spear and the throw-stick reconsidered" in Luft, *Festschrift Kákosy*, 157-169; van Walsem, *Iconography*, 72-74.

77 See Dambach, M. and Wallert, I. (1966). "Das Tilapia-Motiv in der altägyptischen Kunst", *CdÉ* 41(82), 273-294.

78 Säve-Soderbergh, *Egyptian Representations of Hippopotamus Hunting*.

79 Edel, E. (1961). "Zu den Inschriften auf den Jahreszeitenreliefs der 'Weltkammer' aus dem Sonnenheiligtum des Niuserre Part I", *NAWG* 1961, 209-255; Edel, E. (1963). "Zu den Inschriften auf den Jahreszeitenreliefs der 'Weltkammer' aus dem Sonnenheiligtum des Niuserre Part II", *NAWG* 1963, 89-217. See also Assmann, *Search for God*, 55-57.

80 Smith, *HESPOK*, 170.

81 Harpur, *Decoration*, 196.

82 For example, the Leiden Mastaba project. See van Walsem, *Iconography*.

83 ". . . the attention which Egyptian artists of the historical period devoted to recording the animal world excites universal admiration" (Schäfer, *Principles*, 153).

84 Smith, *Art and Architecture*, 17.

capturing the essential and defining qualities of different species,[85] but their need for morphological accuracy had, nevertheless, to be balanced with the demands of the Egyptian method of representation. Like the other elements in wall scenes, such as human figures, inanimate objects, plant forms, and the landscape, animals have been examined by scholars in order to learn how artists, working within these constraints, adapted and manipulated their images.

In his definitive study, *Principles of Egyptian Art*, Heinrich Schäfer drew upon a myriad of examples to illustrate the rules he had detected for the representation of two- and three-dimensional objects. Animal figures were among those offered in support of his fundamental premise that Egyptian art is "based on frontal images".[86] According to this conceptual interpretation, artists composed their representations from mental pictures of what they felt were the most important or characteristic elements of an object. Foreshortening and oblique angles were spurned as these defined views misled the eye about the true form and nature of a depicted object. What mattered was to represent things as the artists knew them to be, rather than how they appeared.

This primary objective affected how the bodies of animals were rendered. Schäfer concluded that the standard side-view by which virtually all quadrupeds were presented in Egyptian two-dimensional art enabled their "essential characteristics" to be seen, such as all four legs, their general body-shape, and distinctive features, such as horns.[87] A purely lateral view of four-footed animals was consequently rare because it misrepresented the number of legs. So, too, bipeds (humans and birds) were rendered so that both limbs were clearly visible.[88] Animal figures also revealed how the artists' need to convey information unambiguously drove the evolution of a "language of forms", for while crocodiles were initially represented in either a fully lateral view or from above with all four legs stretched out, eventually only the former image was acceptable for this particular species and the latter employed exclusively for lizards, to thereby differentiate the two.[89]

The position of physical features such as horns might have presented a representational challenge, but, according to Schäfer's interpretation, the problem of intersecting planes was easily accommodated by creating a composite image that was made up of various frontal views (as the human form was constructed). Nevertheless, he conceded that this method could potentially cause confusion. The frontal view of the hood of a raised cobra, for example, gave the false impression that it was inflated, rather than flared, while the tail position of depicted scorpions suggested that it hung to one side of their body, rather than hovered above it.[90] The animal form could also be adapted in other ways. Artists often took advantage of the longer neck length of some animals to display them with their heads fully reversed (either above their bodies or along their flanks) to convey a turn to the side.[91] For Smith, the head turned back along the flanks was a "definite piece of fore-shortening".[92]

Despite their conformation with the rule of frontal images, it is nevertheless widely accepted that the Egyptian depiction of animals was freer and less complicated than that of humans. For example, it has been asserted that, "the coherence of animal forms in drawing is a simpler matter than that of human beings",[93] that, "Drawing an animal was not regulated by precisely the same standards used for the representation of the human figure",[94] and also that, "il semble qu'ils ont été un peu plus expressifs avec les animaux; peut-être les règles étaient-elles moins rigides pour ceux-ci que pour les humains".[95] In short, it is generally agreed that, "the Egyptian artist was sensitive to

85 Although Groenewegen-Frankfort (*Arrest and Movement*, 41) suggests that the reliefs fail to present animals as "impressive".
86 Schäfer, *Principles*, 91.
87 *ibid.*, 104. For Groenewegen-Frankfort (*Arrest and Movement*, 7-8), the profile view of animals was the result of practical, rather than conceptual, concerns - the product of their symmetrical body shape, which enabled their long axis to run parallel to the surface plane.
88 Schäfer, *Principles*, 104-106.
89 *ibid.*, 150.
90 *ibid.*, 144-145.
91 *ibid.*, 215-216.
92 Smith, *HESPOK*, 174.
93 Groenewegen-Frankfort, *Arrest and Movement*, 41.
94 Davis, *Canonical Tradition*, 39-40.
95 Baud, *Le Caractère du Dessin*, 82.

nature and allowed himself more freedom in representing it".[96] This notion is supported by studies that have revealed that animal figures were not subject to the 'canon of proportions' - the use of squared grids to depict objects in an exact and repeatable manner. This measurement technique was particularly suitable for drafting the static and stiffly dignified postures of human figures (standing, sitting, and kneeling), which were standard elements in formal art, but it was not amenable to the various and lively poses expressed by animals in informal wall scenes.[97]

The location of animal figures within space was also dictated by representational conventions. Like other elements in Egyptian wall scenes, animals were usually firmly attached to the baseline, regardless of their activities or apparent motion. Yet despite this constraint, aspects of depth could still be communicated. For example, dense groupings of animals (such as herds) were depicted as laterally layered, closely overlapping figures. Although the feet of these animals remained in contact with the substrate, Schäfer suggested that the motif, in contrast to vertically layered groups, was intended to convey forward motion.[98] The distribution of animals in the picture field might also offer clues to the spatial layout. For example, Klebs was struck by depictions of goats that feed from the tops of shrubs while situated upon elevated baselines, and concluded that their location revealed that the trees must be perceived as fallen and thus laying away from the viewer.[99] The position of fish and other animals depicted underwater, although giving the appearance of a section through the water column, were nevertheless, for Schäfer, also examples of frontal images viewed from above.[100]

The manner in which animal locomotion was represented has also been explored. Schäfer noted that initially only two forms of quadrupedal gait were possible within the Egyptian method - the walk (in which the legs are parted) and the gallop (in which the front legs move together, as do the back legs). The depiction of these was affected by the pull of the baseline, such that in both motifs, the limbs remained in contact with the substrate. A 'jumping gallop' was later added to the repertoire by the simple modification of elevating the front hooves of a galloping animal.[101] None of these actions was compared by Schäfer with those of living creatures, however Edgerton proposed, after examining sequential photographs of moving animals, that the gallop pose may have been inspired by an observation of flying leaps - distinctive jumps made by some species while running in which the body is elongated and all four legs are lifted off the ground and outstretched.[102] This bounding, elevated run provided the model for the 'flying gallop', a motif that was borrowed from Aegean cultures, and which was incorporated frequently in New Kingdom hunting scenes "for the representation of vigorous motion".[103]

The depiction of avian flight was found to conform to a standard convention in which the wings remained in full view by attaching them to the upper and lower edges of the bird's torso shown in profile. For Schäfer, the placement of the wings and the variable manner in which they are adhered to the body were, despite appearances, unlikely to be attempts at a naturalistic representation of flight.[104] Similarly, Smith concluded that images of flying birds followed two conventions: that noted by Schäfer, and another in which the closer wing was lowered across the body, while the other was folded down behind it.[105]

---

96 Kanawati, *Tomb and Beyond,* 83.

97 Robins, *Proportion and Style*, 23. According to Panofsky, however, squared grids were occasionally employed for the representation of animal figures. See Panofsky, E. (1955). "The history of the theory of human proportions as a reflection of the history of styles" in Panofsky, *Meaning in the Visual Arts*, 60.

98 Schäfer, *Principles*, 187-188.

99 Klebs, L. (1914). "Die Tiefendimension in der Zeichnung des Alten Reiches", *ZÄS* 32, 24. See *Goats* section 7.1 in Chapter 7 for a full discussion of these scenes.

100 Schäfer, *Principles*, 239.

101 *ibid.*, 15.

102 Edgerton, W.F. (1936). "Two notes on the flying gallop", *JAOS* 56(2), 184. See section 4.1.3 in Chapter 4 for further details.

103 Davis, *Canonical Tradition*, 86. See also Kantor, H.J. (1947). "The Aegean and the orient in the second millennium B.C.", *AJA* 51(1), 62-64.

104 He observed, however, that the bodies of flying waterfowl are invariably depicted at an angle, in a posture reminiscent of flushing upward. See Schäfer, *Principles*, 260, n. 3.

105 Smith, *HESPOK*, 129.

Representational constraints did not prevent the inclusion of zoological details. Surviving examples of wall paintings and painted relief reveal that the hue and texture of their body coverings were added to the figures of animals via a deft use of colour and line.[106] Although some images are remarkably accurate in their rendering of species-specific colouring (e.g., the Meidum geese[107]), others show that the pigments chosen for animals could be dictated by convention.[108] It has been suggested, however, that the use of unnatural colours for the depiction of some birds and fish were attempts to convey the reflective qualities of their iridescent feathers[109] and scales.[110] Aspects of the natural history of animals were also expressed via conventions. For example, although the baseline could not be pierced, artists were able to suggest the idea of burrows and cavities in the ground by projecting the landscape upward into a humped shape from which animals were shown either emerging or entering.[111]

Despite their attempts at maintaining a degree of accuracy in their representations of animals,[112] Schäfer noted that artists sometimes made overt mistakes, citing, for example, the depiction of a bee hieroglyph with a beak for stinging. Such errors, however, simply confirmed for him "the role in which thought rather than observation plays in the Egyptian artist's approach".[113]

1.4 THE MISSING ANIMAL

Animals in wall scenes have thus been studied extensively in order to gain insights about ancient Egyptian life and culture, and to deduce some of the fundamental rules of two-dimensional art. Nevertheless, a consistent oversight has been detected in both lines of investigation: namely, the *living* animals upon which the images were originally modelled have not been appreciated fully. This may seem an incongruous statement, as clearly neither the function nor the appearance of depicted fauna could have been described without reference to their living counterparts. Despite this general acknowledgement, however, scholars have been inclined to examine animal images almost exclusively from an anthropocentric viewpoint and indeed, have only considered animals relevant to the extent that they could inform about the lives and activities of the Egyptian people. This human-oriented bias has prevented animal images from being investigated in their own right, independently and objectively, and has therefore limited our understanding of both their meaning and the method of their representation.

Although many studies have analysed depictions of animal-related activities, it is frequently the human participants, rather than the animals, that receive the most thorough treatment. For example, when reviewing calving scenes, Jacques Vandier describes at length the postures of the herdsmen who are in attendance, but reference to the cows is both brief and uncritical.[114] The reader is consequently well informed about the various positions of a staff held by a nearby overseer, but learns little about the animals upon which these scenes are focused.

The physical appearance of depicted animals is often measured against that of their living models, but comparison between their represented behaviour and actual habits is rare. When this does occur, conclusions are often based upon generally held folk beliefs, rather than biological facts. For example, in his description of the 'Battlefield palette', Whitney Davis suggests that a male lion that bites the body of a dead warrior must represent the king. The dead man is seen here as, "the special target of the ruler's leonine power, his ferocity and implacability 'as a lion'".[115] Davis notes that, "It is conceivable that the image should be taken literally at this point . . . as depicting a lion actually mangling the abandoned body of one of the executed enemies he encounters in scavenging. But . . .

106 *ibid.*, 267-270.
107 Goelet, O. (1983). "The migratory geese of Meidum and some Egyptian words for "migratory bird", *BES* 5, 41-48.
108 Smith, *HESPOK*, 257-258; Williams, *Decoration of the Tomb of Per-Neb*, 38-74.
109 For example Keimer, L. (1930). "Quelques hiéroglyphes représentant des oiseaux", *ASAE* 30, 19; Gaillard, C. (1933). "Identification de l'oiseau *Ama* figuré dans une tombe de Béni-Hassan", *BIFAO* 33, 170-174.
110 Keimer, *Ann. Fac. Lett. Univ. Ibrahim* 2(May), 130-131.
111 Schäfer, *Principles*, 126-127.
112 Baines, J. (1985). "Theories and universals of representation: Heinrich Schäfer and Egyptian art", *Art Hist.* 8, 10.
113 Schäfer, *Principles*, 262.
114 Vandier, *Manuel v*, 64-67.
115 Davis, *Masking the Blow*, 139.

the lion is not naturally a scavenger; rather, he attacks and kills his own prey".[116] For his hypothesis to succeed it is not possible for the lion to be picking over the remains of fallen soldiers; rather, to symbolize the ruler, it must be seen to be attacking and killing an enemy. Davis' supposition rests, however, upon a generally held, but incorrect, belief about lion behaviour, for although lions hunt occasionally, they also scavenge whenever the opportunity arises.[117]

Insufficient care is also taken in the terminology used to describe animal depictions; this lack of zoological precision reveals an underlying dismissive attitude towards such images. For example, the terms "gazelle" and "antelope" are frequently used interchangeably when identifying the horned ungulates in wall scenes.[118] Yet both of these labels refer to groups of animals rather than specific species, and while gazelles are a type of antelope, an antelope is not necessarily a gazelle.[119]

Anthropocentric bias is also apparent in studies exploring the Egyptian method of representation. Indeed, the analysis of Egyptian art has always been heavily skewed towards understanding the human form, which Schäfer referred to as "the most important object represented in two dimensions".[120] Animals, like other objects in wall scenes, have clearly furnished examples of the rules of representation, but they have not been afforded the kind of systematic analysis to which human figures and actions have been subjected. In some ways, this is understandable; it is natural for us to be especially drawn to depictions of humans and to be highly motivated to understand their particularly unusual treatment in Egyptian art. I suggest, however, that the amount of attention that has been devoted to animal figures is disproportionately small.[121] In particular, it has failed to match their prevalence in wall scenes. As a result, our understanding of the Egyptian representation of animate beings is incomplete.

It appears, also, that when determining how animals were represented, the assessment by scholars of specific postures and actions (e.g., gait) has been based upon their general beliefs about animals, rather than informed comparison with the movements of living exemplars. The intended meaning of some of the actions exhibited by animal figures may therefore have been misunderstood or even overlooked. Finally, as noted above, it is widely believed that the Egyptians' rendering of animals was freer and not as rule-bound as that of the human form. It is possible, however, that the sheer volume and variety of depicted species has masked our ability to detect patterns of presentation. I suggest that until animal images have been examined in greater depth, such a conclusion cannot be drawn with confidence.

Why have animal images received such superficial treatment? The most likely answer is that scholars have been inhibited by their lack of knowledge about animals generally, and the subjects of Egyptian art in particular. Certainly, few will have had the opportunity to witness firsthand the many African species represented in wall scenes. Unfamiliarity may consequently have rendered the representations of animals less noteworthy than other elements. I suggest also that the method by which animal images have been investigated in the past has itself affected how they have been subsequently perceived and analysed. As noted above, the animals represented in tomb scenes and hieroglyphic signs were identified by comparing their physical appearance with those of contemporary fauna. The exemplars of living animals to which researchers had access consisted of skeletons, mounted specimens, line drawings, colour plates, and photographs. When considering Egyptian depictions of animals, scholars have thus effectively compared like with like - that is, static image with static image. I believe that, as a consequence of this limited and somewhat sterile experience, they have tended to undervalue the animals represented in Egyptian art because they do

[116] *ibid.*, 132.

[117] Estes, *Behavior Guide*, 370. Indeed, long-term field studies have shown that lions acquire most of their food by displacing other species from their kills. Furthermore, it is lionesses, not male lions, which carry out most hunts.

[118] Osborn-Osbornová, *Mammals*, vii.

[119] *ibid.* For a recent example of such inconsistent use in terminology, see Frankfurter, D. (2004). "The binding of antelopes: A Coptic frieze and its Egyptian religious context", *JNES* 63(2), 97-109.

[120] Schäfer, *Principles*, 277.

[121] It is interesting to note that of the 328 figures provided for referral in Schäfer's, *Principles*, only 73 (22%) depict or include the image of an animal.

not respond to them as depictions of dynamic, *living* creatures.[122] This handicap will have been particularly acute for early Egyptologists and art historians, but even though our experience of and exposure to the animal kingdom is now enriched by film, I think it is likely that for most scholars today, animals remain an essentially alien concept.

1.5 BEHAVIOUR AS A KEY TO UNDERSTANDING ANIMAL IMAGES

Their faithful reproduction of the physical appearance of animals reveals close observation of their living models by Egyptian artists. Even if they did not always copy directly from nature, the constant exposure and contact with the animal world that was experienced generally by the Egyptian people will have ensured that most artists were familiar with the characteristics of many species. Attempts to analyse and understand their images without ourselves being fully aware of the animals that inspired them are hence likely to yield incorrect conclusions.

Indeed, the vital nature of animals is thoroughly incorporated into ancient Egyptian images, for they are not presented in wall scenes as rigid, frieze-like objects, but in postures and situations that seem designed to emphasize their movement. In other words, they appear as they do in nature - they *behave*. This is not surprising, for the way that different species conduct themselves is as much a defining characteristic as their appearance. Artists clearly sought to illustrate the morphology of animals as accurately as possible, so it is highly likely that they will have endeavoured equally to reproduce their idiosyncratic movements and habits. What animals *do* in tomb scenes is thus potentially as significant as how they look;[123] this has important implications for the way that their images can and should be analysed.

I propose that the behaviour exhibited by depicted animals provides a key by which to understand more fully both the representational method of Egyptian artists and the intended meaning of their images. This approach forces an awareness of living animals because it focuses upon those aspects that are a direct expression of their animated state.

First, comparison of the behaviour of depicted animals with that of contemporary fauna allows an objective measure by which to judge how artists translated and transformed identifiable dynamic actions into two-dimensional images. The way in which artists chose to convey the discrete movements of particular species may thus be determined with confidence because their images are compared directly to the actual behaviour of living exemplars.

The application of biological information to an evaluation of animal images also ensures a more accurate identification of their actions; indeed, uninformed descriptions based simply upon what depicted animals appear to be doing are likely to yield incorrect or anthropocentric conclusions, especially if images reproduce precise species-specific details. Accordingly, it is clear that the only way to determine objectively what artists were attempting to represent is by reference to the behaviour of live animals. Furthermore, the behavioural features chosen for representation will also inform about the Egyptians' general knowledge of natural history, as well as provide valuable insights about what they noticed and believed about particular species. Behavioural details also have the potential to clarify the interpretation of some depictions that, on face value, appear to be incorrect or anomalous.

This approach treats depicted behaviour as a graphic reference to actual events in the real world, however studies concerned with the symbolic and mythological content of wall scenes have shown that these may sometimes present a distorted version of reality, in which the actions of the participants and the objects chosen for inclusion were affected by the need to convey particular

---

122 In fact, often the only physical contact experienced by Egyptologists with animals is when handling their decidedly static bones and mummified remains!

123 "Although Western scholars frequently focus on the physical form of animals when inspecting the symbolic character of the animal, it is often the ecological status of an animal and even more frequently the nature of its movement, that forms the basis for animal symbolism . . .". See Walens, S. (1987). "Animals" in Eliade, *Encyclopedia of Religion i*, 292.

ideas.[124] Moreover, Egyptian representations have been shown to be conceptual and so do not necessarily correspond to visual impressions of tangible objects. It is also likely that artists reduced many figures to their most essential elements, stripping them of features they deemed irrelevant.[125] It might therefore be argued that these attributes will undermine any attempt to compare real with illustrated behaviour. In fact, an objective assessment of this kind has the potential to reveal precisely where misrepresentations occur - to determine where artists manipulated or modified reality - for either the behaviour exhibited in wall scenes will be appropriate for the species depicted and the contexts in which it occurs, or it will not. Indeed, this scientific approach avoids the problem of the interpreter, "reading into a scene what he wants to deduce from it and of explaining the absence or presence of features which do not tally with his interpretation as artistic incompetence".[126]

Heinrich Schäfer stressed the importance of abandoning our "instinctive sympathy" for Egyptian art in order to understand it properly.[127] I suggest that a focus on behaviour fosters a level of detached objectivity that will allow depictions of animals to be judged critically and dispassionately, independently of their inherent aesthetic appeal.

[124] Robins, *DE* 17, 45-58.

[125] Weeks, K.R. (1979). "Art, word, and the Egyptian world view" in Weeks, *Egyptology and the Social Sciences*, 68.

[126] Groenewegen-Frankfort, *Arrest and Movement*, 11.

[127] Schäfer, *Principles*, 5-6.

CHAPTER 2

THE BEHAVIOURAL ANALYSIS OF ANIMAL IMAGES

2.1 ANIMAL BEHAVIOUR DEFINED

Behaviour may be defined as an animal's active response to its environment, to members of its own or other species, and to its personal requirements. Their actions are driven, in particular, by two primary motivations: the need to survive and the impetus to reproduce. The behavioural repertoire of animals thus consists of a variety of activities that enable them to fulfil these ultimate goals, such as: finding and ingesting food, defending themselves or their territory, choosing and competing for a mate, caring for their young, and communicating with one another. The members of the same species usually carry out these behaviours in a similar ('species-specific') manner. So, for example, all dorcas gazelles mate in an identical fashion. However, separate species not only differ from one another morphologically; they usually express themselves behaviourally in very different ways too. Thus, the sexual behaviour of lions is unlike that of goats.

The behavioural traits that characterise different species have evolved over the course of many hundreds of thousands of years - the result of adaptations and modifications that occurred very slowly in response to sustained physical, social, and environmental pressures. Consequently, the behaviour of wild species today is essentially the same as it was at the dawn of civilization, because substantial changes are unlikely to have emerged in such a short span of (evolutionary) time. During this same period, however, humans have attempted to modify the behaviour of some animals through taming and the process of domestication. During taming, a wild animal is trained to become less aggressive and more tractable; its behaviour is thus muted rather than changed fundamentally. Domestication, on the other hand, involves the selective breeding of animals for particular characteristics that are regarded as useful by humans.[1] Unlike taming, this process imparts a long-term, genetic change that affects both the morphology and behaviour of animals. Domestication can also bring about indirect behavioural changes. For example, domestic fowl that have been bred for meat do not fly, not because they are incapable of flight, but due to their increased weight. Consequently, the habits of domesticated species (e.g., cattle, sheep, dogs, cats, etc.) may differ somewhat from that of their wild ancestors.[2]

When behaving, animals move their bodies in distinctive ways, strike particular poses, vocalize, and/or emit odours. These actions and the contexts in which they occur are the subject of scientific investigation within the field of biology.

2.2 THE STUDY OF ANIMAL BEHAVIOUR

The study of the natural behaviour of animals is known as 'ethology'. Ethologists endeavour to describe and understand what, how, and why animals do the things that they do. Their work involves the detailed examination of different species either in their natural habitats or within captive environments, such as zoos, wildlife parks, and laboratories. Animals are observed in a structured fashion either to further knowledge about a particular species or to answer specific questions. The types and frequencies of particular behaviours are recorded along with the conditions under which they occur (social and environmental), and patterns of activity are noted and analysed.

---

1 Allaby, *Dictionary of Zoology*, 150.

2 Clutton-Brock, J. (1994). "The unnatural world: Behavioural aspects of humans and animals in the process of domestication" in Manning-Serpell, *Animals and Human Society*, 23-35.

In order to describe the stream of activity exhibited by animals, classification schemes have been developed by which to divide their behaviour into natural categories. Scholars sometimes disagree about the divisions, but most schemes include the following broad behavioural themes: contactual (e.g., forming aggregations); ingestive (e.g., eating, drinking); eliminative (i.e., removal of body wastes); sexual (e.g., courtship, copulation); care-giving (e.g., feeding young, incubation); agonistic (e.g., threatening, fighting); allelomimetic (i.e., imitating others); and investigative (i.e., exploration of the environment).[3] These categories provide a framework by which the behavioural repertoires of different species may be described and compared. The results of these systematic studies are published in scientific papers and monographs, and eventually disseminated widely in general works of zoology. Consequently, details of the characteristic behaviour of many species are well understood and readily available.

2.3 ANIMAL BEHAVIOUR IN ANCIENT EGYPTIAN LIFE

The ancient Egyptians also studied closely the behaviour of the animals with which they shared their world. Like modern ethologists, their intense scrutiny was probably motivated by a deep fascination with the different ways that animals express themselves, as well as a desire to understand why they exhibit such strange habits, but their awareness of behaviour was also sustained by more practical concerns.

Knowledge of the behaviour of animals will have been crucial for the purpose of hunting.[4] In order to pursue their prey efficiently, Egyptian hunters needed to be able to predict the movements of animals and to know how they would react in certain situations. For example, some species are nocturnal while others are only found during the day. Some animals rely on olfactory cues to detect predators and so must be stalked from down wind, while others are more sensitive to movements or sounds. Behavioural information may have been used by hunters to trick their prey into accepting their presence or cause them to approach. For example, they may have imitated the sounds made by animals or attempted to dress and move like them.[5] Knowledge about activities, such as the spawning behaviour of fish, breeding behaviour of birds, or the seasonal migration of ungulates will also have helped to determine when some species were most abundant or accessible for hunting.

Egyptian hunters would similarly have relied upon information about animals to avoid being attacked themselves by potential predators. In fact, the Egyptian people generally must have been acutely aware of the habits of many dangerous species that inhabited the Nile Valley, such as hippopotamuses, crocodiles, lions, leopards, and snakes. To protect themselves from becoming inadvertent victims, they will have needed to know where dangerous animals lurked and congregated, how close they could be approached safely, and the situations that caused them to become aggressive.

The domestication of species could not have been accomplished without some understanding of their idiosyncratic social and sexual behaviour. Observation of the natural breeding biology of captive wild animals will have enabled Egyptian pastoralists to learn when mating was likely to take place, as well as the circumstances that either increased or decreased the probability of offspring being produced. As the species that are most readily domesticated are those that live in social groups, herdsmen must also have needed to consider the potentially disruptive effect of dominance hierarchies within their flocks and herds when housing their animals together. Indeed, many husbandry practices were probably developed by the Egyptians in direct response to the behavioural needs and attributes of their animals. For example, the manner in which different

---

3 Scott, J.P. (1956). "The analysis of social organization in animals", *Ecology* 37(2), 213-221.

4 For the importance and application of animal behaviour in tribal societies, see Blurton Jones, N. and Konner, M.J. (1976). "!Kung knowledge of animal behavior (or: The proper study of mankind is animals)" in Lee-De Vore, *Kalahari Hunter-Gatherers*, 325-348.

5 Men depicted on the 'Two Dogs palette' and 'Manchester palette' appear to wear animal masks as they approach game, while those on the 'Hunters palette' wear bushy tails attached to their belts. See Ridley, *Unification of Egypt*, pls. 9, 11, and 12.

species were handled and controlled was likely influenced by their social habits,[6] while the feeding and watering of animals will have been guided by their ingestive requirements. Herdsmen must also have looked for deviations from normal behaviour when monitoring the health of their stock.

Their remarkable qualities - uncanny instincts, extraordinary strength and speed, acute senses, and ability to perform feats of which humans are incapable, such as flight - likely suggested to the Egyptians that animals possessed a special relationship with the divine.[7] Indeed, the behavioural characteristics of some species determined their selection for the representation of specific deities. For example, Thoth, god of writing, was manifested as a sacred ibis because, when foraging for food, this bird appears to scrawl in shallow water with its curved beak. The fierce goddess Sekhmet took the form of an equally ferocious lioness.[8] And Khepri, god of the sun and creation, was represented as a scarab beetle due to its habit of rolling a ball of dung across the landscape in a manner reminiscent of the movement of the sun across the sky.[9] So, too, the symbolic significance of particular species was often derived from behavioural analogies. Thus, the tilapia fish was associated with rebirth and regeneration due to its habit of mouth-brooding - hatching and then spitting forth its young from its mouth - while the unpredictable nature of the hippopotamus rendered it an ideal symbol of chaos.

Artists will have been well acquainted with the movements, postures, sounds, and smells of animals because these attributes were a familiar part of the world they endeavoured to represent in their work. It seems inevitable, then, that they will have drawn upon this behavioural knowledge, either consciously or unconsciously, when illustrating fauna and that, consequently, some subset of these features must be recorded in their art. Thus, if their images are compared with the known behaviour of contemporary species, we may be afforded the opportunity to effectively see what these ancient observers of animals witnessed and experienced.

2.4 A BEHAVIOURAL APPROACH TO ANIMAL IMAGES

The analysis of ancient art from an ethological perspective begins by examining objectively the demeanour of represented animals. This requires inspection of both their *pose* and the *context* in which they are placed.

An animal's pose is evaluated by considering the position of every part of its body: the angle of its head, placement of its mouth, ears, tail, legs, or wings, and the orientation of its torso in space. The combination of all of these elements must be noted in order to arrive at a possible meaning. For example, a distinctive tail position in isolation might suggest a specific behavioural state, but would indicate something quite different if associated with a particular head posture. The gender of depicted figures must also be considered, as the poses adopted by males and females to express some behaviours can differ markedly (e.g., urination), while others may be exhibited by only one sex (e.g., the ritualised courtship displays of males).

The context of a depicted animal consists of its juxtaposition to, and relationship with, other nearby creatures (of the same or different species) or human figures, as well as inanimate objects in its immediate environment. Its general circumstances (i.e., the setting) will also often influence the types of behaviours that might be exhibited. For example, depictions of mating and birth would be expected to occur in pastoral scenes, but also less obvious activities, such as grooming or courtship

6 The nomadic Fulani people of sub-Saharan Africa, for example, exploit the behavioural traits of their cattle to aid their husbandry practices. Herdsmen mimic the aggressive behaviour of dominant bulls in order to establish themselves as the herd leader. Once they achieve this status, their animals can be controlled and moved with relative ease. See Lott, D.F. and Hart, B.L. (1977). "Aggressive domination of cattle by Fulani herdsmen and its relation to aggression in Fulani culture and personality", *Ethos* 5(2), 174-186; Lott, D.F. and Hart, B.L. (1979). "Applied ethology in a nomadic cattle culture", *App. Anim. Ethol.* 5, 309-319.

7 Frankfort, *Ancient Egyptian Religion*, 8-14; Hornung, *Conceptions of God*, 101; Arnold, D. (1995). "An Egyptian bestiary", *BMMA* 52(4), 3-5.

8 Sternberg, H. (1986). "Sachmet" in *LÄ v*, 324.

9 Ward, W.A. (1994). "Beetles in stone", *BA* 57(4), 186-188.

displays. For some animal images, the context might also include the function of the object upon which they appear (e.g., containers, jewellery, etc.).

The pose and context of a depicted animal can usually provide enough clues to identify the general class of behaviour being represented. It is consequently possible to recognize animals engaged in such activities as: walking, running, sleeping, feeding, grooming, urinating, defecating, hunting, threatening, fighting, courting, mating, or nursing. To arrive at a more precise understanding of the behaviour depicted by individual species, however, requires consultation of the ethological literature. Theoretical papers and general references (e.g., animal encyclopaedias[10] and specialized guides devoted to either whole classes or individual species[11]) provide detailed written and visual sketches of the natural history and behaviours expressed by most of the species that are represented in ancient art. By referring to these profiles, it should thus be possible to establish with some confidence what a depicted behaviour was intended to represent and whether it is appropriate for the species exhibiting it, as well as how the artist interpreted its salient features. Recognition of the behaviour depicted may equally aid the identification of animal images for which the species cannot be determined on morphological details alone.

It is important to note that the conceptual nature of Egyptian representations presents both challenges and opportunities for a behavioural approach to ancient animal images. The constraints of their drawing method affected how artists could convey the position and movement of animate beings within two dimensions. Consequently, the depiction of some behaviours will not necessarily match precisely their appearance in nature and may instead include features that reflect an artistic compromise. Furthermore, although behaviour is most often associated with movements of the body, other striking forms of expression are also common among animals, such as vocalizations or odours. As Egyptian artists formed their images from what they knew about their subjects, it is conceivable that they attempted to represent some behavioural characteristics that are not perceived visually, but by other sensory modalities, especially if these were considered an important or defining characteristic of particular animals.[12] The representation of such non-visual attributes might be achieved through the use of symbols or codes, and identified as such in the occurrence of incongruous gestures or features for which a behavioural meaning is not immediately apparent.

2.5 PREVIOUS STUDIES

The benefit of examining the behaviour of ancient animal images has yet to be appreciated fully by scholars, as only a small number of reports have been found to use this approach. However, two European conferences held during the last 30 years, which focussed upon the application of ethology to understanding cultural artefacts, reveal a general interest in this area and illustrate the kind of work that has been carried out to date.

The meaning of palaeolithic animal imagery, found throughout Eurasia on cave walls and small portable objects, has been the subject of considerable debate. These extraordinary images depict a wide range of fauna, including deer, cattle, bison, and horses, in a lively, naturalistic style. Early studies, which were often based upon ethnographic comparisons, concluded that the representations were created for use in hunting rituals and fertility magic,[13] however later work suggested that the combination of animal subjects and their location within the cave environment indicated a symbolic association with spirituality, sexuality, and gender.[14] The lack of "palpable proofs" for any of these magico-religious theories eventually led to the search for a new approach.

---

[10] For example, Grzimek's *Animal Life Encyclopedia,* Cramp and Simmons' *Handbook of the Birds of Europe, the Middle East and North Africa*, or Resh and Cardé's *Encyclopedia of Insects.*

[11] For example, Estes' *Behavior Guide to African Mammals.*

[12] The representation of non-visual phenomena is found in other cultures. See Houston, S. and Taube, K. (2000). "An archaeology of the senses: Perception and cultural expression in ancient Mesoamerica", *CAJ* 10, 261-294; Ouzman, S. (2001). "Seeing is deceiving: Rock art and the non-visual", *World Arch.* 33(2), 237-256.

[13] Reinach, S. (1903). "L'art et la magie à propos des peintures et des gravures de l'âge du renne", *L'Anthropologie* 14, 257-266.

[14] For a summary of this work, see Leroi-Gourhan, A. (1968). "The evolution of paleolithic art", *Sci. Am.* (Feb), 59-70.

It was hoped that an examination of the images from a zoological perspective might yield new insights. This notion was explored in a symposium held in Sigriswil, Switzerland in June 1979, which was devoted to "La contribution de la zoologie et de l'éthologie à l'interprétation de l'art des peuples chasseurs préhistoriques".[15]

Despite the stated concern with ethology, few of the papers that arose from this meeting attempted to analyse in detail the behaviour of specific animal images.[16] Nevertheless, the animals in palaeolithic art were found to fall into four broad behavioural themes: aggressive, sexual, hunting, and everyday activities.[17] Conflict behaviour was found to be a common subject. It was noted, for example, that a high percentage of depicted animals are represented with their tails elevated. This posture often acts as a signal to indicate alarm or a threat. It was concluded, therefore, that this behaviour was recorded frequently because hunters likely witnessed it each time they approached their startled prey.[18] Comparison with the behaviour of modern animals also revealed examples of aggression. The lowered heads and stretched necks of some figures (e.g., bison, rhinos) were found to indicate threat displays, while images in which members of the same species stand head-to-head (e.g., mammoths, bison, and ibex) probably depict fights.[19] Indeed, the benefits of an ethological approach to palaeolithic art had been first suggested by an earlier study in which the demeanour of a stag engraved upon a section of antler discovered in Switzerland was compared to the aggressive displays of reindeer and found to adopt a head-down threat posture, rather than graze passively, as was initially assumed.[20]

Other common activities were also recognized via reference to contemporary species. For example, depictions in which male bulls or reindeer nuzzle the tail of a cow standing before them were found to represent genital sniffing, a process by which the fertility of female ungulates is assessed.[21] Distinctive postures associated with defecation were also noted.[22] In some cases, an ethological approach suggested a new interpretation for well-known images. For example, representations of bison from a cave in Altamira, Spain, had been identified as dead animals because they appear to be collapsed upon the ground. Comparison with living bison, however, revealed that the ancient paintings more likely depict them rolling in the dust, an activity in which they frequently engage.[23]

Herding was represented by large groups of overlapping figures of the same species, but some scenes have also revealed inter-species interactions. Images from France and Germany depict birds either following or at the feet of bison and horses.[24] This is a common occurrence in nature, as birds are often attracted to the insects that are flushed out of the grass by the feet of passing ungulates. Some will also ride upon the backs of ungulates in order to pluck their fur or feed upon skin parasites, an activity that is reproduced in an engraving from Altxerri, Spain.[25]

A few subsequent studies have continued to examine the natural history contained in palaeolithic animal imagery;[26] however, despite the wealth of available material and unresolved questions about

15 The meeting was organised by the Swiss Academy of Human Sciences.

16 See Bandi et al., *La Contribution.*

17 Leroi-Gourhan, A. (1984). "Le réalisme de comportement dans l'art paléolithique d'Europe de l'ouest" in *ibid.*, 75-90.

18 Guthrie, R.D. (1984). "Ethological observations from Palaeolithic art" in Bandi et al., *La Contribution*, 46-47.

19 *ibid.*, 47 and fig. 5.

20 This interpretation was first suggested in 1936, but not confirmed by an ethological study until 1968. For a discussion of this object, see Bandi, H.-G. (1984). "Contribution de la zoologie et de l'éthologie à l'interprétation de l'art des peuples chasseurs préhistoriques", *L'Anthropologie* 88, 565.

21 Guthrie, R.D. (2000). "Paleolithic art as a resource in artiodactl paleobiology" in Vrba-Schaller, *Antelopes, Deer, and Relatives*, 99-101, figs 7.1, 7.2, and 7.3.

22 Camps, G. (1984). "La défécation dans l'art paléolithique" in Bandi et al., *La Contribution*, 251-262.

23 Bandi, *L'Anthropologie* 88, 565, fig. 5 to fig. 8.

24 Guthrie in *Antelopes, Deer, and Relatives*, 102-103, figs. 7.8 and 7.9.

25 *ibid.*

26 Clottes, J., Garner, M. and Maury, G. (1994). "Magdalenian bison in the caves of the Ariège", *Rock Art Res.* 11(1), 58-70; Iakovleva, L. and Pinçon, G. (1998). "The Upper Palaeolithic sculptured ibex of Angles-sur-l'Anglin, France", *OJA* 17(3), 257-268; Guthrie, R.D. (2005). "Paleolithic artists as naturalists" in Guthrie, *Nature of Palaeolithic Art*, 52-112.

the meaning of this art, an ethological perspective has failed to be adopted consistently by prehistorians.[27] This fitful approach is also apparent in studies of the art of Mediterranean and ancient Near Eastern cultures. A meeting held at the University of Liège, Belgium in March 1992 entitled, "L'histoire de la connaissance du comportement animal", considered both artistic and textual evidence from the ancient and medieval past that revealed an awareness of animal behaviour.[28] The depiction of animals in prehistoric Europe, Mesopotamia, and the Greek world were among the subjects discussed.

It was observed, for example, that aspects of their natural behaviour determined the symbolic depiction of animals in Mycenaean art.[29] Species whose behaviour is marked by metamorphosis or hibernation (e.g., butterflies) were used exclusively as funerary symbols to ensure the deceased's transformation and survival in the afterlife. In contrast, images of animals associated with aggression and strength (e.g., lions) or dynamic movement (e.g., swimming dolphins) were usually found in contexts associated with war (such as engravings on daggers). The iconography of hunting dogs in Mesopotamian art was also examined during the meeting,[30] as well as the role of animal behaviour in the practice of Near Eastern prognostication.[31] These investigations supported a general knowledge, and application, of animal behaviour by Near Eastern peoples, a conclusion that had also been reached in an earlier analysis of Mesopotamian milking scenes.[32] This important study revealed that these images, which invariably show milked cows accompanied by their calves, indicate an awareness of the 'milk ejection reflex', a biological response whereby the sight and smell of their calf causes cows to let down their milk. The presence of the young animals was not an artistic flourish, therefore, but evidence of the application of biological knowledge to animal husbandry practices.

The art of Crete has also been subjected to ethological analyses.[33] These studies, uniquely, have attempted to gain a more definitive understanding of particular animal depictions by referring to species-specific descriptions provided in the biological literature. A distinctive characteristic of Minoan art is its focus upon the natural world. This is particularly evident in works from the late period, during which a wide range of terrestrial and marine animals became the subject of painted pottery, small sculptures, and wall frescoes. Some of the most naturalistic and well-preserved paintings come from a Minoan outpost on the nearby island of Thera (Santorini); indeed, recent investigations have confirmed that frescoes from the Theran town of Akrotiri contain unique and detailed observations of animal behaviour.

The 'Spring Fresco', found in 'Building Delta', covers three full walls and consists of an irregular, rocky landscape above which dart seven life-sized birds. Their pointed wings, tail streamers, and distinctive plumage indicate that they are European swallows (*Hirundo rustica*). Three of the birds fly alone, but the remaining swallows form two pairs - one on the west wall, the other on the north. Those on the west wall depict one bird plunging down toward its companion, while the latter, its body inverted, soars upward as though to peck the other's face. Those on the north wall fly headlong toward one another, their wings folded back behind them as their open beaks meet.

These displays have been described variously as "kissing, twittering, and cavorting eagerly in airborne courtship".[34] Two recent studies, however, have sought an explanation in the ornithological literature. The first saw the beak-to-beak contact of the birds as indicative of feeding

---

27 Nor, for that matter, have palaeobiologists and palaeontologists appreciated the value to their work of looking at ancient imagery. See Guthrie in *Antelopes, Deer, and Relatives*, 97.

28 The meeting was held at the University of Liège in Belgium. See Bodson, *L'Histoire.*

29 Laffineur, R. (1993). "Le comportement animal et son interprétation symbolique à l'époque mycénienne" in *ibid.*, 155-167.

30 Trokay, M. (1993). "Le comportement du chien illustré dans l'iconographie mésopotamienne du IVe millénaire" in Bodon, *L'Histoire*, 143-154.

31 Limet, H. (1993). "L'observation des animaux dans les présages en Mésopotamie ancienne" in Bodon, *L'Histoire*, 119-132.

32 Amoroso, E.C. and Jewell, P.A. (1963). "The exploitation of the milk ejection reflex by primitive peoples" in Mourant-Zeuner, *Man and Cattle*, 132-134. This topic is discussed in detail in Chapter 11 of this volume.

33 Masseti, M. (2000). "Did the study of ethology begin in Crete 4,000 years ago?", *Ethol. Ecol. Evol.* 12(1), 89-96.

34 Hollinshead, M.B. (1989). "The swallows and artists of room Delta 2 at Akrotiri, Thera", *AJA* 93, 341.

behaviour.[35] Swallows continue to feed their young for some time after they have left the nest and will often do so while in flight. Given the comparable size of each member of the two pairs, the images were thus viewed as the airborne feeding of adolescent offspring by their adult parents. The second study noted, however, that the tail streamers exhibited by the birds indicate that they are all adults.[36] It instead concluded that the aerial displays represent 'feather fights', ritualised aggressive interactions that take place between European swallows during their nesting period.[37] A feather fight consists of aerial combat in which the birds squabble over a single feather, passing it back and forth between them many times a second, as they tumble downward. These interactions are also often accompanied by vocalizations. The swallows in the 'Spring Fresco' appear to reproduce this noisy display, even though the feathers - the object of their fights - are missing. It has been proposed, however, that holes in the wall above each swallow pair originally held pegs to which real feathers were attached.[38]

In 'Building Beta' at Akrotiri, the walls of Room 1 display an unusual mix of life-sized subjects: six white ungulates arranged in pairs, and two adolescent boys who are boxing with one another. The animals are antelope-like, with long, thin horns (suggesting they are males), markings around their eyes and on their throats and hindquarters, and long tails ending in tufts. These physical characteristics do not match any one species alone, but have features in common with a number of animals, including the Cretan agrimi (a native goat; *Capra aegagrus cretica*[39]) and non-native species, such as oryx. The behaviour expressed by the best-preserved pair, however, has pointed to other possible identifications. The ungulates are oriented in the same direction, with the body of one overlapping the other. The animal in the foreground turns its head back to stare directly at its companion, while both hold their legs stiffly and their tails aloft.

The aggressive behaviour of many antelope species involves ritualised displays in which males attempt to intimidate one another by striking threatening poses. It has been determined that the 'antelope' in Building Beta are similarly engaged, with features of their rigid posturing matching the displays of two African species:[40] Grant's gazelle (*Gazella granti*) and impala (*Aepyceros melampas*). Both possess physical features that are similar to those of the depicted animals. Like these, Grant's gazelle display by standing parallel to one another and then turning their heads to one side.[41] It is more likely, however, that the animals depict impala.[42] Male impala display by first circling one another, then standing motionless with their tails raised and their heads turned to one side, before finally walking together in the same direction. The fresco seems to capture the entire impala sequence by matching their initial head and tail positions, as well as their subsequent parallel walk. Understanding the behaviour of the animals thus provided an explanation for their curious juxtaposition with the pair of boxing boys - all are engaged in a form of ritualised male aggression.[43]

The natural world was also celebrated in 'minor' Minoan arts, such as pottery, figurines, seals, and jewellery. One example is the so-called 'bee pendant', discovered near the palace of Mallia.[44] This curious object has generated much debate regarding the species represented. It consists of two bee-like insects that face one another symmetrically, their heads crowned by a small cage of vertical strands containing a tiny gold bead. The mandibles of each insect are pressed against a small circular object and their legs alight upon a large, granulated disc. Entomologists have suggested that

---

35 *ibid.*, 341-342.
36 Foster, K. (1995). "A flight of swallows", *AJA* 99, 414, n. 15.
37 *ibid.*, 414-415.
38 *ibid.*, 419-420.
39 Porter, R. (1996). "The Cretan wild goat (*Capra aegagrus cretica*) and the Theran "antelopes" in Reese, *Pleistocene and Holocene Fauna*, 295-315. This study stands out for its particularly thorough treatment of the natural history of depicted animals. In his consideration of the Theran paintings, the author has provided full details of the living animals' morphology and behaviour, and emphasizes the need to consult zoological data in order to understand properly the iconography of the agrimi.
40 It has been suggested that exotic species such as these may have been imported into Crete.
41 Masseti, *Ethol. Ecol. Evol.*, 91-92, fig. 5.
42 Morgan, L. (1995). "Of animals and men: The symbolic parallel" in Krzyszkowska, *Klados*, 183-184.
43 *ibid.*, 184; Morgan, L. (1998). "Power of the beast: Human-animal symbolism in Egyptian and Aegean art", *AeUI* 7, 21-22.
44 Demargne, P. (1930). "Bijoux minoens de Malia", *BCH* 54, 411-421 and pl. 19.

they more closely resemble wasps than either hornets or bees.[45] The species was thus tentatively identified as *Polistes*, the common paper wasp, which is found worldwide and nests frequently on man-made structures.[46] This proposal was not confirmed, however, until attention shifted to the behaviour expressed by the insects.[47]

First, it was noted that the disc upon which the pair rest is similar to the nest of *Polistes*, which consists of a circular platform of hexagonal cells sculpted from wood-pulp. After an egg has been deposited into each cell, the compartments are capped, changing their appearance from hexagonal to circular. The granulated disc on the 'bee pendant' is thus not unlike a "miniature model of *Polistes*' single layer paper nest".[48] Next, the small sphere between the insects' mandibles was considered. Various scholars had suggested that this might be a drop of honey, a berry, a dung pellet, or a ball of wood pulp. It was eventually determined, however, that the wasps are engaged in a common activity known as 'trophallaxis', in which a small bolus of liquefied food is passed from one individual to another. While exchanging fluid, the insects move their antennae back and forth vigorously, so rapidly that individual movements of the two feelers become blurred. It was consequently proposed that the curious structure upon the wasps' heads was intended to convey a sense of this rapid movement, the multiple filaments of the cage mimicking the antennae at different points in time, while light catching the gold bead added to the illusion of blurred motion.[49] The intricate behaviour contained within the 'bee pendant' thus not only enabled scholars to confirm the species of wasp depicted, but revealed detailed biological knowledge of these insects by its maker.

The animals in Egyptian art have not previously been subjected to an analysis of this kind. Nevertheless, a general awareness of behaviour is apparent in a few studies. Behavioural criteria have been proposed for the symbolic associations of some species.[50] Behavioural information has also occasionally clarified the meaning of certain scenes and images. For example, Edel suggested that depictions of birds and swimming fish that are found in conjunction with particular inscriptions in the sun-temple of Niuserre might represent the seasonal migrations of birds and mullet,[51] while Keimer identified curious objects emanating from the jaws of depicted hyenas and jackals as evidence of excessive salivation.[52]

Most frequently, however, scholars have alluded to the natural behaviour of animals only when drawing attention to apparent errors, either in the interpretation of particular images or in their rendering by Egyptian artists.[53] These intermittent observations are discussed later in this volume, but one study deserves special mention here for its unique approach. In 1991, Salima Ikram presented the results of an inquiry in which she focussed exclusively upon the depiction of a single behaviour - namely, the mating style of animals in Egyptian wall scenes.[54] The images that she examined were catalogued and described in detail, and then assessed for their biological accuracy by comparison with scientific reports of the behaviour of contemporary species. Her systematic procedure enabled her to conclude that most of the animals in tomb scenes are shown mating in

45 For an argument in favour of bees, despite their appearance in the pendant, see Hood, S. (1976). "The Mallia gold pendant: Wasps or bees?" in Emmison-Stephens, *Tribute*, 59-72.

46 Richards, O.W. (1974). "The Cretan 'hornet pendant", *Antiquity* 48, 22; Davenport, D. and Richards, O.W. (1975). "The Cretan 'hornet", *Antiquity* 49, 212-213.

47 LaFleur, R., Matthews, R.W. and McCorkle Jr, D.B. (1979). "A re-examination of the Mallia insect pendant", *AJA* 83(2), 208-212. However, for an alternate view, see also Kitchell, K.F. (1981). "The Mallia 'wasp' pendant reconsidered", *Antiquity* 55, 9-15.

48 LaFleur et al., *Antiquity* 83(2), 210.

49 *ibid.*, 211.

50 For example, tilapia (see Dambach, M. and Wallert, I. (1966). "Das Tilapia-Motiv in der altägyptischen Kunst", *Cd'É* 41(82), 273-294) and turtles (see Fischer, *Turtles*, 6).

51 Edel, E. (1963). "Zu den Inschriften auf den Jahreszeitenreliefs der 'Weltkammer' aus dem Sonnenheiligtum des Niuserre Part II", *NAWG* 1963, 105-115 and 153-163.

52 Keimer, L. (1955). "Notes de lecture I", *BIFAO* 55, 7-20.

53 See Osborn-Osbornová, *Mammals*, passim.

54 Ikram, S. (1991). "Animal mating motifs in Egyptian funerary representations", *GM* 124, 51-68.

the same manner and that the incorrect mating stance of a few pairs was the result of artists having modelled the sexual behaviour of wild animals upon that of domesticated species.[55]

This study makes it clear that an ethological evaluation of the art of the ancient Egyptians has the potential to provide new insights about both their knowledge of animals, as well as the methods by which their artists endeavoured to represent natural behaviour. The following investigation explores these issues through an examination of the animal-rich tomb reliefs of the Old Kingdom.

[55] I disagree somewhat with this interpretation. See Chapter 10.

CHAPTER 3

AIMS AND METHODS

3.1 SCOPE

The focus of this study is the animal imagery found in Old Kingdom tomb reliefs at Giza and Saqqara. Over eight thousand animal images were examined with the purpose of gathering a representative sample of the types of behaviour of which artists were aware from the 3rd dynasty until the end of the 6th. The rationale for choosing this time frame and these sites is as follows:

Tomb decoration increased dramatically during the Old Kingdom period. In particular, many images from daily life were introduced at this time.[1] These scenes feature a wide range of animal species (both wild and domesticated) that are depicted in many different contexts and engaged in various activities. They are thus a rich source of imagery in which animals might potentially be found to exhibit natural behaviour. As the tombs located at Giza and Saqqara were for the exclusive use of central government officials during the Old Kingdom, the wall reliefs at these two sites exhibit a high level of workmanship and detail, reflecting the wealth and status of their owners. The cemeteries were also used extensively throughout the Old Kingdom period and so provide an ideal opportunity to witness stylistic changes over time, while the large number of tombs at each site ensured that a substantial number of animal images could be located and examined.

It was beyond the scope of this investigation to include a comparison of the animal imagery from Giza and Saqqara with that found in other cemeteries at the capital or outside the Memphite area. Extended analyses of this kind would likely yield further valuable information, however, and provide a different perspective; for example, the art of the provinces may focus on or emphasize animals and behaviour that specifically reflect their different geographical locations. It would equally have been worthwhile to examine the fragmentary remains of wall scenes from royal funerary temples of the Old Kingdom, as many animal motifs found in private tomb decoration undoubtedly originated in these structures.[2] While a full comparison with royal examples was not possible, their influence on non-royal iconography is acknowledged, where applicable, in the following chapters. It is also important to note that this study was not intended to offer a stylistic comparison between images from Giza and Saqqara, nor was its purpose to provide a method for the more accurate dating of tombs. Both of these issues are touched upon in the text where appropriate, but they are peripheral to my primary focus, which was to explore the ethological knowledge of the Egyptian artists as revealed through their representations.

3.2 AIMS

This study had two major objectives:

1) to document the depiction of animal behaviour in wall scenes from private tombs at Giza and Saqqara, and

2) to determine how Egyptian artists in the Old Kingdom translated dynamic animal behaviour into two-dimensional images.

---

1 The repertoire of scenes increased as developments in the shape and construction of tombs resulted in more interior wall space that could then be decorated. See Smith, *Interconnections*, 139; Smith, *HESPOK*, 155.

2 Groenewegen-Frankfort, *Arrest and Movement*, 28-62.

To achieve the first goal, I undertook to examine every extant animal image from Giza and Saqqara - regardless of the species - to assess and record what behaviour, if any, each represented. No assumptions were made about which depicted species would be more or less likely to exhibit some form of behaviour, and so consequently all classes were considered (i.e., mammals, birds, reptiles, amphibians, fish, and invertebrates). Scenes from each tomb and relief fragment were inspected systematically (see section 3.3) to ensure that every depicted animal was located and recorded. The behavioural characteristics of each animal image were evaluated initially on the basis of the species represented, its pose, and the context in which it was placed. Confirmation of the precise behaviour was then made by comparing the image with details of the known repertoire of the depicted species. This last step also enabled a judgment to be made about the accuracy of each depiction.

Once all of the available animal images had been examined and recorded, they were then sorted into eight broad behavioural categories (see section 3.5). The images associated with each of these categories are described and compared to the actions of living species in the following chapters.

The range and types of behaviours found during this study were intended to reveal what Memphite artists noticed about animals, while the care with which these actions are rendered was expected to indicate the level of observation they were afforded. It was anticipated that the subset of activities chosen from the vast range of potential behaviours for inclusion in wall scenes would also provide insights. For example, frequently depicted actions might represent behavioural features that were considered to be 'typical' of certain species, thereby revealing how the Egyptian people possibly categorised some animals.[3] Rare or unique behaviours, on the other hand, might reflect the interests of an individual artist, or, more likely, record independent observations made either by him or the tomb-owner. Incorrect depictions of behaviour were also logged; it was thought that these were likely to represent animals that artists had been unable to witness firsthand, thereby providing indirect information about the accessibility of some species. The behavioural response of some species to their human handlers was noted as it was thought that such interactions might shed light on the Egyptian people's attitude toward animals generally. In short, it was anticipated that this part of the study would provide new insights about the extent of knowledge held by the Egyptians of the Old Kingdom period regarding the natural history of animals.

The second major goal of this investigation was to explore how Egyptian artists transformed expressions of animal behaviour into two dimensions.[4] As noted in Chapter 1, animals have received limited attention in previous analyses of Egyptian art. The potential for animal imagery *uniquely* to contribute to our knowledge of the Egyptian method of representation has consequently been overlooked or at least undervalued by art historians. It is hoped that this study will help to address this imbalance.

The way in which human figures and inanimate objects were depicted in tomb scenes has been examined in detail, revealing much about the Egyptians' idiosyncratic method of two-dimensional representation. Their treatment of static features, such as physical proportions, use of scale, placement of objects in space, and rendering of time, is thus well understood. Many images, however, still defy explanation.[5] The representation of dynamic action presents particular challenges, as it is difficult to assess and analyse the representation of movements for which the only source of information is their depiction in wall scenes. In other words, if the precise actions that artists were attempting to communicate are unknown, how might we begin to understand their method of representing them? Some depicted actions can be guessed at with reasonable certainty (e.g., throwing a spear), while the mechanics of other activities might be obtained via ethnographic comparison with current practices (e.g., agricultural tasks, such as winnowing).[6] The meaning of

3 Weeks, K.R. (1979). "Art, word, and the Egyptian world view" in Weeks, *Egyptology and the Social Sciences*, 59-81.

4 This presents the challenge of looking for natural behaviour in the 'unnatural art' of the Egyptians. See Frandsen, P.J. (1997). "On categorization and metaphorical structuring: Some remarks on Egyptian art and language", *CAJ* 7(1), 72.

5 Schäfer, *Principles*, 144.

6 For example, the representation of a game played by boys in the tombs of *ptḥ-ḥtp II: ṯfj* and *mrrw-k3.jmrj* could be understood only after comparison with a modern Arabic high-jumping competition in which a hurdle, formed by two men seated on the ground and facing one another, is leapt over by a third.

many represented movements, however, has been lost. The depiction of Egyptian dance, for example, is difficult to analyse because we usually do not know how the performers are meant to be arranged in space (side-by-side, in single file, in the round?) or the steps involved.[7] Without known parallels comes the temptation to interpret depicted actions based simply upon what they 'look like'.[8] Yet, given the particular nature of Egyptian art, this is likely to result in misinterpretation of both the meaning of a representation and the method used to produce it.

The behaviour of living animals provides a visual benchmark by which the actions of depicted animals can be compared. As it is possible to view examples today of the kinds of animal behaviour that is represented in tomb scenes, we have the opportunity to assess *precisely* how Egyptian artists attempted to convey specific, known actions within the constraints of their method of representation. The behavioural approach undertaken in this study was thus designed to achieve a clear understanding of depictions of complex movements that might otherwise have remained ambiguous.

As noted in Chapter 1, it has been suggested that the rendering of animals in two dimensions was less constrained than that of human figures, however it is not clear what this opinion is based upon. Egyptian depictions of animals do appear, at first glance, to exhibit a greater freedom of expression than those of humans, but as they have received less attention generally from art historians and have not been afforded the same level of analysis as the human form, it is perhaps presumptuous to assume that their representation was not subject to similar rules and conventions. By examining the Egyptian expression of behaviour, it might therefore be possible to determine if animal images were constrained in similar ways to those of human figures. For example, it has been revealed that most of the animals in tomb scenes are shown mating in an identical fashion - even species for which the depicted posture is incorrect.[9] It is possible that other fundamental behavioural states were represented in a standard format (a single motif), regardless of the true behaviour of individual species. The presence of such conventions would indicate that artists might have reproduced animals that they had not necessarily seen in real life (i.e., their models were derived from copybooks or something similar). Examples of unique depictions of behaviour, on the other hand, would point to direct observations by individual artists.

This generates the intriguing question of whether images of animals potentially drawn directly from life may indeed, by virtue of their lack of prescribed models, be less likely to conform to the rules of representation by which standard motifs were bound. This study thus sought to determine how often unique observations of behaviour were incorporated into tomb scenes and to assess whether their depiction varied substantially from that of more commonly encountered behaviours.

The notion of whether animal images were derived from life or copied from known models also leads to questions about accuracy. Egyptian art was conceptual, an approach that enabled the representation of characteristics of the physical world that would be impossible to recreate using perspective. These included things that cannot necessarily be observed,[10] such as objects that are

---

See Eaton, E.S. (1937). "Two parallels to ancient Egyptian scenes: An Egyptian high-jump", *BMFA* 35, 54-55. See also Samuel, D. (1993). "Ancient Egyptian cereal processing: Beyond the artistic record", *CAJ* 3, 276–283.

7 Indeed, it is often impossible to determine whether an individual is to be perceived as standing still or striding since the leg positions of these two opposite states are nevertheless depicted in an identical manner. See Chapter 4 in this volume and also Groenewegen-Frankfort, *Arrest and Movement*, 10; Wood, W. (1978). "A reconstruction of the reliefs of Hesy-re", *JARCE* 15, 9-24.

8 Schäfer (*Principles*, 146) points out the mistake of interpreting the royal act of 'smiting' based only on how the action appears in two-dimensional representations. In such scenes, the arm holding the weapon is flung back behind the body, suggesting that the king's upper torso was held at right angles to his victim. Statues depicting this act clearly indicate, however, that his upper body faced his opponent and that he raised his mace to the side of his body, with his arm bent and held in the same plane as his shoulders.

9 See Ikram, S. (1991). "Animal mating motifs in Egyptian funerary representations", *GM* 124, 51-68.

10 Schäfer (*Principles*, 90) maintained that although Egyptian imagery was not based on immediate visual impressions, it must nevertheless have begun originally as visual input. While vision tends to be the primary conduit through which artists receive impressions of the world, it is important to note, however, that some are influenced by, and have attempted to communicate, sensations gained through *other* sensory

normally hidden from view (e.g., the contents of containers), but also abstract concepts. Status and importance, for example, could be conveyed graphically through the application of artificial differences in size.[11] Egyptian art can thus potentially include images that are visually inaccurate, but conceptually correct. It was anticipated, therefore, that some animal images were intended to convey the idea of a behaviour, rather than objective reality. As noted in Chapter 2, artists may have endeavoured to reproduce, in some manner, *non-visual* features of behaviour if these were regarded as particularly characteristic of certain animals. They may also have attempted to represent information that was based upon generally held beliefs about animals ('common knowledge'), rather than observed facts. Consistent anomalies in the representation of behaviour were hence studied carefully as it was felt these might nevertheless reveal a logical purpose as well as provide new insights about the conventions applied by Egyptian artists to convey complex information.

3.3 DATA COLLECTION

Data collection proceeded in four stages:

Stage 1
First, all references to animals, animal-related activities or situations in which animals might be incorporated (e.g., agricultural scenes, slaughter scenes, offering scenes, etc.) that are listed in the description of tomb scenes from Giza and Saqqara in Porter and Moss', *Topographical Bibliography of Ancient Egyptian Hieroglyphic Texts, Reliefs and Paintings* (*P-M III*), were noted and the relevant materials collected. Tombs published after *P-M III* (and prior to 2006) were then located via the *Annual Egyptological Bibliography* (*AEB*) and other resources, and added to the corpus. The images that were examined consisted primarily of line drawings in tomb reports, but photographs of wall scenes were sometimes referred to if no other source was available. Photographs were also consulted to check the accuracy of doubtful details in line drawings. Only scenes containing animals were included in the analysis.

Stage 2
Each wall scene was then examined in a systematic manner to ensure that every animal image represented within it could be logged and details of the depicted behaviour recorded on data sheets. Scenes were examined room-by-room (that is, if a tomb contained more than one distinct area, all relevant scenes in one location were recorded first before proceeding to the next). The walls within each room were examined consecutively, disregarding any that lacked depictions of animals.

Animal images were treated as individuals, groups, or clusters. Those depicted entirely separate from others were recorded as individuals. Animals in aggregations with one or more of their own kind ('conspecifics') were treated as a group. A group could be formed by unconstrained animals (e.g., a cow suckling her calf, a pack of donkeys, a flock of birds, etc.) or by animals under the control of humans (e.g., leashed cattle). Clusters consisted of animals of different species that were shown interacting with one another (e.g., a predator and its prey).

The descriptive features of each individual animal image, group, or cluster were recorded, including: the scene type, number of animals, animal type(s), animal behaviour, number of humans present, human behaviour, hieroglyphic label (if any), significant features of the image, and the source of the image. Scene types included: agricultural, pastoral, fording, marsh, fishing, clapnet, desert, hunting, slaughter, offering, domestic, or other (e.g., boat-building). The behaviour of human participants was assigned to one or more general categories: leading, guiding, holding, carrying, pushing, pulling, manipulating, restraining, striking, hunting, and butchering.

Stage 3
During the examination and recording of details for each image, those animals expressing species-specific behaviour were noted (see section 3.5). These images were subsequently scanned and then saved on computer hard-drive.

systems, such as smell, touch, and hearing. See, for example, the art of ancient Mesoamerica or the work of modern artist, Saul Steinberg.

11 *ibid.*, 230-234.

Stage 4
The written records for every animal image were then transferred into a computer database.[12] Details of the tomb from which each image was derived were also appended to its computer record. These included: the name of the tomb-owner, tomb number, tomb location, proposed dates for the tomb, dating sources, *P-M III* citation, bibliographic details for the source of the image, and the position of the image within the tomb. Catalogue or museum inventory numbers were recorded for those derived from relief fragments. The database also offered a facility by which keywords could be attached to individual records. These tags were used subsequently to locate and sort records. Families of descriptive keywords were thus created for the categories: 'Animal species', 'Type of animal behaviour', 'Type of human behaviour', and 'Significant image features'. During data input, appropriate keywords were selected from these lists to describe each scanned image and attached to its record.

A total of 291 tombs and relief fragments were examined (158 from Giza and 133 from Saqqara). This yielded 8,208 records of depicted animals (1,996 from Giza and 6,212 from Saqqara). In approximately half of the tombs that were evaluated, the surviving reliefs showed animals simply standing or walking passively (see Chapter 4 for more details); these were discarded from further analysis. Of the remaining tombs, examples of distinctive, species-specific behaviour were clearly identified in 115 tombs and fragments: 46 from Giza and 69 from Saqqara. These depictions are described in the following chapters.

3.4 DATING

The database record for each animal image included a short list of proposed dates for the tomb or relief fragment from which it had been derived (see section 3.7). Dates were obtained from three sources: *P-M III*, the report or other source from which the image had been obtained, and Yvonne Harpur's, *Decoration in Egyptian Tombs of the Old Kingdom: Studies in Orientation and Context.*

Harpur's dates were chosen as a standard reference by which to sort subsets of images when these were later placed in chronological order. As noted above, the purpose of this study was not to provide new guidelines for the more accurate dating of Old Kingdom tombs. It was felt that Harpur's dates, which had been derived through a careful evaluation of various dating criteria that included features of the wall decoration, were sufficiently reliable to enable the images in this study to be arranged in a sensible and consistent manner. Nevertheless, stylistic changes in the expression of certain behaviours occasionally pointed to an adjustment of some of her suggested tomb dates. This is noted in the following chapters where appropriate. See Appendix 1 for a chronological list of the 115 tombs and fragments that yielded the images of behaviour described in this study.

3.5 BEHAVIOURAL CATEGORIES

The behaviour expressed by each animal image was assigned to one (or more) of eight basic categories:

- Locomotor behaviour - species-specific movement
- Group behaviour - herding, flocking, schooling
- Comfort behaviour - grooming, preening, elimination
- Ingestive behaviour - eating, drinking
- Hunting behaviour - predation
- Aggressive behaviour - threatening, fighting, defending
- Sexual behaviour - courtship, mating
- Parental behaviour - nursing, incubation, protection

These categories are based upon those found in the ethological literature (see Chapter 2), and range from the most fundamental form of behaviour - the expression of movement - through to more complex actions.

---

[12] ImageAXS Pro version 4.0. Note that every animal image was recorded in the database, regardless of whether it expressed a distinctive behaviour or not. This ensured that all images could be located easily in the future should their behavioural significance not become apparent until later.

To determine which behaviour, if any, an animal image represented, I considered: a) the species depicted, b) the pose of the animal, and c) the context in which it was situated. In nature, the behavioural repertoire of individual species is often so idiosyncratic that it can be used for the purpose of identification; as noted in Chapter 2, the term to describe such behaviour is 'species-specific'.[13] The species of depicted animals in wall reliefs thus often suggested, and limited, what types of behaviours they might potentially exhibit. Pose was assessed by noting the torso, ear, tail, head, wing(s) and/or leg positions of animals, along with their proximity to others and to objects in their immediate environment. Interactions with humans were also considered. After this assessment, a preliminary identification was made of the behaviour depicted, which was then checked via consultation of the ethological literature. This consisted of general works of zoology, studies of individual species, and scientific monographs.

When assessing depicted animals, I always began with the assumption that the artist had tried to reproduce their behaviour as accurately as possible. It was accepted, however, that mistakes might have occurred occasionally, either due to confusion about an animal by the draftsman or artistic license. I concluded that an error had been made when, after exhaustive review of the ethological literature, I could find no evidence for such a behaviour, or if recent studies demonstrated unambiguously that the depiction was incorrect. These are indicated, where appropriate, in the text.

3.6 ANALYSIS

The following chapters describe each of the eight behavioural categories in turn. Each chapter begins with a broad definition of the type of behaviour to be considered. The text is then divided into sections, each of which deals with a sub-category of the behaviour in question, and these are further divided by animal species. A chronological list of the tombs containing relevant images appears at the top of each species section. These are identified by the transliterated name of the owner, its behavioural code number(s) (see section 3.7), the location of the tomb, and its proposed date. A summary of the major findings is provided at the end of each of the chapters.

Species are presented in strict taxonomic order: Class, Order, Family, Species. Thus, all of the animals depicting a behaviour from within the Class Mammalia are dealt with first before moving on to those in Class Aves (i.e., birds), then Class Reptilia (e.g. crocodiles), next Amphibia (e.g., frogs), then Pisces (i.e., fish), and finally Class Insecta (e.g., butterflies, grasshoppers). See the end of this chapter for a complete *Species List* of all the animals that are referred to[14] and/or described in the text, arranged taxonomically. Note that the Latin names for each species are provided in the list, but are omitted from the text for the sake of brevity.

The Egyptian depictions of animals are described and discussed in each chapter and are compared with the natural behaviour of the species in question. Descriptions of the mechanics and purpose of each sub-category of behaviour are included in the text where necessary to enable the reader to understand and assess the accuracy of the Egyptian images. I have attempted to keep these as brief as possible, but sufficient biological information has been provided so that readers unfamiliar with such details can gain some understanding of behaviours that may be new to them.

Photographs or drawings illustrating these activities by contemporary animals are also provided for every type of behaviour discussed. These plates are intended to aid comprehension of the text, as well as to act as exemplars by which to compare the Egyptian images. Note that I have not reproduced every pertinent Egyptian depiction of a behaviour, as this would be unmanageable, however the details of every image described or referred to in the text are provided in Appendix 2. The reader is advised to use the behavioural code numbers mentioned in the text (see below) to seek further information about images, as required.

---

13 Immelmann-Beer, *Dictionary of Ethology*, 287.

14 Animals that are only mentioned in passing in the text are marked with an asterisk.

3.7 BEHAVIOURIAL CODE NUMBERS

Each of the images that are described in the following chapters has been assigned a code consisting of two-letters and a number. The letters refer to the behaviour being considered: Locomotor behaviour (LB), Group behaviour (GB), Comfort behaviour (CB), Ingestive behaviour (IB), Hunting behaviour (HB), Aggressive behaviour (AB), Sexual behaviour (SB), and Parental behaviour (PB). The numbers are sequential within each behavioural category. Information about each image can consequently be obtained by noting the code number and then referring to the relevant section in Appendix 2. Note that the order in which the images are listed for each species in Appendix 2 is dictated by the tomb dating system provided by Harpur, *Decoration*.

Some animal images fell into more than one behavioural category. For example, a marsh bird sitting on a nest might be classed as exhibiting simultaneously both aggressive behaviour (threatening approaching predators) and parental behaviour (incubation of eggs). For the sake of clarity, such images are referred to more than once in both the text and Appendix 2, in order to illustrate each behaviour separately. These images have thus received more than one behavioural code number in Appendix 2, according to each of the behaviour types they illustrate.

The following information is provided for each image listed in Appendix 2:

Name
A transliteration of the tomb-owner's name. Honorific titles are not included as these were not relevant to the current study.

Site
Details of the location (Giza or Saqqara), cemetery field, and the tomb number (if any) are presented. The images from Giza and Saqqara are not treated separately, however any stylistic differences observed between the two cemeteries in their depiction of particular behaviours are referred to in the text.

Scene location
The position of the image within the tomb (e.g., chapel, hall, court, door thickness, etc.) is listed, along with the compass direction of the wall upon which it occurs. If the image was derived from a relief fragment held in a museum or other collection, the inventory or catalogue number of the object is provided and the compass direction is listed as 'unknown'.

Scene type
The scene type consists of the general context in which the image is placed, such as: agricultural, pastoral, fording, marsh, fishing, clapnet, desert, hunting, slaughter, offering, domestic, or other.

Source
Abbreviated citation details are provided for *P-M III* and for the publication or other source from which the image was obtained. See the Bibliography for full reference information.

Label
Transliterations and translations are provided for hieroglyphic labels that appear within the register and which either refer directly to the depicted animals (e.g., their name), or to an activity with which they are associated (e.g., 'herding'). Inscriptions that do not match these criteria are listed as 'n/a' (not applicable). Register captions were not considered.

Date
The proposed date provided is usually derived from Harpur, *Decoration*, for which appropriate citation details are offered. The numbering system applied follows Harpur's method, shown below:

IV.1	Snefru	V.1	Userkaf	VI.1	Teti
IV.2	Khufu	V.2	Sahure	VI.2	Pepi I
IV.3	Djedefre	V.3	Neferirkare	VI.3	Merenre
IV.4	Khafre	V.4	Shepseskare	VI.4	Pepi II (years 1-34)
IV.5	Menkaure	V.5	Raneferef	VI.5	Pepi II (years 35-54)
IV.6	Shepseskaf	V.6	Niuserre	VI.6	Pepi II (years 55-85)
		V.7	Menkauhor	VI.7	VI late-VIII
		V.8	Djedkare/Isesi		
		V.9	Unas		

Dates were refined using the letters 'E' (early), 'M' (middle), and 'L' (late).[15] Harpur's date is the only one supplied, except for tombs that have been published or re-examined after her work (i.e., post-1987). In these cases, the date offered in those publications is also provided. If a tomb was not listed by Harpur (a rare occurrence), a proposed date was either provided by *P-M III* or the report from which the image was generated.

3.8. AUTHOR'S NOTE

Every effort was made to examine all surviving material from Giza and Saqqara. Nevertheless, a small number of tombs and fragments had to be excluded either because these were unpublished or were found to contain no depictions of animals. Furthermore, it is inevitable that some images will have been overlooked inadvertently or failed to be included. Such oversights will not adversely affect the primary aim of the study, however, which was to introduce a hitherto unappreciated aspect of Egyptian art. I welcome others to explore in greater depth the ideas contained herein.

15 Harpur, *Decoration*, 33-35.

Class MAMMALIA
Order Perissodactyla (Even-toed ungulates)
Family Equidae
Donkey (*Equus asinus asinus*)
Wild ass (*Equus asinus africanus*)
Order Artiodactyla (Odd-toed ungulates)
Family Bovidae
Cattle (*Bos taurus*)
Zebu (*Bos indicus*)*
Addax (*Addax nasomaculatus*)
Scimitar oryx (*Oryx dammah*)
Hartebeest (*Alcelaphus buselaphus*)
Impala (*Aepyceros melampus*)*
Dorcas gazelle (*Gazella dorcas*)
Sömmering's gazelle (*Gazella sömmeringi*)*
Dik dik (*Modoqua* sp.)*
Nubian ibex (*Capra ibex nubiana*)
Goat (*Capra hircus*)
Sheep (*Ovis aries*)
Barbary sheep (*Ammotragus lervia*)
Family Hippopatamidae
Hippopotamus (*Hippopotamus amphibius*)
Family Cervidae
Dama deer (*Dama mesopotamica*)*
Order Carnivora
Family Felidae
Lion (*Panthera leo*)
Leopard (*Panthera pardus*)
Caracal (*Caracal caracal*)
Wild cat (*Felis silvestris*)
Family Hyaenidae
Striped hyena (*Hyaena hyaena*)
Spotted hyena (*Crocuta crocuta*)*
Family Viverridae
Egyptian mongoose (*Herpestes ichneumon*)
Common genet (*Genetta genetta*)
Large-spotted genet (*Genetta tigrina*)*
Family Canidae
Domestic dog (*Canis familiaris*)
Golden jackal (*Canis aureus*)
Black-backed jackal (*Canis mesomelas*)*
African wild dog (*Lycaon pictus*)*
Red fox (*Vulpes vulpes*)
Rüppell's sand fox (*Vulpes rueppelli*)*
Family Mustelidae
Common otter (*Lutra lutra*)*
Spotted-neck otter (*Lutra maculicollis*)
Libyan striped weasel (*Poecilictis libyca*)
Ratel (*Mellivora capensis*)*
Zorilla (*Ictonyx striatus*)*
Order Rodentia
Family Dipodidae
Lesser Egyptian jerboa (*Jaculus jaculus*)
Order Lagomorpha
Family Leporidae
Cape hare (*Lepus capensis*)
Order Primates
Family Cercopithecidae
Hamadryas baboon (*Papio hamadryas*)
Anubis baboon (*Papio cynocephalus anubis*)
Vervet monkey (*Cercopithecus aethiops*)
Order Insectivora
Family Erinaceidae
Desert hedgehog (*Paraechinus* sp.)
Long-eared hedgehog (*Hemiechinus auritus*)
Family Soricida
Greater musk shrew (*Crocidura flavescens*

Class AVES
Order Struthiformes
Family Struthionidae
Ostrich (*Struthio camelus*)
Order Pelecaniformes
Family Pelecanidae
White pelican (*Pelecanus onocrotalus*)
Family Phalacrocoracidae
Reed cormorant (*Phalacrocorax africanus*)
Great cormorant (*Phalacrocorax carbo*)*
Shag (*Phalacrocorax aristotelis*)*
Order Ciconiiformes
Family Ardeidae
Grey heron (*Ardea cinerea*)
Great blue heron (*Ardea herodias*)*
Little egret (*Egretta garzetta*)
Bittern (*Botaurus stellaris*)
Family Threskiornithidae
Eurasian spoonbill (*Platalea leucorodia*)
Glossy ibis (*Plegadis falcinellus*)*
Sacred ibis (*Threskiornis aethiopicus*)
Order Anseriformes
Family Anatidae
Egyptian goose (*Alopochen aegyptiacus*)
Bean goose (*Anser fabalis*)*
Greylag goose (*Anser anser*)
White-fronted goose (*Anser albifrons*)*
Lesser white-fronted goose (*Anser erythropus*)*
Barnacle goose (*Branta leucopsis*)*
Brent goose (*Branta bernicla*)*
Red-breasted goose (*Branta ruficollis*)*
Spur-winged goose (*Plectopterus gambensis*)*
European wigeon (*Anas penelope*)*
Green-winged teal (*Anas crecca*)*
Mallard (*Anas platyrhynchos*)*
Pintail (*Anas acuta*)
Common shelduck (*Tadorna tadorna*)*
Ruddy shelduck (*Tadorna ferruginea*)*
Pochard (*Nyroca ferina*)*
Tufted duck (*Aythya fugilula*)*
Order Falconiformes
Family Accipitridae
Black kite (*Milvus migrans*)

Order Galliformes
  Family Phasianidae
    Common quail (*Coturnix coturnix*)
Order Gruiformes
  Family Gruidae
    Common crane (*Grus grus*)
    Demoiselle crane (*Anthropoides virgo*)
  Family Rallidae
    Purple gallinule (*Porphyrio porphyrio*)
Order Charadriiformes
  Family Charadriidae
    Lapwing (*Vanellus vanellus*)
  Family Laridae
    Seagull (*Larus* sp.)*
Order Columbiformes
  Family Columbidae
    Pigeon (*Columba* sp.)
Order Coraciiformes
  Family Alcedinidae
    Kingfisher (*Alcedo atthis*)*
    Pied kingfisher (*Ceryle rudis*)
  Family Upupidae
    Hoopoe (*Upupa epops*)
Order Passeriformes
  Family Oriolidae
    Golden oriole (*Oriolus oriolus*)

Class REPTILIA
Order Crocodylia
  Family Crocodylidae
    Nile crocodile (*Crocodylu niloticus*)
Order Squamata
  Family Varanidae
    Nile monitor (*Varanus niloticus*)*
Order Testudinae
  Family Trionychidae
    Nile soft-shell turtle (*Trionyx triunguis*)
  Family Testudinidae
    Egyptian tortoise (*Testudo kleinmanni*)*

Class AMPHIBIA
Order Anura
  Family Ranidae
    Frog sp.
  Family Bufonidae
    Damietta toad (*Bufo kassasii*)*

Class PISCES
Order Cypriniformes
  Family Cichlidae
    Tilapia (*Tilapia* sp.)
Order Mugiliformes
  Family Mugilidae
    Grey mullet (*Mugil* sp.)
Order Siluriformes
  Family Mochokidae
    African catfish (*Synodontis schall*)
    Upside-down catfish (*Brachysynodontis batensoda*)
  Family Clariidae
    Walking catfish (*Clarias* sp.)
Order Anguilliformes
  Family Anguillidae
    European eel (*Anguilla anguilla*)

Class INSECTA
Order Hymenoptera
  Family Apidae
    Honey bee (*Apis mellifera*)
  Family Vespidae
    Yellowjacket wasp (*Vespula germanica*)
    Hornet (*Vespa* sp.)*
Order Lepidoptera
  Family Nymphalidae
    Monarch butterfly (*Danaus chrysippus*)
  Family Pterophoridae
    Plume moth sp.
Order Odonata
  Infraorder Anisoptera
    Dragonfly sp.
Order Orthoptera
  Family Acrididae
    Egyptian grasshopper (*Anacridium aegyptium*)
    Desert locust (*Schistocerca gregaria*)
Order Diptera
  Family Muscidae
    Housefly (*Musca domestica*)*
    Stable fly (*Stomoxys* sp.)*
Order Mantodea
  Family Mantidae
    Praying mantis (*Mantis religiosa*)

CHAPTER 4

LOCOMOTOR BEHAVIOUR

DEFINITION

Locomotor behaviour is comprised of the different methods by which animals travel through the environment (e.g., walk, run, fly, swim, etc.), and the postures and movements associated with achieving and experiencing rest. How animals move is dictated by their physical structure (whether they are a quadruped or biped, the size and shape of their body, length and articulation of their limbs, shape of their feet, etc.), the type of habitat in which they live (e.g., terrestrial, aquatic, etc.), and their pattern of activity (diurnal or nocturnal).

4.1 MOVEMENT

4.1.1 Walking/Standing

Approximately half of the animals examined during this study were found to exhibit no apparent behaviour other than walking or standing still, reflecting the common occurrence of these general activities. Both bipedal and quadrupedal animals were each depicted in this manner using a standard motif. As it is neither feasible, nor necessary, to reproduce all of these images here, only a subset of representative examples are presented and referred to below.

Quadrupeds

nj-ꜥnḫ-ẖnmw/ẖnmw-ḥtp (LB1: Saqqara, V.6L-7)
jtj-sn (LB2: Giza, V.6-8?)
nḫt.kꜣ.j (LB3: Giza, V.6-9?)
snḏm-jb: jntj (LB4: Giza, V.8M-L)
sḫm-ꜥnḫ-ptḥ (LB5: Saqqara, V.8-9)
ṯjj (LB6-7: Saqqara, V.8-9)
nfr-jrt-n.f (LB8: Saqqara, V.8-9)
ꜣḫtj-ḥtp (LB9: Saqqara, V.8L-9E)
ptḥ-ḥtp II: ṯfj (LB10-11: Saqqara, V.9M-L)
Anon (LB12: Saqqara, VI)
sḫm-kꜣ.j (LB13: Giza, V.9-VI.1?)
sḫm-ꜥnḫ-ptḥ (LB14: Giza, V.9-VI.1?)
nfr I: jdw (LB15: Giza, VI.1)
mrrw-kꜣ.j: mrj (LB16-17: Saqqara, VI.1M-L)
ḥzj (LB18: Saqqara, VI.1L-2E)
jnw-mnw (LB19: Saqqara, VI.2E)
mrjj-ttj: mrj (LB20: Saqqara, VI.2M-L)
mḥw (LB21: Saqqara, VI.2M-3?)

Quadrupeds in Memphite tomb scenes are represented either standing still or walking via a single pose (e.g., LB1 to LB11, LB13, LB14, LB16, LB18 to LB20). This consists of both the forelegs and the hind legs being parted, with the more distant member of each pair always advancing first, regardless of the direction in which the animal faces.[1] Furthermore, all four legs are usually held stiffly[2] and all four feet are shown in contact with the ground at once. The pose is applied to all quadrupeds, regardless of size or species. Many of the animals depicted in this posture are engaged in activities that suggest they should be perceived as walking (e.g., transporting produce - LB7; see fig. 4-1), however tethered and caged quadrupeds are also shown in the same attitude, even though they would be largely incapable of forward movement (e.g., LB3 and LB10; see fig. 4-2). It is consequently difficult to be sure in many cases whether an artist intended to depict a moving or stationary animal.[3]

1 This rule is also exhibited by human figures. See Erman, *Ägypten*, 532-534.
2 Although the hock (ankle joint) of each hind leg is usually represented, giving these limbs a slight bend.
3 The interchangeable meaning of this pose has been noted elsewhere. See Edgerton, W.F. (1936). "Two notes on the flying gallop", *JAOS* 56(2), 182-183. Schäfer (*Principles*, 293) also makes the same point regarding human figures.

When compared to the sequence of leg movements exhibited in nature by walking four-legged animals (cf. pl. 4A), it can be seen that the Egyptian pose is ambiguous. When moving forward slowly, quadrupeds always set down their feet in the following order: left foreleg, right hind leg, right foreleg, left hind leg.[4] This pattern of footfalls allows them to keep at least three feet on the ground at one time, creating a triangle of support upon which to maintain their balance as they move forward.[5] As each limb is advanced, the joints of the affected leg are flexed so that it is bent and lifted off the ground before being swung forward. Once placed upon the substrate again, the limb acts to support the trunk as it is thrust forward. Quadrupeds thus always have three, but never four, feet in contact with the ground simultaneously when walking slowly.[6] When standing still, however, they keep all four feet on the ground. Each pair of legs is either placed side-by-side or parted slightly. The limbs of large quadrupeds are usually held straight when standing still, while those of smaller animals tend to be slightly bent.[7]

By showing quadrupeds with straight legs and, more importantly, all four feet upon the ground at once, Egyptian artists have produced an image that more closely resembles a stationary animal than one that is moving. The joints of the depicted animals' legs do not bend as they would when advancing their limbs forward. Furthermore, the Egyptian motif, in which either both left limbs or both right limbs advance simultaneously, does not correspond to the natural pattern of walking quadrupedal movement. Nevertheless, it is clear that in many scenes these seemingly stationary animals are to be perceived as walking (e.g., when presented to the deceased, ploughing, transporting produce, etc.).

The key to determining which of these two states (moving or standing still) the artists intended might possibly be found in the contexts in which the animals occur. A tethered quadruped in this pose would therefore be understood to be stationary, while an unfettered animal would be seen to be walking. Yet it seems that many of the depicted animals could equally be doing either. The Egyptians themselves must also have struggled to ascertain which of the two states was represented in some scenes, an ambiguity that artists were perhaps content to endure. This would be at odds, however, with the overwhelming inclination in their art to convey what they understood about the world as clearly as their method of representation would allow.

Some authors have suggested that the pose represents neither walking nor standing, but was simply intended to provide an unimpeded view of all four legs.[8] The artistic dictates of symmetry may also have influenced the treatment of animal limbs by compelling draftsmen to present each side of the body as equal opposites. I would like to propose a different interpretation of the standing/walking pose: namely, that it was intended to illustrate *both* states simultaneously. The motif clearly contains elements of each activity: the limbs are advanced, expressing forward movement, but they are straight and the feet held flat upon the ground, indicating a state of inertia. If it is accepted as neither a depiction of movement nor stasis *per se*, but a composite of the two, then the pose can instead be seen to encompass a rolling *sequence* of motion: standing still followed by walking, or vice versa. This explanation for the otherwise ambiguous stance is in accord with a tendency, found often during this study, to present two apparently conflicting actions within the same scene in order to represent two different time-frames (for example, see section 4.1.3). This interpretation could apply equally to the 'striding' pose of human figures.

It is interesting to note that only seven images examined during this study were found to break the conventional rule of showing the limbs furthermost from the viewer advancing first.[9] An Egyptian mongoose (LB140) and a dorcas gazelle (see fig. 10-13 in Chapter 10) are depicted with their closer foreleg flung forward, a hamadryas baboon (LB12), two vervet monkeys (LB17 and LB21; see fig. 4-3) and another Egyptian mongoose (LB109) move their closer hind legs first, and a vervet in the tomb of *nfr I: jdw* (LB15) advances both limbs closest to the viewer. Schäfer noted that a small

4 Alexander, *Exploring Biomechanics*, 32.
5 Alexander, *Locomotion*, 82-85.
6 Gray, *Animal Locomotion*, 280.
7 Alexander, *Locomotion*, 104-107.
8 Edgerton, *JAOS* 56(2), 182.
9 See also sections 11.2.1 and 11.3.1 in Chapter 11 for a possible further examples.

number of human figures are represented with the nearest foot forward and suggested that the novel pose was introduced deliberately to emphasize their movement.[10] Curiously, most of the errant animals appear to be climbing - the mongooses are ascending papyrus stalks, one of the vervets is teetering upon the head of its handler, while the others scale the rigging and sails of ships. The location of the leashed baboon is more difficult to discern, but it is nevertheless elevated above and behind its handler who twists around to brandish a baton at it. These images may reflect an attempt by a few artists to produce a more realistic depiction of moving animals or, as Schäfer proposed, the rule has been broken to highlight movements associated specifically with climbing.

Bipeds

ṯjj (LB22: Saqqara, V.8-9)
ptḥ-ḥtp II: ṯfj (LB23: Saqqara, V.9M-L)
ʿnḫ-m-ʿ-ḥr: zzj (LB24: Saqqara, VI.1L-2E)
ḥzj (LB25-26a,b: Saqqara, VI.1L-2E)

Bidepal animal figures, consisting largely of free-moving birds, occur in a variety of contexts in wall scenes. In nature, birds that spend most of their time in trees, hop when they move across a substrate, but those that nest and/or forage for food on the ground (i.e., are more terrestrial) usually walk.[11] The most common bird species shown moving upon the substrate in wall scenes from Giza and Saqqara are storks, cranes, herons, sacred ibis, geese, ducks, and pigeons, all of which spend a considerable amount of time on the ground even though they are capable of flight. These animals are depicted frequently either moving forward (e.g., LB23; see fig. 4-4) or standing still (e.g., when tethered - LB24; see fig. 4-5). Deciding between these two states can be difficult, however, as, like quadrupeds, both are usually represented in an identical manner, employing a single standardised pose in which the birds (regardless of species) are rendered with their legs parted, the limb furthermost from the viewer advancing first, both legs held straight or slightly bent, and both sets of toes flattened against the substrate.

The structure of bird hind limbs reveals much about avian terrestrial locomotion. Their legs consist of three parts: The uppermost segment, the 'femur', is quite short and positioned deep within the trunk. The femur articulates with the 'tibia', a long bone that extends from the bird's body down to its ankle. At the ankle, the tibia intersects with the 'tarso-metatarsus', a long, slender foot bone from which the toes extend. The tarso-metatarsus and the toes are usually covered with small scales, while above the ankle the leg is feathered.[12] The long length of the tarsus-metatarsus gives the impression that it forms part of the leg, but it is actually a segment of the foot. Consequently, the joint where the tibia meets the tarsus-metatarsus can be mistaken for the knee, when, in fact, it is the ankle. When birds walk, therefore, they appear to bend each leg backwards, but they are actually flexing their ankles (see pl. 4B). When standing still, small birds will often bend their legs slightly while larger birds hold their limbs relatively straight. The structure of their legs also requires that birds stand and walk upon their toes. During terrestrial movement, the toes are first flexed and pursed as the foot is lifted, and then extended again halfway through the step as the foot approaches the ground (see pl. 4B).

The majority of birds shown standing upon the substrate in tomb scenes reveal little evidence of forward movement (although for rare exceptions, see below). The leading limb invariably inclines forward from the bird's body without bending, while the following leg either bends very slightly at the ankle or not at all. Although parted as when walking, the legs of these bird images nevertheless lack the ankle bend associated with locomotion. Furthermore, both sets of toes are always depicted flat against the substrate. This further suggests that the birds are not advancing forward but are, instead, stationary. Nevertheless, the widely spaced legs of the animals, like those of quadrupeds imply that they are moving, as when birds stand still they usually place their feet together directly beneath their body.[13] The images therefore have elements in common with both walking and standing still.

---

10 Schäfer, *Principles*, 293.
11 Burton, *Bird Behaviour*, 25.
12 Dorst, *Life of Birds i*, 36-38.
13 Alexander, *Locomotion*, 104.

A small number of reliefs were found to deviate from the standard bipedal pose. Purple gallinules depicted in two tombs (LB26a,b and LB198; see fig. 4-6) flex their advancing limbs abruptly at the ankle, capturing precisely the apparent backward bend of the leg that occurs when birds take a step. Gallinules possess extremely long toes (see pl. 4B), a feature that enables them to walk upon floating vegetation and to cling to, and ascend, stalks and reeds. Their locomotor behaviour is thus a defining attribute of the species - a feature so intimately connected to purple gallinules that a few artists were compelled to manipulate the bipedal motif in order to reproduce the birds' actual leg movements.

The advancing leg of a small bird depicted in a marsh scene in the tomb of *nj-kꜣw-jzzj* (LB87) is both bent at the ankle and lifted above the substrate. A crest on its head identifies it as an adult lapwing, but its rounded shape and lack of a tail suggest that it is a chick.[14] The animal stands upon a papyrus stalk beside a genet that is attacking egret nestlings; it thus appears to kick its raised leg defensively towards the predator.[15]

The tomb of *nj-kꜣw-jzzj* also contains images of geese in which the toes have been rendered in a unique fashion (see figs. 7-40 and 7-41 in Chapter 7). The birds do not hold their feet flat, but instead gather their toes together so that only the tips make contact with the ground. As noted above, birds purse their toes reflexively to lift them clear of the substrate as they step forward (cf. pl. 4B). The reliefs are thus a remarkable attempt to reproduce a visually accurate depiction of walking by geese. This is most apparent in the image of a bird that has paused mid-step, holding its advanced foot flat while gathering the toes of its following foot (see fig. 7-41, right). Two birds, however, are shown with both sets of toes pursed (see figs. 7-40 and 7-41). I suggest that these images were intended to convey a sense of on-going movement in which each foot is lifted rapidly one after the other (cf. the toes of waterbirds in section 4.1.6).

4.1.2 Balking

Animals under human control will sometimes refuse to advance and so must be coaxed or forced to move forward. Balking is often provoked by objects in their path that the animal fears.

Representative species

nfr-mꜣꜥt (LB27: Giza, IV.2-4)
mrs-ꜥnḫ III (LB28: Giza, IV.6)
kꜣ-pw-nswt: kꜣj (LB29: Giza, V.1-5)
jj-mrjj (LB30: Giza, V.3L)
rꜥ-špss (LB31-32: Saqqara, V.8M)
ṯjj (LB33: Saqqara, V.8-9)
nbt (LB34-37: Saqqara, V.9)
sšm-nfr IV (LB38: Giza, V.9-VI.1)
jꜣzn (LB39-40: Giza, VI.1M-2M)
ptḥ-ḥtp (LB41: Saqqara, VI.2? or LV-VIE?)

Quadrupeds are occasionally represented in a stance that conveys unambiguously to the viewer that they are stationary. In this pose, both forelegs are held together, preventing forward movement, while the hind legs are usually separated (although see below). The posture is adopted primarily by ungulates, but has also been observed in depictions of dogs, hippos, and donkeys. The pose occurs most frequently in scenes where humans are manipulating large and small ungulates, often when they are forcing them to move against their will. Indeed, many animals will lock their limbs and use them as a brace when pulling back (see pl. 4C). Balking is particularly common in offering scenes in which animals are sometimes reluctant to walk forward for presentation to the deceased. A subset of these reliefs, in which bearers intervene to make the animals proceed, has been chosen to illustrate this posture.

---

14 Kanawati-Abder-Raziq, *Nikauisesi*, 40.

15 Adult lapwings will sometimes attempt to distract and then draw a predator away from their nests by feigning injury - flapping and dragging their wings as though they are broken, and limping to suggest they are lame. The image may possibly allude to this distinctive display, however chicks do not exhibit the behaviour. See Brom, T.G. (1983). "Lapwing (*Vanellus vanellus*)" in Cramp-Simmons, *Handbook iii*, 262.

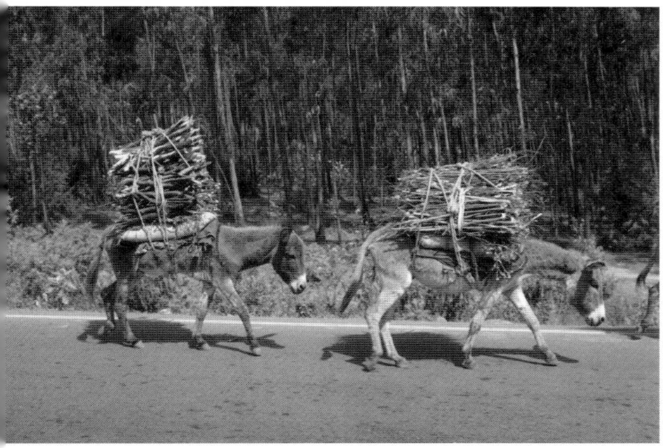

Pl. 4A Quadrupedal walking gait
© iStockphoto.com/Klaas Lingbeek-van Kranen

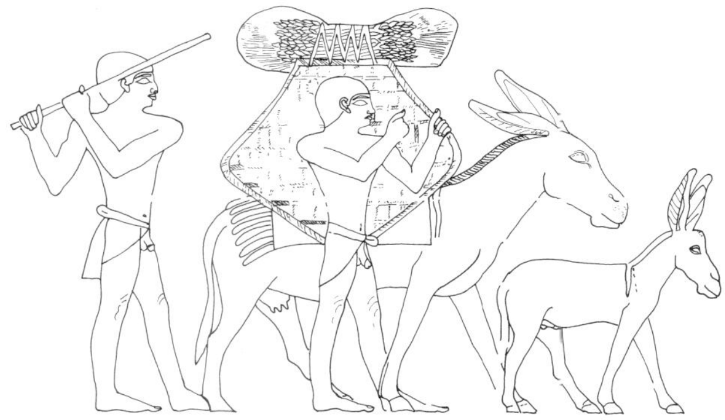

Fig. 4-1 (LB7)

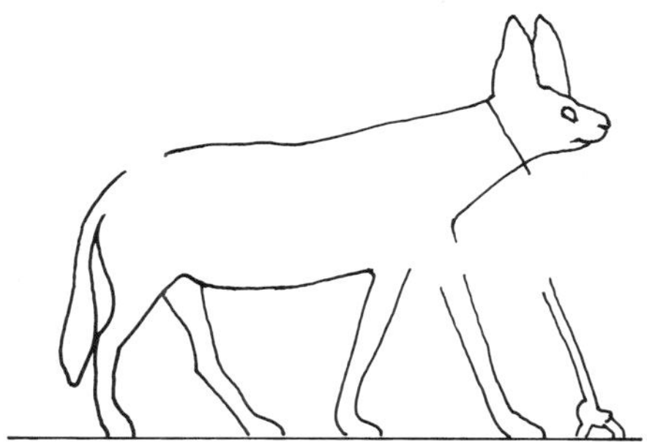

Fig. 4-2 (LB3)

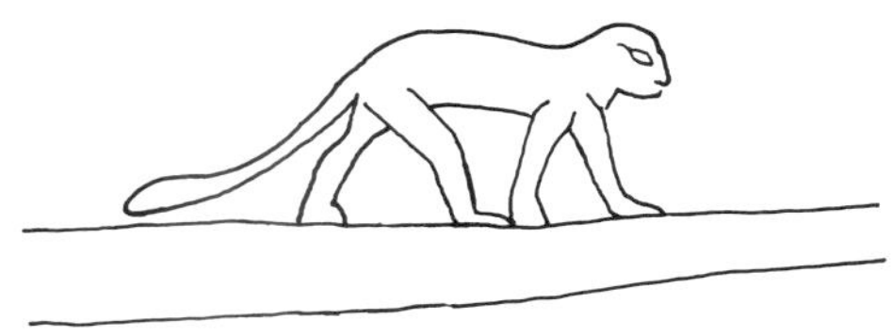

Fig. 4-3 (LB17)

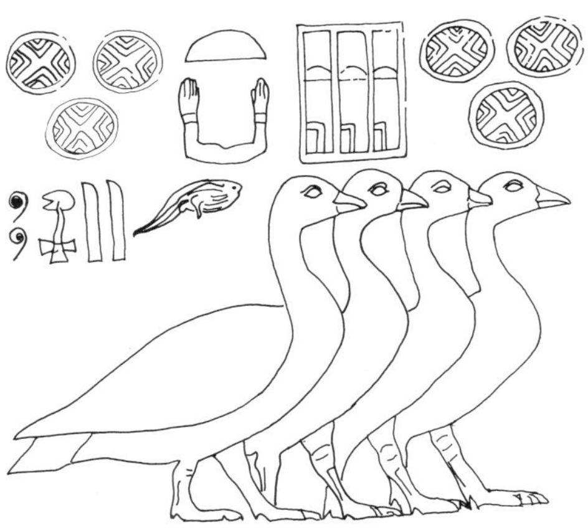

Fig. 4-4 (LB23)

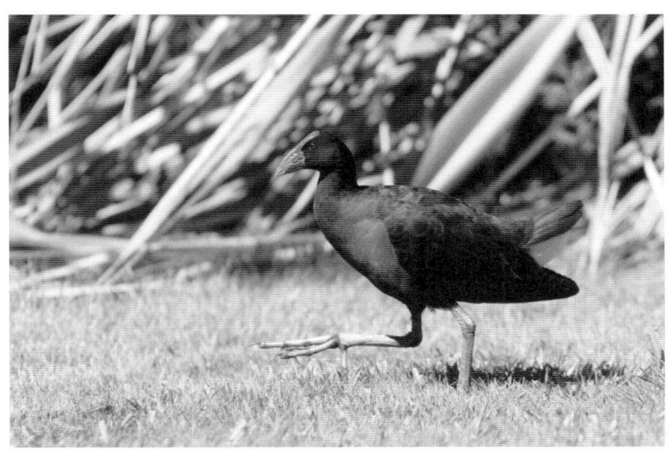

Pl. 4B Bipedal walking gait
© iStockphoto.com/Harry Kolenbrander

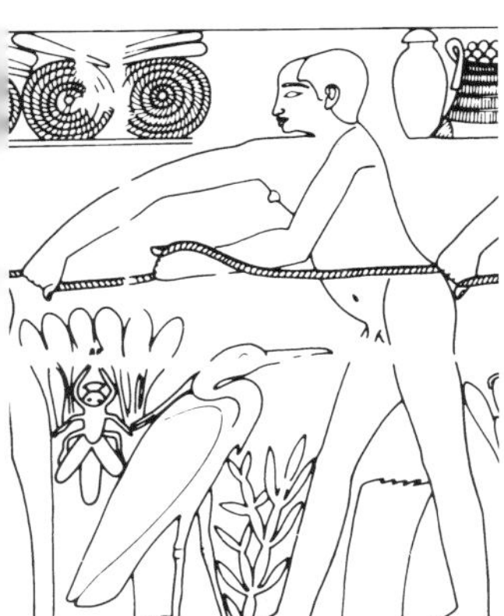

Fig. 4-5 (LB24)

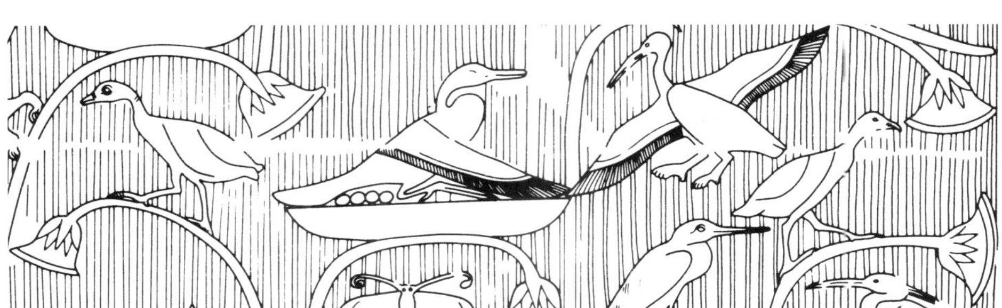

Fig. 4-6 (LB26a,b)

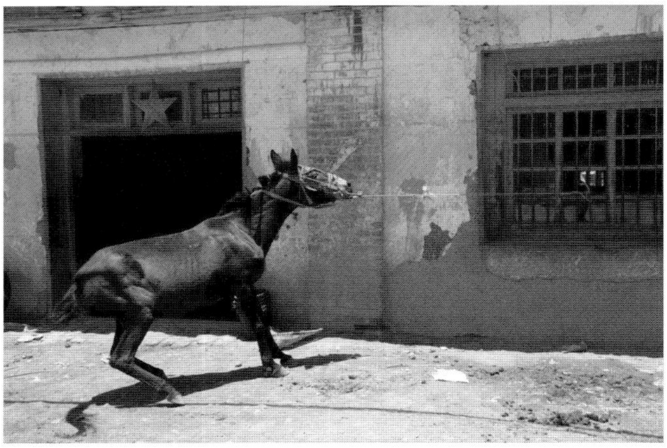

Pl. 4C Balking quadruped
© iStockphoto.com/Harald Tjøstheim

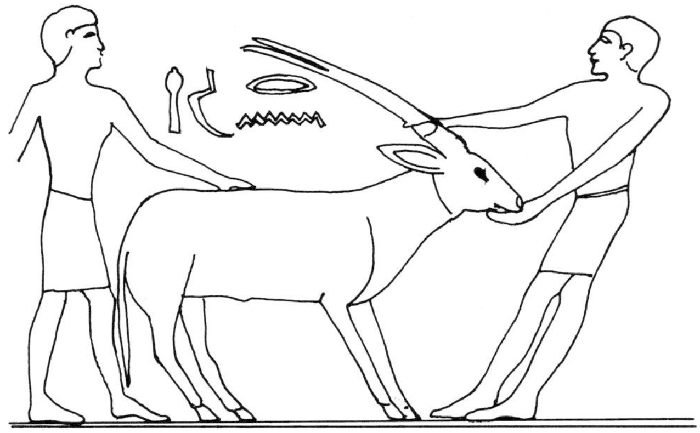

Fig. 4-7 (LB27)

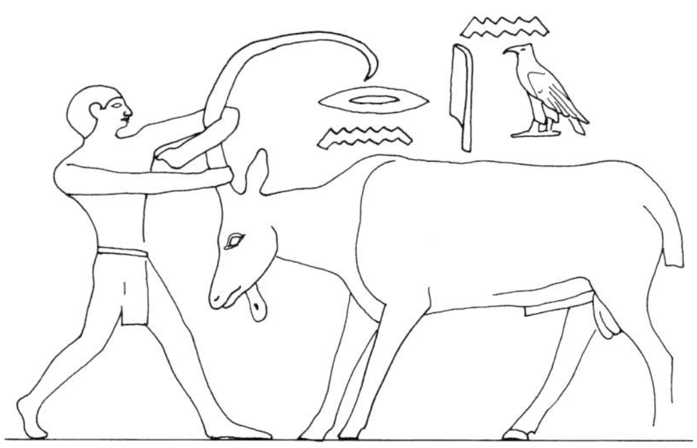

Fig. 4-8 (LB29)

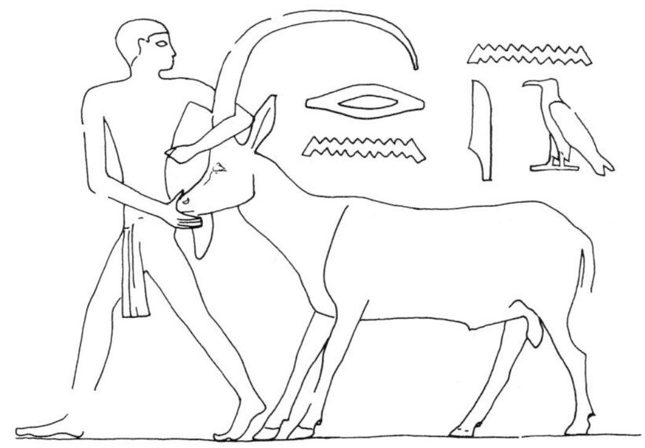

Fig. 4-9 (LB30)

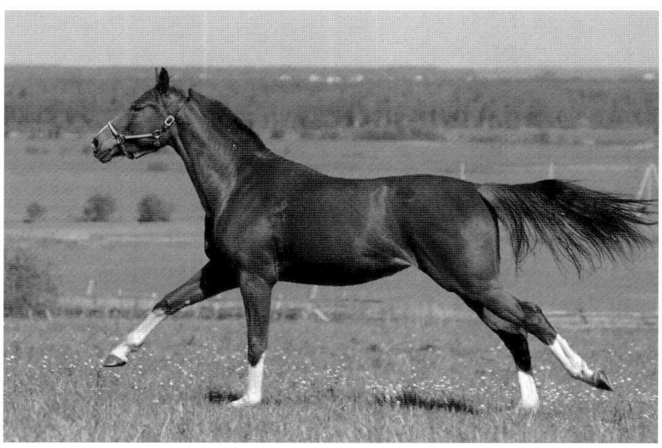

Pl. 4D Horse gallop
© iStockphoto.com/Kseniya Abramova

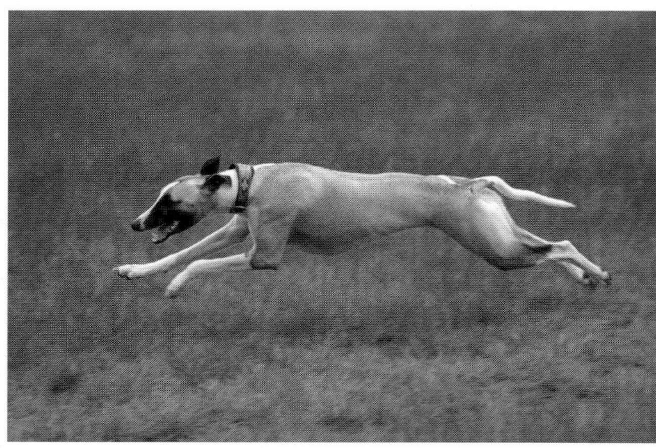

Pl. 4E Jumping gallop
© iStockphoto.com/Tamara Bauer

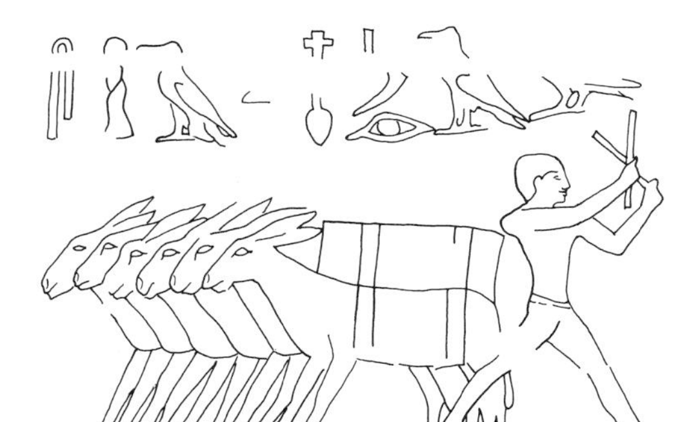

Fig. 4-10 (LB51)

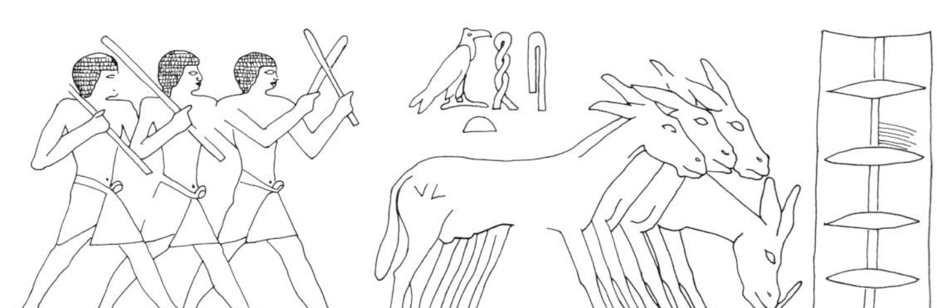

Fig. 4-11 (LB55)

The animals' refusal to move is often made more apparent by the behaviour of their human handlers who attempt strenuously to drag them forward. An image from the early tomb of *nfr-m3ʿt* is particularly striking (LB27; see fig. 4-7). In this example, an oryx has locked both its forelegs and its hind legs. An attendant stands in front of it and leans back precariously, his own legs mimicking the braced forelegs of the animal as he uses all of his weight to shift his stubborn charge. Although the inclined and paired limbs of the oryx present a clear picture of resistance, most depictions of balking animals show them with their hind legs parted (e.g., LB28 to LB30, LB33 to LB36, LB38 to LB41; see figs. 4-8 and 4-9). This allowed the stance to be differentiated from that of running animals in which both pairs of legs are usually held together (see section 4.1.3). Indeed, a balking ibex that adopts this latter pose in the tomb of *rʿ-špss* (LB31) is not convincing, even though its handler appears to pull it vigorously.

In some images, the animals stretch their necks upward while they are being towed, but ibex in the tombs of *k3-pw-nswt: k3j* (LB29; see fig. 4-8) and *ṯjj* (LB33) and a bull in the tomb of *sšm-nfr IV* (LB38) lower their heads in response to their treatment. For most ungulates, especially bulls, this head position indicates a threat (see Chapter 9). Their posture thus further indicates that they are challenging the herdsmen.

In many of the remaining examples, the attendants do not face their charges or struggle to control them. Instead, they walk forward while reaching back to hold or hook the animals' horns with one arm and grasp their muzzles with their free hands (e.g., LB28, LB30, LB40, and LB41; see fig. 4-9). I propose that in these cases, the men are not trying to move the animals forward, but rather, causing them to stand still. A technique used in the present day for controlling cattle involves standing to one side of an animal, grasping the closest ear or horn with one hand, and then pinching the septum nasi between a finger and thumb on the other hand. Cattle usually react to this nasal discomfort by pulling back their head and refusing to move.[16] It is unknown whether this technique could be used successfully with other ungulates, but note that the thumb of each handler is indeed positioned directly above the animals' nostrils (see figs. 4-7 and 4-9). Their balking postures would thus represent their attempted withdrawal from the herdsmen's handholds. This method would have enabled the Egyptians to control the movement of an animal with relative ease - releasing their hold when they wished it to proceed and pinching the skin when it was required to stop. This simple technique may have been well known in Egypt, as in many of the offering scenes examined during this study, handlers frequently guide desert ungulates by simultaneously grasping their horns and muzzle rather than using a leash.[17] Curiously, this handling method is rare in depictions of cattle.

4.1.3 Galloping

Animals exhibit two forms of gallop: cattle, donkeys, and other large ungulates produce a 'horse gallop' (pl. 4D), while a 'jumping gallop' (pl. 4E) is performed by rodents, carnivores, and pigs. During a jumping gallop, animals give flying leaps in which the trunk is stretched, the limbs are extended, and all four legs are lifted off the ground simultaneously. The horse gallop of large ungulates does not include flying leaps as their more rigid spines prevent this.[18] The horse gallop does include a moment when the animal is suspended above the ground, but during this phase all four legs are bent. When the legs are extended, at least one foot is always in contact with the ground (pl. 4D).

Depictions of donkeys have been chosen to describe the representation of galloping in Memphite tomb scenes.

16 Ewbank, R. (1968). "The behavior of animals in restraint" in Fox, *Abnormal Behavior*, 167-168.

17 The use of this hand-hold was found in 161 images from Giza and Saqqara. These include: scimitar oryx (72), Nubian ibex (37), addax (15), dorcas gazelle (15), unidentified ungulates (9), cattle (7), hartebeest (3), barbary sheep (1), donkeys (1), and deer (1).

18 Nickel et al., *Locomotor System*, 461-465.

Donkeys

jj-mrjj (LB42: Giza, V.3L)
nj-ʿnḫ-ẖnmw/ẖnmw-ḥtp (LB43: Saqqara, V.6L-7)
pḥn-wj-kȝ.j (LB44: Saqqara, V.6-8E)
ḥtp-ḥr-ȝḫtj (LB45: Saqqara, V.6-8E)
jrj-n-kȝ-ptḥ (LB46: Saqqara, V.6-9)
zȝ-jb (LB47: Giza, V.8)
sḫm-ʿnḫ-ptḥ (LB48: Saqqara, V.8-9)
ṯjj (LB49: Saqqara, V.8-9)
nfr-jrt-n.f (LB50: Saqqara, V.8-9)
kȝ.j-m-nfrt (LB51: Saqqara, V.8-9?)
ȝḫtj-ḥtp (LB52: Saqqara, V.8L-9E)
nj-kȝw-ḥr (LB53: Saqqara, V.9)
ẖnwt (LB54: Saqqara, V.9)
sḫm-kȝ.j (LB55: Giza, V.9-VI.1?)
kȝ.j-m-nfrt (LB56: Giza, V-VI)
mrrw-kȝ.j: mrj (LB57: Saqqara, VI.1M-L)
zšzšt: jdwt (LB58: Saqqara, V.9 r/u - VI.1-2E?)
mṯṯj (LB59: Saqqara, VI.2?)
mḥw (LB60: Saqqara, VI.2M-3?)
kȝ-ḥj.f (LB61: Giza, VI.2M-3?)

Twenty tombs dating from the early 5th until the mid-6th dynasty were found to display scenes in which small herds of donkeys return to the field to collect bundles of barley and emmer wheat for threshing (e.g., see figs. 4-10 and 4-11). A gang of herdsmen wielding sticks chases each group. Many of the images are simply labelled *sḥȝt*, 'donkey pack', but the captions to some scenes indicate that the herds are moving, as they refer to the animals as being 'driven' (LB45 and LB60) or 'descending' to the sheaves (LB44, LB46, LB52, and LB61). Indeed, Junker describes those in the tomb of *kȝ-ḥj.f* (LB61) as galloping.[19] The herds are depicted using 'lateral layering' in which the body of each individual partially overlaps that of the next (see Chapter 5). Schäfer suggested that this motif was employed "for groups moving forwards",[20] but in most of these scenes, the animals' motion is also emphasized by their widely spaced legs, in which each pair of fore- and hind limbs are held closely together.[21] This stance contrasts greatly with that adopted by donkeys when encumbered by heavy loads and clearly walking slowly (cf. fig. 4-1). The splayed legs and raised back heels of the men, who are also presented as layered groups, similarly suggest swift movement (see fig. 4-11).

Donkeys produce a horse gallop when running and can achieve speeds of up to 50 km per hour.[22] When they gallop, they appear to advance both forelegs or both hind legs simultaneously before moving the other pair. In fact, each member of a pair is bent, lifted, and then set down one after the other, but at such a high speed and for such a short period of time that their movements seem to be synchronous.[23]

The pose of the depicted donkeys is biologically inaccurate. Their rigid legs lack the flexion associated with movement of their limbs, and, unlike galloping animals, all of their feet remain in contact with the ground. The close pairing of their fore- and hind legs is also unnatural. Nevertheless, their widely spaced legs give the impression that their torsos are extended, resulting in a posture that resembles a phase of the horse gallop in which the body is fully elongated and both sets of legs are flung outward (cf. pl. 4D). Furthermore, in most of the examples, the forelegs of the donkeys are more inclined than their hind legs (e.g., figs. 4-10 and 4-11). They consequently appear to be stretching forward, imparting a sense of momentum to the group. Thus, although their legs and feet positions are incorrect, the animals do exhibit some characteristics associated with galloping.

The running pose is found frequently in other contexts, notably hunting scenes in which panicked animals are chased by dogs. William Edgerton tried to determine how this pose acquired its meaning and suggested that it might have been influenced by observations of flying leaps made by animals such as dogs, during which the legs are held similarly (see pl. 4E).[24] It may be significant then, that in the earliest examples of the running pose found during this study, which occur in a

19 Junker, *Giza vi*, 144.
20 Schäfer, *Principles*, 188.
21 Three of the returning donkey packs were not represented in this manner (LB53, LB57, and LB60), but are instead shown with their legs in a conventional walking pose.
22 Nowak, *Walker's Mammals ii*, 1011.
23 Gray, *Animal Locomotion*, 275.
24 Edgerton, *JAOS* 56(2), 184. Dogs, like other carnivores, are capable of producing a jumping gallop.

hunting scene in the 4th dynasty tomb of *mṯn*, both desert ungulates *and* the dogs that pursue them adopt this stance (see fig. 8-24 in Chapter 8). Whatever its origins, however, the sharp contrast between this conventional pose and that of standing/walking signals immediately to the viewer that an animal is behaving differently, while the context in which it occurs usually conveys the need for swift movement.

It should be noted that despite their leg postures, the behaviour of some of the depicted donkeys seems to indicate that they are not running. In these examples, one of the animals bends forward deeply so that its muzzle hovers above the substrate (LB46, LB47, LB50, LB59, and LB61), and in the Giza tomb of *sḫm-kꜣ.j* (LB55; see fig. 4-11), a member of the herd even appears to browse on foliage (see Chapter 7). This posture/behaviour is at odds with rapid movement. When interpreting the scene in the tomb of *kꜣ-ḥj.f* (LB61), Junker proposed that it is a composite, showing both the approach and arrival of the herd in the one image (possibly due to space constraints).[25] This explanation allows two antithetical behaviours (running and bending/browsing) to be represented, because two different time-frames are being expressed simultaneously (approach and arrival). The correct perception of such an image, however, depends upon the viewer having some knowledge of the normal behaviour of these animals.

4.1.4 Jumping

When jumping or hopping, the body is lifted above the substrate and propelled forward for some distance. The method by which different species launch themselves into the air is dependent upon their physical structure and muscular capabilities.

Jerboas

mṯn (LB62: Saqqara, IV.1-2)
nj-ꜥnḫ-ẖnmw/ẖnmw-ḥtp (LB63: Saqqara, V.6L-7)
ptḥ-ḥtp II: ṯfj (LB64: Saqqara, V.9M-L)
Anon (LB65: Saqqara, VI)
jnw-mnw (LB66-67: Saqqara, VI.2E)

Jerboas are small desert rodents with short forelimbs, extremely long hind legs and feet, and a long, tufted tail. They are most active at night and hide during the day in excavated burrows. Jerboas move by hopping on their powerful hind legs and holding out their tail for balance (see pl. 4F).[26] On rare occasions, however, they will hop slowly on all four legs.[27] They are able to jump to a height of approximately one metre and can cover a distance of between one and three metres when moving at high speed.[28]

Six depictions of jerboas were found in desert/hunting scenes situated in five Saqqara tombs. These images probably represent the lesser Egyptian jerboa, a species that is found frequently in desert habitats.[29] The earliest example, located in the 4th dynasty tomb of *mṯn* (LB62),[30] provides an accurate depiction of the hopping movement of these animals. The jerboa stands erect on its hind legs and holds its tail aloft; its body is also inclined slightly, suggesting forward motion. The animals in the remaining images also elevate their tails, indicating that they, too, are moving.

In two scenes, however, the jerboas lean forward so that their forepaws touch the ground. Both of these examples feature the animals' burrows, which are rendered as humped shapes emerging out of the substrate.[31] In one, the jerboa faces its stylised den and thus seems to be entering it (LB64;

25 "In Wirklichkeit dürften die Tiere nicht mehr laufen, denn sie sind schon bei den Garben angekommen; man hat also das Herannahen mit dem Eintreffen in einem Bilde verschmolzen". See Junker, *Giza vi*, 144.
26 Hoath, *Field Guide*, 190-191.
27 Nowak, *Walker's Mammals ii*, 1329.
28 *ibid.*
29 Hoath, *Field Guide*, 194.
30 This image thus predates the first known occurrence of a jerboa in a royal tomb, namely that of the 5th dynasty king Sahure at Abu Sir. See Borchardt, *Sahure ii*, pl. 17.
31 As noted by Schäfer (*Principles*, 124, 126-127), holes and hollows never pierce the baseline in Egyptian representations, but are instead rendered as distinct shapes placed on top of it.

see fig. 4-13),[32] while in the other it appears to be leaving (LB63). The burrows of lesser Egyptian jerboas consist of a hole at ground level, which opens to a slanted passage roughly one metre long.[33] I suggest, therefore, that the artists responsible for these scenes have tried to indicate the animals' movements associated with entry and exit from the den by causing them to bend over and use their forepaws, precisely as they would be required to do in order to hop slowly through the narrow entrance.

Although jerboas move their hind legs in unison when hopping, the artists responsible for the images in the tombs of *mtn* (LB62) and *ptḥ-ḥtp II: ṯfj* (LB64; see fig. 4-13) nevertheless adhered to the rule of the leg furthermost from the viewer advancing first. In the remaining images, however, the animals' hind limbs have been reduced to a single leg to emphasize their unusual gait (e.g., LB66 and LB67; see fig. 4-12). The jerboa that appears to float far above the substrate in the tomb of *jnw-mnw* (fig. 4-12) is possibly exhibiting a high, springing jump that is produced occasionally by this species.

Fish

mrrw-k3.j: mrj (LB68-69: Saqqara, VI.1M-L) *zšzšt: jdwt* (LB70: Saqqara, V.9 r/u - VI.1-2E?)

Two roughly contemporary tomb scenes from Saqqara depict fish leaping out of hand-held nets as they are hauled out of the water. Each dip-net is filled to capacity with tiny fish of various kinds.[34] Bony projections emanating from the heads of the jumping fish in the tomb of *mrrw-k3.j: mrj* (LB68 and LB69; see fig. 7-30 in Chapter 7), reveal that these are catfish (probably African catfish),[35] while the fins of the animal in the tomb of *zšzšt: jdwt* (LB70: see fig. 4-14) indicate that it is a mullet.[36] These species appear regularly in fishing and marsh scenes.

Both catfish and mullet jump frequently. Grey mullet are capable of clearing the surface of the water by more than three times their body length (cf. pl. 4G).[37] They occasionally leap to avoid predators, but will also do so in hypoxic (oxygen depleted) water to force air into a specialized organ at the back of their throats to increase their oxygen intake.[38] Indeed, the lower the oxygen level of the water, the more mullet will jump. The animal depicted in the tomb of *zšzšt: jdwt* might therefore be leaping in response to its capture or to the reduction in oxygen caused by its removal from the water. That the Egyptians were aware of this striking behaviour by mullet is revealed by one of its names, *sṯpw*, 'jumper'.[39]

Catfish have also been observed to escape during dip-netting in precisely the same manner as those depicted in the tomb of *mrrw-k3.j: mrj*.[40] The reason why these fish jump is less clear, but predator avoidance or oxygen depletion might equally account for their behaviour.

4.1.5 Climbing

Climbing is defined here as the ascent of an animal above the substrate while clasping a vertical surface. Some species are unable to climb due to an inability to support their body weight or because their feet lack appropriate modifications for maintaining an adequate grip (e.g., claws).[41] Species that climb do so for the purpose of feeding, hunting, and/or avoiding predators.

---

32 The forelegs of this animal are too long. The hind legs of some species of jerboas are up to four times longer than their forelimbs. See Hoath, *Field Guide*, 190.

33 Osborn-Helmy, *Contemporary Land Mammals*, 350.

34 Brewer and Friedman (*Fish and Fishing*, 38) suggest that hand nets were used to catch small to medium-sized fish.

35 *ibid.*, 67-69.

36 *ibid.*, 72-73.

37 Francesconi, *Striped Mullet*, 1.

38 Hoese, H.D. (1985). "Jumping mullet: The internal diving bell hypotheses", *Env. Biol. Fishes* 13(4), 309-314.

39 Edel, E. (1980). "*sṯpw* 'Springer' als Bezeichnung der Mugiliden: Der älteste Beleg für die Anlage von Mugilidenteichen zur Vorratshaltung", *Or* 49, 204-207.

40 Schwedler et al., *Raising Catfish*, 8.

41 Alexander, *Locomotion*, 107-110.

Goats

Goats are very agile animals and often climb into the branches of trees and shrubs to browse on leaves. See Chapter 7 for a detailed discussion of the climbing behaviour of this species.

Common genets

nb.j-m-ꜣḫtj (LB71: Giza, IV.6-V.1)
nj-ꜥnḫ-ẖnmw/ẖnmw-ḥtp (LB72-73: Saqqara, V.6L-7)
ḥtp-ḥr-ꜣḫtj (LB74: Saqqara, V.6-8E)
jtj-sn (LB75: Giza, V.6-8?)
jrj-n-kꜣ-ptḥ (LB76: Saqqara, V.6-9)
rꜥ-špss (LB77: Saqqara, V.8M)
snḏm-jb: jntj (LB78-79: Giza, V.8M-L)
ṯjj (LB80-81: Saqqara, V.8-9)
kꜣ.j-m-nfrt (LB82: Saqqara, V.8-9?)
ꜣḫtj-ḥtp (LB83-84: Saqqara, V.8L-9E)
ḫnwt (LB85: Saqqara, V.9)
jj-nfrt: šꜣ.n.f (LB86: Saqqara, V.9)
nj-kꜣw-jzzj (LB87-88: Saqqara, VI.1M)
mrrw-kꜣ.j: mrj (LB89: Saqqara, VI.1M-L)
zšzšt: jdwt (LB90: Saqqara, V.9 r/u - VI.1-2E?)
sꜥnḫ-w(j)-ptḥ (LB91-92: Saqqara, VI.1L)
ḥzj (LB93-95: Saqqara, VI.1L-2E)
mrjj-rꜥ-mrj-ptḥ-ꜥnḫ: nḫbw (LB96: Giza, VI.2)
jnw-mnw (LB97: Saqqara, VI.2E)
mḥw (LB98-101: Saqqara, VI.2M-3?)
jj-nfrt (LB102-103: Giza, VI.3-4E?)
mrjj-rꜥ-nfr: ḳꜣr (LB104: Giza, VI.4)
ꜣḫtj-ḥtp (LB105: Saqqara, VI.5?)

Common genets[42] were identified in marsh scenes in 23 tombs dating from the late 4th until the late 6th dynasty, where they are frequently depicted beside a similar small mammal, the Egyptian mongoose (see *Egyptian mongooses* below).[43] Common genets are characterised by a lithe body, short legs, long striped tail, spotted or striped coat, a sharply pointed muzzle, and pricked ears (see pl. 4H).[44] They rarely inhabit open marshland, but are abundant in vegetated areas beside watercourses;[45] indeed, access to cover is an essential requirement wherever they exist. They are primarily carnivorous and hunt both on and above the substrate; however, most of their prey is caught at ground level.[46]

In each wall scene, the genets climb a single papyrus stalk and advance toward the nests of birds (LB71 to LB105; see fig. 4-15). In the tomb of *ḥzj* (LB94; see fig. 4-16), however, one animal descends headfirst while carrying its prey. All of the animals walk in the customary quadrupedal manner in which the limbs furthermost from the viewer advance first.

Common genets can climb with great agility due to sharp, retractile claws on their fore- and hind paws with which they grip the substrate. They can ascend and descend vertical surfaces inclined up to 90°;[47] unlike less accomplished climbers, they can descend headfirst.[48] When walking, genets keep their body low, their spine horizontal, and their tail elevated. Tail elevation is particularly important when they are climbing as this helps them to shift and maintain their centre of balance.

---

42 It has been suggested that the animals depicted in tomb scenes may instead be 'large-spotted genets', however, this species is not known in Egypt and is today only found south of the Sahara. Common and large-spotted genets are morphologically very similar. See Aufrère, S.H. (1998). "La loutre, le chat, la genette et l'ichneumon, hôtes du fourré de papyrus: Présages, prédateurs des marécages et croyances funéraires", *DE* 41, 9.

43 Artists were inconsistent in their rendering of common genets and Egyptian mongooses, making it difficult at times to differentiate between images of them. I have largely concurred with the authors of the relevant tomb reports in their identification of anomalous animals, but sometimes offer a different interpretation (see LB images in Appendix 2). I concluded that a genet was intended if the animal featured pricked ears, and a mongoose if this feature was lacking. If the head region of the animal was damaged and only the tail remained, the image was usually not considered. Genets elevate their tails above the ground when they walk, whereas Egyptian mongooses drag theirs – a feature that is sometimes noted correctly in wall scenes – but, as images of genets with dragging tails and mongooses with elevated tails also occur, I felt that I could not identify the species on this detail alone. For the tail positions of genets and mongooses, see Taylor, M.E. (1970). "Locomotion in some East African viverrids", *J. Mamm.* 51(1), 44.

44 Haltenorth-Diller, *Field Guide*, 188.

45 Larivière, S. and Calzada, J. (2001). "Genetta genetta", *Mam. Sp.* (680), 3.

46 Schliemann, H. (1990). "Viverrids" in *Grzimek's Encyclopedia iii*, 518.

47 Wemmer, *Comparative Ethology*, 13.

48 Estes, *Behavior Guide*, 281.

On rough, wide surfaces (such as a tree-trunk), they can ascend using a walking gait by embedding their curved claws and gripping with their toes. The animals in wall scenes, however, climb slender stalks of papyrus, a feat that many authors have suggested is impossible (see *Egyptian mongooses* below). Ascending narrow surfaces involves an entirely different gait for the genet. In this situation, the animal grips with both of its forepaws before lifting its hindquarters and slowly advancing its hind legs. Once the hind legs have been brought close to the forepaws (such that the animal arches its back), the genet releases its grip and quickly moves its forelegs yet higher.[49] It is thus possible for genets to ascend an object as smooth and narrow as a stalk of papyrus, but their depiction in wall scenes, which show them employing a walking gait, is incorrect.

Egyptian mongooses

nb.j-m-ȝḫtj (LB106: Giza, IV.6-V.1)
nj-ʿnḫ-ẖnmw/ẖnmw-ḥtp (LB107: Saqqara, V.6L-7)
ȝḫtj-ḥtp (LB108: Saqqara, V.6-8E)
ẖnmw-ḥtp (LB109: Saqqara, V.6-8E?)
kȝ.j-dwȝ (LB110-111: Giza, V.6-9?)
rʿ-špss (LB112-113: Saqqara, V.8M)
snḏm-jb: jntj (LB114-115: Giza, V.8M-L)
ṯjj (LB116: Saqqara, V.8-9)
nfr-jrt-n.f (LB117: Saqqara, V.8-9)
kȝ.j-m-nfrt (LB118: Saqqara, V.8-9?)
ȝḫtj-ḥtp (LB119-120: Saqqara, V.8L-9E)
nbt (LB121: Saqqara, V.9)
jj-nfrt: šȝ.n.f (LB122: Saqqara, V.9)
nj-kȝw-jzzj (LB123-124: Saqqara, VI.1M)
mrrw-kȝ.j: mrj (LB125-127: Saqqara, VI.1M-L)
zšzšt: jdwt (LB128: Saqqara, V.9 r/u - VI.1-2E?)
jȝzn (LB129: Giza, VI.1M-2M)
sʿnḫ-w(j)-ptḥ (LB130-131: Saqqara, VI.1L)
ḥzj (LB132-133: Saqqara, VI.1L-2E)
mṯṯj (LB134: Saqqara, VI.2?)
mrrj (LB135: Saqqara, VI.2E?)
mḥw (LB136-139: Saqqara, VI.2M-3?)
mrjj-rʿ-nfr: ḳȝr (LB140: Giza, VI.4)
ȝḫtj-ḥtp (LB141: Saqqara, VI.5?)

The general morphology of Egyptian mongooses (also known as 'ichneumons') closely resembles that of common genets. Both are small, short-limbed, long-tailed carnivorous mammals, however Egyptian mongooses are larger and heavier than genets, their grey coat is unmarked, and their ears are small and sit tightly against their head (see pl. 4I).[50] When they walk, they drag their tail upon the ground. They inhabit dense thickets near water, such as riverine forests and reed beds, and rarely emerge into open areas. They are most active during the day ('diurnal'), but are nevertheless rarely seen, as they prefer to stay in deep cover.[51]

Egyptian mongooses were identified in 24 Memphite tombs dating from the late 4th until the late 6th dynasty. Like common genets, they are depicted ascending papyrus stalks and approaching the nests of birds (LB106 to LB141; see fig. 4-17); in one example, found in the tomb of *mrrw-kȝ.j: mrj* (LB126; see fig. 8-15 in Chapter 8), a mongoose descends headfirst. The animals adopt a walking stance in which the furthermost legs advance first (although see LB109 and LB140).

It has been noted by a number of scholars that the behaviour of both the mongooses and genets in marsh scenes is improbable as a single stalk of papyrus would be unable to support their weight.[52] A detail that has been overlooked with regard to Egyptian mongooses, however, is that these

49 Wemmer, *Comparative Ethology*, 13-14.

50 Haltenorth-Diller, *Field Guide*, 201-202; Estes, *Behavior Guide*, 298.

51 *ibid.*, 299.

52 For example, "The ichneumon is shown in a number of tombs . . . . Usually it can be found in a small cameo scene where it is eagerly climbing a papyrus stalk, which bends under its weight, in search of a bird's nest. Reality and artistic licence almost certainly merge here: the conventions of Egyptian art required a baseline on which to place individual figures, and while in most cases these were straight horizontal lines separating registers, the baseline is in these cases provided by a single papyrus stalk. The weight of the animal makes such a feat somewhat unlikely in real life" (Malek, *The Cat*, 32-33); and "Two stealthy common genets and an Egyptian mongoose (or ichneumon) are eagerly creeping up papyrus stems that merely bend under their weight, searching for eggs or unguarded fledglings. These particular details are, of course, quite imaginative, and purely the product of artistic license, since under no circumstances whatsoever could a slender papyrus stem support the weight of a living genet or mongoose. For that matter, this is also equally true for the many nests that are depicted precariously balanced on top of fragile papyrus blossoms" (Houlihan, P.F. (1996). "A guide to the wildlife represented in the great swampland scene in the offering-chapel of Ti (no. 60) at Saqqara", *GM* 155, 20).

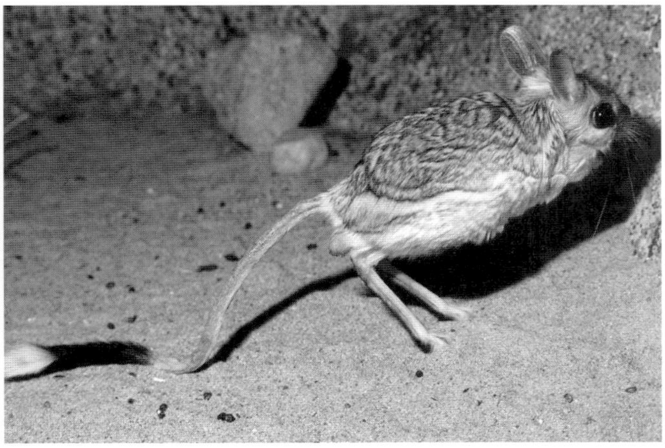

Pl. 4F Hopping jerboa
© ardea.com/Pat Morris

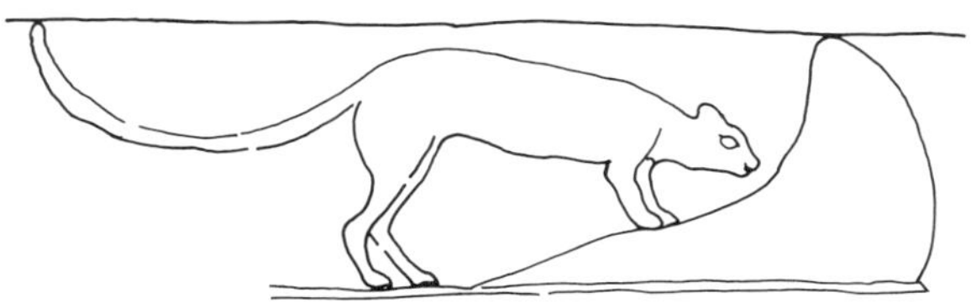

Fig. 4-13 (LB64)

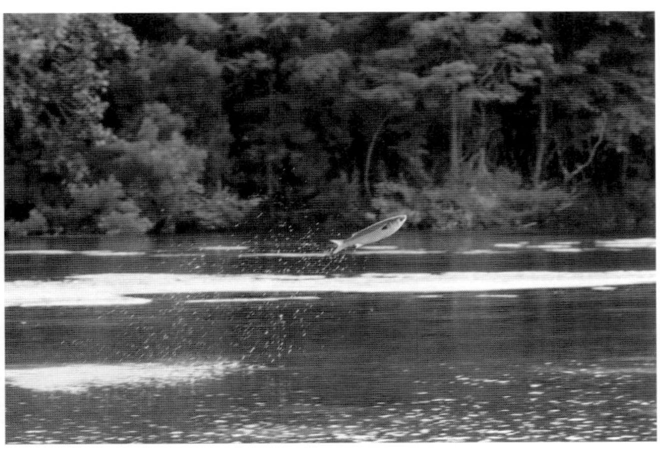

Pl. 4G Mullet jumping
© iStockphoto.com/Tony Campbell

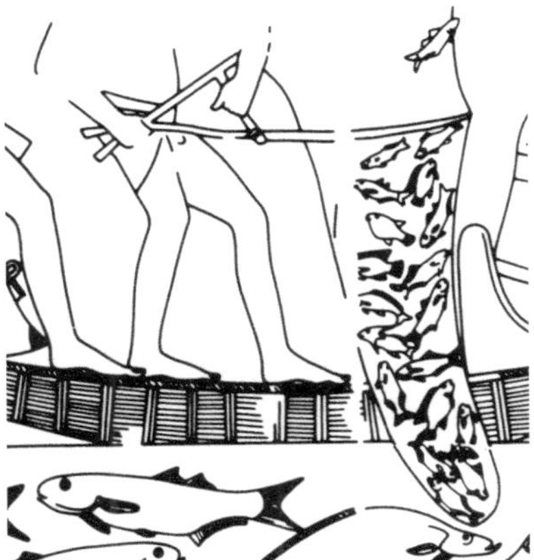

Fig. 4-14 (LB70)

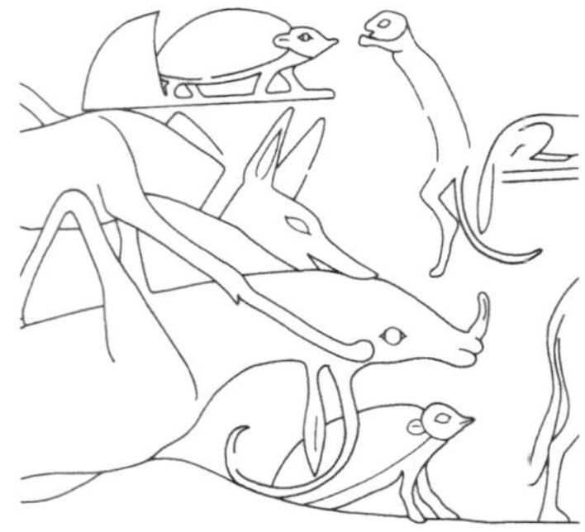

Fig. 4-12 (LB66)

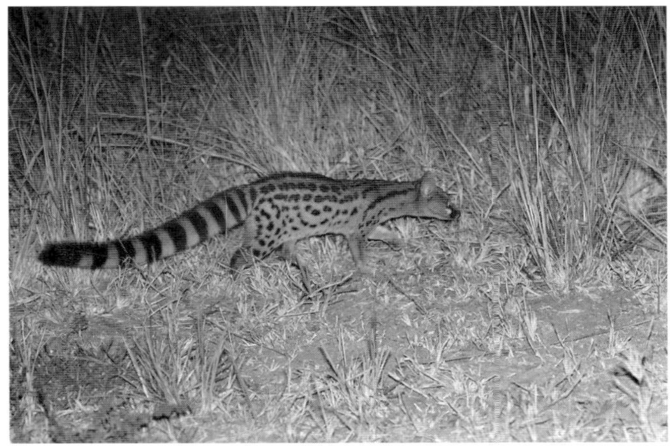

Pl. 4H Common genet
© ardea.com/M. Watson

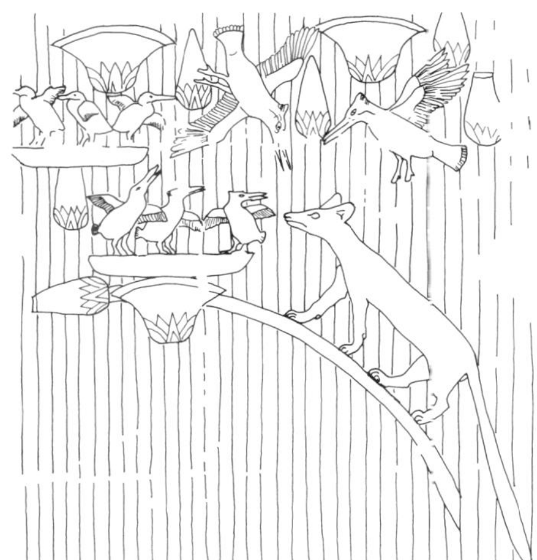

Fig. 4-15 (LB81)

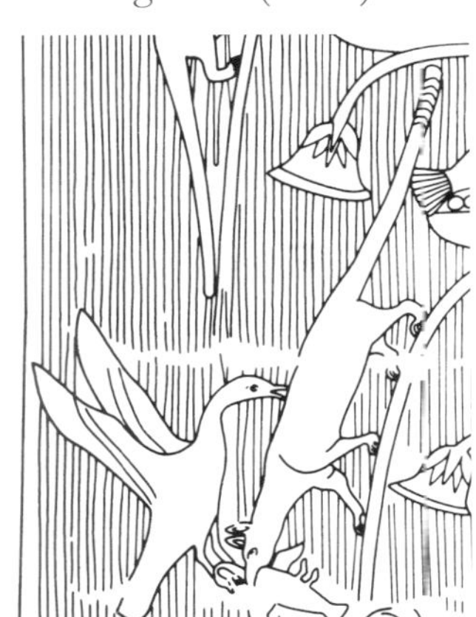

Fig. 4-16 (LB94)

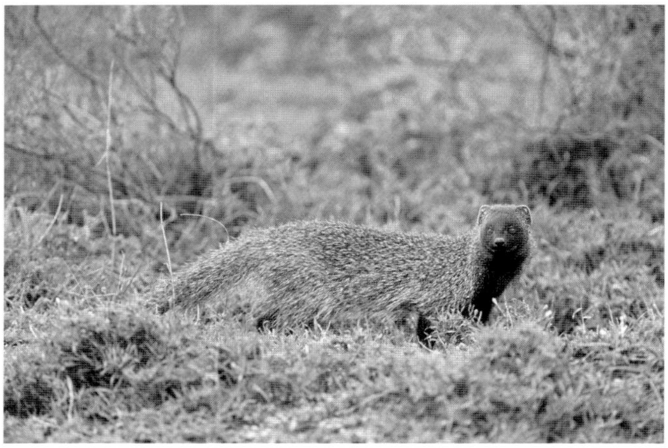

Pl. 4I Egyptian mongoose

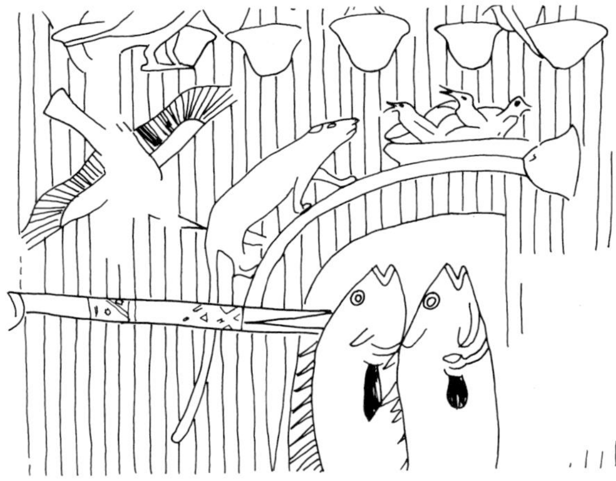

Fig. 4-17 (LB117)

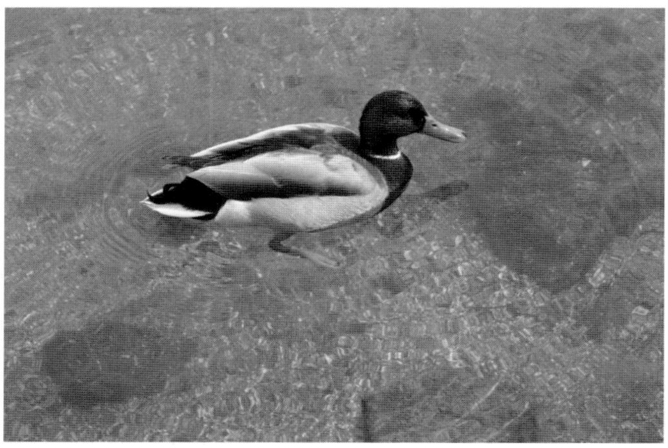

Pl. 4J Duck swimming

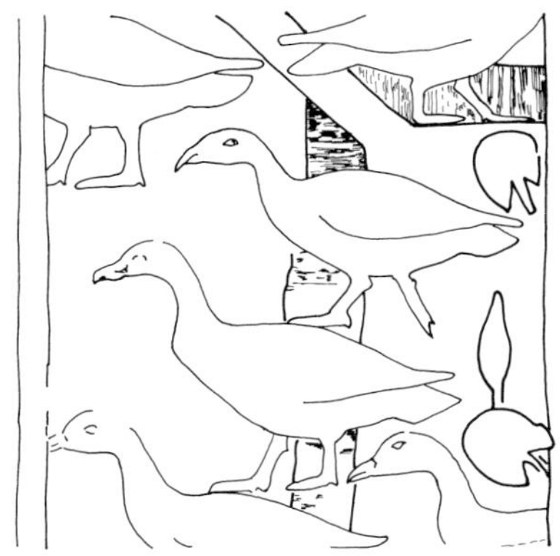

Fig. 4-18 (LB143)

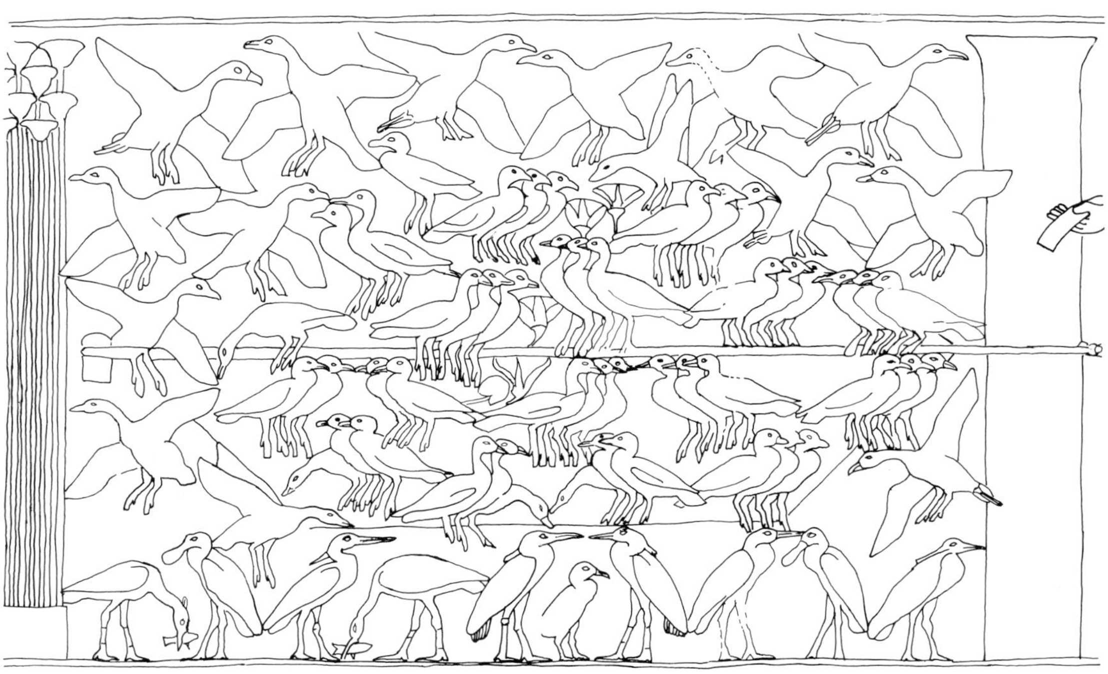

Fig. 4-19 (LB142)

animals are strictly terrestrial and do not climb at all.[53] Although they hunt for prey as shown in wall reliefs, it has been observed that, "but for a few ground-nesting species, birds would normally be beyond (their) reach".[54]

This information calls into question our understanding of what marsh scenes were intended to represent and highlights the importance of considering the biological aspects of Egyptian depictions of animals before drawing conclusions about their accuracy. I suggest, therefore, that in order to evaluate properly the behaviour of the animals shown climbing in papyrus thickets, we should begin by learning more about the plant upon which they are shown moving and the environment in which they are represented.

Papyrus (*Cyperus papyrus*) is a perennial, aquatic plant found along the banks of streams and in shallow lakes.[55] The stalks grow in large, dense thickets, along with other aquatic plants, such as reeds (e.g., *Phragmites* sp. etc.). New stems, which consist of a tough outer rind surrounding a water-laden pith, emerge from submerged, horizontal roots embedded in the mud. Apart from a few leaves at its base, each stalk tapers smoothly upward to a height of three to four metres to culminate in a single, drooping, tassel-like flower or 'umbel'. The stems reach their full height in just one year, after which they begin to bend and break, before eventually collapsing into the water below. The mass of roots and fallen stalks form a floating mat, one to two metres thick, which can be trodden upon.[56]

Papyrus thickets thus consist of three distinct layers: the accumulation of broken and decomposing stalks at the bottom, then a tangle of vertical and bent stalks, and finally the umbels at the top.[57] The umbels form a dense canopy that prevents sunlight from reaching the lower levels; visibility within the thicket is consequently poor, but the shady conditions provide ideal cover for birds that like to hide away from the eyes of potential predators. These species are difficult to see, but make their presence known through their calls and songs.[58] A few smaller varieties (e.g., warblers) are able to nest within the delicate umbels,[59] but most swamp birds place their nests amongst the collapsed stems. Many birds hunt for insects within the thicket during the day or roost there at night. The most common animals to be found in this type of habitat, apart from birds, are insects and spiders, snakes, mice, water rats and a variety of shrews.[60] These last two prey upon the eggs and young of nesting birds.

The genets and mongooses in marsh scenes hunt exclusively for birds (see Chapter 8), most of which are depicted in bowl-shaped nests that are each placed upon a single, curving papyrus stalk. The unlikelihood of this physical arrangement has been noted by Patrick Houlihan.[61] In reality, the nesting behaviour of the depicted species varies greatly. Some build large platform nests in trees (e.g., sacred ibis, grey herons), however many of the species shown (e.g., lapwings, purple gallinules, ducks, geese, and egrets) do not elevate their nests above the ground, but instead position them either directly upon the substrate or low down, within dense vegetation.[62] Furthermore, hoopoes[63] and pied kingfishers,[64] which are similarly depicted, are 'hole-nesters' (species that hide and

---

53 Estes, *Behavior Guide*, 299.

54 Kingdon, *East African Mammals iii*, 180.

55 For the importance of papyrus and its representation in Egyptian art and architecture, see "Papyrussäulen" in Borchardt, *Aegyptische Pflanzensäule*, 25-43.

56 O'Casey, *Making of Papyrus*, 3-11; Britton, P.L. (1978). "Seasonality, density and diversity of birds of a papyrus swamp in western Kenya", *Ibis* 120(4), 450; Vande Weghe, J. (1981). "L'avifaune des papyraies au Rawanda et au Burundi", *Gerfaut* 71(4), 492-498; Thompson, K. and Hamilton, A.C. (1983). "Peatlands and swamps of the African continent" in Gore, *Mires 4B*, 334-335; Parkinson-Quirke, *Papyrus*, 10.

57 Vande Weghe, *Gerfaut* 71(4), 496.

58 Someren, V. (1956). "East African birds", *Fieldiana* 38, 324.

59 Britton, *Ibis* 120(4), 458-460.

60 Speight, M.C.D. and Blackith, R.E. (1983). "The animals" in Gore, *Mires 4A*, 349-365.

61 Houlihan, *GM* 155, 20.

62 Burton, *Bird Behaviour*, 166-170.

63 Someren, *Fieldiana* 38, 198-200.

64 *ibid.*, 176.

incubate their eggs in cavities and excavated tunnels, rather than out in the open; see Chapters 8 and 9 for more details).

I suggest that the properties of papyrus, the nature of thickets, and the avifauna found within them call for a different interpretation of the behaviour of both mongooses and genets in marsh scenes, one that it is based upon, and supported by, the Egyptian method of representation. Lack of perspective in their art freed Egyptian draftsmen from the constraints of the strictly visual. This enabled them to depict the information contained within an image, rather than what they saw. For example, hidden objects, such as the contents of boxes, could be made apparent to the viewer by rendering the sides of the container invisible,[65] or by drawing the items on top or to one side of it.[66] Far away objects, undiminished in size or occluded by structures in the foreground, could be implied by the simple rule of moving them higher up within the register.[67] The demands of perspective did not need to be considered or accounted for because what we see in Egyptian art is not simply a record of what the artists witnessed visually, but what they knew.

I propose, therefore, that the predators in marsh scenes are not meant to be perceived as climbing *up* stalks, but instead, passing *through* the papyrus thickets at, or near, ground-level.[68] If the rule, 'higher up equals further away', is applied to these scenes,[69] then it quickly becomes apparent that the animals we see in the wall of vegetation above the boating tomb-owner are actually meant to be viewed as some distance from him, hidden deep within the thicket. In short, the landscape must be perceived as unfolding away from the viewer. The predators can consequently be seen to advance towards their prey through - and into - the thicket along bending horizontal - not vertical - stalks.[70] The nesting birds should equally be regarded as remote and surrounded by vegetation, their nest cups placed on curved stalks to suggest the flattened vegetation upon which these structures would be situated in nature. These scenes thus reveal to the viewer what the artists knew, or believed, lay hidden within these impenetrable stands, a representational liberty that could even allow the behaviour of hole-nesting kingfishers and hoopoes to be displayed.

4.1.6 Swimming

Movement through the water can be achieved using different techniques, including undulation of the body, jet propulsion, or rowing movements of the limbs.[71] The bodies of animals that spend all or much of their lives in water are often modified for this purpose, giving rise to such features as a water-proof outer layer, streamlined shape, webbed feet, flippers, etc.

---

65 'False transparencies'. See Schäfer, *Principles*, 121-128.

66 *ibid.*, 166-172.

67 *ibid.*, 189-198.

68 Keimer states that he observed a cat traversing a stand of papyrus in a manner similar to that of the depicted predators. The stalks bent as the animal passed over them rapidly, but did not break because they were packed so closely together. This observation convinced him that the images of genets and mongooses in marsh scenes represent a moment during their movement through a thicket, in which individual stalks bend briefly as they pass over them. He seems not to have known, however, that Egyptian mongooses do not climb. See Keimer, L. (1936). "Sur quelques représentations de caméléon de l'ancienne Égypte", *BIFAO* 36, 86-87.

69 Schäfer (*Principles*, 186-189) noted that one method by which artists implied distance was by overlapping figures in the vertical plane ('vertical layering'), but that this technique did not occur in Egyptian art until the Middle Kingdom. Distance could also be conveyed by 'loose groups' (*Principles*, 189-198), in which isolated objects placed higher in the picture field were to be perceived as further away. The late occurrence of vertical layering might indicate that the second method is also missing from reliefs of the Old Kingdom, but Schäfer believed that this form of representation was so fundamental, that it is likely to have arisen earlier. Its first occurrence cannot be determined, however, due to the difficulty in distinguishing between compositions in which isolated objects are unrelated and those in which they are ordered spatially. Consequently, he suggested that it was "a matter of personal opinion whether a given picture should be taken as a view over a wide area or not" (*Principles*, 193).

70 Aufrère (*DE* 41, 8) suggests similarly that the animals are depicted upon individual stalks to represent the effect of their passage upon the vegetation as they walk through the thicket.

71 Alexander, *Locomotion*, 13.

Birds

ptḥ-ḥtp I (LB142: Saqqara, V.8M-L)

ṯjj (LB143: Saqqara, V.8-9)

Waterfowl appear in a range of contexts in wall reliefs, including offering, pastoral, marsh, and clapnet scenes. In some of these situations, they are provided with an opportunity to swim, either in man-made pools or natural waterways. The foot actions performed by waterbirds during swimming differ from those exhibited when walking upon a hard substrate. Artists noticed this distinction and incorporated it into their representations of avian aquatic behaviour. Only two examples are described here, but they have been chosen specifically because they illustrate, within the same scene, the two different foot positions.[72] A clapnet scene in the tomb of *ptḥ-ḥtp I* (LB142; see fig. 4-19) and a poultry-yard scene in the tomb of *ṯjj* (LB143) reveal that while birds on dry land are depicted in the standard bipedal pose (see *Bipeds* above) with their feet flat upon the substrate, those that are situated in ponds purse their toes. This foot position matches the natural behaviour of swimming waterfowl.

Swimming birds paddle with alternating strokes, one foot pushing the water behind as the other returns.[73] Webbing between their toes increases the surface area of each foot. As the foot is thrust back, the membranes spread out to provide a broader surface with which to push against the water.[74] As the foot is brought forward, however, the toes are pursed and the webbing folded to reduce the amount of resistance (see pl. 4J), enabling the bird to glide quickly and easily through the water.

The distinction that artists made between land and water foot poses is particularly clear in the depiction of a goose emerging from a pond in the *ṯjj* poultry-yard scene (LB143; see fig. 4-18), for while the toes of its foot on land are flat, those still in the water are pursed. Flocks of swimming birds are occasionally depicted with flat feet, but these images are not common.[75] Thus, although the submerged feet of birds are usually not visible from a distance, artists at Giza and Saqqara nevertheless chose to reveal this hidden feature and use it as a marker to ensure that the viewer understood clearly the animals' aquatic behaviour.

Nile crocodiles

ḥtp-ḥr-ȝḫtj (LB144: Saqqara, V.6-8E)
snḏm-jb: jntj (LB145-146: Giza, V.8M-L)
ṯjj (LB147: Saqqara, V.8-9)
sšm-nfr IV (LB148-149: Giza, V.9-VI.1)
ȝḫtj-mrw-nswt (LB150: Giza, V.9-VI.1)
nj-mȝʿt-rʿ [*jzzj-mr-nṯr*] (LB151: Giza, V.9-VI.1)
kȝ-gm-nj: mmj (LB152a,b: Saqqara, VI.1M)
ḥzj (LB153-154: Saqqara, VI.1L-2E)
jj-nfrt (LB155: Giza, VI.3-4E?)

Nile crocodiles do not occur regularly in tomb reliefs until the mid- to late 5th dynasty. Thereafter, they are depicted frequently in fording and marsh scenes, and occasionally in other contexts, such as harpooning, boating, and fishing activities. In the majority of images, the animals are shown deep underwater, resting upon the riverbed (e.g., LB146; see fig. 4-20, left, and examples in Chapters 5 and 8). In a subset of tombs, however, they are suspended above the substrate and thus appear to be swimming (e.g., LB146; see fig. 4-20, right). In early scenes, the crocodiles hover just above the river bottom (e.g., LB144, LB145, and LB147; see fig. 4-21), but in later images they are positioned closer to the water surface (see fig. 8-37 in Chapter 8). Their bodies are sometimes held horizontally, but most are inclined upward slightly (LB146, LB151 to LB154; see fig. 4-20, right). In the tomb of *ȝḫtj-mrw-nswt*, the heads of two crocodiles actually emerge from the water (LB150; see fig. 4-22). One of these animals has its jaws agape, a pose that is also exhibited by four other

72 For more examples, the reader is referred to clapnet scenes described in *Confidence decoys* in Chapter 5.
73 Leahy, *Companion*, 697.
74 *ibid.*, 408.
75 Thirty-five of the 42 clapnet scenes examined during this study were sufficiently well preserved to determine the shape of the birds' feet. Birds with flat toes were found in only six images. The remaining scenes all show waterfowl with pursed toes, the earliest example of which occurs in the early 5th dynasty tomb *wr-jr.n-ptḥ* at Saqqara.

swimming crocodiles (LB148, LB152a, LB152b, and LB153; see fig 8-36 in Chapter 8). This behaviour is not usually displayed by crocodiles resting upon the riverbed, except for those that are eating (see Chapter 7) or about to bite the newborn young of hippopotamuses (see Chapter 11).

Unlike images of other quadrupeds, the depicted crocodiles do not display both pairs of legs. Instead, only the limbs closest to the viewer are portrayed.[76] In nature, Nile crocodiles position their legs quite differently depending upon the speed at which they are moving, and whether they are on land or in the water.[77] When they are at rest or basking on land, their fore- and hind limbs are bent sharply and their ventral surface lies flat upon the substrate (see pl. 4K). When moving at slow speeds, they usually 'belly crawl' by holding their stomach and tail on or slightly above the ground and pulling themselves along by placing their splayed legs in an alternating diagonal sequence.[78]

The limbs of crocodilians play a lesser role when they swim. Trailing their legs back against their body to reduce drag, they move forward by sweeping their tail from side to side in a sinusoidal pattern and pointing their head in the direction they wish to go (see pl. 4L).[79] At slow speeds, however, and when hanging motionless in the water, their limbs are held out from their body for stability and to allow them to correct their direction. They position themselves in the water by adjusting the amount of air in their lungs. This enables them to either float at the water surface with only their head or nostrils apparent (see pl. 8Q in Chapter 8) or to submerge completely to either swim or walk and rest upon the riverbed (where they can stay for up to 45 minutes or longer if necessary).[80] Crocodiles are also capable of raising their head and body far above the surface of deep water by undulating their tail at speed and pushing themselves upward. In shallow water, they can also use both their feet and tail to lunge forward suddenly on to the shore. Both of these movements are used to catch prey.

Crocodiles depicted directly upon the riverbed bend their legs abruptly and flatten their bellies against the substrate (e.g., see fig. 4-20, left). This matches both the resting posture of these animals and their low-slung belly crawl. The Egyptians may have observed crocodiles lying in repose below them when boating over clear water; however, it is perhaps more feasible to suggest that this underwater pose was probably generated from those seen on land (cf. pl. 4K).

In scenes of apparent swimming, the animals adopt an identical posture - that is, their legs are bent in a resting position, rather than back against the sides of their body. This would seem to indicate that artists were uncertain how crocodiles swim and so simply elevated images of resting reptiles in order to convey this behaviour. The gaping jaws of many of the swimming animals may also reveal that their pose is land-based as Nile crocodiles frequently open their mouth in this manner when basking (see pl. 4K) or lying in shallow water, to aid thermoregulation.[81]

Nevertheless, small details suggest that some artists were familiar with the aquatic movements of crocodiles. In the tomb of *sndm-jb: jntj* (LB146; see fig. 4-20, right), the pointed tip of the reptile's tail curves downward languidly in manner that might suggest its sinuous movement while swimming (cf. pl. 4L); note that the tail of the resting crocodile with which it shares the scene does not display this feature. So, too, in the tomb of *ꜣḫtj-mrw-nswt* (LB150; see fig. 4-22, right), the tail of one of the crocodiles is curved and lifted even though its body rests upon the substrate. This animal displays the classic crocodilian pose in which just the top of its head and nostrils break the

76 Schäfer (*Principles*, 105) believed that a fully lateral view is provided for crocodiles because they are short-legged.

77 Cott, H.B. (1961). "Scientific results of an inquiry into the ecology and economic status of the Nile crocodile (*Crocodilus niloticus*) in Uganda and Northern Rhodesia", *Trans. Zool. Soc. Lond.* 29, 229-232.

78 The structure of their ankle joints enables crocodilians to rotate their feet so that they can point forwards as well as out from the sides of their body. Other reptiles, such as lizards, do not possess this ankle structure and so are unable to direct their feet forward.

79 Cott, *Trans. Zool. Soc. Lond.* 29, 232; Mazzotti, F. (1989). "Structure and function" in Ross, *Crocodiles*, 43.

80 Cott, *Trans. Zool. Soc. Lond.* 29, 233-235. See also Seebacher, F., Franklin, C.E. and Read, M. (2005). "Diving behaviour of a reptile (*Crocodylus johnstoni*) in the wild: Interactions with heart rate and body temperature", *Phys. Biochem. Zool.* 78(1), 1-8.

81 It should be noted, however, that crocodiles are able to open their mouth while underwater without risk of drowning due to a 'palatal valve' in their throat, which, when shut, prevents water from entering their trachea.

surface of the water - a posture that is associated with both swimming and floating (pl. 8Q). The head of the other reptile in this scene (LB150; see fig. 4-22, left) appears to burst from the band of water; it is tempting to see in this highly unusual representation the upward thrust exhibited by Nile crocodiles when attacking.

The Egyptians were undoubtedly aware that a crocodile on the move is more dangerous and a more immediate threat than one that is resting in deep underwater. The tendency for swimming crocodiles to be displayed with open jaws may in fact point to an association drawn by the Egyptians between movement and an imminent strike; indeed, each gaping reptile approaches a possible prey item - namely, fish (LB153), hippos (LB148, LB152b; see fig. 8-36 in Chapter 8), and cattle (LB150 and LB152a; see fig. 4-22). The greater perceived risk posed to the deceased within the magical world of the tomb environment by depictions of swimming crocodiles would thus explain the lower incidence of such scenes compared to those of reptiles in repose.[82]

Nile soft-shell turtles

ḥzj (LB156: Saqqara, VI.1L-2E) *mḥw* (LB157-158: Saqqara, VI.2M-3?)

Images of turtles were located in only two tombs during this study, both dating to the early 6th dynasty. This small number reflects the relative rarity of such animals in wall scenes generally throughout the pharaonic period.[83] Their more frequent representation in other media, such as palettes and amulets, has allowed the species to be identified as the Nile soft-shell.

Soft-shell turtles are so-named because their low and slightly domed carapace is covered with leathery skin, rather than horny shields. Their most distinctive feature, however, is their nostrils, which are long and tapered to form a type of proboscis. They are able to absorb oxygen from the surrounding water through both their skin and throat. This enables them to stay submerged for hours at a time, often buried in silt where they lie in wait to strike at passing prey. When they do come up to the surface to breathe they simply project their nostrils out of the water, enabling them to remain completely hidden (see pl. 4M).[84] Combined with their nocturnal habits, Nile soft-shell turtles are consequently a most secretive and elusive species.

Images in both tombs show the turtles floating in the water as boating parties glide by. On the north wall of room II in the tomb of *mḥw* (LB158; see fig. 4-24), the animal is depicted in a vertical position such that its nose intersects with the water surface. In another scene in the same tomb (LB157; see fig. 4-25) and in the earlier tomb of *ḥzj* (LB156; see fig. 4-23), turtles are also presented vertically, but at an incline. Their heads have not been preserved, but their bodies indicate that the animals' snouts were also originally located just beneath the water surface. I suggest that the decidedly upright orientation of the three turtles is intended to represent their curious tendency to lurk in the shallows while breathing via their protruding nostrils.

The *ḥzj* and *mḥw* scenes are predated by a relief fragment from the sun-temple of Niuserre that also displays a pair of soft-shells.[85] Unfortunately, it is impossible to determine the orientation of these animals, but later depictions of turtles usually show them swimming horizontally.[86] The upright orientation of the *ḥzj* turtle may thus record a unique observation of their behaviour, copied subsequently in the tomb of *mḥw*, but then no longer repeated.

82 Nile crocodiles were found in 32 tombs during this study: 22 from Saqqara and 10 from Giza. Nine of these tombs display swimming crocodiles: 4 from Saqqara and 5 from Giza. Thus, half of the Giza tombs in which crocodiles appear show them in a swimming pose, indicating that the motif was especially popular at this location.

83 Fischer, *Turtles*, 9.

84 *ibid.*, 5-6; Obst, F.J. (1998). "Turtles and tortoises" in Cogger-Zweifel, *Reptiles and Amphibians*, 117.

85 Wenig, *Jahreszeitenreliefs*, fig. 10.

86 See Fischer, *Turtles*, fig. 4.

Fish

nj-mȝʿt-rʿ [*jzzj-mr-nṯr*] (LB159: Giza, V.9-VI.1) *ḥzj* (LB160: Saqqara, VI.1L-2E)

Fish are able to swim via the action of small muscle movements that run down the length of their body. The expansion and contraction of these muscles alternate on either side of their torso, bending the animal's body first one way, then the other. At the end of each cascade of movements, their tail is pulled back and forth, propelling the fish forward. The muscular contractions cause some fish, such as eels, to undulate their bodies in a sinusoidal fashion as they swim (see pl. 4N), but in most other species, only the tail is seen to move.[87]

Fish appear frequently, and abundantly, in wall scenes in one of three conditions - dead, caught, or free-swimming. Images of free-swimming fish in marsh scenes (including the 'mound of water') were examined during this study for evidence of species-specific locomotor behaviour. This exercise revealed that the animals are presented in a uniform manner (although see below). Due to the identical nature and large number of these images, only two representative scenes (one from Giza and one from Saqqqara) are considered here.

Most fish species are depicted in a side-view; walking catfish, uniquely, are shown with their bodies in profile, but their heads as though seen from above, such that both eyes are represented[88] (see (c) in fig. 4-27). Fish are largely represented horizontally, but with a slight rise or fall as though to suggest their movement through the water.[89] Upside-down catfish are frequently inverted (see (a) in fig. 4-27), which is a characteristic behaviour of this species (see Chapter 7 for more details). The orientation and depth of other species may also have behavioural significance, as some Nile fishes, such as walking catfish, are 'air gulpers' that must come to the surface frequently to breathe.[90] Logging the precise location of individual species was beyond the scope of the present study, but may be worthy of further investigation.

Each fish is usually depicted as a separate entity, unimpeded by the bodies of others. Also, members of the same species do not often appear next to one another in great numbers. In the Saqqara tomb of *ḥzj* (LB160; see (d) in fig. 4-27), however, both of these rules are broken in an apparent depiction of schooling (see Chapter 5).

---

[87] Biewener, *Animal Locomotion*, 106-108.

[88] This special treatment of *Clarias* catfish is curious. Brewer and Friedman (*Fish and Fishing*, 61) observed that it is difficult to keep the head of this species in a profile view when positioning a dead specimen. This might account for its pose, however I suggest instead that a solution lies in a behavioural feature that is exhibited only by *Clarias* fish. Walking catfish possess a specialised breathing organ that enables them to survive out of water. Furthermore, as their common name implies, this species can move over land for some distance by flexing their bodies and 'walking' on their pectoral spines. This behaviour allows them to travel to a new water source if their home pond dries up or becomes contaminated after heavy rain (see Johnels, A.G. (1957). "The mode of terrestrial locomotion in Clarias", *Oikos* 8(2), 122-129). For example, Brewer and Friedman (*Fish and Fishing*, 60) noted that a column of c. 30 *Clarias lazera* (a species found in the Nile), has been observed walking from a swamp to a new location in a river, covering a distance of around 200 metres in an hour. They also report that Nile fishermen today believe that *Clarias* species leave the river to feed on vegetation. I propose that the ancient Egyptians witnessed this locomotor behaviour and that consequently, this species alone is depicted with its head as though seen from above because this is *precisely* how it would have been viewed as it walked over land. The pose also helps to differentiate this extraordinary species from African and upside-down catfishes with which it often shares wall scenes.

[89] It is difficult to understand the artists' intentions with regard to the underwater view. Schäfer (*Principles*, 239-240) maintained that, despite their appearance, underwater scenes are not cross-sections, but frontal views seen from above. To make his point, he compared such images to depictions of gaming boards and their pieces in which the horizontal plane is clearly lifted. While this interpretation is consistent with the Egyptians' general method of representation, I believe that, because an apparent flat plane of water *also* has depth (which a gaming board patently does not), it is possible that this area was treated differently. The position of fish within the water column, for example, which are often inclined upward or downward, seem intended to suggest a sense of depth.

[90] Brewer-Friedman, *Fish and Fishing*, 60.

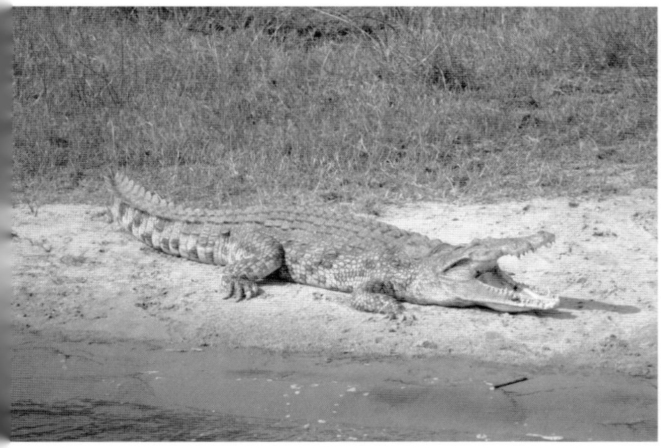

Pl. 4K Nile crocodile resting
© iStockphoto.com/Frank van den Bergh

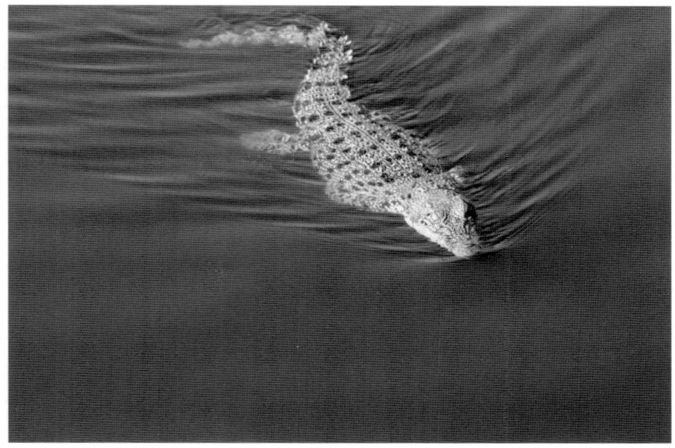

Pl. 4L Nile crocodile swimming
© iStockphoto.com/George Clerk

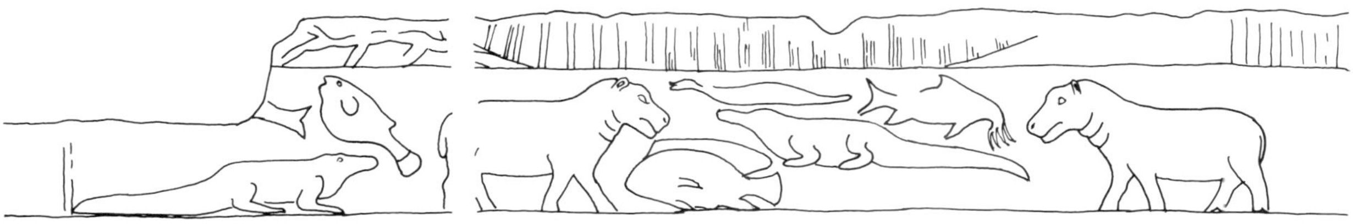

Fig. 4-20 (LB146)

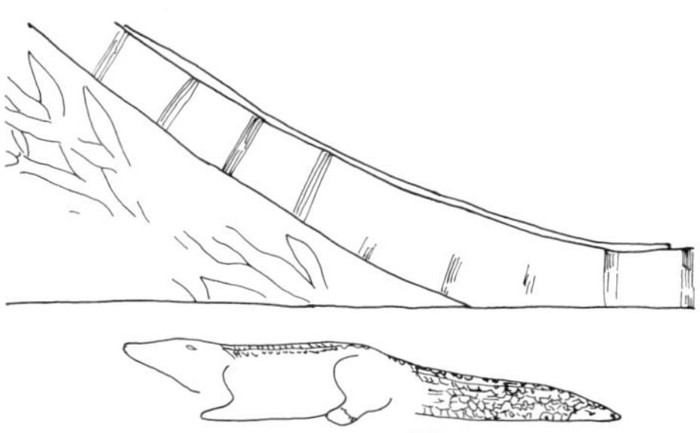

Fig. 4-21 (LB144)

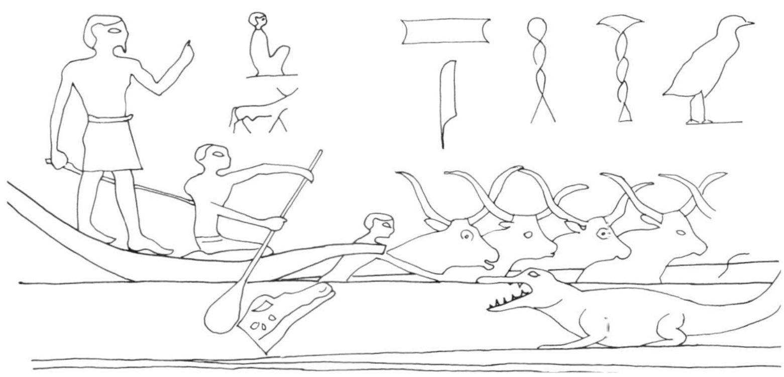

Fig. 4-22 (LB150)

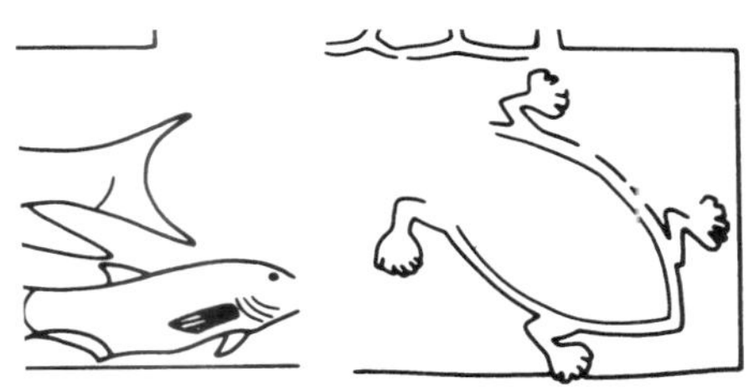

Fig. 4-23 (LB156)

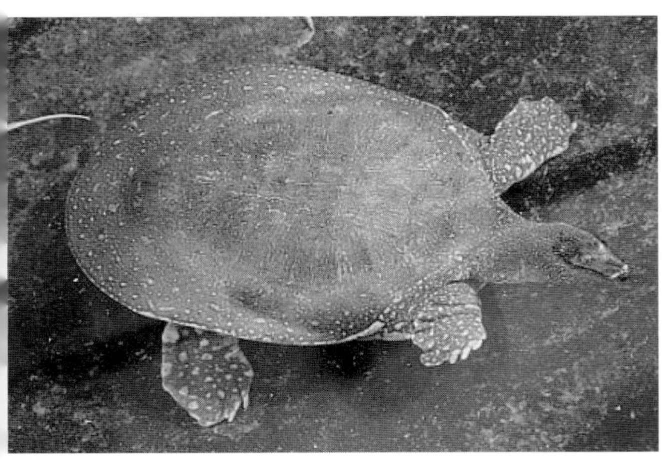

Pl. 4M Nile soft-shell turtle
© Andreas S. Hennig
Reproduced with permission

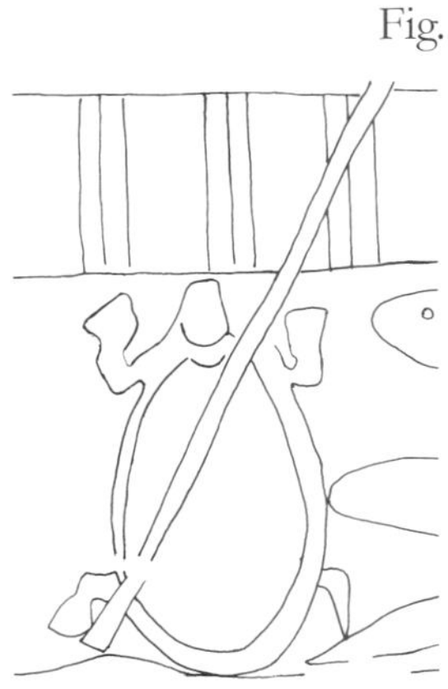

Fig. 4-24 (LB158)

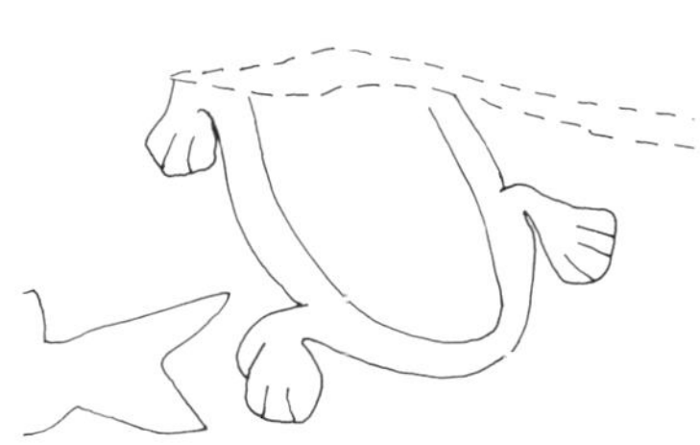

Fig. 4-25 (LB157)

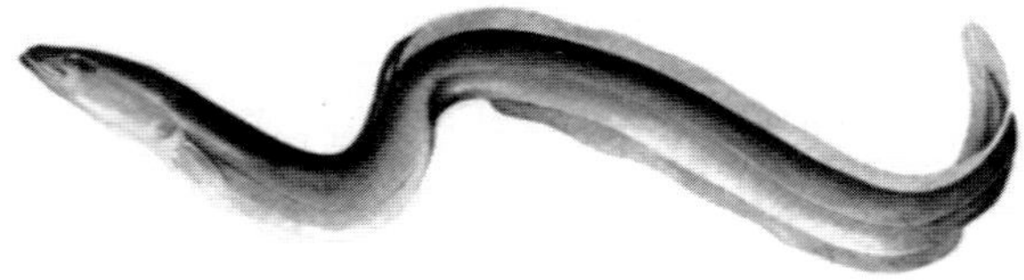

Pl. 4N Eel swimming
© Demetris Mina
Reproduced with permission

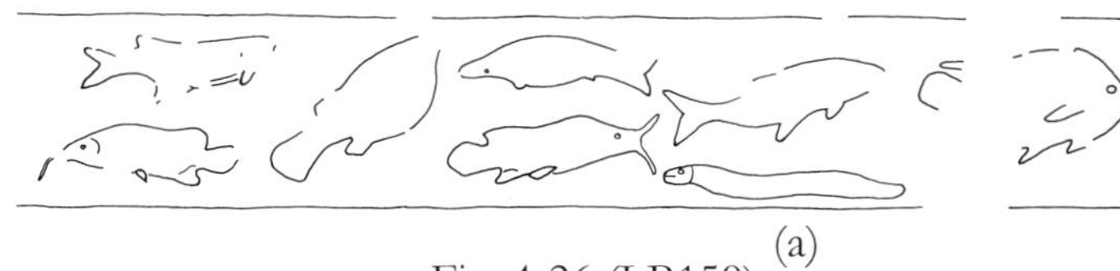

Fig. 4-26 (LB159)

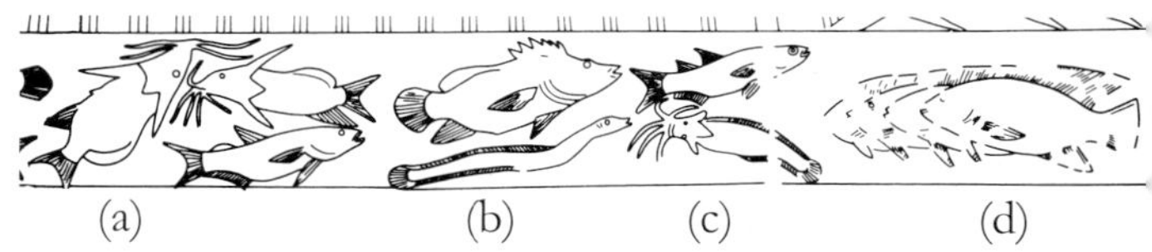

Fig. 4-27 (LB160)

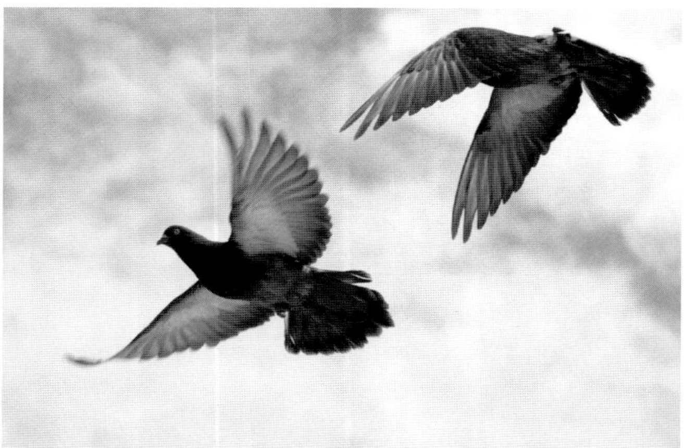

Pl. 4O Bird flight
© iStockphoto.com/Sue Mack

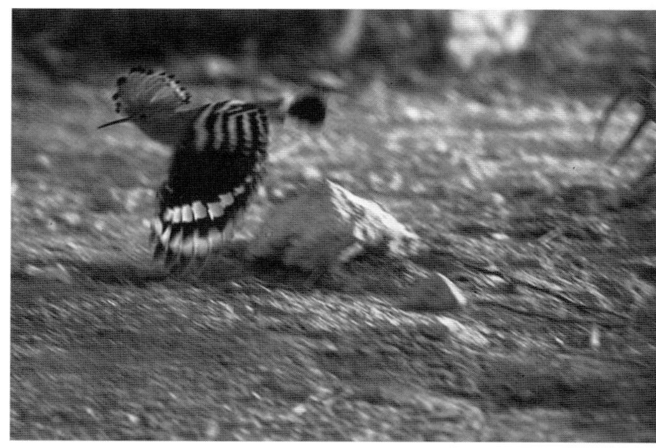

Pl. 4P Hoopoe flight
© G. Gerra and S. Sommazzi – www.justbirds.it
Reproduced with permission

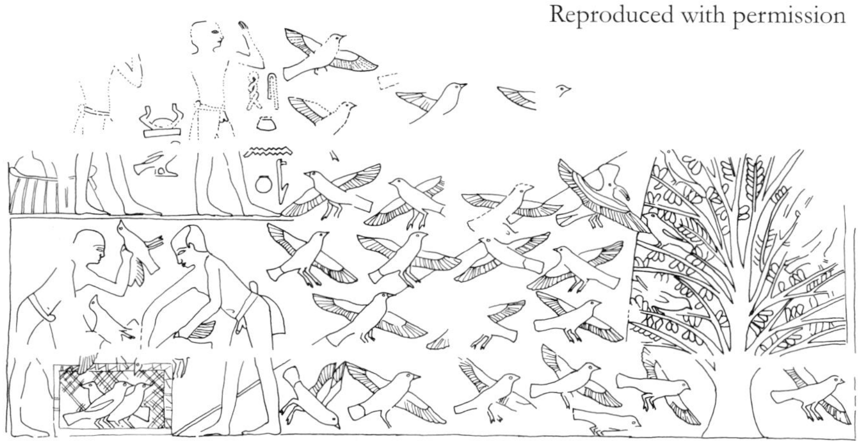

Fig. 4-28 (LB167)

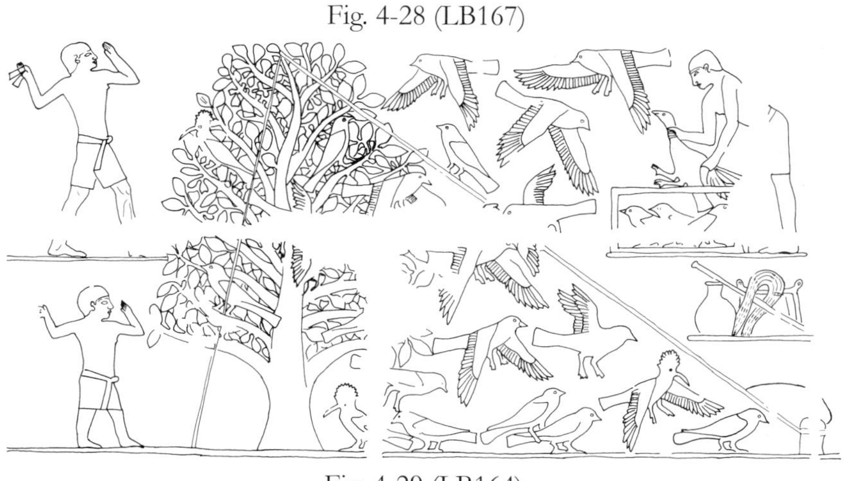

Fig. 4-29 (LB164)

No distinctive tail or fin postures were detected by which a sense of individual motion might be implied or understood. This is not surprising as the horizontal, side-to-side movement of their tails combined with the streamlined shape of their bodies renders the depiction of fish locomotion in two dimensions particularly challenging. Eels, however, were often found to display a slight bend roughly halfway down the side of their body (see (a) in fig. 4-26 and (b) in fig. 4-27), possibly in an effort to show their undulating, snake-like movement when swimming (cf. pl. 4N).

4.1.7 Flying

Natural powered flight is achieved via the shape and movement of wings, which together generate lift aerodynamically. Both birds and insects in wall scenes express this behaviour. The hovering and diving movements of some animals, such as the pied kingfisher, are particularly distinctive and tied intimately to their hunting behaviour. This is treated in Chapter 8.

Birds

jj-mrjj (LB161: Giza, V.3L)
nj-ʿnḫ-ẖnmw/ẖnmw-ḥtp (LB162: Saqqara, V.6L-7)
ḥtp-ḥr-ꜣḫtj (LB163: Saqqara, V.6-8E)
ꜣḫtj-ḥtp (LB164: Saqqara, V.6-8E)
rʿ-šps (LB165: Saqqara, V.8M)
ṯjj (LB166: Saqqara, V.8-9)
nfr-jrt-n.f (LB167: Saqqara, V.8-9)
mrrw-kꜣ.j: mrj (LB168: Saqqara, VI.1M-L)
nj-ʿnḫ-nswt (LB169: Saqqara, VI.1-2E)
ḥzj (LB170: Saqqara, VI.1L-2E)
ʿnḫ-m-ʿ-ḥr: zzj (LB171: Saqqara, VI.1L-2E)

Flying birds enliven many tomb reliefs, particularly in clapnet, agricultural, and marsh scenes. To illustrate the manner in which the Egyptians depicted avian flight, birds at wing in netting scenes are considered here (LB161 to LB171).

Birds are able to rise (achieve 'lift') due to the aerodynamic shape of their wings, but to move through the air they require some form of propulsion. They achieve this by flapping their wings. The wing-beats of flying birds are extremely complex, with both the angle of each wing and its feathers adjusting in subtle and interdependent ways during each stroke (see pl. 4O).[91] To the unaided eye, however, bird flight appears to be quite simple - first the wings are lifted above the body and then they descend to push the air down and back. The Egyptian depiction of flight can thus only be compared to this apparent motion.

Examination of the 11 netting scenes from Giza and Saqqara reveals immediately that the birds hold their wings in a variety of postures. The most common arrangement, however, consists of each wing outstretched and attached to the outline of the bird's torso, which is rendered in profile (e.g., LB167; see fig. 4-28). This is at odds with Heinrich Schäfer's assertion that the way in which the animals' wings were adhered to the body changed in the 4th dynasty so that thereafter, "one wing grows out of the body itself without a dividing line, while the other emerges from the clearly marked outline".[92] This pose does occur (LB161, LB163 and LB166), but it is less common. Flapping is also conveyed in tomb reliefs by the wings held above the body (e.g., LB162 to LB164, LB167, LB170, and LB171; see fig. 4-29) or by depressing the nearer wing while keeping the other horizontal (e.g., LB164 and LB170; see fig. 4-29). Uniquely, in the tomb of *ṯjj* (LB166), a number of birds are depicted with only one wing, possibly in an attempt to show horizontal flight. These non-standard wing postures inject variety and impart a sense of fluttering movement to each scene. Nevertheless, a bird with outstretched wings was the predominant motif by which flight was represented. By adjusting the angle of the torso while in this posture, artists could also produce convincing depictions of birds that appear to dive and soar.

Morphological details have occasionally been included that help the viewer to determine the animals' orientation in space. In the tomb of *nfr-jrt-n.f*, the wings of a hoopoe (LB167; see fig. 4-28, left of the tree) have been rendered as a continuous mantle. The breast feathers of this species

91 Leahy, *Companion*, 281-289; Butler, P.J. (1991). "Movement" in Brooke-Birkhead, *Ornithology*, 53-56.
92 Schäfer, *Principles*, 260.

are a uniform colour, but both their wings and back are marked with black and white stripes.[93] During flight, the bands on the back become more visible so that the birds appear to have a set of contiguous stripes running from one wing to the other (see pl. 4P). The *nfr-jrt-n.f* artist has incorporated this unusual feature in his depiction, enabling the hoopoe to be perceived as flying away from the viewer.[94]

One feature that is shared by all the depicted birds is the position of their legs, which, regardless of their wing posture or orientation, dangle beneath them with their toes pursed. In nature, however, these are usually lifted and either held close to the body during flight or trail back behind to help maintain a streamlined shape and reduce drag. Their legs only hang down in the manner shown for a short time after take off and when landing. Perhaps the importance of presenting the birds' bodies in their entirety overrode the need for biological accuracy, or representational constraints (such as the need to reveal the limb furthermost from the viewer advancing first), prevented artists from lifting the animals' legs into a more natural position.

Butterflies

sḫm-kꜣ-rꜥ (LB172: Giza, V.2)
nfr/kꜣ-ḥꜣ.j (LB173: Saqqara, V.6)
nj-ꜥnḫ-ḫnmw/ḫnmw-ḥtp (LB174: Saqqara, V.6L-7)
ḥtp-ḥr-ꜣḫtj (LB175: Saqqara, V.6-8E)
ꜣḫtj-ḥtp (LB176: Saqqara, V.6-8E)
jtj-sn (LB177: Giza, V.6-8?)
jrj-n-kꜣ-ptḥ (LB178: Saqqara, V.6-9)
snḏm-jb: jntj (LB179: Giza, V.8M-L)
sḫm-ꜥnḫ-ptḥ (LB180: (Saqqara, V.8-9)
ṯjj (LB181: Saqqara, V.8-9)
kꜣ.j-m-nfrt (LB182: Saqqara, V.8-9?)
ꜣḫtj-ḥtp (LB183: Saqqara, V.8L-9E)
sḫntjw/nfr-sšm-ptḥ (LB184: Saqqara, V.9)
ḫnwt (LB185: Saqqara, V.9)
nbt (LB186: Saqqara, V.9)
jj-nfrt: šꜣ.n.f (LB187: Saqqara, V.9)
ptḥ-ḥtp II: ṯfj (LB188-189: Saqqara, V.9M-L)
nj-mꜣꜥt-rꜥ [*jzzj-mr-nṯr*] (LB190: Giza, V.9-VI.1)
ḥtpt (LB191: Giza, V-VI)
kꜣ.j-m-ꜥnḫ (LB192-194: Giza, VI)
kꜣ-gm-nj: mmj (LB195: Saqqara, VI.1M)
nj-kꜣw-jzzj (LB196: Saqqara, VI.1M)
mrrw-kꜣ.j: mrj (LB197: Saqqara, VI.1M-L)
nj-ꜥnḫ-nswt (LB198: Saqqara, VI.1-2E)
zšzšt: jdwt (LB199: Saqqara, V.9 r/u – VI.1-2E?)
jꜣzn (LB200: Giza, VI.1M-2M)
sꜥnḫ-w(j)-ptḥ (LB201: Saqqara, VI.1L)
ḥzj (LB202-205: Saqqara, VI.1L-2E)
ꜥnḫ-m-ꜥ-ḥr: zzj (LB206: Saqqara, VI.1L-2E)
mrjj-rꜥ-mrj-ptḥ-ꜥnḫ: nḫbw (LB207: Giza, VI.2)
mḥw (LB208-210: Saqqara, VI.2M-3?)
wr-nw (LB211: Saqqara, VI.3-4?)
ꜣḫtj-ḥtp (LB212: Saqqara, VI.5?)

Eighty-four depictions of butterflies were located in 33 tombs during this study; at least one example from each tomb is compared here (LB172 to LB212).[95]

Butterflies and moths possess two pairs of wings (forewings and hind wings) that are attached to either side of their thorax. During flight, however, each fore- and hind wing is coupled together so that they move as one (see pl. 4Q). The antennae of moths are usually feathery and plume-like, while those of butterflies are delicate, with a small knob on each tip.[96]

---

93 Roselaar, C.S. (1985). "Hoopoe (*Upupa epops*)" in Cramp-Simmons, *Handbook iv*, 786.

94 No other species in this or any other netting scene are depicted in this manner; clearly, the distinctive markings of hoopoes in particular allowed this wing treatment to be applied. This detail has also been noted in flying hoopoes in marsh scenes in the tombs of *ṯjj* and *mḥw*.

95 For the depiction of butterflies in Egyptian art, see Keimer, L. (1934). "Pendeloques en forme d'insectes faisant partie de colliers Égyptiens D. Pendeloques-amulettes en forme de papillons", *ASAE* 34, 177[107]-213[143]; Keimer, L. (1937). "Pendeloques en forme d'insectes faisant partie de colliers Égyptiens E. Notes additionelles (suite)", *ASAE* 37, 143[170]-172[199]; Verhoeven, U. (1986). "Schmetterling", *LÄ v*, 663-664; Lopez-Moncet, A. and Aufrère, S. (1999). "Les papillons ("Monarques") du tombeau de Khnoumhotep II à Béni-Hasan (Moyen Empire, XII Dynastie)" in Aufrère, *Encyclopédie Religieuse*, 265-277. Harpur (*Decoration*, 189-190) states that the earliest representation of a butterfly occurs in the mid-5th dynasty tomb of *nfr/kꜣ.ḥꜣ.j*, but I place the earliest image in the early 5th dynasty tomb of *sḫm-kꜣ-rꜥ*. Harpur rules out this depiction as examination of a photograph of the relief failed to convince her of its presence.

96 Dalton, *Borne on the Wind*, 38-40.

Examination of tomb scenes reveals two basic designs for butterflies: in one, the insects' wings are rendered as two large, unsegmented lobes (LB172 to LB182, LB185, LB190? to LB194, and LB200; see figs. 4-30 to 4-32), while in the other, all four wings are represented (LB183, LB184, LB187, LB188, LB195 to LB199, LB201, LB205 to LB207, LB210 to LB212; see figs. 4-33 to 4-35). It is possible that the two-winged images were originally painted so as to indicate four wings and that these details have since disappeared. Nevertheless, there is a general tendency for reliefs of butterflies with unsegmented wings to occur prior to the late 5th dynasty, while those with divided wings appear from the reign of Unas.[97] The presence of two-winged butterflies in some tombs may indicate an earlier date for their decoration than is currently accepted.[98]

The wing movements of most flying insects are too rapid to be perceived by human eyes, but the flight of large butterflies and moths can be observed because their wing-flaps are comparatively slow.[99] Butterflies are able to fly at abrupt angles (loop and dive), but they most often travel horizontally in a slightly meandering fashion. During flight, their legs hang beneath them. The majority of depictions from Giza and Saqqara show butterflies in mid-air, but holding their wings as though seen from above, rather than the side. They thus do not present a biologically accurate depiction of the insects' flight.

In five tombs, however, butterflies are shown in profile (LB186, LB189, LB198, LB202 to LB204, LB208, and LB209; see figs. 4-36 to 4-38). Like four-winged butterflies, these images first occur in the late 5th dynasty, during the reign of Unas.[100] In two scenes, a perspective view of the insects' large forewings is provided (LB186 and LB198; see fig. 4-36), while in the others, a side view of one forewing (above the thorax) and one hind wing (above the abdomen) is displayed. It should be noted that a few of these insects (LB189, LB208, and LB209; see fig. 4-38) are very similar to Egyptian images of bees or wasps (see *Wasps* below), which are always shown from the side.[101] The legs of many of the profile butterflies are also represented like those of depicted bees in which the feet are fused to form a type of baseline (see LB189, LB203, LB208, and LB209; see figs. 4-37 and 4-38). Nevertheless, the size and shape of their wings are more indicative of butterflies or moths; the presence of the hind wing in some images also differs from that of bees, in which only two wings are depicted (emerging from the thorax).

Schäfer pointed out that Egyptian artists settled upon, and standardised, particular forms for both animate and inanimate objects in order to communicate clearly and unambiguously the things they were intended to signify (e.g., crocodiles are always represented in profile, while lizards, which are

---

97 The earliest example of a four-winged butterfly found during this study was in the Saqqara tomb of *ȝḫtj-ḥtp* (LB183), which is dated to the late Djedkare/Isesi – early Unas period.

98 The tomb of *kȝ.j-m-ʿnḫ* (G4561) has been widely dated to the 6th dynasty (for example, see Junker, *Giza iv*, 4; Baer, *Rank and Title*, 141; Smith, *HESPOK*, 206; Bolshakov, *Man and his Double*, 119-120), however Kanawati (*Kaiemankh*, 15-18) argues that architectural and artistic details point, instead, to the late 5th dynasty (probably the reign of Djedkare/Isesi). The two-winged butterflies displayed in the tomb's hall (LB192 and LB193) and burial chamber (LB194) support Kanawati's conclusion and indicate a date no later than early Unas. See also Woods, A. (in press). "Contribution to a controversy: A date for the tomb of *kȝ(=i)-m-ʿnḫ* at Giza", *JEA* 95.
A two-winged butterfly also appears in the Giza tomb of *jȝzn* (LB200), which Harpur (*Decoration*, 36-37) has dated tentatively to between the reigns of Teti and Pepi I based largely upon architectural details. She suggests, however, that a late 5th dynasty date cannot be ruled out. Indeed, Simpson (*HESPOK*, 197) placed *jȝzn* in the second half of the 5th dynasty due to strong similarities with the tomb of *jj-mrjj* (G6020), both in the treatment of its walls and its scene content. The *jȝzn* artist seems also to have been influenced by the decoration in the Giza tombs of *kȝ.j-m-nfrt* (LG63) and *kȝpj* (G2091), which are dated by Harpur to V-VI and V.9-VI, respectively. None of these earlier tombs includes the image of a butterfly. I suggest that the presence of the two-winged butterfly in the tomb of *jȝzn* either indicates that the artist copied the motif from yet another earlier tomb, or that the tomb itself must date from the late 5th dynasty, rather than the 6th.

99 Dalton, *Borne on the Wind*, 26-27.

100 Smith (*HESPOK*, 319, n. 1) cites only three examples of butterflies in this orientation - one from the Middle Kingdom and two from the New Kingdom period. Harpur (*Decoration*, 202) describes one of two profile butterflies found in the tomb of *mḥw* (LB208), but states that no other depictions of this type have been preserved from the Old Kingdom.

101 Schäfer, *Principles*, 150.

morphologically similar, are shown from above).[102] As the side-view was intimately associated with the idea of stinging insects, I suggest that this probably prevented the depiction of butterflies in profile from becoming more widespread in wall scenes.

Some reliefs appear to show insects that have landed, as they are depicted against tall shrubs or stands of reeds (LB197, LB202, LB204 to LB206, and LB210). Moths and butterflies hold their wings differently when at rest – butterflies close their wings in a vertical position above their thorax, while moths open their wings fully and lay them out flat.[103] Most of the resting insects would thus appear to illustrate moths. In the tomb of *ḥzj*, however, two of these animals hold their wings vertically (LB202 and LB204; see fig. 4-39), indicating clearly that they are butterflies. Many of the insects also display terminating knobs on their antennae, identifying them as butterflies. Markings on the wings of some (see fig. 4-31) reveal that they may depict the Monarch butterfly.[104]

Dragonflies/Grasshoppers

k3.j-m-nfrt (LB213: Saqqara, V.8-9?)
3ḫtj-ḥtp (LB214-215: Saqqara, V.8L-9E)
jj-nfrt: š3.n.f (LB216-217: Saqqara, V.9)
3ḫtj-mrw-nswt (LB218-219: Giza, V.9-VI.1)
k3-gm-nj: mmj (LB220: Saqqara, VI.1M)
nj-k3w-jzzj (LB221-224: Saqqara, VI.1M)
j3zn (LB225: Giza, VI.1M-2M)
sʿnḫ-w(j)-ptḥ (LB226: Saqqara, VI.1L)
ḥzj (LB227: Saqqara, VI.1L-2E)
mrjj-rʿ-mrj-ptḥ-ʿnḫ: nḫbw (LB228: Giza, VI.2)
mrrj (LB229: Saqqara, VI.2E?)
mḥw (LB230-232: Saqqara, VI.2M-3?)

In the late 5th dynasty, a new insect form joined that of butterflies in marsh scenes (LB213 to LB232). Like butterflies, they are depicted flying both above and within papyrus thickets and are represented as though viewed from above. Each of them possesses at least two pairs of elongated wings and most are characterized by long antennae and large hind legs that dangle on either side of their abdomen (LB213 to LB216, LB220, LB222 to LB224, LB226, LB227, LB230 to LB232; see figs. 4-40 to 4-42). These creatures are often described as dragonflies,[105] but the length of their antennae and hind legs (which are diminutive in species of Odonata) renders this identification unlikely. Twenty depictions of the animal located in 12 tombs were found during this study.

Keimer identified the flying insects as grasshoppers,[106] but the only reason he offered for this belief is, "car en plusieurs cas ce même insecte est figuré au repos à côté de l'insecte volant".[107] On the basis of later images and textual evidence, he proposed that the species represented is the desert locust.[108] Desert locusts are short-horned grasshoppers that are found in semi-arid and desert areas. They are usually solitary, but are one of a small number of species that will form vast swarms if their living conditions become overcrowded. Like other grasshoppers, they possess two sets of wings - a set of narrow, leathery forewings and a pair of broad, transparent hind wings that only becomes apparent when they fly (see pl. 4R). When swarming, locusts may fly for many hours; grasshoppers and non-swarming locusts can do so only for short periods.[109]

The insects depicted in tomb scenes show some similarities to flying locusts, particularly in the length of their antennae and hind limbs, however their wings differ markedly, for while the hind wings of locusts are a different shape from their forewings (see pl. 4R), the Egyptian images consistently show each pair as identical to one another. The flying insects also bear little resemblance to depicted grasshoppers at rest. These are invariably represented with stout, elliptical

[102] *ibid.*
[103] Dalton, *Borne on the Wind*, 38.
[104] Larsen, *Butterflies*, 8.
[105] von Bissing, *Gem-ni-kai i*, 6; Harpur, *Decoration*, 196, n. 147; Houlihan, *Animal World*, 194 and fig. 131.
[106] Keimer, L. (1932). "Pendeloques en forme d'insectes faisant partie de colliers égyptiens B. Pendeloques et pièces de colliers en forme de sauterelles", *ASAE* 32, 129[43]-150[64].
[107] *ibid.*, 136[50].
[108] Keimer, L. (1933). "Pendeloques en forme d'insectes faisant partie de colliers égyptiens II. Textes pouvent expliquer l'emploi des pendeloques ou des pièces de colliers (amulettes, etc.) en forme de sauterelles", *ASAE* 33, 125[93]-127[95].
[109] Cressman, K. (1998). "Monitoring desert locusts in the Middle East: An overview", *YF&ESB* 103, 125.

bodies and straight, erect antennae. The bodies of the flying insects are often long and narrow, with the thorax and abdomen clearly differentiated (e.g., LB216, LB223, and LB229). The antennae of many are long and languid, with some ending in discs. A few of the insects also have three sets of wings (LB227, LB230, and LB232; see fig. 4-42).

Contrary to Keimer's claim, only one of the insects in question was found beside a grasshopper at rest. This image occurs in the tomb of *kꜣ-gm-nj: mmj* (LB220; see fig. 4-41) - the same example that Keimer cites to make his case (his fig. 31). Furthermore, this is the only tomb in which *both* a grasshopper at rest and a flying insect were found among the reliefs. None of the remaining 11 tombs that depict flying insects also include images of seated grasshoppers. These were found in only six tombs during this study,[110] and date from the late 5th dynasty. They thus overlap in time with those of flying insects, yet apart from that of *kꜣ-gm-nj: mmj*, none is found in the same tomb as one.

I believe that the differently shaped wings of locusts, the limited morphological resemblance between the flying insects and seated grasshoppers, and the lack of overlap between these two image types in individual tombs indicate that the creatures in question do not represent grasshoppers/locusts. Determining which species they do depict is beyond the scope of the current study, but I suggest that some examples may represent a form of butterfly. A few have exaggerated discs on the tips of their antennae, a feature found frequently in depictions of butterflies (cf. the insect and butterfly in LB224). In particular, their physical attributes are very similar to 'plume moths'. In these species, the wings are divided into two or more long, narrow lobes, each of which resembles a feather (pl. 4S).[111] They also possess long antennae and extremely long hind legs. Most of the depicted insects closely resemble these moths, especially those that have been rendered with three pairs of wings. Finally, a small number of the depictions, in which neither the antennae nor the hind legs are apparent (LB217 to LB219, and LB225), may indeed represent dragonflies.

Wasps

nj-ꜥnḫ-ẖnmw/ẖnmw-ḥtp (LB233-234: Saqqara, V.6L-7)

Two unique scenes in the tomb of *nj-ꜥnḫ-ẖnmw/ẖnmw-ḥtp* depict insects that have been attracted to gutted fish (LB233 and LB234). In one vignette, three creatures hover directly above the worktable of a man as he slices open a fish (LB233; see fig. 4-43), while in the other, two insects have been drawn to the smell of cooking and drying flesh (LB234; see fig. 4-44). Although the images are damaged and somewhat sketchy, examination reveals that the insects possess two narrow, elongated wings[112] and long antennae. This last feature reveals that they are not flies (in which the antennae are extremely short), but either bees or wasps. Indeed, the shape of the insects matches closely that of the *bjt* bee, the royal symbol of Lower Egypt.[113] As noted above, however, the legs of bees are usually depicted such that their feet form a baseline; the legs of the insects represented here dangle freely. Also, bees are nectar-feeders and so would unlikely be drawn to fish.

I suggest that the context in which these insects are represented indicates that they are probably yellowjacket wasps (see pl. 4T), as this species, which is native to North Africa, is particularly attracted to, and feeds upon, meat.[114] Fish is used frequently as a bait to poison this species[115] – a

---

110 These are: *rꜥ-špss* (LS16), *sšm-nfr IV* (LG53), *mrrw-kꜣ.j: mrj* (ntn), *ꜥnḫ-m-ꜥ-ḥr: zzj* (ntn), *kꜣ-gm-nj: mmj* (LS10), and *mṯṯj* (ntn).

111 See members of the family Pterophoridae or the family Alucitidae. Resh-Cardé, *Insects*, 655.

112 The number of wings might indicate that they are houseflies (Diptera sp.), as these insects possess only one functional pair. However, although bees and wasps (Hymenoptera sp.) are fitted with two sets of wings, these are usually hooked together to create two single, broad areas. Consequently, when flying, they appear to have only two wings. See Dalton, *Borne on the Wind*, 44-45, 46-47.

113 The *bjt* gylph is used as a determinative for the names of various stinging insects, such as bees, hornets, and wasps. See *Wb i*, 434. Harpur (*Decoration*, 193, n. 144), citing Schäfer (*Principles*, 150), also suggests that the shape of the depicted insects is more akin to that of bees or wasps, as flies in Egyptian art are normally represented from above, rather than in profile.

114 Spurr, E.B. (1995). "Protein bait preferences of wasps (*Vespula vulgaris* and *V. germanica*) at Mt Thomas, Canterbury, New Zealand", *NZ J. Zool.* 22, 281-289.

necessary procedure because yellowjackets, which have a painful sting, are often found in large numbers around human activities.[116] The cleaning and laying out of fish carcasses will thus have provided a most effective lure.

4.2 REST

Most living organisms require a form of rest after long periods of activity. Some species also pause in extreme physical conditions (e.g., high temperatures, limited water supply, etc.) in order to preserve energy. Many animals rest simply by becoming motionless while remaining upright, but others change their posture by lowering their body and bending their limbs to sit or lie down.[117]

4.2.1 Sitting

When sitting, the hind legs are bent and the rump is placed upon the substrate. In quadrupeds, the forelegs are usually are held straight, with the feet resting upon the ground to support the upper body.[118]

Hippopotamuses

dbḥnj (LB235: Giza, IV.5)
pr-sn (LB236: Saqqara, V.2-3)
ȝḫtj-ḥtp (LB237: Saqqara, V.6-8E)
kȝ.j-m-nfrt (LB238: Giza, V.6-8E?)
ṯjj (LB239: Saqqara, V.8-9)
ḫnwt (LB240: Saqqara, V.9)
nbt (LB241: Saqqara, V.9)
sšm-nfr IV (LB242-243: Giza, V.9-VI.1)
Anon (LB244: Saqqara, V-VI)
kȝ-gm-nj: mmj (LB245-247: Saqqara, VI.1M)
mrrw-kȝ.j: mrj (LB248: Saqqara, VI.1M-L)
zšzšt: jdwt (LB249-250: Saqqara, V.9 r/u - VI.1-2E?)
sʿnḫ-w(j)-ptḥ (LB251: Saqqara, VI.1L)
mḥw (LB252: Saqqara,VI.2M-3?)
jj-nfrt (LB253: Giza, VI.3-4E?)
nb-kȝw-ḥr: jdw (LB254: Saqqara, V.9 r/u - VI.7)

Due to their bulk and short limbs, hippopotamuses are unable to lie down or rise in one smooth movement, but must instead sit briefly during each transition (see pl. 4U). They descend by first lowering themselves onto their haunches with their hind legs extended beneath them and their forelegs held straight, before finally reclining. When rising, they sit once again, pushing back on to their rump before pulling themselves up by their forelegs.[119]

Hippos sit frequently in wall scenes (beginning in the late 4th dynasty), often when they are engaged in other activities, such as 'yawning' (LB235 to LB254) (see Chapter 9 for details). In each example, the animals' hind legs are folded under them while their forelimbs are braced (e.g., LB241; see fig. 4-45). The foreleg furthermost from the viewer invariably advances first, either to conform to the rules of Egyptian representation or possibly to imply that the animals are rising from a recumbent position by reaching out with their forelimbs.

Birds

Birds sit while roosting and also when they are incubating eggs. This behaviour is dealt with in Chapters 9 and 11.

115 *ibid.*
116 D'Adamo, P. and Lozada, M. (2005). "Conspecific and food attraction in the wasp *Vespula germanica* (Hymenoptera Vespidae), and their possible contributions to control", *Ann. Entom. Soc. Am.* 98(2), 236.
117 Fraser-Broom, *Farm Animal Behaviour*, 135-137.
118 Nickel et al., *Locomotor System*, 451-452.
119 Estes, *Behavior Guide*, 224.

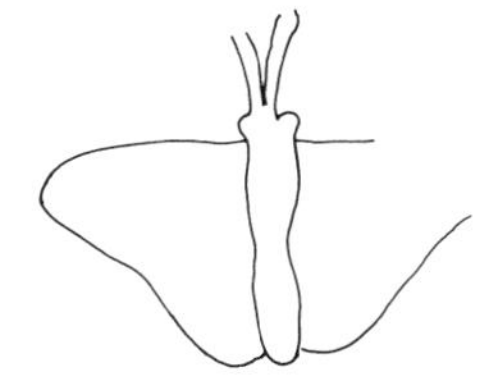

Fig. 4-30 (LB174)

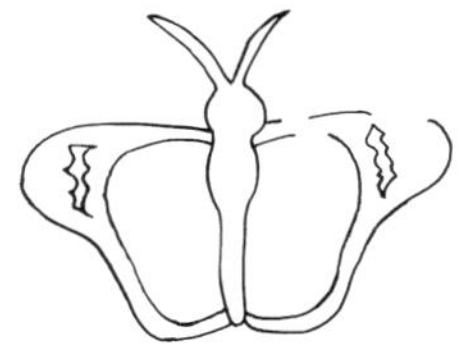

Fig. 4-31 (LB181)

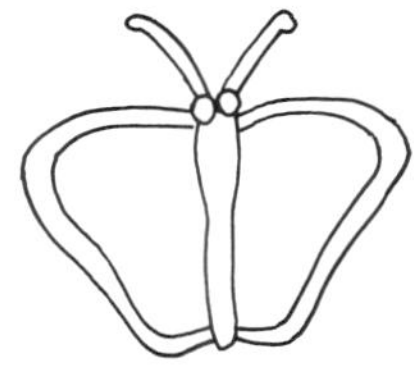

Fig. 4-32 (LB182)

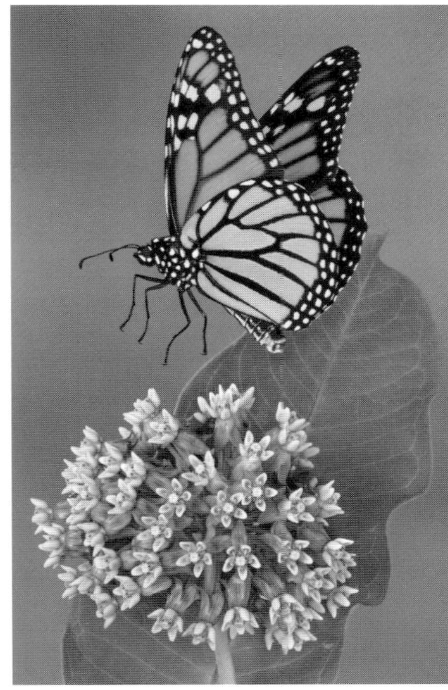

Pl. 4Q Butterfly flight
© iStockphoto.com/Cathy Keifer

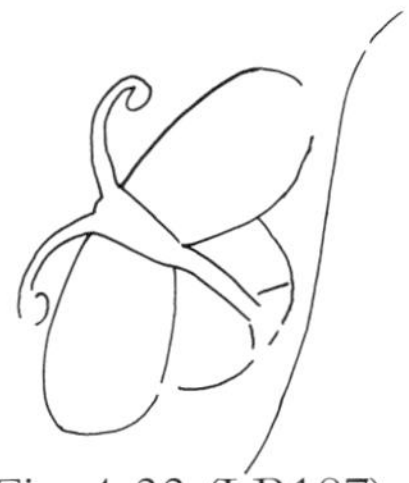

Fig. 4-33 (LB187)

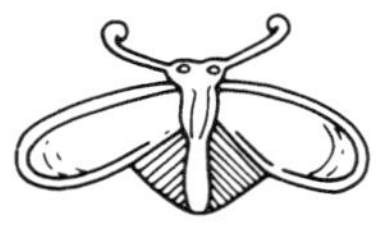

Fig. 4-34 (LB196)

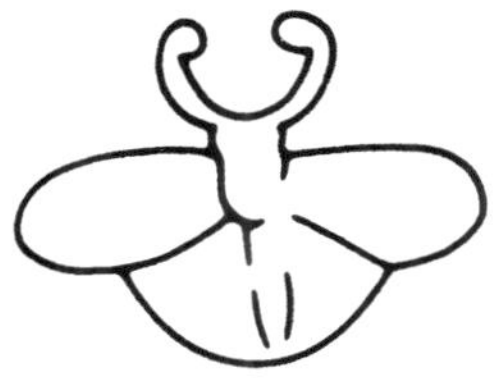

Fig. 4-35 (LB199)

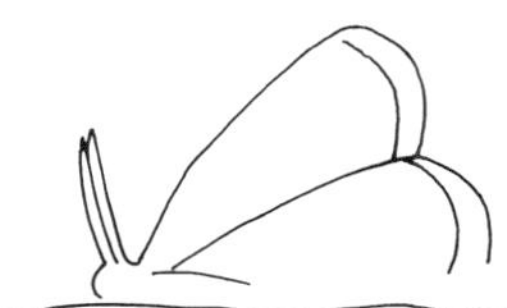

Fig. 4-36 (LB186)

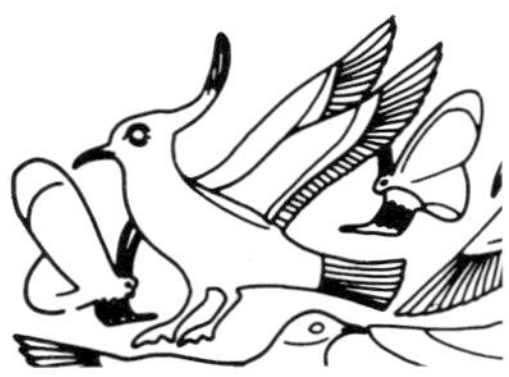

Fig. 4-37 (LB203)

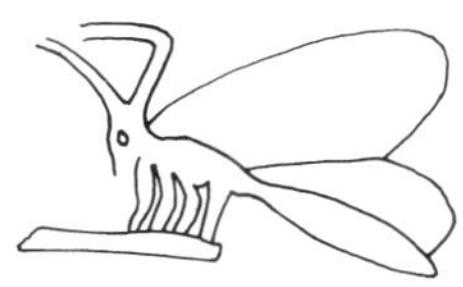

Fig. 4-38 (LB209)

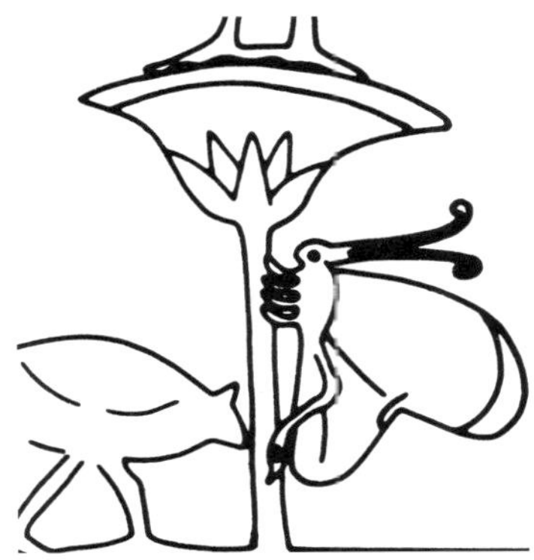

Fig. 4-39 (LB202)

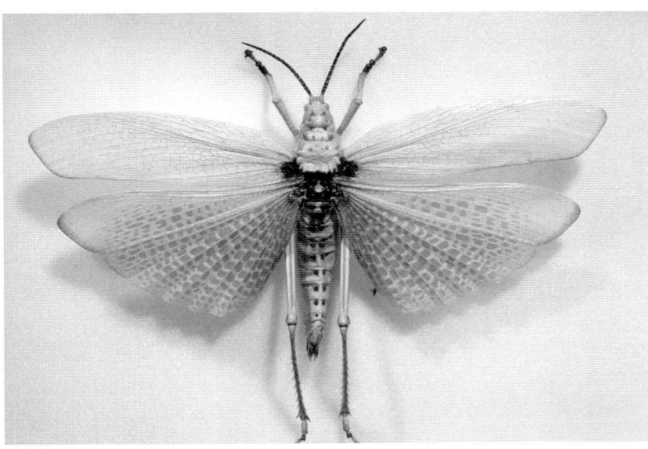

Pl. 4R Locust wing structure
© iStockphoto com/Tony Francischiello

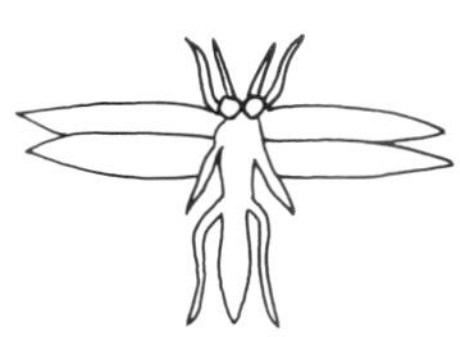

Fig. 4-40 (LB213)

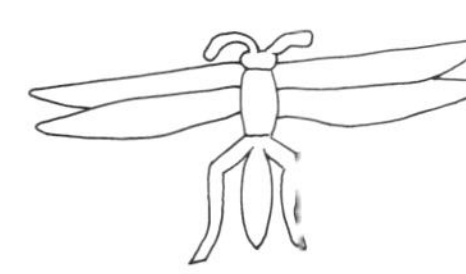

Fig. 4-41 (LB220)

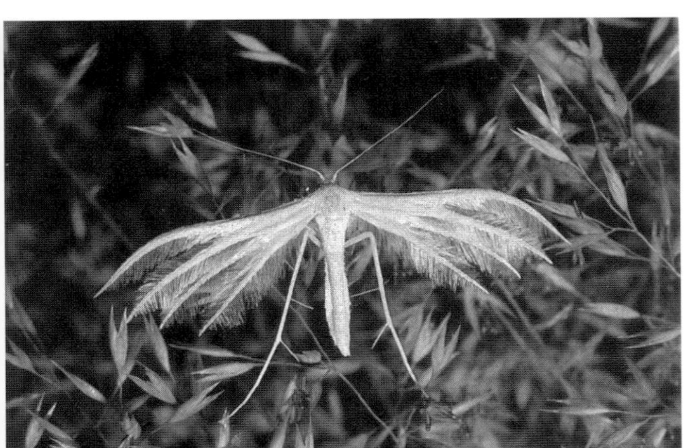

Pl. 4S Plume moth
© Yves Lanceau
Reproduced with permission

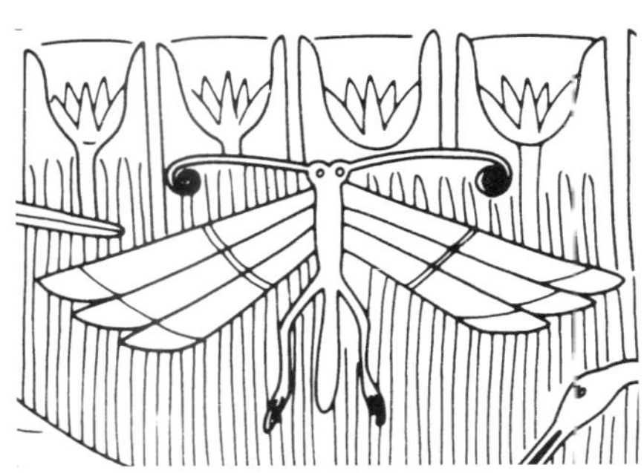

Fig. 4-42 (LB227)

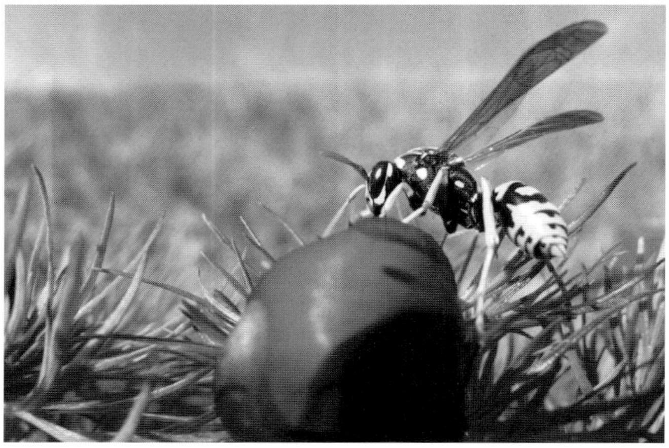

Pl. 4T Yellowjacket wasp

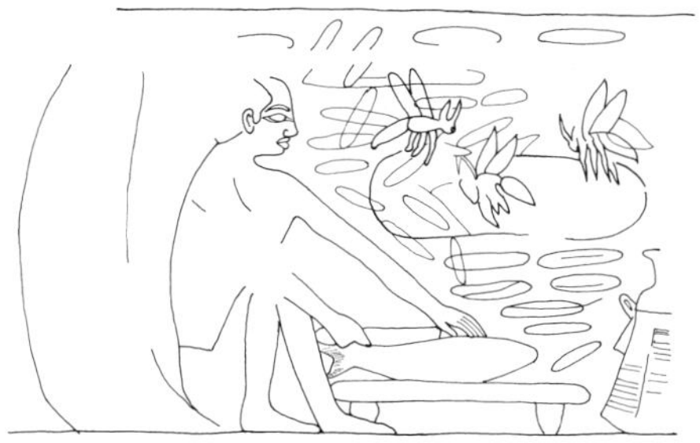

Fig. 4-43 (LB233)

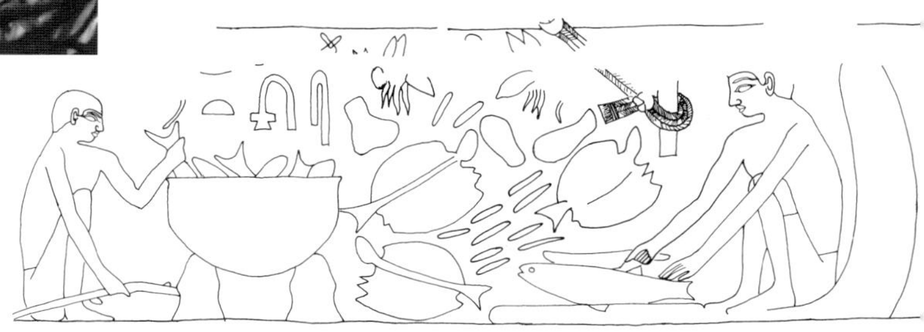

Fig. 4-44 (LB234)

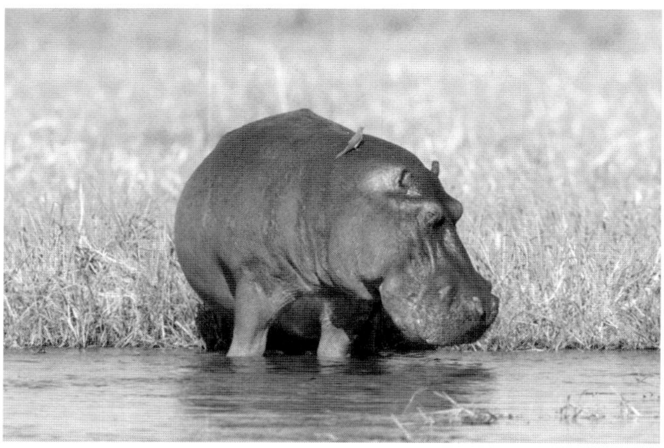

Pl. 4U Hippopotamus sitting

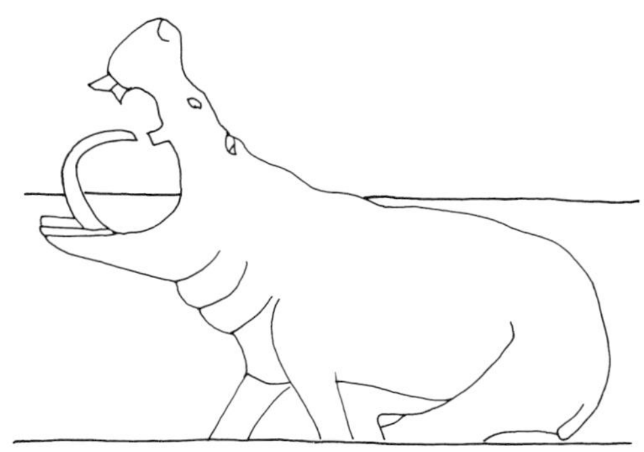

Fig. 4-45 (LB241)

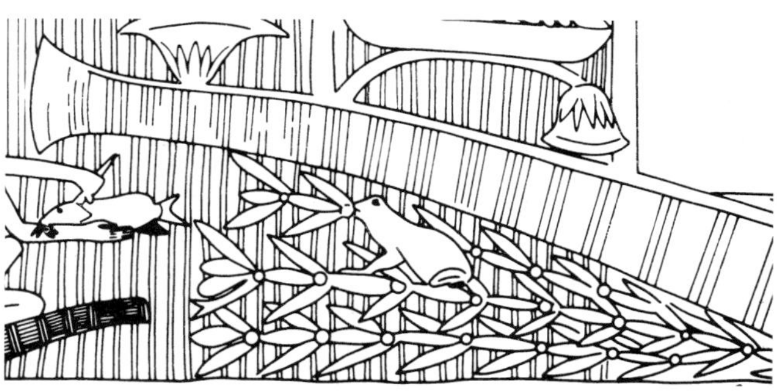

Fig. 4-46 (LB273)

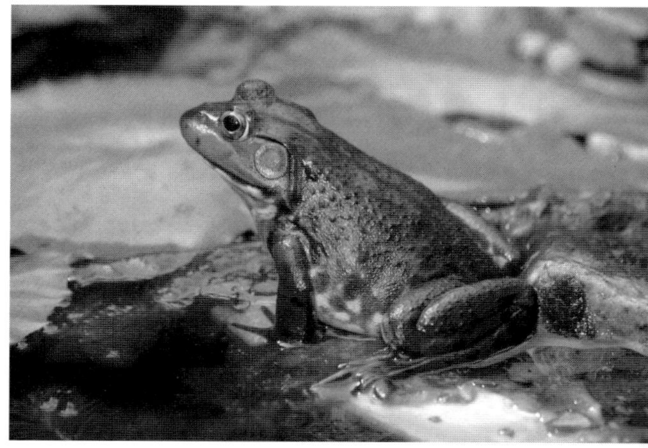

Pl. 4V Frog sitting

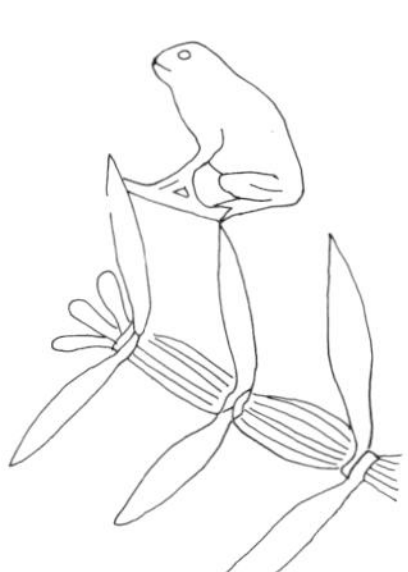

Fig. 4-47 (LB269)

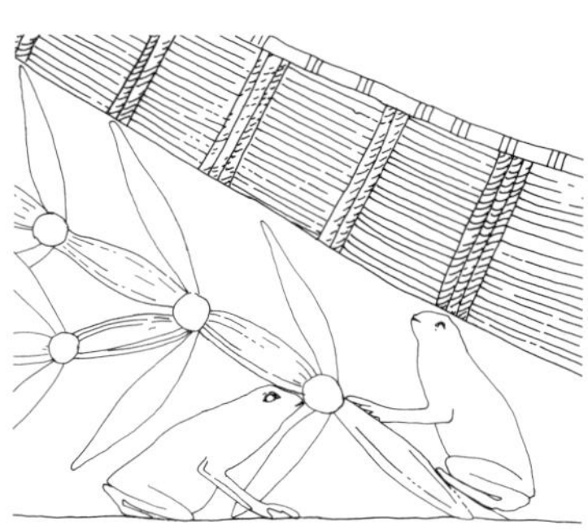

Fig. 4-48 (LB266)

Frogs

nj-ʿnḫ-ẖnmw/ẖnmw-ḥtp (LB255: Saqqara, V.6L-7)
rʿ-šps (LB256: Saqqara, V.8M)
nfr-jrt-n.f (LB257: Saqqara, V.8-9)
nbt (LB258: Saqqara, V.9)
jj-nfrt: šȝ.n.f (LB259: Saqqara, V.9)
ptḥ-ḥtp II: ṯfj (LB260: Saqqara, V.9M-L)
sšm-nfr IV (LB261: Giza, V.9-VI.1)
kȝ.j-m-ʿnḫ (LB262: Giza, VI)
kȝ-gm-nj: mmj (LB263-267: Saqqara, VI.1M)
nj-kȝw-jzzj (LB268: Saqqara, VI.1M)
mrrw-kȝ.j: mrj (LB269-272: Saqqara, VI.1M-L)
ḥzj (LB273-274: Saqqara, VI.1L-2E)
mrrj (LB275: Saqqara, VI.2E?)
mḥw (LB276: Saqqara, VI.2M-3?)
jj-nfrt (LB277: Giza, VI.3-4E?)
wr-nw (LB278: Saqqara, VI.3-4?)

Seated frogs were located in 16 marsh scenes. The earliest image occurs in the mid-5th dynasty tomb of *nj-ʿnḫ-ẖnmw/ẖnmw-ḥtp* (LB255), but the motif does not seem to have become popular until the late 5th and early 6th dynasties. In most examples, the animals' hind legs are folded beneath them as they sit squarely upon their haunches, but a few lean forward and lift their rumps slightly (e.g., LB259, LB261, LB265, LB269, LB270, and LB277). Frogs at rest adopt both of these postures (see pl. 4V).

In the tomb of *ptḥ-ḥtp II: ṯfj* (LB260), a small amphibian perches upon an emerging lotus bud, but in the remaining scenes, the animals are usually found on leafy foliage situated beneath the raised stern or prow of the deceased's canoe (e.g., LB273; see fig. 4-46). The plants amongst which they sit have been identified as pondweed, *Potamogeton* sp.[120] Contrary to its bush-like appearance in tomb scenes, pondweed does not grow upright, but instead spreads out horizontally and floats upon the water surface. Some of the depicted frogs sit on the outer branches of this plant, while others are deep within its foliage. Given the growth habits of pondweed, however, each of these animals should in fact be viewed as sitting *upon* its floating leaves, not nestled among them.[121] Their limbs are usually well supported by the foliage beneath them, but in the tomb of *mrrw-kȝ.j: mrj* (LB269; see fig. 4-47), the amphibian appears to sit upon a baseline strung between two leaf tips. In other images, the frogs are adjacent to the plant, rather than within it, and are positioned directly upon the line that indicates the water surface (LB258, LB263, and LB265; see fig. 4-48). I suggest that these examples are an attempt to convey the true location of the animals on the horizontal plane of water (supported by the floating foliage).

The depiction of pondweed and the frogs seated upon it indicate that this feature should be viewed as unfolding away from the viewer.[122] These images thus add further support to the proposition, made earlier in this chapter, that the nature of papyrus and the behaviour of the animals within thickets indicate that the marsh scene as a whole is best perceived in this manner.

4.2.2 Recumbent

When quadrupeds are recumbent, they lie upon their sternum and bring their hind legs forward. Their forelegs may either be bent back beneath their chest or stretched forward. Some animals bend their spine in an arc when lying down, such that their head may be carried around to the side of their body.[123]

---

120 It is unclear whether the plant depicted is *Potamogeton lucens*, *P. schweinfurthii*, or *P. crispus*. See Keimer, L. (1927). "Le *Potamogeton lucens* L. dans l'Égypte ancienne", *REA* 1, 182-197; Keimer, L. (1929). "Nouvelles recherches au sujet du *Potamogeton lucens* L. dans l'Égypte ancienne et remarques sur l'ornementation des hippopotames en faience du Moyen Empire", *REA* 2, 210-253; Keimer, L. (1931). "Notes additionelles", *REA* 3, 36-41; and more recently, Servajean, F. (1999). "Du singulier à l'universal: Le *Potamogeton* dans les scènes cynégétiques des marais", in Aufrère, *Encyclopédie Religieuse*, 249-264. For an alternative interpretation, see Beaux, N. (1988). "The representation of *Polygonum senegalense* Meisn. in ancient Egyptian reliefs and paintings", *JEA* 74, 248-252.

121 This point has also been made by Harpur (*Decoration*, 186, n. 136). Unfortunately, the frogs are not rendered in sufficient detail to identify the species with certainty. It is interesting to note, however, that a newly discovered Egyptian species, the Damietta toad, has been described as highly aquatic and is often found amongst dense floating vegetation. See Saleh, *Amphibians and Reptiles*, 209.

122 Indeed, Keimer (*REA* 1, 193) reminds the reader "que les dessinateurs égyptiens projetaient souvent ainsi en hauteur des objets qui se trouvaient, en réalité, dans une position horizontale."

123 Fraser-Broom, *Farm Animal Behaviour*, 136.

Dogs, hares, and ungulates adopt a recumbent pose in wall scenes. The images considered here are those for which the position has a behavioural significance beyond mere rest.

Cattle

bw-nfr (LB279: Giza, IV.6-V.1?)
jrj-n-k3-ptḥ (LB280: Saqqara, V.6-9)
rˁ-špss (LB281: Saqqara, V.8M)
sndm-jb: jntj (LB282: Giza, V.8M-L)
ṯjj (LB283: Saqqara, V.8-9)
nfr-jrt-n.f (LB284: Saqqara, V.8-9)
k3.j-m-nfrt (LB285: Saqqara, V.8-9?)
ptḥ-ḥtp II: ṯfj (LB286: Saqqara, V.9M-L)
nj-m3ˁt-rˁ [jzzj-mr-nṯr] (LB287: Giza, V.9-VI.1)
k3.j-m-nfrt (LB288: Giza, V-VI)
k3-gm-nj: mmj (LB289: Saqqara, VI.1M)
nj-k3w-jzzj (LB290: Saqqara, VI.1M)
mrrw-k3.j: mrj (LB291: Saqqara, VI.1M-L)
ˁnḫ-m-ˁ-ḥr: zzj (LB292: Saqqara, VI.1L-2E)
ptḥ-ḥtp (LB293: Saqqara, VI.2? or LV-VIE?)
mrrj (LB294: Saqqara, VI.2E?)
mrw: ttj-snb (LB295: Saqqara, VI.2E-M?)
mḥw (LB296: Saqqara, VI.2M-3?)

Recumbent cattle are presented regularly in wall scenes. In a subset of these, they are depicted while being attended to by herdsmen. Labels accompanying the images, and the behaviour of the men, suggest that the animals are being fed while prone. Eighteen tombs were found to display images of recumbent feeding; one example from each tomb is provided here for discussion.

All of the depicted animals lie upon their sternums (LB279 to LB296). In each example, only one hind leg is apparent and this is invariably brought forward, along their flanks. The tip of their tail either peeks out from behind this leg (e.g., LB283; see figs. 4-49 and 4-51) or sweeps across it (e.g., LB288: see fig. 4-50). Rendering of the animals' forelegs is more variable. Usually, only one limb is shown, bending back and under the body (e.g., fig. 4-49). Occasionally, however, both forelimbs are displayed such that while one leg is flexed, the other may be flung forward (LB279, LB285, LB286, LB288, and LB291; see figs. 4-50 and 4-51). This latter limb is often furthermost from the viewer, but in the tomb of *ptḥ-ḥtp II: ṯfj* (LB286), the closer leg is extended.

In each scene, a kneeling herdsman holds one hand to the mouth of his charge and a short leash around the animal's neck, or attached to its lower jaw, in the other. In some examples, a food item can be seen (LB283, LB285, LB289; see fig. 4-49 and Chapter 7 for further details). In others, the men appear to cup the lower lips (LB279, LB281, LB286, LB287, LB289, LB292, and LB293) or touch the muzzles of the cattle (LB282 and LB288; see fig. 4-50). In a small number of images, they insert their hands deep inside the animals' mouths (LB290, LB291, LB294, and LB296; see fig. 4-51). In five examples (LB279, LB285, LB287, LB288, and LB293), their activities are labelled, *wš3* 'Fattening' (e.g., fig. 4-50).

Cattle normally feed while standing (see Chapter 7). In order to digest their food, however, they must cease grazing to ruminate.[124] During this process, boluses of slightly chewed food are regurgitated from the first stomach compartment (the 'rumen'), masticated again, and then re-swallowed. Rumination can take place while the animal is standing quietly, but cattle show a marked preference for lying down, resting upon their sternum with their forelegs bent back and under their chest and their hind legs brought forward (see pl. 4W). Rumination occurs approximately 15 to 20 times a day, with each bout lasting from around two minutes up to an hour.[125]

Thus, although the animals considered here appear to be being fed, their posture is actually more indicative of rumination. When discussing similar scenes, Jacques Vandier suggested that the recumbent cattle are being fattened through force-feeding[126] – i.e., by pushing food down their oesophagus. Images in which men place their hands deep into the mouths of their charges would certainly seem to suggest this. Indeed, the term *wš3* is associated frequently with depictions of geese that are clearly being fed against their will.[127] For Vandier, then, the cattle are being made to lie down

[124] Estes, *Behavior Guide*, 4-5.
[125] Hafez, E.S.E., Schein, M.W. and Ewbank, R. (1969). "The behaviour of cattle" in Hafez, *Behaviour of Domestic Animals*, 248.
[126] Vandier, *Manuel v*, 80-83.
[127] Montet (*Scènes*, 111-112) translates *wš3* as 'cramming'.

in order to eat the "repas forcé" of bread and foliage offered to them by their seated handlers.[128] He suggests that the animals' muzzles are cupped in order to prevent them from opening their mouths and rejecting the fodder.[129] It is widely accepted that the Egyptians fed some of their cattle forcibly.[130]

There are a number of problems with this interpretation of the scenes. First, Vandier implies that the cattle are being held in position during this uncomfortable procedure by the short leash held by their handlers, yet it does not seem possible that this alone would have prevented such powerful animals from standing if they so desired. He also does not explain why the animals actually need to be recumbent during this procedure. The most serious flaw to Vandier's theory, however, is that it is not possible to force-feed a ruminant without endangering its health. In particular, over-loading of the rumen can cause potentially life-threatening conditions in cattle, such as bowel obstruction and bloat.[131] It is consequently inconceivable that Egyptian herdsmen would have put their animals at risk by forcing food upon them in this manner. Furthermore, such extreme measures will have been totally unnecessary for, given sufficiently palatable food, most animals will continue to eat without encouragement. Even under these circumstances, however, ruminants would still be required to pause periodically to ruminate.

To understand the meaning of these curious scenes, it is necessary to reduce them to their fundamental elements. First, the label *wš3* clearly indicates that the purpose of the depicted activities was to fatten the animals; the images must therefore contain elements of the fattening process. There are just two basic requirements to make an animal gain weight: it must eat a great deal of nutritious food and its movements must be reduced. Both of these elements are contained within each scene.

First, I suggest that the quality of the food being offered to the animals is conveyed by the fact that it is delivered by hand. Thus the presence and individual attention of the kneeling herdsmen inform the viewer of the animals' special feeding regimen. The scenes either show hand-feeding directly - that is, a food item being held to the animal's mouth - or it is implied when the herdsmen either touch the lips of their charge, or indeed, place their hands inside its mouth. Note that I do not believe that in agricultural practice herdsmen actually placed food directly into the mouths of their cattle (for, as noted above, this would be unnecessary), but that their presence in these scenes is a visual device by which to convey to the viewer the special foods and treatment to which these fattened animals had access.

The animals' restricted movement is suggested to the viewer in three ways: by their recumbency, by the sedentary nature of their handlers, and by the shortness of their leash, which implies that they cannot roam far. Finally, while the possibility that Egyptian herdsmen were able to induce their cattle to eat while lying down cannot be ruled out, I think it more likely that the images depict a merging of time in which feeding and then subsequent rumination have been depicted within the same image.

Ungulate calves

rˁ-m-k3.j (LB297: Saqqara, V.6? r/u - V.6-7)
nj-ˁnḫ-ẖnmw/ẖnmw-ḥtp (LB298-301: Saqqara, V.6L-7)
3ḫtj-ḥtp (LB302: Saqqara, V.8L-9E)
ptḥ-ḥtp II: *ṯfj* (LB303: Saqqara, V.9M-L)
sšm-nfr IV (LB304-305: Giza, V.9-VI.1)
Anon (LB306: Saqqara, VI)
mrrw-k3.j: mrj (LB307: Saqqara, VI.1M-L)
sˁnḫ-w(j)-ptḥ (LB308-310: Saqqara, VI.1L)
mrjj-ttj: mrj (LB311-319: Saqqara, VI.2M-L)

Desert hunting scenes located in nine tombs at Saqqara were found to display small recumbent ungulates on undulating sub-registers, frequently situated beside small shrubs and bushes. They thus

128 Vandier, *Manuel v*, 80.
129 *ibid.*, 78.
130 For example, *ibid.*, 80; Brewer et al., *Domestic Plants and Animals*, 86; Houlihan, *Animal World*, 13-14; Germond, *Egyptian Bestiary*, 56.
131 Thear-Fraser, *Complete Book*, 191.

appear to be secreted in the background, away from the main action. The posture of many of the animals supports this conclusion and also suggests two possible explanations for their behaviour.

The first is that the ungulates are resting during the heat of the day. The images depict three species: dorcas gazelles[132] (LB297, LB298, LB300 to LB305, LB308, LB309, LB312, LB313, LB315, LB316, and LB318; see fig. 4-52 and fig. 8-25 in Chapter 8), Nubian ibex (LB319), and small animals with nub horns that are probably scimitar oryx calves (LB299, LB306, LB307, LB310, LB311, LB314, and LB317; see fig. 4-53).[133] During the summer months, dorcas gazelle rest at midday by either standing or lying down in the shade of available trees or bushes.[134] Nubian ibex[135] and scimitar oryx[136] also pause during the hottest hours, and will sometimes lie down in small scrapes that they dig with their hooves. The postures of the depicted animals is thus in accord with resting. It is curious, however, that only oryx calves are shown engaged in this behaviour and never adults.

An alternative explanation is that the images depict young animals that are hiding in response to the hunters and dogs with which they share each scene. The behaviour of young ungulates after birth differs among species. Many are 'followers', animals that trail after their mothers as soon as they are able to stand, but some species are 'hiders' in which the young are left alone and concealed in the undergrowth for anywhere from a few days to many weeks. The calves lie completely immobile while hidden in order to avoid detection by predators and only emerge from their hiding places when their mothers return to nurse them.[137] When lying prone, hiders adopt a particular posture in which they stretch their head and neck forward upon the substrate and flatten their body so as to "produce the lowest possible profile" (see pl. 4X).[138] The hiding response is particularly strong in newborns, and wanes as they age.

---

132 The earliest example, which was found in the tomb of *rꜥ-m-kꜣ.j* (LB297), dates to the reign of Niuserre. Dorcas gazelles adopting similar postures also occur in three fragments from the sun-temple of this king. In each image, the animals are located within the main register, rather than on separate baselines. In one relief, the gazelle lies upon a small rise beside a tree and adopts a posture identical to that found in private tombs (Edel, E. (1961). "Zu den Inschriften auf den Jahreszeitenreliefs der 'Weltkammer' aus dem Sonnenheiligtum des Niuserre", *NAWG* 1961, fig. 7, section 251). In another example, a gazelle again lies upon a hummock beside a small shrub, but extends one foreleg and elevates its head (Edel, E. (1963). "Zu den Inschriften auf den Jahreszeitenreliefs der 'Weltkammer' aus dem Sonnenheiligtum des Niuserre II", *NAWG* 1963, fig. 4, section 705), while in a third fragment, the animal lies with its head lifted and its forelegs placed under its sternum (Edel-Wenig, *Jahreszeitenreliefs Ne-user-re*, pl. 17, section 804). All of these examples are predated by the depiction of a recumbent gazelle in the mortuary temple of Sahure, which, while lying upon a sub-register in a hunting scene, curls its head and neck around to rest its muzzle upon its hind leg (see Borchardt, *Sahure ii*, pl. 17). For a recent examination of these and other depictions of dorcas gazelle, see Strandberg, *Gazelle*, passim.

133 Depictions of calves bearing nub horns begin to occur in offering and post-hunting scenes from the mid-5th dynasty. Authors differ in their identification of these animals with some suggesting that they are young scimitar oryx, while others believe they are hartebeest. Indeed, the newborn calves of these species are very similar morphologically, as both possess slender legs, long tails, nub horns, and large prominent ears. In an offering scene in room III of the tomb of *ṯjj* (Wild, *Ti iii*, pl. 166), calves carried by two adjacent bearers are labelled respectively, *jnjt sšꜣ*, 'Bringing a hartebeest' and *jnjt mꜣ-ḥḏ*, 'Bringing an oryx', even though the animals are virtually identical (only the length of their small horns and the width of their muzzles differ slightly). In room IX of the tomb of *zšzšt: jdwt* (Kanawati-Abder-Raziq, *Iynefert*, pl. 71), however, a carried calf is labelled, *bḥz n mꜣ-ḥḏ*, 'calf of a scimitar oryx'. Furthermore, many of the animals possess a distinctive marking around their eye, a feature that is often found in depictions of adult oryx. I suggest, therefore, that most of the images in question probably represent scimitar oryx calves, matching the greater number of unambiguous depictions of this species that were found during the current study, compared to hartebeest (59 images of hartebeest vs 445 of oryx).

134 Yom-Tov, Y. (1995). "Gazella dorcas", *Mam. Sp.* (491), 4.

135 Hoath, *Field Guide*, 157

136 *ibid.*, 150.

137 Estes, *Behavior Guide*, 17.

138 Lent, P.C. (1974). "Mother-infant relationships in ungulates" in Geist-Walther, *Behaviour of Ungulates i*, 28.

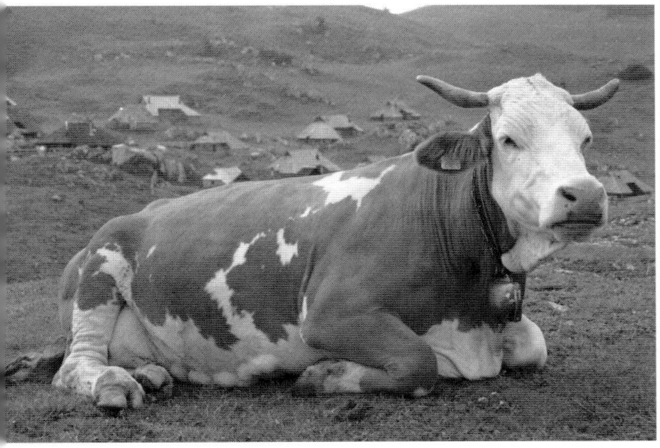

Pl. 4W Recumbent cow

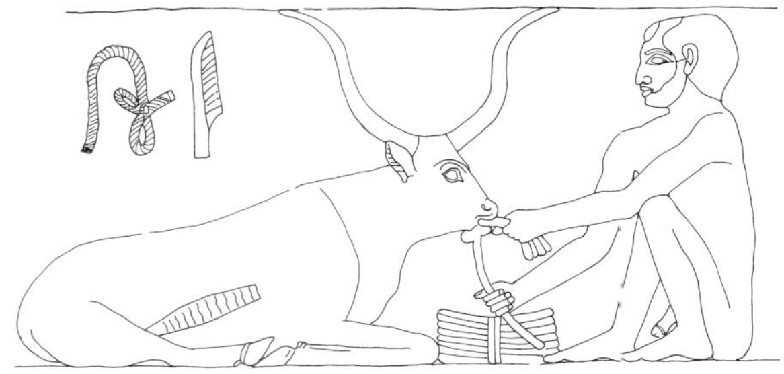

Fig. 4-49 (LB283)

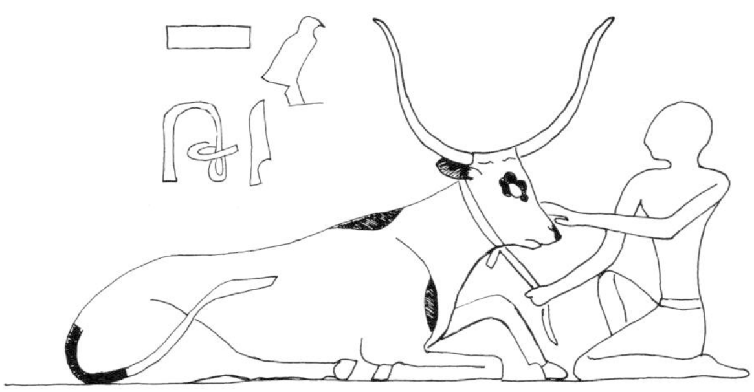

Fig. 4-50 (LB288)

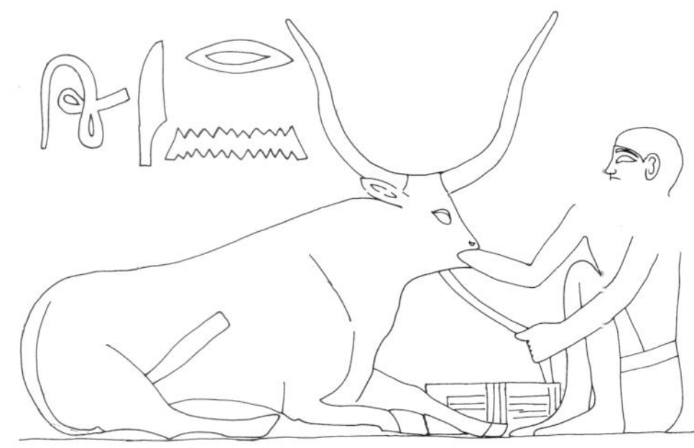

Fig. 4-51 (LB291)

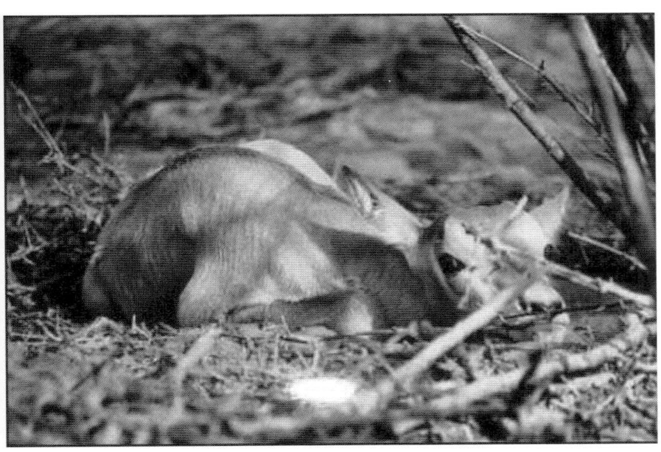

Pl. 4X Recumbent dorcas gazelle calf

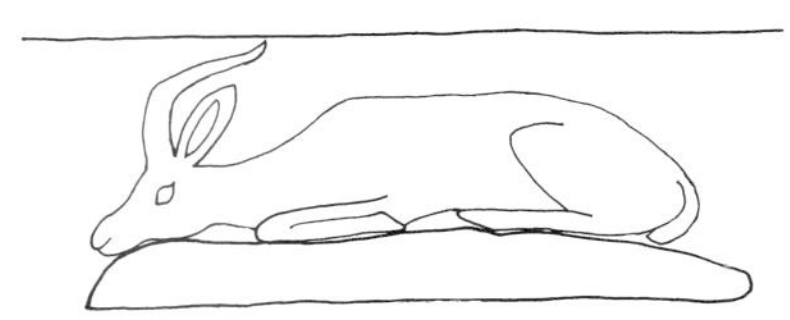

Fig. 4-52 (LB304)

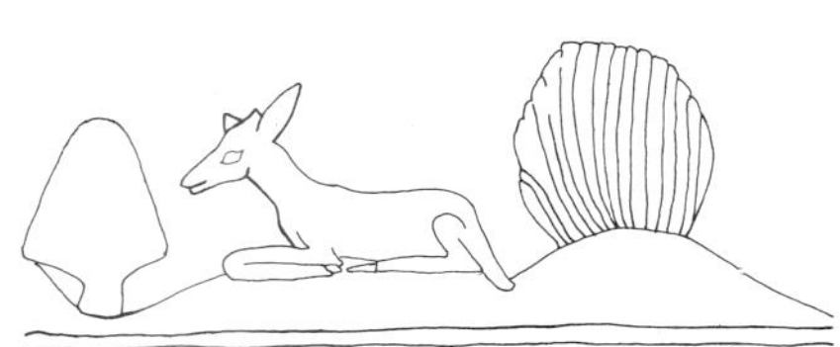

Fig. 4-53 (LB307)

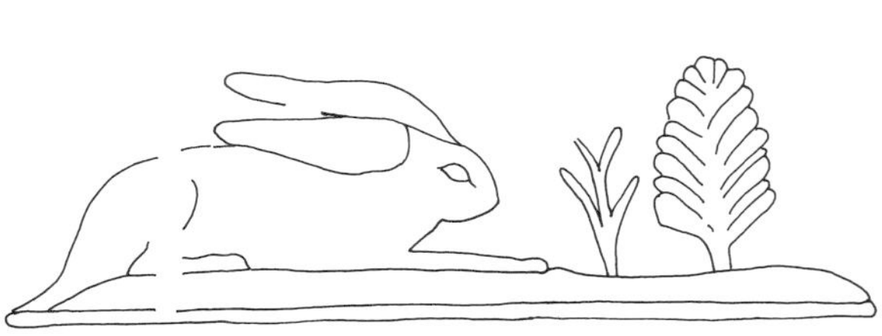

Fig. 4-54 (LB320)

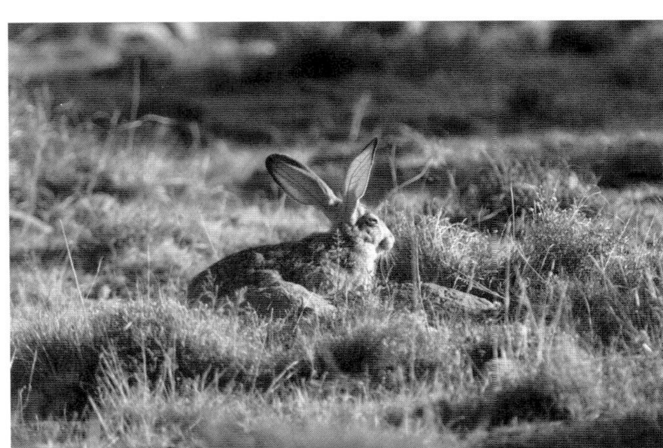

Pl. 4Y Cape hare sitting in a 'form'

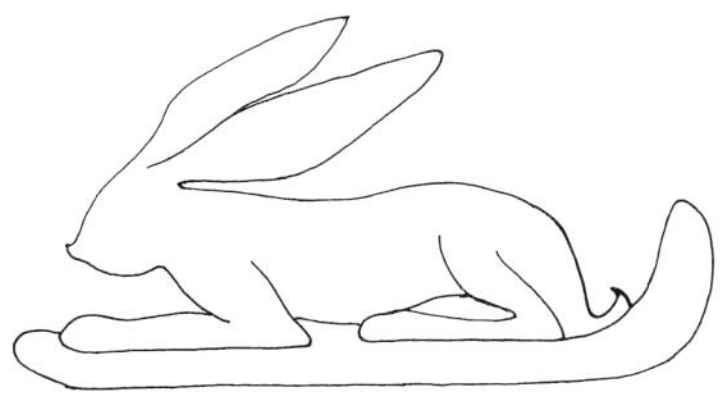

Fig. 4-55 (LB322)

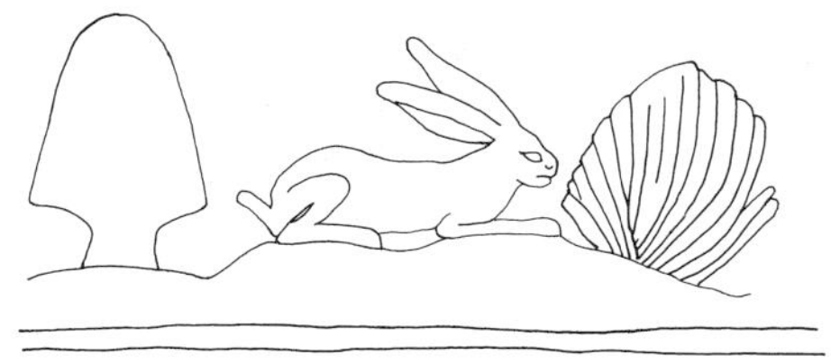

Fig. 4-56 (LB325)

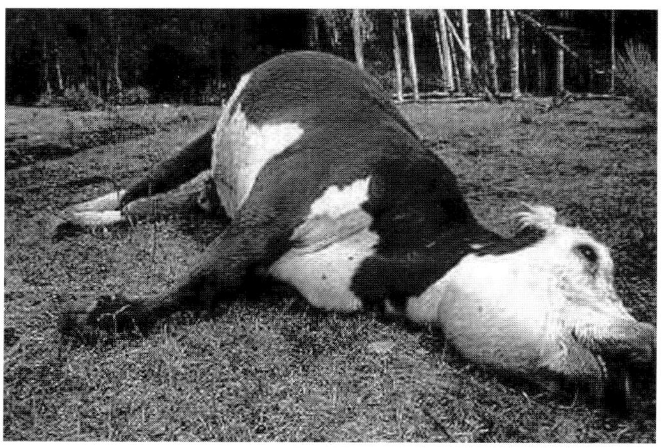

Pl. 4Z Dead cow
© Western Watersheds Project
Reproduced with permission

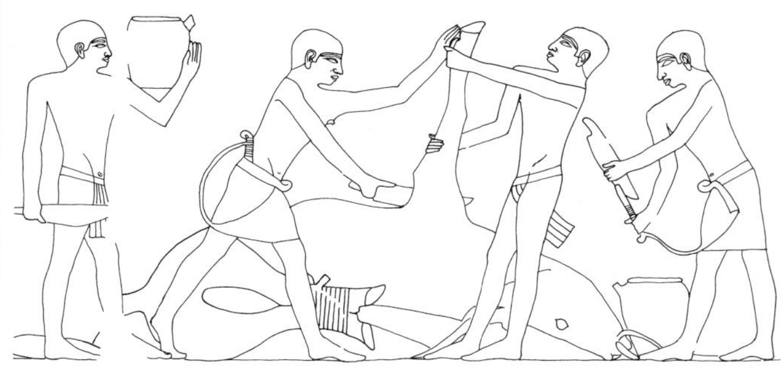

Fig. 4-57 (LB326)

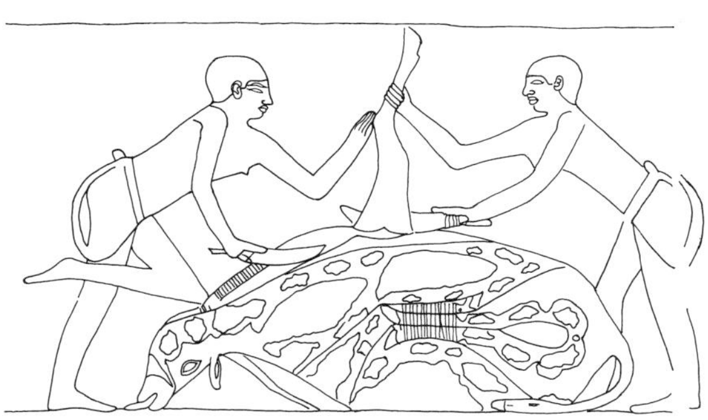

Fig. 4-58 (LB327)

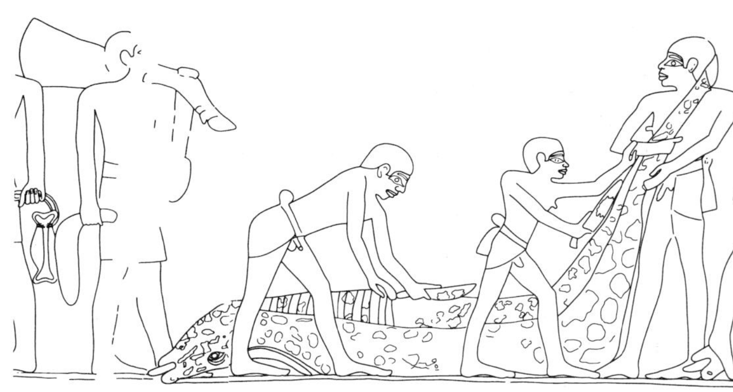

Fig. 4-59 (LB328)

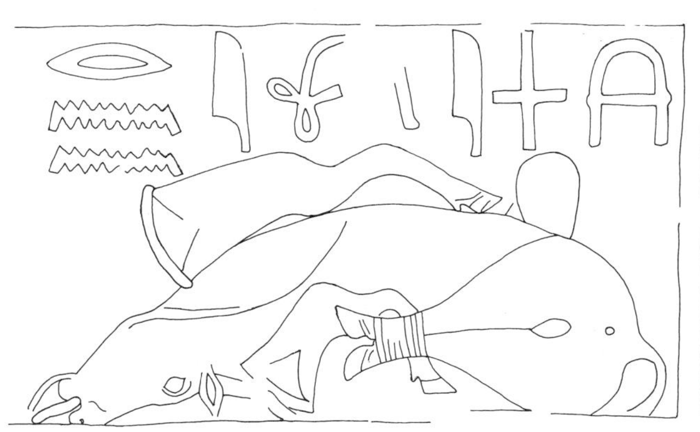

Fig. 4-60 (LB329)

Dorcas gazelle[139] and scimitar oryx[140] are both species in which the young hide for the first few weeks of life. The posture and location of the recumbent gazelles in wall scenes matches precisely that of hider calves, as they are depicted with both their fore- and hind legs folded under them and their head stretched forward so that it rests upon the substrate (cf. fig. 4-52 with pl. 4X).[141] It might be argued however, that these animals cannot be calves as they each exhibit full-size horns. I suggest that this adult feature has been included simply as a species marker, to enable the viewer to clearly identify them as dorcas gazelles.

The depicted oryx calves adopt a somewhat different posture, as they are not fully prone, but instead elevate their heads (see fig. 4-53). Nevertheless, their behaviour is also consistent with hiding. Nubian ibex are 'followers' and consequently do not hide their young.[142] I suggest, therefore, that the single example of a recumbent ibex that occurs in the mid-6th dynasty tomb of *mrjj-ttj: mrj* (LB319) is either a mistake or depicts an adult at rest.[143]

Hider calves will maintain their position even if a predator approaches. The posture can also be induced in response to disturbances nearby. The hunting context in which the recumbent animals are found is thus appropriate for such a reaction, and one that is more conducive to hiding than resting behaviour. Hunters will likely have discovered such secreted calves during desert expeditions. As the young do not move, even when approached by predators, they will have been caught easily. As noted by Osborn and Osbornová, this probably explains their frequent occurrence in scenes in which large numbers of diminutive ungulates are shown being transported in boxes and bundles.[144]

Cape hares

rʿ-m-k3.j (LB320: Saqqara, V.6? r/u - V.6-7)
nj-ʿnḫ-ẖnmw/ẖnmw-ḥtp (LB321-322: Saqqara, V.6L-7)
pḥn-wj-k3.j (LB323: Saqqara, V.6-8E)
ṯjj (LB324: Saqqara, V.8-9)
mrrw-k3.j: mrj (LB325: Saqqara, VI.1M-L)

Cape hares are large rodent-like mammals with long, elongated ears and large hind feet. They are found in deserts and open savannas where they forage at night and rest during the day. They do not use a burrow for this latter purpose, but instead scrape a shallow depression in the soil (called a 'form') and then lie down within it (see pl. 4Y). When reclining, their ears, which are usually erect, are held flat against their back; in this posture they are well camouflaged, as the animals' coat colour allows them to blend in with their desert surroundings.[145]

Five images of recumbent hares were found during this study (LB320 to LB325).[146] Each animal squats upon the ground with its hind legs folded under it, its forelegs outstretched, and its head elevated. All of the reliefs occur on sub-registers in hunting/desert scenes, suggesting that the animals, like young prostrate ungulates (see above), are to be perceived as 'hidden' in the background. The scenes reveal an awareness of the resting behaviour of this species, as each hare lies upon a low mound - the artists' response to the challenge of conveying an indentation in the soil. The earliest example, located in the tomb of *rʿ-m-k3.j* (LB320; see fig. 4-54) depicts a hare reclining upon a long, thin ridge surmounting a short baseline. The animal's ears stream across its back in accordance with its state of repose; in later images, however, many hares hold their ears in a more alert posture (e.g., LB322 and LB325; see fig. 4-55). The shape of the scrape is also quite varied, reflecting the different approaches taken by artists to the representation of this distinct area. In the tomb of *nj-ʿnḫ-ẖnmw/ẖnmw-ḥtp* (LB322; see fig. 4-55), for example, the resting place is

139 Yom-Tov, *Mam. Sp.* (491), 4.
140 Estes, *Behavior Guide*, 132.
141 Åsa Standberg (*Gazelles*, 27-28) has also noted this behaviour in hunting scenes.
142 Ralls, K., Lundrigan, B. and Kranz, K. (1987). "Mother-young relationships in captive ungulates: Spatial and temporal patterns", *Zoo Biol.* 6, 14.
143 The inspiration for this motif may have come from elsewhere. For example, Labrousse and Moussa (*Complexe Funéraire*, 42, fig. 48) describe a recumbent ibex that is depicted in the funerary complex of Unas.
144 Osborn-Osbornová, *Mammals*, 177.
145 Hoath, *Field Guide*, 218-220.
146 For an overview of the role of Cape hares in Egyptian art and culture, see Stoof, *Hasendarstellungen*.

presented as a thin slab with one upturned edge, while in the tomb of *ṯjj* (LB324), it takes the form of a thick platter with two elongated lips. The raised mound with a slight indentation at the crest that appears in the tomb of *mrrw-kꜣ.j: mrj* (LB325; see fig. 4-56) is a particularly realistic depiction; the addition of small shrubs on either side of the scrape is also accurate (see also fig. 4-54), as Cape hares are often found hiding amongst clumps of vegetation.[147]

4.2.3 Death

Death is sometimes associated with particular physical and postural changes that can differentiate it visually from a living state.[148] Contexts in which dead animals occur in wall reliefs include offering, hunting, and slaughter scenes.

Cattle

sḫm-ꜥnḫ-ptḥ (LB326: Saqqara, V.8-9)
ṯtw: kꜣ.j-nswt (LB327: Giza, VI)
jrw-kꜣ-ptḥ: ḫnw (LB328: Saqqara, VI)
ḫntj-kꜣ.j: jḫḫj (LB329: Saqqara, VI.2M)

In images depicting the slaughter of cattle and other ungulates, the death of many of the victims can be verified due to the dismembered state of their bodies, but scholars have been unable to agree whether animals are represented alive or dead during the most important part of the slaughtering process: the ritual severing of the foreleg.[149] It is beyond the scope of the current work to examine this issue in detail, but one behavioural observation must be made in passing.

During this study, 553 slaughter scenes in 145 tombs were examined. I present here just one example of the most commonly represented stages of the butchering process: that is, cutting of the foreleg (LB326; see fig. 4-57), removal of the leg after severing (LB327; see fig. 4-58), the division of the carcass (LB328; see fig. 4-59), and presentation of the prepared body as an offering to the deceased (LB329; see fig. 4-60). While the bulls in the first two images might conceivably still be alive, the remaining two are clearly not. Note, however, that while the animals' bodies are rendered in a range of postures, one feature remains constant in each: their tongues are extended so that the tip touches the ground. This feature was observed in the majority of slaughter scenes, beginning with the earliest example, found in the early 4th dynasty tomb of *mṯn*.[150] Two hundred and seven images of foreleg severing were investigated. In these, the animal's tongue was found to be clearly visible in 124 scenes (60%), unable to be determined due to poor preservation of the relief in 50 examples, and not depicted at all in only 33. It is important to note that when represented in such scenes, the protruded tongue always makes contact with the ground - either along its entire length or, most frequently, by the tip - regardless of the animal's orientation.

Eggebrecht noted that during butchering, the muscles holding an animal's tongue in position are usually severed when the throat is cut, which causes the organ to slip out of the muzzle (cf. pl. 4Z and also Colour Plate D).[151] He felt, consequently, that the presence of this feature in slaughter scenes provided definitive proof that a depicted animal was dead. I agree with this, but would add the proviso that the organ must be in contact with the ground for this state to be confirmed. As will be shown in subsequent chapters, animals frequently expose their tongues in wall scenes in situations where they are clearly alive, but in these depictions, the organ does not touch the substrate.

[147] Osborn-Helmy, *Contemporary Land Mammals*, 90.

[148] Barber, *Vampires, Burial, and Death*, 102-119; Iserson, K.V. (2002). "Rigor mortis and other postmortem changes" in Kastenbaum, *Encyclopedia of Death ii*, 721-723.

[149] See Weigall, A.E.P.B. (1915). "An ancient Egyptian funeral ceremony", *JEA* 2, 10-12; Junker, *Giza iii*, 229-231; Vandier, *Manuel v*, 136-138; Ikram, *Choice Cuts*, 50-51; Gordon-Schwabe, *Quick and the Dead*, 77, n. 7.

[150] See *LD ii*, pl. 4 [left]. I note, however, that this example is predated by a scene in the late 3rd to early 4th dynasty Meidum tomb of *rꜥ-ḥtp* (no. 6) in which an inverted bull, whose throat has been clearly cut, protrudes its tongue prominently as two butchers set to work. See Harpur, *Nefermaat and Rahotep*, 101, fig. 94 and pls. 44 and 46.

[151] Eggebrecht, *Schlachtungsbräuche*, 70. Ikram (*Choice Cuts*, 50) has also noted this feature.

SUMMARY

- Different forms and speeds of locomotion are communicated in wall scenes by simple adjustments of the legs and feet of animal figures – parted legs with feet flat convey a sequence of walking and standing, paired forelegs combined with parted hind legs suggest stasis, while the pairing of both fore- and hind legs impart a sense of speed. Flattened toes are generally associated with terrestrial locomotion, while pursed toes (in depictions of birds) indicate aquatic or aerial movement.
- The inclusion of antithetical behaviours in the same scene - such as running and grazing by donkeys or feeding and rumination by cattle - suggests a merging of two time-frames within the one image.
- A number of species-specific behaviours were found represented, such as bipedal hopping by jerboas, underwater breathing by Nile soft-shell turtles, attraction to meat by yellowjacket wasps, hiding by ungulate calves, and resting behaviour by Cape hares.
- The natural behaviour of many of the animals depicted in papyrus thickets (common genets, Egyptian mongooses, nesting birds) indicate that the marsh scene is best understood as folding away from the viewer. This is supported by the position of frogs within apparent branches of pondweed, a plant that actually grows horizontally across the water surface.

CHAPTER 5

GROUP BEHAVIOUR

DEFINITION

Social animals live in contact with other members of their own species. This close association allows the organization of individuals into groups, from breeding pairs and family units up to larger aggregations, such as schools, flocks, and herds. Group members often influence, and are influenced by, the behaviour of others ('social facilitation'); there is thus a tendency for groups to engage in activities in a synchronized manner.

Group living has many benefits, the most significant of which is that members improve their ability to detect and defend themselves against predators while simultaneously reducing the probability that they will fall victim to one. Social species find isolation stressful and will endeavour to rejoin their group if they become separated.

5.1 HERDING

A herd is an aggregation of mammals of the same species; however, the term is most commonly applied to grazing ungulates. Both wild and domesticated animals form herds; these may range in size from a few dozen to many hundreds of animals, depending on the species, time of year, and in the case of livestock, farm practices.

Donkeys, cattle, sheep, and goats are depicted frequently in wall reliefs in unleashed groups of three or more. Herds of scimitar oryx, Nubian ibex, or dorcas gazelle are much less common, but they occur in a few desert and offering scenes, particularly from the late 5th and early 6th dynasties; unusually, these wild ungulates are sometimes represented in mixed species groups.

Artists commonly depict herds using 'lateral layering', a motif in which the body of each individual partially overlaps that of the next.[1] Heinrich Schäfer proposed that this method was employed for groups that were moving forward;[2] indeed, one of the defining attributes of a herd is the cohesion of its members when in motion. Laterally layered groups featuring many individuals often produce a convincing impression of a massing of animals, but the regimented composition did not allow artists to easily communicate the intrinsic disorder that can characterize large herds. Sometimes, the uniformity is broken by one or more individuals that lower their heads or turn back in the opposite direction. Occasionally, though, a more realistic representation of a herd was attempted by showing the bodies of the animals in an irregular manner such that their individual poses, and the degree to which their bodies overlapped within the group, was varied.

Among the many images of wild and domesticated animal groups that are depicted in tomb reliefs were found two scene types that exhibit a particular knowledge of the idiosyncratic herding behaviour of some species. These examples also illustrate the artists' response to the challenge of representing the structure and movement of animal groups.

---

1 Schäfer, *Principles*, 178-185.

2 *ibid.*, 188.

Sheep

mrs-ꜥnḫ III (GB1: Giza, IV.6)
nb.j-m-ꜣḫtj (GB2: Giza, IV.6-V.1)
jj-mrjj (GB3: Giza, V.3L)
wr-jr.n-ptḥ (GB4: Saqqara, V.3-5?)
nfr-bꜣw-ptḥ (GB5-6: Giza, V.6)
nfr/kꜣ-ḥꜣ.j (GB7: Saqqara, V.6)
nj-ꜥnḫ-ẖnmw/ẖnmw-ḥtp (GB8: Saqqara, V.6L-7)
ḥtp-ḥr-ꜣḫtj (GB9: Saqqara, V.6-8E)
ꜣḫtj-ḥtp (GB10: Saqqara, V.6-8E)
jrj-n-kꜣ-ptḥ (GB11: Saqqara, V.6-9)
zꜣ-jb (GB12: Giza, V.8)
snḏm-jb: jntj (GB13: Giza, V.8M-L)
sḫm-ꜥnḫ-ptḥ (GB14: Saqqara, V.8-9)
ṯjj (GB15: Saqqara, V.8-9)
nfr-jrt-n.f (GB16: Saqqara, V.8-9)
kꜣ.j-m-nfrt (GB17: Saqqara, V.8-9?)
nbt (GB18: Saqqara, V.9)
nfr I (GB19: Giza, V.9-VI.1)
sḫm-kꜣ.j (GB20: Giza, V.9-VI.1?)
kꜣ.j-m-ḥzt (GB21: Saqqara, V.9-VI.1?)
kꜣ.j-m-nfrt (GB22: Giza, V-VI)
jꜣzn (GB23: Giza, VI.1M-2M)
mḥw (GB24: Saqqara, VI.2M-3?)
kꜣ-ḥj.f (GB25: Giza, VI.2M-3?)
jj-nfrt (GB26: Giza, VI.3-4E?)
nb-kꜣw-ḥr: jdw (GB27: Saqqara, V.9 r/u VI.7)

Agricultural scenes reveal that some tasks could be carried out by more than one animal species (e.g., threshing, which was executed by either donkeys or cattle), however the trampling in of sown seed appears to have been performed exclusively by sheep. While the predominance of these animals in treading scenes may simply be due to artistic preference, I propose that it represents an established farm practice by the Egyptians, which was based upon the particular herding behaviour of this species.

Thirty-two treading scenes were examined during this study. In five reliefs, however, only the distinctive posture of the accompanying shepherds indicates that such a scene was once present, as images of the sheep have been lost.[3] These are not considered here. The remaining examples all follow a similar pattern: A flock of sheep, averaging between four and seven individuals, advances towards a field-hand carrying a seed bag (e.g., GB1; see fig. 5-1).[4] The sower walks away from the group (although see GB1, GB13, and GB21), but turns back to hold out one hand towards the lead animal. In many scenes, the sheep places its muzzle in his palm (GB2, GB4, GB7, GB11, GB14, GB15, GB17, GB19, and GB21; see fig. 5-3); seed falling from the shepherd's hand in the tomb of *kꜣ.j-m-nfrt* (GB17; see fig. 5-2) indicates clearly that the lead animal is being fed (see Chapter 7).

The flock is often represented in a stylised manner using lateral layering, however this method is abandoned in some scenes to instead show the group moving in a more erratic fashion. Some animals lower their heads (GB1 to GB8, GB10, GB15, GB16, GB19 to GB26; see fig. 5-1), or turn in the opposite direction (figs. GB1, GB8, GB16, GB22, and GB24; see fig. 5-1, middle). In four examples, one animal rears up to place its forelegs over the back of a neighbour (GB1, GB16, GB21, and GB24; see fig. 5-1), and in the tomb of *sḫm-ꜥnḫ-ptḥ* (GB14; see fig. 5-3), a ewe lifts her hind legs off the ground and plummets forward over the back of a nearby companion. In every scene, the flock is followed by a group of men who wield whips above their heads and hold other implements in their lowered hands. The men appear to be running, as many raise the heel of their back foot and incline their body forward. Some also hold a free hand above their head (GB1; see fig. 5-1) or to their mouth (GB3 and GB5), suggesting that they may be calling out.

The herding instinct in sheep is particularly well developed. In natural conditions, they form small groups of around four to seven individuals[5] that are often related to one another. Although farmers usually maintain very large flocks, the animals within these nevertheless continue to form small, discrete clusters, each with its own home range.[6] Sheep stay close to one another when grazing and

---

3 These reliefs are found in the following tombs: *snḏm-jb: mḥj* (G2378=LG26) from Giza, and *sḫntjw/nfr-sšm-ptḥ* (ntn), *ḫnwt* (ntn), *ptḥ-ḥtp: jj-n-ꜥnḫ* (ntn) and *ḥzj* (ntn) from Saqqara.

4 Some flocks are headed by two sower/coaxers. For a discussion of the role of the sower/coaxer, see Vandier, *Manuel vi*, 18.

5 Kilgour-Dalton, *Livestock Behaviour*, 72.

6 Rutter, S.M. (2002). "Behaviour of sheep and goats" in Jensen, *Ethology*, 149.

orient themselves so that they can keep their neighbours in view.[7] Their desire to stay together is so strong that the flock will tend to follow any member that happens to move off in a purposeful manner. They react to alarming situations (such as the presence of a predator) by fleeing. Unlike other ungulates, however, which are inclined to disperse in the face of danger (e.g., goats and cattle), sheep crowd together and herd ever more tightly the more frightened they become.[8] Farmers have taken advantage of this fear response by using dogs, which the sheep perceive as predators, to move and control large groups (see pl. 5A).

I believe that Egyptian farmers were aware that sheep bunch when they are alarmed, and that they employed this knowledge in their agricultural practices.[9] In short, I suggest that the shepherds who follow the animals in treading scenes are deliberately scaring them to make them form a compact group. Note that they appear to pursue the sheep at speed, cracking their whips, waving their arms, and calling out. Such behaviour will undoubtedly have caused the flock great fright, inducing them to both flee and herd closely together. This conclusion is supported by studies that have shown that sheep perceive humans as potential predators,[10] and that the use of noise-making devices, such as rattles, can be extremely effective in encouraging them to move.[11]

Their constrained method of representation did not offer artists a straightforward means by which to depict a compact group. As noted above, most of the flocks display lateral layering, but some artists rejected this in favour of a less orderly and, consequently, more visually accurate composition. In particular, flocks that include animals that rear up (GB1, GB16, GB21, and GB24; fig. 5-1) and fall forward (GB14; see fig. 5-3) appear to reproduce a specific behaviour given by sheep in response to close contact and alarm. When panicked, sheep will sometimes leap in the air as they attempt to escape the perceived threat.[12] Furthermore, when pressed together, animals at the centre of the flock may become so cramped that they will attempt to break free by jumping up. I propose, therefore, that the rearing and falling sheep in treading scenes are intended to impart a leaping motion.[13] The highly unusual pose of the ewe in the tomb of *sḫm-ʿnḫ-ptḥ* (GB14; see fig. 5-3), clearly suggests an animal that is tumbling downward - its ungainly movement imparting a roiling momentum to the otherwise regimented group. The injection of these postures into treading scenes is thus both an accurate portrayal of the animals' response to danger and an ingenious way to represent, indirectly, their agitated flocking behaviour.

The purpose of the running shepherds was thus to make the animals bunch and to maintain, through their continued scare tactics, the flock's tight formation as it proceeded across the field[14] The direction in which the animals advanced, however, was controlled by the sower. By enticing one of the sheep with a handful of seeds as he walked along the furrows, he will have created a leader behind which other members of the flock would follow. This animal was probably chosen at

---

7 Fisher, A. and Matthews, L. (2001). "The social behaviour of sheep" in Keeling-Gonyou, *Social Behaviour*, 218-220.

8 *ibid.*, fig. 8.3; Hafez, E.S.E., Cairns, R.B., Hulet, C.V. and Scott, J.P. (1969). "The behaviour of sheep and goats" in Hafez, *Behaviour of Domestic Animals*, 332; Müller, *Sheep*, 57.

9 It is intriguing to note that the depicted groups frequently consist of around four to seven animals, which matches precisely the size of naturally occurring flocks. Egyptian farmers may have discovered, through trial and error, that this group size was easier to work with. See Kilgour-Dalton, *Livestock Behaviour*, 72.

10 Bouissou, M.F. and Vandenheede, M. (1995). "Fear reactions of domestic sheep confronted with either a human or a human-like model", *Beh. Proc.* 34, 81-92; Hargreaves, A.L. and Hutson, G.D. (1997). "Handling systems for sheep", *Livest. Prod. Sci.* 49(2), 127.

11 Kilgour-Dalton, *Livestock Behaviour*, 77.

12 Gill, *Applied Sheep Behaviour*, 24

13 As two of the rearing animals are rams (GB1 and GB24; see fig. 5-1), their apparent mounting behaviour has been described elsewhere as sexual (see Ikram, S. (1991). "Animal mating motifs in Egyptian funerary representations", *GM* 124, 51-52), but the forward movement of the flock would have largely prevented this, as ewes usually stand still during mating. It is also important to note that one of the rearing sheep appears to be a female (GB16).

14 The men may have chased the sheep as a tight group, but it is more likely that they stood beside one another in an arc, to encircle the rear and sides of the advancing flock. The latter seems to be implied in scenes where shepherds with whips appear within the flock (GB1; see fig. 5-1), and both in front and behind the sheep (GB3 and GB22).

random. It is interesting to note, however, that in modern farm practice, individual sheep are sometimes trained, through the use of food rewards, to lead flocks on command through areas where they might otherwise be reluctant to go.[15]

The animals were driven over ploughed fields in order to embed the sown seed in the soil with the action of their feet. While any hoofed creature might have accomplished this to some degree, sheep will have been the most effective and efficient species to employ because of their particular tendency to stay in a group. This behaviour will have allowed the shepherds to move the animals as a controlled, discrete unit directly over the path of the seed. In contrast, other herding animals that might have been used for this purpose (e.g., cattle or goats) will have been more inclined to form loose and unruly amalgamations that walked over the seed in a haphazard fashion. Furthermore, the cohesion of the sheep flock will have kept their many hooves concentrated within a defined area to push the seed deep into the substrate while simultaneously kneading and compacting the soil around it.[16]

Treading scenes from Giza and Saqqara thus exhibit a practical knowledge of the idiosyncratic social behaviour of sheep and a creative artistic solution to the problem of imparting their tight flocking response to fear.

Mixed species

ptḥ-ḥtp II: ṯfj (GB28: Saqqara, V.9M-L)
sšm-nfr IV (GB29-30: Giza, V.9-VI.1)
nj-mꜣꜥt-rꜥ [*jzzj-mr-nṯr*] (GB31: Giza, V.9-VI.1)
mrrw-kꜣ.j: mrj (GB32: Saqqara, VI.1M-L)
jnw-mnw (GB33: Saqqara, VI.2E)
mrjj-ttj: mrj (GB34-37: Saqqara, VI.2M-L)
wr-nw (GB38: Saqqara, VI.3-4?)

Seven tombs dating from the end of the 5th until the mid-6th dynasty show ungulates of different species running together as cohesive groups. All of the mixed herds appear in desert hunting scenes in which animals are being attacked: pursued and captured by men, bitten and pulled down by dogs, or overpowered by lions (see Chapter 8). Members of some of the mixed herds are themselves lassoed, grasped, or chased (e.g., GB28, GB29, GB32, and GB35; see fig. 5-5), and, in the tomb of *mrjj-ttj: mrj*, dogs maul animals at the rear of two groups (GB36 and GB37). Each cluster consists of a combination of animals drawn from the following species: wild cattle, addax, scimitar oryx, hartebeest, dorcas gazelle, Sömmering's gazelle, Nubian ibex, and dama deer. One herd is also accompanied by a jerboa (GB33; see fig. 5-4).

All of the depicted ungulates live in herds or small family groups, however no species, including those pictured, would attempt to join, or be accepted in, the herd of another. Social animals associate readily with those of their own kind, but they do not normally mix with members of other species. Thus the images, as presented, do not match the natural herding behaviour of animals. Nevertheless, the hunting context in which the mixed ungulates appear may offer a solution.

The response of many animals to the sudden presence of a predator is to flee (pl. 5B).[17] In herding animals, such as antelope species and deer, this can result in a stampede in which group members run *en masse* in different directions as they try to escape the threat. Under these circumstances, the social boundaries between herds break down as individuals disperse and overlap with the members

---

15 Kilgour-Dalton, *Livestock Behaviour*, 79-80.

16 The special attributes of their treading behaviour have been incorporated into a device known as a 'sheep's foot roller' - a large metal cylinder covered by blunt spikes - which is used in construction to compact soil. As the roller moves, the spikes enter the ground and compress the dirt at some depth, but agitate the topsoil as they are extracted. A common problem in agriculture is compaction of the soil in pastures that is caused by the hooves of grazing flocks of sheep. See Edmond, D.B. (1974). "Effects of sheep treading on measured pasture yield and physical conditions of four soils", *NZ J. Exp. Ag.* 2, 39-43; Drewry, J.J. and Paton, R.J. (2005). "Effects of sheep treading on soil physical properties and pasture yield of newly sown pastures", *NZ J. Ag. Res.* 48, 39-46.

17 Grandin, T. and Deesing, M.J. (1998). "Genetics and behavior during handling, restraint, and herding" in Grandin, *Genetics*, 114.

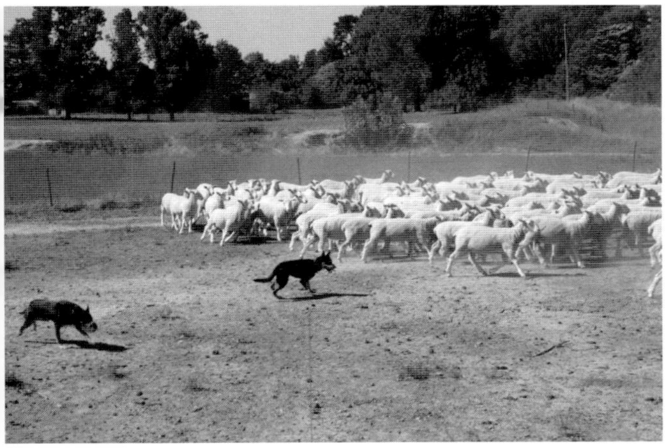

Pl. 5A Sheep bunching
© iStockphoto.com/John Simmons

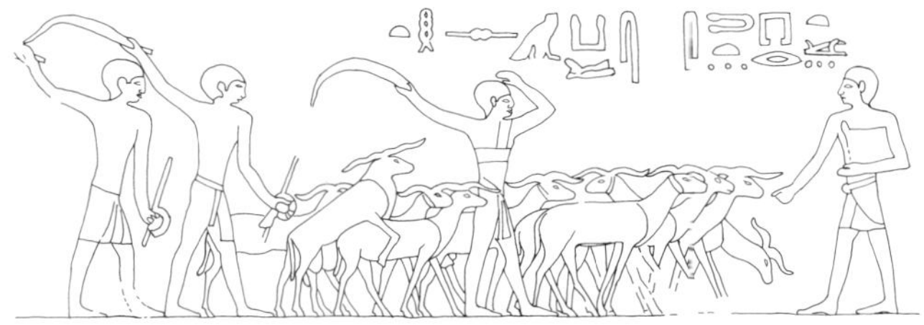

Fig. 5-1 (GB1)

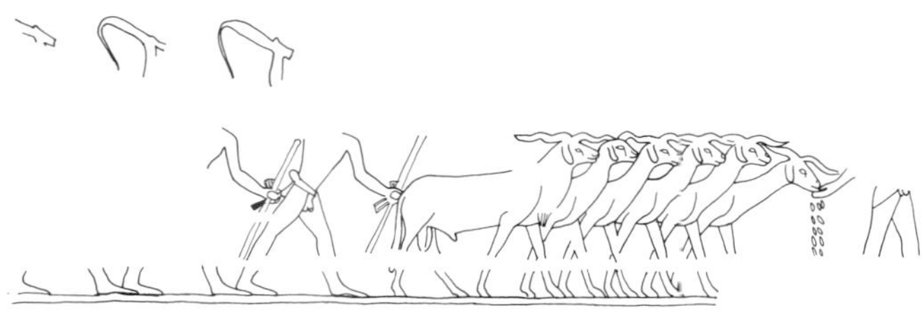

Fig. 5-2 (GB17)

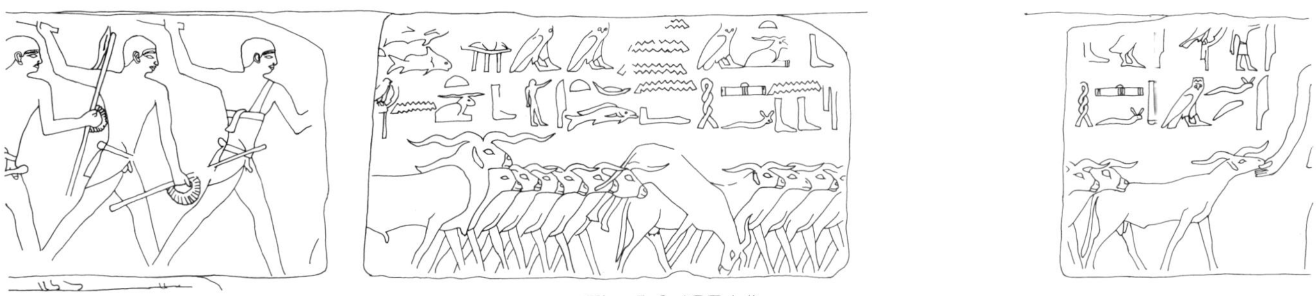

Fig. 5-3 (GB14)

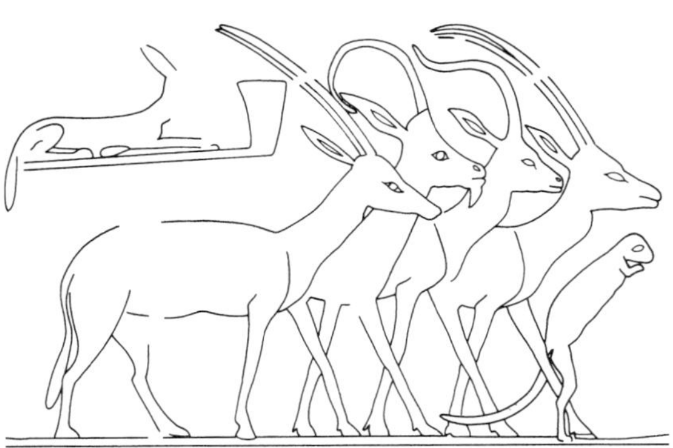

Fig. 5-4 (GB33)

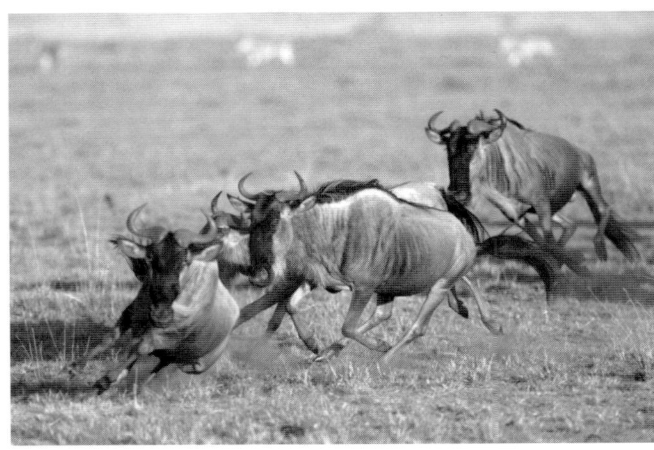

Pl. 5B Fleeing from predators
© iStockphoto.com/Graeme Purdy

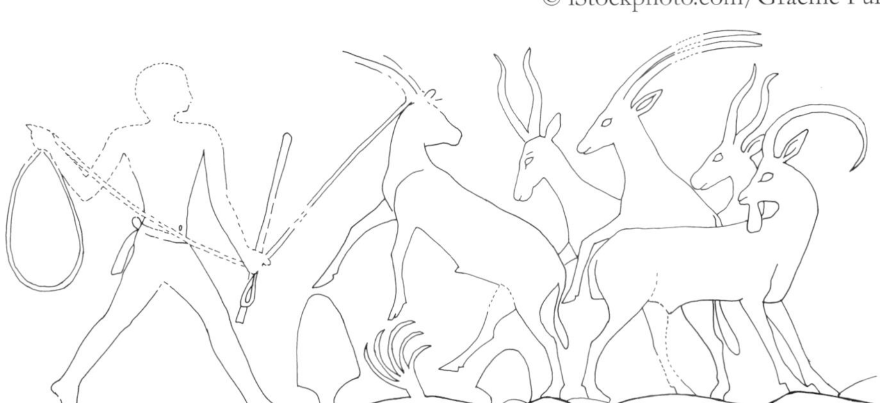

Fig. 5-5 (GB29)

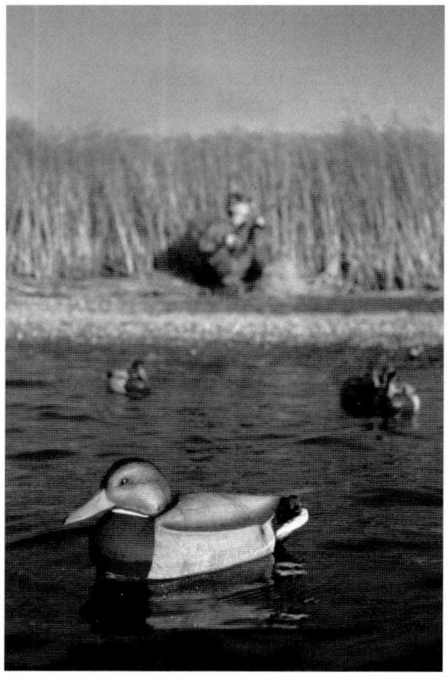

Pl. 5C Artificial decoys
© iStockphoto.com/Ryan Howe

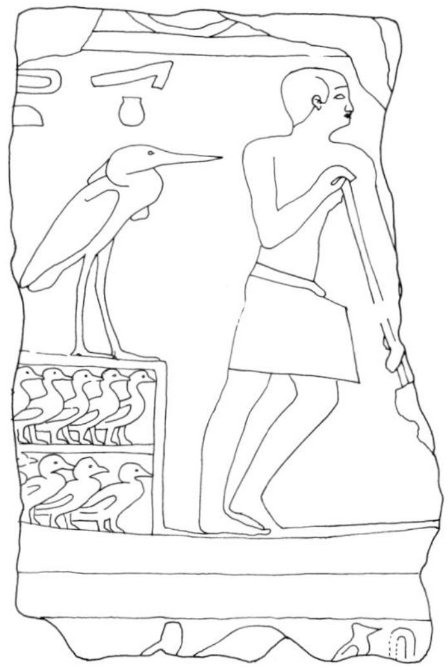

Fig. 5-6 (GB39)

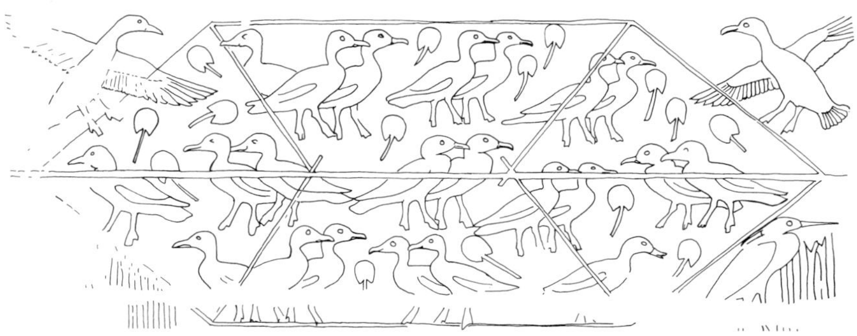

Fig. 5-7 (GB43)

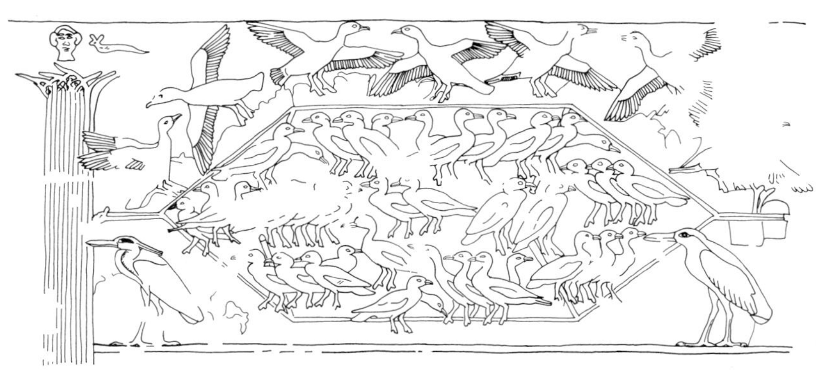

Fig. 5-8 (GB47)

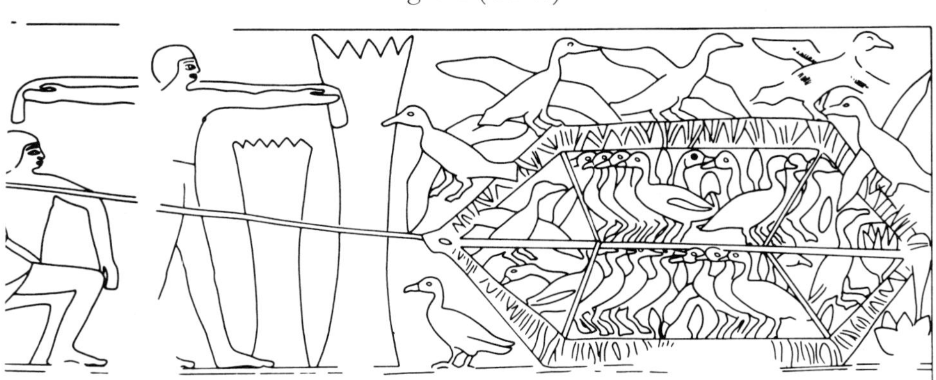

Fig. 5-9 (GB58)

of other groups. The mixed ungulates in wall scenes may thus be exhibiting an anti-predator response to the hunters, dogs, and lions that are represented nearby. Despite the lack of a clear running pose (see Chapter 4) in some images, I suggest that the artists have attempted to convey the intermingling of panicked herds by depicting a heterogenous group of animals whose disparity is emphasized by their different horn shapes.[18] The mix of species in each depicted herd is not the result of observation, but individual choice by the artists, driven only by the need to produce an image of confusion and disarray. Indeed, some of the animals would never be found running together naturally because they inhabit different ecological environments (e.g., ibex, which are found in mountainous regions are depicted beside scimitar oryx, which are a desert species).

Other details point to a panicked reaction by the animals. For example, a rearing dorcas gazelle depicted in the tomb of *sšm-nfr IV* (GB30), may be 'stotting'[19] - a high, stiff-legged jump that is given by this species when running away from predators. So, too, the apparent inter-species mounting of a dorcas gazelle upon a Sömmerings gazelle in the tomb of *nj-m3ˁt-rˁ [jzzj-mr-ntr]* (GB31),[20] may instead depict a bounding leap - as might images in which animals straddle the hindquarters of their neighbours (GB29 and GB38; see fig. 5-5 and cf. rearing sheep in treading scenes). The jerboa leading the mixed herd in the tomb of *jnw-mnw* (GB33; see fig. 5-4) also points to a stampede as these animals are known to leave their burrows and then run about erratically in response to outside disturbances.[21] The rodent's addition to the group, although somewhat humorous, is thus in accord with the general theme of a fear-driven mixing of species.

Finally, it is also worth considering another, yet complementary, explanation for these herds: that they are composed of exemplars of different species that have been caught and held in hunting parks. All of the images appear within a desert context, but two are clearly situated in fenced areas (GB32 and GB33). The intermingling of the animals would thus be the result of not only their panic, but also their close confinement. Indeed, the apparent mounting behaviour by some may suggest their extreme proximity (see *Sheep* above).

5.2 FLOCKING

Many bird species form aggregations called flocks. Some species spend their entire lives in close association with their own kind, while others only come together in groups at particular times and/or for specific purposes, such as defence, feeding, or when migrating.[22] Flocks are occasionally represented in wall reliefs as loose amalgamations (e.g., cranes in pastoral scenes), but more frequently, they are rendered as tight groups of laterally layered birds of the same species.

Confidence decoys

k3.j-wˁb (GB39: Giza, IV.2-3)
rˁ-m-k3.j (GB40: Saqqara, V.6? r/u - V.6-7)
nj-ˁnḫ-ẖnmw/ḫnmw-ḥtp (GB41: Saqqara, V.6L-7)
ḥtp-ḥr-3ḫtj (GB42: Saqqara, V.6-8E)
3ḫtj-ḥtp (GB43: Saqqara, V.6-8E)
jrj-n-k3-ptḥ (GB44: Saqqara, V.6-9)
rˁ-špss (GB45: Saqqara, V.8M)
ptḥ-ḥtp I (GB46: Saqqara, V.8M-L)
ṯjj (GB47-49: Saqqara, V.8-9)
nfr-jrt-n.f (GB50: Saqqara, V.8-9)
ẖww-wr (GB51: Giza, V.8-9M)
ẖnwt (GB52: Saqqara, V.9)
jj-nfrt: š3.n.f (GB53: Saqqara, V.9)
ptḥ-ḥtp II: ṯfj (GB54: Saqqara, V.9M-L)
sḫm-k3.j (GB55: Giza, V.9-VI.1?)
Anon (GB56: Giza, V-VI)
k3-gm-nj: mmj (GB57: Saqqara, VI.1M)
sˁnḫ-w(j)-ptḥ (GB58: Saqqara, VI.1L)
ḥzj (GB59: Saqqara, VI.1L-2E)
ˁnḫ-m-ˁ-ḥr: zzj (GB60-62: Saqqara, VI.1L-2E)
wḏ3-ḥ3-ttj: ššj (GB63-64: Saqqara, VI.1L-2E)
mṯtj (GB65: Saqqara, VI.2?)
mḥw (GB66-69: Saqqara, VI.2M-3?)

18 Schäfer (*Principles*, 181) noted that, "Only very rarely are entirely different animals or objects arranged in layers", but that "sixth dynasty artists love to unite three animals with different horns in one layered group".
19 Yom-Tov, Y. (1995). "Gazella dorcas", *Mam. Sp.* (491), 4.
20 Strandberg (*Gazelles*, 80) interprets the composition of this scene differently.
21 This behaviour has been noted in the Lesser Egyptian jerboa. See Osborn-Helmy, *Contemporary Land Mammals*, 349.
22 Leahy, *Companion*, 291-293.

Thirty-four tombs inspected during this study were found to include reliefs of men capturing birds with a clapnet. In each scene, a hexagonal-shaped net is laid out in an area of open water (defined by either an oval line or vegetation around the perimeter). The space within the clapnet is usually filled to capacity with swimming and feeding waterfowl (see Chapters 4 and 7) that seem unaware of their precarious situation. The tightly packed birds are usually clumped into groups of two or more overlapping figures, a visual ploy that emphasizes the crowded space and the abundance of animals. More than one species is frequently represented within each net, as the bird groups often vary in body size and neck length (e.g., GB40, GB47, GB50, GB53, and GB55; see fig. 5-8). Most are clearly ducks and geese, but uniquely in the tomb of *mḥw*, a clapnet contains four cranes (GB67). Men who wait to pull each trap shut are depicted nearby, often behind a 'blind' of papyrus.

In approximately half of the scenes examined, a grey heron was found standing prominently beside the clapnet, staring in the direction of the hunters (e.g., see figs. 5-7 and 5-8, left); this subset of images is considered here. Occasionally, a second heron, or a bird that is sometimes identified as a bittern (e.g., GB44, GB47, GB63, and GB66; see fig. 5-8, right), stands in an obvious manner at the opposite end of the clapnet. It has been suggested that all of these vigilant birds were tame decoys employed to attract waterfowl to the net.[23] A special relationship between the herons and fowlers is certainly indicated in early images in which they travel with the men as they canoe home with their captured fowl (GB39, GB41, GB42, and GB45);[24] the birds stand calmly upon crates containing ducks (e.g., GB39; see fig. 5-6), thereby establishing a link between the decoys and their victims. In later scenes, the herons are also often restrained by a short leash tied to their legs (GB48, GB61, and GB62; see fig. 4-5 in Chapter 4) or tail-feathers (GB65), indicating that they belong to, and are valued by, the fowlers.

Hunters have used decoys to attract flocks of waterfowl for hundreds of years.[25] These are most often models (pl. 5C) - replicas of the hunted birds made from wood or other materials and usually painted to resemble particular species - but live animals can also be used as bait.[26] Floating artificial decoys are set out in large numbers in open waters to entice migrating waterbirds to land in the vicinity. The technique works because ducks and geese are highly gregarious and perceive safety in numbers. The size of the artificial flock assures approaching birds that the area is safe to land and thus draws them within range of waiting hunters. Live decoys, however, are used to lure birds directly into prepared nets and traps. The decoys are usually caged and then positioned either within or close to the trap so that approaching birds will become ensnared. The decoys are usually members of the targeted species and so are approached by others for the purpose of social interaction. Both live and artificial birds are often labelled 'confidence' decoys, as their primary function is to lull wary waterfowl into a false sense of security.

The bitterns and grey herons in clapnet scenes appear to be used as live decoys, however, rather than attracting members of their own species, these birds were clearly intended to lure ducks and geese to the location of the net. Indeed, fowlers in modern Egypt are reported to use herons for this purpose[27] and hunters in North America often place a full-sized model of a great blue heron near their spread of duck decoys as it is believed that the sight of this bird increases the chance that waterfowl will approach.[28]

Grey herons are solitary for most of their lives. When hunting for fish, they often stand motionless in shallow water as they wait for a potential prey item to swim within striking range (see *Herons* in Chapter 7). They are found in a variety of habitats, including fields, lakes, marshes, rivers and

---

23 Houlihan, *Birds*, 13-15 and 21.

24 The 4th dynasty tomb chapel of *jtt* at Meidum (no. 16) features one of the earliest examples of a clapnet scene. In the panel directly above this image, a grey heron is depicted on the stern of a fowler's canoe. See Petrie, *Medum*, pl. 24; Harpur, *Nefermaat and Rahotep*, fig. 86 and pl. 34.

25 Baldassarre-Bolen, *Waterfowl Ecology*, 4-6; Tuohy, D.R. and Napton, L.K. (1986). "Duck decoys from Lovelock Cave, Nevada, dated by ^{14}C accelerator mass spectrometry", *Am. Antiq.* 51(4), 813-816.

26 Evrard, J.O. and Bacon, B.R. (1998). "Duck trapping success and mortality using four trap designs", *NABB* 23(4), 110-114.

27 Houlihan, *Birds*, 15.

28 Sisson, D. (1995). "Take a heron hunting", *FS* 100, 44.

streams, but they prefer wide-open surroundings.[29] They are cautious and suspicious birds that are sensitive to disturbance; consequently, they usually choose quiet places in which to forage. The presence of a grey heron would thus be an indication to passing waterfowl that the area was safe as it would be unlikely to stay if threatened. Herons are also very adaptable birds that can be tamed[30] and are capable of learning,[31] so it would seem feasible that they could be trained to stand calmly beside a clapnet.

In contrast, it is difficult to understand why bitterns would be chosen to act as decoys. Bitterns are extremely shy and secretive creatures that prefer to stay in dense cover. If they are disturbed, they sleek their feathers and stand immobile, a behaviour that, in combination with their cryptic colouration, enables them to blend in with their surroundings. They are rarely seen out in the open and are difficult to locate from a distance.[32] It is unlikely then, that bitterns would make effective decoys, as waterfowl would be unlikely to recognize members of this species and react to their presence. I propose, therefore, that their depiction is an artistic invention intended to inject some variety into clapnet scenes. The comparatively small number of images featuring bitterns would tend to support this notion.

Other species have also been identified as potential decoys. Spoonbills, for example, are often depicted in the waters surrounding clapnets (GB46, GB50, and GB59),[33] and in the Giza tomb of *sḫm-kꜣ.j* (GB55), a member of this species stands, along with a grey heron, in the same conspicuous position that is adopted by bitterns. Nevertheless, spoonbills, along with other birds that often line the lower edge of the net in later scenes (e.g., ducks, cranes, and ibis) are perhaps best viewed as simply illustrating the avifauna of the marshes. A more convincing case can be made for a duck decoy in the tomb of *sʿnḫ-w(j)-ptḥ* (GB58; see fig. 5-9). The bird stands before a clapnet filled with members of the same species and stares at the fowlers nearby, mimicking precisely the conspicuous behaviour of grey herons. As noted above, decoy birds from the targeted species make particularly effective lures.

5.3 SCHOOLING

When feeding, fish often aggregate in loose groups ('shoals') in which the members are randomly oriented to one another. As they begin to move, however, they may form a 'school' in which they turn to face in the same direction and swim together in a synchronized manner (see pl. 5D). A school is defined as a group of three or more fish in which each individual maintains a precise distance from its neighbours (those above and below, as well as to the sides) and adjusts its speed and course in harmony with those around it.[34]

Although schooling behaviour is most easily observed underwater, it can also be seen from above (e.g., from a boat) in clear, relatively shallow water.

Fish

ḥzj (GB70-72: Saqqara, VI.1L-2E)

Marsh and fishing scenes provide the observer with a view beneath the water surface. The underwater section[35] is often rendered as a very shallow band, constraining the way in which aquatic animals and vegetation could be depicted. The representation of fish, for example, tends to follow a rigid pattern. The low water level often prevented artists from presenting more than one

---

[29] Cramp, S. and Simmons, K.E.L (1977). "Grey heron (*Ardea cinerea*)" in Cramp-Simmons, *Handbook i*, 303.

[30] The Mohana people, who live on Lake Manchar in Pakistan, treat many waterbirds as members of their families. Each Mohana owns one special heron or egret, known as a 'Kong', which acts as a guard for its owner. See Pelletier, F.-X. (1999). "The bird people of Manchar", *Geo Aust.* 21, 68-73.

[31] McKilligan, *Herons*, 38.

[32] Cramp, S. and Simmons, K.E.L. (1977). "Bittern (*Botaurus stellaris*)" in Cramp-Simmons, *Handbook i*, 247.

[33] Houlihan, *Birds*, 34.

[34] Partridge, B.L. (1982). "The structure and function of fish schools", *Sci. Am.* 246(6), 90-99.

[35] See n. 89 in Chapter 4.

fish at a time, resulting in a chain-like composition. The bodies of the animals never overlap, even when they are situated in cramped spaces (such as within dip-nets; see fig. 4-14 in Chapter 4). Furthermore, members of the same species rarely appear next to one another. In fact, artists seem to have deliberately mixed the different types, perhaps to achieve a more varied visual effect.[36] This is also apparent in scenes where the represented water depth is increased slightly (such as in the 'mound of water', or within some seine fishing-nets), in which the orientation of their bodies seems to be entirely random (although see *Catfish* in Chapter 7). The depiction of fish in Memphite reliefs thus seems largely to have been dictated by artistic, rather than biological, considerations.

Images found in the tomb of *ḥzj* are consequently unique and reveal a special interest by the artist in piscine behaviour. In each of two marsh scenes that grace the southern wall of the tomb's portico, groups of tilapia[37] are depicted in the waters below the deceased's canoe - three fish occur in the scene to the east of the entrance (GB70; see fig. 5-10), and four in the western scene (GB71; see fig. 5-11). Groups of four tilapia and three mullet[38] are also represented in a seine-net depicted on the east wall of the portico (GB72; see fig. 5-12). In each case, the bodies of the fish overlap one another horizontally as they advance.

I propose that the unusual arrangement of these fish was intended to show them moving in schools. The artist has again taken advantage of the lateral layering method used to represent groups of animals; however, here the regimented order of the motif has, either by chance or design, resulted in an extremely accurate depiction of the polarized body placement of schooling fish (cf. pl 5D). Schäfer's proposal that lateral layering was intended to convey forward movement seems especially likely in the *ḥzj* images, as one of the most striking features of schooling behaviour is the fluid manner in which the members of the group travel in unison.

Tilapia and mullet are both schooling fish. The young of many tilapia species stay together as a group after hatching and often join other schools of juveniles as they mature. Nile tilapia (one of the most common species in the Nile river) are reported to form very substantial schools (up to 100 metres^2) for the purposes of feeding and protection against predators.[39] Grey mullet also travel in large groups (especially during their breeding season), enabling fishermen to catch them with relative ease.[40] The *ḥzj* reliefs clearly do not reproduce the large groups that these species can attain, but it is tempting to see in the trios of tilapia (fig. 5-10) and mullet (fig. 5-12, right) a possible symbolic allusion to plurality/abundance.[41]

The schooling behaviour by these and other Nile species will have been well known to Egyptian fishermen. It is surprising, then, that this common and captivating behaviour was shunned by all but the artist responsible for the tomb of *ḥzj*.[42]

5.4 SEPARATION

Isolation can cause great anxiety in social species. Separated individuals often exhibit signs of stress by performing particular actions, emitting distress vocalizations, and making concerted efforts to rejoin their social group. The effect of separation upon mothers and their progeny may be particularly acute.

36 The juxtaposition of some species may have a symbolic meaning. Brewer and Friedman (*Fish and Fishing*, 77-79) suggest that the combination of a tilapia (which are abundant in the Delta) and a lates (which are common in the south) in bident fishing scenes may allude to the unification of Upper and Lower Egypt.

37 *ibid.*, 76-79.

38 *ibid.*, 72-73.

39 Philippart, J.-C. and Ruwet, J.-C. (1982). "Ecology and distribution of tilapias" in Pullin-Lowe-McConnell, *Tilapias*, 54.

40 Brewer-Friedman, *Fish and Fishing*, 72.

41 For the association between the number three and plurality, see Wilkinson, *Symbol and Magic*, 131-133.

42 In contrast, the reproductive behaviour of tilapia was of great interest to artists. Tilapia were employed frequently as a decorative motif due to the symbolic association drawn between the mouth-brooding behaviour of some species and the concept of rebirth. Some images appear to show the fish 'ingesting' their eggs by grouping them around a central sphere. See Dambach, M. and Wallert, I. (1966). "Das Tilapia-Motiv in der altägyptischen Kunst", *CdÉ* 41, 273-294.

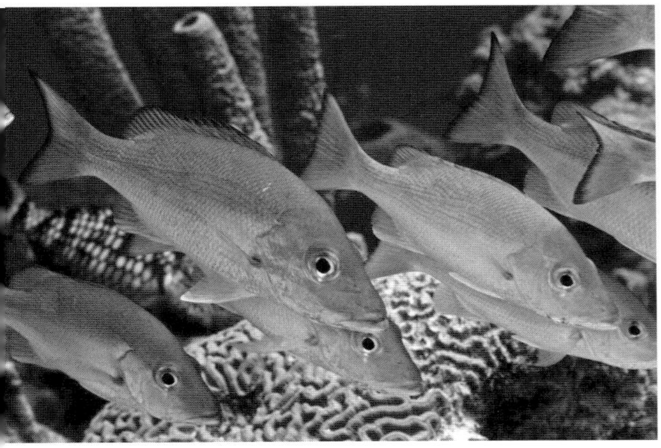

Pl. 5D Schooling fish
© iStockphoto.com/John Anderson

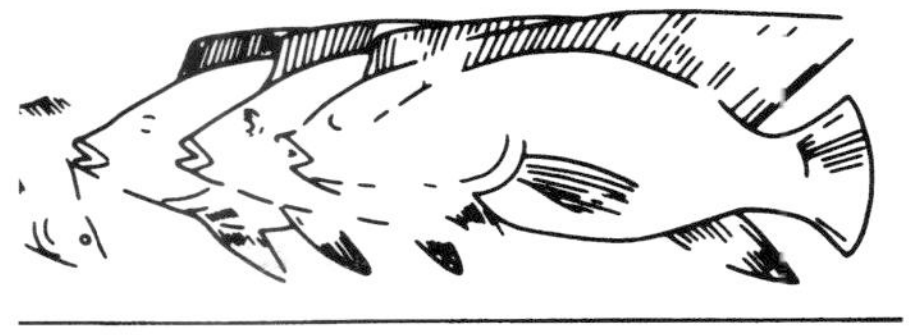

Fig. 5-10 (GB70)

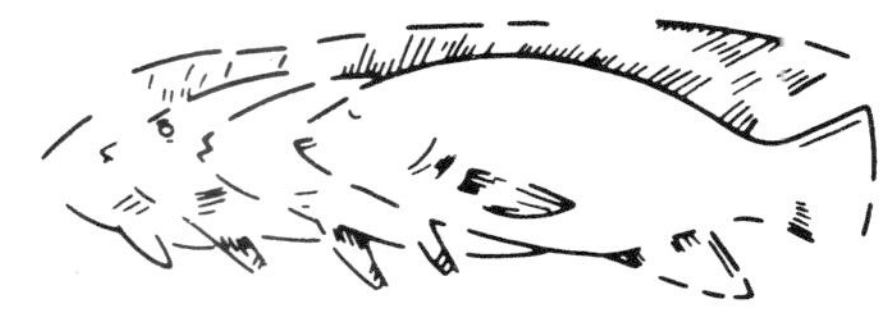

Fig. 5-11 (GB71)

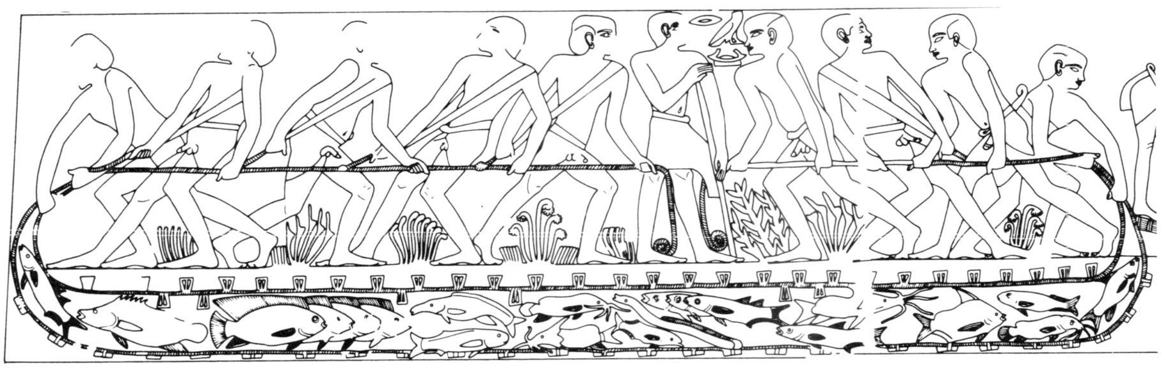

Fig. 5-12 (GB72)

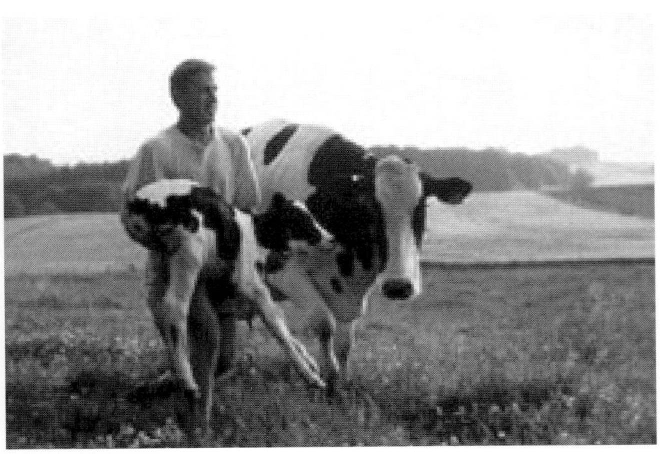

Pl. 5E Cow following her calf
© European Communities, 1995-2008
Reproduced with permission

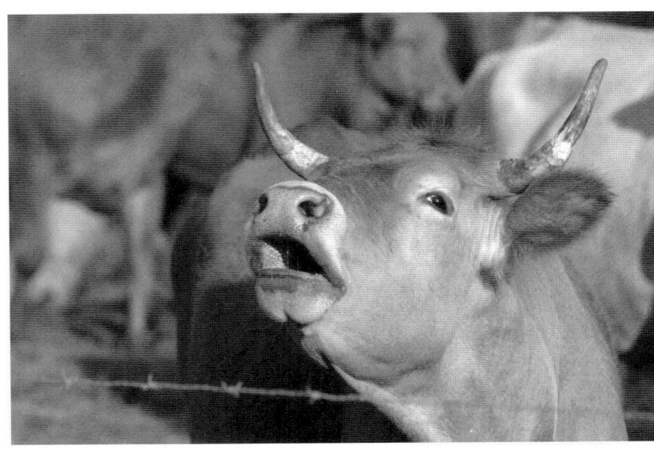

Pl. 5F Cow distress calling
© iStockphoto.com/Steve Simzer

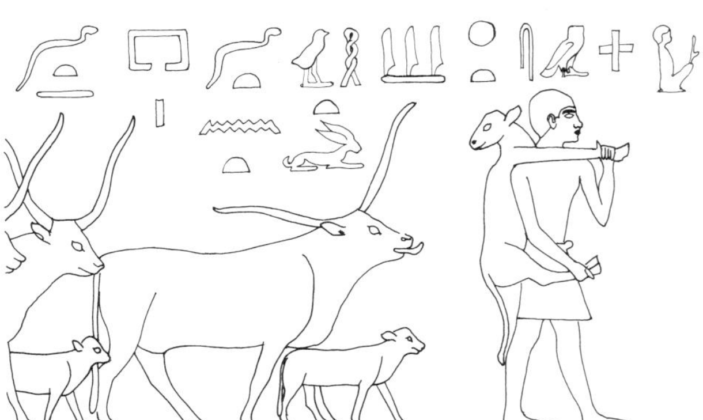

Fig. 5-13 (GB74)

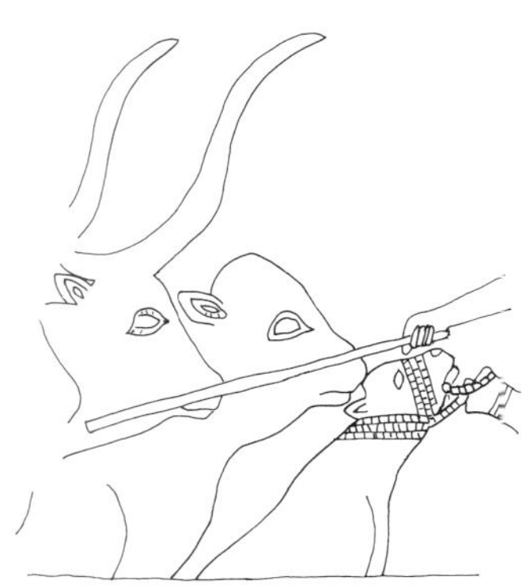

Fig. 5-14 (GB78)

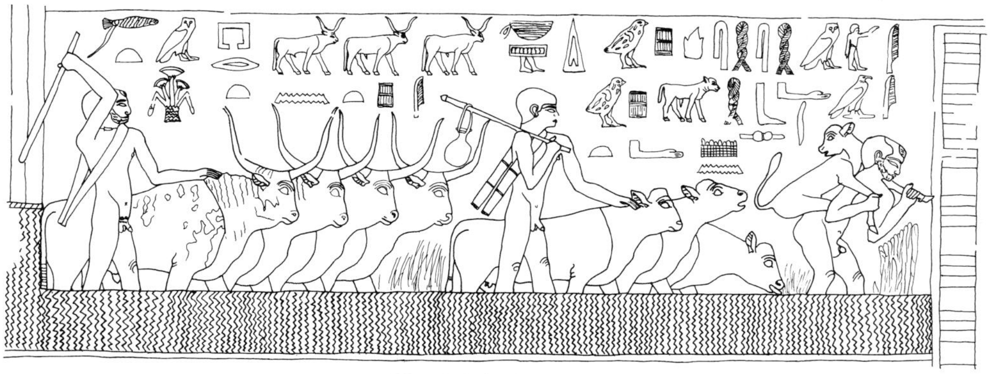

Fig. 5-15 (GB83)

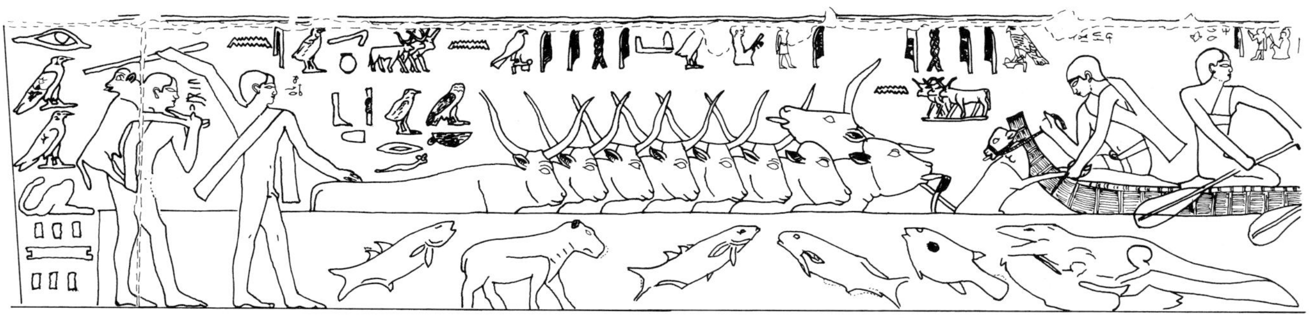

Fig. 5-16 (GB93)

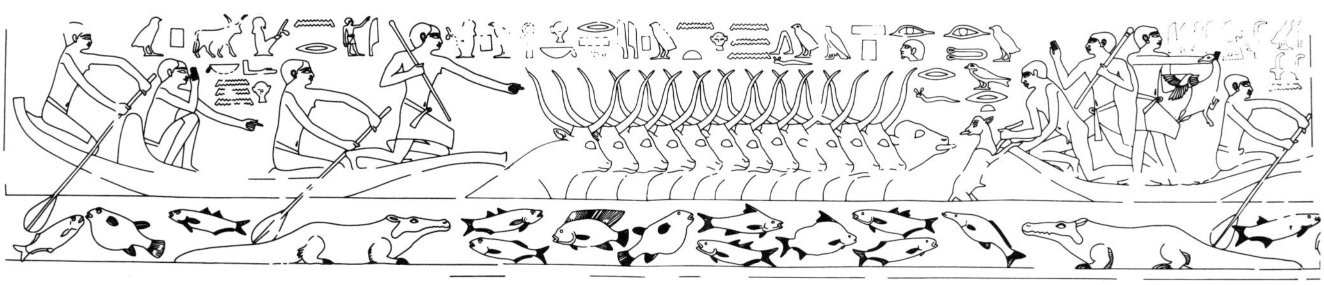

Fig. 5-17 (GB97)

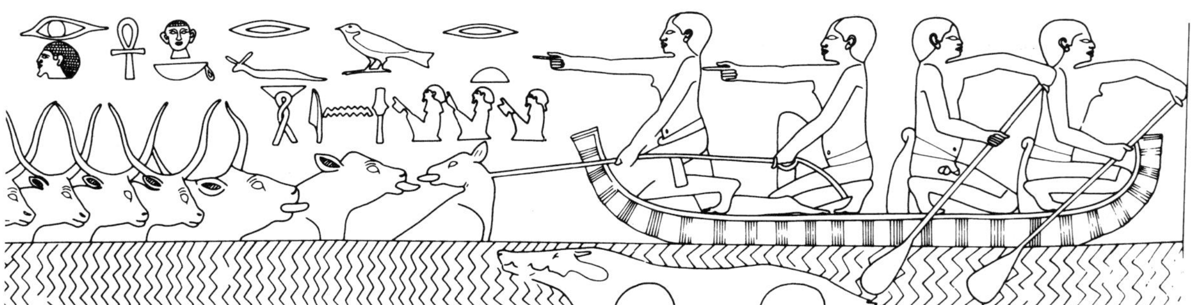

Fig. 5-18 (GB101)

Cattle

s3bw (GB73: Saqqara, V?)
ṯntj (GB74: Giza, V.1-5)
nfr-msdr-ḫwfw (GB75: Giza, V.2-8, possibly V.9)
ḥtp-ḥr-3ḫtj (GB76: Saqqara, V.6-8E)
3ḫtj-ḥtp (GB77: Saqqara, V.6-8E)
k3.j-m-nfrt (GB78: Giza, V.6-8E?)
z3-jb (GB79: Giza, V.8)
k3.j-m-rḥw (GB80: Saqqara, V.8)
rʿ-špss (GB81: Saqqara, V.8M)
ṯjj (GB82-83: Saqqara, V.8-9)
3ḫtj-ḥtp (GB84-85: Saqqara, V.8L-9E)
snḏm-jb: mḥj (GB86: Giza, V.9)
nj-m3ʿt-rʿ (GB87: Giza, V.9)
ḫnwt (GB88-89: Saqqara, V.9)
ptḥ-ḥtp II: ṯfj (GB90: Saqqara, V.9M-L)
nj-m3ʿt-rʿ [*jzzj-mr-nṯr*] (GB91: Giza, V.9-VI.1)
k3-gm-nj: mmj (GB92-94: Saqqara, VI.1M)
mrrw-k3.j: mrj (GB95: Saqqara, VI.1M-L)
nj-ʿnḫ-nswt (GB96: Saqqara, VI.1-2E)
zšzšt: jdwt (GB97: Saqqara, V.9 r/u - VI.1-2E?)
j3zn (GB98: Giza, VI.1M-2M)
ḥzj (GB99-100: Saqqara, VI.1L-2E)
ʿnḫ-m-ʿ-ḥr: zzj (GB101: Saqqara, VI.1L-2E)
mḥw (GB102-104: Saqqara, VI.2M-3?)

A large number of scenes dating from the early 5th dynasty onward depict a calf that has been separated from its herd and moved by stockmen. In most examples, a group of cattle follows closely behind, often headed by an adult that lifts its head to peer directly at the manhandled calf. This particular interest suggests that, in each case, the lead animal represents the calf's mother.

In one scene type, separated calves are carried on the backs of trudging herdsmen who frequently grasp the animals' forelegs with one hand and their hind legs with the other (GB74, GB75, GB77, GB79, GB83 to GB85, GB87, GB88, GB92, GB96, GB98, GB100, and GB102; see figs. 5-13, 5-16, and also fig. 7-55 in Chapter 7). Offering bearers also hold calves on their backs in three scenes (GB73, GB94, and GB104). The small animals, without exception, turn their heads back to look behind them. In images where a herd follows, the calves thus seem to gaze anxiously at their mothers; in a few scenes, however, the object of their attention is not clear (e.g., GB93; see fig. 5-16, far left). Calves are transported in this manner both on land and through shallow water (cf. fig. 5-13 and fig. 5-15).

Separated calves also play a pivotal role in deep-water fording scenes from the mid-5th dynasty onward. In each example, the young animals are suspended over the side of a canoe containing herdsmen. In earlier scenes, the calves are held in position by a rope halter and/or collar attached to a leash (GB76, GB78, GB80, and GB82; see fig. 5-14). In the late 5th dynasty tomb of *rʿ-špss* (GB81), however, a herdsman holds the calf overboard by its foreleg; thereafter, such physical handling becomes more common.

When held by ropes alone, only the heads and shoulders of the calves are usually apparent above the water surface, but when herdsmen seize them, more of their torso is gradually made visible. At first, the handled animals can only be seen from the belly up (GB81, GB86, and GB90), but in later scenes they are elevated so that their entire body is displayed - including their hind legs, even when these are still immersed (GB99 and GB103). Calves that are held by hand turn their heads back to gaze behind them (e.g., GB93 and GB97; figs. 5-16 and 5-17), while animals that are dragged through the water via ropes usually look forward (although see GB101; fig. 5-18).

When analysing the curious text associated with a shallow-water fording scene in the tomb of *ṯjj* (GB83; see fig. 5-15), René van Walsem proposed that the calf has been picked up and "swung backwards around" the herdsman's shoulders because its slow movement through the water was impeding the herd's progress.[43] Yet, depictions also show calves being transported in this manner on land (GB73, GB74, GB77, GB84, GB87, GB88, GB94, GB98, GB100, GB102, and GB104; see fig. 5-13).[44] An alternative explanation is that the young animals have been seized, not because they have caused the herd to stop moving, but to *make* the adults move. This motivation can also account for the treatment of calves in deep-water fording scenes. Clues are found in the natural herding behaviour of cattle.

---

43 van Walsem, R. (1998). "The caption to a cattle-fording scene in a tomb at Saqqara and its implications for the Seh/Sinnbild discussion on Egyptian iconography" in Clarysse et al., *Egyptian Religion ii*, 1469-1485.

44 A dorcas gazelle is also transported in this way in an offering scene in the late 5th dynasty Saqqara tomb of *jj-nfrt: š3.n.f.* See Kanawati-Abder-Raziq, *Iynefert*, pl. 44 [a].

Cattle are very sociable animals. Herd members stay close to one another during both day and night and recognise each other based upon visual, vocal, and olfactory cues. They produce a range of calls that are associated with different behaviours and situations. Physical contact via grooming (e.g., see sections 6.1.1 and 6.1.3 in Chapter 6) occurs frequently, and serves to strengthen social bonds.[45] Herd members monitor one another constantly and often pattern their own behaviour upon that of others. They find social isolation extremely stressful and vocalize a great deal if they become separated from their herd.[46]

The bond between cows and their newborn calves is particularly strong (see Chapter 11); studies have indicated that separation for even short periods of time is distressing for both.[47] If her calf is physically removed in her presence, a cow will make every effort to be reunited with it and will follow it closely (see pl. 5E). In these circumstances, the mother utters a specific vocalization - a high frequency distress call that is designed to travel over distance.[48] Calls emitted by the cow usually have little effect on herd members, but *this* vocalization causes other cattle (especially cows with young) to approach her.[49]

The stockmen in tomb scenes appear to have taken advantage of the maternal instinct of cows, and the cohesive social behaviour of cattle, to drive their herds. By removing a calf (especially a newborn) from its mother, the men will have induced her to begin moving and calling anxiously. This behaviour will have alerted the rest of the herd, causing them to approach and then follow the calling cow. By this method, the herdsmen transporting the calf could keep the herd together *and* control the direction in which it advanced, enabling them to move the animals long distances and to lead them through shallow- or deep-water streams with relative ease.

The vocal behaviour of the cows is a crucial part of this process, for it is their distress calls that draw the rest of the herd towards them. This behaviour is also represented in separation scenes. Cattle vocalizations range from quiet contact sounds through to louder calls associated with hunger, fear, and stress. Softer calls are uttered with the muzzle closed, but when giving louder vocalizations their mouth is opened with the lips covering the teeth. The tongue is usually either barely visible or not seen at all (see pl. 5F and Colour Plate B). In the tomb of *ṯjj* (GB83; see fig. 5-15), both the mother and carried calf open their mouths and thus appear to be vocalizing, however the animals in other scenes behave differently. In an early image located in the Giza tomb of *ṯntj* (GB74; see fig. 5-13), the cow lifts her head and protrudes her tongue prominently as she gazes at the hoisted calf ahead of her. A cow again extends her tongue in the late 5th dynasty tomb of *ꜣḫtj-ḥtp* (GB85); thereafter, this feature occurs reliably in both shallow- and deep-water fording scenes (GB90, GB93, GB95, GB96, GB97, GB101, GB102, and GB103; see figs. 5-16 to 5-18). In each case, the tongue fills the entire mouth cavity and extends well beyond the animals' snouts. They are thus not simply showing their tongues as they open their mouths, but are depicted as poking them out from between their lips. I suggest that, although the tongues of vocalizing cattle are not normally visible, artists have added this exaggerated attribute to their depictions in order to convey unambiguously to the viewer the acute distress calling of the segregated animals - thereby representing one of the most important and striking aspects of the separation process.

---

45 For detailed overviews of the social behaviour of cattle, see Phillips, *Cattle Behaviour*, 33-74; Bouissou, M.-F., Boissy, A., Le Neindre, P. and Veissier, I. (2001). "The social behaviour of cattle" in Keeling-Gonyou, *Social Behaviour*, 113-145.

46 Boissy, A. and Le Neindre, P. (1997). "Behavioral, cardiac and cortisol responses to brief peer separation and reunion in cattle", *Phys. Beh.* 61(5), 696; Watts, J.M. and Stookey, J.M. (2000). "Vocal behaviour in cattle: The animal's commentary on its biological processes and welfare", *App. Anim. Beh. Sci.* 67(1-2), 23.

47 Weary, D.M. and Chua, B. (2000). "Effects of early separation on the dairy cow and calf: 1. Separation at 6 h, 1 day and 4 days after birth", *App. Anim. Beh. Sci.* 69(3), 177-188; Flower, F.C. and Weary, D.M. (2001). "Effects of early separation on the dairy cow and calf: 2. Separation at 1 day and 2 weeks after birth", *App. Anim. Beh. Sci.* 70(4), 275-284.

48 The 'men*e*nh' call. See Kiley, M. (1972). "The vocalizations of ungulates, their causation and function", *Z. Tierpsychol.* 31, 193; Fraser-Broom, *Farm Animal Behaviour*, 219. For a visual representation ('spectrogram') of this call, see Weary, D.W. and Fraser, D. (2002). "Social and reproductive behaviour" in Jensen, *Ethology*, fig. 5.1.

49 *Pers. comm.* Prof. Joseph M. Stookey, Dept. of Large Animal Clinical Sciences, Univ. of Saskatchewan, Canada.

Separated calves are also shown with protruded tongues, but this feature is introduced somewhat later than for cows. The earliest example occurs in the late 5th dynasty deep-water fording scene of *ptḥ-ḥtp II: ṯfj* (GB90). Thereafter, this feature is included reliably in reliefs in which herdsmen dangle calves from canoes (GB93, GB95, GB97, and GB101; see figs. 5-16 to 5-18). Animals that are carried on the backs of herdsmen are not depicted with extended tongues until the early 6th dynasty; however, once this detail is introduced, the motif is repeated frequently (GB92, GB94, GB96, GB102, and GB104; see fig. 5-16, far left). It is important to note that in many reliefs the calves' tongues are greatly enlarged and out of proportion to the size of their heads.

The calves' protruded tongues indicate that they, too, are calling. Studies have revealed that calves will vocalize in response to both social isolation and physical discomfort. Newborns are usually silent as they are being separated from their mothers,[50] but will begin distress calling once they have been fully isolated for around 18 hours.[51] Calves are segregated in this manner during weaning; their increased distress calling is thus prompted as much by the pain of hunger (due to their reduced access to milk) as separation anxiety.[52] Cows respond to these distress vocalizations by orienting towards the sound and then approaching.[53] Indeed, stockman today will sometimes imitate the calls of a newborn calf in order to lure a recently-calved cow out from a herd.[54] This orienting behaviour may be apparent in a ploughing scene in the tomb of *ḫnwt* (GB89), in which yoked cows appear to follow a vocalizing calf.

Calves with protruded tongues are first depicted in scenes in which they are held over the sides of boats. This context suggests that the Egyptians were aware that a calf would be likely to vocalize loudly in response to the discomfort of being grasped roughly and then immersed in cold water.[55] The calf's calls will have ensured that its mother remained focussed upon it, that she would continue to approach and call to it, and that the herd would continue to follow her. Immersion of the calves was thus a deliberate ploy designed to make them vocalize so as to provoke the herd into crossing deep-water streams.

It is interesting to note that this vocal symbol is not found in early examples of calves slung upon the backs of herdsmen - matching the typically silent response of newborns when first removed from their mother. Calves in later examples of this scene, however, *do* extend their tongues. This would suggest that the motif was copied from deep-water fording scenes, but it is also possible that artists began to include this feature in order to emphasize the animals' anxiety and general discomfort. Isolated calves in other contexts, such as offering, boating, and pastoral scenes, are depicted frequently with protruded tongues, illustrating their distressed vocal response to both the social separation and physical treatment inflicted upon them by their human handlers.

The calling behaviour of the cows and their young has been noted by other scholars. Macramallah, for example, says of the deep-water fording scene in the tomb of *zšzšt: jdwt* (GB97; fig. 5-17) that the calf "se retourne vers sa mère en meuglant", to which she responds similarly.[56] Germond also states that the calf in the tomb of *kꜣ-gm-nj: mmj* (GB93; fig. 5-16), "calls for its mother".[57] And when interpreting the caption to the shallow-water scene in the tomb of *ṯjj* (GB83; fig. 5-15), both Erman[58] and Altenmüller[59] suggest that the enigmatic term *jꜣ* might refer to the vocal behaviour of the calf. To my knowledge, however, the protruded tongue has not been recognized as a device by which Egyptian artists represented animal sounds.

---

50 Thomas, T.J., Weary, D.M. and Appleby, M.C. (2001). "Newborn and 5-week-old calves vocalize in response to milk deprivation", *App. Anim. Beh. Sci.* 74(3), 171. Separated newborns are probably less inclined to call than their mothers to avoid alerting nearby predators. See Barfield, C.H., Tang-Martinez, Z. and Trainer, J.M. (1994). "Domestic calves (*Bos taurus*) recognize their own mother by auditory cues", *Ethology* 97, 263.

51 Flower-Weary, *App. Anim. Beh. Sci.* 70(4), 280.

52 Thomas et al., *App. Anim. Beh. Sci.* 74(3), 171.

53 Marchant-Forde, J.N., Marchant-Forde, R.M. and Weary, D. (2002). "Responses of dairy cows and calves to each other's vocalisations after early separation", *App. Anim. Beh. Sci.* 78(1), 26.

54 Ewbank, R. (1968). "The behavior of animals in restraint" in Fox, *Abnormal Behavior*, 173.

55 *Pers. comm.* Prof. Joseph M. Stookey, Dept. of Large Animal Clinical Sciences, Univ. of Saskatchewan, Canada.

56 Macramallah, *Idout*, 16.

57 Germond, *Egyptian Bestiary*, 13, fig. 8.

58 Erman, *Reden*, 28-29.

59 Altenmüller, H. (1989). "Kälberhirte und Schafhirte: Bemerkungen zur Rückkehr des Grabherrn", *SAK* 16, 6.

Finally, I propose that the separated calves are deliberately held in such a way that their mothers could see them. Indeed, the visual contact between the cows and their young seems to be emphasized by the calves' tendency to look behind them. Note that carried calves are not held in the arms of herdsmen, but on their backs, enabling the animals' mothers to see them clearly as they are borne away. Similarly, ferried calves do not travel in the canoes, but are dropped directly in front of the advancing herds; the gradual elevation of their bodies in the water (see above) also suggests that they are situated increasingly in a more discernible position.

It would be difficult for a man to carry a domesticated calf in the manner shown, as these weigh approximately 40 kilograms at birth; in farm practice, they are usually roped and led, but very young animals are sometimes transported a short distance while cradled in both arms (see pl. 5E). The young of primitive cattle, however, which may approximate ancient Egyptian breeds, are roughly half this weight.[60] It would thus be feasible for an animal of this size to be carried on the back of a herdsman or hung from a boat, locations that would keep it fully in sight of the lead cow.

The visual importance of the calf is confirmed by the caption to a deep-water fording scene in the tomb of *k3-gm-nj: mmj* (GB93; see fig. 5-16). In this image, the calf is identified with the son of the cow goddess, Hathor, as it is labelled, *jḥy n k3w,* 'Ihy of the cattle'.[61] A herdsman walking at the rear of the herd shouts ahead to the men who hold the calf over the side of their canoe, imploring them to, "Give (or show) Ihy of the cattle to them. Behold! They turn against him"! Indeed, one of the animals at the head of the line has turned away from the canoe and is heading back to shore; it appears to be panicking, as its head is lifted and its protruded tongue suggests that it is vocalizing. The herdsman's request thus shows clearly that the Egyptians viewed the presence of the calf (both visual and aural) as a key element in keeping their travelling herds together and moving forward.[62]

SUMMARY

- Wall reliefs from Giza and Saqqara indicate that the Egyptians understood how to manipulate the social behaviour of animals for their own purposes. Treading scenes reveal their awareness of the effect of fear on the grouping response of sheep and the application of this knowledge in their agricultural practices. Fording scenes show how stockmen exploited the gregarious nature of cattle, and the maternal and vocal behaviour of cows and calves, to control and move their herds efficiently. Clapnet scenes point to an understanding of interspecies relationships and the flocking behaviour of waterfowl.
- Lateral layering is the most common method by which herds, flocks, and schools are represented in wall scenes, however some artists adapted this highly stylised motif to better convey the characteristic disorder of animated animal groups by adjusting the body, head, and leg positions of individual members. The unusual postures of sheep in treading scenes may be intended to represent their leaping in response to fear and close physical proximity.
- As Schäfer proposed, the application of lateral layering seems also to have been used to imply forward motion, as movement is often a primary feature of the depicted animal groups: sheep are chased across sown fields, flocks of waterfowl swim together calmly, fish align with one another as they glide underwater, and herds of cattle are driven over land and through streams.
- The depiction of cattle with unnaturally protruded tongues in fording scenes reveals a graphic device by which artists were able to visualize the vocal behaviour of animals.

[60] Roy, *The Calf i*, 17-27.

[61] Ihy, who was associated with sistrums and music, is usually represented as a young boy with the side-lock of youth; depictions of him as a calf are less frequent. It is beyond the scope of the current study to explore the possible mythological significance of his presence in the *k3-gm-nj: mmj* scene. See *Wb iii*, 125-126; Rundle-Clark, *Myth and Symbol*, 87-89; Wilkinson, *Complete Gods*, 132-133.

[62] Firth and Gunn (*Teti Pyr. ii*, 114) translate this inscription as "Show Ihy to the cattle that I may keep them together". This alternate interpretation nevertheless implies that the calf was viewed as essential for maintaining the integrity of the herd.

CHAPTER 6

COMFORT BEHAVIOUR

DEFINITION

Comfort behaviour includes any actions that help an animal stay in good condition and free of irritation, such as grooming (by mammals), preening (by birds), scratching and stretching. Most animals devote a large part of their daily activity to such behaviour.[1] Elimination (defecation and urination) is also included in this category.

6.1 GROOMING

Grooming consists of movements directed toward the body surface that are intended to remove ectoparasites (lice, ticks, etc.), relieve skin irritation, and clean and condition the coat. An animal can groom itself (autogrooming) and may also groom, or be groomed by, others (allogrooming).[2] In mammals that lack opposable thumbs (i.e., non-primates), these activities are carried out with the tongue (licking), the teeth (nibbling, scraping), or feet, hooves or horns (scratching), and may also involve contact with external objects (rubbing).

6.1.1 Licking

Licking is defined as the movement of the tip or surface of the tongue over an object. During grooming, licking is performed to remove matter (dirt, ectoparasites, etc.) from among the hairs of the coat and to spread saliva over the body surface.

Twelve examples of licking by animals were detected in tomb scenes, all located at Saqqara. Most of these depict grooming by cattle. Only one other species, a dorcas gazelle, is represented autogrooming.

Cattle

rˁ-špss (CB1: Saqqara, V.8M)
snḏm-jb: jntj (CB2: Giza, V.8M-L)
ṯjj (CB2a: Saqqara, V.8-9)
nbt (CB3: Saqqara,V.9)
k3-gm-nj: mmj (CB4-5: Saqqara, VI.1M)
wḏ3-ḥ3-ttj: ššj (CB6-7: Saqqara, VI.1L-2E)
mṯtj (CB8: Saqqara, VI.2?)
ḫntj-k3.j: jḥḥj (CB9-11: Saqqara, VI.2M)

Licking is the primary method by which cattle groom.[3] All accessible parts of the body are licked during autogrooming, but when grooming others, it is largely restricted to the head and neck region.[4] Family members often perform allogrooming, as this activity helps to maintain and strengthen the bonds between them. Cows lick their calves frequently.

The tomb of *nbt* clearly displays the image of an ox that has twisted its head back to lick its shoulder with its protruded tongue (CB3; see fig. 6-1). Comparison with live cattle (pl. 6A) reveals that this representation is extremely accurate. The posture of an ox in the earlier tomb of *rˁ-špss* (CB1; see fig. 6-2) is very similar to this scene such that, although the animal does not reveal its tongue, it

1 Fraser-Broom, *Farm Animal Behaviour*, 99.
2 Allaby, *Dictionary of Zoology*, 205.
3 Hart, B.L., Hart, L.A., Mooring, M.S. and Olubayo, R. (1992). "Biological basis of grooming behaviour in antelope: The body-size, vigilance and habitat principles", *Anim. Behav.* 44(4), 615.
4 Phillips, *Cattle Behaviour*, 41.

seems likely that licking was also intended. Offering scenes from the tombs of *sndm-jb: jntj* (CB2), *tjj* (CB2a), *k3-gm-nj: mmj* (CB4 and CB5), *wd3-ḥ3-ttj: ššj [nfr-sšm-ptḥ]* (CB6 and CB7), and *ḫntj-k3.j: jḫḫj* (CB9 to CB11) also appear to show cattle autogrooming as in each case a leashed bull turns its head back sharply along its flank while simultaneously protruding its tongue (e.g. CB11; see fig. 6-3). While these examples undoubtedly give the appearance of licking, it is equally possible, however, that they are instead intended to depict cattle calling while turning their heads back or to the side. This ambiguity renders their identification as grooming less convincing.

Only one example of allogrooming was found. This occurs in the early 6th dynasty tomb of *mttj* (CB8), and depicts a cow that licks the head of her gambolling calf while she is being milked. This intimate interaction reflects the importance of the calf in the milking process, as cows will not usually 'let-down' their milk unless they sense that their young are present (see Chapter 11).

Dorcas gazelle

tjj (CB12: Saqqara, V.8-9)

In the tomb of *tjj*, a tethered dorcas gazelle grooms itself by elevating a hind leg, passing it under its body, and then turning back its head to lick its lifted limb (CB12; see fig. 6-4). Although animals often adopt awkward postures when attempting to groom inaccessible parts of their body, that shown by the gazelle is physically impossible. The contortion here has been caused by the image conforming to the artistic rule of the limb furthermost from the viewer advancing first. The behaviour exhibited is also incorrect for, unlike cattle, gazelle do not groom themselves by licking, but instead rake their lower incisors through their coat in long, upward strokes (see pl. 6B).[5] It is possible that the artist was unaware of how dorcas gazelle actually groom and so reproduced the method employed by cattle and other domestic animals. I suggest, however, that the protruded tongue may have been added to the scene to remove any ambiguity about what the animal is doing. The technique is wrong, but the message is nevertheless correct: the gazelle is grooming itself.

6.1.2 Scratching

Scratching consists of scraping a hoof, paw, or horn a number of times through the hairs of the coat or across the body surface. This action can dislodge ectoparasites and relieve skin irritation.

Cattle

ntr-wsr (CB13: Saqqara, V.6L-7)
rˁ-špss (CB14: Saqqara, V.8M)
jj-nfrt: š3.n.f (CB15: Saqqara, V.9)

Three depictions of scratching were found, all located at Saqqara. In examples from the tombs of *ntr-wsr* (CB13; see fig. 6-5) and *jj-nfrt: š3.n.f* (CB15; see fig. 6-7), a leashed ox raises a hind leg and turns its head back to pass its hoof across the bridge of its muzzle. In the tomb of *rˁ-špss* (CB14; see fig. 6-6), a leashed bull again lifts a hind leg and also turns back its head, but scratches its flank. In each scene, a herdsman struggles to control the animals by pulling on their leash. In the case of *ntr-wsr*, the ox also protrudes its tongue. It is unlikely that the animal is both licking and scratching at the same time; an alternative explanation is that the protruded tongue here indicates that it is vocalizing its displeasure at the handler's attempted interruption to its grooming (see section 9.4 in Chapter 9).

Cattle scratch either by lifting a hind leg and then scraping their hoof across their flank or other body part (see pl. 6C), or by twisting their head back and applying the tip of a horn to the irritated region. All three images appear to conform to this behaviour, however the posture of the bull in the tomb of *rˁ-špss* may depict another action: kicking. When under attack from painful biting insects, such as stable flies, cattle often respond by lifting and holding a leg next to their body, or by kicking

[5] Hart et al., *Anim. Behav.* 44(4), 615.

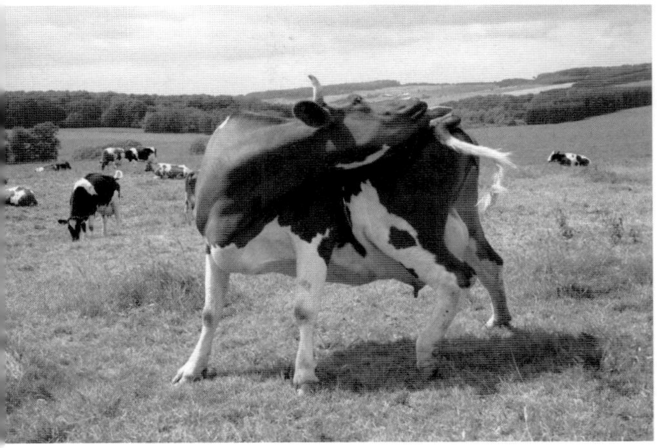

Pl. 6A Cow licking
© iStockphoto.com/Alexander Briel Perez

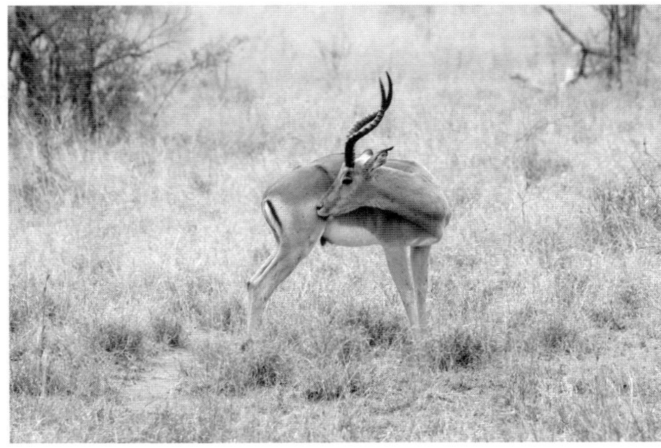

Pl. 6B Gazelle grooming
© iStockphoto.com/Darrell Fraser

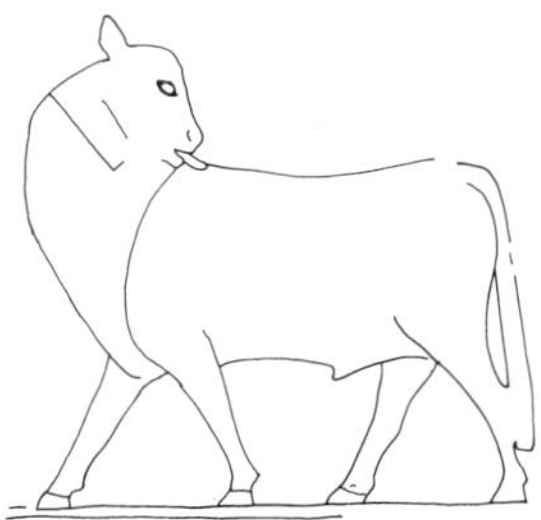

Fig. 6-1 (CB3)

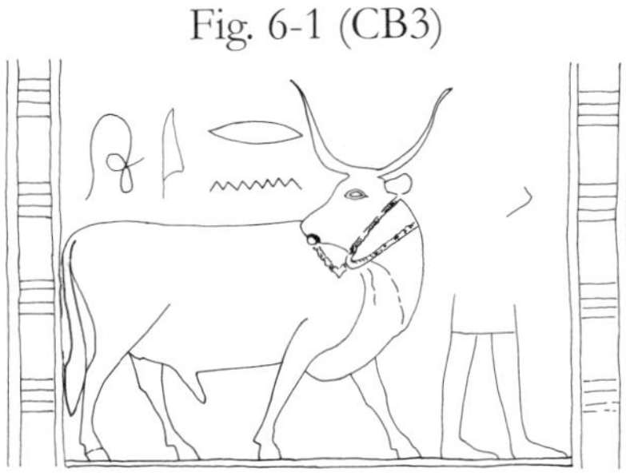

Fig. 6-2 (CB1)

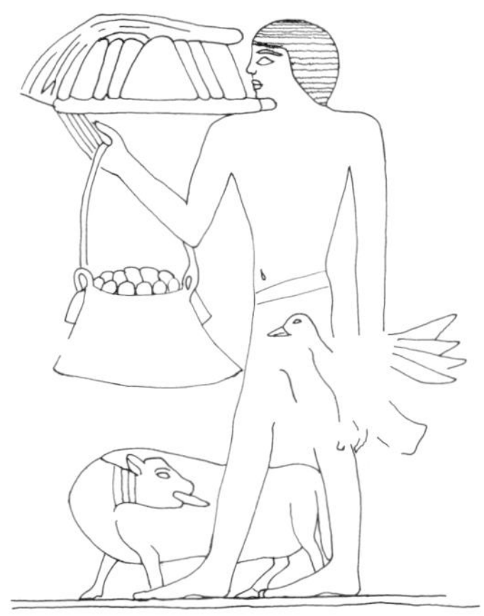

Fig. 6-3 (CB11)

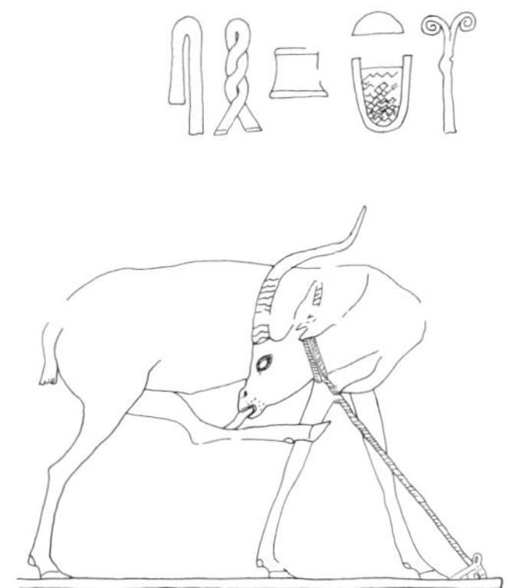

Fig. 6-4 (CB12)

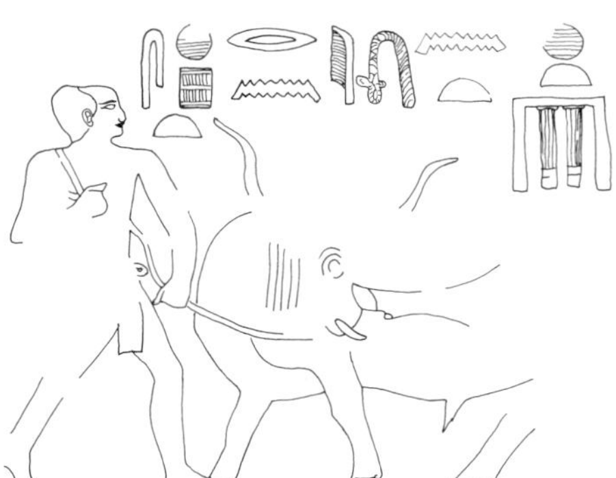

Fig. 6-5 (CB13)

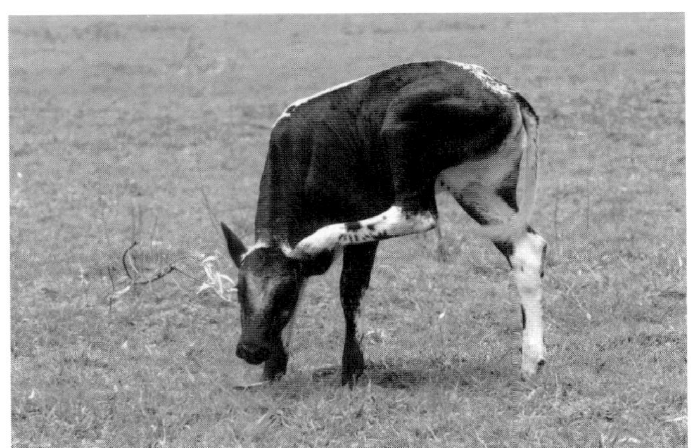

Pl. 6C Cow scratching
© iStockphoto.com/Mile Stone

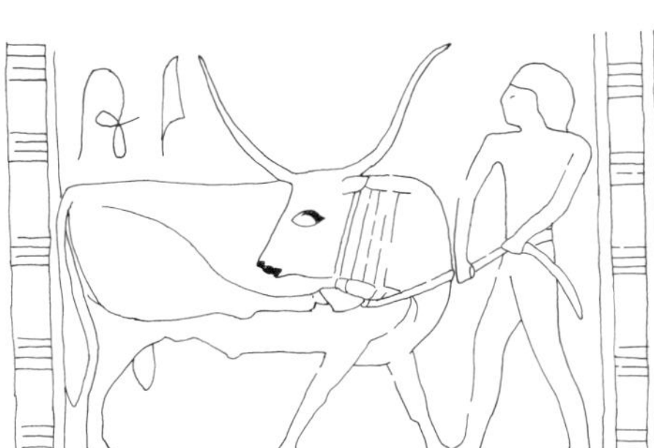

Fig. 6-6 (CB14)

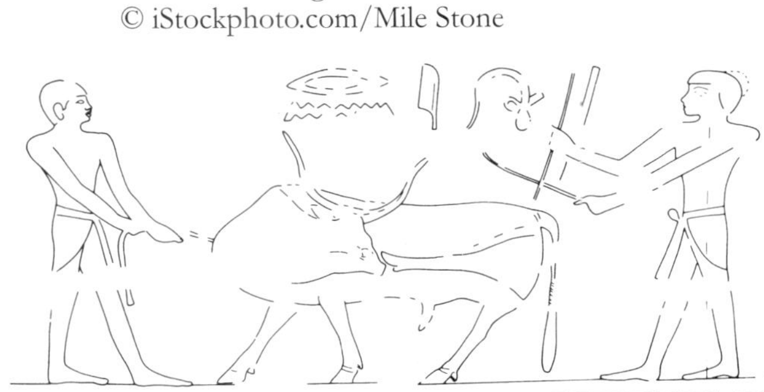

Fig. 6-7 (CB15)

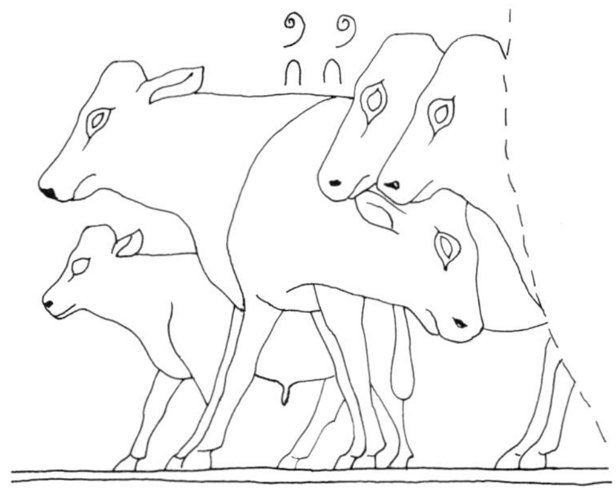

Fig. 6-8 (CB16)

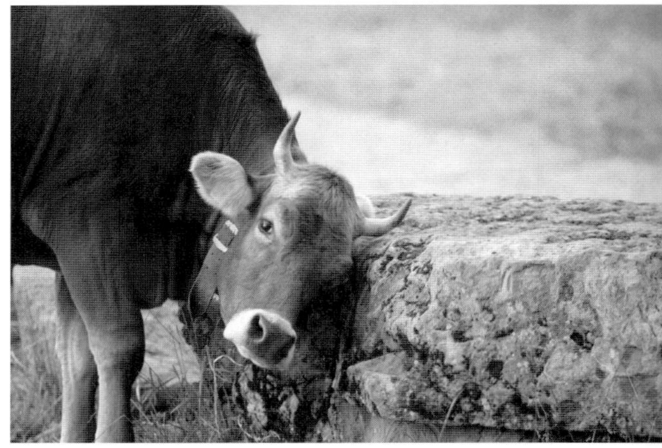

Pl. 6D Cow rubbing
© iStockphoto.com/Bernd Klumpp

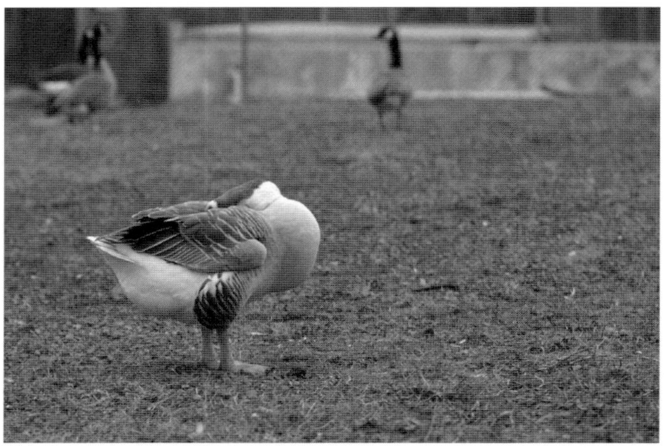

Pl. 6E 'Headless' goose
© iStockphoto.com/Jason Barnes

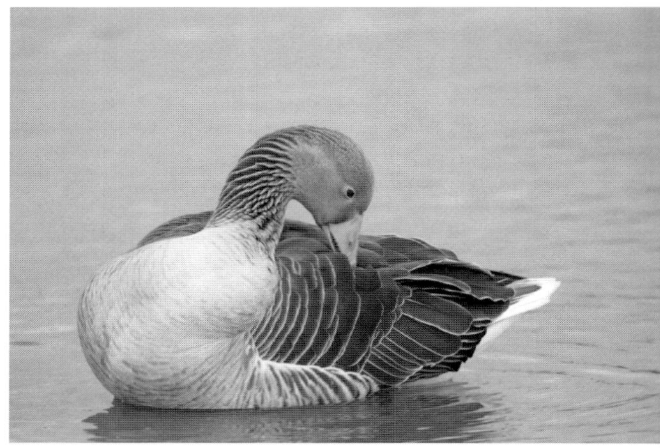

Pl. 6F Duck preening back
© iStockphoto.com/James O'Neill

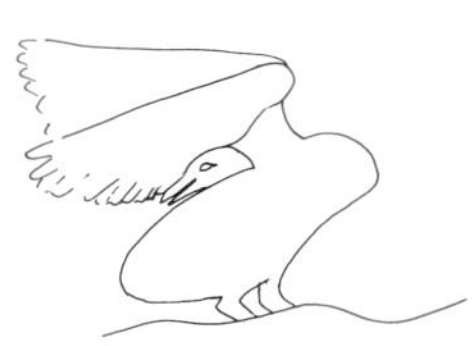

Fig. 6-9 (CB17)

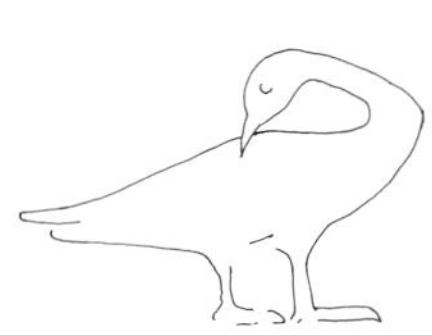

Fig. 6-10 (CB19)

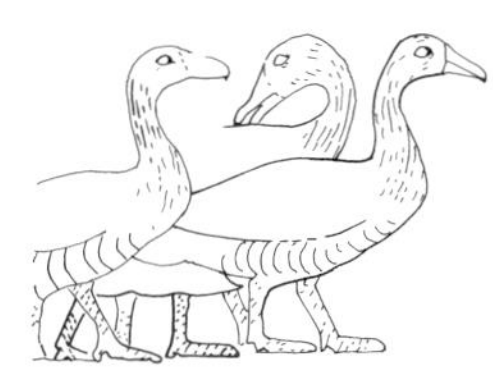

Fig. 6-11 (CB21)

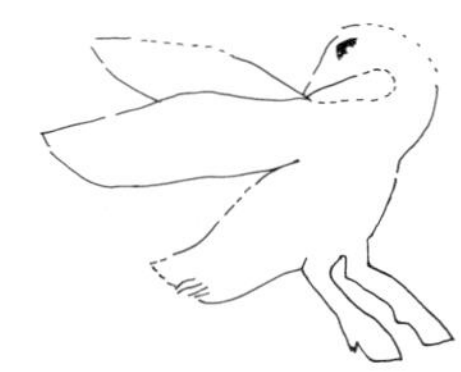

Fig. 6-12 (CB22)

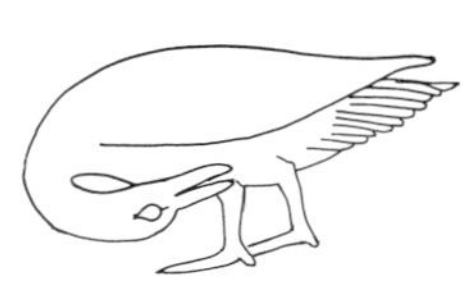

Fig. 6-13 (CB23)

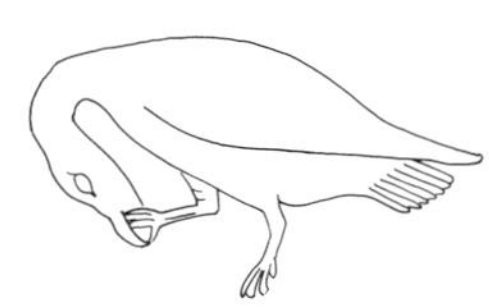

Fig. 6-14 (CB23)

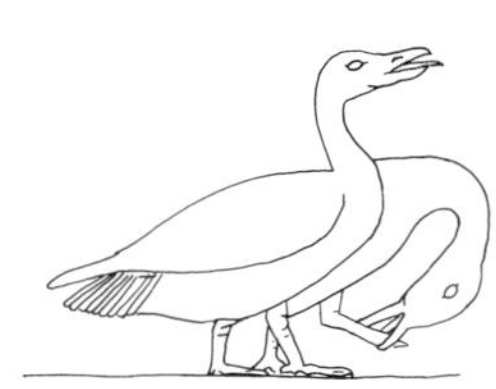

Fig. 6-15 (CB25)

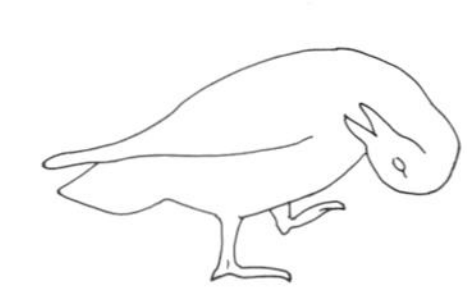

Fig. 6-16 (CB24)

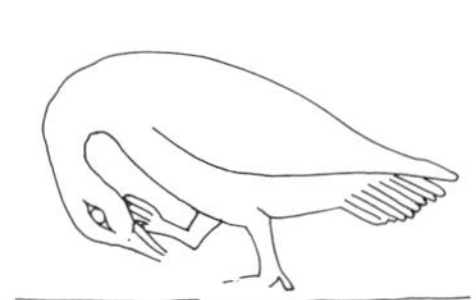

Fig. 6-17 (CB24)

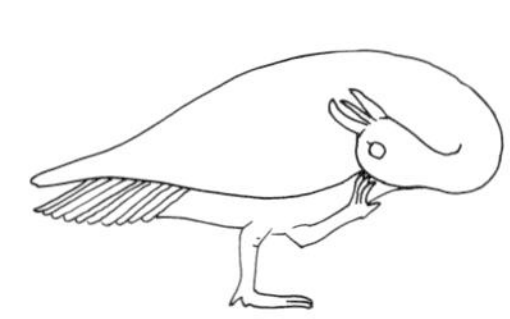

Fig. 6-18 (CB24)

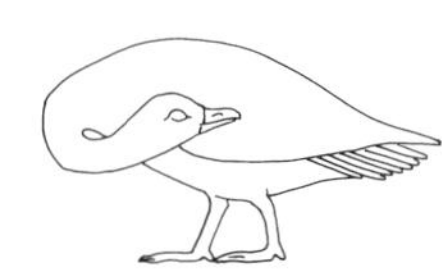

Fig. 6-19 (CB24)

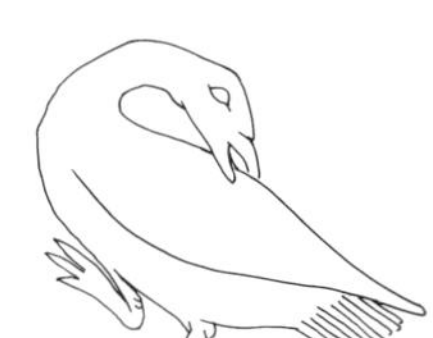

Fig. 6-20 (CB24)

at their flanks (cf. *Cows* in section 11.2.1 in Chapter 11).[6] They may also toss their head violently when bitten; this might explain why the animal in the *rʿ-špss* scene also looks back.

6.1.3 Rubbing

Rubbing involves pushing part of the body (often the head) repeatedly back and forth across an external object (such as a tree stump or wall; see pl. 6D), or against the body of another animal. Rubbing can relieve irritation, but also aids in the removal of dead skin, ectoparasites, and, in some species, remnants of their seasonal coat. Rubbing is a leisurely behaviour that is performed by animals during periods of inactivity.

Cattle

rʿ-ḫʿ.f-ʿnḫ (CB16: Giza, V.6?)

One example of apparent rubbing was found in an offering scene from the tomb of *rʿ-ḫʿ.f-ʿnḫ* at Giza (CB16; see fig. 6-8). In this depiction, an ox turns back and appears to hook its head under the lower jaws of two of its herd-mates in a movement that suggests that it is pushing its brow against them. It should be noted that a number of other images of herding cattle were found to include animals that hold their heads close to one another in a manner reminiscent of rubbing, however, as these scenes depict animals in the process of moving (e.g., fording), it seems unlikely that rubbing (which is performed while largely stationary) was intended.

6.2 PREENING

Preening is a grooming behaviour performed by birds to both clean and oil their feathers.[7] These require constant maintenance to keep them in good condition; consequently, all bird species devote a great deal of time each day to preening.[8] Each feather consists of a central shaft ('rachis'), from which branch delicate, lateral 'barbs'. Tiny projections on these structures ('barbules'), enable each barb to intersect with the next, giving the feather a smooth, flexible, cohesive structure.[9] During preening, the bill or beak is passed through the plumage, drawing individual feathers between the mandibles to clean and reset the barbs. The feathers are also coated with a protective, waxy substance which is secreted from a preen gland at the base of the tail and applied with the bill or beak (a process known as 'oiling').

The flexibility of their light skeleton enables birds to bend in such a way that they can reach most of their plumage during preening: the breast, back ('saddle'), over and under the wings, and the tail feathers. Dirt and ectoparasites are removed from the head by scratching; most birds do this by reaching a foot up under a wing, but a few species (notably passerines) drop one wing and pass a foot over the top of it.[10] Irritations on the legs and feet may be relieved by nibbling with the beak or bill.[11] Birds will also preen the head and neck region of their mate during courtship and to maintain their pairbond (a behaviour known as 'allopreening').[12] If water is available, many birds will begin a preening session by bathing, wetting all the feathers by splashing vigorously or, in some species, plunging underwater briefly.[13] At the end of a preening bout, birds often fluff all their feathers convulsively and then flap their wings.[14]

---

6 Stable flies are common around barns and animal pens. Cattle find them extremely irritating. See Dougherty, C.T., Knapp, F.W., Burrus, P.B., Willis, D.C. and Cornelius, P.L. (1995). "Behavior of grazing cattle exposed to small populations of stable flies (*Stomoxys calcitrans* L.)", *App. Anim. Beh. Sci.* 42(4), 232.

7 Allaby, *Dictionary of Zoology*, 375.

8 Gill, *Ornithology*, 77.

9 *ibid.*, 65-68.

10 Simmons, K.E.L. (1957). "The taxonomic significance of the head-scratching methods of birds", *Ibis* 99, 178-181; Simmons, K.E.L. (1964). "Feather maintenance" in Thomson, *Dictionary of Birds*, 281-282.

11 McKinney, F. (1965). "The comfort movements of Anatidae", *Behaviour* 25(1), 154.

12 Burton, *Bird Behaviour*, 148.

13 *ibid.*, 30-33.

14 *ibid.*, 33.

6.2.1 Preening movements

During oiling and preening, birds adopt a variety of positions as they attempt to reach different parts of their body. Common movements include: reaching back to preen the saddle and vent (see pls. 6E and 6F), straightening the legs and bending the head forward deeply to preen the belly, and stretching the neck up and tucking in the bill to access the neck and breast feathers (see pl. 6G).[15] Two types of bill movement occur during preening: either nibbling, in which the mandibles open to bite repeatedly at the feathers, or stroking movements, in which the bill remains closed while it is passed rapidly across the feathers to smooth them.[16] Preen oil is often applied to the plumage by stroking.

Preening movements were found in 10 tombs, all located at Saqqara. The earliest image comes from the reign of Niuserre, but the remaining examples date from the late 5th dynasty on. The action is performed exclusively by waterfowl and cranes, and most examples occur in overstocked poultry-yards or scenes depicting force-feeding. The inclusion of preening among the large numbers of birds that congregate in scenes of this kind is correct; preening is a social activity and its expression by one animal often causes others to behave similarly. Pools of water are included in many poultry-yard scenes, but not in those featuring force-feeding.

Birds

nfr/k3-ḥ3.j (CB17: Saqqara, V.6)
ṯjj (CB18-21: Saqqara, V.8-9)
3ḫtj-ḥtp (CB22: Saqqara, V.8L-9E)
k3-gm-nj: mmj (CB23-26: Saqqara, VI.1M)
nj-k3w-jzzj (CB27: Saqqara, VI.1M)
mrrw-k3.j: mrj (CB28-33: Saqqara, VI.1M-L)
ḥzj (CB34: Saqqara, VI.1L-2E)
mrrj (CB35: Saqqara, VI.2E?)
mrjj-ttj: mrj (CB36: Saqqara, VI.2M-L)
mḥw (CB37: Saqqara, VI.2M-3?)

The earliest example of preening comes from the tomb of *nfr/k3-ḥ3.j* (CB17; see fig. 6-9), and appears in a scene in which captured waterfowl are being loaded into crates. The image is highly stylised, consisting of little more than a globular body from which two wings project abruptly. Nevertheless, it is quite accurate, as birds will sometimes lift their wings to gain better access to their saddle and tail feathers and, when twisted fully backward, often appear headless (cf. pl. 6E). The artist has undoubtedly attempted to reproduce this impression as, at first glance, the bird's head is quite difficult to locate.[17]

Preening does not occur again until the late 5th dynasty tomb of *ṯjj*. Among scenes of force-feeding in the portico (CB18 to CB20) and pillared hall (CB21), a small number of geese and pintail ducks appear to twist their necks to probe at their saddle feathers (e.g., in CB 18, CB19, and CB20; see fig. 6-10 and pl. 6F). One bird nibbles at its plumage (CB21; see fig. 6-11), while the remaining birds preen with closed bills. One bird also looks back, but may possibly preen one wing rather than its saddle (CB19). This posture, in which the head is turned fully backward, is unambiguous and easy to differentiate from other activities in the poultry-yard, such as eating and wing-flapping (see below). Back preening subsequently becomes a standard motif by which artists illustrated this behaviour.

Back preening occurs again in a unique example from the tomb of *3ḫtj-ḥtp* (CB22; see fig. 6-12). In a marsh scene, a bird while in mid-flight, turns its head so that its beak touches its saddle. The position of the head is thus identical to the preening geese in the tomb of *ṯjj*. Although birds occasionally perform acrobatic movements in flight, the posture displayed by the bird in this tomb would have compromised its ability to fly. Thus, it seems that the artist has made a clumsy mistake.

---

15 Simmons in Thomson, *Dictionary of Birds*, 280-281.

16 *ibid.*

17 Moussa and Altenmüller (*Nefer and Ka-hay*, 23) believe that the bird is endeavouring to "clean her feathers". It should be noted that the posture of this animal strongly resembles those of birds whose wings have been overlapped after capture (that is, one wing placed over the other), to stop them from flying away. Unlike the bird in the tomb of *nfr/k3-ḥ3.j*, however, the heads of captured birds are usually held upright, not placed under their wings. Captured birds also sit, while the bird in question is standing. For similarities between the posture of captured waterfowl and the *rḫj.t*-bird, see Nibbi, *Lapwings and Libyans*, 11-17.

Although rare, however, some species do preen while in flight.[18] This can range from simple shaking and scratching of the head, to preening of the breast feathers. Preening of the back feathers while flying has not been observed. It is nevertheless intriguing to consider whether the artist who produced the *3ḫtj-ḥtp* bird chose to represent a rare glimpse of in-flight preening with a pose that was visually incorrect, but conceptually accurate.

In the tomb of *k3-gm-nj: mmj* (CB23 to CB26), the range of preening movements broadens. Geese in poultry-yards and force-feeding scenes are represented with an attention to detail that suggests that they were observed closely in life. Back preening continues to be the most common action to be depicted (with both closed and open bills), but to this is now added: preening of the wings and flanks (CB23, CB24, and CB26; see fig. 6-13), nibbling and probing of the feet (CB23, CB24, and CB25; see figs. 6-14 and 6-15), scratching of the head and bill (CB24; see figs. 6-16 to 6-18), and stroking the bill along the sides of the body (CB24; see fig. 6-19). The birds bend in a pleasing variety of postures, sometimes lifting a foot while straining to reach inaccessible body parts (e.g., CB23; see fig. 6-20). The representations of scratching and foot pecking are particularly striking. When tending to their feet, the birds either lean over to probe at them while standing or lift one foot and hold it as it is nibbled with the bill (e.g., see figs. 6-14 and 6-15). The artist must have been influenced by a chance observation of this behaviour, as in life it is not expressed as often as other preening movements.[19] All of the poses reflect an acute awareness of the way in which these animals actually move, even when adjustments were made to ensure that they remained within the boundaries of the Egyptian method of representation. For example, when scratching their heads, the geese twist their necks upside-down along their flanks and raise one foot from (correctly) under one wing.[20] Their head is scratched, however, with the foot furthermost from the viewer, requiring it to pass under their body in an unnatural manner (e.g., fig. 6-18). Yet, rather than detracting from the overall impression, the resulting distortion simply adds to the naturally contorted posture of the birds.

One common preening movement, however, is missing from the tomb of *k3-gm-nj: mmj* - preening the breast feathers (see pl. 6G). This action first appears in the roughly contemporary tomb of *nj-k3w-jzzj* (CB27; see fig. 6-21, left), where the artist has taken advantage of the extremely long neck of a common crane to produce this distinctive pose. Breast-feather preening then occurs a number of times in the tomb of *mrrw-k3.j: mrj*, expressed by both long-necked and short-necked birds (e.g., CB28, CB29, CB31 left, and CB33; see fig. 6-22). The repertoire of movements here is less adventurous than that of the tomb of *k3-gm-nj: mmj*, but uniquely it includes two possible examples of allopreening. In chamber A6, a duck turns back and touches, with its open beak, the head of another that appears to be actively soliciting the attention (CB29; see fig. 6-23),[21] however this apparent allopreening may simply be an artefact of the close proximity of the figures. It is also important to note that, although many bird species engage in social grooming, it is not a common behaviour among ducks.[22] More convincing is an example from the same wall, in which a common crane standing in a small flock, bends forward and lifts its head back and up to touch the face of another with its beak (CB31; see fig. 6-25, right). The focus on the region of the head in both examples supports the notion that allopreening is being depicted.

In the tomb of *ḥzj*, a new posture is introduced in which geese turn back to probe at their feathers beneath one raised wing (CB34; see fig. 6-24).[23] Preening under the wings is a common movement among birds,[24] but this motif may have originated in the nearby tomb of *k3-gm-nj: mmj* where a goose also lifts one wing while apparently turning back its head (CB26). Damage to part of this

---

18 Goodwin, R.E. (1959). "Records of flight preening and related aerial activities in birds, particularly the black tern", *Auk* 76(October), 521-523; Michaelis, H.J. (1976). "Sich im Fluge putzende Lachmöwe, *Larus ridibundus*", *BV* 21(3/4), 304.

19 McKinney, *Behaviour* 25(1), 154.

20 *ibid.*, 152.

21 The pursed feet of the recipient of this action are difficult to explain in this context. See Chapter 4 for a discussion of this foot pose.

22 McKinney, *Behaviour* 25(1), 165-166.

23 It is interesting to note that this posture closely resembles the bird performing in-flight preening in the tomb of *3ḫtj-ḥtp*.

24 In some duck species, this movement is also part of a ritualised, pre-copulation display. See McKinney, *Behaviour* 25(1), 167-168.

relief, however, makes it difficult to determine precisely what the animal is doing. It is also possible that this posture is simply a variation of the back preening motif.

In the tomb of *mrrj*, two birds are again shown nibbling their feet (CB35). It seems likely, though, that this motif was copied from the tomb of *k3-gm-nj: mmj*, which, like that of *ḥzj*, is situated nearby.[25] The depictions lack the detail and finesse of those of *k3-gm-nj: mmj*, as does the single example of breast-feather preening from this tomb, which is little more than a sketch. This also describes the treatment of preening in the tomb of *mrjj-ttj: mrj* (CB36). Although this site features many depictions of preening the breast, the birds are rendered in a clumsy fashion and the repetition of the motif suggests that the artist had little direct knowledge of this avian behaviour. This is particularly evident where the artist attempted to introduce a new movement - preening the breast while simultaneously flapping the wings, both forwards (fig. 6-26) and backwards (fig. 6-27). To my knowledge, the concurrent expression of these two movements is physically impossible and not part of the preening repertoire of any bird species. It is thus an incorrect amalgamation of two independent poses and not a reflection of reality.

Finally, a pintail duck appears to preen its saddle feathers in a pastoral scene from the tomb of *mḥw*, however the animal's wings are crossed and the position of its legs suggests that it has collapsed (CB37; see fig. 6-28). As it is situated among the belongings of a herdsman, it seems more likely that the bird has been killed. Consequently, its head position is probably due to the limp state of its body (cf. the bodies of birds after their necks have been wrung in offering scenes).

6.2.2 Wing-flapping

A large number of depictions show waterfowl flapping their wings while standing on the ground (i.e., these are not movements associated with flight). Some birds stretch out their wings on either side of their bodies, while others hold them back and above their torsos. These two poses possibly capture the movement of real wings through space, from their initial upward thrust (see pl. 6H), to when they are brought forward and down in a sweeping motion.

Birds generally adopt an upright posture while wing-flapping. If this is done with sufficient vigour, some are able to achieve lift (see Chapter 4), such that only their toes stay in contact with the ground. Wing-flapping may also occur on water. Usually two to six flapping movements occur in a bout, but birds will occasionally wing-flap for longer periods.

Many birds mark the end of preening by flapping their wings elaborately. As the images in question often occur in contexts in which other birds are shown tending their feathers (see above), it seems likely that this action simply illustrates the last phase of preening. It may nevertheless be worthwhile to consider other behavioural explanations.

Birds

spd-ḥtp (CB38: Saqqara, V.6L-9?)
ṯjj (CB39-42: Saqqara, V.8-9)
k3-gm-nj: mmj (CB43-45: Saqqara, VI.1M)
ḥzj (CB46: Saqqara, VI.1L-2E)
wḏ3-ḥ3-ttj: ššj (CB47: Saqqara, VI.1L-2E)
mṯtj (CB48: Saqqara, VI.2?)
mrrj (CB49-50: Saqqara, VI.2E?)
mrjj-ttj: mrj (CB51: Saqqara, VI.2M-L)

In the tomb of *mrrj* (CB50), a duck (?) flaps its wings vigorously while in a pool situated in the centre of a poultry-yard. It appears to be bathing. Most bird species will wash their feathers when given the opportunity, but the method used by land and water species varies. Some bathe while standing in shallow water, but waterfowl do so when afloat - first dipping their head and shoulders and then beating their wings against the water so that it splashes over their body.[26] The bird in the tomb of *mrrj* undoubtedly expresses this common behaviour. It is curious that no other examples of bathing were detected in poultry-yard scenes from either Giza or Saqqara, even though pools of

25 See Map LII in *P-M III*.
26 Simmons in Thomson, *Dictionary of Birds*, 278-279.

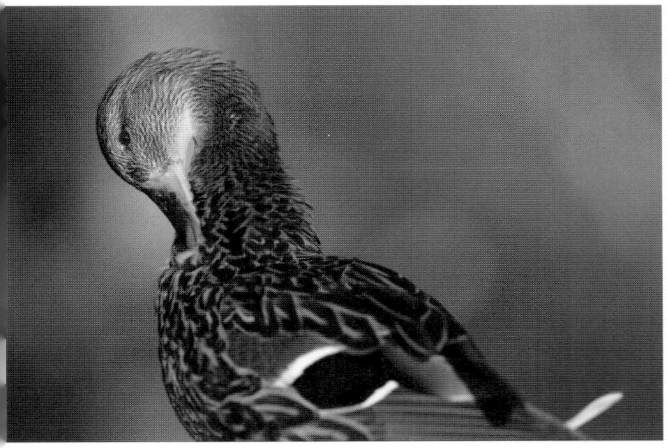

Pl. 6G Duck preening breast
© iStockphoto com/David Birkbeck

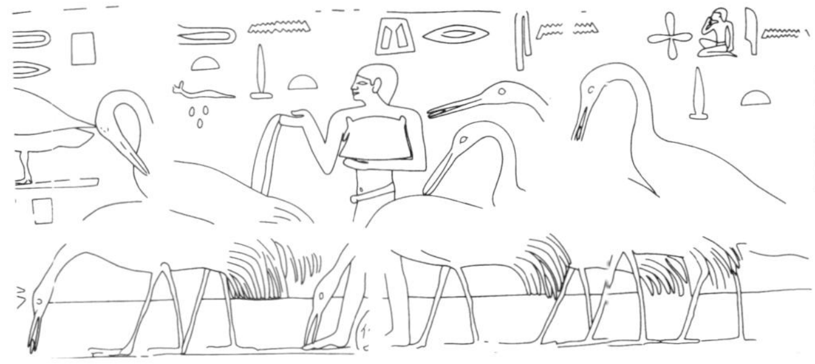

Fig. 6-21 (CB27)

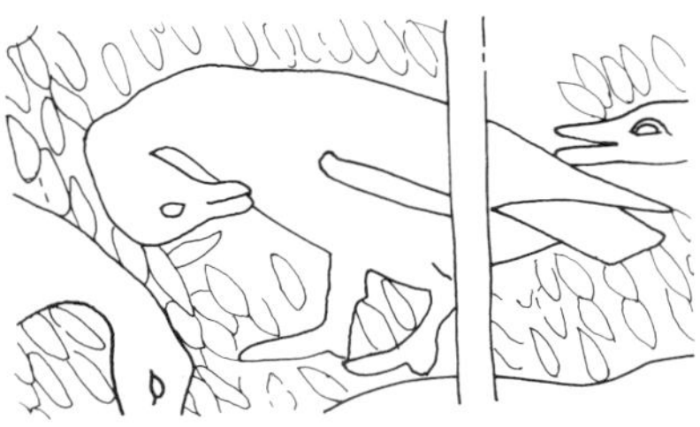

Fig. 6-22 (CB28)

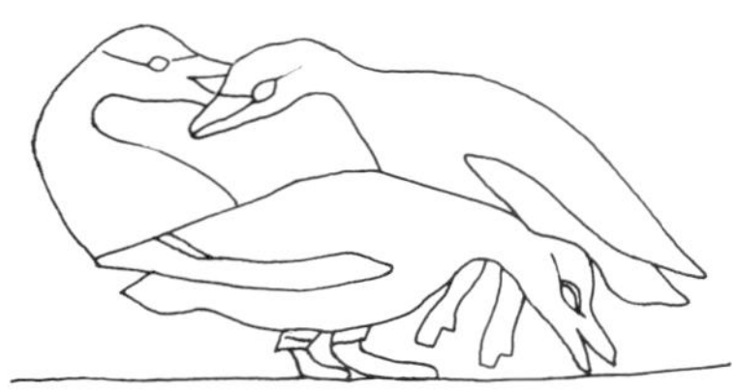

Fig. 6-23 (CB29)

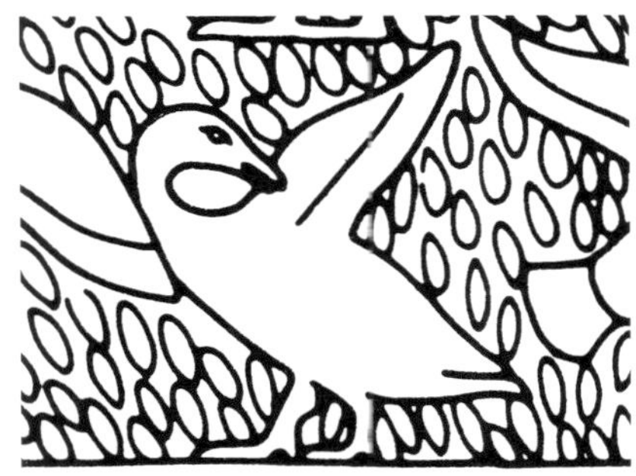

Fig. 6-24 (CB34)

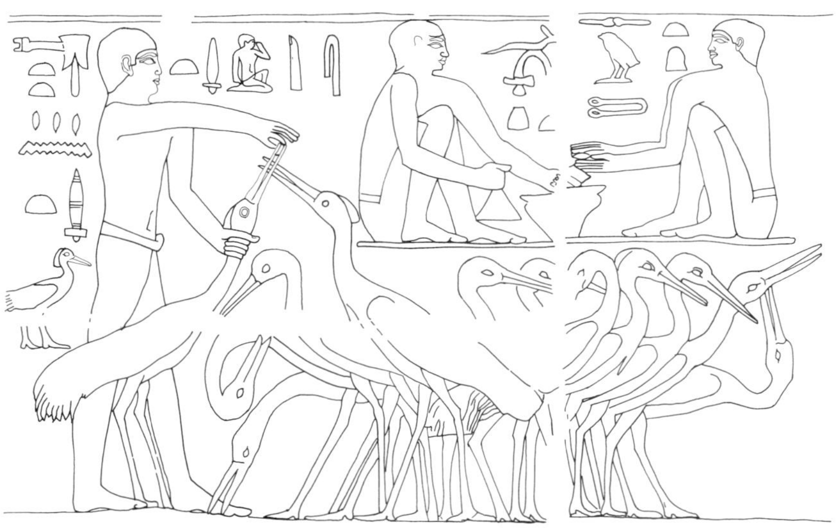

Fig. 6-25 (CB31)

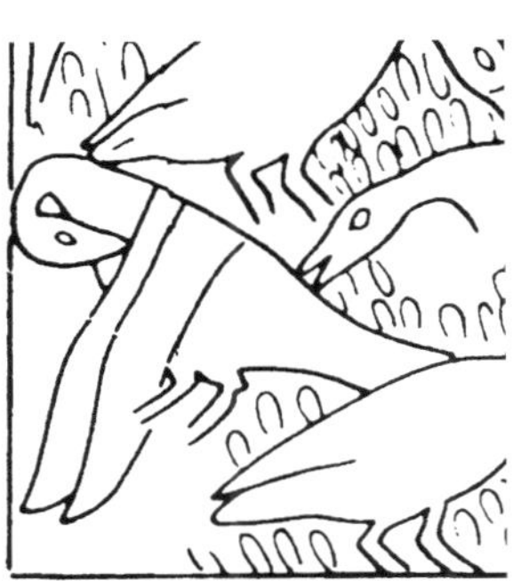

Fig. 6-26 (CB36)

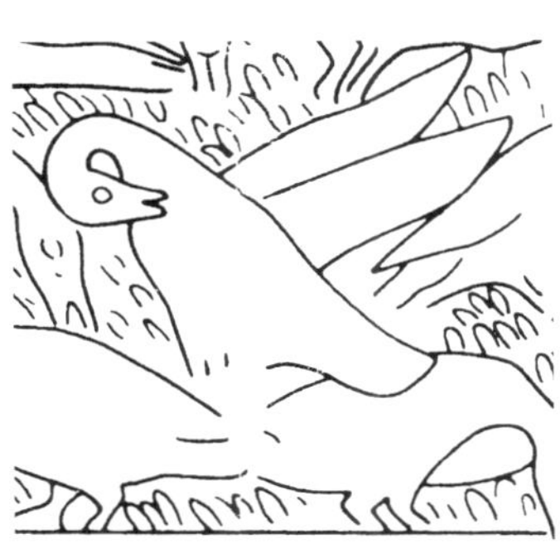

Fig. 6-27 (CB36)

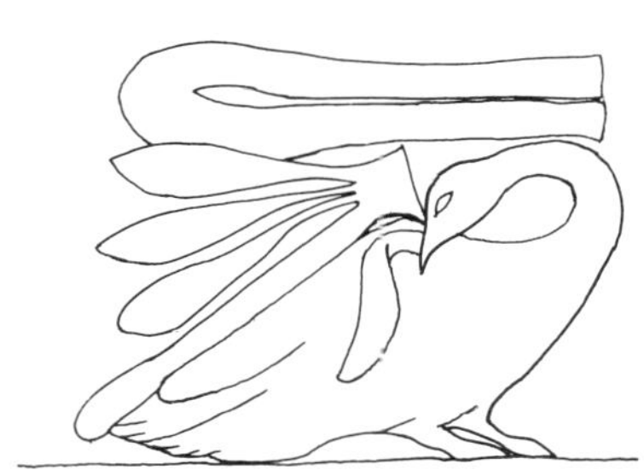

Fig. 6-28 (CB37)

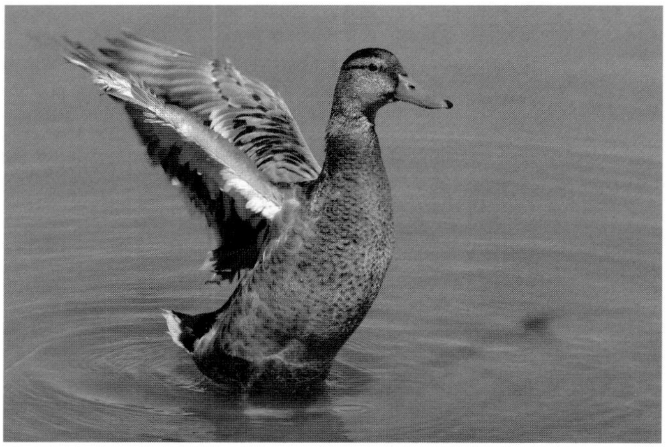

Pl. 6H Wing-flapping
© iStockphoto.com/Terry Alexander

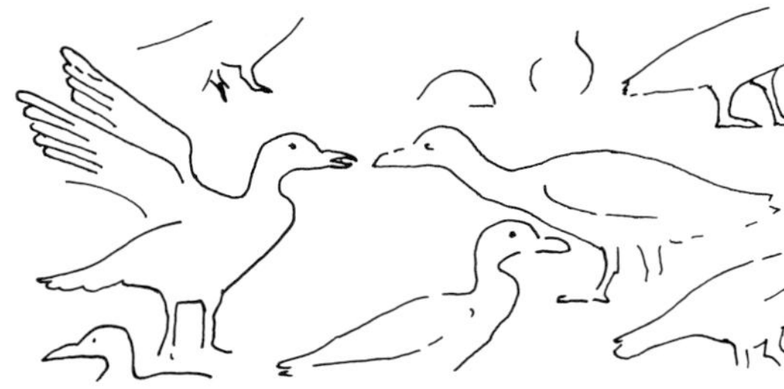

Fig. 6-29 (CB40)

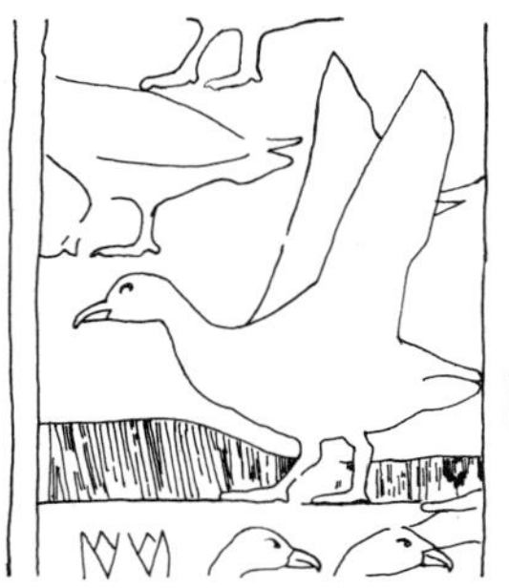

Fig. 6-30 (CB41)

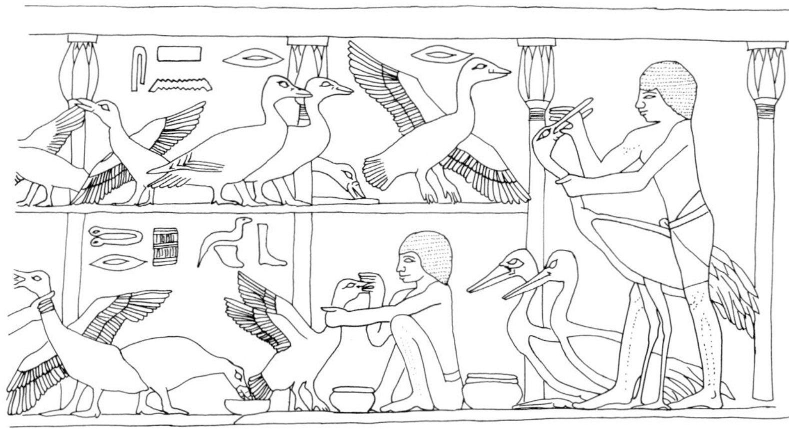

Fig. 6-31 (CB38)

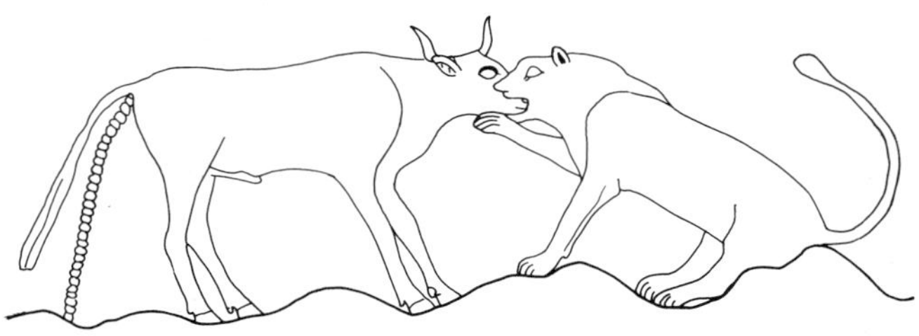

Fig. 6-32 (CB52)

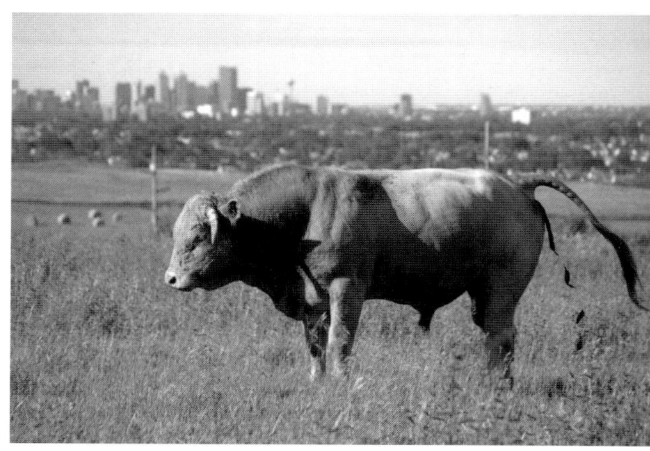

Pl. 6I Bull defecating
© iStockphoto.com/Wesley Pohl

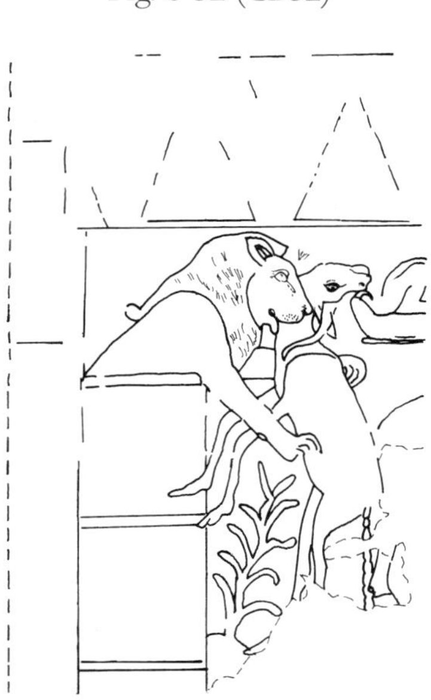

Fig. 6-33 (CB59)

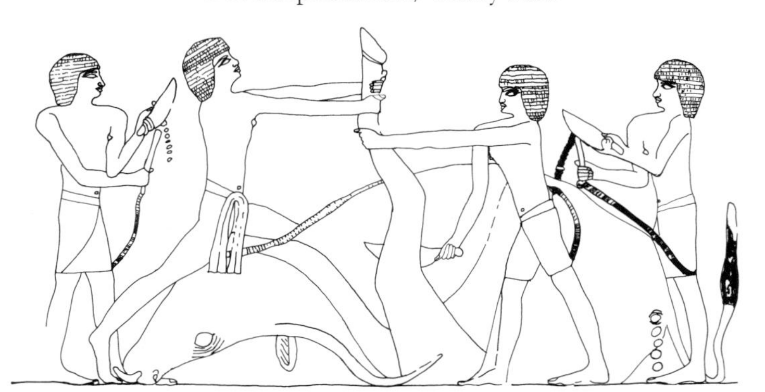

Fig. 6-34 (CB55)

water are often present (i.e., in the tombs of *ṯjj*, *k3-gm-nj: mmj*, *ḥzj*, *wḏ3-ḥ3-ttj: ššj* [*nfr-sšm-ptḥ*], *mrrj*, and *mrjj-ttj: mrj*).[27]

Bathing frequently precedes preening and oiling. Before preening can commence, however, birds must first dry their feathers. This is done by opening their wings and waving them back and forth to remove water droplets.[28] Rather than signalling the end of a preening bout, then, many images of flapping birds may actually indicate the *first* stage of the process. Feather drying is certainly suggested in an example from the tomb of *ṯjj*. Among feeding geese in a large, heavily populated poultry-yard, a single bird flaps its wings (CB41; see fig. 6-30). It does so, however, as it stands upon the banks of a central pool, giving the impression that it has just emerged from the water and is now drying off.[29]

Feather drying is a feasible explanation for wing-flapping in poultry-yards where water is present, but this cannot account for its occurrence in other contexts, such as force-feeding scenes. Waterfowl will often flap in response to disturbances in their immediate environment (including those caused by humans)[30] and so it is possible that the birds in fattening scenes are reacting to the repeated capture of their flock-mates nearby. An example from the tomb of *mṯtj* (CB48) gives this impression. In some waterfowl, wing-flapping may also occur during or after aggressive interactions between birds.[31] Indeed, a goose flapping its wings in the tomb of *ṯjj* (CB40; see fig. 6-29) appears to be reacting to the behaviour of a flock-mate that stretches its neck forward in a distinctive threat display (see *Geese* in section 9.1.1 in Chapter 9), while those in the tomb of *spd-ḥtp* (CB38; see fig. 6-31, lower panel) flap in response to aggressive treatment by their human handlers.

Birds will usually stretch their legs and wings after sleeping or sitting, slowly extending each appendage to its full extent.[32] Flapping the wings can similarly serve to increase the blood supply to cramped flight muscles.[33] It has been noted that birds in captive populations will often wing-flap a number of times after periods of physical restraint, both to settle their ruffled plumage and to stretch out their wings. Flapping birds in force-feeding scenes may consequently be recovering from recent handling.

It is important to recall that both poultry-yard and force-feeding scenes depict the Egyptians' attempts to fatten fowl for the table. The birds in each situation appear to roam about freely - seemingly able to fly away if they wished - yet none show physical evidence of wing-clipping or pinioning, which would render them flightless.[34] It is entirely possible, then, that their increased weight kept them from flying and that wing-flapping consequently depicts their fruitless attempts to take flight. Some seem to achieve a small amount of lift as they are clearly shown standing on their toes. This is particularly evident in examples from the tombs of *spd-ḥtp* (CB38; see fig. 6-31, upper panel), *k3-gm-nj: mmj* (CB43 and CB44), *ḥzj* (CB46), and *mrjj-ttj: mrj* (CB51). Finally, it has been observed that captive waterfowl will engage in prolonged wing-flapping just prior to feeding, and often flap while eating.[35] This behaviour would thus be appropriate in any scene in which birds are being fed.

---

27 It is important to note that birds in clapnet scenes occasionally appear to flap their wings while swimming, holding them both fully outstretched and above their bodies. I have not considered these to be depictions of bathing, as it is not clear whether the birds are behaving freely or reacting to the closure of the net. The flapping wings in these situations may instead indicate birds in flight, flushing upward.

28 Simmons in Thomson, *Dictionary of Birds*, 279-280.

29 The undulating, irregular banks of the pool are remarkably realistic, quite unlike the precise rectangular structures found in later examples of this scene.

30 McKinney, *Behaviour* 25(1), 141-142.

31 *ibid.*, 145.

32 *ibid.*, 149-152; see also Simmons in Thomson, *Dictionary of Birds*, 284-285.

33 McKinney, *Behaviour* 25(1), 143.

34 It may not have been necessary to stop captive waterfowl from escaping. Geese and ducks are not inclined to leave areas where they are safe and where food is provided regularly (*Pers. comm.* Cathy Alexander, Bird Department, Taronga Zoo, Sydney). Both wing-clipping (in which the primary flight feathers of a wing are cut) and pinioning (in which the outer portion of a wing is amputated) change the shape of the wings. None of the images from Giza or Saqqara, in which birds are depicted with outstretched wings, show evidence of these procedures.

35 McKinney, *Behaviour* 25(1), 142.

It will be apparent from the discussion above that it is impossible to pinpoint the precise reason for the occurrence of wing-flapping in poultry and force-feeding scenes, but such movements are undoubtedly a natural part of the comfort repertoire of waterfowl and so their inclusion in these situations must reflect a general awareness of their behaviour. It is interesting to note, however, that large birds, such as cranes, which also appear in poultry-yard and force-feeding scenes, are never shown wing-flapping.

6.3 ELIMINATION

Animal species differ greatly in the postures they adopt when defecating and urinating, the location where eliminative behaviour is performed, and the content and shape of the faecal matter produced.[36] Males and females within species often display different postures when urinating, but adopt a similar stance when defecating (see pl. 6I).

Animals that live in semi-permanent structures (nests, dens, etc.) are often able to withhold elimination until they are away from their living quarters, but species that move constantly, such as ungulates, usually defecate and urinate indiscriminately.[37] Some species deposit their faeces in prominent locations to serve as territorial markers, while others use communal latrines that can also communicate olfactory information about group members. The size, shape, and amount of faecal material produced depends primarily upon the size and age of the individual and the type of food it has ingested.[38] The consistency of faeces may also be affected by illness and emotional state (e.g., anxiety or fear). Finally, the frequency with which animals perform eliminative behaviour can be influenced by social and territorial requirements, but is largely dependent upon their access to food and water: animals that live in desert environments urinate and defecate infrequently, while those with adequate food and water do so many times a day.

It appears that Old Kingdom artists were reluctant to illustrate this extremely common aspect of animal life, as elimination is depicted in only five tombs, all of which date from the limited period of the late 5th dynasty to the reign of Teti.[39] One dorcas gazelle and seven cattle are shown defecating, however none appear to be doing so as part of their normal eliminative behaviour. Instead, four of the animals are defecating in response to an attack by a predator, while the remaining four images show animals clearly defecating after death.

Cattle

ptḥ-ḥtp II: ṯfj (CB52: Saqqara, V.9M-L)
ḥtp-k3.j: tp-k3.j (CB53-56: Giza, V.9-VI.1)
nj-m3ˁt-rˁ [*jzzj-mr-nṯr*] (CB57: Giza, V.9-VI.1)
mrrw-k3.j: mrj (CB58: Saqqara, VI.1M-L)

The first example that depicts defecating cattle occurs in the tomb of *ptḥ-ḥtp II: ṯfi* (CB52: see fig. 6-32). Here, a lion attacks a bull, clamping its jaws over its muzzle (see also Chapter 8). In response, the bull has locked its fore- and hind legs and pulls back, away from its aggressor. A chain of faecal pellets emerges from below its tail, which is held out abruptly from its body. Cattle usually raise their tails in this manner while defecating to avoid contaminating their coats (cf. pl. 6I), however they do not normally produce distinct pellets.[40] It may be possible, then, that the artist was attempting to represent a more liquid form of faeces produced by nervous cattle, to emphasize the bull's likely fear. In an almost identical scene from the later tomb of *mrrw-k3.j: mrj* (CB58; see fig. 8-3 in Chapter 8), in which a bull is also attacked by a lion, the chain of pellets is replaced by two parallel lines, clearly representing a stream of faeces.

In a poorly preserved scene from the Giza tomb of *nj-m3ˁt-rˁ* [*jzzj-mr-nṯr*] (CB57), only the back, hind legs, and tail of an unidentified ungulate remain. The inclination of the hind legs and the angle

---

36 Chame, M. (2003). "Terrestrial mammal feces: A morphometric summary and description", *MIOC, Rio de Janeiro* 98(Suppl. I), 71-94.
37 Grier, *Biology*, 166.
38 Chame, *MIOC, Rio de Janeiro* 98, 71-72.
39 In the Middle Kingdom, however, bulls are occasionally depicted defecating while fighting with one another.
40 The high water content of cattle faeces normally results in large cakes of dung. See *ibid.*, 74 (group vi).

of the tail are virtually identical to those of the bull defecating while under attack by a lion in the tomb of *ptḥ-ḥtp II*: *ṯfi*. Consequently, it seems likely that this animal is responding similarly. That other animals on the same register are also being attacked (by hunting dogs) supports this notion.

Finally, four examples from the tomb of *ḥtp-k3.j: tp-k3.j* show cattle defecating after death. In each case, a small chain of faecal pellets descends from the anus of each ungulate while it is being slaughtered (CB53 to CB56; see fig. 6-34). Such a response is not unexpected, as complete relaxation of the muscles after death ('primary flaccidity') can sometimes result in spontaneous urination or defecation.[41]

Dorcas gazelle

ṯjj (CB59: Saqqara, V.8-9)

The earliest example of defecation found in this study comes from the tomb of *ṯjj* (CB59; see fig. 6-33) and depicts a lion dragging the body of a dorcas gazelle over a fence (see also Chapter 8). The predator has bitten the gazelle's throat, and so it is likely that if the animal is not yet dead, it is close to death. It hangs limply from the jaws of the lion, its head lolling backward and its legs dangling in mid-air as a rope-like stream of faeces descends from its body. Dorcas gazelle normally expel a shower of small, dry pellets.[42] The chain of faeces is thus probably intended to represent the rapid succession and motion of the pellets as they emerge and fall away from the animal's body (cf. the depiction of falling seeds during sowing in sheep treading scenes).[43] As the gazelle is defecating while under attack, however, it is also possible that this stream of pellets was intended to suggest a liquid state brought on by its fear. It may equally be defecating after death.

SUMMARY

- Comfort activities occur most frequently in tomb scenes from Saqqara dating from the late 5th to the early 6th dynasty.
- The majority of examples in this behavioural category depict domesticated species (cattle, geese, and ducks). Most animals spend a great deal of time keeping themselves comfortable, but artists would undoubtedly have had a greater opportunity to witness this activity in species that were kept in captivity. Such animals would have been relatively easy to observe, but also sufficiently at ease in the presence of humans to allow the expression of relaxed behaviour.
- Grooming behaviour is portrayed accurately for cattle, but not for the single example of a dorcas gazelle, which is shown licking. This may be the result of lack of familiarity with the animal, but could equally be an artistic device by which to convey the act of grooming unambiguously. Only one clear case of allogrooming was noted.
- The range of depicted preening movements broadens over time and appears to reflect close observation of bird behaviour. Only one unequivocal example of allopreening was detected.
- The artist responsible for the preening images in the tomb of *k3-gm-nj: mmj* stands out. Representations of poultry from the site reveal that this individual took great pleasure in the contortions that birds adopt as they preen and was at pains to reproduce their full repertoire as accurately as possible. Generally, however, the proportion of tombs from Giza and Saqqara showing comfort behaviour is quite small and so it seems that most artists were either not aware of these activities or did not deem them worthy of reproduction in wall scenes.
- Elimination behaviour appears rarely in wall scenes. The examples found were all associated with dead or dying animals.

41 Iserson, K.V. (2003). "Rigor mortis and other postmortem changes" in Kastenbaum, *Encyclopedia of Death ii*, 721.

42 Faeces with low water content are produced in the form of pellets; the water content of dorcas gazelle faeces is very low (similar to that of camels). See Wilson, *Ecophysiology*, 80-84.

43 Pellets of incense are also shown in this manner. See Schäfer, *Principles*, 228.

CHAPTER 7

INGESTIVE BEHAVIOUR

DEFINITION

Ingestion refers to the intake of food and fluids. The way that animals do this is dictated by their body structure (e.g., size, shape, dentition, digestive anatomy, etc.), energy requirements (e.g., type of metabolism), what they ingest, and the availability of food and water. The diets of different species vary considerably, as does the frequency with which they must feed and drink.

7.1 FEEDING

Feeding consists of all behaviours associated with the acquisition, manipulation, and ingestion of food. Animals can be classified into one of three broad categories according to the type of food they ingest: *carnivores* (flesh-eaters), *herbivores* (plant-eaters), or *omnivores* (both plants and meat are consumed). Carnivores can include *insectivores* (insect-eaters), *piscivores* (fish-eaters), and *scavengers*. Herbivores may also be divided into: *frugivores* (fruit-eaters), *granivores* (seed-eaters) and nectar-feeders. The type of food ingested affects greatly how it is obtained, handled, chewed, and swallowed by an animal.

Feeding behaviour was represented by 21 species in tomb scenes from Giza and Saqqara. Note that predatory behaviour, in which some carnivore species obtain food by catching, killing, and consuming other animals, will be treated separately in the following chapter.

Donkeys

rˁ-ḫˁ.f-ˁnḫ (IB1-2: Giza, V.6?)
rˁ-m-k3.j (IB3: Saqqara, V.6? r/u - V.6-7)
nj-ˁnḫ-ẖnmw/ẖnmw-ḥtp (IB4-5: Saqqara, V.6L-7)
pḥn-wj-k3.j (IB6: Saqqara, V.6-8E)
ḥtp-ḥr-3ḫtj (IB7: Saqqara, V.6-8E)
3ḫtj-ḥtp (IB8: Saqqara, V.6-8E)
jrj-n-k3-ptḥ (IB9: Saqqara, V.6-9)
k3.j-m-rḥw (IB10: Saqqara, V.8)
snḏm-jb: jntj (IB11: Giza, V.8M-L)
ptḥ-ḥtp I (IB12: Saqqara, V.8M-L)
sḫm-ˁnḫ-ptḥ (IB13: Saqqara, V.8-9)
ṯjj (IB14: Saqqara, V.8-9)
nfr-jrt-n.f (IB15: Saqqara, V.8-9)
k3.j-m-nfrt (IB16: Saqqara, V.8-9?)
ḫww-wr (IB17-18: Giza, V.8-9M)
snḏm-jb: mḥj (IB19: Giza, V.9)
nj-k3w-ḥr (IB20: Saqqara, V.9)
3ḫtj-mrw-nswt (IB21: Giza, V.9-VI)
ḥtp-k3.j: tp-k3.j (IB22: Giza, V.9-VI.1)
sḫm-k3.j (IB23: Giza, V.9-VI.1?)
mrrw-k3.j: mrj (IB24: Saqqara, VI.1M-L)
ˁnḫ-m-ˁ-ḥr: zzj (IB25: Saqqara, VI.1L-2E)
k3-ḥj.f (IB26: Giza, VI.2M-3?)

Domestic donkeys graze upon a wide range of plants, including shrubs, forbs, and grasses.[1] As they possess both upper and lower incisors (cf. *Sheep* and *Goats* below), and a strong and mobile upper lip, they are able to bite plants very close to the ground (see pl. 7A). They can ingest an adequate amount of grass if they are able to graze for long periods (c. 6-7 hours per day), but if they are required to work, their diet must be supplemented with other foods. These can include straw, hay, and grains, but also the bi-products of threshing and cereal milling (i.e., chaff).[2]

Twenty-six examples of feeding donkeys were found in 23 tombs dating from the mid-5th to the early 6th dynasty. Most of these images occur in threshing or transportation scenes. As

1 Moehlman, P.D. (1998). "Behavioral patterns and communication in feral asses (*Equus africanus*)". *App. Anim. Beh. Sci.* 60(2-3), 132.
2 Aganga, A.A., Letso, M. and Aganga, A.O. (2000). "Feeding donkeys", *LRRD* 12(2), 1.

donkeys graze by bringing their mouths in close contact with food items, only depicted animals whose muzzles actually touch foliage, containers, or the ground have been considered.[3]

Donkeys (and also cattle) are frequently shown moving in small herds upon threshing floors, a procedure that allowed, through the weight and action of the animals' feet, for ears of grain to be removed from their stems and seeds to break free from their husks. In line with the constraints of Egyptian representation, these flat, circular arenas are rendered as long, low rectangles upon, or behind, which the animals move.[4] While most herd members usually maintain an erect position, some lean well forward so that their mouths touch the platform. In earlier depictions, the animals' muzzles are often hidden behind the floor (IB2, IB6, IB8, and IB16; see fig. 7-2), or occasionally hang in front of it (IB3, IB7, IB14, and IB21). Later, however, their muzzles more commonly rest upon the top of the platform (IB9 to IB13, IB15, IB17, IB19, IB22, IB24 to IB26; see fig. 7-1). Although in each of these examples no food is shown in the donkeys' mouths, both their behaviour and the context in which it occurs indicate that feeding is depicted.[5] As noted earlier, donkeys will eat straw, grain, and also the husks of seeds (chaff), all of which would be available to them upon the threshing floor. Some of the depicted animals open their mouths as though to bite (IB3, IB7, IB9, IB11, IB21?, IB25, and IB26). Feeding is also suggested by an image from the tomb of *ḥtp-k3.j: tp-k3.j* (IB22; see fig. 7-1), in which a donkey protrudes its tongue so that it touches the platform. Although the tongue is usually not involved until after donkeys have obtained a mouthful, this biologically inaccurate detail manages, nevertheless, to link the animal's mouth unambiguously with the source of food.

Feeding donkeys also appear in scenes in which sacks of wheat sheaves are transported. In an example in the tomb of *sḫm-k3.j* at Giza (IB23; see fig. 4-11 in Chapter 4), the animals have made their delivery and are being hurried back to the field by men who chase them brandishing sticks. Despite their apparent haste, a donkey at the head of the pack bends its head down so that its muzzle touches a plant. The contact between the animal's mouth and potential food suggests strongly that it is feeding. As was noted in Chapter 4, because it is not physically possible for donkeys to run and graze simultaneously, this scene must represent a contraction of time and space, in which the two conflicting behaviours indicate the beginning of their journey (running) and its end (eating).[6]

Finally, three images depict donkeys that feed while carrying men upon their backs.[7] The later of these occurs in the Giza tomb of *ḫww-wr* (IB18; see fig. 7-3) and shows a pair of donkeys, one of which feeds by leaning forward and burying its muzzle in a small shrub. This behaviour suggests that the animals are not moving.[8] The two remaining images are found in the earlier tomb complex of *nj-ꜥnḫ-ẖnmw/ẖnmw-ḥtp* (IB4 and IB5). Each scene depicts a pair of donkeys, one of which places its mouth close to a bundle of greenery offered by a man who walks in advance of the animals. As no contact occurs between the foliage and the donkeys' mouths, eating is not represented; instead, I suggest that these images depict their *anticipation* of feeding. In contrast to the *ḫww-wr* image, then, it is likely that these animals are moving, driven forward by their constant attempt to reach the leaves that are kept tantalizingly out of reach.

3 Some depictions show donkeys leaning well forward so that their muzzles hover close to the ground. These were possibly copied from images of feeding animals, but a food source has failed to be included.

4 Siebels, R. (2000). "Agricultural scenes" in Donovan-McCorquodale, *Egyptian Art*, 61-62.

5 Simpson (*Sekhem-ankh-ptah*, 16) believes that the donkeys in the tomb of *sḫm-ꜥnḫ-ptḥ* (IB13) are leaning over to eat.

6 For a discussion of the representation of time in Egyptian art, see Bochi, P.A. (1994). "Images of time in ancient Egyptian art", *JARCE* 31, 55-62; Bochi, P.A. (2003). "Time in the art of ancient Egypt: From ideological concept to visual construct", *KronoScope* 3(1), 51-82.

7 For a discussion of riding donkeys, see Houlihan, P.F. (2002). "Some instances of humor associated with animal riding in ancient Egypt", *GM* 190, 35-45.

8 Alternatively, the scene may again show a contraction of time, representing both travel and arrival.

Cattle

nfr-msḏr-ḫwfw (IB27: Giza, V.2-8, possibly V.9)
nj-ʿnḫ-ẖnmw/ẖnmw-ḥtp (IB28-31: Saqqara, V.6L-7)
ȝḫtj-ḥtp (IB32: Saqqara, V.6-8E)
jtj-sn (IB33-35: Giza, V.6-8?)
zȝ-jb (IB36: Giza, V.8)
ṯjj (IB37-38: Saqqara, V.8-9)
kȝ.j-m-nfrt (IB39: Saqqara, V.8-9?)
kȝ-gm-nj: mmj (IB40-42: Saqqara, VI.1M)
ʿnḫ-m-ʿ-ḥr: zzj (IB43-44: Saqqara, VI.1L-2E)
jnw-mnw (IB44a: Saqqara, VI.2E)
ẖntj-kȝ.j: jḫḫj (IB45: Saqqara, VI.2M)
mḥw (IB46-47: Saqqara, VI.2M-3?)

Twenty-two examples of feeding cattle, dating from the early 5th to the mid-6th dynasty, were detected in 12 tombs. The animals are shown eating in different contexts, including from troughs, while on the threshing-floor, fording, and recumbent.

Cattle graze primarily on grass, but in agricultural practice their diet may also be supplemented by hay, silage (fermented plant material), and grains. Regardless of what type of food they ingest, they use their highly mobile tongues extensively when they feed. When grazing upon foliage, their tongue is extended and wrapped around leaves or grass stems to hold them in place while they are then bitten or torn off by jerking their head (see pl. 7B). If feeding on grain from a trough, particles are first scooped with the tongue and then sucked into the mouth.[9] After a sufficient amount of food has been swallowed, they must pause to ruminate. This essential activity typically takes place while the animal is recumbent (see Chapter 4).[10]

Only one animal has been depicted grazing upon grass. In the mid-6th dynasty tomb of *jnw-mnw* (IB44a; see fig. 7-4), a cow leans forward deeply to place her protruded tongue upon a stylised clump of vegetation, illustrating correctly the manner in which cattle manipulate foliage prior to ingesting it. Three other animals, however, are clearly shown ingesting swamp plants. In the tomb complex of *nj-ʿnḫ-ẖnmw/ẖnmw-ḥtp* (IB31; see fig. 7-6), an animal at the head of a herd of fording cattle places its muzzle in the midst of a tall stand of reeds (*Phragmites* sp.) and protrudes its tongue so that it touches the foliage. In another fording scene in the tomb of *ȝḫtj-ḥtp* (IB32; see fig. 7-7), an animal in the centre of the herd bites and pulls at a waterweed or water lily as it passes by. And in a unique example in the tomb of *zȝ-jb* (IB36; see fig. 7-5), a recumbent ox appears to ingest a stalk of papyrus (*Cyperus papyrus*) that emerges from a thicket and enters its open mouth.

The method of feeding is correct in all of these examples, but the plants ingested by the animals require further comment. Cattle do not commonly graze upon papyrus as it grows in water (in streams, rivers, and irrigation canals), making it difficult to access. It is a highly nutritious food (higher in protein than grass),[11] however, and if given the opportunity, they will ingest it readily. Immature phragmites reeds are much favoured by cattle[12] and are considered by farmers to be a high-quality forage for livestock. Cattle will also eat a variety of floating water plants, including water lilies.

During the Egyptian dry season, when vegetation for grazing was reduced in the Nile Valley, herdsmen were forced to move their cattle to the Delta where they had access to grass. Pollen spectra found in core samples taken from this region have indicated, however, that although the local flora did include dry-ground shrubs and weeds, the most predominant species were papyrus, phragmites reeds, and bulrushes (*Typha* sp.).[13] It would thus appear that when grazing in the Delta, cattle were as likely to have ingested these swamp species as grass. The images of feeding cattle found in this study, two of which are labelled, *prt m mḥt* 'Coming from the marshes' (IB31 and

[9] Hafez, E.S., Schein, M.W. and Ewbank, R. (1969). "The behaviour of cattle" in Hafez, *Behaviour of Domestic Animals*, 236.
[10] *ibid.*, 248
[11] Muthuri, F.M. and Kinyamario, J.I. (1989). "Nutritive value of papyrus (*Cyperus papyrus*, Cyperaceae), a tropical emergent macrophyte", *Econ. Bot.* 43(1), 23-30; Brewer et al., *Domestic Plants and Animals*, 40-41.
[12] Van Deursen, E.J.M. and Drost, H.J. (1990). "Defoliation and treading by cattle of reed *Phragmites australis*", *J. App. Ecol.* 27, 284-297.
[13] Saad, S.I. and Sami, S. (1967). "Studies of pollen and spores content of Nile Delta deposits (Berenbal region)", *PS* 9, 500.

IB32; see fig. 7-7), support the notion that swamp plants probably formed a substantial part of their diet when they were in this region.

Cattle feeding from troughs are depicted in only one tomb – that of *jtj-sn* (IB33 to IB35; see fig. 7-8). In each case, the animals bend their heads and protrude their tongues so that they touch a tall manger. This behaviour matches closely the feeding method used by cattle when ingesting small particles, such as grains or chopped fodder.

Feeding cattle also appear in two threshing-floor scenes (IB29 and IB38). In both examples, the animals lean forward deeply and rest their muzzles on top of a low structure representing the threshing arena; as both grains and chaff would be present on this floor, it seems likely that they are feeding. The earlier of the two examples (IB29) dates from the mid-late 5th dynasty. The position of the animal's muzzle on top of the arena at this early date is somewhat at odds with similar examples of donkeys feeding (see above), where it does not occur reliably until later in the 5th dynasty.

Four examples of hand-feeding were found in which the proffered food is clearly visible and held to the mouth of the animal.[14] In the tomb complex of *nj-ꜥnḫ-ẖnmw/ẖnmw-ḥtp* (IB28), one member of a ploughing pair is enticed to walk forward by a herdsman who offers it a bundle of reeds or grasses. This matches precisely two similar scenes from the same tomb in which donkeys are also encouraged to move by a handler offering food (see above; IB4 and IB5). In other examples of hand-feeding, the animals are immobile. In both the tomb of *ṯjj* (IB37; see fig. 4-49 in Chapter 4) and that of *kꜣ.j-m-nfrt* (IB39), a recumbent bull or ox bites a stalk of foliage offered to it by a seated herdsman. In the final example, a calf bites at a ball of bread or dough as it kneels before a handler (IB27; see fig. 7-9).[15] It has been suggested that bread dough was used by the Egyptians to supplement the diet of cattle during the dry season to provide valuable proteins and minerals.[16]

Mineral deficiency may also explain the behaviour of seven calves and one horned bull(?) (IB40 to IB47; see fig. 7-10) that either lick the ground or touch it with their muzzles in offering scenes. It is possible that the contact between the cattle's mouths and the substrate was intended only to introduce some postural variety in the repetitive lines of animals found in such scenes. It is worth noting, however, that many herbivores are known to engage in 'geophagia', the eating or licking of dirt, if their diet is deficient in salts and minerals;[17] indeed earth-eating by cattle is widespread in Africa.[18] Newly-weaned calves are also particularly inclined to lick and nibble any available surface.[19] The damaged image of a calf in the tomb complex of *nj-ꜥnḫ-ẖnmw/ẖnmw-ḥtp* (IB30) also places its muzzle on the ground, but the presence of its tethered companion nearby suggests that here this posture may simply be the result of its struggling.

Addax

mrrw-kꜣ.j: mrj (IB48: Saqqara, VI.1M-L)

A single representation of a feeding addax was found in the tomb of *mrrw-kꜣ.j: mrj* at Saqqara (IB48; see fig. 7-11). The collared animal is tethered to a short rope and stands before a low manger. Although no food is visible, the addax stretches its neck forward and protrudes its tongue so that it touches the upper surface of the trough. This is an accurate depiction, as all antelope species, including addax, possess long, prehensile tongues that they use to select and pull food items into their mouths.[20] Addax inhabit arid areas and graze primarily upon coarse desert grasses

---

14 See Chapter 4 for a discussion of recumbent, hand-fed cattle in which food items are not apparent.

15 Although adults usually stand while feeding, calves are known to occasionally graze while lying prone. See Hafez et al. in Hafez, *Behaviour of Domestic Animals*, 236.

16 Brewer et al., *Domestic Plants and Animals*, 86.

17 Fraser-Broom, *Farm Animal Behaviour*, 320; Smith, J.W., Adebowale, E.A., Ogundola, F.I., Taiwo, A.A., Akpavie, S.O., Larbi, A. and Jabbar, M.A. (2000). "Influence of minerals on the aetiology of geophagia in periurban dairy cattle in the derived savannah of Nigeria", *Trop. Anim. Health Prod.* 32(5), 315-327.

18 Hafez et al. in Hafez, *Behaviour of Domestic Animals*, 246.

19 Fraser-Broom, *Farm Animal Behaviour*, 319.

20 Spinage, *Antelopes*, 19.

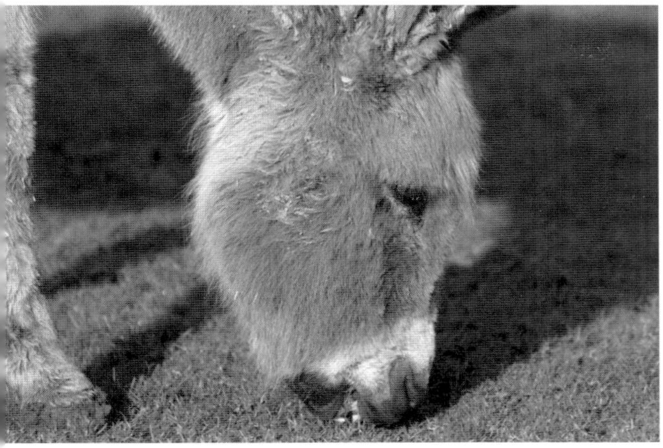

Pl. 7A Donkey grazing
© iStockphoto.com/Gary Forsyth

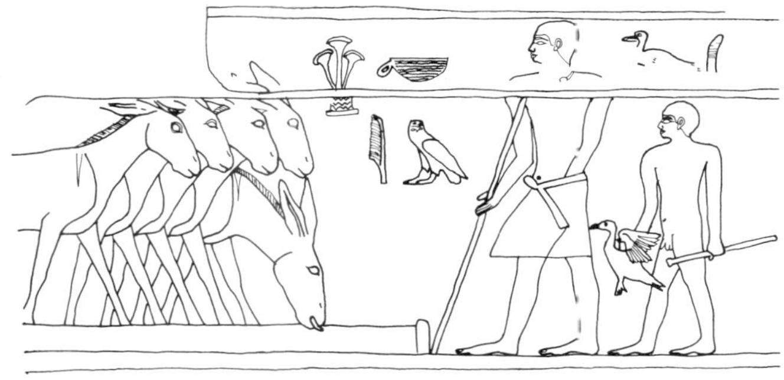

Fig. 7-1 (IB22)

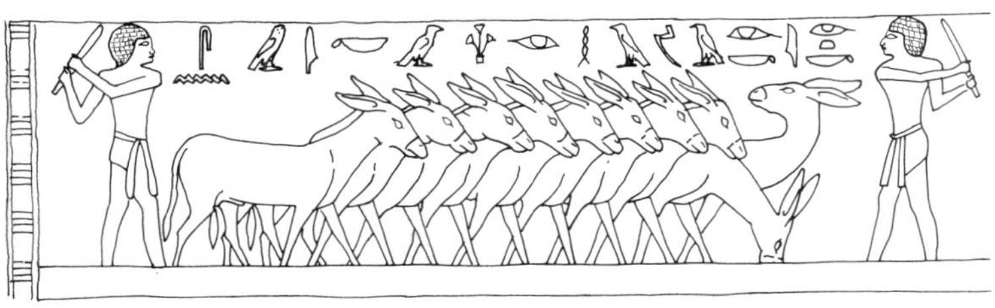

Fig. 7-2 (IB6)

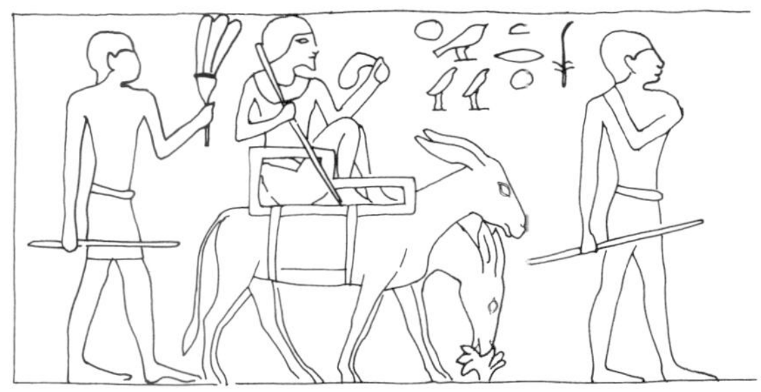

Fig. 7-3 (IB18)

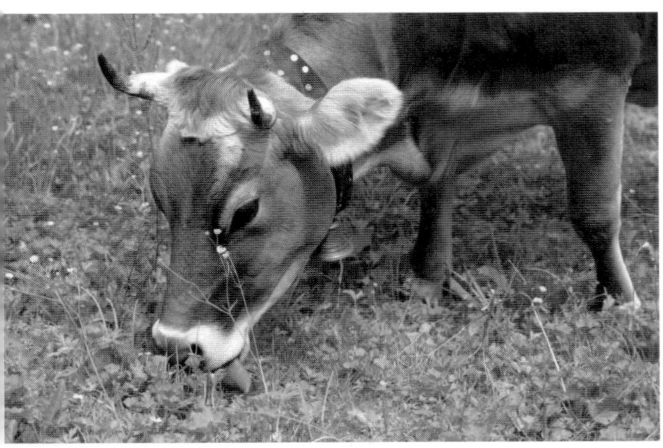

Pl. 7B Cow grazing
© iStockphoto.com/Bettina Ritter

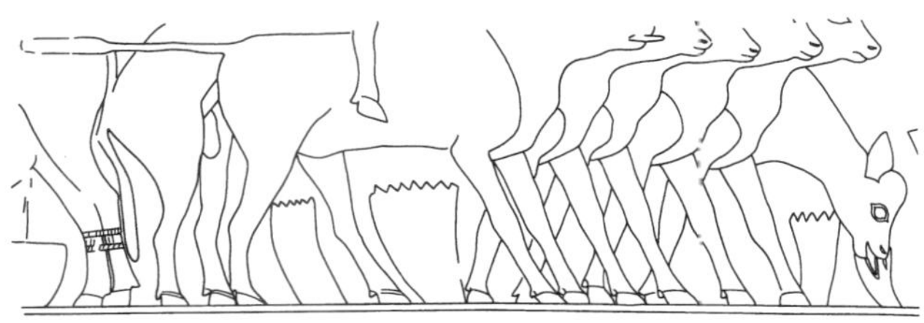

Fig. 7-4 (IB44a)

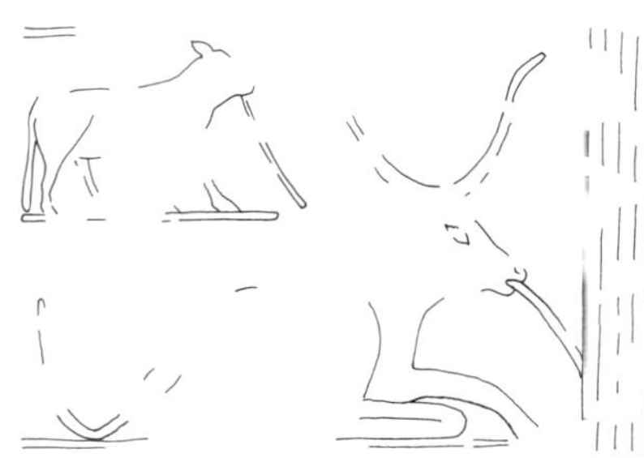

Fig. 7-5 (IB36)

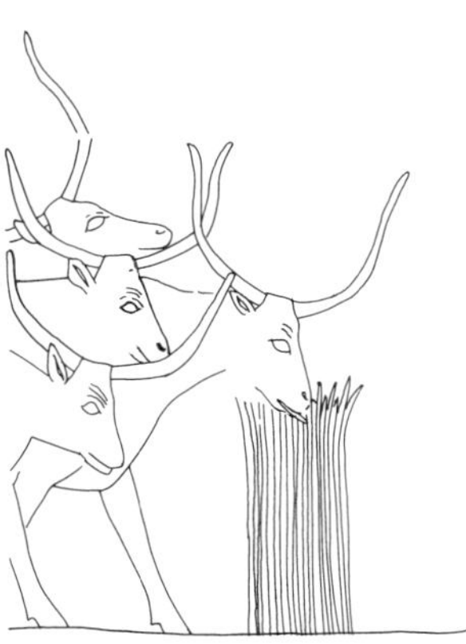

Fig. 7-6 (IB31)

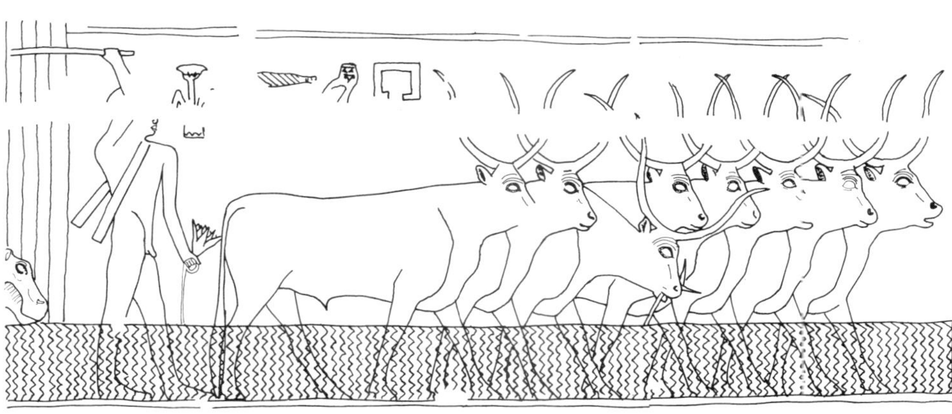

Fig. 7-7 (IB32)

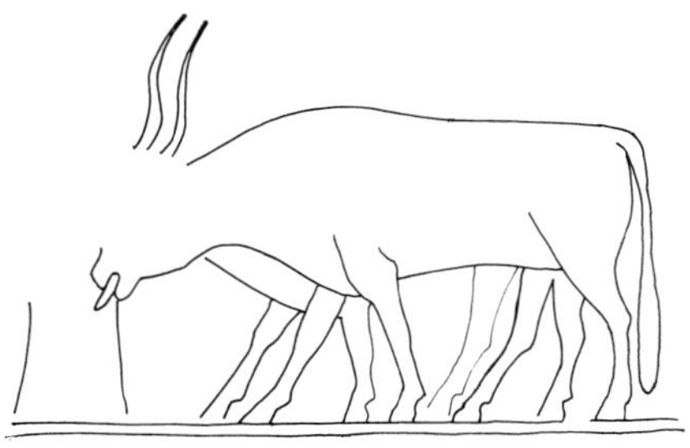

Fig. 7-8 (IB34)

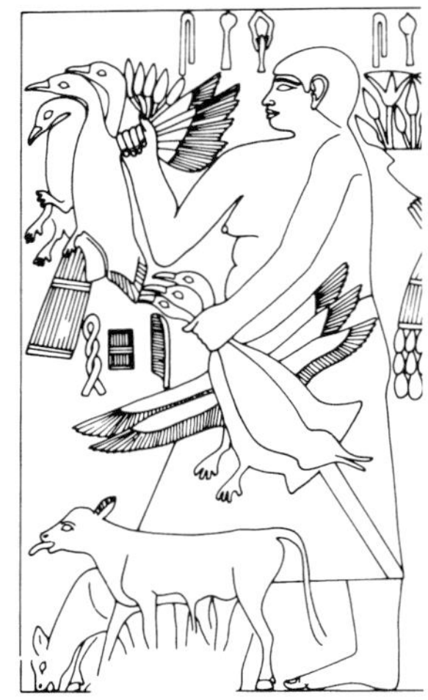

Fig. 7-10 (IB44)

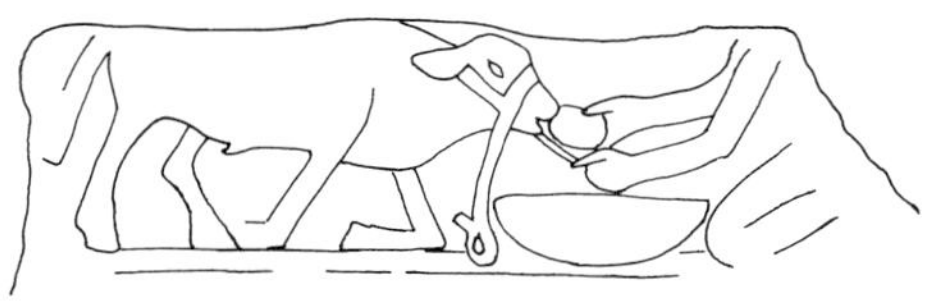

Fig. 7-9 (IB27)

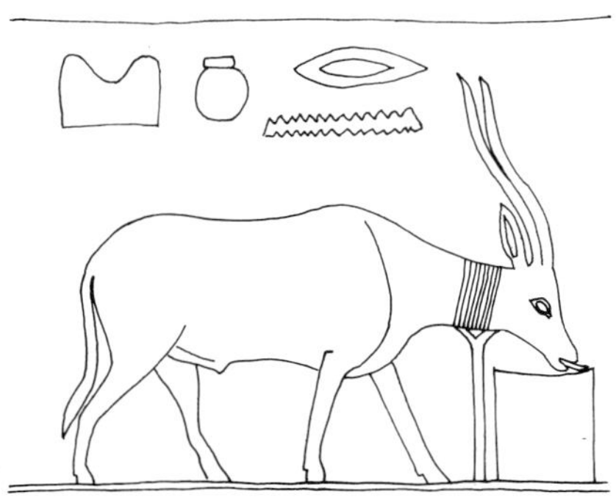

Fig. 7-11 (IB48)

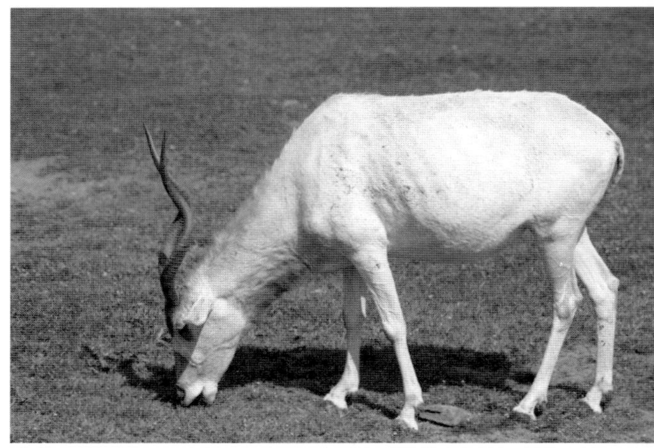

Pl. 7C Addax grazing
© iStockphoto.com/Jan Gottwald

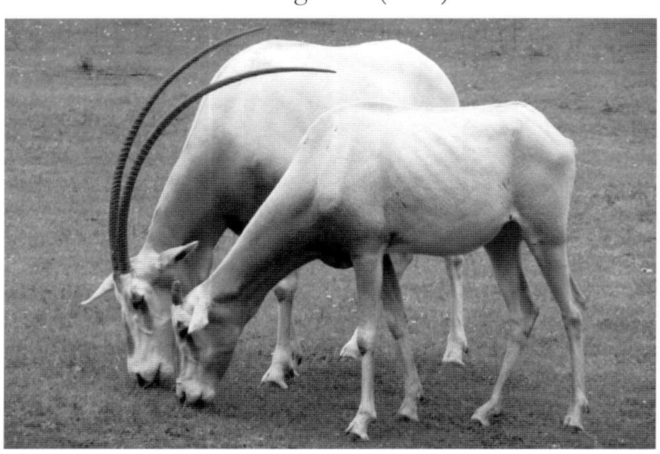

Pl. 7D Scimitar oryx grazing
© iStockphoto.com/Dariusz Urbanczyk

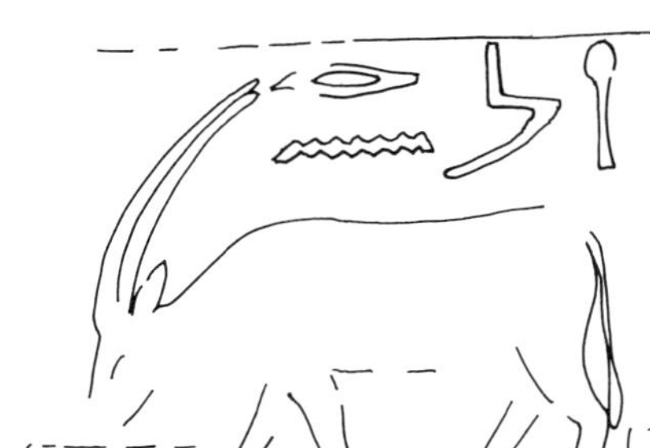

Fig. 7-12 (IB51)

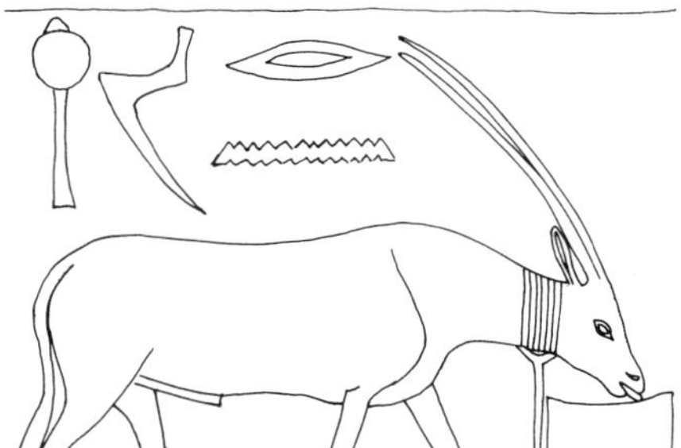

Fig. 7-13 (IB53)

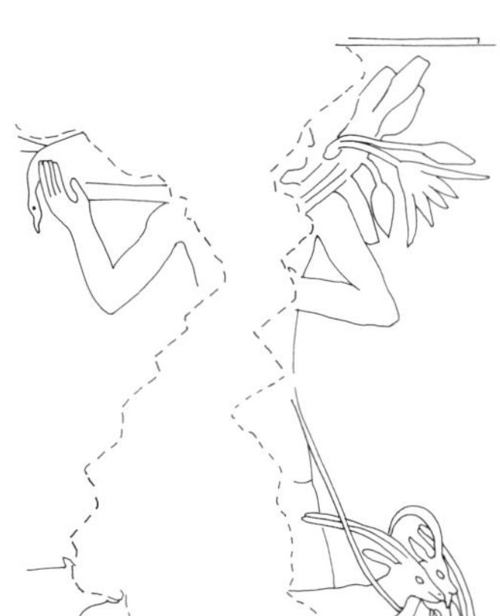

Fig. 7-14 (IB54)

(see pl. 7C), however they will also travel great distances to feed upon new shoots that emerge after rain. The captive addax depicted in the tomb *mrrw-k3.j: mrj* may thus have been fed a mixture of hay and fresh grasses.

Scimitar oryx

ṯjj (IB49-52: Saqqara, V.8-9)
mrrw-k3.j: mrj (IB53: Saqqara, VI.1M-L)
ptḥ-špss II (IB54: Saqqara, VI.3-4)

Three tombs were found to include depictions of captive scimitar oryx while eating. In the tomb of *ṯjj* (IB49 to IB52; see fig. 7-12) and that of *mrrw-k3.j: mrj* (IB53; see fig. 7-13), the animals feed from troughs. In each case, the oryx stretch their heads forward and down, but in the tomb of *mrrw-k3.j: mrj*, sufficient detail remains to show that the animal also protrudes its tongue, with which it touches the top of the manger. As noted above (see *Addax*), antelope species use their highly mobile tongues when feeding and so this depiction is feasible. Like addax, oryx are desert herbivores that ingest perennial grasses, herbs, and shrubs (see pl. 7D). In captivity today, they are usually fed hay and grasses.

The final depiction appears to show an oryx licking the ground (IB54; see fig. 7-14). The drawing is somewhat clumsy, however, and may not accurately reproduce the original bas-relief as the tongue seems too wide for the animal's mouth. Nevertheless, an ingestive behaviour may be considered for, as noted above (see *Cattle*), herbivores will sometimes lick the substrate or eat dirt when their diet is deficient in essential salts and minerals.

Dorcas gazelle

mrrw-k3.j: mrj (IB55: Saqqara, VI.1M-L)
ꜥnḫ-m-ꜥ-ḥr: zzj (IB56-59: Saqqara, VI.1L-2E)
mḥw (IB60: Saqqara, VI.2M-3?)

Only one image was found to clearly represent a dorcas gazelle while feeding. In the tomb of *mrrw-k3.j: mrj* (IB55; see fig. 7-15), a tethered gazelle touches the upper surface of a food trough with its long and tapered tongue, suggesting that it is pulling foliage into its mouth. Dorcas gazelle graze upon a variety of grasses and bushes, and are particularly fond of the leaves, flowers, and seeds of acacia trees (*Acacia* sp.), which they will sometimes browse by standing on their hind legs.[21]

The remaining images (IB56 to IB60; see fig. 7-16) depict small groups of gazelle being presented in offering scenes in which one member leans forward to either lick the substrate or touch it with its muzzle. Dorcas gazelle do not engage in geophagy as frequently as other species as they are able to tolerate extremely arid conditions and are less inclined than other animals to lose salts and minerals. During the summer months, however, when very little foliage is available, this species will sometimes unearth and eat bulbs;[22] they consequently appear to mouth at the dirt (cf. pl. 7E). The Egyptian images possibly represent this curious behaviour, but it seems unlikely in the context of offering scenes.

Nubian ibex

ṯjj (IB61: Saqqara, V.8-9)
mrrw-k3.j: mrj (IB62: Saqqara, VI.1M-L)

Two images from Saqqara depict Nubian ibex feeding. The earlier of the two scenes is poorly preserved (IB61; see fig. 7-17), but the presence of scimitar oryx feeding from mangers on the same register (IB49 to IB52; cf. fig. 7-12) suggests that this animal may be similarly engaged. In a later image from the tomb of *mrrw-k3.j: mrj* (IB62; see fig. 7-18), a tethered ibex stands before a manger and protrudes its tongue so that it touches a rectangular trough before it. Although no food items are drawn, its behaviour is consistent with feeding, as ibex use their tongues to pull food

21 Yom-Tov, Y. (1995). "Gazella dorcas", *Mam. Sp.* (491), 4.

22 Satz, D. and Ward, D. (2000). "Responding to a three-pronged attack: Desert lilies subject to herbivory by dorcas gazelle", *Plant Ecol.* 148, 127-138.

items into their mouths. Nubian ibex inhabit cliffs and rocky terrain; their diet consists mainly of mountain vegetation (see pl. 7F), but during the spring, they will descend into wadis to graze upon new desert grasses. Ibex were thus likely fed grasses in captivity.

Goats

nb.j-m-ꜣḫtj (IB63: Giza, IV.6-V.1)
nfr/kꜣ-ḥꜣ.j (IB63a-67: Saqqara, V.6)
nj-ꜥnḫ-ẖnmw/ẖnmw-ḥtp (IB68: Saqqara, V.6L-7)
ḥtp-ḥr-ꜣḫtj (IB69-72: Saqqara, V.6-8E)
ꜣḫtj-ḥtp (IB73: Saqqara, V.6-8E)
ꜣḫtj-mrw-nswt (IB74: Giza, V.9-VI.1)
mrrw-kꜣ.j: mrj (IB75-78: Saqqara, VI.1M-L)
mrrj (IB79: Saqqara, VI.2E?)
nb-kꜣw-ḥr: jdw (IB80-83: Saqqara, V.9 r/u VI.7)

Most ruminants graze almost exclusively on grass, but goats browse ('nibble') on a wide range of plant foods, including weeds, shoots, shrubs, leaves, bark, and tree branches (see pl. 7G). They bite at foliage by gripping it between their lower incisors and a plate in their upper jaw and then tearing. Goats are extremely agile animals; they climb well and can scale fences up to a height of 1.5 m.[23] They often browse on the leaves and bark of shrubs by standing on their hind legs, and will climb into the canopies of trees - jumping from branch to branch - to eat foliage that would otherwise be out of reach (see pl. 7H). Long-legged breeds will bend their forelegs and kneel to graze, if necessary.[24]

Nine tombs were found to accurately reproduce the browsing behaviour of goats. The tomb of *ḥtp-ḥr-ꜣḫtj* includes animals nibbling on both short and tall grasses (IB69 to IB71; see fig. 7-19), but the remaining examples all depict goats feeding on trees. The earliest of these occurs in the 4th dynasty Giza tomb of *nb.j-m-ꜣḫtj* (IB63), but few details can be discerned due to its fragmentary state. The mid-5th dynasty tomb of *nfr/kꜣ-ḥꜣ.j*, however, displays five separate scenes of browsing. Four of these images show the animals standing on their hind legs on either side of unidentified bushy trees (IB64 to IB67), their forelegs resting upon lower branches as they stretch their heads forward and up to reach the foliage. In one scene (IB66; see fig. 7-20), a buck steps on the back of his companion in his eagerness to reach the leaves,[25] and in another (IB65), a kid takes advantage of its rearing mother's exposed belly to suckle from her (see fig. 11-18 in Chapter 11). Herders assist the goats by using hooked poles to bring the topmost branches down within reach of their animals (see fig. 7-20). This tool-use is repeated in the subsequent tomb of *nj-ꜥnḫ-ẖnmw/ẖnmw-ḥtp* (IB68). As goats can climb, however, such help would not always have been necessary. This behaviour is acknowledged in a delightful scene from the Giza tomb of *ꜣḫtj-mrw-nswt* (IB74; see fig. 7-21) in which goats not only rear up on the ground, but also scramble amongst the branches of two trees. One animal appears to either protrude its tongue or take a leaf into its mouth from an area of branch where the foliage has already been stripped.

Other depictions of climbing are more difficult to interpret. Examples from the tombs of *nfr/kꜣ-ḥꜣ.j* (IB63a and IB66; see fig. 7-20), *ḥtp-ḥr-ꜣḫtj* (IB72), *ꜣḫtj-ḥtp* (IB73), and *nb-kꜣw-ḥr: jdw* (IB80, IB82, and IB83; see fig. 7-22), show goats nibbling at the upper branches of trees while they stand or kneel upon separate registers or baselines. As they are not positioned within the canopies, it is not clear whether they should be perceived as climbing. When discussing similar scenes, Luise Klebs noted that they sometimes occur in conjunction with wood-chopping activities and suggested that the images might actually represent goats browsing from trees *after* they had fallen.[26]

---

23 Thear-Fraser, *Complete Book*, 82.

24 Kilgour-Dalton, *Livestock Behaviour*, 85.

25 Davies et al. (*Saqqara Tombs i*, 13) describe a pair of rearing goats in the tomb of *mrrj* (IB79, left) as 'copulating', but goats do not mate in this manner (see Chapter 10). It is more likely that they are stepping upon one another in their attempts to reach the foliage.

26 See Klebs, L. (1914). "Die Tiefendimension in der Zeichnung des alten Reiches", *ZÄS* 32, 24. The 6th dynasty scenes discussed by Klebs occur in tomb No. 2 (*LD ii*, pl. 108) and tomb No. 14 (*LD ii*, pl. 111) at Zawyet el-Amwat. In tomb No. 2, wood-chopping occurs on the register directly below that of the browsing goats, while in tomb No. 14, felling appears on the same register as the feeding animals. It should be noted that Harpur (*Decoration*, 110-111, fig. 201) states that the dynasty V.2 tomb of *sḫm-kꜣ-rꜥ* at Giza is the earliest in which a combination of browsing goats and tree-felling occurs, however the fragmentary image that she interprets as a feeding goat is labelled, *jdt msj* 'Cow/female giving birth' (see

According to this argument, the trees should be perceived as lying away from the viewer, and the goats, on separate baselines, as standing or lying down on the ground beside the felled timber. Heinrich Schäfer noted, however, that felling trees is not always associated with goats feeding on sub-registers. He was also disturbed by the lack of any indication on the trees that they had been cut down. He felt, instead, that goats feeding from upper registers were simply an artistic attempt to produce an aesthetically pleasing image - to accommodate the height of trees projecting into upper registers by placing goats symmetrically on either side of the canopy. After witnessing a photograph of goats climbing in a tree in Morocco, however, Schäfer acknowledged that some depictions might also represent this behaviour.[27]

Two examples from the tomb of *nb-k3w-ḥr: jdw* (IB80 and IB82; see fig. 7-22) appear at first glance to support Klebs' interpretation as they depict goats both on sub-registers and within the canopies of trees that are in the process of being felled. In both cases, a woodchopper hurls his axe against the trunk of a tree, which leans to the right as goats feed on the canopy from an upper register and/or individual baselines. Two goats are also pictured without ground-lines and 'float' within and to the right of the foliage (IB80; also IB83; see fig. 7-22). The behaviour of these animals, in particular, would seem to indicate that both the tree and the goats are situated on the ground.

I propose an alternative explanation. Rather than representing depth, I suggest that the *nb-k3w-ḥr: jdw* scenes may instead depict a contraction of time in which separate temporal events are presented within the same image. The order of these events might therefore be: first the trees are felled, and then the goats feed upon them (thus agreeing, temporally at least, with Klebs). The position of the goats both within the canopy and on sub-registers would thus represent their activities on the ground. An alternative sequence is also possible, however: first the goats feed in the canopies, and then the trees are cut down. This latter interpretation views the animals browsing from sub-registers as climbing, but with their images removed from the canopy for clarity and also, as Schäfer suggested, to create a balanced composition. 'Floating' goats, according to this explanation, would represent animals jumping among the tree branches.

It is interesting to note that, along with those from the tomb of *nb-k3w-ḥr: jdw*, images of tree-cutting also occur on the same wall as the browsing goats in the tomb of *nfr/k3-ḥ3.j* and on the same register as those in the tomb complex of *nj-ʿnḫ-ẖnmw/ẖnmw-ḥtp*. An association between woodchopping and goats browsing has also been noted in a number of later tombs outside of the Memphite area.[28] The concurrence probably indicates that goats were allowed access to trees that had been slated for cutting. This would make both economic and practical sense as, by allowing the animals to remove much of the foliage and bark from trees before and also after felling, the wood-cutters' task of stripping the fallen logs would have been reduced. The trees upon which the goats are shown grazing cannot be identified with certainty, but possible species include tamarisk, sycamore, and acacia. The last, which was used extensively for woodworking in ancient Egypt, is also commonly used as a goat fodder in parts of Africa today.[29]

Finally, in the tomb of *mrrw-k3.j: mrj* (IB75 to IB78), four screw-horned goats are depicted being hand-fed. Each recumbent goat is labelled, *rn wnḏw* 'young goat' and is tended by a herdsman who squats before his charge and holds stalks of foliage or bread dough for them to feed upon.[30] Two of the goats appear to either bite or rest their muzzle upon the wrist of their handler (IB77 and IB78). This may simply be the result of clumsy placement by the artist. As goats are very inquisitive

Hassan, *Giza iv*, fig. 57). The animal stands with its head drooped and, although it is located beside a small shrub, it faces in the opposite direction and does not make contact with it. Consequently, I have not considered this image here.

27 Schäfer, *Principles*, 253-255 and fig. 273.

28 Vandier, *Manuel v*, 86-92.

29 Komwihangilo, D.M., Goromela, E.H. and Bwire, J.M.N. (1995). "Indigenous knowledge in utilization of local trees and shrubs for sustainable livestock production in central Tanzania", *LRRD* 6(3), 1-6.

30 Two similarly labelled goats also occur in the tomb of *mrrj*, but although herdsmen place their hands to the animals' mouths, no food is depicted (see Davies et al., *Saqqara i*, pl. 6).

animals, however, and will often investigate unfamiliar objects by nibbling them, the scenes may also represent exploratory behaviour.[31]

Sheep

nb.j-m-3ḫtj (IB84: Giza, IV.6-V.1)
wr-jr.n-ptḥ (IB84a: Saqqara, V.3-5?)
nfr/k3-ḥ3.j (IB85: Saqqara, V.6)
ḥtp-ḥr-3ḫtj (IB86: Saqqara, V.6-8E)
3ḫtj-ḥtp (IB87: Saqqara, V.6-8E)
jrj-n-k3-ptḥ (IB88-89: Saqqara,V.6-9)
sḫm-ʿnḫ-ptḥ (IB90: Saqqara, V.8-9)
ṯjj (IB91: Saqqara, V.8-9)
nfr-jrt-n.f (IB92: Saqqara, V.8-9)
k3.j-m-nfrt (IB93: Saqqara, V.8-9?)
nfr I (IB94: Giza, V.9-VI.1)
sḫm-k3.j (IB95: Giza, V.9-VI.1?)
k3.j-m-ḥzt (IB96: Saqqara, V.9-VI.1?)
mḥw (IB96a: Saqqara, VI.2M-3?)

Feeding animals are sometimes included among depictions of sheep treading seed into newly ploughed fields (see Chapter 5). Two types of behaviour occur: Some sheep lean well forward so that their muzzles touch plants, grains that have fallen on to the ground, or the ground itself,[32] while others place their snout upon the hand of a herdsman who, while sowing, offers them seeds. Sometimes, both of these behaviours occur in the same scene, expressed by different animals within the one flock.

Sheep are grazing animals that feed primarily upon grasses and other plants. If only poor pastures are available, their diet can be supplemented with hay, straw, and grains. Sheep use their upper lips and lower incisors to eat (see pl. 7I). Unlike cattle, which pull leaves into their mouths with their tongues, sheep (like goats) bite at foliage by pressing it between their bottom teeth and a horny plate in their upper jaw.[33] Consequently, their tongues do not protrude during grazing as they do in cattle.[34] Their upper lip is split and extremely manipulable, enabling them to nibble grass very close to the ground and to graze selectively, avoiding some plants in favour of others.

In treading scenes, sheep are driven across ploughed fields. Therefore, they do not graze as they would normally as, apart from the odd tussock, the only food available is seeds - either those that have already been thrown onto the ground, or those offered by a herdsman from a seed bag. In order to pick up small grains from either the ground or a hand, sheep would need to place their muzzles directly on top of the food and then bite, using their lips to guide the seeds into their mouths (cf. pl. 7I). The earliest example of feeding, found in the tomb of *nb.j-m-3ḫtj* (IB84; see fig. 7-23), appears to show both grazing and hand-feeding, for while one sheep at the head of the line rests its closed muzzle in the palm of the sower, another leans forward deeply to examine the ground. This scene is essentially reproduced in the subsequent tombs of *wr-jr.n-ptḥ* and *nfr/k3-ḥ3.j*, although in these, the second animal bends over to either place its muzzle upon a tuft of grass (IB84a; see fig. 7-24) or a small pile of fallen grain (IB85; see fig. 7-25). In later tombs, only one of these two behaviours is usually included: in that of *ḥtp-ḥr-3ḫtj* (IB86)[35], *nfr I* (IB94), and *k3.j-m-ḥzt* (IB96),[36] only hand-feeding is depicted, while in that of *3ḫtj-ḥtp* (IB87), *nfr-jrt-n.f* (IB92), *sḫm-k3.j* (IB95), and (possibly) *mḥw* (IB96a), one or more sheep lean forward to apparently feed on the ground.

---

31 It is also possible that these scenes represent a contraction of time, similar to that proposed for recumbent cattle in Chapter 4.

32 Animals that simply bend their heads to peer at the ground have not been considered here.

33 Müller, *Sheep*, 7.

34 Hafez, E.S.E., Cairns, R.B., Hulet, C.V. and Scott, J.P. (1969). "The behaviour of sheep and goats" in Hafez, *Behaviour of Domestic Animals*, 302.

35 Mohr (*Hetep-her-akhti*, 80) suggests that the sower is a child offering salt. Although sheep require salt in their diet, similarities with comparable scenes suggest that he is instead holding seeds.

36 Note that McFarlane (*Kaiemheset*, 31 and pl. 44) identifies these animals as donkeys, however the posture of the herdsman to the right of the group, who bends his left arm while holding out his right, resembles closely that of hand-feeding sowers in sheep treading scenes.

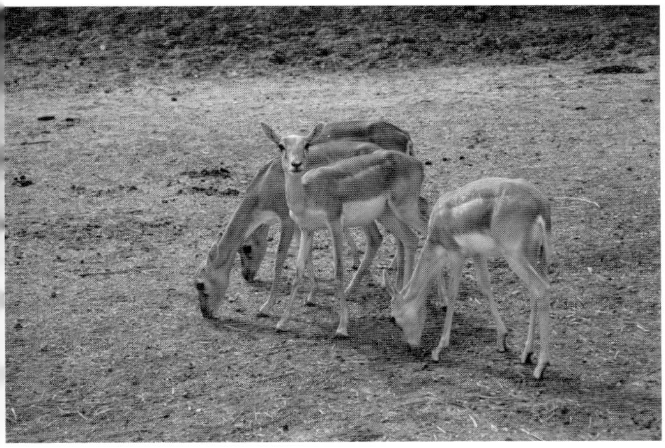

Pl. 7E Gazelle grazing
© iStockphoto.com/Charles Gibson

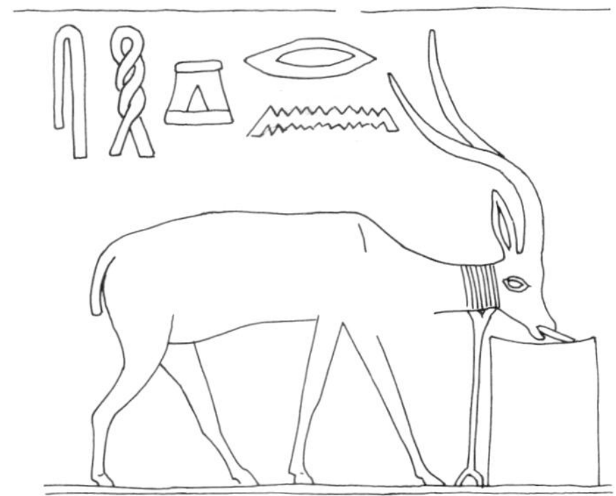

Fig. 7-15 (IB55)

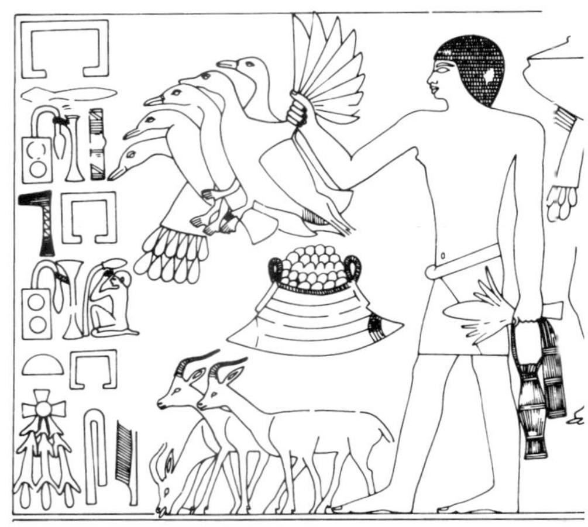

Fig. 7-16 (IB56)

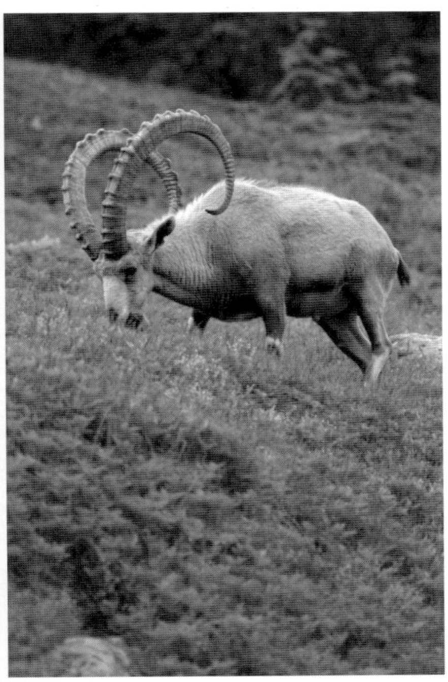

Pl. 7F Nubian ibex grazing
© iStockphoto.com/Stefan Ekernas

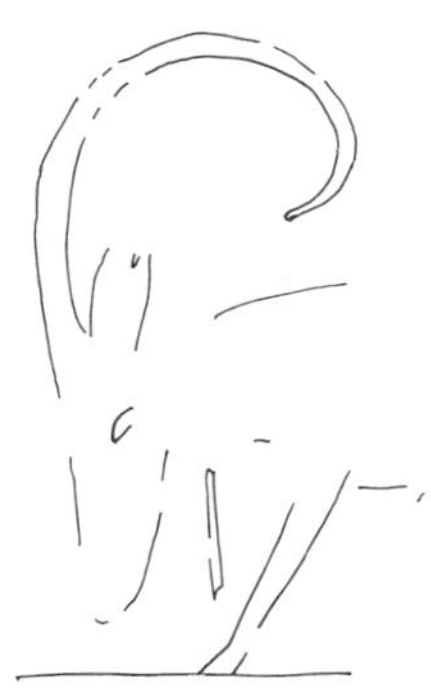

Fig. 7-17 (IB61)

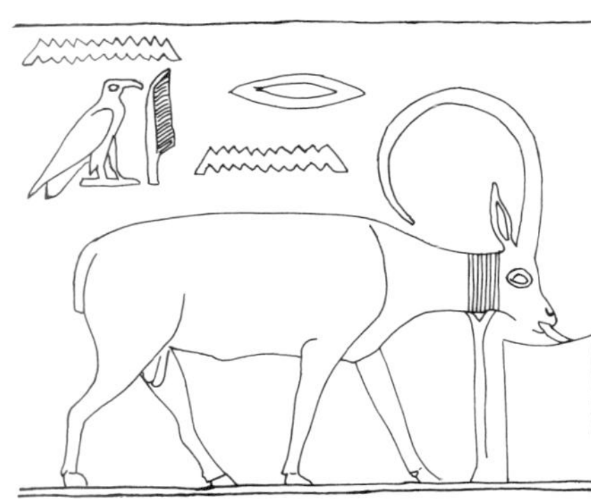

Fig. 7-18 (IB62)

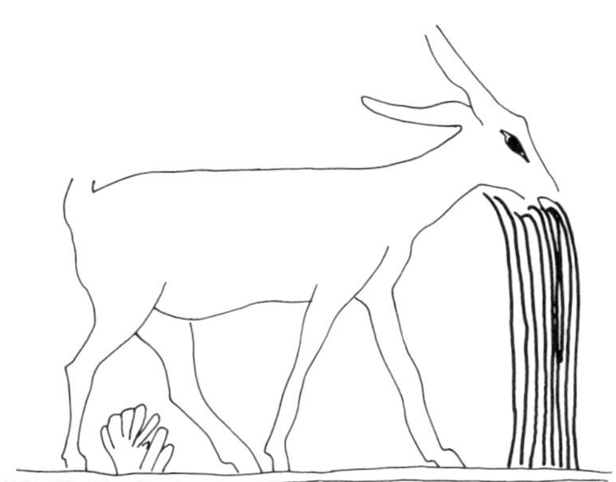

Fig. 7-19 (IB70)

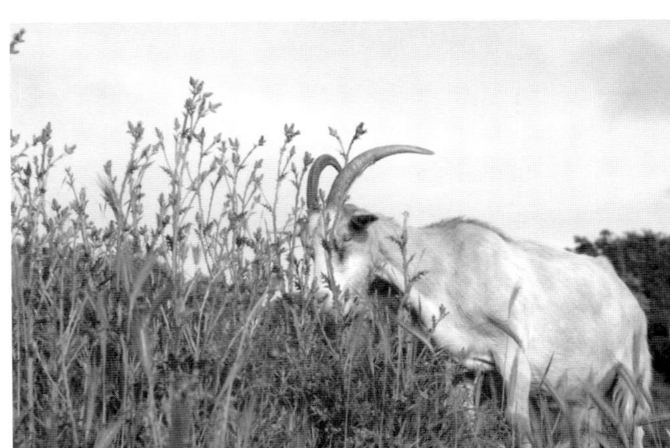

Pl. 7G Goat browsing
© iStockphoto.com/Eliandric

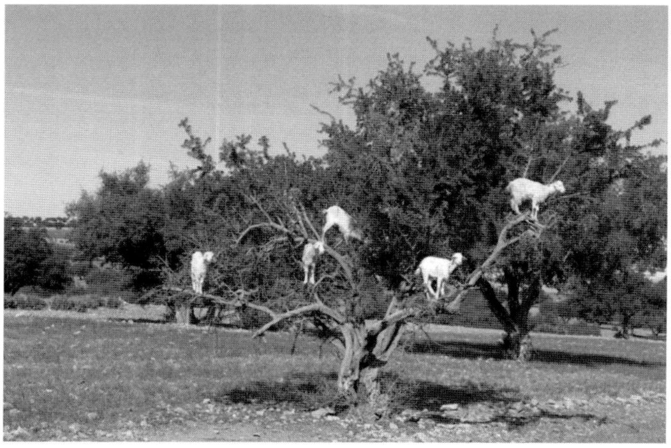

Pl. 7H Goats browsing in a tree
© iStockphoto.com/Mosaik Photography

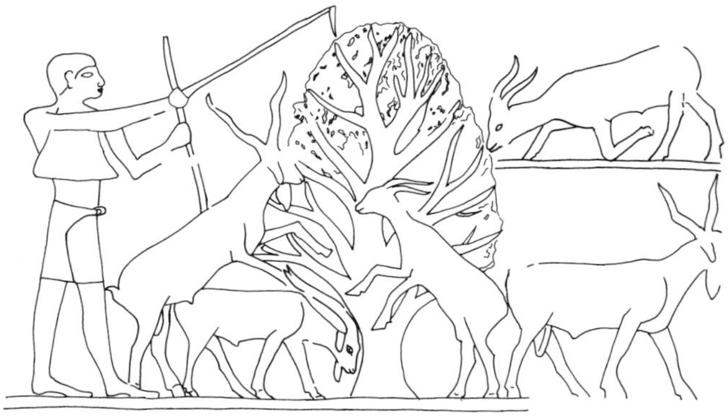

Fig. 7-20 (IB66)

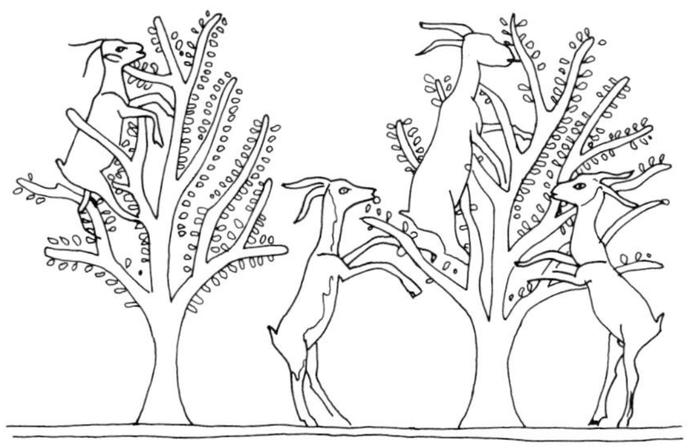

Fig. 7-21 (IB74)

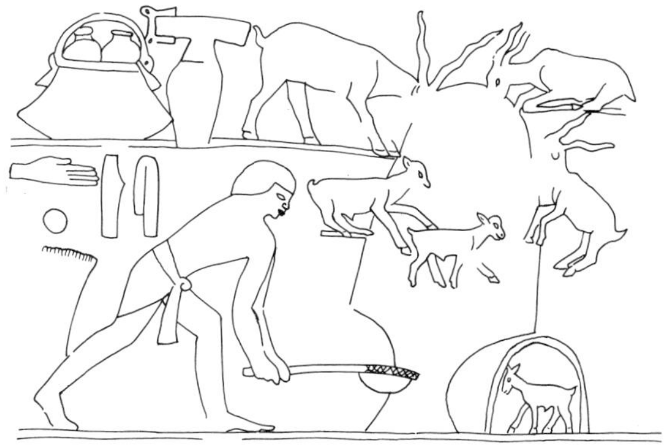

Fig. 7-22 (IB80)

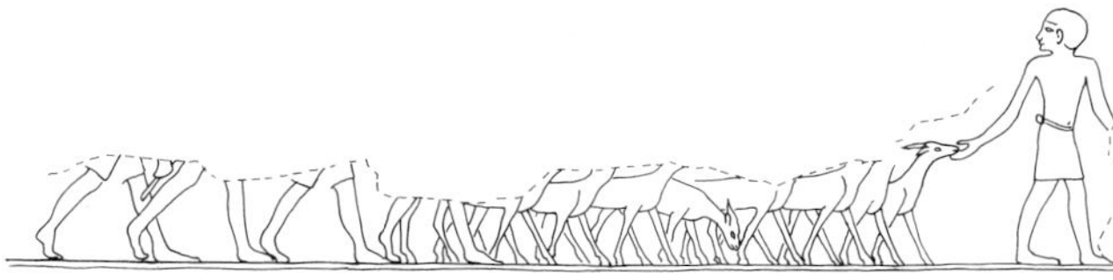

Fig. 7-23 (IB84)

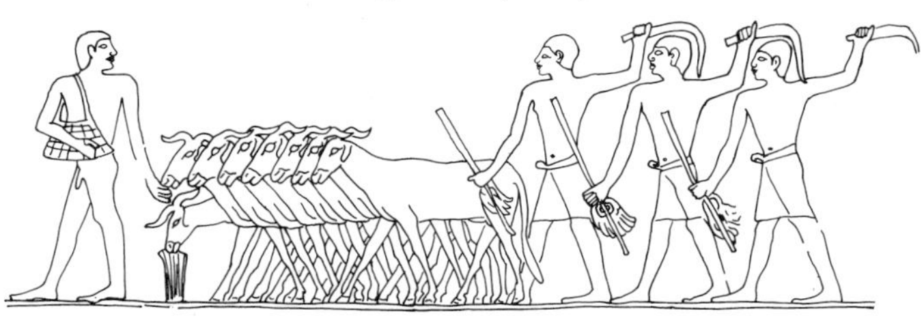

Fig. 7-24 (IB84a)

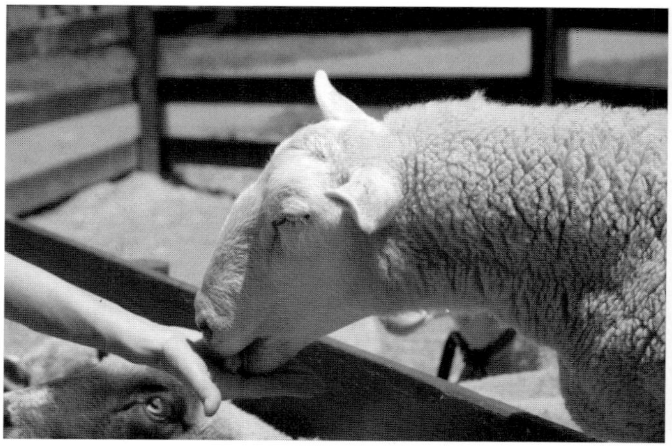

Pl. 7I Sheep eating from hand
© iStockphoto.com/Mariada Design

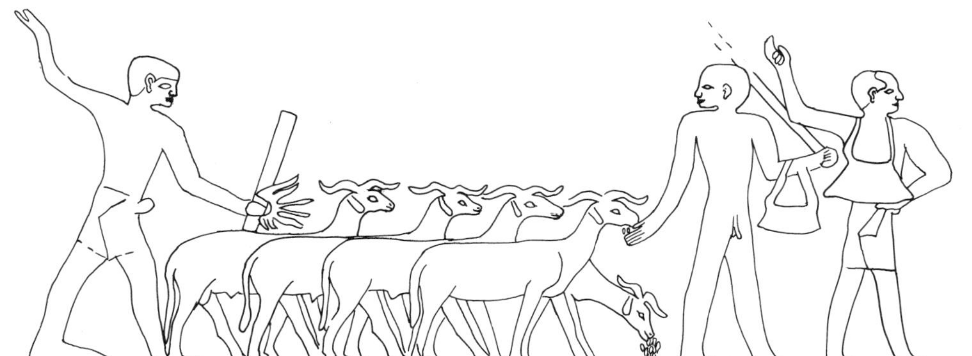

Fig. 7-25 (IB85)

The tomb of *jrj-n-k3-ptḥ* shows feeding in both a seed treading (IB88) and a threshing scene (IB89).[37] In the latter, one animal bends its head deeply, but as the image is badly preserved, it cannot be determined whether it is ingesting grains from the threshing-floor. In the treading scene, the mouth of the sheep is open, as though biting at the sower's hand, a detail that is also repeated in the later tomb of *sḫm-ʿnḫ-ptḥ* (IB90; see fig. 5-3). Here, though, the muzzle is positioned in such a way that the space between the sower's thumb and fingers gives the impression that the animal's tongue is protruding. This was possibly the artist's intention. Feeding sheep do not appear to open their mouths in any other examples, but many of the scenes are too poorly preserved to verify this detail. In the tomb of *ṯjj* (IB91), however, the animal's upper jaw projects beyond its lower in an accurate representation of the dexterous action of their lips. Finally, it should be noted that although no grains are ever depicted in the palms of the sowers, in the late 5th dynasty tomb of *k3.j-m-nfrt* (IB93; see fig. 5-2), a shower of seeds falls directly from the herdsman's open hand, as though knocked aside by the animal's searching muzzle.

Egyptian mongooses

mrrw-k3.j: mrj (IB97: Saqqara, VI.1M-L) *ḥzj* (IB98: Saqqara, VI.1L-2E)

Although many examples can be found of Egyptian mongooses hunting for prey (see Chapter 8), only two reliefs were found to illustrate them while feeding. Both of these occur in marsh scenes from Saqqara and each shows the animals eating a fish.

The earlier of the two images comes from the tomb of *mrrw-k3.j: mrj* (IB97; see fig. 7-26) and depicts a large mongoose lying beside a water course against a backdrop of papyrus stalks, its elongated body tilted upward slightly as it places both of its forepaws on what appears to be a mullet. The mongoose bites at the fish's head, just below its pectoral fin. There is much disagreement about the identification of this animal, with many scholars suggesting that based upon physical features, it is instead an otter (*Lutra* sp.).[38] Furthermore, whereas mongooses are usually represented hunting for birds, this animal eats a fish, the primary food of otters. The creature is also situated close to the water's edge, rather than up in the thicket, the usual location for depicted mongooses.

It should be noted, however, that the morphological attributes that the *mrrw-k3.j: mrj* image shares with otters are equally characteristic of Egyptian mongooses - namely, a lithe body, short limbs, a long tapered tail, and a short muzzle with small, flat ears that sit tightly against the skull. Fragmentary skeletal remains of spotted-necked otters have been identified at a late Palaeolithic site at Makhadma in middle Egypt and the bones of an unidentified otter species dating from the same era have been found at Catfish Cave in Egyptian Nubia.[39] Nevertheless, no evidence exists that indicates unambiguously that any otter species lived in the Nile during the historic period (despite an assertion made by Herodotus).[40]

If behavioural, rather than physical, features of the two species are compared, it becomes apparent that the *mrrw-k3.j: mrj* animal is more likely to be a mongoose than an otter. As others have

37 The depiction of sheep in threshing scenes is unusual. During this study, only one other example was located - in chamber A13 of the tomb of *mrrw-k3.j: mrj*. See Duell, *Mereruka ii*, pl. 169.

38 For example, see Keimer, L. (1942). "Quelques représentations rarissimes de mustélidés conservées sur des bas-reliefs de l'ancien empire", *Études* 4, 1-10; Boessneck, *Tierwelt*, 51. For a detailed discussion, see Houlihan, P.F. (1996). "On Herodotus (II, 72) and the question of the existence of the otter (*Lutra* sp.) in ancient Egypt", *GM* 153, 45-53 and Evans, L. (2010). "Otter or mongoose? Chewing over the evidence" in Woods et al., *Egyptian Culture and Society*, 119-129. Some authors maintain that other accepted representations of Egyptian mongooses are also otters. See Aufrère, S.H. (1998). "La loutre, le chat, la genette et l'ichneumon, hôtes du fourré de papyrus: Présages, prédateurs des marécages et croyances funéraires", *DE* 41, 7-28.

39 van Neer, W. (2002). "Archaeozoological evidence for the former presence of spotted-necked otter (*Lutra maculicollis*) in Egypt", *Bonn. Zool. Beitr.* 50(3), 199-209.

40 "There are also otters in the river, and Egyptians hold these sacred too" (II, 76). See Blanco-Tolbert Roberts, *Herodotus The Histories*, 97.

noted,[41] like otters, mongooses also include fish in their diet;[42] they swim well and have even been found raiding fish traps.[43] After catching a fish, however, the behaviour of the two species differs. Spotted-necked otters reliably eat their prey tail-first, and then discard the head (especially when the fish is large).[44] Egyptian mongooses, on the other hand, begin with a bite to the neck or the head, and then eat their prey headfirst (see pl. 7J).[45] The behaviour of the *mrrw-k3.j: mrj* animal, which clearly focuses its attention on the head region of the mullet, is thus more characteristic of mongooses, than spotted-necked otters.

A second fish-eating animal, which appears in the tomb of *ḥzj* (IB98; see fig. 7-27), is also a mongoose and not an otter.[46] Like the *mrrw-k3.j: mrj* animal, this creature feeds upon a fish headfirst, which it holds in place with its forepaws. The image may have been copied from that in the earlier tomb, but here the artist has provided more mongoose-like features. This time the animal is situated within the thicket,[47] where mongooses are usually depicted (although it lies down upon a papyrus stalk, rather than walks along it). The marsh scene in which the image occurs also includes another mongoose and two genets hunting for birds in the usual manner. The animal in question bears a strong physical resemblance to the mongoose; as genets and mongooses frequently occur in pairs in marsh scenes, I suggest therefore that the fish-eating animal is the intended 'mate' of the other.

Vervet monkeys

k3.j-nfr (IB99: Giza, V.1-3)
jj-nfrt: š3.n.f (IB100: Saqqara, V.9)
nj-k3w-jzzj (IB101-102: Saqqara, VI.1M)
ʿnḫ-m-ʿ-ḥr: zzj (IB103-104: Saqqara, VI.1L-2E)
mḥw (IB105: Saqqara, VI.2M-3?)

Seven depictions were found of vervet monkeys eating. This behaviour was identified based upon the position of the animals' arms and hands, and the apparent availability of food items. In all but one example, the monkeys squat upon the head or shoulders of a human caretaker (often a dwarf), thereby freeing their hands. In each case, one paw is held to their mouth while the other sometimes touches their handler's head, possibly to keep their balance. In accordance with artistic convention, the raised hand is always furthermost from the viewer. In five of the seven examples, an object held in the raised hand is brought to the animals' mouths. Most of the monkeys are leashed to collars around their necks (see figs. 7-28 and 7-29), although the vervet in the early 5th dynasty tomb of *k3.j-nfr* (IB99) also wears a belt around its waist.[48] A final example depicts a vervet sitting under the chair of the tomb-owner, *jj-nfrt: š3.n.f* and his wife (IB100), holding a small bowl in one hand.[49]

Vervet monkeys are omnivores and include fruits, leaves, seeds, flowers, grasses, invertebrates, and occasionally even small vertebrates in their diet. They are also extremely dexterous and use their

---

41 Houlihan, *GM* 153, 47.

42 Haltenorth-Diller, *Field Guide*, 202; Ben-Yaacov, R. and Yom-Tov, Y. (1983). "On the biology of the Egyptian mongoose, *Herpestes ichneumon*, in Israel", *Z. Säugetier.* 48, 43; Kingdon, *East African Mammals iii Part A*, 180.

43 Estes, *Behavior Guide*, 299.

44 *ibid.*, 440; Kingdon, *East African Mammals iii Part A*, 112; Larivière, S. (2002). "Lutra maculicollis", *Mam. Sp.* (712), 4.

45 Estes, *Behavior Guide*, 300; Ben-Yaacov-Yom-Tov, *Z. Säugetier.* 48, 43.

46 Kanawati and Abder-Raziq (*Hesi*, 28) identify this animal as an Egyptian mongoose. Bouvier-Closse (2005) has suggested recently that it is instead an otter, however this identification is based entirely upon its perceived morphological similarity to the animal in the tomb of *mrrw-k3.j: mrj*, which the author accepts as an otter. See Bouvier-Closse, K. (2005). "Une nouvelle représentation de loutre", *GM* 204, 21-36 and also Evans in Woods et al., *Egyptian Culture and Society*.

47 It is important to note that, unlike otters, Egyptian mongooses usually take their prey into dense cover prior to feeding. See Ben-Yaacov-Yom-Tov, *Z. Säugetier.* 48, 39.

48 Early depictions of vervet monkeys from Giza and Saqqara show them leashed to a belt around their waist. Then, in the late 4th to early 5th dynasty, leashing to a collar became more common. A few later images of monkeys show them wearing both a collar and a belt (e.g., that in the tomb of *nw-nṯr* at Giza). During the New Kingdom, both baboons and vervet monkeys were again leashed to belts.

49 Monkeys are often depicted under the chairs of women. See Osborn-Osbornová, *Mammals*, 40.

hands constantly when feeding. Different postures are adopted depending on what food is being obtained. For example, when foraging for termites, which are found on the ground, they stand quadrupedally,[50] but when eating leaves and fruits, they often sit in order to free their hands for reaching and holding items (see pl. 7K). Fruits with inedible skins may be peeled first, which can require both hands,[51] but most plant foods are usually picked up and placed in the mouth with one hand, without manipulating them in any way.[52]

It is impossible to determine what each of the depicted monkeys is ingesting, but the rounded shape of the items they hold suggest that they may be fruits, of which vervets are particularly fond. The small objects in a bowl held by a vervet in the tomb of *nj-k3w-jzzj* (IB101; see fig. 7-28) have this appearance, as do those grasped by the *jj-nfrt: š3.n.f* monkey (IB100). The items in a basket held by a dwarf in the tomb of *ʿnḫ-m-ʿ-ḥr: zzj* (IB103; see fig. 7-29) have been identified as doum nuts,[53] but, as the apple-sized fruits of the doum palm are considerably larger than the depicted objects, this would seem unlikely.[54] A doum nut is a feasible explanation, however, for a large spherical object held to the mouth of a monkey in the tomb of *mḥw* (IB105). Although nothing can be seen in the hand of the *k3.j-nfr* monkey (IB99), and the raised hands of vervets in the tomb of *jj-nfrt: š3.n.f* (IB100) and on a fragment from the tomb of *ʿnḫ-m-ʿ-ḥr: zzj* (IB104) are not preserved, similarities with comparable images suggest that they also depict feeding vervets. All of the Egyptian images thus represent accurately the behaviour of these primates when eating small food items.

Cormorants

mrrw-k3.j: mrj (IB106: Saqqara, VI.1M-L)

A scene from the tomb of *mrrw-k3.j: mrj* (IB106; see fig. 7-30) presents a unique representation of the feeding behaviour of cormorants. In the waters beneath a pair of fishermen's canoes, three birds are depicted standing upon the riverbed, one of which has a fish clenched between its mandibles. Patrick Houlihan, who identified this last animal as a cormorant based upon its general morphology, states that the bird is "near the water's edge",[55] but it is clearly *underwater* as a small fish swims before it and it is surrounded by water lilies (*Nymphaea* sp.). The image thus correctly reproduces the feeding method of cormorants as these birds hunt for fish and crustaceans by submerging fully and then pursuing their prey underwater.[56] The birds swim by moving their feet (not their wings), and use their long, stiff tail as a rudder.[57] Large prey are usually brought to the surface prior to swallowing (pl. 7L), but small fish are sometimes ingested while the bird is still submerged.[58] The Egyptian depiction appears to show the moment of capture, although the cormorant stands upon the riverbed, rather than swims; possibly the artist was unsure how these birds actually caught their prey after they disappeared below the water surface.

The bird situated immediately behind the first may also be a cormorant as it, too, stands underwater and its general morphology matches its companion. It also appears to be foraging, as it bends forward to touch its partially open bill upon the substrate. Although some cormorant species do feed on bottom-dwelling fish, it seems more likely that the artist simply did not know how these birds hunted underwater and so reproduced the terrestrial feeding method of more familiar species (see *Sacred ibis* and *Ducks* below).

---

50 Harrison, K.E. and Byrne, R.W. (2000). "Hand preferences in unimanual and bimanual feeding by wild vervet monkeys (*Cercopithecus aethiops*)", *J. Comp. Psychol.* 114(1), 15.

51 *ibid.*

52 Estes, *Behavior Guide*, 502.

53 Osborn-Osbornová, *Mammals of Ancient Egypt*, fig. 4-35, 40.

54 For a description of doum fruit, see Brewer et al., *Domestic Plants and Animals*, 50.

55 Houlihan, *Birds*, 8.

56 The Egyptians' knowledge of the swimming behaviour of cormorants is reflected in their name for them: *ʿg3w* (or *ʿg3.t*), 'the one that sinks, submerges'. See Wolterman, C. (1991-1992). "On the names of birds and hieroglyphic sign-list G22, G35 and H3", *JEOL* 32, 125-127.

57 Thomson, *Dictionary of Birds*, 796.

58 *ibid.*, 151.

Three species of cormorant have been reported in Egypt: the great cormorant, the shag, and the reed (or long-tailed) cormorant.[59] The first two species are still found in Egypt today, but the reed cormorant, although once prevalent in the Delta and Faiyum, has not been sighted in the country since 1875.[60] As no trace of colour remains on the *mrrw-kꜣ.j: mrj* bas-reliefs, and the images lack any specific plumage details, it would seem impossible to determine what species of cormorant is depicted, however a large white pelican,[61] which stands to the right of these two birds, may offer a clue.

Unlike cormorants, pelicans do not swim underwater and so its presence beside the other fully submerged birds is incorrect. Nevertheless, their method of feeding *does* occur below the water surface. Pelicans catch fish by tipping their body forward and plunging their head underwater (see pl. 7M) in order to scoop prey into a large pouch that hangs below their bill. After they bring their head back to the surface, the pouch is drained by raising their bill vertically; any fish caught within it are then swallowed. Pelicans frequently hunt in groups, swimming beside one another in a coordinated fashion to herd fish.[62]

It has been observed that reed cormorants often forage wherever white pelicans congregate, following them closely and even positioning themselves within the flocks to snatch any fish that might fall as the birds drain their pouches.[63] I suggest that the combination of cormorants with a pelican, a bird that appears rarely in Egyptian art,[64] indicates that the images in the tomb of *mrrw-kꜣ.j: mrj* depict reed cormorants.[65]

Although reed cormorants usually forage alone, they occasionally form flocks of up to forty individuals in order to hunt cooperatively - diving synchronously to drive shoals of fish into shallow water.[66] White pelicans will similarly work together to herd fish. As both of these species appear in a scene in which fishermen are hauling in abundant catches, it is intriguing to ask whether the ancient Egyptians knew about the birds' herding behaviour and possibly took advantage of it to aid their own fishing activities. It is also tempting to see in the juxtaposition of the pelican and the fishermen a possible allusion to both the shape and function of the bird's swollen pouch and the men's pendulous hand-nets.

Herons

ptḥ-ḥtp I (IB107-108: Saqqara, V.8M-L)
kꜣ-gm-nj: mmj (IB109: Saqqara, VI.1M)
ḥzj (IB110: Saqqara, VI.1L-2E)

Although herons appear frequently in tomb scenes, only four images were found to depict them while feeding. These birds hunt and eat a wide range of prey, but their main food source is fish. They hunt primarily in shallow water by either standing motionless or walking slowly, waiting for a fish to swim within reach. Prey are caught by a rapid downward thrust of the beak to either impale it, or grasp it between their mandibles (pl. 7N).[67] Fish with spines (such as catfish) are usually swallowed headfirst.[68]

---

59 Houlihan, *Birds*, 7.

60 Cramp, S. (1977). "Long-tailed cormorant (*Phalacrocorax africanus*)" in Cramp-Simmons, *Handbook i*, 220.

61 Cramp, S. (1977). "White pelican (*Pelecanus onocrotalus*)" in *ibid.*, 227-233.

62 *ibid.*, 230.

63 Whitfield, A.K. and Blaber, S.J.M. (1979). "Feeding ecology of piscivorous birds at Lake St Lucia, Part 3: Swimming birds", *Ostrich* 50(1), 17.

64 During this study, pelicans were also located in a weir-fishing scene in the tomb of *nj-ꜥnḫ-ẖnmw/ẖnmw-ḥtp* at Saqqara. For a description of pelicans in Egyptian art, see Houlihan, *Birds*, 10-13.

65 Houlihan suggested that a nesting bird on the north wall of the chapel of *ṯjj* may also be a reed cormorant, as "this species is a former breeding resident of the Faiyum". He felt, however, that such a conclusion was nevertheless too speculative. See Houlihan, P.F. (1996). "A guide to the wildlife represented in the great swampland scene in the offering-chapel of Ti (no. 60) at Saqqara", *GM* 155, 35.

66 Burton, *Bird Behaviour*, 67.

67 Hancock-Kushlan, *Herons*, 47.

68 McKilligan, *Herons*, 40.

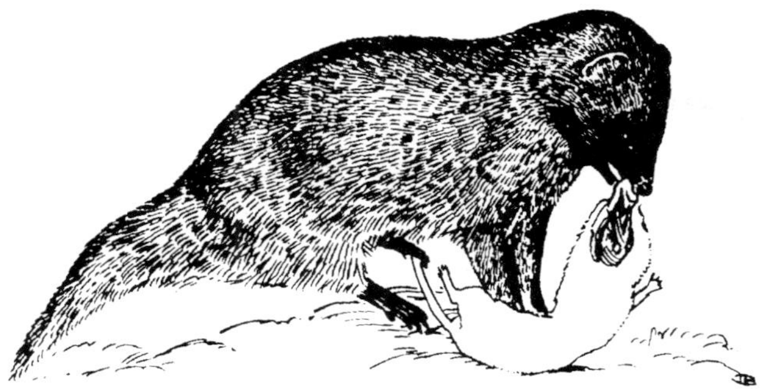

Pl. 7J Egyptian mongoose feeding
Reprinted from: Estes, *Behavior Guide*, fig 19.18
By permission of © University of California Press

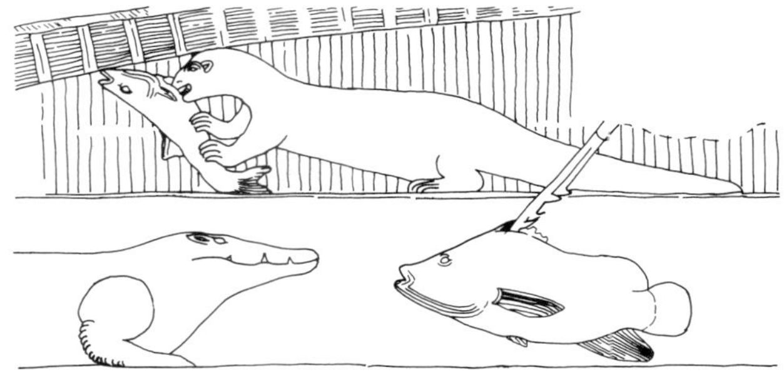

Fig. 7-26 (IB97)

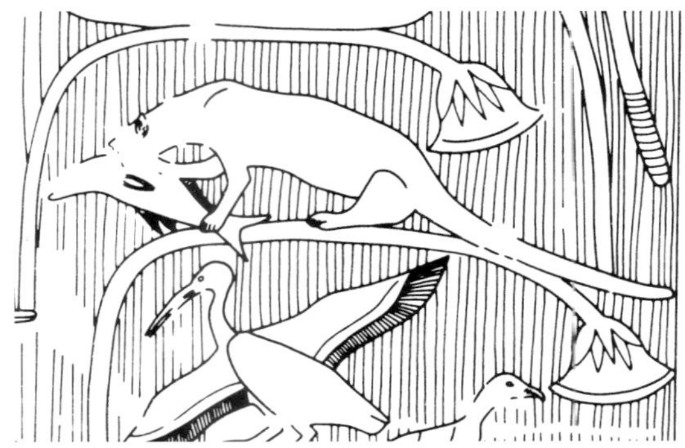

Fig. 7-27 (IB98)

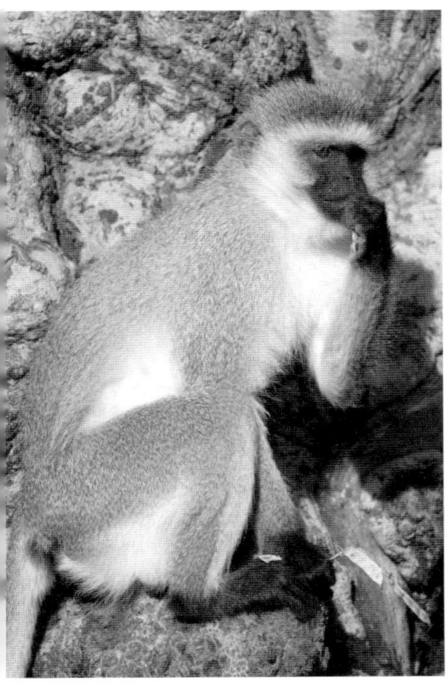

Pl. 7K Vervet monkey eating
© iStockphoto.com/Sebastien Burel

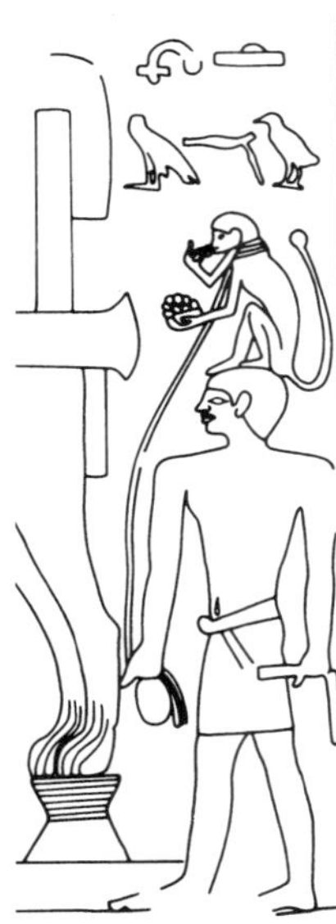

Fig. 7-28 (IB101)

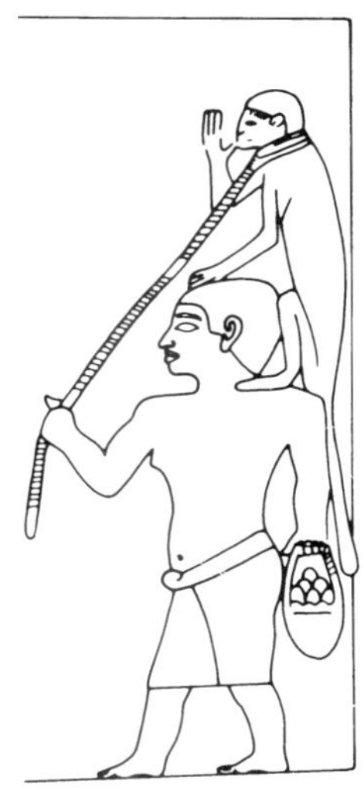

Fig. 7-29 (IB103)

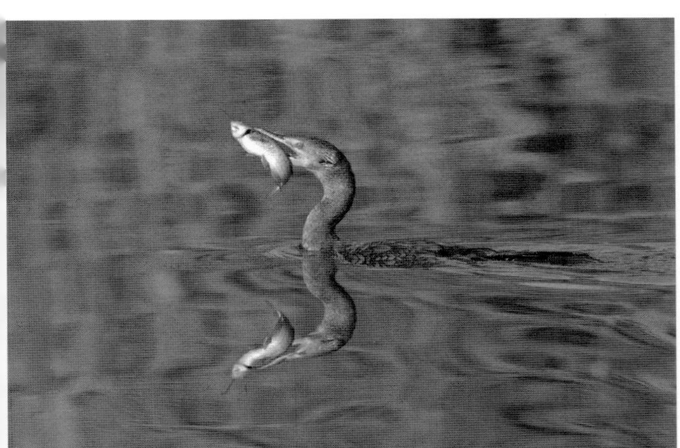

Pl. 7L Cormorant with fish
© iStockphoto.com/Stephen Muskie

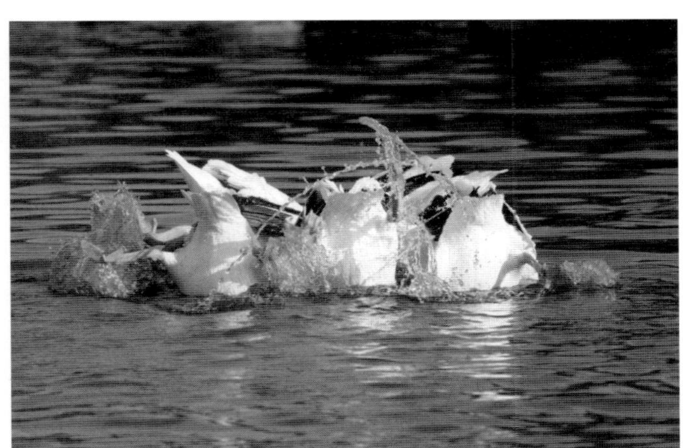

Pl. 7M Pelicans diving
© iStockphoto.com/Igor Karon

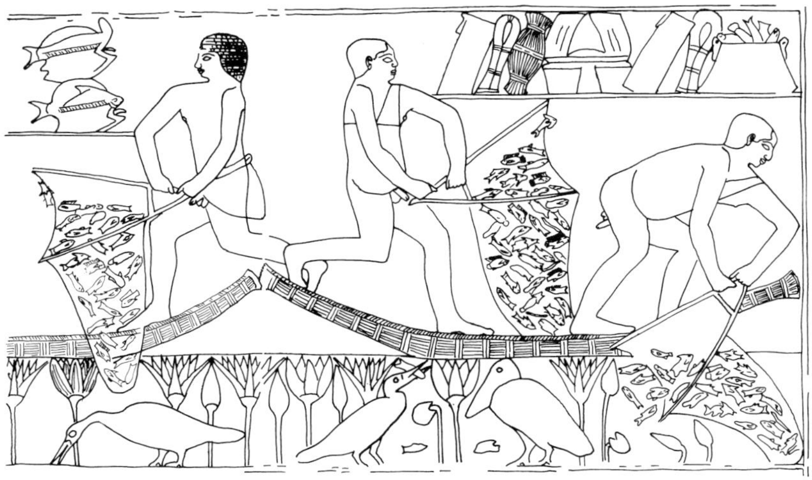

Fig. 7-30 (IB106)

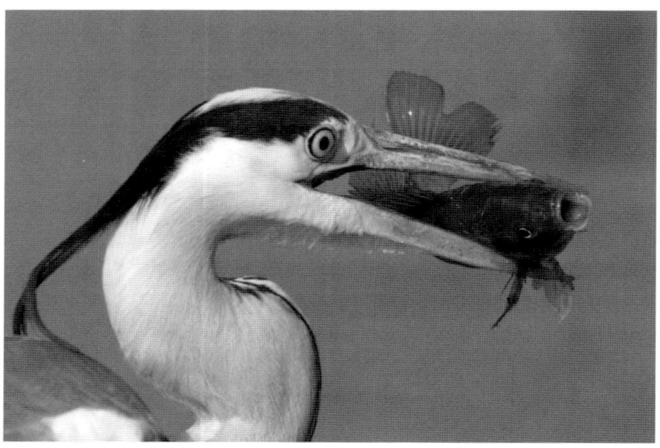

Pl. 7N Heron swallowing fish
© iStockphoto.com/Hilton Kotze

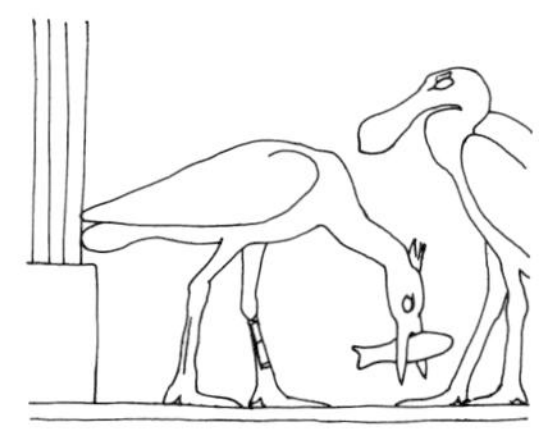

Fig. 7-31 (IB107)

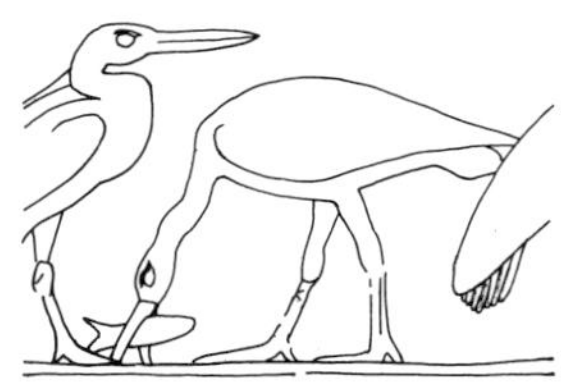

Fig. 7-32 (IB108)

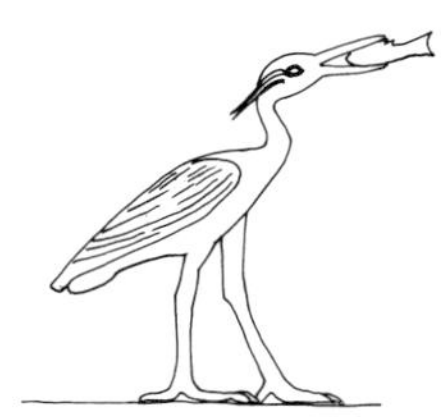

Fig. 7-33 (IB109)

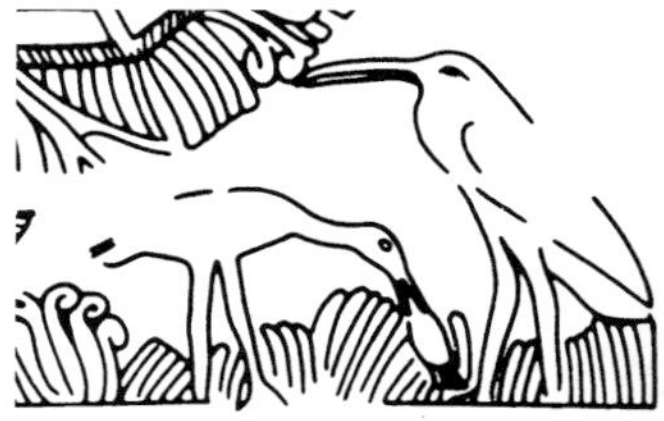

Fig. 7-34 (IB110)

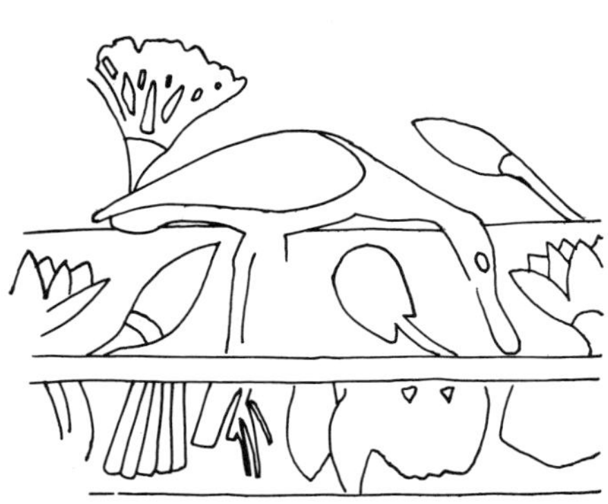

Fig. 7-35 (IB111)

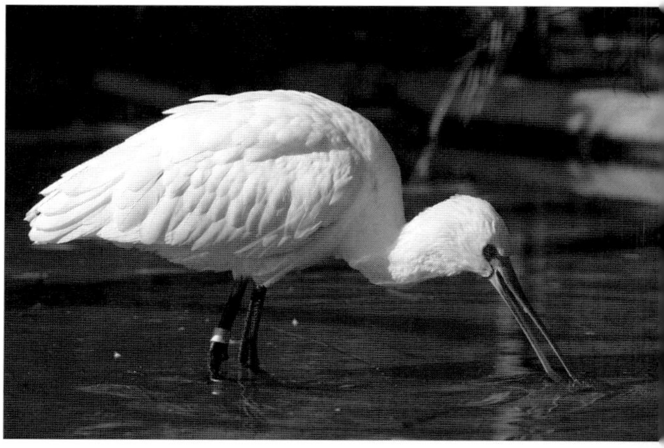

Pl. 7O Spoonbill foraging
© iStockphoto.com/Jan Gottwald

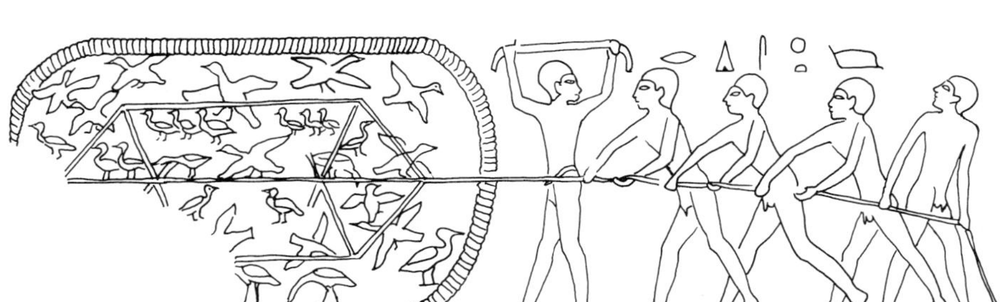

Fig. 7-36 (IB113)

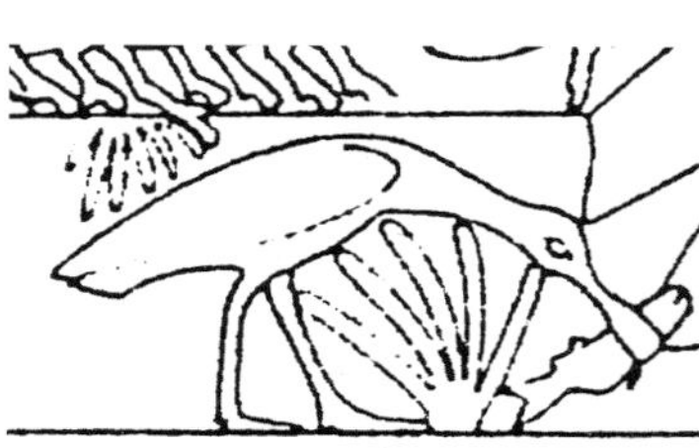

Fig. 7-37 (IB113)

The earliest depictions of feeding herons, which occur in the tomb of *ptḥ-ḥtp I*, show the birds manipulating fish prior to swallowing (IB107 and IB108; see figs. 7-31 and 7-32). Each fish is held between the birds' mandibles, suggesting that they have just been caught. The remaining images, which date from the early 6th dynasty, depict the swallowing of prey (IB109 and IB110; see figs. 7-33 and 7-34). In each example, the birds grasp an object with a forked tail in their open beak, clearly indicating that they are swallowing a fish headfirst.

Two of the birds are grey herons, as revealed by their characteristic head plumes (IB107 and IB109). The bird featured in the tomb of *ḥzj* (IB110; see fig. 7-34) is somewhat ambiguous and may instead represent an egret, a species that also preys upon fish. Finally, it should be noted that although none of the birds appear to be standing in water, all of them occur in clapnet scenes. They are thus immediately adjacent to a water source.

A heron eating a fish also forms the hieroglyphic determinative: Gardiner sign list G51, in the word, *ḥ3m* 'catch fish'.[69] In this study, G51 was found in inscriptions accompanying five fishing scenes.[70] The determinative also displays a change in the behaviour of the bird over time, for while the first three texts (all dating from the mid-5th dynasty) each depict the heron biting the body of a fish, the latest examples (dating from the late 5th and early 6th dynasties) show it ingesting its prey headfirst. The dates for these inscriptions indicate that the motif of a heron eating a fish possibly occurred first in texts before being incorporated in tomb scenes.

Spoonbills

nfr-jrt-n.f (IB111: Saqqara, V.8-9)
ḫww-wr (IB112: Saqqara, V.8-9)
wḏ3-ḥ3-ttj: ššj (IB113: Saqqara, VI.1L-2E)

Three images, all in clapnet scenes, were found to depict spoonbills feeding. Two of the birds bend forward so that their distinctive, spatula-shaped bills (for which the species is named)[71] touch the bottom of a stretch of water (figs. 7-35 and 7-36), while a third clasps a small fish (fig. 7-37).

Eurasian spoonbills (pl. 7O) feed on fish, crustaceans, molluscs, and insects, which they find by plunging their bill into shallow water until the tip reaches the substrate and then sweeping it back and forth.[72] Prey items are detected by feel, not sight.[73] The representations in the tombs of *nfr-jrt-n.f* (IB111; see fig. 7-35) and *ḫww-wr* (IB112; see fig. 7-36) show spoonbills foraging correctly upon the river-bottom. In the *nfr-jrt-n.f* example, however, the bird's head is fully submerged so that it peers downward through the water. Spoonbills will immerse the entire length of their beak when foraging, but as they only hunt in shallow water and do not use their eyes to find food their head usually stays above the surface (see pl. 7O). The *nfr-jrt-n.f* image is thus incorrect.

Sacred ibis

ḥzj (IB114: Saqqara, VI.1L-2E)
ꜥnḫ-m-ꜥ-ḥr: zzj (IB115: Saqqara, VI.1L-2E)

Sacred ibis search for invertebrates (their primary prey), either by wading in shallow pools and probing the mud with their beaks, or by pecking for insects on dry land situated next to water (see pl. 7P).[74] This latter method matches closely a depiction in the tomb of *ḥzj* (IB114; see fig. 7-38) in

69 The example offered by Gardiner (*Egyptian Grammar*, 473) for this sign shows a long-necked bird leaning over and pecking at a fish below its feet, which was derived from the 12th dynasty Theban tomb of *znt* and *jnt.f-jkr* (TT60). He proposed that it represents an egret, but evidence from this study suggest that it is more likely to be a heron.

70 G51 could be clearly identified in inscriptions in the tombs of *nj-ꜥnḫ-ẖnmw/ẖnmw-ḥtp* (ntn), *3ḫtj-ḥtp* (ntn), *ḥtp-ḥr-3ḫtj* (D60), *sḫm-k3.j* (G1029), and *mḥw* (ntn). This sign is also found in a large number of offering-lists in the item, *ḥ3mw*.

71 The flattened shape of the bird's bill was also the inspiration for its Egyptian name of *p3ḳ*, the root of which is 'to be thin, flat (i.e., beaten metal)'. See Wolterman, *JEOL* 32, 127-129.

72 Olney, P.J.S. (1977). "Spoonbill (*Platalea leucorodia*)" in Cramp-Simmons, *Handbook i*, 355.

73 Hancock et al., *Storks, Ibises and Spoonbills*, 256.

74 Roselaar, C.S. (1977). "Sacred ibis (*Threskiornis aethiopicus*)" in Cramp-Simmons, *Handbook i*, 349.

which an ibis, while standing beside a pool over which a clapnet has been deployed, bends forward to touch the tip of its beak upon the ground. I propose that a pair of birds in a clapnet scene from the tomb of *ʿnḫ-m-ʿ-ḥr: zzj* (IB115; see fig. 7-39) are also sacred ibis. Although their beaks do not curve as strongly as other depictions of this species in Egyptian art, their general morphology is similar to ibis and the method of foraging portrayed by one of them is identical to that of the bird in the tomb of *ḥzj*.

Geese

spd-ḥtp (IB116: Saqqara, V.6L-9?)
ṯjj (IB117-120: Saqqara, V.8-9)
k3-gm-nj: mmj (IB121-124: Saqqara, VI.1M)
nj-k3w-jzzj (IB125-126: Saqqara, VI.1M)
mrrw-k3.j: mrj (IB127-129: Saqqara, VI.1M-L)
ḥzj (IB130: Saqqara, VI.1L-2E)
wḏ3-ḥ3-ttj: ššj (IB131: Saqqara, VI.1L-2E)
k3.j-jrr (IB132: Saqqara, VI.2?)
mrrj (IB133-134: Saqqara, VI.2E?)
mrjj-ttj: mrj (IB135-136: Saqqara, VI.2M-L)

Both poultry-yard and force-feeding scenes include many images of geese eating naturally. At least nine species of goose were known in ancient Egypt,[75] but it has often proved impossible to identify individual types in representations due to a lack of sufficient morphological details. The feeding behaviour expressed by all geese is largely similar, however: Most species graze on land by bending their necks and biting rapidly at vegetation as they walk forward (see pl. 7Q). They eat a variety of plant materials including grasses, seeds, roots, and cereal grains. Some species, such as greylag geese, also feed while swimming by immersing their foreparts ('upending') and plucking submerged plants.

The images found in this study show geese in their characteristic head-down grazing posture (IB116 to IB136).[76] Many appear to touch the substrate with their bill, while others simply bend forward, often with their beak ajar. Most of the birds do not hold anything between their open mandibles. Nevertheless, as the contexts in which this behaviour occurs are reliably associated with the presence of food, it is likely that feeding is depicted. The position of the birds' legs usually does not indicate whether they are moving forward as they would be when grazing. However in a unique example from the tomb of *nj-k3w-jzzj* (IB125 and IB126; see figs. 7-40 and 7-41), three geese purse their toes, a reflex action that is performed by walking birds when they lift their feet to take a step (see Chapter 4 for further details). The grazing behaviour of these birds is also confirmed by the presence of seeds in their beaks as well as the accompanying inscription, *wnm jt jn r ṯrp* 'Feeding on grain by the *r*-geese and the *ṯrp*-geese'.

Many poultry-yard scenes display a pool of water in the centre of the enclosure. In three tombs (IB123, IB130, and IB135), a small number of geese that either swim or wade in these ponds are shown leaning forward. Although geese usually 'up-end' fully when foraging in water, it is nevertheless possible that these images are intended to depict aquatic feeding.

Finally, in a reference to the voracious appetite of these birds, the tomb of *mrrw-k3.j: mrj* (IB129) displays the image of a goose in a force-feeding scene, which steals items from a tray of prepared food. This amusing motif was later copied in the tomb of his son, *mrjj-ttj: mrj* (IB136).

75 These include: the barnacle goose, bean goose, brent goose, Egyptian goose, greylag goose, red-breasted goose, spur-winged goose, white-fronted goose, and the lesser white-fronted goose. Some of these species are migratory and only transient visitors to Egypt. Their terminology for such birds reveals awareness by the Egyptians of their migratory habits. See Goelet, O. (1983). "The migratory geese of Meidum and some Egyptian words for "migratory bird", *BES* 5, 41-60.

76 It is interesting to note that Gardiner sign list G52 - a determinative in the word, *snm* 'feed'- represents a goose in precisely this posture. See Gardiner, *Egyptian Grammar*, 473.

Ducks

nb.j-m-ꜣḫtj (IB137: Giza, IV.6-V.1)
wr-jr.n-ptḥ (IB138: Saqqara, V.3-5?)
nfr/kꜣ-ḥꜣ.j (IB139: Saqqara, V.6)
rʿ-ḫʿ.f-ʿnḫ (IB140: Giza, V.6?)
rʿ-m-kꜣ.j (IB141: Saqqara, V.6? r/u - V.6-7)
nj-ʿnḫ-ẖnmw/ẖnmw-ḥtp (IB142: Saqqara, V.6L-7)
ḥtp-ḥr-ꜣḫtj (IB143: Saqqara, V.6-8E)
jrj-n-kꜣ-ptḥ (IB144: Saqqara, V.6-9)
spd-ḥtp (IB145: Saqqara, V.6L-9?)
ptḥ-ḥtp I (IB146: Saqqara, V.8M-L)
ṯjj (IB147: Saqqara, V.8-9)
nfr-jrt-n.f (IB148: Saqqara, V.8-9)
kꜣ.j-m-nfrt (IB149: Saqqara, V.8-9?)
ꜣḫtj-ḥtp (IB150: Saqqara, V.8L-9E)
ptḥ-ḥtp II: ṯfj (IB151: Saqqara, V.9M-L)
sḫm-kꜣ.j (IB152: Giza, V.9-VI.1?)
Anon (IB153: Giza, V-VI)
kꜣ-gm-nj: mmj (IB154: Saqqara, VI.1M)
sʿnḫ-w(j)-ptḥ (IB155: Saqqara, VI.1L)
ḥzj (IB156: Saqqara, VI.1L-2E)
ʿnḫ-m-ʿ-ḥr: zzj (IB157: Saqqara, VI.1L-2E)
mḥw (IB158-159: Saqqara, VI.2M-3?)

Ducks feed primarily upon water plants and aquatic insects that they locate by either swimming or wading through shallow water and 'dabbling' their bills along the water surface – i.e., opening and closing their mandibles rapidly to filter out edible items (pl. 7R).[77]

Many species of duck have been identified in Egyptian art,[78] and they are represented frequently in scenes at Giza and Saqqara. Twenty-two tombs were found to depict ducks in a feeding posture, all of them in clapnet scenes (IB137 to IB159). Reconstructions of this method of capturing waterfowl suggest that the net was probably laid out across the water surface and submerged slightly to enable ducks attracted to the open area to swim (see Chapter 5).[79] In clapnet scenes, ducks both within the net area and in the waters immediately adjacent, often bend their heads forward deeply, as though dabbling. This can be seen quite clearly in the earliest example from the tomb of *nb.j-m-ꜣḫtj* (IB137), where a bird within the net leans forward to possibly touch its beak upon the surface of the water. In subsequent examples, the pursed toes of the ducks indicate clearly that they are swimming while in this posture (e.g., IB141, IB143, and IB152), while others appear to dabble while walking in shallow water (e.g., IB144, IB149, and IB153; see fig. 7-42).

Other images of 'feeding' ducks consist of little more than a head and neck that emerges and bends forward from a clump of its swimming companions (e.g., IB148). Although a feeding posture cannot be ruled out entirely, it seems more likely that these schematic representations are an artistic flourish, intended only to fill all the available spaces within the clapnet and expand the number of birds present.

Quail

sḫm-kꜣ.j (IB160: Giza, V.9-VI.1?)
mrrw-kꜣ.j: mrj (IB161-163: Saqqara, VI.1M-L)
ḥzj (IB164-166: Saqqara, VI.1L-2E)
mḥw (IB167: Saqqara, VI.2M-3?)

Four tombs were found to depict quail that adopt a feeding posture. Quail are small granivorous birds that live and breed in grassland and cereal fields where they are able to feed upon both seeds and insects. The images from Giza and Saqqara all appear in harvest scenes, a context in which an abundance of grain would be available for the birds to ingest. Quail are nevertheless rare in such scenes.

The earliest example occurs in the Giza tomb of *sḫm-kꜣ.j* (IB160) and shows a long-legged bird striding amongst crops in a field.[80] Although it stands near a workman, it is possible that it should

[77] McKinney, F. (1969). "The behaviour of ducks" in Hafez, *Behaviour of Domestic Animals*, 595-596.

[78] These include: the common shelduck, European wigeon, green-winged teal, mallard, pintail, ruddy shelduck, pochard, and tufted duck. See Moreau, R.E. (1930). "The birds of ancient Egypt" in Meinertzhagen, *Nicoll's Birds*, 64; Houlihan, *Birds*, 65-74.

[79] For an attempted reconstruction of the clapnet, see Dunham, D. (1937). "Two parallels to ancient Egyptian scenes", *BMFA* 35(210), 50-54.

[80] The bird's legs are much longer than those of quail. Nevertheless, as only quail occur in harvest scenes, it seems likely that it is intended to represent this species. Harpur (*Decoration*, 215) suggests that the tomb of

be perceived as hidden amongst the stubble and uncut stalks. In later images, however, the quail do not stand erect, but instead squat low on the ground with their legs folded beneath them and their heads tilted forward (IB161 to IB167). One animal in the tomb of *mrrw-kȝ.j: mrj* (IB163; see fig. 7-43) sits as it nibbles ears of barley, while another on the same register pecks at the ground as men make their way across a field, flushing and capturing the small birds in a net strung between them (IB162; see fig. 7-44). The apparent nonchalance of this quail while its companions fly upward suggests that the men have not yet reached it and it is still feeding calmly amongst the stubble.[81] A similar scene in the tomb of *ḥzj* (IB166) was probably copied from that of *mrrw-kȝ.j: mrj*, but here the feeding quail's beak does not make contact with the substrate.

The seated posture of these feeding birds is intriguing. Recent studies have shown that ground-foraging birds such as quail are myopic.[82] In order for them to discriminate seeds and other edible objects on the ground, they must bring both their eyes to bear, requiring them to lean well forward (see pl. 7S). If a large amount of seed is concentrated in a small area, such as in a cereal field, the birds do not need to walk continuously in order to forage. They may thus remain and feed at the one location, resting comfortably on their haunches as they lean forward to peck. The seated quail images from Saqqara would seem to suggest that the Egyptians were aware of this distinctive feeding posture, even though quail are notoriously difficult to observe in the wild.

Cranes

ṯjj (IB168-169: Saqqara, V.8-9)
nj-kȝw-jzzj (IB170: Saqqara, VI.1M)
mrrw-kȝ.j: mrj (IB171-173: Saqqara, VI.1M-L)
wḏȝ-ḥȝ-ttj: ššj (IB174: Saqqara, VI.1L-2E)
mrjj-ttj: mrj (IB175-177: Saqqara, VI.2M-L)

Five tombs depict small flocks of cranes in which one or more members are shown feeding. The flocks consist primarily of common cranes, but some also include demoiselle cranes, which can be identified by their distinctive head crests.

Bill length plays an important role in determining both how and what cranes eat. The common crane uses its short bill for grazing in a goose-like manner, walking with its head lowered, pecking and probing at the ground (cf. pl. 7T). When an edible object is found, it is swallowed with a characteristic backward movement of the neck. The common crane ingests a wide range of foods, but is particularly fond of grain and so frequently forages in croplands. Demoiselle cranes also graze in a similar manner.

Many Egyptian depictions show common cranes leaning over to peck either at the ground (IB168, IB172, IB174, and IB176; see fig. 7-45) or at grain that has been thrown down by handlers for them to ingest (IB170, IB171, and IB177; see fig. 6-21 in Chapter 6). Some can be seen to hold a seed within their mandibles shortly after they have plucked it from the ground or as they tip back their head and are about to swallow it. Scenes in the tombs of *ṯjj* (IB171), *mrrw-kȝ.j: mrj* (IB172 and IB173), and *mrjj-ttj: mrj* (IB175 and IB176; see fig. 7-46) also show both common and demoiselle cranes being force-fed. In all but one of these examples, the procedure is interrupted by another bird, which pecks at the beak of its immobile companion. Common cranes are known to be highly aggressive when feeding and will engage in 'kleptoparasitism' - stealing food from their companions

sḫm-kȝ.j may date to the early 6th dynasty and that this image was copied from that of *mrrw-kȝ.j: mrj*. Smith (*HESPOK*, 197) also believes that this scene must have been copied from elsewhere.

81 Two methods for catching quail have been noted in modern Egypt. In one, a net is first laid out over a field, covering it completely, before quail are flushed upward to catch them in the mesh. The other method consists of dragging a weighted net across a field to achieve the same result. Scenes of catching quail in the tombs of *mrrw-kȝ.j: mrj* and *ḥzj* may show the latter method, as the hunters appear to be running, which would not be necessary if the net were simply laid over the top of the feeding birds. See Goodman-Meininger, *Birds of Egypt*, 91; also Vandier, *Manuel v*, 318-320.

82 Hodos, W. and Erichsen J.T. (1990). "Lower-field myopia in birds: An adaptation that keeps the ground in focus", *Vis. Res.* 30(5), 653-657.

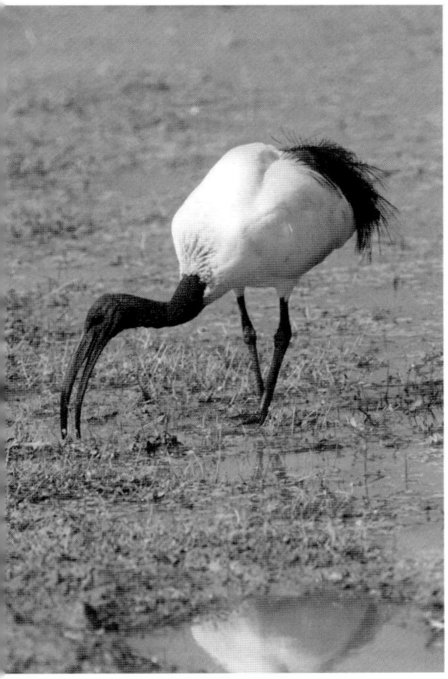

Pl. 7P Sacred ibis foraging
© ardeacom/Alan J.S. Weaving

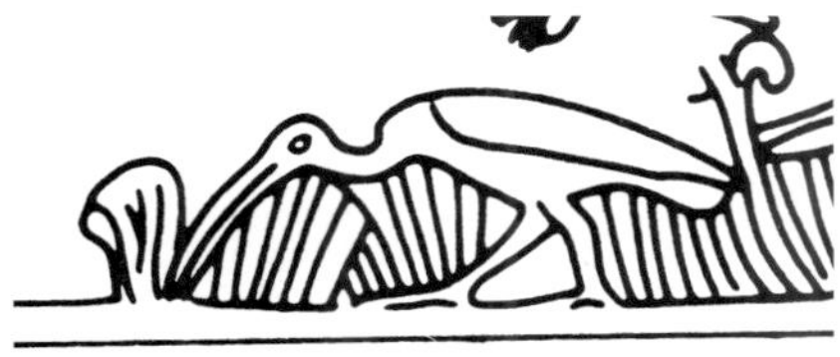

Fig. 7-38 (IB114)

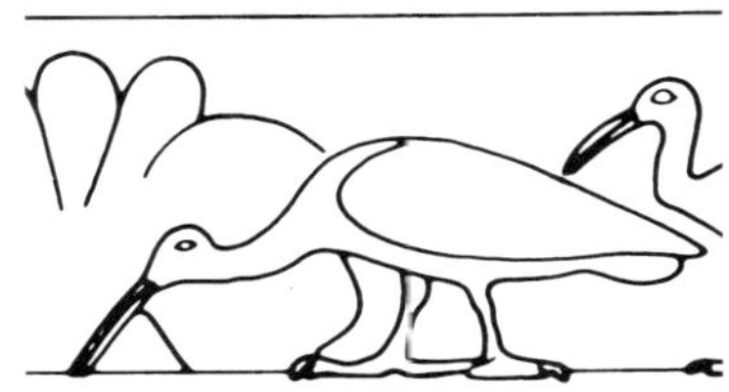

Fig. 7-39 (IB115)

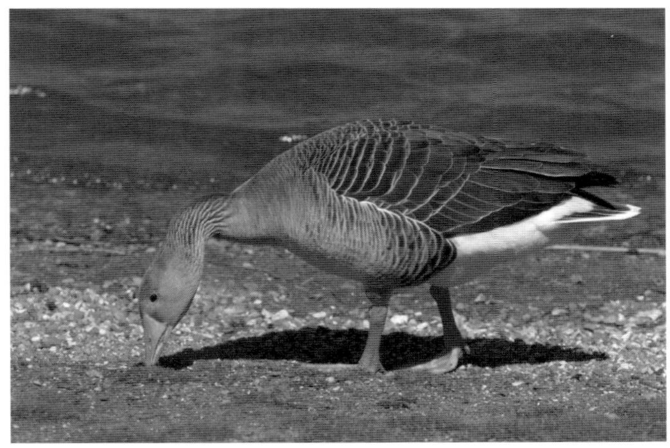

Pl. 7Q Goose grazing
© iStockphoto.com/Jerome Whittingham

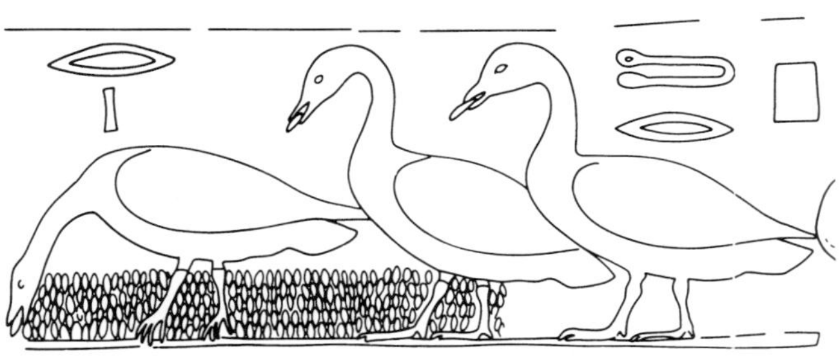

Fig. 7-40 (IB125)

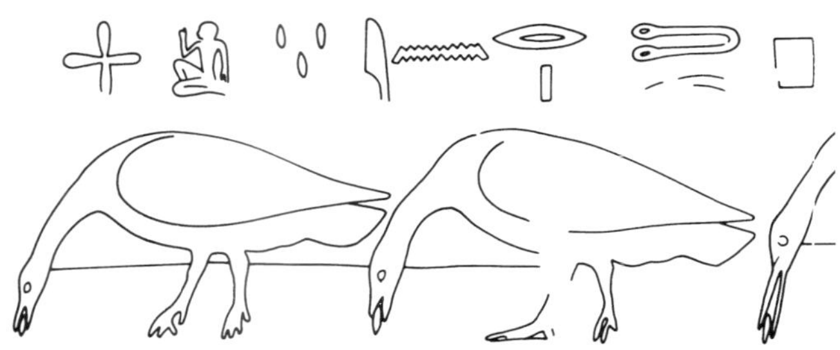

Fig. 7-41 (IB126)

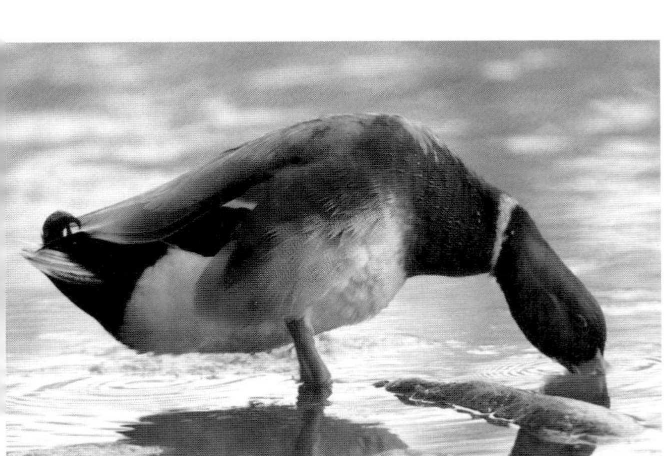

Pl. 7R Duck foraging
© iStockphoto.com/Bernd Lang

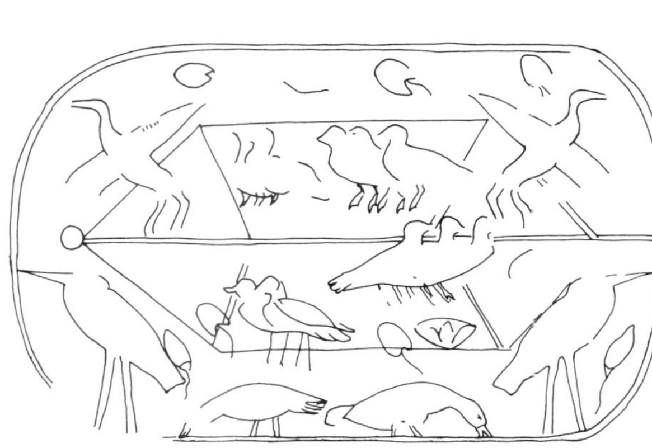

Fig. 7-42 (IB144)

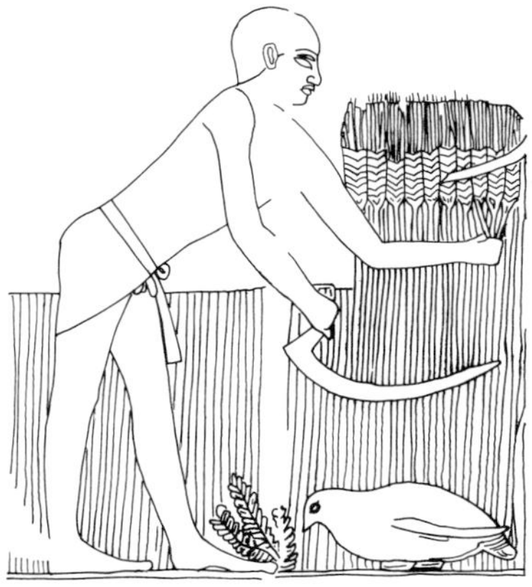

Fig. 7-43 (IB163)

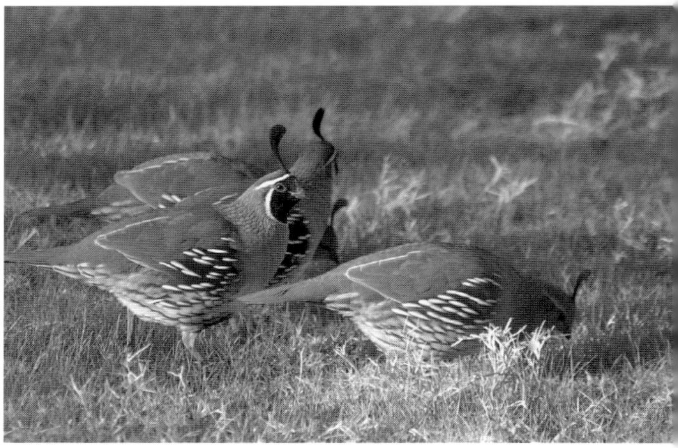

Pl. 7S Quail feeding
© iStockphoto.com/Anton Foltin

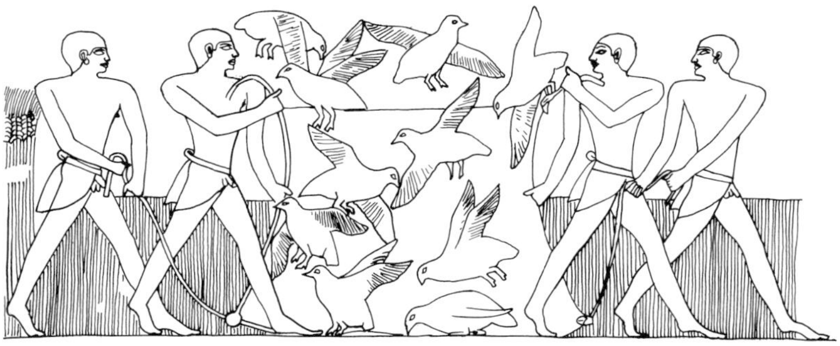

Fig. 7-44 (IB162)

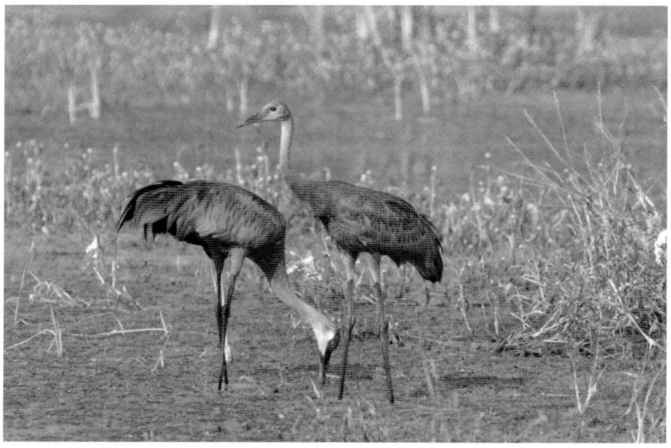

Pl. 7T Cranes foraging
© iStockphoto.com/Walter Spina

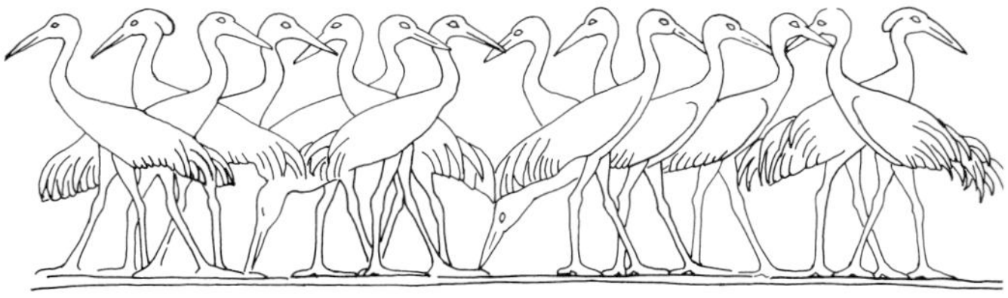

Fig. 7-45 (IB168)

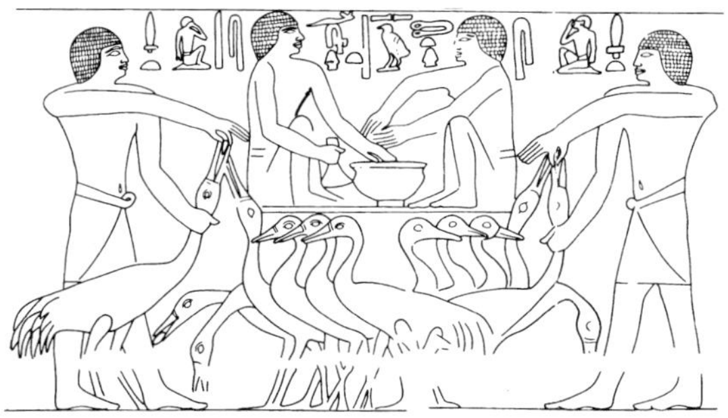

Fig. 7-46 (IB176)

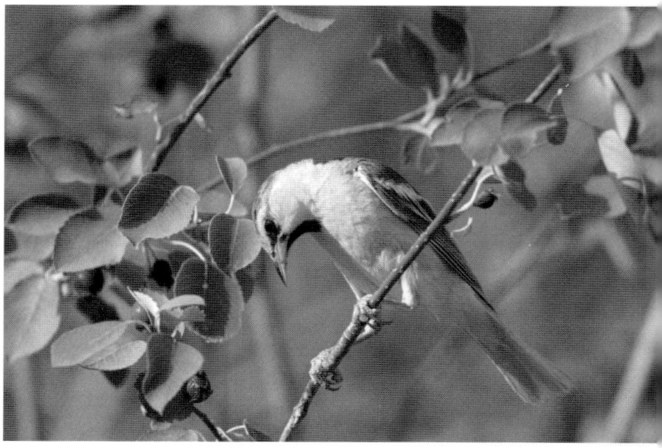

Pl. 7U Oriole in a fruit tree
© iStockphoto.com/Frank Leung

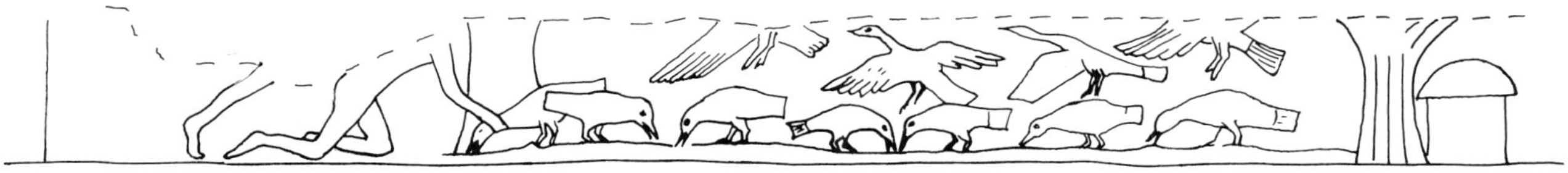

Fig. 7-47 (IB178)

whenever the opportunity presents itself.[83] It is possible, then, that the unrestrained cranes in force-feeding scenes are trying to take the food that is placed in their companions' beaks.

Finally, some images appear to show cranes pecking at the toes of either their companions or those of their human handlers (IB168, IB170, IB171, IB172, and IB176; see fig. 6-21 in Chapter 6 and fig. 7-45). Although the placement of the birds' beaks may be accidental, the repetition of this detail in different scenes suggests that it may be a deliberate attempt by artists to present a humorous image.

Golden orioles

rˁ-špss (IB178: Saqqara, V.8M)

Eleven scenes of netting birds were examined during this study, but only one of these was found to include images of feeding. Scenes of this kind show the capture of birds in orchards, either in spring-traps laid out below the trees or by flushing them into nets placed over the canopies (see figs. 4-28 and 4-29 in Chapter 4).[84] The birds in these scenes have been identified as golden orioles based upon morphological features and a few examples where remaining paint has indicated yellow plumage[85] - a distinctive feature of this species. Hoopoes are also found occasionally among the flocks.

Golden orioles feed upon both insects and vegetation (mainly seeds and fruits). They are a shy and intensely arboreal species, spending most of their time above the ground, feeding among the treetops (see pl. 7U).[86] On very rare occasions, however, they will forage for insects at ground level.

It is believed that the Egyptians netted these birds because they stole important fruit crops.[87] It is surprising, then, that only one netting scene - that of *rˁ-špss* (IB178; see fig. 7-47) - yielded images of feeding birds. The relief is damaged and the available line drawing is very schematic, but the scene appears to show two men attempting to catch a flock of birds at the base of a tree, six of which adopt a feeding posture by bending forward deeply to touch their beaks upon the substrate. A small box-like object, which possibly represents a trap, is drawn to the right of the tree-trunk. Curiously, then, rather than stealing fruit up in the canopy, the orioles are thus shown feeding upon the ground, something they rarely do. This suggests that the animals represented in the *rˁ-špss* scene may be a different species.

Nile crocodiles

ẖnm-ntj (IB179: Giza, VI.1)
kꜣ-gm-nj: mmj (IB179a-180: Saqqara, VI.1M)
mrrj (IB181: Saqqara, VI.2E?)
mḥw (IB182-183: Saqqara, VI.2M-3)

Six Memphite images were found to depict Nile crocodiles feeding (IB179 to IB183). Each crocodile is situated underwater, resting upon the riverbed with its jaws agape as it attempts to swallow a large fish (see figs. 7-48 and 7-49). Although only one fish appears within each animal's open mouth, the artists have in all but one example (IB183), drawn other fishes bunched around their jaws, giving the impression that each crocodile is trying to cram in more than one.

Nile crocodiles eat a wide range of prey, including fish. Two methods are used to catch their prey: in shallow in-shore waters, they lie submerged and motionless with their jaws open, waiting to snap at any unsuspecting creature that comes within reach, while in deeper water they will actively pursue fish.[88] Crocodiles are able to open their mouths for long periods when underwater due to a valve in

83 Stillman, R.A., Bautista, L.M., Alonso, J.C. and Alonso, J.A. (2002). "Modelling state-dependent interference in common cranes", *J. Anim. Ecol.* 71, 874-882.
84 Vandier, *Manuel v*, 313-318.
85 Gaillard, C. (1929). "Sur deux oiseaux figures dans les tombeaux de Béni-Hassan", *Kemi* 2, pl. 2.
86 Roselaar, C.S. (1993). "Golden oriole (*Oriolus oriolus*)" in Cramp-Simmons, *Handbook vii*, 420.
87 Houlihan, *Birds*, 130-131.
88 Cott, H.B. (1961). "Scientific results of an inquiry into the ecology and economic status of the Nile crocodile (*Crocodilus niloticus*) in Uganda and Northern Rhodesia", *Trans. Zool. Soc. Lond.* 29, 302-303.

their throat that closes off their respiratory system, however all food must be brought to the surface before it can be swallowed. When eating a fish, crocodiles lift their head out of the water and then manipulate their victim, tossing it in their mouth until it can be swallowed headfirst (cf. pl. 7V).[89] This orientation prevents sharp spines on the fish from injuring their throat and gullet as it is swallowed.

All six depictions appear to show crocodiles swallowing their prey incorrectly underwater. Alternatively, these images may be intended to show them catching their prey prior to swallowing, using the passive capture method described above. Four of the depicted crocodiles swallow their fish correctly headfirst (IB179, IB179a, IB181 to IB183; see fig. 7-48), while one is shown biting its prey on the tail (IB180; see fig. 7-49). This suggests strongly that the Egyptians were aware of how crocodiles manipulate fish prior to eating and thus supports the notion that these images show them in the act of swallowing, rather than catching their prey. It would seem, then, that behaviour viewed at the water surface has been depicted underwater (although see below).

The species of fish being swallowed can be tentatively identified based upon features of their fins. Some of the crocodiles appear to ingest 'walking catfish', as revealed by their long dorsal and anal fins, and rounded caudal (tail) fins[90] (IB179,[91] IB179a, IB181, and IB182; see fig. 7-48). These fish possess sharp spines on their pectoral fins, which would explain why they are depicted being eaten headfirst. Another crocodile (IB183) clearly eats a tilapia, indicated by its flared caudal fin and large, pointed dorsal and anal fins.[92] These latter fins are heavily spined. The only fish to be shown eaten tail-first occurs in the tomb of *k3-gm-nj: mmj* (IB180; see fig. 7-49). The elongated shape of its body and small mouth suggest that it is a grey mullet.[93] This species has two dorsal fins, one of which (the anterior) is made up of 3-5 hard spines.[94] Perhaps the mullet's narrow shape allowed the crocodile to swallow this species tail-first without the spines catching in its throat?

Walking catfish, tilapia, and mullet are all preyed upon by Nile crocodiles.[95] These species also prefer to inhabit shallow water. As described above, when hunting in-shore fish, Nile crocodiles use a passive, open-mouthed, 'sit and wait' hunting style. The depicted crocodiles, which rest upon the riverbed and bite at shallow water fish, thus appear to represent this capture method.

The images therefore possess features that match *both* the capture and swallowing methods of Nile crocodiles. These behaviours are sequential, however, and occur in different locations: first capture, which takes place underwater, and then swallowing, which occurs above the water surface. In order to account for these two activities appearing in the same image, I suggest that the artists may have attempted to depict the entire sequence in the one scene.

Catfish

snḏm-jb: jntj (IB184: Giza, V.8M-L)
ṯjj (IB185: Saqqara, V.8-9)
ẖnwt (IB186: Saqqara, V.9)
Anon (IB187: Saqqara, late V-VI)
k3.j-m-ʿnḫ (IB188: Giza, VI)
k3-gm-nj: mmj (IB189-190: Saqqara, VI.1M)
mrrw-k3.j: mrj (IB191-192: Saqqara, VI.1M-L)
sʿnḫ-w(j)-ptḥ (IB193: Saqqara, VI.1L)
ḥzj (IB194-196: Saqqara, VI.1L-2E)
jnw-mnw (IB196a: Saqqara, VI.2E)
mḥw (IB197-201: Saqqara, VI.2M-3?)
wr-nw (IB202: Saqqara, VI.3-4?)
3ẖtj-ḥtp (IB203: Saqqara, VI.5?)

---

89 *ibid.*, 303; Whitfield, A.K. and Blaber, S.J.M. (1979). "Predation on striped mullet (*Mugil cephalus*) by *Crocodylus niloticus* at St. Lucia, South Africa", *Copeia* 2, 269.

90 Brewer-Friedman, *Fish and Fishing*, 60-63.

91 Brovarski (*Senedjemib Complex Part I*, 119) also suggests that the crocodile in the tomb of *ẖnm-ntj* is ingesting a catfish.

92 Brewer-Friedman, *Fish and Fishing*, 76-79.

93 *ibid.*, 72-73.

94 Boulenger (*Fishes of the Nile*, 429) notes the interesting observation by Strabo that mullet are often found near *Schall* catfish which "by means of their strong spines, defend them against the crocodiles". I do not know if modern zoologists have noted this behaviour.

95 Cott, *Trans. Zool. Soc. Lond.* 29, 294-295; Whitfield-Blaber, *Copeia* 2, 266-269.

Two species of *Synodontis* catfish have been identified in Egyptian wall scenes: 'African catfish'[96] and 'upside-down catfish'.[97] These fish are immediately recognizable due to distinctive bony plates that project from their heads and whisker-like 'barbels' below their snouts. The behaviour of the two is fundamentally different, however, for unlike African catfish and other fish species, upside-down catfish spend most of their time inverted (pl. 7W).[98] Although many catfish species feed on bottom-dwelling organisms and detritus, the upside-down catfish ingests plankton, small insects, and other food items that are found at, or near, the water surface, which it reaches by inverting its body and skimming with its ventrally located mouth. This action is assisted by its long barbels, which spread out to scoop matter towards it. The fish is easily identified in wall scenes because artists have reproduced its distinctive feeding behaviour of swimming inverted.[99]

Examples of upside-down catfish swimming freely were found in 13 tombs. It should be noted that catfish trapped in seine fishing-nets were not included in this sample as it was felt that the position of their bodies within the water was more likely to be the result of the crowded space available than an expression of their natural behaviour.

Fourteen examples from eight tombs depict upside-down catfish swimming in the waters of marsh and fishing scenes. The earliest of these occurs in the tomb of *ṯjj* (IB185), and shows an inverted catfish deep underwater. Its barbels do not touch the water surface; nor do those of subsequent late 5th dynasty images from the tomb of *ẖnwt* (IB186) and an anonymous individual (IB187). In the early 6th dynasty tomb of *kꜣ-gm-nj: mmj*, however, a remarkable depiction from room I shows the barbels of an upside-down catfish actually piercing the plane of water (IB189; see fig. 7-50), while in a scene from the pillared hall, the 'whiskers' of another fish fan out just below the surface (IB190), precisely as they do in nature (see pl. 7W). This latter behaviour is repeated in the tombs of *mrrw-kꜣ.j: mrj* (IB192), *ḥzj* (IB95 and IB196; see fig. 7-51), and *mḥw* (IB197 and IB198), while in a later example from the tomb of *wr-nw* (IB202; see fig. 7-52), a catfish again appears to push its barbels out of the water.[100] Thus over time, free-swimming upside-down catfish are depicted increasingly closer to the water surface, possibly to draw attention to their idiosyncratic feeding method. Although the images of catfish in the tombs of *kꜣ-gm-nj: mmj* (fig. 7-50) and *wr-nw* (fig. 7-51), which hold their barbels above water-level, are biologically inaccurate, they nevertheless serve to emphasize yet more strongly the highly unusual behaviour of this particular fish species.

It is interesting to contemplate why artists chose to draw these fish closer to the water surface in the 6th dynasty than they did in the 5th dynasty. A recent study has revealed that upside-down catfish spend more time swimming at the surface when they are in deoxygenated waters. Diffusion from the atmosphere produces a thin micro-layer at the water surface that is well-oxygenated, regardless of how hypoxic the water below may be.[101] Experiments have shown that although the fish do indeed feed at the water surface, they will also take advantage of the air available in this micro-layer when their environment becomes oxygen deficient. The shallow water in papyrus swamps is generally low in oxygen, and this decreases further during the dry season when the water

---

[96] For the importance of this fish in the diet of the ancient Egyptians, see Luff, R.M. and Bailey, G.N. (2000). "Analysis of size changes and incremental growth structures in African catfish *Synodontis schall* (Schall) from Tell El-Amarna, Middle Egypt", *J. Arch. Sci.* 27(9), 821-835; Luff, R.M. and Bailey, G. (2000). "The aquatic basis of ancient civilisations: The case of *Synodontis schall* and the Nile Valley" in Bailey et al., *Human Ecodynamics*, 100-113.

[97] This fish was known formerly as *Synodontis batensoda,* but was recently re-named *Brachysynodontis batensoda* ('short *Synodontis*'). For details about *S. batensoda*, see Brewer-Friedman, *Fish and Fishing*, 67-69.

[98] A closely related species, *Hemisynodontis membranaceus* (formerly *Synodontis membranaceus*) also swims upside-down. See *ibid.*, 68; Boulenger, *Fishes of the Nile*, 382.

[99] Brewer-Friedman (*Fish and Fishing*, 69) point out that the shape of the two species is often identical in wall scenes, however, as only *B. batensoda* swims upside-down and *S. schall* never does, it seems likely that any representation of a catfish that is shown inverted must be *B. batensoda*. Only catfish depicted in this orientation are considered here.

[100] This feature may indicate that the tomb of *wr-nw* is closer in date to that of *kꜣ-gm-nj: mmj*, rather than the late 6th dynasty (VI.3-4?), as tentatively suggested by Harpur.

[101] Chapman, L.J., Kaufman, L. and Chapman, C.A. (1994). "Why swim upside down?: A comparative study of two Mochokid catfishes", *Copeia* 1994(1), 130-135.

level drops.[102] Swamps are also affected by seasonal flood cycles, with low floods resulting in more hypoxic water. Recent geoscientific studies have suggested that the annual Nile flood decreased gradually throughout the Old Kingdom period, to reach its lowest level during the reign of Pepi II (c. 2160 BC).[103] Perhaps, then, the increasing tendency to depict upside-down catfish at the water surface in the 6th dynasty mimics what artists were seeing in nature - that is, as the Nile became more hypoxic, these fish spent more of their time inverted and consequently were observed more often.

Finally, seven examples were also found of upside-down catfish swimming in the 'mound of water', in which the tomb-owner spears fish with a bident. In an example from the late 5th dynasty Giza tomb of *snḏm-jb: jntj* (IB184), a catfish can be seen in the centre of the mound, but in tombs of the 6th dynasty, they are consistently located on the edge of the column of water that is furthermost from the tomb-owner (cf. IB191, IB193, IB194, IB196a, and possibly IB203; see fig. 7-53).[104] It is interesting to consider whether the positioning of the fish in these later depictions offers some clue about the orientation and interpretation of the 'mound'.[105] If free-swimming upside-down catfish in the 6th dynasty are usually shown just below the water, then, as they appear reliably on the far side of the mound, this must *also* represent the surface. It may be significant, therefore, that, like depictions of the water surface in horizontal underwater scenes, the far side of most of the mounds in question is drawn with a straight line, distinguishing it from that on the other, which is usually curved (see fig. 7-53).

7.2 DRINKING

All animals require fluid intake to survive. Animals differ in the amount of water they require for survival, as well as in the frequency with which they need to drink. Some animals, particularly desert-adapted species, are able to cope with long periods of dehydration. Most species, however, require a constant supply of fluids to function normally. Animals drink in a variety of ways: lapping with the tongue, gulping and tipping back the head or sucking fluid up into the mouth.

Drinking was represented by five species (cattle, dogs, ducks, geese, and cranes) in tomb scenes from Giza and Saqqara.

Cattle

nb.j-m-ꜣḫtj (IB204: Giza, IV.6-V.1)
Anon (IB205: Saqqara, V)
nfr/kꜣ-ḥꜣ.j (IB206-208: Saqqara, V.6)
nj-ꜥnḫ-ẖnmw/ẖnmw-ḥtp (IB209-210: Saqqara, V.6L-7)
kꜣ.j-dwꜣ (IB211: Giza, V.6-9?)
rꜥ-špss (IB212-213: Saqqara, V.8M)
sḫm-ꜥnḫ-ptḥ (IB214: Saqqara,V.8-9)
ṯjj (IB215-216: Saqqara, V.8-9)
nfr-jrt-n.f (IB217: Saqqara, V.8-9)
ꜣḫtj-ḥtp (IB218: Saqqara, V.8L-9E)
nbt (IB219: Saqqara, V.9)
sḫm-ꜥnḫ-ptḥ (IB220: Giza, V.9-VI.1?)
mrrw-kꜣ.j: mrj (IB221: Saqqara, VI.1M-L)
nj-ꜥnḫ-nswt (IB222: Saqqara, VI.1-2E)
jꜣzn (IB223: Giza, VI.1M-2M)
mrjj-rꜥ-nfr: kꜣr (IB224: Giza, VI.4)

Cattle do not drink by lapping with their tongues, but instead, dip their muzzles into water and then suck the fluid into their mouths (see pl. 7X).[106] They usually drink during the day and rarely do so at night. Cattle housed in barns, where feed is abundant, tend to drink more often and consume more water than those that are grazed on open pastures. As they are domesticated animals, water is usually provided for them, either in large drinking troughs or prepared waterholes.

[102] Chapman, L.J., Chapman, C.A. and Crisman, T.L. (1998). "Limnological observations of a papyrus swamp in Uganda: Implications for fish faunal structure and diversity", *Verh. Internat. Verein. Limnol.* 26, 1821-1826.
[103] Stanley, J.-D., Krom, M.D., Cliff, R.A. and Woodward, J.C. (2003). "Nile flow failure at the end of the Old Kingdom, Egypt: Strontium isotopic and petrologic evidence", *Geoarchaeology* 18(3), 395-402.
[104] A catfish is also positioned in the centre of the mound of water in the tomb of *kꜣ.j-m-ꜥnḫ* (IB188), thus hinting at a date for this structure earlier than that of the 6th dynasty. For a discussion of the evidence supporting a late 5th dynasty date for this tomb, see Kanawati, *Kaiemankh*, 15-21. See also Chapter 4, n. 98.
[105] For a discussion of this feature, see Mohr, *Hetep-her-akhti*, 68-71.
[106] Hafez et al. in Hafez, *Behaviour of Domestic Animals*, 251; Fraser-Broom, *Farm Animal Behaviour*, 93.

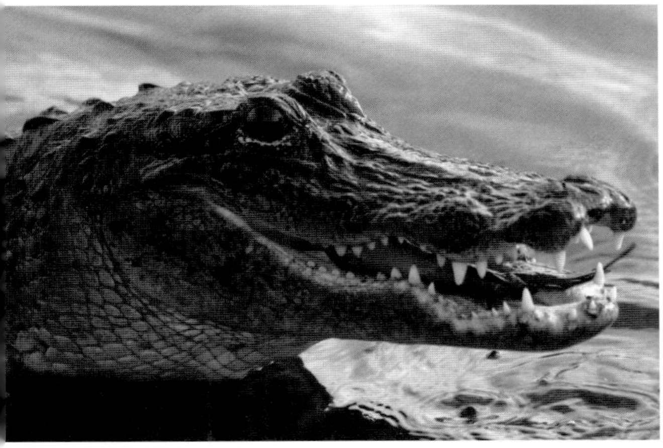

Pl. 7V Alligator eating a fish headfirst

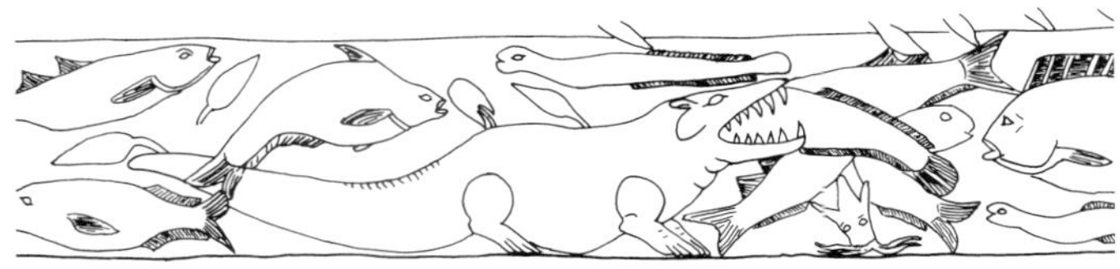

Fig. 7-48 (IB179a)

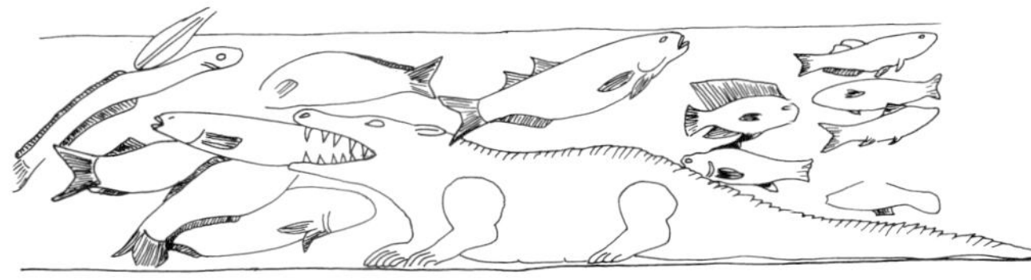

Fig. 7-49 (IB180)

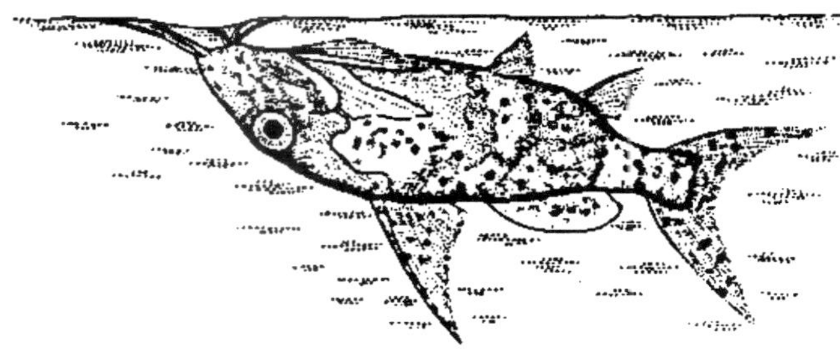

Pl. 7W Upside down catfish at water surface
Reprinted from: Chapman, L.J. et al., *Copeia* 1994(1), fig. 2

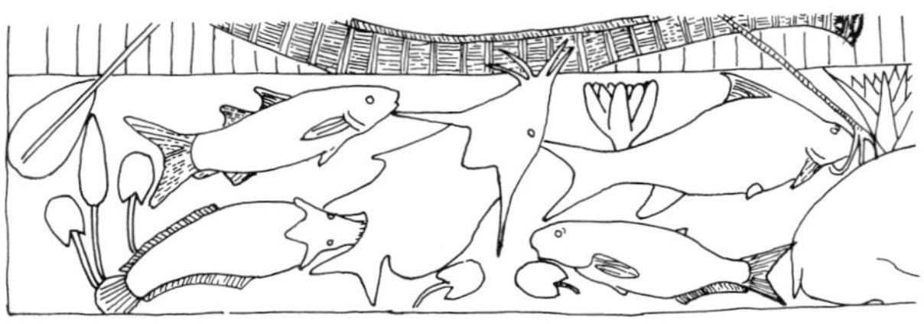

Fig. 7-50 (IB189)

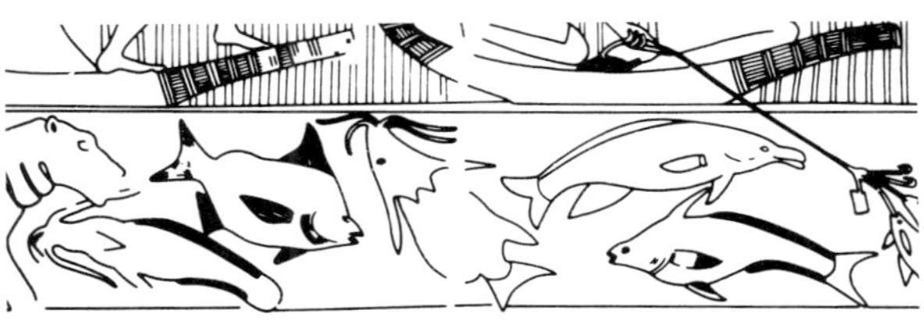

Fig. 7-51 (IB195)

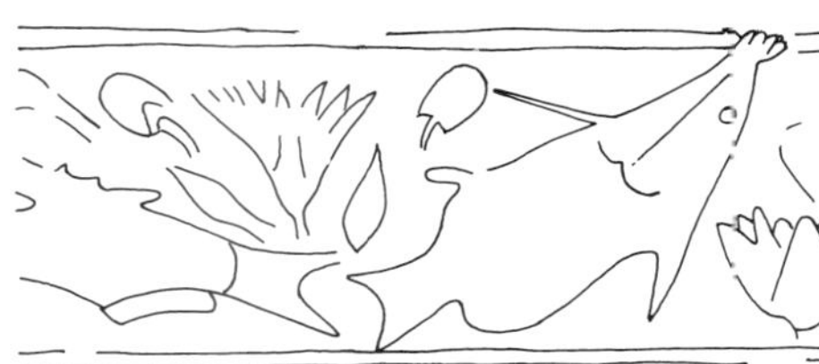

Fig. 7-52 (IB202)

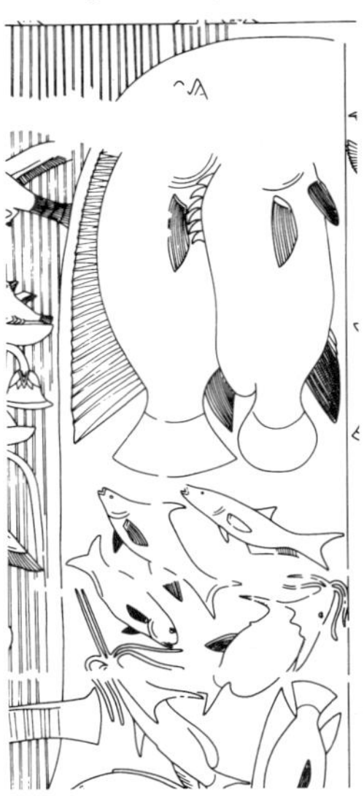

Fig. 7-53 (IB194)

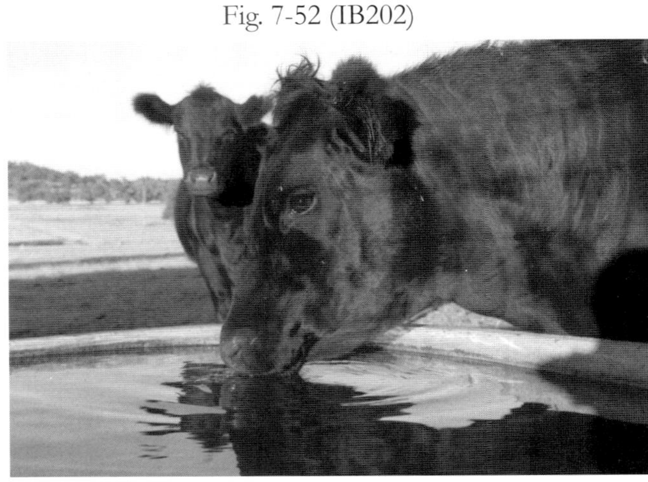

Pl. 7X Cow drinking

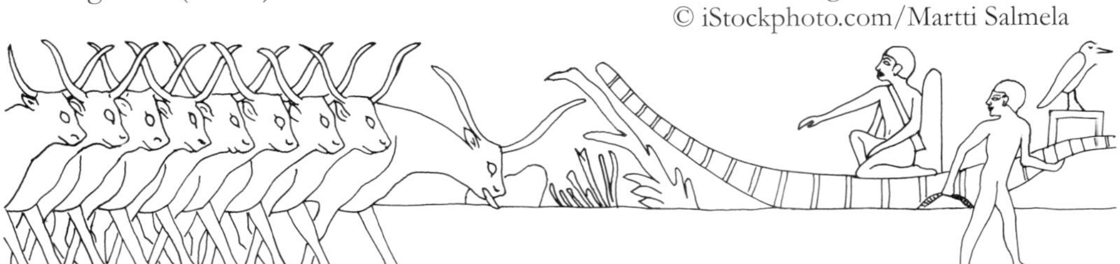

Fig. 7-54 (IB204)

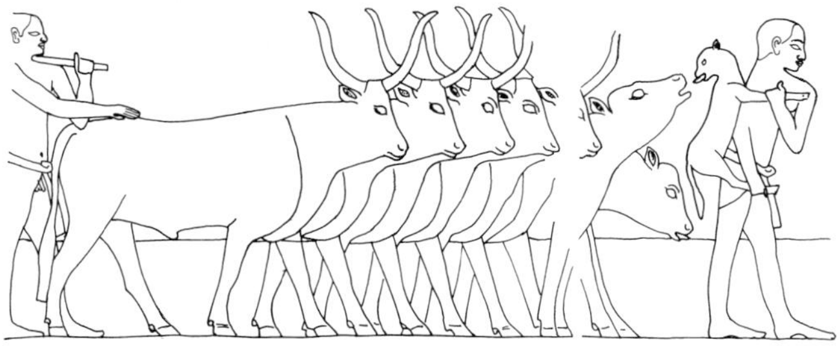

Fig. 7-55 (IB222)

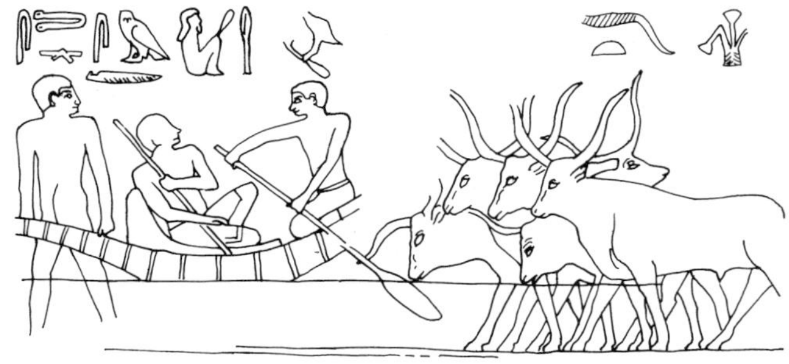

Fig. 7-56 (IB205)

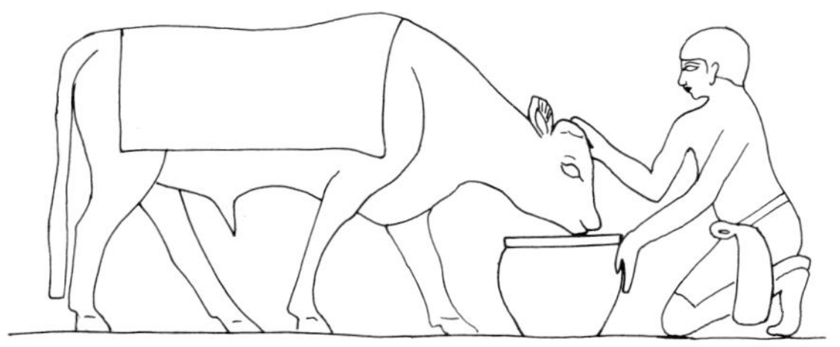

Fig. 7-57 (IB208)

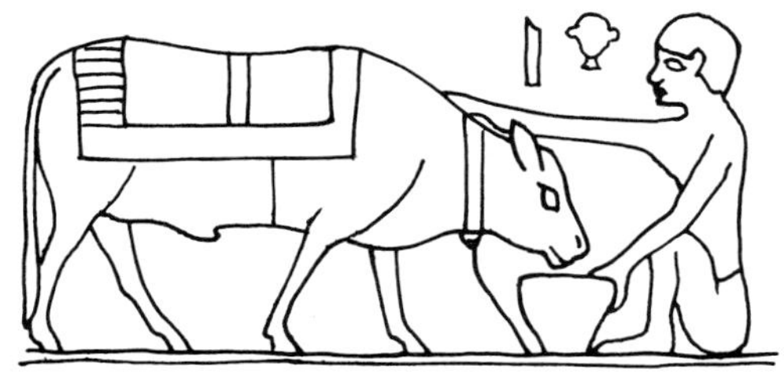

Fig. 7-58 (IB212)

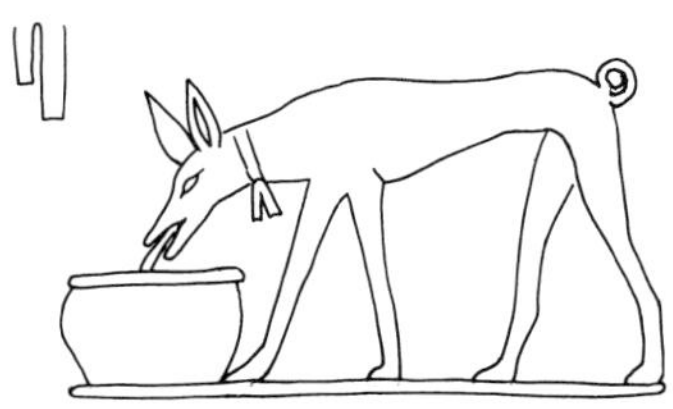

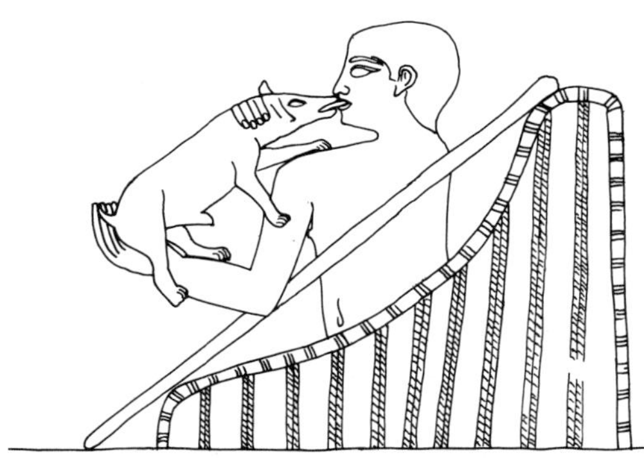

Fig. 7-59 (IB225)

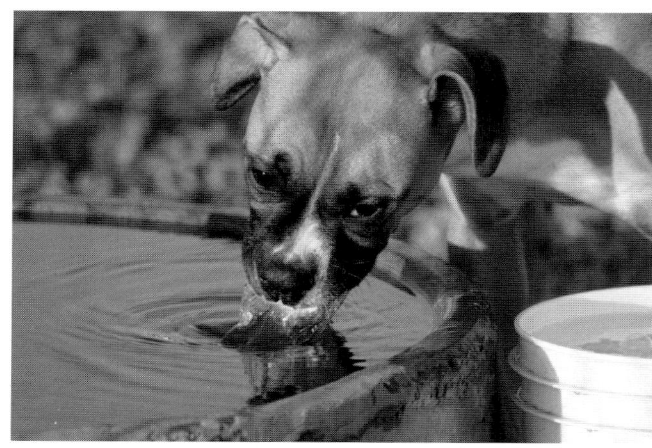

Pl. 7Y Dog drinking
© iStockphoto.com/Chris Johnson

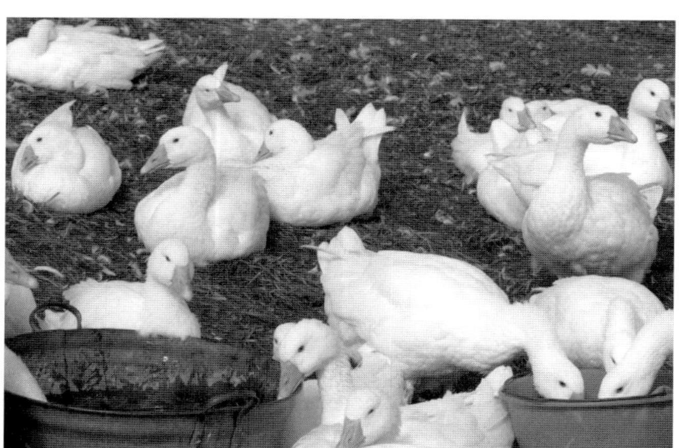

Pl. 7Z Geese drinking
© iStockphoto.com/Hanne Melbye-Hansen

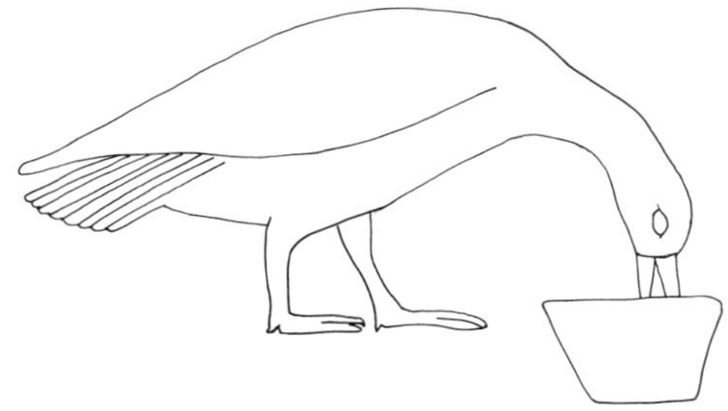

Fig. 7-60 (IB232)

Twenty-one examples of drinking by cattle were found in 16 Memphite tombs and fragments. Of these, seven depict animals drinking as they ford natural streams, while the remaining 14 show cattle imbibing from bowls provided by their handlers.

The earliest example occurs at Giza in the late 4th dynasty tomb of *nb.j-m-3ḫtj* (IB204; see fig. 7-54). Here, a herd of nine cattle fords a shallow stream, following a small boat containing a man who gestures toward the animals with his right arm.[107] The animal at the head of the line leans forward and breaks the surface of the water with the tapered tip of its tongue, which protrudes well beyond its open mouth. It thus appears to lap the water, something that cattle do not do when they drink. Three later depictions of fording cattle, all from Saqqara, are very similar to this example. In the late 5th dynasty tomb of *ṯjj* (IB216; see fig. 5-15 in Chapter 5), in which a herd of three animals follows a herdsman with a calf upon his back, the cow at the head of the line also leans forward and touches, with its protruded tongue, the surface of the stream through which they are walking. So too, in an almost identical depiction from the contemporary tomb of *3ḫtj-ḥtp* (IB218), the lead cow also protrudes its tongue, but fails to reach the surface of the water; nonetheless, it seems likely that this was the artist's intention. Finally, in a fragment from the early 6th dynasty tomb of *nj-ʿnḫ-nswt* (IB222; see fig. 7-55), which also depicts cattle fording a stream while following a herdsman carrying a calf, an animal at the head of the line touches the water surface with its tongue.

Other representations indicate that some artists were aware that cattle do not lap while drinking, by showing, instead, only their muzzles in contact with water. In a fragment from an anonymous 5th dynasty Saqqara tomb (IB205; see fig. 7-56), two members of a fording herd lean well forward so that the ends of their muzzles touch the water surface. Cattle are also shown in a similar manner in fording scenes from the mid-late 5th dynasty tomb of *nj-ʿnḫ-ẖnmw/ẖnmw-ḥtp* (IB210) and that of *rʿ-šps* (IB213). Although these last examples are biologically correct, we cannot be certain that a depiction of drinking was actually intended; the placement of the animals' heads may simply be accidental. Curiously, the biologically incorrect examples, in which cattle protrude their tongues into water, seem *less* ambiguous as the direct contact between the animals' mouths and fluid suggests, at once, that they must be drinking. It is not possible to say, however, whether this mistake (and its ensuing impression), was intentional or not. It is also important to note that these differences may simply be the result of copying. Three of the four incorrect examples show cattle following a herdsman carrying a calf upon his back, while all of the biologically correct depictions do not include this feature, suggesting that copying from earlier tombs might also explain the apparent difference in behavioural awareness (e.g., the scenes from the tombs of *ṯjj* and *3ḫtj-ḥtp* are almost identical).

The remaining images all depict cattle that appear to drink from deep bowls placed on the ground before them. Although the sex of the animals is often hard to determine, none of them appear to be female. In most examples, a herdsman stands or squats nearby. Four of the animals are horned, while the remaining ten are all hornless. One of the horned cattle (IB214) provides confirmation of their behaviour as it is clearly labelled, *zwr r(n) jw3* 'Young ox drinking'.[108] The animal appears to lap at the fluid, as its tongue seems to protrude from its open mouth, but damage to this area of the relief makes this difficult to confirm. Subsequent examples of horned cattle show two with their muzzles in contact with the surface of the container (IB220 and IB221), while another holds its head above a broad-rimmed bowl (IB224). It may be argued that this last animal is not drinking, but provision of a container for cattle is sufficiently unique in wall scenes that, when compared with the images discussed above, this interpretation seems feasible. While all four horned animals are standing, an ox in the tomb of *mrrw-k3.j: mrj* bends its forelegs and kneels in order to access its bowl. Cattle will kneel in this manner if a desirable food item is beyond their reach,[109] and also when suffering from lameness. It is also possible that the animal is depicted in the process of either getting up or lying down.[110]

107 For the meaning and significance of hand signals in fording scenes, see Dominicus, *Gesten und Gebärden*, 131-143.

108 Simpson, *Sekhem-ankh-ptah*, 9.

109 Hafez et al. in Hafez, *Behaviour of Domestic Animals*, 236.

110 Vandier, *Manuel v*, 77.

The ten hornless cattle generate some interesting questions. In all but two of these images, a herdsman sits before his charge, and stretches out one arm to touch it gently. There is some suggestion of a stylistic change in this action, for while earlier examples show the herdsmen touching the top of the animals' heads (IB206, IB208, IB209, and IB211; see fig. 7-57), in later images the hand position shifts further back to touch their shoulders (IB212, IB215, and IB217; see fig. 7-58 and fig. 11-20 in Chapter 11). Both of these hand positions draw the viewer's attention to the animals' lack of horns.

It is widely believed that images of hornless cattle represent a distinct breed in ancient Egypt, different from long-horned (*ng3w*) and short-horned (*wndw*) cattle.[111] The breed is frequently referred to as *ḥry-ḏbˁ*, because cattle with this label are invariably depicted without horns.[112] The term *ḥry-ḏbˁ* has been translated variously as 'That which has a pointed head' in reference to the animals' dome-shaped crowns,[113] and 'That which is on the finger',[114] but Hans Goedicke offered an alternative definition: 'Finger-suckled calf'.[115] This new interpretation was derived from an inscription found in the tomb of *mrs-ˁnḫ III* (Giza, IV.6): *jtj bḥs šdjj ḥry-ḏbˁ*, which he translated as 'Pulling the calf which was suckled upon the finger'.

The meaning of this expression was found in modern farm practice. After they are separated from their mothers, calves must be taught how to drink from a container. This is done by taking advantage of their readiness to suck on any teat-like object, such as a finger. Once it has latched on to a digit, the hand is slowly lowered into a bucket of milk (or substitute). By this method, the calf eventually learns the association between the container and being fed.[116] Goedicke believed that *ḥry-ḏbˁ* thus refers to this weaning method and that the term denoted "the young calf reared in this way as 'the one upon the finger'".[117] This finding suggests that cattle labelled *ḥry-ḏbˁ* must consequently be calves, and, as animals so-labelled are always hornless, by extension, many images of hornless cattle may also represent calves. Depictions from Giza and Saqqara of hornless cattle drinking from bowls have a number of features that suggest that they are indeed calves.[118]

One of the images of drinking cattle is clearly labelled *ḥry-ḏbˁ* (IB212; see fig. 7-58). This example, which appears in the late 5th dynasty tomb of *rˁ-špss*, shows a small animal wearing a blanket, leaning well-forward so that its muzzle hovers above the rim of a deep bowl proffered by its handler. The herdsman does not touch the animal's mouth, but holds the bowl steady with one hand while patting his charge with the other. In an earlier, but damaged, relief from the tomb of *k3.j-dw3* (IB211), a handler does seem to hold his hand to the muzzle of his charge, but as the bowl is missing in this example, it can not be treated as an unambiguous example of drinking. In a later example from the tomb of *j3zn* (IB223), however, a herdsman undoubtedly places one hand on the mouth of a young animal whose head is suspended above a bowl. Unfortunately, the placement of his fingers is unclear. Nevertheless, these three examples (particularly the *ḥry-ḏbˁ* from the tomb of *rˁ-špss*) may indicate that calves were indeed weaned by 'hand'.

The animal in the tomb of *j3zn*, like that of *rˁ-špss*, also wears a blanket upon its back. In fact, five of the ten hornless animals drinking from bowls have received this special treatment. In modern farm practice, calves are sometimes fitted with a blanket of cloth or wool during extremely cold weather or, more commonly, when they are ill. Adult cattle are not usually provided with a cover. It is possible then, that cattle shown wearing blankets in wall scenes may indicate that they are in fact

111 For example, see Paton, *Animals of Ancient Egypt*, E.3 (d) 'Hornless Cattle'; Brewer et al., *Domestic Plants and Animals*, 84.

112 *Wb iii* (136, 6) defines *ḥry-ḏbˁ* as 'hornloses Rind'. It should be noted, however, that depictions of cattle labelled *jw3*, *ng*, or *k3* are also occasionally shown without horns.

113 Andersson, E. (1912). "La dénomination égyptienne des boeufs sans cornes", *Sphinx* 16, 145-164.

114 Montet, *Scènes*, 139.

115 Goedicke, H. (1957). "*ḥry-ḏbˁ*, 'Suckled calf'", *RdÉ* 11, 159-161.

116 This is the method that is used by farmers today. It is a slow process, taking anywhere from two days to a week before the calf is feeding completely on its own. See Thear-Fraser, *Complete Book*, 178. Goedicke presents a slightly different method whereby the calf is encouraged to suck upon the fingers of a hand held upon the surface of a bucket of milk.

117 Goedicke, *RdÉ* 11, 160.

118 Vandier (*Manuel v*, 79) identifies five hornless animals drinking from bowls as calves.

calves.[119] It is intriguing to note that recent research has revealed that both beef and dairy calves will gain weight at a significantly faster rate if reared for the first few months of life wearing a blanket.[120] Possibly the ancient Egyptian farmers and pastoralists were similarly aware of the benefits of this simple husbandry procedure.

The youth of the hornless animals is also suggested by a final feature: an umbilical cord hanging from their belly. This is particularly noticeable in an example from the tomb of *ṯjj* (IB215; see fig. 11-20 in Chapter 11), which shows a very small calf with a rope-like object stretching down to a tuft of grass between its legs. At least 15 cm of umbilical cord are usually left hanging from a calf's belly after birth. This slowly shrivels during the following two weeks before ultimately falling off. The presence of a cord in Egyptian depictions of calves thus gives us a reliable indicator of their age (i.e., they can be no more than two weeks old). Other images may also show this feature. For example, the cattle in the tombs of *nfr/k3-ḥ3.j* (IB206 to IB208; see fig. 7-57), *nj-ʿnḫ-ẖnmw/ẖnmw-ḥtp* (IB209), and *j3zn* (IB223) each appear to have what may be an umbilical cord hanging from their ventral surface. These protuberances are far shorter than that on the *ṯjj* calf, however, and so they may instead depict the penis sheath, which is located next to the naval and can be of considerable length in adult bulls.

The general impression presented by the hornless cattle drinking from bowls is that they depict calves. This individualized attention likely represents the process of weaning in which calves learned to drink instead of suckle. Douglas Brewer and colleagues suggest that, based on modern ethnographic evidence, milk production by ancient Egyptian cows was relatively low.[121] Weaning will have enabled herdsmen to halt unlimited access by calves to their mothers' milk, thereby freeing more of it for human consumption (for more details about milk production, see Chapter 11).[122] It is not surprising, then, that this essential husbandry practice was deemed worthy of reproduction in tomb scenes, along with other important agricultural procedures.

Dogs

k3-gm-nj: mmj (IB225: Saqqara, VI.1M)

Two examples were found of dogs drinking, both located in the same scene on the northern wall of the pillared hall of *k3-gm-nj: mmj* (IB225; see fig. 7-59). The first of these depicts a female, adult dog with its tongue protruding from its open mouth so that the tip touches the rim of a large bowl set before it. The hieroglyphic label that accompanies this image is eroded, but the sign *s* (Gardiner sign list S29) is still apparent. Firth and Gunn suggested that this label was "probably *s[nmt]* 'feeding animals'", as they believed that the dog was "eating from a jar".[123] While such an interpretation is undoubtedly feasible, I suggest that similarities between this image and those of cattle imbibing from bowls render it more likely that the dog is drinking.[124]

It should be noted that the artist has particularly emphasized the animal's tongue. Dogs use their tongues both for eating and for drinking. Unlike other carnivores, however, which lap at water by curling their tongues upwards and then scooping it into their mouths, dogs bend the tip of theirs downward to draw the fluid in from underneath (pl. 7Y). Their tongue is thus particularly apparent

[119] Sixty-four examples of animals wearing blankets were found during this study; of these, 39 depict cattle. Four of these animals are horned, but the remaining 35 (90%) are all hornless. There is thus a strong association between special care (i.e., provision of a blanket) and the lack of horns.

[120] See Loy, T.W., Schroeder, J.W., Lardy, G.P., Wallace, G.T., Zimmerman, M., Rose, M., Dhuyvetter, J. and Slanger, W.D. (1999). "Effect of wool/polyester blankets on performance and gain of newborn dairy and beef calves", *J. Dairy Sci.* 82(Suppl. 1), 843.

[121] Brewer et al., *Domestic Plants and Animals*, 85.

[122] A domesticated cow can continue to produce milk for seven to 10 months after her calf has been weaned. See Thear-Fraser, *Complete Book*, 180.

[123] Firth-Gunn, *Teti Pyr. i*, 114.

[124] The hieroglyphic label might thus be *s[wr]*, 'Drinking'. While later renditions of this word begin with Gardiner sign list S29 (i.e., *swr*), earlier spellings appear with Gardiner sign list O34 (i.e., *zwr*). The hieroglyphic label accompanying IB225 is possibly an early example of this later spelling. See *Wb iii*, 428-429.

during this awkward method, which also forces them to gulp noisily - features that may have attracted *k3-gm-nj: mmj*'s artist. I suggest, furthermore, that the dog is drinking milk.

Activities associated with milk characterize the entire register in which the dog appears as, further to the left, a cow is shown being milked by a herdsman while another is lassoed as she suckles her calf (see fig. 11-21 in Chapter 11). This interpretation is also supported by a second image appearing just below the first that depicts a seated man holding a puppy to his face (see fig. 7-59). This small animal has often been confused for a piglet due to the shape of its muzzle, tightly curled tail, and small pendulous ears.[125] The first two features are shared with the adult dog in the sub-register above and so, as it also has paws and not hooves, it is undoubtedly a puppy.[126]

It has been proposed that the man is squirting fluid from his mouth into that of the dog, as the two are connected by an elongated structure that emerges from their lips.[127] It seems likely, then, that the herdsman is attempting to wean the puppy in this complicated fashion using milk taken from a jar offered by a colleague who stands before him. I propose, however, that in accordance with other depictions of animals drinking in wall scenes, the puppy is instead protruding its tongue to lick the fluid directly from the man's mouth. The depictions of both dog and puppy thus mirror one another: they are both lapping milk.

Birds

spd-ḥtp (IB226: Saqqara, V.6L-9?)
ṯjj (IB227-230: Saqqara, V.8-9)
k3-gm-nj: mmj (IB231-232: Saqqara, VI.1M)
mrrw-k3.j: mrj (IB233: Saqqara, VI.1M-L)
mrjj-ttj: mrj (IB234: Saqqara, VI.2M-L)

In force-feeding scenes from Saqqara, ducks, geese, and cranes are depicted both restrained (while undergoing the procedure), and roaming freely, either before or after being fed. In a few scenes, unrestrained birds are sometimes shown leaning over small bowls placed upon the ground. I propose that these containers are water bowls provided specifically for birds that have just been fed.

Most bird species drink by first taking a sip and then throwing back their heads to allow the fluid to travel down their throats.[128] All birds must drink to replace body fluids lost during respiration and excretion, but some species have extra requirements. Ducks need water to digest their food. They must consequently drink frequently (cf. pl. 7Z) and will even place food in water prior to ingesting it. Adult ducks will become ill if deprived of water while feeding, and ducklings can die. It is important that captive ducks are provided with water containers that are of sufficient depth to allow their entire bill to be immersed.

In force-feeding scenes, birds are shown being held by the throat while small food items are placed into their beaks. As the intention was to fatten them, it is assumed that a large quantity of food (i.e., more than one item) was pushed into each animal during one sitting. The need for water immediately after such a procedure would have been essential. It is likely, then, that the occurrence of bowls in force-feeding scenes was to provide water for birds that had recently been fed manually, to aid the passage of dry feed through their digestive tracts. The birds in the tombs of *spd-ḥtp* (IB226; see fig. 6-31 in Chapter 6), *ṯjj* (IB227 to IB230), *k3-gm-nj: mmj* (IB231 and IB232; see fig. 7-60), and *mrrw-k3.j: mrj* (IB233) should thus be seen as sipping at water by bending forward deeply to place the tips of their bills on to the surface of the proposed water bowls. A goose in the tomb of *ṯjj* (IB229) also elevates its bill, as though tipping the fluid down its throat.

[125] This mistake has been repeated many times even though Firth and Gunn (*Teti Pyr. i*, 114) originally, and correctly, identified it as a "young dog". For example, see Handoussa, T. (1986). "Le chien d'agrément en ancienne Égypte", *GM* 89, 29, n. 50; Janssen-Janssen, *Egyptian Household Animals*, 33; Ikram, *Choice Cuts*, 30 and 305; Germond, *Egyptian Bestiary*, fig. 62.

[126] Some authors have re-drawn the image so that, instead of paws, it is shown with hooves! See Janssen-Janssen, *Egyptian Household Animals*, fig. 27.

[127] Harpur, Y. (2005). "Recording in the Saqqara tomb-chapel of Kagemni: A summary of work, opinions and ideas", *ASAE* 79, 119.

[128] Burton, *Bird Behaviour*, 72.

Finally, in the tomb of *mrjj-ttj: mrj*, a number of geese wandering freely in a poultry-yard are shown bending forward so that their bills enter a pool of water in which other birds swim (IB234). They thus appear to be drinking directly from the pond. It should be noted, however, that many other birds in the immediate environment adopt a similar posture in order to peck at seeds on the ground. It is possible, then, that the connection between the birds' bills and water is accidental and simply due to clumsy placement of the animals by the artist. No other examples of drinking in poultry-yard scenes were detected.[129]

SUMMARY

- A wide range of species are depicted eating and drinking in Memphite tomb scenes.
- Cattle, donkeys, and sheep are rarely shown grazing naturally; rather, they are usually depicted feeding as a result of human intervention or activity, such as while threshing, treading seeds, ploughing, or transporting items. Some are fed by hand. Goats are shown browsing in a natural manner, but only while under human control.
- The depiction of wild animals feeding often reveals an awareness of species-specific behaviour: for example, underwater foraging by cormorants or the tendency for Nile crocodiles and some waterbirds to ingest fish headfirst.
- Some animals protrude their tongues while feeding or drinking, such that the tip either touches a potential food item or is immersed in water. This feature accurately reproduces the natural behaviour of some animals (e.g., feeding by cattle), but is occasionally included incorrectly (e.g., feeding by donkeys, drinking by cattle). Regardless of its biological accuracy, however, animals with extended tongues were found to clearly convey the idea of ingestion due to the unambiguous visual link between their mouths and food or water.

[129] In poultry-scenes in both the tomb of *ṯjj* (portico, south wall) and the tomb of *wḏꜣ-ḥꜣ-ttj: ššj* [*nfr-sšm-ptḥ*] (room II, east wall), however, a bird is depicted leaning over so that its bill touches the upper edge of the central pool. As in each case the bill does not enter the water, I have not included these as examples of drinking.

CHAPTER 8

HUNTING BEHAVIOUR

DEFINITION

Predation is defined as an interaction between species in which one animal obtains food by hunting, killing, and consuming another.[1] Predators are found in all animal classes (mammals, birds, reptiles, amphibians, fish, and invertebrates). Some predators (e.g., insectivores, piscivores, etc.) prey exclusively on one animal group, but most catch and eat a variety of species. Predators that rely on stealth and surprise to catch their prey are often solitary. Social species usually hunt in groups to out-manoeuvre and chase down their prey, often in a coordinated fashion. Many animals also prey indirectly by scavenging from the kills made by other predators.

8.1 PREDATION

Lions

nj-ʿnḫ-ẖnmw/ẖnmw-ḥtp (HB1-2: Saqqara, V.6L-7)
ṯjj (HB3: Saqqara, V.8-9)
ꜣḫtj-ḥtp (HB4: Saqqara, V.8L-9E)
ptḥ-ḥtp II: ṯfj (HB5: Saqqara,V.9M-L)
sšm-nfr IV (HB6: Giza, V.9-VI.1)
nj-mꜣʿt-rʿ [*jzzj-mr-nṯr*] (HB6a: Giza, V.9-VI.1)
Anon (HB7: Saqqara, V-VI)
mrrw-kꜣ.j: mrj (HB8: Saqqara, VI.1M-L)
mrjj-ttj: mrj (HB9-11: Saqqara, VI.2M-L)
wr-nw (HB12: Saqqara, VI.3-4?)
ḥʿ-bꜣw-ẖnm (HB13: Saqqara, VI.4)

African lions are accomplished and adaptable predators. They hunt and eat a variety of animal prey (particularly ungulates), but they will scavenge readily if the opportunity arises. Lions usually search for prey alone, but they are also known to hunt cooperatively in pairs or small family groups.[2] A single lion can bring down an animal that is twice, and sometimes four times, its own weight (e.g., buffaloes).[3] Lionesses hunt more frequently than males;[4] however, due to their greater size and strength, male lions are able to attack larger prey than females.

Once a potential prey animal has been located, the lion will endeavour to get as close as possible to it without being detected. It advances slowly in a stalking posture with its body held low and tense and its head stretched forward.[5] Once it is close enough, it then rushes at its target. Medium-sized prey are chased from behind and brought down by a slap with one paw to their hindquarters, or by clutching them on their haunches with both paws and dragging them to the ground. Lions attack large prey by rearing up and throwing their body over that of their quarry and then using both their strength and weight to pull it over (see pl. 8A); the neck, shoulders, or back of the victim are also bitten to help the predator maintain its grip. Shortly after the animal falls, the lion either strangles it to death by biting its neck or throat, or slowly suffocates it by clamping its jaws over the prey's mouth and nostrils (see pl 8B).[6] Animals are occasionally consumed where they are killed, but usually their bodies are moved some distance away before feeding begins.[7] Small animals (up to 40 kg) are lifted and carried (see pl. 8C); large prey are either dragged or pulled.[8] Hunting is more frequent at night than during the day.

---

1 Allaby, *Dictionary of Zoology*, 375.
2 Estes, *Behavior Guide*, 372.
3 *ibid.*, 370.
4 Eighty-five to 90 percent of hunts are carried out by females. See *ibid.*, 372-373.
5 Rudnai, *Social Life of the Lion*, 87.
6 Schaller, *Serengeti Lion*, 263-267; Rudnai, *Social Life of the Lion*, 89-91; Estes, *Behavior Guide*, 373.
7 Schaller, *Serengeti Lion*, 267-268; Rudnai, *Social Life of the Lion*, 91.
8 *ibid.*, 91-92.

Thirteen reliefs from 10 tombs were found to depict the predatory behaviour of lions. The scenes, which date from the mid-5th to early 6th dynasty, can be divided into three motifs, each of which represents a different part of the lions' hunting sequence. These are: 1) the attack, 2) suffocation of prey, and 3) moving of prey. All of the depicted lions have a mane (either a full or partial ruff) indicating that they are males.

Two images show animals being knocked to the ground. The first of these, which appears in the Giza tomb of *sšm-nfr IV* (HB6; see fig. 8-1), depicts a lion that has leapt upon the back of a large bull, causing its hind legs to collapse. The presence of both forepaws on one side of the victim's body give the lion an awkward appearance, but the artist's depiction of the predator's jaws sinking into the bull's shoulder is surprisingly accurate (cf. pl. 8A). In a later scene from the tomb of *wr-nw* (HB12; see fig. 8-2), the artist has successfully placed the lion's forelegs on either side of a fallen oryx upon whose back it has sprung. The predator's tail is erect and curved over its body. Although this can indicate a general sense of excitement, lions usually hold their tails straight and largely horizontal when chasing prey;[9] the depicted posture may thus be due to lack of space at the end of the register. Both scenes show a remarkable awareness of the lions' method of attacking large prey, particularly the use of their forepaws to bring down their quarry. This knowledge can possibly be traced back to the predynastic period, as objects such as the Gebel el-Tarif and Gebel el-Arak knives depict lions that rear up and grab the haunches of their victims.[10]

Their art also reveals that the Egyptians possessed a clear understanding of the lions' ability to suffocate prey. Three scenes from the Saqqara tombs of *ptḥ-ḥtp II: ṯfj* (HB5; see fig. 6-32 in Chapter 6), *mrrw-k3.j: mrj* (HB8; see fig. 8-3), and *mrjj-ttj: mrj* (HB11; see fig. 8-21), depict lions engaged in this behaviour while attacking bulls.[11] A number of idiosyncratic details have been included in all three examples, indicating that they must be the result of close observation (cf. pl. 8B). First, the jaws of the lions are clamped tightly over the muzzles of the bulls, precisely as they are when smothering prey. Each lion 'sits' upon the substrate. As the suffocation of prey can take many minutes,[12] lions will often lie upon the ground to help maintain their mouth-hold and to apply their dead weight to keeping their victim still. The predators in the tombs of *mrrw-k3.j: mrj* and his son are in a seated posture, but the lion pictured in the earlier tomb of *ptḥ-ḥtp II: ṯfj* is shown with outstretched hind legs, as though to indicate that it is fully recumbent (cf. pl. 8B). Each of the depicted animals also lifts one paw to the muzzle of its prey, imitating identical paw movements made by lions to keep their victims still (cf. pl. 8B). Finally, the lions hold their tails in a highly erect position, indicating a heightened sense of excitement.

Other examples of the suffocating muzzle-bite can be seen in Egyptian art, but its meaning has rarely been interpreted correctly.[13] One of the earliest depictions of this behaviour can be seen on the predynastic Oxford palette (reverse) in which two lions press their muzzles against those of two gazelles.[14] One lion also places a forepaw on the body of its prey. These interactions have been described as "fraternising",[15] "biting in a fashion that almost resembles kissing",[16] and

---

9 Schaller, *Serengeti Lion*, 103.

10 It should be noted, however, that these objects, or at least their motifs, may be of Mesopotamian origin. For a discussion of Near Eastern influences on early Egyptian iconography, see Davis, *Masking the Blow*, 22-27.

11 A badly preserved image in the tomb of *nj-m3ʿt-rʿ* [*jzzj-mr-ntr*] (HB6a) may also have originally depicted this behaviour, as it appears to show a large animal defecating and stretching away from an unidentified aggressor. Roth (*Cem. Pal. Att.*, 132-133) described it as such based upon its similarity to the relief in the tomb of *ptḥ-ḥtp II: ṯfj*.

12 Rudnai, *Social Life of the Lion*, 89. Estes (*Behavior Guide*, 373) notes that the strangulation of prey can take up to 13 minutes.

13 Kanawati and Abder-Raziq (*Meryteti*, 25), however, state that the lion in the tomb of *mrjj-ttj: mrj* bites the muzzle of a wild bull "in order to suffocate it".

14 Fischer has pointed out that none of the animals on the Oxford palette actually bites their prey, but that, like the lions, simply press their muzzles against their victims; he suggests that this was an artistic convention that was eventually replaced after the 1st dynasty by a bite with open jaws. Nevertheless, only the lions on the palette abut their mouths against those of their prey. See Fischer, H.G. (1958). "A fragment of late predynastic Egyptian relief from the eastern Delta", *AA* 21(1), 75-77.

15 Schott, S. (1952). "Kulturprobleme der frühzeit Ägyptens", *MDOG* 84, 8; Ridley, *Unification of Egypt*, 27.

16 Davis, *Masking the Blow*, 84.

"metaphorical rather than literal".[17] On the contrary, they are highly accurate, if somewhat stylised, representations of the suffocation of prey by lions.

The remaining images consist of a curious motif that appears to depict the moving of prey animals (HB1 to HB4, HB7, HB9, and HB10; see figs. 8-4 and 8-5; also fig. 8-9).[18] The predators occur in scenes in which wild animals confined in an enclosure are hunted by men and dogs. Netting fences represented as long (sometimes hatched) strips, run down either side of the areas within which the trapped animals are pursued. At the top of each barrier, the heavily-maned head and forequarters of a lion emerges from, and reaches 'over', the fence to bite an ungulate upon the throat and grasp its body with its forepaws. The prey animals in the earliest examples are scimitar oryx, but in all subsequent images, dorcas gazelle are depicted. The bodies of the prey animals hang lifelessly: their heads are flung back or loll forward and their limbs droop languidly. In one example, the lion has managed to lift a gazelle off the substrate (HB7; see fig. 8-5). In a fragment from the tomb of *ꜣḫtj-ḥtp* (HB4), the lion is no longer preserved, but a single paw and the gazelle's posture attest to its presence. In the tomb of *ṯjj* (HB3; see fig. 6-33 in Chapter 6), the gazelle defecates as its body is hoisted, while in an example from the tomb of *mrjj-ttj: mrj* (HB10), the victim protrudes its tongue, either indicating a loss of muscle control associated with death or a cry of pain.

The tombs of *nj-ꜥnḫ-ẖnmw/ẖnmw-ḥtp* (HB1 and HB2; see figs. 8-4 and 8-9) and *mrjj-ttj: mrj* (HB9 and HB10) feature a lion atop each fence, suggesting that the remaining examples may also have originally formed part of pairs. This is supported by a cylinder seal found in the mid-6th dynasty tomb of *ḥꜥ-bꜣw-ẖnm* (HB13) that depicts the bodies of gazelle suspended over crude fences,[19] on either side of which stand 'dual' lions.

I propose that the motif represents the removal of prey animals from hunting enclosures by lions (cf. pl. 8C). Lions are often attracted to areas where potential prey are concentrated, such as corrals holding cattle,[20] and so a well-stocked hunting park would provide an irresistible target. Wall scenes reveal that netting fences surrounded these enclosures. It is impossible to determine how high this fencing may have been, but as it was probably intended to be a temporary structure, I suggest that it was of only moderate height. Lions are capable of jumping three metres high from a standing start and can leap up to 12 metres horizontally.[21] They are also known to enter fenced enclosures ('bomas') in many parts of Africa where they frequently attack and kill livestock.[22] It is entirely feasible then, that lions may have raided hunting parks by jumping over the temporary fencing, killing or seizing an ungulate by a bite to the throat, and then leaping back across the barrier with their victim to consume it elsewhere. As noted above, the prey animals in the earliest scenes (HB1 and HB2; see figs. 8-4 and 8-9) are scimitar oryx, which can weigh up to c. 180 kilograms; male lions can drag animals of this size, but it is unlikely that they could lift one over a fence. Dorcas gazelle, however, which are the most frequent prey in these scenes, weigh no more than 20 kilograms and could be carried easily, particularly by male lions.

Support for this interpretation of the motif was found in the tomb of *ꜣḫtj-mrw-nswt* in which men are depicted filling granaries.[23] The lower parts of the men's bodies are hidden so that only their head, arms, and chest emerge from the top of the tall structures as they reach out to accept basketfuls of grain from their colleagues standing below. Although Smith believed that the men are standing *inside* the granaries,[24] they are in fact reaching *over* the roofs of these sealed structures in order to pour the grain into small apertures at the top.

---

17 *ibid.*

18 Smith (*HESPOK*, 192) also describes these images. He mention one further example, which I have not been able to view, namely Cairo Museum 8.6.23.1, which was found by Quibell.

19 Jequier (*Monument Funéraire iii*, 67) believed that the gazelle are hanging over the vertical bodies of fish.

20 Peterhans, J.C.K. and Gnoske, T. (2001). "The science of 'man-eating' among lions *Panthera leo* with a reconstruction of the natural history of the Man-eaters of Tsavo", *J. East Afr. Nat. Hist.* 90, 12-17.

21 Burton-Burton, *International Wildlife*, 1452.

22 Patterson, B.D., Kasiki, S.M., Selempo, E. and Kays, R.W. (2004). "Livestock predation by lions (*Panthera leo*) and other carnivores on ranches neighboring Tsavo National Parks, Kenya", *Biol. Cons.* 119(4), 507-516.

23 See Smith, *HESPOK*, fig. 229.

24 *ibid.*, 347.

The lions atop fences encapsulate, in one concentrated image, all the essential elements of a raid on a hunting park: the manipulation of the prey, the throat bite, the lifting of the body, and the barrier. This is also achieved without jeopardizing the integrity of the hunt scene as a whole. If the lions had been depicted literally - that is, actually jumping the barrier – then their bodies and that of their prey would have broken the enclosure lines, creating a visually untidy image and leaving the scene potentially open to misinterpretation (is the lion jumping over the fence or breaking through it?).[25] The presence of two lions, one on either side of the enclosure, helps to maintain a balanced composition; it is interesting to note, however, that young male lions that have yet to establish a pride, often form pairs ('coalitions')[26] in which they work together to bring down prey.[27]

Leopards

nj-mꜣꜥt-rꜥ [*jzzj-mr-nṯr*] (HB14: Giza, V.9-VI.1)

A large cat that is depicted in a fragmentary hunting scene in the Giza tomb of *nj-mꜣꜥt-rꜥ* [*jzzj-mr-nṯr*] (HB14; see fig. 8-6) closely resembles a leopard in both morphology and posture.[28] The leopard appears in the midst of a pack of domestic hunting dogs that are killing animals. One dog immediately behind it has brought down a large ungulate, while ahead of the cat a pair of dogs are busy attacking a small, hoofed animal whose hind legs have collapsed. The leopard holds its head below the level of its hunched shoulders, a posture that is adopted by cat species when stalking prey (see pl. 8D). Comparison with other depictions of leopards reveals that Egyptian artists systematically represented this species with its head held in this manner in order to differentiate it from cheetahs which, although physically very similar to leopards, are almost always shown with an excessively long neck and their head elevated above shoulder level.[29] Despite the possibility that the artist has simply produced the standard 'leopard motif', I suggest that the scene in the tomb of *nj-mꜣꜥt-rꜥ* [*jzzj-mr-nṯr*] shows this cat in the act of stalking.

Leopards hunt primarily at night and rely on the cover of darkness to creep as close as possible to their intended victims. Large animals are seized by the throat and strangled, while smaller prey are killed quickly by a bite to the back of the neck.[30] They are opportunistic hunters that prey upon a range of vertebrates, including reptiles, fish, and mammals. They will also take advantage of kills made by other animals and will steal prey from cheetahs, hyenas, and any carnivore smaller than themselves.[31] Unlike lions, they frequently hunt and eat other predators, such as cheetahs and lion cubs.[32] In particular, it has been noted that they are especially prone to attack jackals and domestic dogs, with some leopards actually preferring canine flesh to other sources of meat.[33]

The leopard in the tomb of *nj-mꜣꜥt-rꜥ* [*jzzj-mr-nṯr*] advances toward two domestic dogs while adopting a stalking posture (cf. pl. 8D). I suggest that the image represents the leopard either as it is about to steal the hunting dogs' prey or actively stalking the dogs themselves. The scene is thus tense with anticipation as the leopard, slinking behind the preoccupied and unsuspecting canids, prepares to pounce. It is interesting to note that in the 3rd dynasty tomb of *nfr-mꜣꜥt* at Meidum (no. 16), a stalking leopard is featured in a register directly below a dog that attacks a small, collapsed ungulate.[34] The postures of the dog and its prey are very similar to those pictured before the leopard in the *nj-mꜣꜥt-rꜥ* [*jzzj-mr-nṯr*] scene. The presence of a hunter crouched near the Meidum

---

25 In later hunting scenes, however, the fence is sometimes 'broken' by a gap to accommodate the passage of arrows sent by hunters standing outside the park (e.g., in the Middle Kingdom tomb of *znt* and *jnt.f-jkr* (TT60) at Thebes. See Davies, *Antefoker*, pl. 7).

26 Estes, *Behavior Guide*, 371.

27 Bygott, D., Bertram, B. and Hanby, J. (1979). "Male lions in large coalitions gain reproductive advantages", *Nature* 282(20), 839-841.

28 Roth (*Cem. Pal. Att.*, 132) describes it as a "striding panther or another large cat".

29 Jèquier, G. (1913). "La panthère dans l'ancienne Égypte", *RES* 4, 353-372; Osborn-Osbornová, *Mammals*, 119.

30 Nowak, *Walker's Mammals i*, 828.

31 Apps, *Wild Ways*, 73.

32 Bothma, *Larger Carnivores*, 75.

33 Ewer, *The Carnivores*, 210; Apps, *Wild Ways*, 72; Bothma, *Larger Carnivores*, 75.

34 Petrie, *Medum*, pl. 17, lower left; Harpur, *Nefermaat and Rahotep*, 62, fig. 74 and pl. 5.

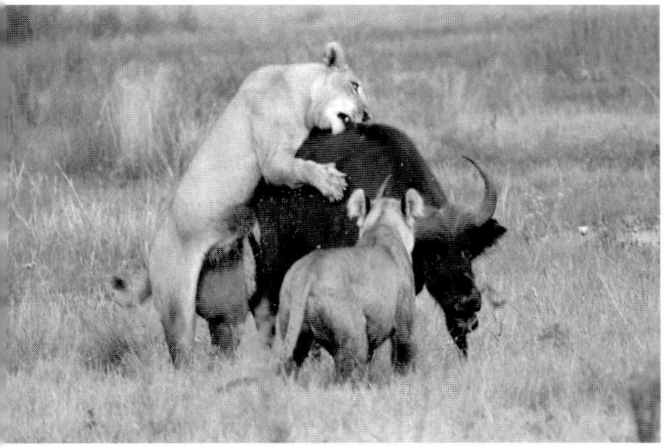

Pl. 8A Lions attacking
© iStockphoto.com/Peter Malsbury

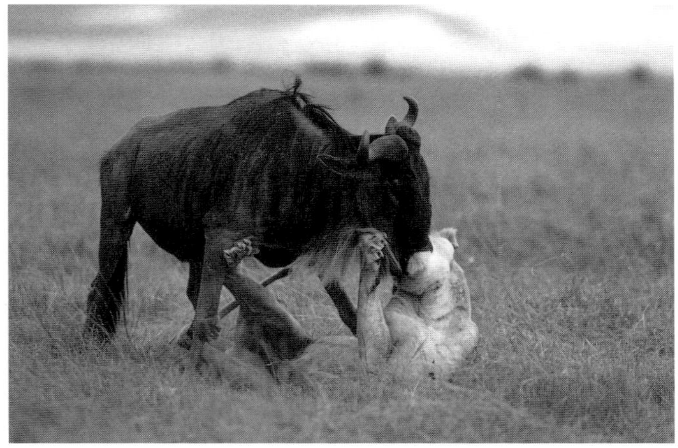

Pl. 8B Lion suffocating prey
© Richard Matthews and Samantha Purdy
Reproduced with permission

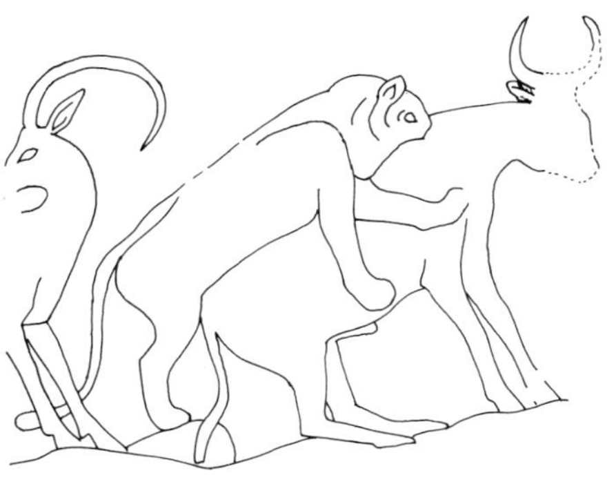

Fig. 8-1 (HB6)

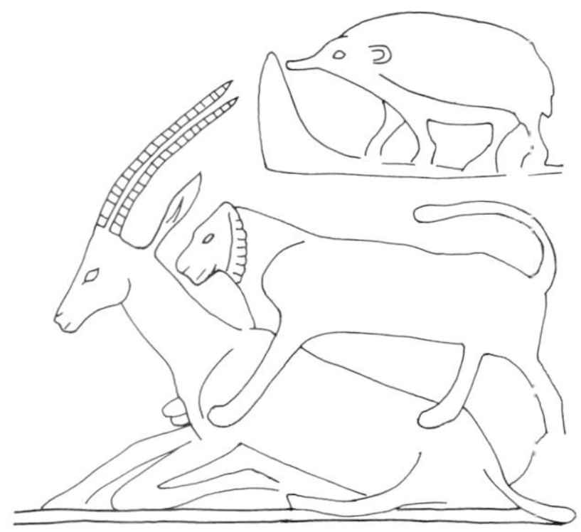

Fig. 8-2 (HB12)

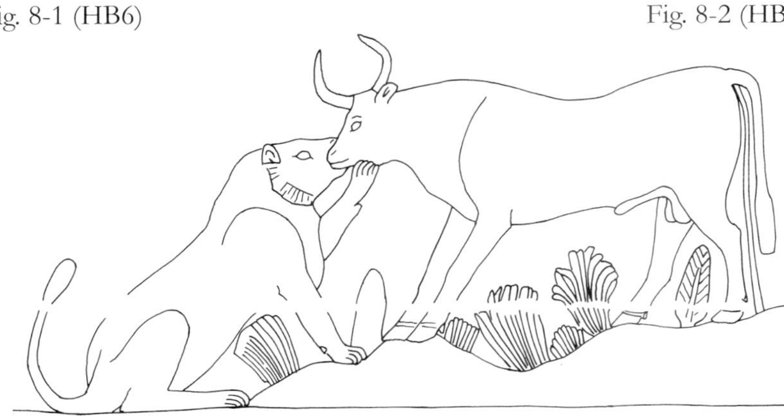

Fig. 8-3 (HB8)

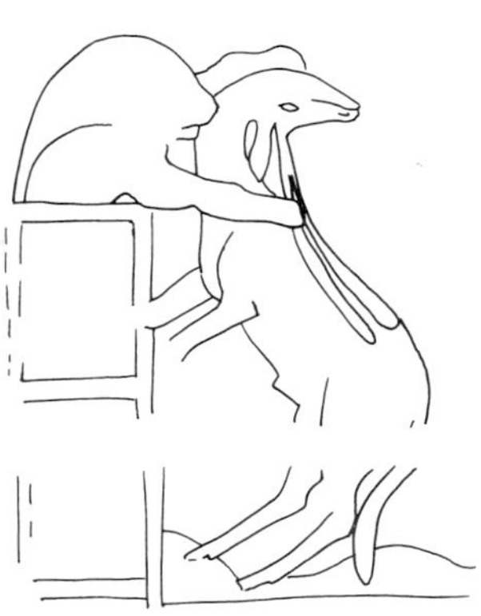

Fig. 8-4 (HB1)

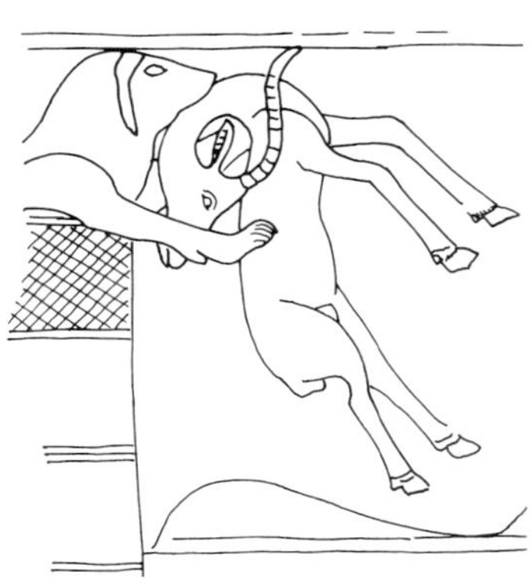

Fig. 8-5 (HB7)

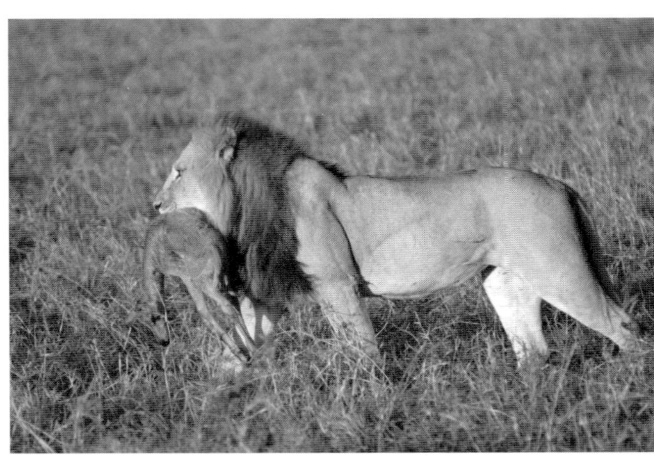

Pl. 8C Lion moving prey
© iStockphoto.com/Alain Couillard

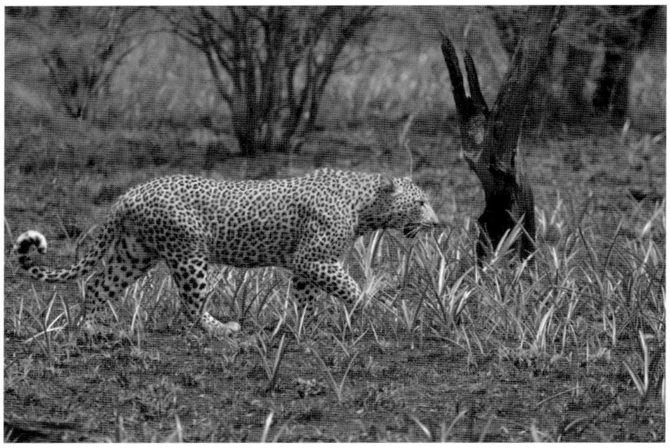

Pl. 8D Leopard stalking
© iStockphoto.com/Benjamin Tupper

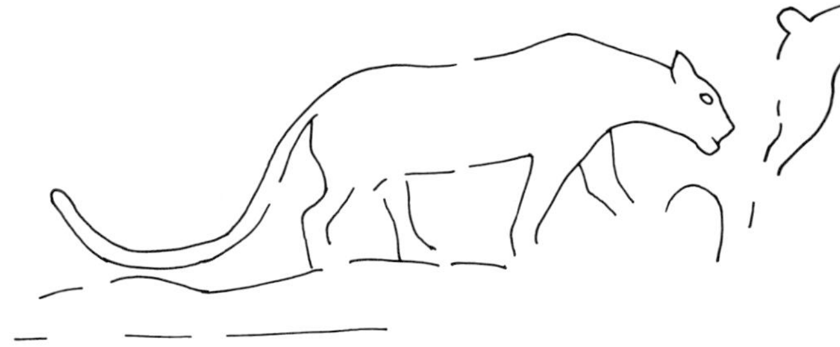

Fig. 8-6 (HB14)

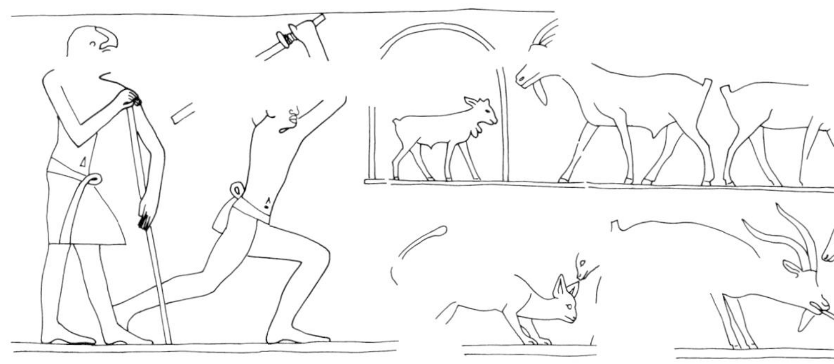

Fig. 8-7 (HB15)

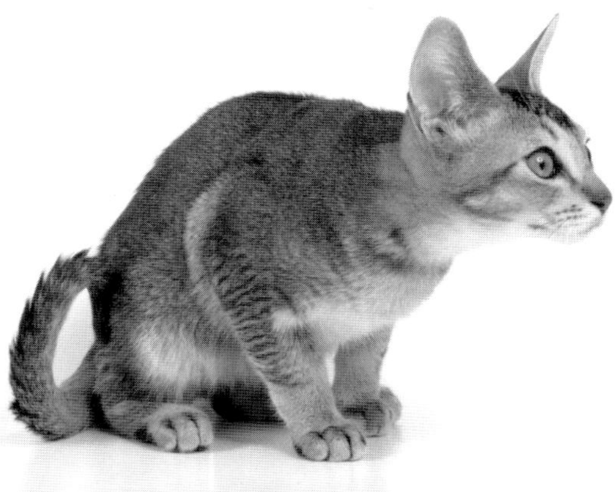

Pl. 8E Cat staring
© iStockphoto.com/Vladimir Suponev

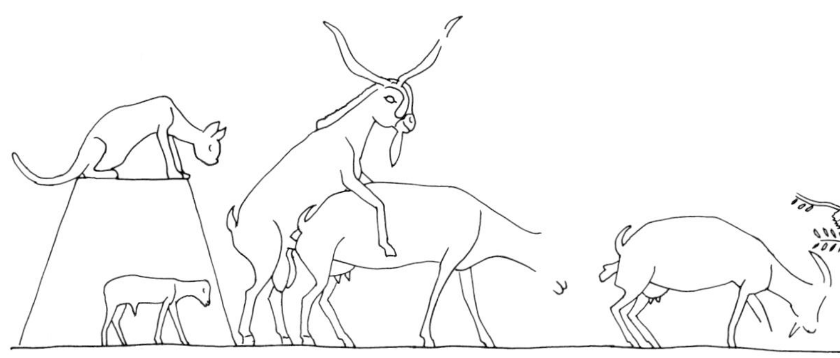

Fig. 8-8 (HB16)

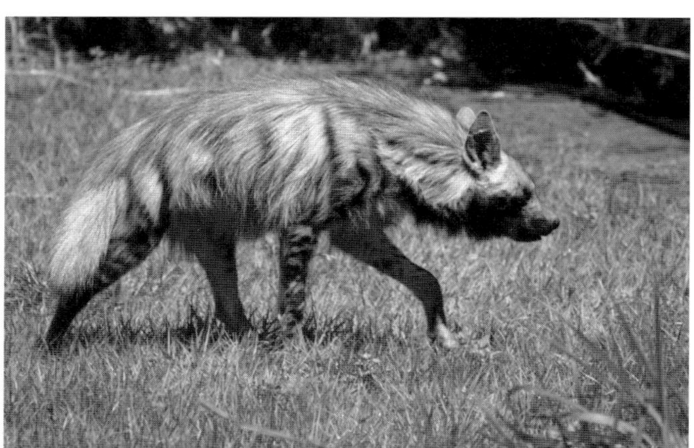

Pl. 8F Striped hyena foraging
© iStockphoto.com/Dawn Nichols

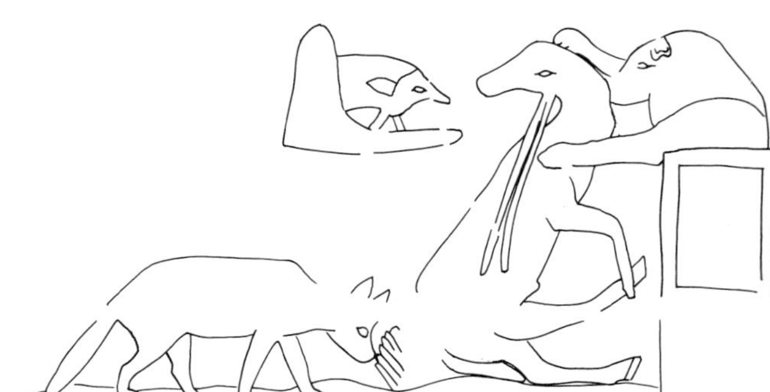

Fig. 8-9 (HB17)

leopard has prompted some to suggest that it is tame.[35] Although images of leashed leopards may indicate that the Egyptians kept some in captivity,[36] I believe that those appearing in hunting scenes represent wild animals that have been attracted to the possibility of a meal presented by both the dogs and their prey. The simultaneous appearance of leopards and dogs in hunting scenes is thus not arbitrary, but instead reveals awareness by the Egyptians of a particular, idiosyncratic aspect of leopard predatory behaviour.

Wild cats

nj-ꜥnḫ-ẖnmw/ẖnmw-ḥtp (HB15: Saqqara, V.6L-7) *ꜣḫtj-mrw-nswt* (HB16: Giza, V.9-VI.1)

In a pastoral scene in the tomb of *nj-ꜥnḫ-ẖnmw/ẖnmw-ḥtp* (HB15; see fig. 8-7), an animal that has been identified as a 'wild cat' threatens a goat.[37] African wild cats are medium-sized, nocturnal carnivores that prey upon a range of animals, including small mammals, birds, reptiles, and insects.[38] Their method of hunting is similar to that of most felines. Victims are first stalked until they are within striking range. The cats then pounce upon and grab their prey with their clawed forepaws, dispatching it with either a bite to the throat or by mauling the back of the neck, chest, and head. Wild cats can kill animals up to the size of a rabbit, and are also capable of taking the calves of small antelope species. They have been reported to kill young sheep and goats,[39] but others suggest that this is unlikely.[40] Hunting usually takes place at night.

The Egyptian relief shows a wild cat standing directly behind a goat giving birth. The predator's head, which is lowered in a stalking posture (see *Lions* and *Leopards* above), is held just below that of the emerging kid. The cat's legs are spread, indicating that it is moving quickly (see Chapter 4), and its tail is erect and arched. Cats usually hold their tails out horizontally from their bodies when they are hunting, however during the final rush at their prey, it is sometimes elevated. The wild cat's posture (head, tail, and legs) and the behaviour of a nearby herdsman (who is about to strike the predator with a stick), suggest that it is advancing toward the newborn kid and about to pounce. Contrary to current belief then, this image supports the notion that wild cats may indeed prey upon young goats.

A later scene from the tomb of *ꜣḫtj-mrw-nswt* at Giza (HB16; see fig. 8-8) also represents the predation of goats. The image depicts a large cat seated upon an elevated structure, which lowers its head to peer menacingly at a small herd. The animal's line of sight indicates that it is not the mating animals immediately before it that have caught its attention, but rather, a female giving birth further afield. The cat's pricked ears, braced forelegs, and erect tail indicate that it is in an alert and agitated state. I suggest that, like the feline in the tomb of *nj-ꜥnḫ-ẖnmw/ẖnmw-ḥtp*, it is preparing to strike the emerging newborn kid.

Cats are able to hear sounds at very high frequencies and, when hunting, they often first become aware of the presence of their prey via acoustic cues. When they detect the sounds of a possible victim, they orient both their outer ears (pinnae) and head towards the source. They then stare in the direction of the sound (a behaviour known as 'interest for a locus'; cf. pl. 8E), waiting to detect any signs of movement (to which their eyes are particularly sensitive).[41] The scene in the tomb of *ꜣḫtj-mrw-nswt* appears to depict this orienting behaviour. Goats vocalize a great deal when giving birth.[42] It is likely that the wild cat has detected these sounds and then dropped its head to peer in the direction of their source, which is both beyond and below where it is seated.

---

35 Osborn-Osbornová, *Mammals*, 120.

36 For example, a leashed leopard and vervet monkey are led by a dwarf on a relief fragment from the early 6th dynasty tomb of *nj-ꜥnḫ-nswt* (Kofler-Truniger Coll. no. K 415 A). See Müller, *Ägyptische Kunstwerke*, pl. A90.

37 " . . . lauert am Schwanzende eine Wildkatze auf das neugeborene Zicklein". See Moussa-Altenmüller, *Nianchchnum*, 73.

38 Nowak, *Walker's Mammals i*, 801.

39 Shortridge, *Mammals of South West Africa i*, 93.

40 Smithers, *Mammals of Botswana*, 124.

41 Turner, D.C. and Meister, O. (1988). "Hunting behaviour of the domestic cat" in Turner-Bateson, *Domestic Cat*, 112-113.

42 "It is not uncommon for goats to scream as the kids are being born . . .". See Thear-Fraser, *Complete Book*, 90. Note that the goat in fig. 8-7 protrudes its tongue dramatically as she gives birth. See Chapter 11 for more details.

African wild cats are the ancestors of the domesticated cat,[43] but full domestication did not occur until well after the Old Kingdom period.[44] The images from the tombs of *nj-ʿnḫ-ẖnmw/ẖnmw-ḥtp* and *ꜣḫtj-mrw-nswt* reveal, however, that the Egyptians were becoming familiar with the behaviour of this species as early as the mid-5th dynasty.[45]

Striped hyenas

nj-ʿnḫ-ẖnmw/ẖnmw-ḥtp (HB17: Saqqara, V.6L-7)

The tomb of *nj-ʿnḫ-ẖnmw/ẖnmw-ḥtp* features a striped hyena (HB17; see fig. 8-9)[46] that approaches the body of a scimitar oryx as it is hauled over a fence by a lion (see *Lions* above). Striped hyenas are usually depicted in an alert posture with their head erect - a pose that emphasizes the characteristic sloping back of this species.[47] Although space on the register would have allowed the artist to show the animal in this manner, the hyena instead drops its head well below the level of its shoulders in a pose akin to stalking.

East African striped hyenas[48] are largely solitary animals that forage alone and at night. They occasionally hunt small animals (see *Golden jackals* below),[49] but obtain most of their food through scavenging. Faecal analyses have revealed that they ingest vegetable matter, insects, reptiles, birds, and both large and small mammals. Food from large mammals is often obtained by feeding from the carcasses of animals that have been killed by other carnivores. The hyena's powerful jaws, teeth, and digestive system enable it to chew and digest carrion and the remains of kills that other predators leave behind, such as skin and bones.[50] It has been reported that they obtain some meat by driving leopards from their kills, but more frequently it is they who are chased by other carnivores from food they have acquired.[51] They are attracted to the remains of kills made by lions, but usually avoid contact with this species.[52]

43 "The domestic cat is descended primarily from the wild cat of Africa and extreme southwestern Asia, *Felis sylvestris libyca*". See Nowak, *Walker's Mammals i*, 801.

44 Malek, *The Cat*, 45ff.

45 Two other images were located during this study that may also depict African wild cats. The first appears on the north wall of the mid-5th dynasty chapel of *rʿ-ḫʿ.f-ʿnḫ* (G7948=LG75)(see *LD Ergänz.*, pl. 28 [c]). The animal, seated at the base of a papyrus thicket with an erect, tufted tail, appears in a marsh scene, however wild cats are more commonly found in dry habitats. The second image satisfies this environmental criterion, as it occurs in a fragmentary desert scene in the late 5th dynasty tomb of *nj-mꜣʿt-rʿ* [*jzzj-mr-nṯr*] (G2097) at Giza (see Roth, *Cem. Pal. Att.*, pl. 189). The small animal stands behind an ungulate that may be defecating in response to a lion attack (HB6a). Both its size and the length of its pricked ears resemble those of a wild cat. Its tail, however, is extremely short, indicating that it may instead represent a caracal. Roth (*ibid.*, 133) suggests that it is "perhaps a member of the cat family".

46 The animal's tail is rather too long, but this is a mistake that Egyptian artists repeated frequently.

47 For examples, see Osborn-Osbornová, *Mammals*, 97-104.

48 Five subspecies of striped hyena have been identified: *Hyaena hyaena barbara* in North and West Africa; *H. h. syriaca* in Syria and Asia Minor; *H. h. hyaena* in India; *H. h. sultana* in Arabia; and *H. h. dubbah* in North and East Africa. The northern subspecies (*barbara*, *syriaca*, and *hyaena*) are considerably larger than those in the south (*sultana* and *dubbah*) and so are more inclined to attack and kill large prey. The most prevalent subspecies found in Egypt is *H. h. dubbah*, but it is believed that both *H. h. sultana* and the bigger and more aggressive *H. h. syriaca* may also occur there. See Rieger, I. (1979). "A review of the biology of striped hyaenas, *Hyaena hyaena* (Linné, 1758)", *Säugetier. Mitteil.* 27, 81-95; Rieger, I. (1981). "Hyaena hyaena", *Mam. Sp.* (150), 2; Hoath, *Field Guide*, 92.

49 Ewer, *The Carnivores*, 204; Kruuk, H. (1976). "Feeding and social behaviour of the striped hyaena (*Hyaena vulgaris* Demarest)", *EAWJ* 14, 102-103; Leakey, L.N., Milledge, S.A.H., Leakey, S.M., Edung, J., Haynes, P., Kiptoo, D.K. and McGeorge, A. (1999). "Diet of striped hyaena in northern Kenya", *AJE* 37(3), 314-326.

50 Estes, *Behavior Guide*, 324 and 329.

51 *ibid.*, 330-331.

52 Kruuk (*EAWJ* 14, 107) observed that striped hyenas in Kenya stayed at least 50 metres away from lions. A recent review by Ikram (2001) suggests that striped hyenas are capable of scaring lions from their kills; however, this has only been reported for leopard kills. Only spotted hyenas, which, unlike striped hyenas, are a social species, are able to form groups large enough to intimidate lions. The author attributes to striped hyenas a number of other behaviours that are only exhibited by spotted hyenas (e.g., hunting in

When foraging, striped hyenas hold their head at, or slightly below, shoulder level to sniff for food in holes, shrubs, and under rocks (see pl. 8F).[53] The head of the animal in the tomb *nj-ꜥnḫ-ẖnmw/ẖnmw-ḥtp* is considerably lower than this, but may represent this simple searching behaviour. The proximity of the hyena to the body of the oryx could also indicate that it has detected the lion's kill and is approaching in anticipation of later access to the carcass. By depicting the animal in a posture resembling stalking, the image conveys very effectively the stealth with which the hyena would have to undertake such a dangerous exercise.

Common genets

nb.j-m-ꜣḫtj (HB18: Giza, IV.6-V.1)
nj-ꜥnḫ-ẖnmw/ẖnmw-ḥtp (HB19-20: Saqqara, V.6L-7)
ḥtp-ḥr-ꜣḫtj (HB21: Saqqara, V.6-8E)
jtj-sn (HB22: Giza, V.6-8?)
jrj-n-kꜣ-ptḥ (HB23: Saqqara, V.6-9)
*rꜥ-šps*s (HB24: Saqqara, V.8M)
snḏm-jb: jntj (HB25-26: Giza, V.8M-L)
ṯjj (HB27-28: Saqqara, V.8-9)
kꜣ.j-m-nfrt (HB29: Saqqara, V.8-9?)
ꜣḫtj-ḥtp (HB30-31: Saqqara, V.8L-9E)
ẖnwt (HB32: Saqqara, V.9)
jj-nfrt: šꜣ.n.f (HB33: Saqqara, V.9)
nj-kꜣw-jzzj (HB34-35: Saqqara, VI.1M)
mrrw-kꜣ.j: mrj (HB36: Saqqara, VI.1M-L)
zšzšt: jdwt (HB37: Saqqara, V.9 r/u - VI.1-2E?)
sꜥnḫ-w(j)-ptḥ (HB38-39: Saqqara, VI.1L)
ḥzj (HB40-42: Saqqara, VI.1L-2E)
mrjj-rꜥ-mrj-ptḥ-ꜥnḫ: nḫbw (HB43: Giza, VI.2)
jnw-mnw (HB43a: Saqqara, VI.2E)
mḥw (HB44-47: Saqqara, VI.2M-3?)
jj-nfrt (HB48-49: Giza, VI.3-4E?)
mrjj-rꜥ-nfr: ḳꜣr (HB50: Giza, VI.4)
ꜣḫtj-ḥtp (HB51: Saqqara, VI.5?)

Genets engaged in predation are depicted in 23 tombs dating from the late 4th until the late 6th dynasty, indicating that this was a popular motif throughout the Old Kingdom period.[54] Each cat-like carnivore appears among papyrus thickets in marsh scenes, often alongside Egyptian mongooses.[55]

Genets are solitary predators that hunt by night (especially just after dark and before dawn) and sleep in cover during the day. Long, sharp claws on each of their paws enable them to climb effortlessly; they are thus able to hunt animals in trees that would be safe from other predators. Nevertheless, they obtain most of their prey when on the ground (see pl. 8G). As they hunt in the dark, they detect their prey primarily by smell.[56] Their method of hunting consists of ambushing prey, either by lying in hiding or actively stalking their victim before pouncing upon it. They often use their paws to hold and subdue their prey before dispatching it with a quick bite to the head. Feeding also begins at the head;[57] indeed, frequently only the head or breast is eaten. They hunt and eat a wide range of foods including small mammals (particularly rodents), birds and their eggs, lizards, snakes, frogs, and insects.[58] Although they are small animals (c. 2-3 kg in weight), genets will attack adult birds up to the size of a chicken. Nestlings are taken easily.

The genets in tomb scenes pursue birds; no other prey type is depicted. The species of birds that are attacked can be tentatively identified in 27 of the 35 scenes: ducks or geese (14), pied kingfishers (11), and herons/egrets (2). In three images dating from the mid- to late 5th dynasty, the carnivores approach adult ducks or geese seated upon nests (HB22, HB23, and HB25), possibly with the intention of attacking the birds, but more likely in pursuit of their eggs. Nestlings form the genets' prey in all of the remaining examples (although see below).

groups, intolerance between females, extreme aggression, etc.). See Ikram, S. (2001). "The iconography of the hyena in ancient Egyptian art", *MDAIK* 57, 127-140.

53 Kruuk, *EAWJ* 14, 98-99.

54 The possible religious significance of the predatory behaviour of genets and other viverrids in marsh scenes is discussed at length in Aufrère, S.H. (1998). "La loutre, le chat, la genette et l'ichneumon, hôtes du fourré de papyrus: Présages, prédateurs des marécages et croyances funéraires", *DE* 41, 7-28.

55 See Chapter 4, n. 43 for a discussion of the ways in which common genets may be differentiated from Egyptian mongooses.

56 Larivière, S. and Calzada, J. (2001). "Genetta genetta", *Mam. Sp.* (680), 4.

57 *ibid.*

58 *ibid.*, 3.

Prior to the late 5th dynasty, the genets approach nests without making physical contact with the birds. Some advance from a short distance away (e.g., HB18 and HB25; see fig. 9-25 in Chapter 9), but most are depicted within easy striking distance of their intended victims (see fig. 4-15 in Chapter 4). From the late 5th dynasty onward, however, the carnivores are more frequently depicted in the act of biting nestlings and pulling their bodies from their nests. This occurs in 14 examples, beginning with that in the tomb of *ꜣḫtj-ḥtp* (HB30; see fig. 8-10), in which a young pied kingfisher is bitten upon one wing and lifted away from its siblings.[59] The wings of nestlings are also grasped in later images from the tombs of *ḥzj* (HB42), *mrjj-rꜥ-mrj-ptḥ-ꜥnḫ: nḫbw* (HB43), *jnw-mnw* (HB43a), and *mḥw* (HB44). As noted above, genets usually kill by biting the head. This locus of attack is found in examples from the tombs of *sꜥnḫ-w(j)-ptḥ* (HB39; see fig. 8-12), *mḥw* (HB45), and (possibly) *mrjj-rꜥ-nfr: kꜣr* (HB50), while in others, the nestlings are grabbed by the throat (e.g., HB34, HB35, and HB37). None of the genets use their paws to hold their prey. Nevertheless, the animals' method of killing is largely accurate.

A unique image in the tomb of *mḥw* depicts a genet attacking an adult bird (HB46; see fig. 8-13). The carnivore bites the duck(?) on one wing and appears to pluck it from the air mid-flight. The bird hangs in the predator's jaws in essentially the same manner as nestlings that have been similarly bitten and lifted from their nests. Earlier scenes in the tombs of *nj-kꜣw-jzzj* (HB68; see fig. 9-32 in Chapter 9) and *ḥzj* (HB76; see fig. 8-18), in which Egyptian mongooses throttle adult geese, may have inspired this image (see *Egyptian mongooses* below).[60]

Each genet 'climbs' a single papyrus stalk (for a discussion of this behaviour, see Chapter 4). In most scenes, they advance along the same stem that supports the nest of their prey. In some images, however, the nest is situated on a stalk or umbel that is above or adjacent to that of the genet. In four of these examples (HB31, HB34, HB39, and HB51; see figs. 8-11 and 8-12), the genet looks back over its shoulder towards the nestlings as though it had almost passed by them, but then suddenly detected their presence. This suggests that the nests should perhaps be perceived as hidden from view. In other images (HB29, HB40, and HB43), the genet turns its head away from the nestlings as it comes under attack from their parents (see *Birds vs Predators* in Chapter 9). Finally, a genet in the tomb of *ḥzj* is shown descending headfirst from the thicket while carrying a nestling in its jaws (HB41; see fig. 4-16 in Chapter 4). This image may have been copied from that in the tomb of *mrrw-kꜣ.j: mrj* (HB70; see fig. 8-15), in which an Egyptian mongoose travels down a papyrus stalk with a captured nestling. Genets usually descend headfirst when climbing; as noted in Chapter 4, Egyptian mongooses are not arboreal.

The bird species that the genets attack are all found in marsh environments: herons usually position their nests in trees, while ducks, geese, and egrets all nest at ground level and often within reed beds. Pied kingfishers, however, incubate and rear their young in holes that they excavate in riverbanks. Thus, both the depiction of pied kingfisher chicks sitting on nests and, consequently, the genets' behaviour towards them, are improbable.

An explanation for these anomalous images may possibly be found in the parental behaviour of pied kingfishers. When juvenile kingfishers first leave their burrow, their parents continue to feed them for several days until they are able to fly and also hunt for themselves (see *Pied kingfishers* below).[61] The chicks usually stay close to the nest-hole during this vulnerable time and may thus have been witnessed by the Egyptians. The depictions of 'nesting' pied kingfishers hunted by genets may possibly allude to this brief period during which the juveniles are still cared for by their parents and, due to their inability to fly competently, are particularly subject to predation (for further discussion, see Chapter 9). The nests upon which the chicks are situated are thus a signal to the viewer of their immaturity and continued dependence upon their parents.

---

59 Note that a genet pictured in the mid-5th dynasty tomb of *nj-ꜥnḫ-ẖnmw/ẖnmw-ḥtp* (HB19) appears to touch the outstretched wing of a nestling with its muzzle, but the bird is not bitten and remains standing in the nest.

60 For more details, see Evans, L. (2010). "Otter or mongoose? Chewing over the evidence" in Woods et al., *Egyptian Culture and Society*, 119-129.

61 Roselaar, C.S. (1985). "Pied kingfisher (*Ceryle rudis*)" in Cramp-Simmons, *Handbook iv*, 728.

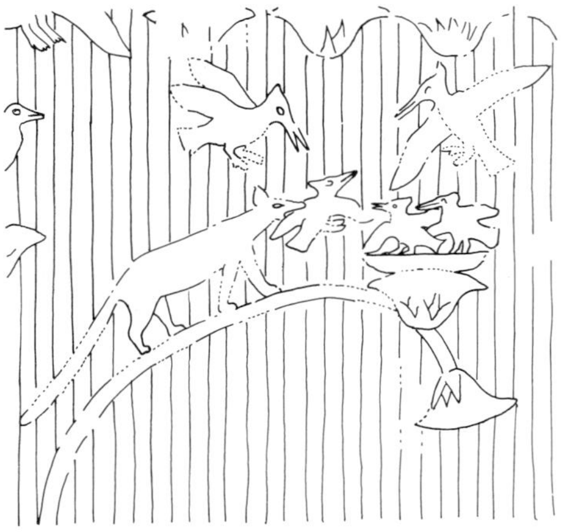

Fig. 8-10 (HB30)

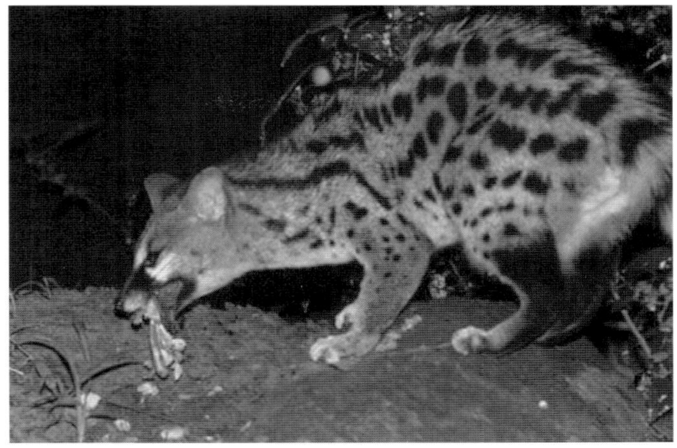

Pl. 8G Common genet feeding
© Howard Norfolk
Reproduced with permission

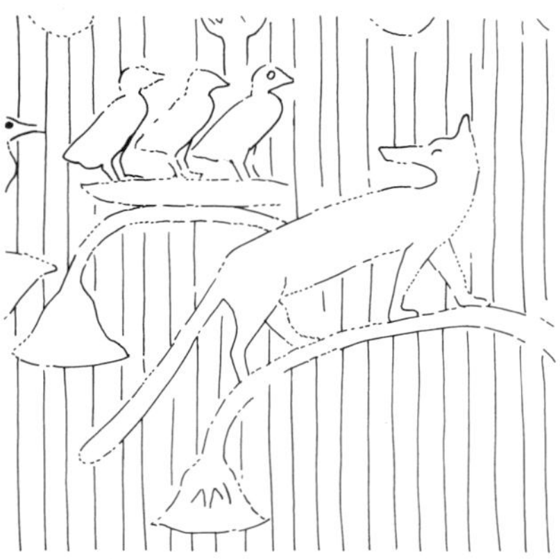

Fig. 8-11 (HB31)

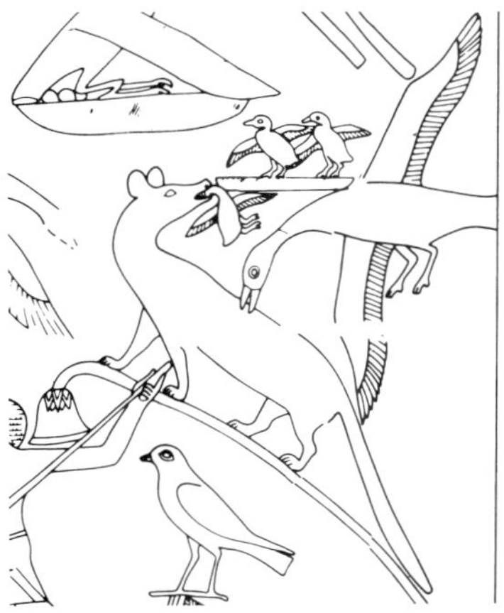

Fig. 8-12 (HB39)

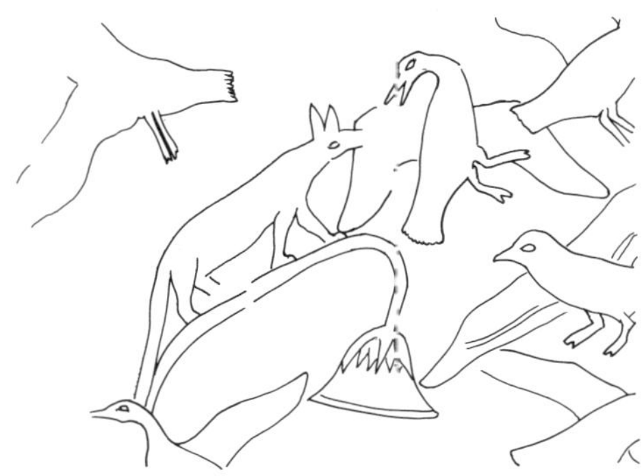

Fig. 8-13 (HB46)

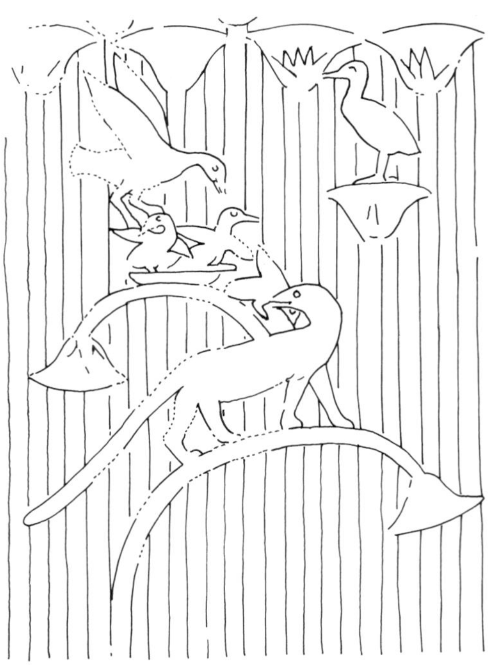

Fig. 8-14 (HB63)

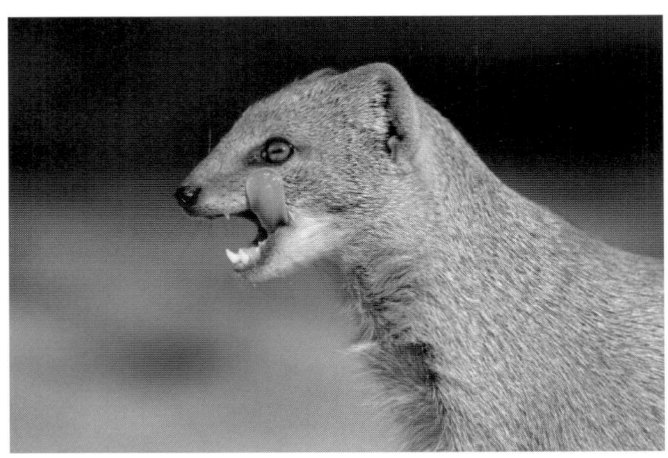

Pl. 8H Mongoose teeth
© iStockphoto.com/Nico Smit

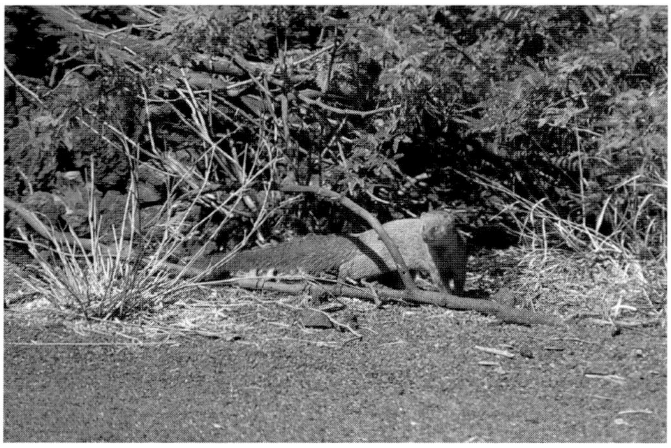

Pl. 8I Mongoose moving into cover
© iStockphoto.com/Kathleen Reidnauer

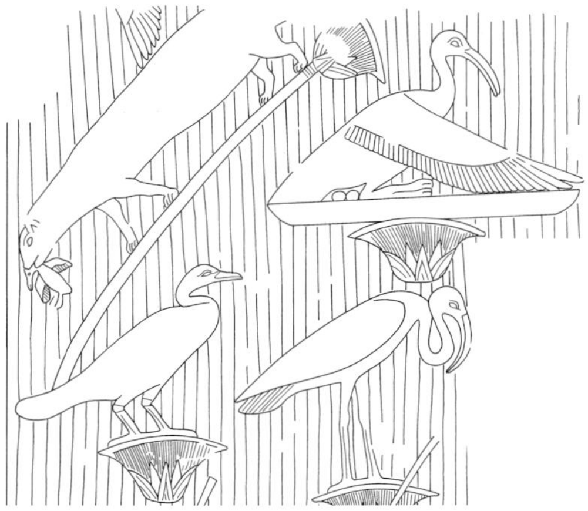

Fig. 8-15 (HB70)

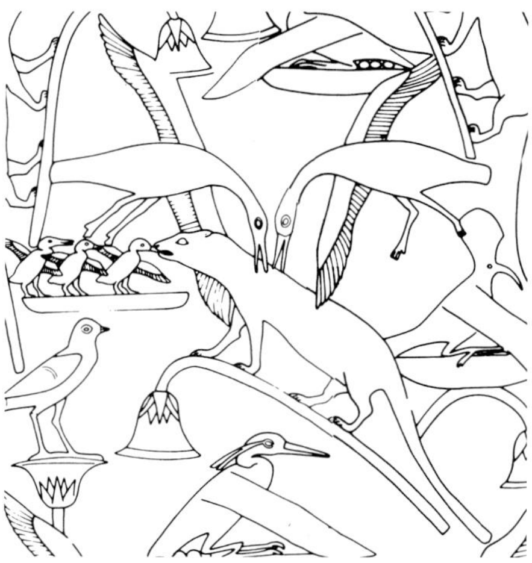

Fig. 8-16 (HB75)

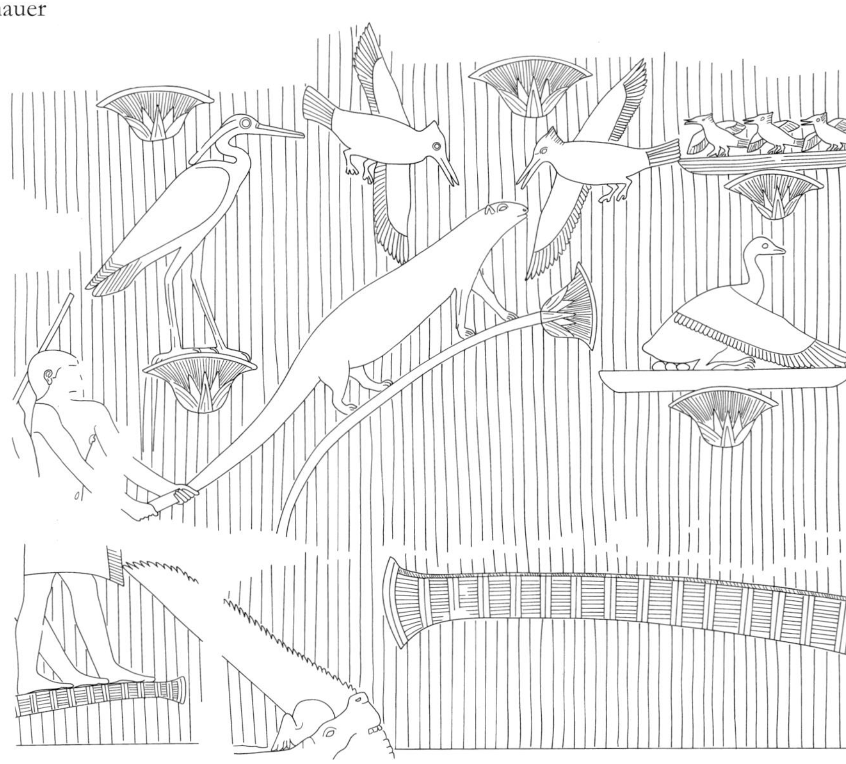

Fig. 8-17 (HB71)

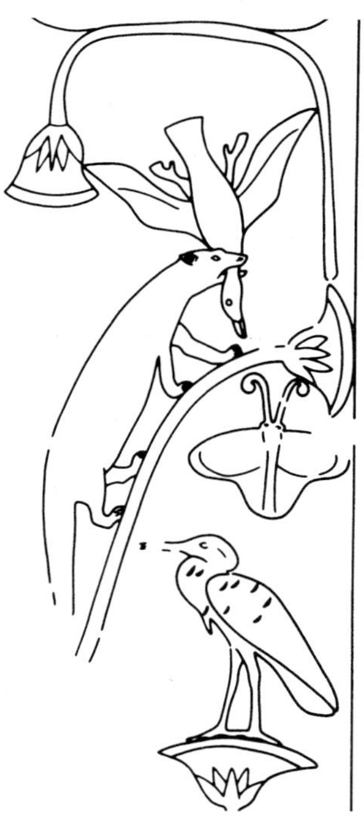

Fig. 8-18 (HB76)

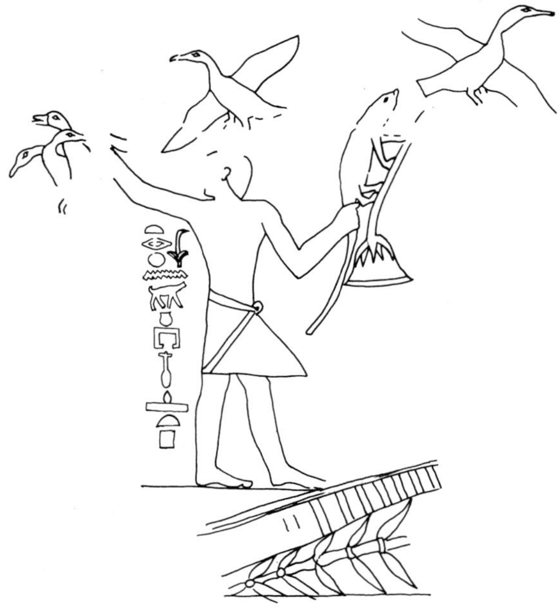

Fig. 8-19 (HB82)

Seven wall scenes clearly depict a pair of genets hunting within the same papyrus thicket[62] (*ṯjj*: HB27 and HB28; *ꜣḫtj-ḥtp*: HB30 and HB31 - see figs. 8-10 and 8-11; *nj-kꜣw-jzzj*: HB34 and HB35; *sʿnḫ-w(j)-ptḥ*: HB38 and HB39; *ḥzj*: HB40 and HB41; *mḥw*: HB45 and HB46; and *jj-nfrt*: HB48 and HB49). It should be noted that common genets live alone for most of their lives. The territories of males and females sometimes overlap[63] and pairs have been sighted occasionally in the wild, but interactions between individuals are a rare occurrence.[64] The depiction of two genets in close proximity is thus either incorrect and the result of artistic license, or an indication that marsh scenes actually represent a very wide section of the physical environment.

Egyptian mongooses

nb.j-m-ꜣḫtj (HB52: Giza, IV.6-V.1)
nj-ʿnḫ-ẖnmw/ẖnmw-ḥtp (HB52a: Saqqara, V.6L-7)
ꜣḫtj-ḥtp (HB53: Saqqara, V.6-8E)
ẖnmw-ḥtp (HB54: Saqqara, V.6-8E?)
kꜣ.j-dwꜣ (HB55-56: Giza, V.6-9?)
rʿ-špss (HB57: Saqqara, V.8M)
snḏm-jb: jntj (HB58-59: Giza, V.8M-L)
ṯjj (HB60: Saqqara, V.8-9)
nfr-jrt-n.f (HB61: Saqqara, V.8-9)
kꜣ.j-m-nfrt (HB62: Saqqara, V.8-9?)
ꜣḫtj-ḥtp (HB63-64: Saqqara, V.8L-9E)
nbt (HB65: Saqqara, V.9)
jj-nfrt: šꜣ.n.f (HB66: Saqqara, V.9)
nj-kꜣw-jzzj (HB67-68: Saqqara, VI.1M)
mrrw-kꜣ.j: mrj (HB69-71: Saqqara, VI.1M-L)
zšzšt: jdwt (HB72: Saqqara, V.9 r/u - VI.1-2E?)
jꜣzn (HB73: Giza, VI.1M-2M)
sʿnḫ-w(j)-ptḥ (HB74-75: Saqqara, VI.1L)
ḥzj (HB76-77: Saqqara, VI.1L-2E)
mṯṯj (HB78: Saqqara, VI.2?)
mrrj (HB79: Saqqara, VI.2E?)
mḥw (HB80-83: Saqqara, VI.2M-3?)
mrjj-rʿ-nfr: ḳꜣr (HB84: Giza, VI.4)
ꜣḫtj-ḥtp (HB85: Saqqara, VI.5?)

Egyptian mongooses are small, semi-social carnivores that prey upon a range of species, including insects, fish, reptiles, ground-nesting birds, eggs, and small rodents.[65] Prey are detected by using their acute sense of smell, or opportunistically as they wander through the undergrowth. Their victims are grabbed in their jaws (see pl. 8H) after a sudden rush. Large animals are caught and killed by biting the neck, while small prey are dispatched by a crushing bite to the head (see pl. 7J in Chapter 7).[66] Food items are usually taken into dense cover to be eaten (cf. pl. 8I).[67] Unlike common genets, with which they are frequently represented, Egyptian mongooses usually do not climb and so must hunt on the ground (see Chapter 4).[68]

Thirty-five images of predatory mongooses were identified in marsh scenes from 24 tombs.[69] The images share many similarities with those of genets. Birds are the most commonly depicted prey (although see *Egyptian mongooses* in Chapter 7), the species of which can be roughly identified in 27 scenes: ducks or geese (16), pied kingfishers (7), and egrets (4). These numbers match closely the depicted prey of common genets. As noted above, ducks, geese, and egrets are ground-nesting birds and so these species would all be within reach for the terrestrial mongoose; the inaccessibility of pied kingfisher nestlings has already been discussed above (see *Genets*). In two examples (HB56 and HB59), however, the predators approach empty nests. The futility of their behaviour was perhaps intended to be a humorous comment or possibly a reference to the somewhat random nature of the mongooses' hunting style.

Egyptian mongooses are not depicted biting their prey until the late 5th dynasty. Prior to this period, the predators only approach their intended victims, either by traversing the same papyrus stalk

62 It has been noted that often one Egyptian mongoose may be pictured with two genets, but that the opposite combination never occurs. See Aufrère, *DE* 41, 9.
63 Fuller, T.K., Biknevicius, A.R. and Kat, P.W. (1990). "Movements and behavior of large spotted genets (*Genetta maculata* Gray 1830) near Elmenteita, Kenya (Mammalia Viverridae)", *Trop. Biol.* 3(1), 18.
64 Estes, *Behavior Guide*, 286.
65 *ibid.*, 299.
66 *ibid.*, 299-300.
67 Ben-Yaacov, R. and Yom-Tov, Y. (1983). "On the biology of the Egyptian mongoose, *Herpestes ichneumon*, in Israel", *Z. Säugetier.* 48, 39.
68 Kingdon, *East African Mammals iii Part A*, 180; Estes, *Behavior Guide*, 299.
69 For the possible religious significance of mongooses, see Brunner-Traut, E. (1965). "Spitzmaus und Ichneumon als Tiere des Sonnengottes", *NAWG* 7, 123-163.

upon which the nest of their prey is perched or on an adjacent shoot. In two examples, a mongoose approaches an adult bird incubating eggs (HB52 and HB79). Usually, however, the nests that they advance upon contain only young birds. In four examples, the mongooses turn their head to one side as though they have detected the birds' presence only after they had passed by (HB63, HB69, HB78, and HB79). The earliest of these scenes (HB63; see fig. 8-14) includes the first representation of a bird being attacked by a mongoose. This occurs in the tomb of *ȝḫtj-ḥtp*, which also features the first example of a genet biting its prey (HB30; see fig. 8-10). Although Egyptian mongooses usually dispatch small animals with a bite to the head, the nestling here has been seized by the back of the neck. The necks or throats of nestlings are also attacked in all but one of the subsequent depictions; in the tomb of *sʿnḫ-w(j)-ptḥ* (HB75; see fig. 8-16), a mongoose bites a gosling upon the beak. In contrast to images of genets, none of the nestlings are bitten upon their wings.[70] A single example depicts one of the predators descending a papyrus stalk headfirst while holding a nestling in its jaws (HB70; see fig. 8-15); as noted above, Egyptian mongooses do not climb (see Chapter 4).

In some scenes, the object of the predators' attention is unclear (i.e., no nestlings are in their immediate vicinity; HB52, HB72, HB81), but their presence nonetheless appears to provoke an angry response from adult birds, which hover nearby. Their panic may be understood after examining two scenes in which mongooses attack adult birds. In both of these examples, Egyptian geese are seized by the throat: in the tomb of *ḥzj*, a goose is plucked from the air mid-flight (HB76; see fig. 8-18), while in the earlier tomb of *nj-kȝw-jzzj*, the bird is mauled moments after its capture (HB68; see fig. 9-32 in Chapter 9). The images show correctly the mongoose's method of killing larger prey, but it seems unlikely that they would be capable of tackling an Egyptian goose, which can weigh up to two and a half kilograms (Egyptian mongooses weigh c. 2-4 kg).[71] Nevertheless, it may be possible that the mongooses' speed and agility in capturing prey, combined with the element of surprise afforded by their tendency to hide in the undergrowth, could enable them to seize an unsuspecting goose, should it fly too close.

The secretive nature of these predators may also offer a clue to understanding two curious images in which men, standing in canoes, hold the animals by their tails (HB71 and HB82). It has been suggested that these are tame mongooses sent into the papyrus thicket to flush birds for the tomb owner to hunt with his throw-stick.[72] Indeed, in the tomb of *mrrw-kȝ.j: mrj* (HB71; see fig. 8-17), the mongoose has attracted the attention of pied kingfishers, and in the tomb of *mḥw* (HB82; see fig. 8-19), Egyptian geese swirl about the predator's head. Some mongoose species, including the Egyptian mongoose, can apparently be tamed and have been kept as pets.[73] They are reported to be capable of learning their name and coming when called. The tomb reliefs in question may thus indicate that the mongooses have been sent out into the thicket to serve their human owners (their relationship with whom is indicated by the holding of their tails). It is equally possible, however, that the reverse is true: that the flushing birds have alerted the men to the presence of the mongooses hidden in the impenetrable thicket, allowing them to locate and catch by the tail the otherwise elusive animals as they meander through the undergrowth.

Golden jackals

ȝḫtj-mrw-nswt (HB86: Giza, V.9-VI.1) *mrjj-ttj: mrj* (HB87-88: Saqqara, VI.2M-L)

Golden jackals are small, nocturnal, dog-like omnivores that are most frequently found in mated pairs or small family groups. Their diet is extremely varied, including fruits, invertebrates, reptiles, birds, and mammals. They scavenge from kills made by larger predators, carrion, and human garbage dumps, but also hunt small prey animals, such as rodents, hares, and gazelle calves (see pl.

70 The muzzle of an advancing mongoose in HB74 appears to touch the tip of a nestling's outspread wing, but does not bite it. The artist here has created a dramatic visual focal point by causing the predator's snout, the nestling's wing, and the adult's beak to all converge.

71 Haltenorth-Diller, *Field Guide*, 202; Estes, *Behavior Guide*, 298.

72 Malek, *The Cat*, 33; Osborn-Osbornová, *Mammals*, 94-95. Altenmüller (*Mehu*, 100) suggests that the animal shown in fig. 8-19 is slipping down the papyrus stalk to which it is clasped.

73 Hinton-Dunn, *Mongooses*, 91-92.

8J).[74] They have been reported to cause damage to agricultural crops, and to kill poultry and very young or sick sheep and goats.[75] Golden jackals, like many canids, hoard surplus food by digging a hole and burying it.[76]

In the late 5th dynasty Giza tomb of *3ḫtj-mrw-nswt* (HB86; see fig. 8-20), a jackal lunges at a newborn calf as its mother strains to deliver it.[77] Like the *Wild cats* described above, the predator rushes toward its intended victim with its tail held out from its body. It does not approach in a hunched stalking posture, but instead tilts its head up so that it is in line with that of the calf; its posture suggests that it is about to bite the newborn's head. A recent study in Israel has reported that golden jackals do indeed prey upon calves and that frequently the face and tongue of newborns are attacked as they emerge from their mothers.[78] The image in the tomb of *3ḫtj-mrw-nswt* is thus entirely feasible. The scene may also have been inspired by an earlier image in the Saqqara tomb of *3ḫtj-ḥtp* (HB97), in which a dog menaces a goat as she gives birth (see *Dogs* below).

I suggest that a group of small canids depicted on a sub-register in the tomb of *mrjj-ttj: mrj* (HB88; see fig. 8-21) are not striped hyenas (as identified elsewhere),[79] but jackals. First, the animals lack the sloping backs of striped hyenas, a feature that is usually emphasized in Egyptian depictions. Like jackals, the muzzles of the figures are sharply pointed whereas hyenas are always represented with blunt snouts.[80] The three canids also possess exceedingly long tails that drop below the register line, a characteristic that is found frequently in depictions of the reclining jackal-like figure of Anubis.[81] It is also worth noting that, in life, striped hyenas are not normally found in social groups.

This alternative identification may be supported by the group's proximity to a lion attacking a bull in the register below. Despite being strongly territorial, golden jackals will occasionally gather in large numbers around food sources, such as garbage dumps or recent animal kills;[82] this behaviour is especially likely in areas where food is scarce.[83] Furthermore, although they are usually intimidated by large predators, they have nevertheless been observed to steal meat directly from lion kills while the cats are still feeding.[84]

---

74 Estes, *Behavior Guide*, 399.

75 Jhala, Y.V. and Moehlman, P.D. (1990). "Golden jackal, *Canis aureus*" in Sillero-Zubiri et. al., *Canids*, 158-159.

76 There is much disagreement concerning which canid species represented Anubis, god of cemeteries and embalming. The animal is often thought to resemble a golden jackal, but the combination of its physical features suggests that it could equally represent a fox, a wolf, or a domestic dog. These species differ morphologically, but as they are all members of the *Canid* family they share many behavioural traits. One of these is food caching - the burying of surplus meat. Studies have shown that all canids bury prey in exactly the same manner. I have proposed that the Anubis animal does not represent a golden jackal, or any one particular species, but that it is instead an amalgamation or synthesis of *all* the canids with which the Egyptians were familiar, based upon their shared tendency to bury flesh. This behaviour will have rendered any canid an appropriate symbol for a god associated with the preparation and safe burial of the dead. See Evans, L. (2008). "The Anubis animal: A behavioural solution?", *GM* 216, 17-24. For a summary of the evidence supporting different canids as the prototype for the Anubis animal, see Osborn-Osbornová, *Mammals*, 75-79.

77 The animal's lack of a collar, long tail, and short legs indicate that it is probably a wild canid rather than a domesticated dog. Smith (*HESPOK*, 340) identified it as a jackal.

78 Yom-Tov, Y., Ashkenazi, S. and Viner, O. (1995). "Cattle predation by the golden jackal, *Canis aareus* in the Golan Heights, Israel", *Biol. Cons.* 73(1), 19-22.

79 Kanawati-Abder-Raziq, *Meryteti*, 25.

80 Osborn-Osbornová, *Mammal*, 97-104.

81 An animal morphologically identical to these appears on a relief fragment from the late 5th dynasty funerary complex of Unas. It has been identified simply as "un canidé". See Labrousse-Moussa, *Complexe Funéraire du Roi Ounas*, 47 and Doc. 45, fig. 59. Note, however, that the tails of live jackals are quite short (c. 20-30 cm).

82 Estes, *Behavior Guide*, 399.

83 van Lawick-Goodall, *Innocent Killers*, 134.

84 *ibid.*, 120. It is interesting to note that black-backed jackals, a closely-related East African species found to the south of Egypt, will become excited and begin to howl when they detect a lion kill in progress. This behaviour draws others to the site. The milling group then lingers nearby, waiting for the lions to become sated and leave the carcass for them to pick over. See Grzimek, B. (1990). "Jackal behavior in the wild" in *Grzimek's Encyclopedia iv*, 112.

Another canid figure, which has also been identified as a hyena,[85] appears in the same panel as the jackal group (HB87; see fig. 8-22). The animal rears up and bites an ibex directly upon the muzzle, wrenching its head and causing it to collapse. I am uncertain what predator species the artist intended to represent here. It bears a strong resemblance to the foraging jackals, but due to its unusual posture, a striped hyena cannot be ruled out. Neither striped hyenas nor golden jackals normally hunt animals as large as ibex, however, unless these are weak or incapacitated. I know of no other depictions of striped hyenas attacking animals; indeed, they are usually presented in submissive roles – being chased, wounded, or led. The method of attack is also curious, for while lions are often depicted biting the muzzles of their prey (see *Lions* above), canids are rarely shown engaging in this behaviour (see *Dogs* below). I suggest that the image may have been derived from an earlier hunting scene in the royal tomb of Sahure in which a striped hyena faces an ibex, but does not attack it.[86] The *mrjj-ttj: mrj* artist has modified the motif, however, so that the hyena or jackal bites the muzzle of its adversary in the same manner as the lion attacking a bull with which it shares the register (cf. fig. 8-21). The composition also resembles closely that of a dog in a lower register on the same wall (HB120; see *Dogs* below), which bites the muzzle of a collapsed dorcas gazelle.

Dogs

mṯn (HB89-91: Saqqara, IV.1-2)
rꜥ-m-kꜣ.j (HB92-93: Saqqara, V.6? r/u - V.6-7)
nj-ꜥnḫ-ẖnmw/ẖnmw-ḥtp (HB94: Saqqara, V.6L-7)
pḥn-wj-kꜣ.j (HB95-96: Saqqara, V.6-8E)
ꜣḫtj-ḥtp (HB97: Saqqara, V.6-8E)
ꜣḫtj-ḥtp (HB98: Saqqara, V.8L-9)
ptḥ-ḥtp II: ṯfj (HB99-103: Saqqara, V.9M-L)
sšm-nfr IV (HB104: Giza, V.9-VI.1)
ꜣḫtj-mrw-nswt (HB105: Giza, V.9-VI.1)
nj-mꜣꜥt-rꜥ [*jzzj-mr-nṯr*](HB106-111: Giza, V.9-VI.1)
Anon (HB112: Saqqara, V-VI)
Anon (HB113: Saqqara, VI)
mrrw-kꜣ.j: mrj (HB114-116: Saqqara, VI.1M-L)
sꜥnḫ-w(j)-ptḥ (HB117: Saqqara, VI.1L)
jnw-mnw (HB118-118a: Saqqara, VI.2E)
mrjj-ttj: mrj (HB119-122: Saqqara, VI.2M-L)

Sixteen Memphite tombs ranging in date from the early 4th to the early 6th dynasty were found to depict domestic dogs in the act of hunting. All of the dogs resemble the *ṯsm* morph with a slender body, long legs, pricked ears, and a tightly curled tail.[87] The lack of obvious genitals suggests that the majority are female. Most of the animals wear a cravat collar tied around their necks.

In one example, a male dog watches the impending birth of a goat kid (HB97). A herdsman nearby prepares to strike the animal, suggesting that he is trying to stop its approach. The predatory instincts of domestic dogs, particularly those used in hunting, can be triggered suddenly in certain circumstances, making them somewhat unpredictable. The sounds and smells associated with the goat's labour would very likely have stirred the dog's hunting impulse, prompting the herdsman to anticipate a possible attack.

In the remaining images, dogs pursue prey under the direction of hunters. Their predatory behaviour is thus not for the purpose of ingesting the prey themselves, but to provide meat for their human owners. Hunting with dogs is an ancient practice in which humans have taken advantage of the canine's superior senses (smell and hearing) and greater speed to locate, track, and catch game. In one form of hunting known as 'coursing', hounds are trained to respond to the sight of fleeing animals by running after them. If they are pursuing small game, they usually hunt in pairs, but if the prey is large and can fight back, a pack of dogs will be sent out. Coursing hounds usually kill the game after they have caught it. There is much disagreement, however, about how this is done. Hunters maintain that hounds usually dispatch their prey quickly with a bite to the neck, severing the cervical vertebrae (cf. pl. 8K), however examination of the bodies of animals killed by hunting dogs has revealed wounds on the abdomen, thorax, and, in particular, the hindquarters.

85 Kanawati-Abder-Raziq, *Meryteti*, 25.
86 Borchardt, *Sahure ii*, pl. 17.
87 See Osborn-Osbornová, *Mammals*, 60-64.

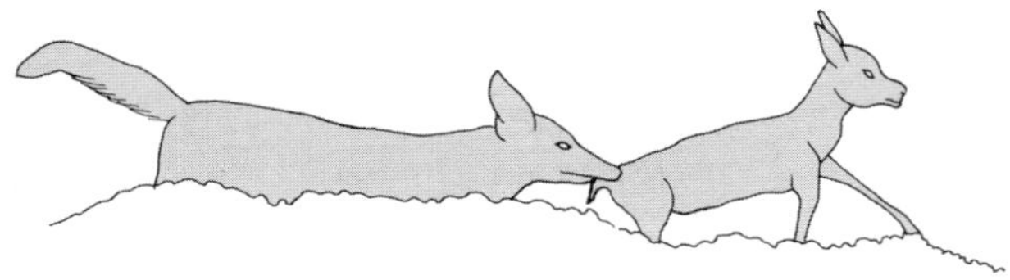

Pl. 8J Golden jackal pursuing a gazelle calf
After: Van Lawick-Goodall, *Innocent Killers*, pl. 11

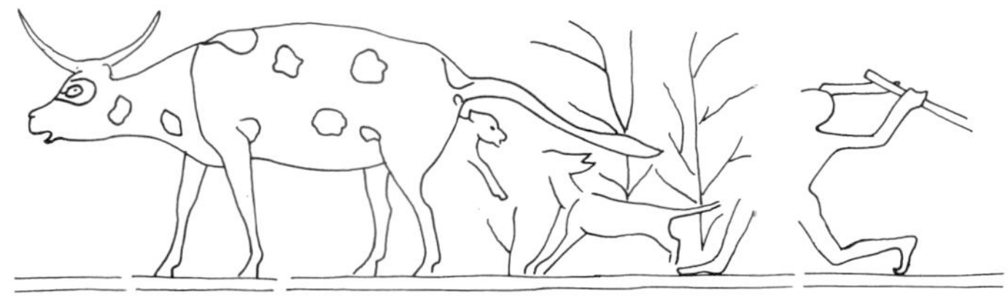

Fig. 8-20 (HB86)

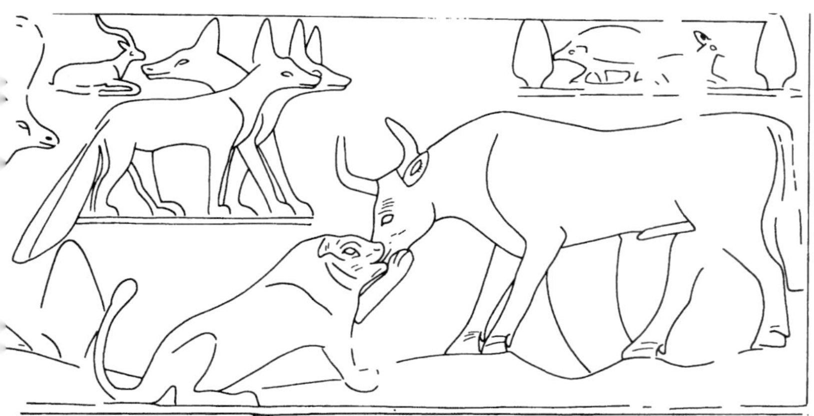

Fig. 8-21 (HB88)

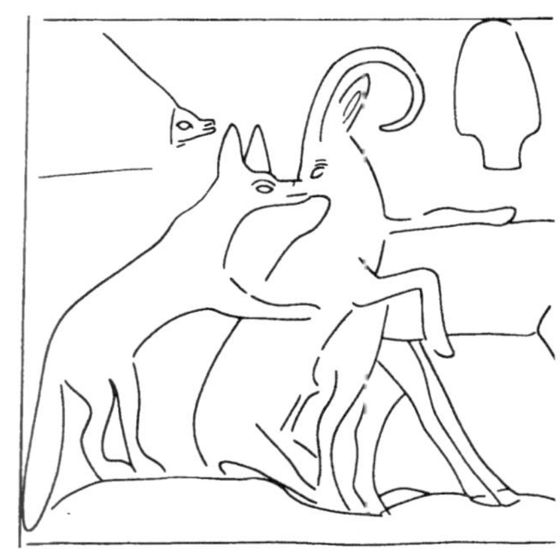

Fig. 8-22 (HB37)

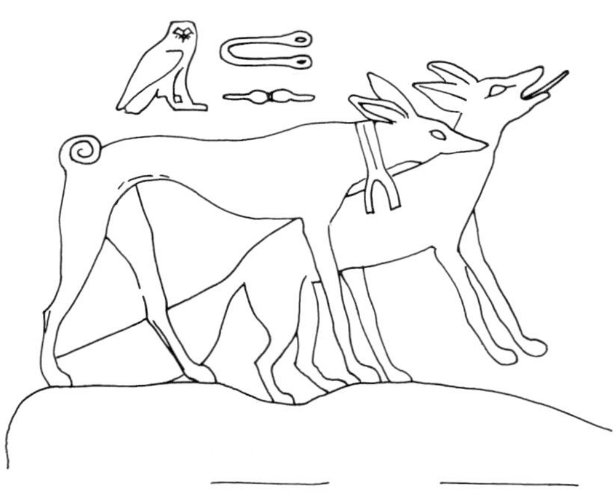

Fig. 8-23 (HB92)

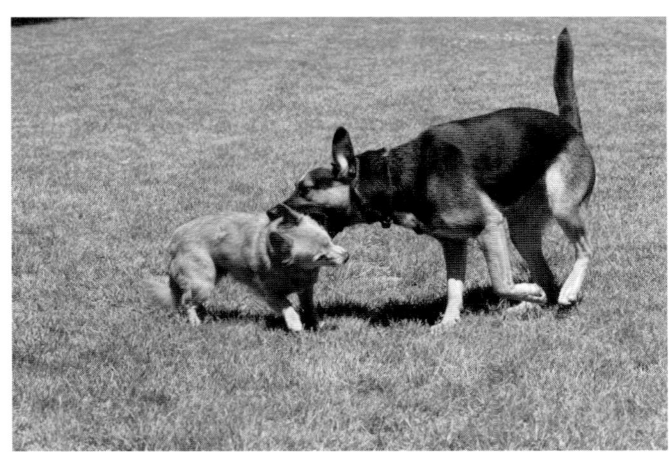

Pl. 8K Dog biting neck
© iStockphoto.com/Joop Snijder

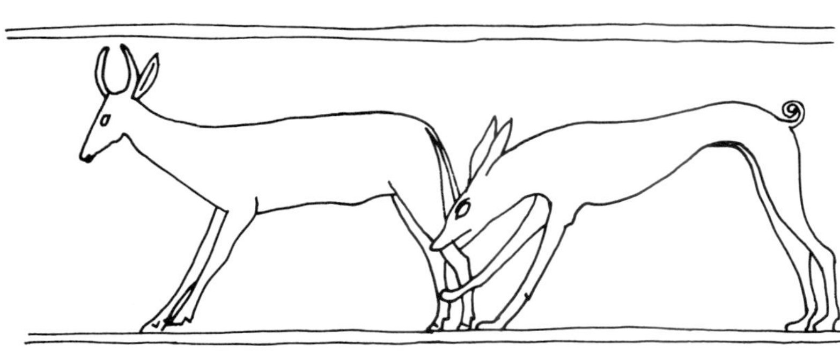

Fig. 8-24 (HB90)

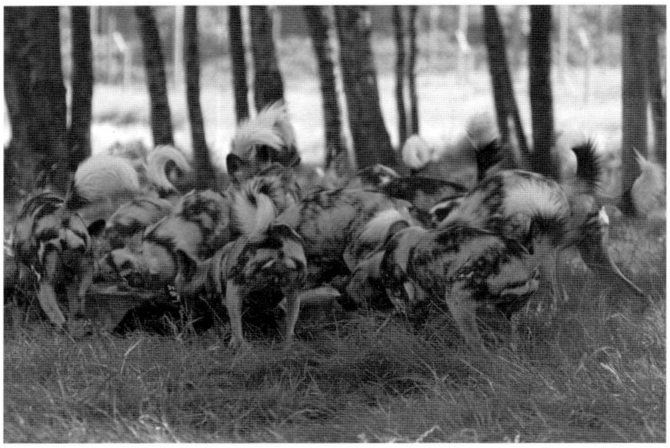

Pl. 8L Hunting dogs tearing apart prey

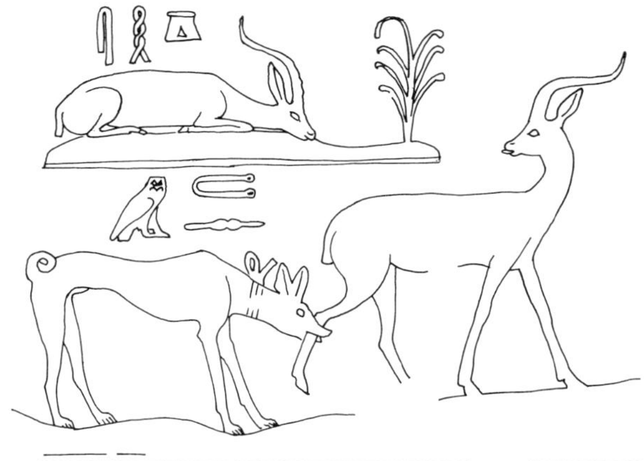

Fig. 8-25 (HB93)

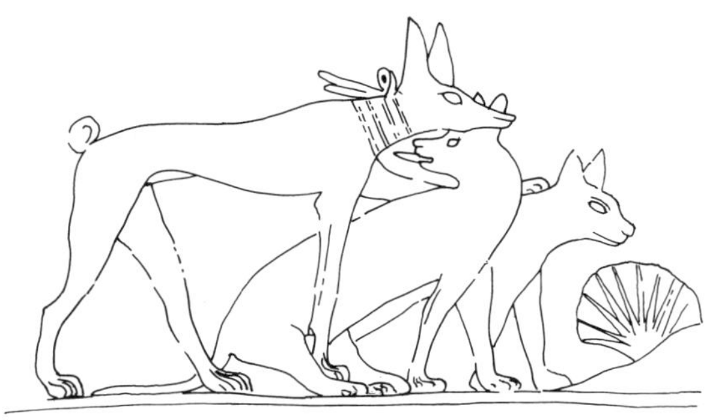

Fig. 8-26 (HB99)

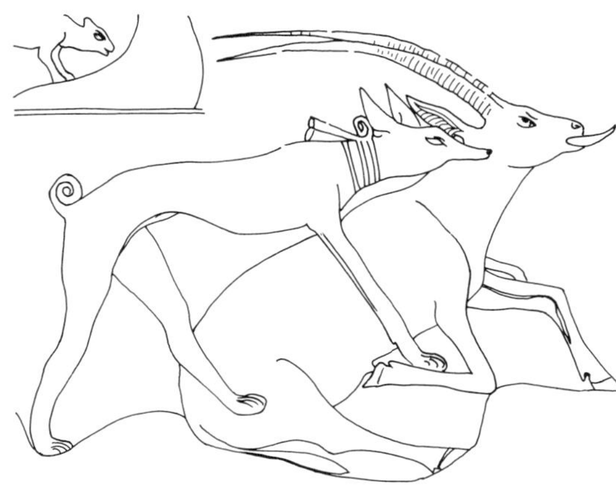

Fig. 8-27 (HB103)

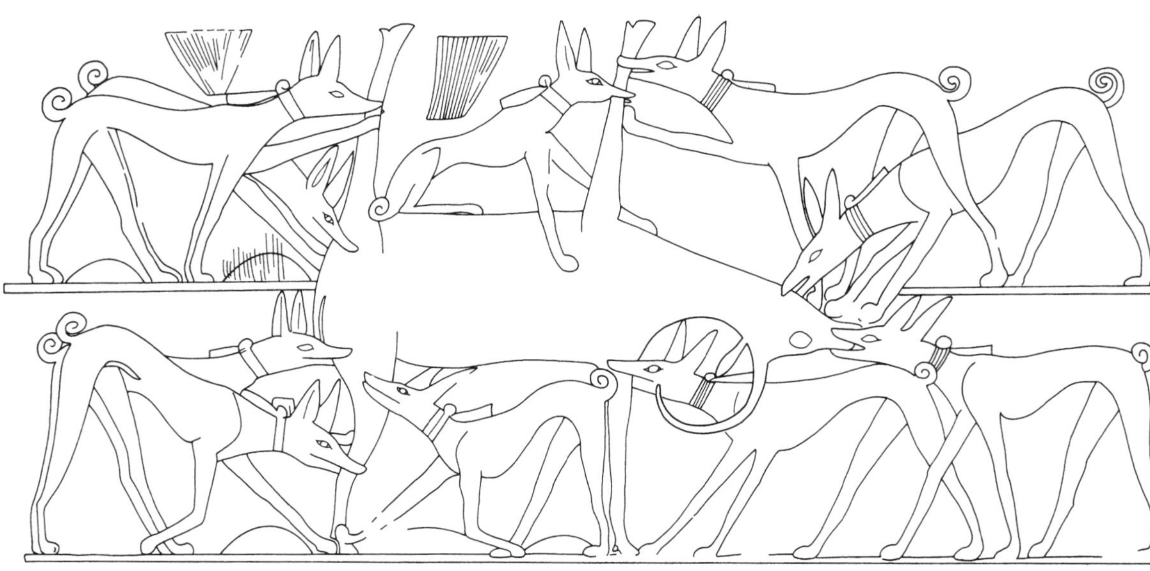

Fig. 8-28 (HB118)

In desert hunting scenes, men are sometimes shown kneeling beside their dogs and pointing, as though drawing the animals' attention to the game running before them.[88] The species that are attacked are: Nubian ibex (10); scimitar oryx (5); unidentified animals (6); dorcas gazelle (4); red fox or golden jackal (4); hartebeest (1); addax (1); Sömmerings gazelle (1); striped hyena (1); goat (1); and Cape hare (1). In most of the 35 examples, a single hound bites the prey animal, but in two scenes, a pair of dogs makes the kill, and three images feature a pack of nine or ten dogs attacking an ibex. In seven images, beginning with the earliest scenes in the tomb of *mṯn*, prey are bitten on their hind legs as they attempt to flee (HB89 to HB91, HB93, HB101, HB104, and HB117; see figs. 8-24 and 8-25). Bites are also made to the tail or rump of running animals (HB111). The ears (HB99 and HB119; see fig. 8-26), head (HB120), jaw (HB119), and muzzle (HB114, HB118, and HB121) are the focus of attack for some, while in two examples, the prey has been turned upside down (perhaps to indicate a tossing motion?) and the bite delivered to their abdomens (HB106 and HB113). The majority of bites, however, are made to the neck or throat region (see figs. 8-23 and 8-27), supporting the assertion made by modern hunters that hounds kill in this manner. Prey animals are bitten while they are chased and after their attackers have brought them down to the ground. Some animals lay sprawled upon their backs (HB96, HB98, HB100, HB102, HB108, HB118a, and HB119). Others 'sit' upon the ground with their hind legs stretched out before them (HB103, HB107, HB116, and HB121; see fig. 8-27) or, while continuing to stand, bend their legs in abrupt angles (HB105, HB115, and HB122),[89] giving the impression that we are witnessing the very moment of their collapse. The dogs often straddle their fallen victims (see fig. 8-23) or stand upon them as they deliver their bites (see fig. 8-27).

It has been proposed that striped hyenas seated beside a hunting dog in the tomb of *ptḥ-ḥtp II: ṯfj* were tame (HB99; see fig. 8-26).[90] The dog bites the ear of one of the hyenas while placing a foot on its outstretched hind legs and pawing the back of its companion. I suggest that the extended back legs of the first hyena indicate that it is not sitting normally, but has collapsed. It has responded to the dog's behaviour by turning its head and protruding its tongue, an action that may signify that it is in pain (see below).[91] Directly in front of this trio, another dog stands upon and bites an oryx, which also protrudes its tongue while twisting its head (HB100). The striped hyenas are thus being attacked in precisely the same manner as all the other prey in this relief. Their proximity to both the hound and the hunt do not indicate that they were trained for the chase or have a good rapport with the dog, as suggested by Jacques Vandier,[92] but simply that they are prey.

Ibex, scimitar oryx, and dorcas gazelle are the most commonly depicted prey animals. Nubian ibex are quite small compared to other ibexes (weight: 25-70 kg; height 65-75 cm), and dorcas gazelle are diminutive ungulates (weight: 15-20 kg; height: 55-65 cm). It seems feasible, therefore, that a single, fit dog could manage to attack animals of this size, however it is difficult to accept that one dog alone could bring down other prey species, such as scimitar oryx (weight: 135-180 kg; height: 110-125 cm), which are considerably larger and armed with dangerous horns. Large animals, however, were probably chased to the point of exhaustion by the dogs prior to their attack.[93]

---

88 Pointing hunters are depicted in the tombs of *nj-ꜥnḫ-ẖnmw/ẖnmw-ḥtp* (ntn), *ptḥ-ḥtp II: ṯfj* (D64), *nj-mꜣꜥt-rꜥ* [*jzzj-mr-nṯr*] (G2097), *sꜥnḫ-w(j)-ptḥ* (ntn), and *mrjj-ttj: mrj* (ntn). In three of these tombs, the hunters grab their dogs with their free hand, possibly to prevent them from beginning the chase too soon. A similar motif may occur in the funerary reliefs of Unas. See Labrousse-Moussa, *Complexe Funéraire au Roi Ounas,* Doc. 45, fig. 59.

89 The oryxes in HB115 and HB122 each appear to have broken a hind leg, as these limbs are bent forward at a 90° angle.

90 Vandier, *Manuel iv*, 797.

91 Osborn and Osbornová (*Mammals*, 101-102) suggest that the hyena licks the dog's throat, but note that in the reproduction provided by Quibell (*Ptah-hetep*, pl. 32), the animal's tongue does not make contact with the dog (see fig. 8-26).

92 " . . . et deux hyènes domestiques, dressées pour la chasse, en même temps que les chiens avec lesquels . . . elles semblent faire très bon ménage". See Vandier, *Manuel iv*, 797.

93 Unlike other animals in ancient Egypt, dogs were frequently provided with names. I believe that hunting dogs were named specifically so that hunters could direct individual members of a pack. Malek (*The Cat*, 51) proposed that one such name, *ḥmw-mꜣ*, 'Steering-oar of the lion' refers to the animal's tail. I suggest that it instead describes a hunting dog's ability to stay with its prey during the chase - remaining at its heels through every twist and turn, in much the same way as a steering-oar moves in concert with a boat.

Three reliefs depict groups of dogs that grab an ibex from all sides, pulling its body taut (HB114, HB118, and HB120; see fig. 8-28). Foxhounds squabbling over prey will sometimes stretch the body between them in such a manner, and African wild dogs, which hunt in large packs, frequently tear the bodies of their victims apart when all the participants grab their share simultaneously (see pl. 8L). It is likely that the ibex represented in these examples have been given to the packs for them to feed upon, as modern hunters sometimes share part of the kill with their hounds.[94] The subsequent frenzied attack has been conveyed very effectively by the variety of postures adopted by the dogs as they jostle and pull, and by allowing the body of the ibex to be extended across two registers. Many of the dogs display 'balking' poses (forelegs together, hind legs separated or both sets of legs held together and inclined; see Chapter 4), which impart a strong sense of tugging and 'pulling back' away from the victim's body.

Eleven images show the dogs' prey protruding their tongues (HB92, HB93, HB95, HB99 to HB101, HB103, HB109, HB115, HB116, and HB122; see figs. 8-23, 8-25, 8-26, and 8-27). In the tomb of *rˁ-m-k3.j* (HB92; fig. 8-23), for example, a fox that is grabbed upon the neck opens its mouth and stretches its tongue well forward, while further along on the same register, a dorcas gazelle responds to a bite on its hind leg with a very small protrusion of its tongue (HB93; fig. 8-25). In the tomb of *pḥn-wj-k3.j* (HB95), the tongue of a jackal issues far from its mouth as it is bitten on its neck. Ibex and oryx in many reliefs stretch their heads forward and protrude their tongues as they collapse to the ground (see fig. 8-27). The animals' behaviour gives the strong impression that they are crying out, either in pain or fear. Many animals that do not normally vocalize will nevertheless call when they are alarmed or distressed.[95] There is thus frequently an auditory component to predation events. It appears that some Egyptian artists considered this to be an important feature of predatory interactions and found a way to represent it in their depictions.

Hedgehogs

pḥn-wj-k3.j (HB123: Saqqara, V.6-8E)
ptḥ-ḥtp II: ṯfj (HB124: Saqqara, V.9M-L)
Anon (HB125: Saqqara, VI)

Three reliefs were found to depict hedgehogs catching prey. The most unambiguous image, which occurs in the tomb of *ptḥ-ḥtp II: ṯfj* (HB124: see fig. 8-30) shows a small hedgehog emerging from a stylised burrow[96] carrying a scorpion or locust in its mouth. A similar depiction can be seen on a later, somewhat eroded relief fragment (HB125), which shows a hedgehog emerging from a burrow (travelling left) with a small narrow object in its mouth. The earliest of the images depicts a hedgehog standing freely with a peg-like shape in its jaws (HB123; see fig. 8-29). I suggest that this object was also intended to represent an insect or other prey item and that it has been rendered incorrectly, either by the original artist or the modern copyist.

Desert hedgehogs and long-eared hedgehogs are both found in Egypt today.[97] It is impossible to determine which of these is represented in tomb scenes, as the two species are very similar morphologically,[98] however, as their hunting behaviour and diets are comparable, the general accuracy of the Egyptian depictions can nevertheless be assessed.

For the names of dogs, see Janssen, J.M.A. (1958). "Über Hundenamen im pharaonischen Ägypten", *MDAIK* 16, 176-182; Fischer, H.G. (1961). "A supplement to Janssen's list of dogs' names", *JEA* 47, 152-153; Fischer, H.G. (1978). "More ancient Egyptian names of dogs and other animals", *MMJ* 12 (Suppl.), 173-178; Vachala, B. (2002). "Zwei Hundenamen aus Abusir", *GM* 190, 83-85.

94 This has also been proposed by Vandier, *Manuel iv*, 800.

95 Dorcas gazelle, for example, bleat loudly when they are in danger or in pain. See Yom-Tov, Y. (1995). "Gazella dorcas", *Mam. Sp.* (491), 5.

96 For the Egyptian depiction of animal burrows, see Schäfer, *Principles,* 124 and 126-127.

97 Long-eared hedgehogs are more common in Egypt than desert hedgehogs. They are usually found in densely vegetated areas around human settlements, particularly in gardens, while desert hedgehogs prefer drier habitats. See Hoath, *Field Guide*, 25-29.

98 Although long-eared hedgehogs are smaller than desert hedgehogs, their ears are much larger and project well above their spines. The arrangement of the spines on their heads also differs. See *ibid.*, pl. 1.

The diet of desert and long-eared hedgehogs consists primarily of invertebrates (see pl. 8M), but they also feed upon eggs and small vertebrate animals. Both species rest during the day, but emerge from their burrows at dusk to hunt for prey at night, which they detect using their acute senses of smell and hearing. After capture, prey are immobilized and killed by locking their jaws and crushing. The Egyptian depictions thus show these animals hunting correctly. Furthermore, as two of the hedgehogs are depicted leaving burrows, I suggest that the images also show them pursuing prey nocturnally.[99]

Sacred ibis

mḥw (HB126: Saqqara, VI.2M-3?)

The combination of a sacred ibis and a butterfly atop a papyrus thicket in the tomb of *mḥw* (HB126; fig. 8-31) may represent an act of predation. The butterfly appears to be in flight as it is depicted above the foliage with its wings spread. The bird approaches the over-sized insect and touches it upon the head with the tip of its curved beak. The juxtaposition of the bird and butterfly may be accidental, but it should be noted that although sacred ibis obtain most of their food by probing mud and soil with their beaks (see Chapter 7), they will occasionally catch insects and other creatures (cf. pl. 8N) by running forward with their wings partially open and snapping at them.[100] Although the ibis in the *mḥw* relief does not appear to run, a predatory interpretation may nevertheless be feasible.

Pied kingfishers

nfr/k3-ḥ3.j (HB127: Saqqara, V.6)
nj-ʿnḫ-ẖnmw/ẖnmw-ḥtp (HB128: Saqqara, V.6L-7)
ṯjj (HB129: Saqqara, V.8-9)
3ḫtj-ḥtp (HB130-131: Saqqara, V.8L-9E)
k3.j-m-ʿnḫ (HB132: Giza, VI)
ḥzj (HB133: Saqqara, VI.1L-2E)

Pied kingfishers can be identified in tomb scenes based upon morphological cues, such as their sharp beaks and prominent head crests.[101] They are also distinguished by their predatory behaviour, which is reproduced in six tombs at Giza and Saqqara.[102]

Pied kingfishers prey upon fish, crustaceans, frogs, and aquatic insects.[103] When hunting, they either perch motionless on a branch above water or fly for short periods and then hover approximately two to ten metres above the water surface (see pl. 8O), waiting (in both cases) until a fish or other aquatic animal comes into view below. Once they have detected their prey, they then dive downward in stages. Pulling their wings in (see pl. 8P), they plummet headfirst before plunging into the water to seize their prey in their beak. Small fish are eaten on the wing, but larger prey are carried to a branch or other perch and then beaten to death before being swallowed.[104] Both the hover and the dive of pied kingfishers are reproduced in tomb scenes.[105]

Hovering is a very conspicuous behaviour, as kingfishers must flap their wings at a fast rate to maintain the same position in the air. While sweeping their wings back and forth, their body is

[99] During this study, 25 depictions were found of hedgehogs appearing in a natural setting. Seventeen of these are shown in association with a stylised burrow. Thirteen images depict the animals emerging from the burrow (facing out), while only four show them entering. I suggest that this indicates that the Egyptians were well aware of their nocturnal behaviour.

[100] Hancock et al., *Storks, Ibises and Spoonbills*, 214.

[101] See Gaillard, C. (1931). "Quelques représentations du Martin-Pêcheur Pie (*Ceryle rudis*), sur les monuments de l'Égypte ancienne", *BIFAO* 30(1), 249-271.

[102] The depiction of this distinctive behaviour in Egyptian art has been noted elsewhere. The earliest example occurs in the early 5th dynasty mortuary temple of Userkaf. See *ibid.*, 265-266; Houlihan, *Birds*, 114-116.

[103] Roselaar in Cramp-Simmons, *Handbook iv*, 725.

[104] Whitfield, A.K. and Blaber, S.J.M. (1978). "Feeding ecology of piscivorous birds at Lake St Lucia, Part 1: Diving birds", *Ostrich* 49(4), 190. A relief fragment from the mortuary temple of king Sahure displays a pied kingfisher holding a fish in its beak while perched within a papyrus thicket. See Borchardt, *Sahure ii*, pl. 16.

[105] These behaviours are also represented in scenes where kingfishers are attempting to both feed their young and protect them from approaching predators. Only flight manoeuvres associated with predation have been considered in this chapter, while the remaining images are discussed in Chapters 9 and 11.

usually held upright and their head and beak tilted downward (see pl. 8O) so that they can study the waters below. The birds are able to hold their heads perfectly still while their bodies move, allowing them to keep their prey (which is also often moving) in focus.[106]

Hovering for the purpose of predation is reproduced in five images dating from the mid-5th to early 6th dynasty (HB127, HB128, HB130, HB131, and HB133; see figs. 8-32 and 8-34). In each example, the birds are depicted with their wings spread, their bodies upright, and their heads bent down.[107] The combination of frontal (body and wings) and profile (head) views, so characteristic of Egyptian two-dimensional art, captures precisely the planes of the birds' bodies when engaged in this particular action. Three of the birds fly above papyrus thickets, while that in the tomb of *ḥzj* (HB133; fig. 8-34) appears amongst the wall of vegetation. Fish teeming in the waters below each bird are undoubtedly their intended prey. The image in the tomb of *ḥzj* is somewhat distorted, as it gives the impression that the bird's body is facing away from the viewer. I suggest that here the artist may have attempted to show the transition from hovering to diving by bending the bird's wings in towards its body (see below) and swinging its legs to one side.

Two scenes were found to represent pied kingfishers diving while hunting (HB129 and HB132; see fig. 8-33).[108] In both cases, the birds are depicted upside down with their wings outstretched and slightly bent, and their heads in profile. Studies have shown that the position of their wings during the dive is affected by the depth of the kingfishers' prey in the water: if it is close to the surface, the birds' wings are extended as they dive, but when pursuing deeper prey, their wings are held closer to their body.[109] It has also been noted that the angle of their dive tends to be steeper (more vertical) for deep prey.[110] The Egyptian images thus combine incorrectly the wingspread associated with hunting shallow prey with the near-vertical dive employed for deep prey. Nevertheless, they convey effectively the drama of this part of the kingfishers' predatory sequence.

Nile crocodiles

ḥzjj-rʿ (HB134: Saqqara, III)
k3.j-m-nfrt (HB135: Giza, V.6-8E?)
k3.j-m-rḥw (HB136: Saqqara, V.8)
ṯjj (HB137: Saqqara, V.8-9)
snḏm-jb: mḥj (HB138: Giza, V.9)
ptḥ-ḥtp II: ṯfj (HB139: Saqqara, V.9M-L)
sšm-nfr IV (HB140-141: Giza, V.9-VI.1)
3ḫtj-mrw-nswt (HB142: Giza, V.9-VI.1)
nj-m3ʿt-rʿ [jzzj-mr-nṯr] (HB143: Giza, V.9-VI.1)
jrw-k3-ptḥ: ḫnw (HB144: Saqqara, VI)
k3-gm-nj: mmj (HB145-148: Saqqara, VI.1M)
mrrw-k3.j: mrj (HB149-150: Saqqara, VI.1M-L)
zšzšt: jdwt (HB151-153: Saqqara, V.9 r/u - VI.1-2E?)
ḥzj (HB154: Saqqara, VI.1L-2E)
ʿnḫ-m-ʿ-ḥr: zzj (HB155-156: Saqqara, VI.1L-2E)
mṯṯj (HB157: Saqqara, VI.2?)
mrrj (HB158: Saqqara, VI.2E?)
mḥw (HB159: Saqqara, VI.2M-3?)

Nile crocodiles are among the most aggressive of all crocodilians. They hunt and eat a wide range of prey: insects, turtles, fish, other reptiles, birds, and mammals (including humans).[111] They hunt primarily at night and rest during the day, either on land or underwater (see Chapter 4). Their method of attack changes somewhat depending on the prey type and where the kill takes place (either on land or in water), but they always approach silently and without warning, and are rarely seen by their victims until it is too late.[112] When hunting, they frequently position themselves in the water so that only their eyes and nostrils are visible, hanging motionless as they wait to locate potential victims (see pl. 8Q). From this semi-submerged and well-camouflaged position, they can

---

106 Schwab, I.R. and Hart, N.S. (2004). "Cover illustration: Halcyon days", *Brit. J. Ophthalmol.* 88, 613.

107 Although the image in the tomb of *nj-ʿnḫ-ḫnmw/ḫnmw-ḥtp* (HB128) is somewhat damaged, the characteristic position of the bird's beak clearly indicates that it is a hovering kingfisher.

108 Note that pied kingfishers, as well as other bird species, can also adopt an inverted posture in wall scenes when mobbing predators that threaten their chicks. See *Birds vs predators* in section 9.3 in Chapter 9.

109 Katzir, G., Labinger, Z. and Benjamini, Y. (2003). "Prey choice in pied kingfishers (*Ceryle rudis* L.): Prey size and prey depth", *Bird Behav.* 15(2), 48.

110 Schwab-Hart, *Brit. J. Ophthalmol.* 88, 613.

111 Cott, H.B. (1961). "Scientific results of an inquiry into the ecology and economic status of the Nile crocodile (*Crocodilus niloticus*) in Uganda and Northern Rhodesia", *Trans. Zool. Soc. Lond.* 29, 283-297.

112 Pooley, A.C., Hines, T. and Shield, J. (1989). "Attacks on humans" in Ross, *Crocodiles*, 177.

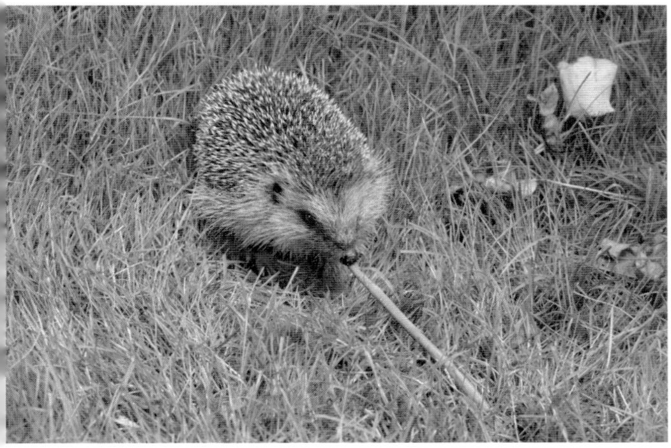

Pl. 8M Hedgehog with worm
© ardea.com/Brian Bevan

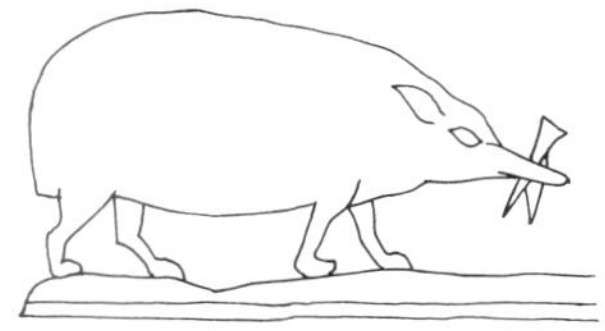

Fig. 8-29 (HB123)

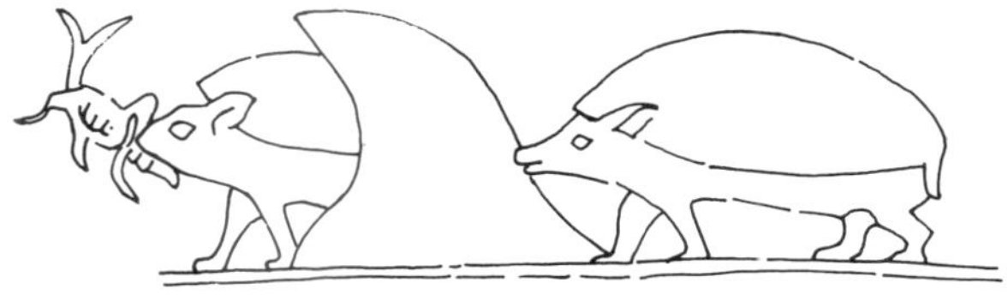

Fig. 8-30 (HB124)

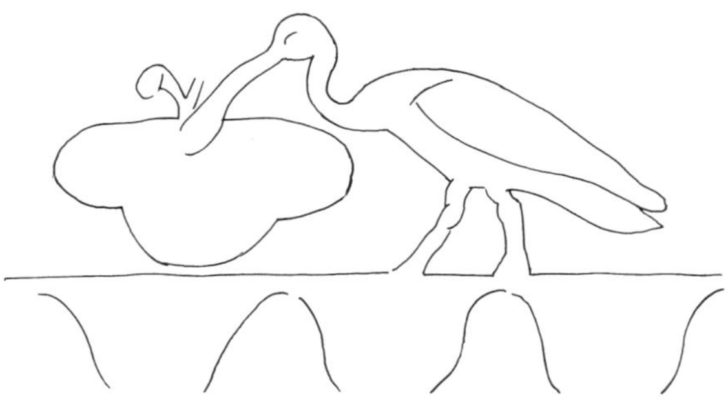

Fig. 8-31 (HB126)

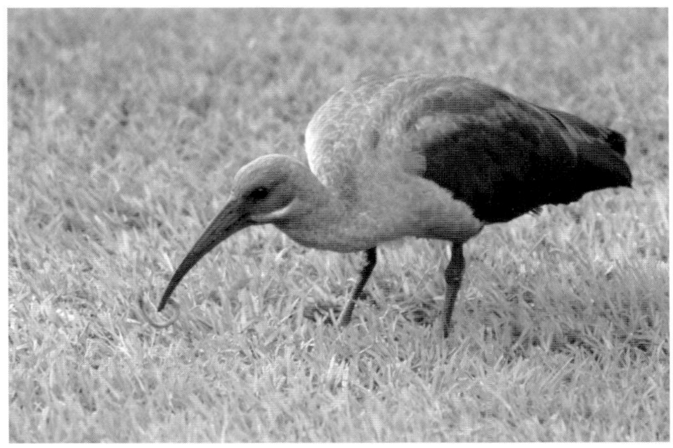

Pl. 8N Ibis with worm
© iStockphoto.com/Mark Atkins

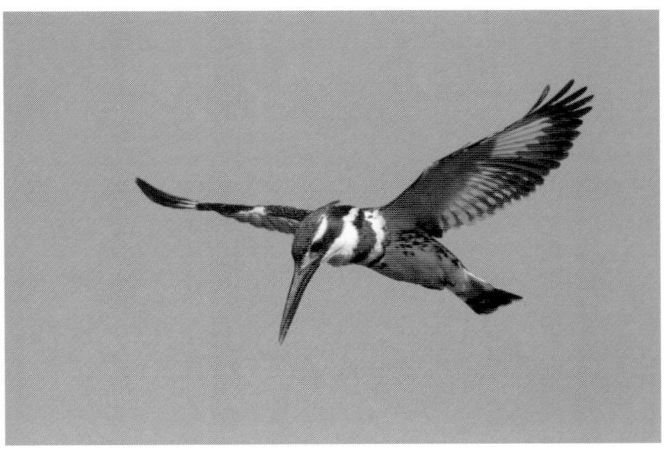

Pl. 8O Pied kingfisher hovering
© iStockphoto.com/Stephen Earle

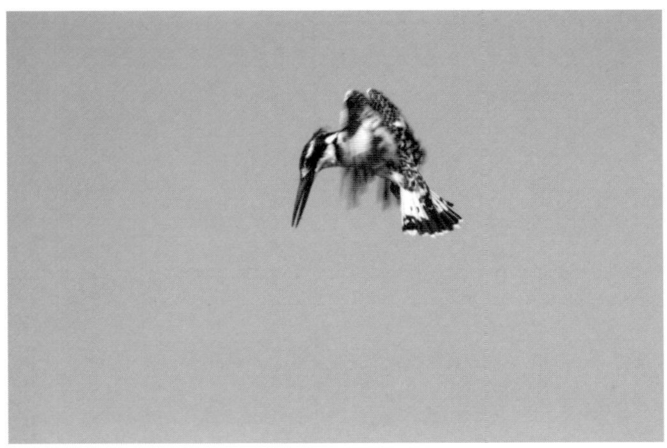

Pl. 8P Pied kingfisher beginning to dive
© iStockphoto.com/David Garry

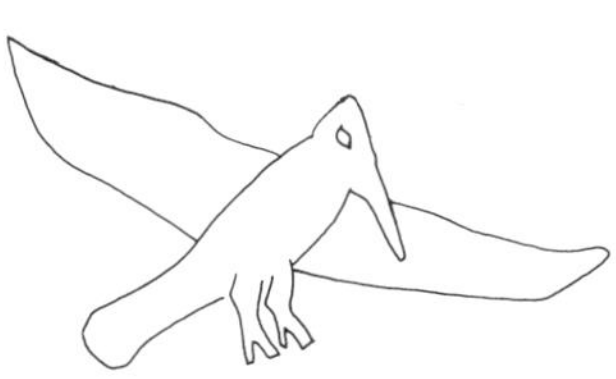

Fig. 8-32 (HB127)

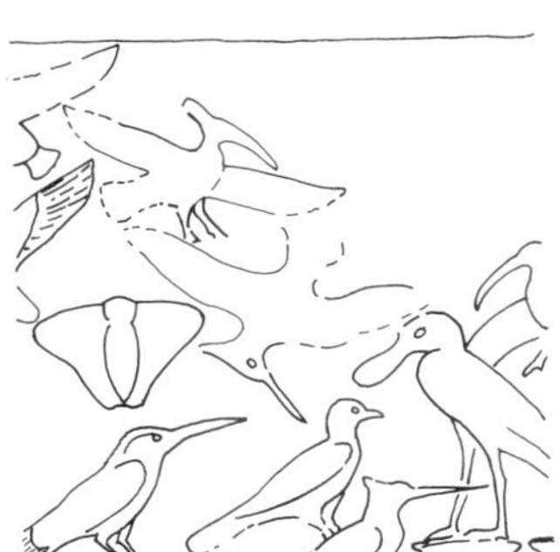

Fig. 8-33 (HB132)

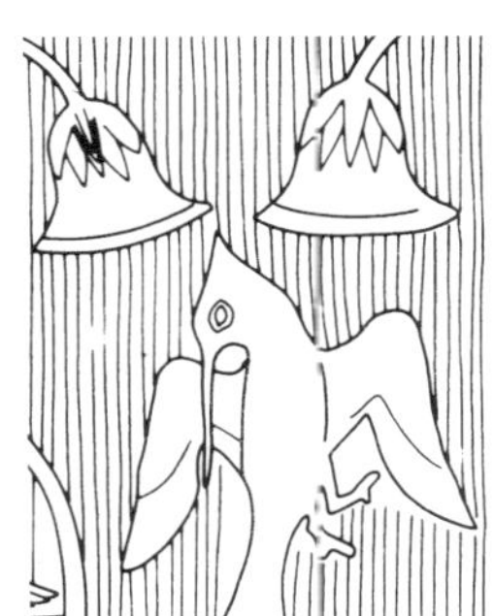

Fig. 8-34 (HB133)

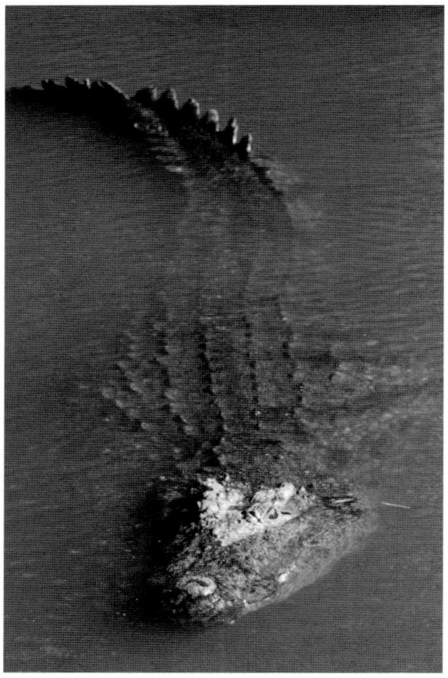

Pl. 8Q Semi-submerged crocodile

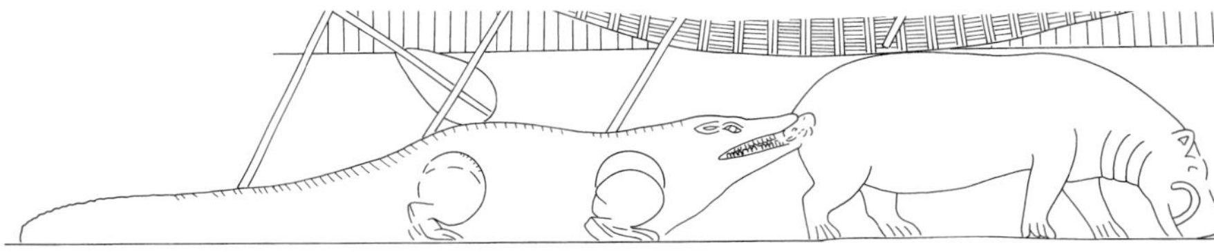

Fig. 8-35 (HB150)

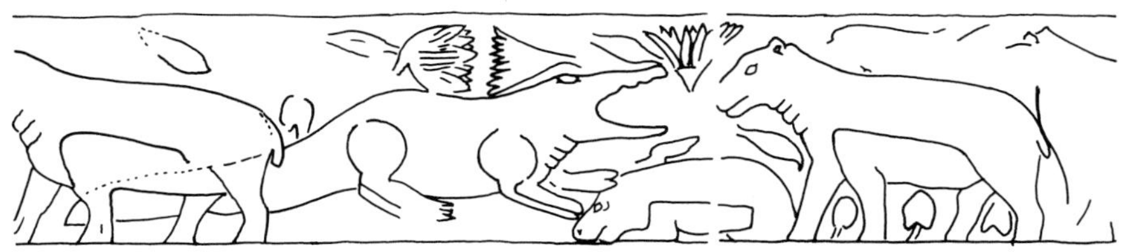

Fig. 8-36 (HB147)

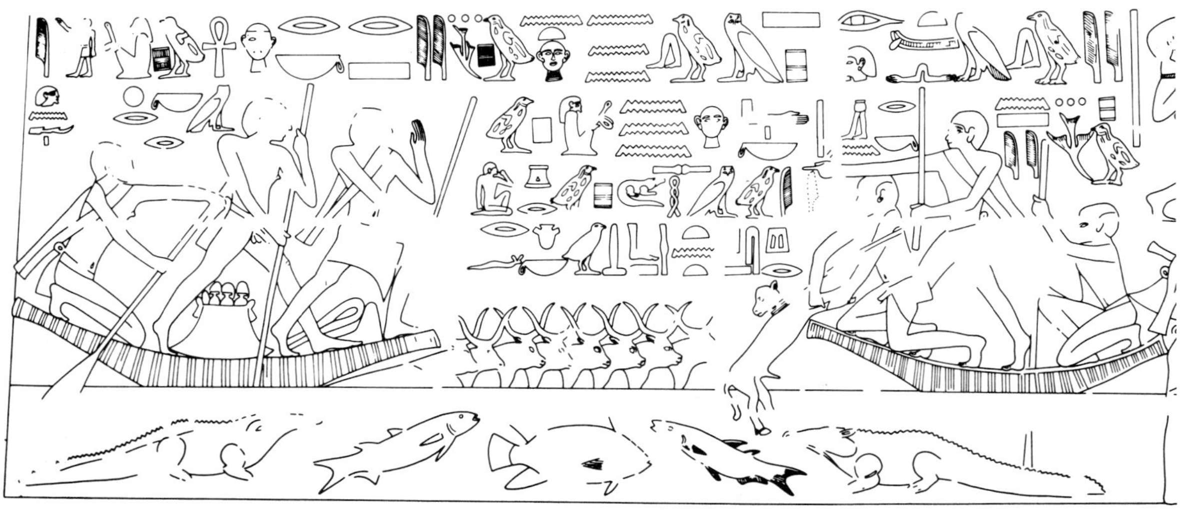

Fig. 8-37 (HB154)

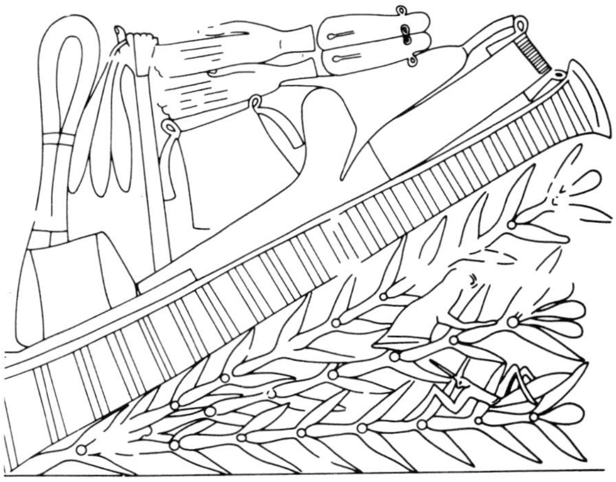

Fig. 8-38 (HB160)

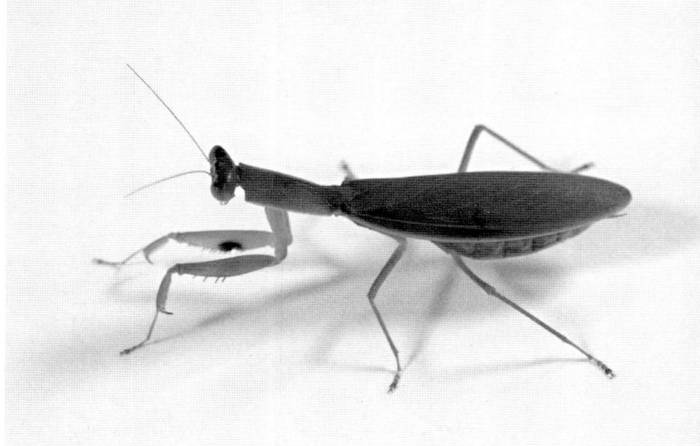

Pl. 8R Praying mantis

launch themselves with astonishing speed directly out of the water at their unsuspecting victims. Prey are usually seized by the muzzle or leg and then dragged underwater and drowned. Animals at waterholes and river-crossings are particularly vulnerable to attack.[113]

In tomb scenes, Nile crocodiles exhibit predatory behaviour towards both cattle and newborn hippopotamus calves. Crocodiles attack cattle frequently in many parts of Africa today and are responsible for substantial stock losses,[114] but they rarely prey upon healthy adult hippos as these are considerably larger than the reptiles and are equipped with long tusks that are capable of inflicting fatal injuries (see Chapter 9). They will attack injured hippos, however, and also juveniles that stray too far from adults. Hippopotamus calves are also taken, but these are usually safe from predation, as their mothers guard them fiercely and will pursue and attack any crocodile that comes too close. In order to snatch a calf from its mother without being injured, crocodiles need to rely on both stealth and speed.[115] The remains of cattle and both adult and newborn hippos have been recovered from the stomachs of Nile crocodiles.[116]

Eight tomb scenes depict crocodiles standing behind hippopotamus cows that are giving birth (HB138, possibly HB145 and HB148, HB150, HB151, HB152, HB156, and HB158). In each case, the reptiles are situated so that their jaws either linger near or below the hippos' tails, from whence emerge their newborn calves, or they bite at the foetus (HB150 and HB158; see fig. 8-35). The crocodiles thus seem to be waiting to snap at the young after they have been fully expelled or to pluck them directly from the birth canal (see Chapter 11).

Hippo cows leave their herd just prior to calving and do not rejoin the group until some weeks later. They are thus isolated during the birth, which takes place either on land or in shallow water. Zoo observations suggest that parturition lasts approximately 3 hours.[117] Crocodiles consequently have ample time to detect an impending birth and draw closer to the cow. Calves are extremely vulnerable immediately after birth, particularly those born underwater, as they must be helped to the surface by their mother to take their first breath. A crocodile would nevertheless need to attack quickly before the cow became aware of, and react to, its presence. The Egyptian images thus convey the speed with which the crocodiles were required to strike by placing the predators' jaws directly below, and sometimes touching, the still unborn prey.[118]

In another depiction of an interaction between a crocodile and a hippopotamus, the hippo again fails to see the predator. In a scene from the tomb of *sšm-nfr IV* (HB140), a crocodile approaches an adult hippopotamus standing beside a calf. The close proximity between the small animal and the larger suggests that the latter is probably its mother. The crocodile is elevated above the riverbed and thus may be represented swimming (see Chapter 4). Its open jaws indicate that it is about to attack either the adult hippo, whose back is turned away from the predator, or the calf that, due to the adult's lack of awareness, wanders directly toward the crocodile. A similar drama also unfolds in the later tomb of *k3-gm-nj: mmj* (HB147; see fig. 8-36), in which a swimming crocodile again faces an adult hippopotamus with a calf. The calf lies prostrate upon the riverbed directly below the gaping jaws of the predator. Here, however, the adult is oriented towards the crocodile and so must be aware of the imminent danger.

The importance of vigilance and sight are also stressed in scenes involving cattle. Sixteen images were found to depict herds of cattle fording canals in which one or more crocodiles are submerged.

113 Pooley, A.C. (1989). "Feeding and food habits" in *ibid.*, 86.

114 Mulonga, S., Suich, H. and Murphy, C. (2003). "The conflict continues: Human wildlife conflict and livelihoods in Caprivi", *DEA Research Discussion Paper* 59, 12.

115 In southern Africa, hippopotamus births peak during the summer rainy season, which runs from November until April. This coincides with the period during which male Nile crocodiles are most aggressive because warmer water temperatures at this time of year enable the cold-blooded reptiles to be more active. The rainy season in East Africa lasts from April until June.

116 Cott, *Trans. Zool. Soc. Lond.* 29, 291-292; Guggisberg, *Crocodiles*, 91.

117 Senior, M. and Tong, E.H. (1963). "Parturition in a hippopotamus", *Proc. Zool. Soc. Lond.* 140, 57-59.

118 It is important to note, however, that it is unlikely that a crocodile would be able to take a calf before it had been fully expelled.

The water is deep as often only the heads and shoulders of the animals are visible; the slope of their backs suggests that they are swimming. Some examples feature only one crocodile, but in the tomb of *ṯjj* (HB137) and thereafter, two crocodiles are commonly depicted (see fig. 8-37 and also fig. 5-17 in Chapter 5). Nile crocodiles are usually solitary hunters and so the appearance of two threatening reptiles is curious. It should be noted, however, that male Nile crocodiles hold and defend territories - stretches of shoreline and their adjoining waters – from which they exclude other crocodiles. They patrol their territory regularly by swimming back and forth along their claimed section of shoreline.[119] It is feasible, then, that a fording herd could pass through the territory of a crocodile on one side of a stream and then through that of another male on the opposite side. Nevertheless, it seems more likely that the artists have situated a crocodile at either end of the line of cattle simply to present a symmetrically balanced composition and to exaggerate the threat posed to the herd.

The earliest fording scene comes from a painted wall relief in the 3rd dynasty tomb of *ḥzjj-rˁ* (HB134). The image is poorly preserved, but originally depicted four oxen and two herdsmen that are threatened by a single crocodile resting upon the riverbed. Like most of the crocodiles that appear subsequently, it adopts an alert posture with its head raised. In the 6th dynasty tomb of *mḥw*, however, one crocodile lays its head upon the substrate as though sleeping (HB159, left). Nile crocodiles exhibit both of these postures frequently when they are basking on land; the artists have thus incorporated these easily observed, terrestrial postures in their depictions of the submerged reptiles. Crocodiles in the fording scenes of *ṯjj* (HB137, left), *sšm-nfr IV* (HB141), *nj-m3ˁt-rˁ* [*jzzj-mr-nṯr*] (HB143), *k3-gm-nj: mmj* (HB146), and *ḥzj* (HB154; see fig. 8-37) are suspended above the riverbed, possibly to suggest that they are swimming (see Chapter 4). The elevated crocodiles usually face the oncoming herd, as though drawn to the sounds of their approach. Indeed, studies have shown that crocodiles use sound to locate their prey (see below), and are particularly attracted to splashing noises made by animals struggling in the water.[120]

In all but two of the fording scenes, the crocodiles are situated beneath the canoes of the herdsmen, but in the tombs of *ptḥ-ḥtp II: ṯfj* (HB139) and *3ḫtj-mrw-nswt* (HB142; see fig. 4-22 in Chapter 4), the predators are depicted directly beside the cattle. The latter scene deserves particular mention for the accuracy with which it depicts the behaviour of two crocodiles. In characteristically cryptic fashion, one animal has only its eyes and nostrils visible above the water surface while its body rests upon the riverbed (cf. pl. 8Q).[121] The artist has presented a cross-section of the water column around the predator, allowing the viewer to see both above and below simultaneously and thus to appreciate fully the imminent threat posed by its open jaws. Another crocodile in the same scene appears to raise its entire head above the water next to a young man who swims across the river with the herd. As noted in Chapter 4, crocodiles often lift their bodies out of the water when attacking and so the image may be intended to convey an impending strike on the boy.[122] Thus the artist has incorporated both a vertical and a horizontal view within the same stretch of water in an apparent attempt to present two aspects of crocodilian predatory behaviour – the stealthy approach and the attack.

The cattle in each scene appear to be oblivious to the presence of the predators, which is perhaps not surprising since most of the crocodiles are fully submerged. In the tombs of *k3-gm-nj: mmj* (HB146; see fig. 5-16 in Chapter 5) and *mḥw* (HB158), however, one member of the herd seems to panic as it has turned to face the opposite direction, away from the crocodile. Cattle have excellent senses of smell and hearing, but they depend primarily upon vision to detect threats. The position of their eyes provides them with a panoramic view of their surroundings, but their ability to focus and depth perception are poor.[123] They are consequently very sensitive to contrasts in their environment and will often balk at shadows. If the dark shape of a moving crocodile were detected in the waters below, they would undoubtedly refuse to advance and may attempt to flee.

---

119 Modha, M.L. (1967). "The ecology of the Nile crocodile (*Crocodylus niloticus* Laurenti) on Central Island, Lake Rudolf", *EAWJ* 5(August), 79-80.

120 Pooley et al. in Ross, *Crocodile*, 176-177.

121 As noted in Chapter 4, this animal may be swimming.

122 Crocodiles also lift their snouts out of the water in a similar fashion during courtship and aggressive interactions, however the context in which the image occurs suggests that a predatory function is more likely here.

123 Phillips, *Cattle Behaviour*, 1-5.

The inscriptions that accompany fording scenes from the late 5th dynasty onward often refer to the crocodile as coming 'blind-of-head' (HB136, HB137, HB149, HB153 to HB155, and HB159; see fig. 8-37 and also fig. 5-17 in Chapter 5). These texts, combined with magic hand gestures performed by the herdsmen,[124] were intended to have the apotropaic function of preventing the reptiles from attacking the cattle as they waded through the water. The enigmatic phrase, 'blind-of-head', seems to allude to their predatory behaviour. A clue to its meaning is revealed by Herodotus (II, 68) who, when describing the animals of Egypt, stated that the crocodile "cannot see in the water, but in the open air it is the most-sharp-sighted of all".[125] This would suggest that the Egyptians might have believed that *submerged* crocodiles were unable to see their prey. The desire that the crocodile come 'blind-of-head' may thus have expressed a hope that it would remain fully underwater and consequently unable to locate the herd visually. Remarkably, a recent study examining the visual acuity of six crocodilian species found that while Nile crocodiles have excellent vision above the water surface, their ability to focus is severely impaired when submerged. It has been concluded that, when hunting underwater, they must consequently rely upon senses other than vision (e.g., hearing).[126] The Egyptians were therefore correct about the reptiles' limited ability to see in the water, but they were mistaken if they believed that they, and their animals, were safe from attack from crocodiles lurking in the depths.

Praying mantids

ḥzj (HB160: Saqqara, VI.1L-2E)

In marsh scenes of the late 5th and early 6th dynasties, frogs, butterflies, and grasshoppers are often depicted amongst foliage situated in the void beneath the stern of the deceased's canoe. In the tomb of *ḥzj*, however, this area was found to contain a unique depiction of a praying mantis (HB160; see fig. 8-38, lower right).[127]

Praying mantids are large, carnivorous insects and masters of camouflage, which they achieve both through their appearance and their behaviour. All mantids have a triangular-shaped head with long antennae, a slender thorax, and an elongated abdomen (pl. 8R). They are often the same colour as the environment they inhabit, and the wings of some species can also take the form of the leaves or flowers they hide among. These morphological features camouflage them so well that they are able to blend in with their surroundings. This effective concealment is crucial to their method of hunting. Mantids are 'sit and wait' predators; perched on a branch or leaf, they wait motionless until a possible victim comes within range. They then strike with lightning speed by reaching out with their pincer-like forelegs to catch and hold their prey. When at rest, mantids raise their forelegs in a manner resembling prayer - a habit that has given rise to their name.[128] They hunt and eat a variety of insects, as well as other small prey, such as tree frogs, lizards, and small birds.[129]

The insect in the tomb of *ḥzj* has been captured precisely as it would appear in nature. Sitting amid stylised leaves of pondweed (*Potamogeton lucens*), it is depicted with its forelegs raised in the characteristic 'prayer' position, its triangular-shaped head cocked to one side, and its abdomen tilted upward to mimic the surrounding foliage. Unlike representations of grasshoppers, the mantid image is very difficult to make out against the surrounding foliage. This suggests that the artist knew that these animals blend in with their environment to avoid detection and positioned his image intentionally so that its appearance in the relief was obscured. It seems likely, then, that the artist would also have been aware of their predatory habits. It is consequently intriguing to consider whether the mantid image had a practical function, which was to magically 'eat' any harmful insects that might enter the tomb.

124 Dominicus, *Gesten und Gebärden*, 131-143.
125 Blanco-Tolbert Roberts, *Herodotus The Histories*, 96.
126 Fleishman, L.J., Howland, H.C., Howland, M.J., Rand, A.S. and Davenport, M.L. (1988). "Crocodiles don't focus underwater", *J. Comp. Physiol. A* 163(4), 441-443.
127 For further details, see Evans, L. (2004). "The praying mantis in ancient Egypt", *BACE* 15, 7-18.
128 Hurd, L.E. (2003). "Mantodea (Praying mantids)" in Resh-Cardé, *Insects*, 675-677.
129 Mazer, C.L. (2004). "Mantodea" in *Grzimek's Animal Life Encyclopedia iii*, 180.

SUMMARY

- The representation of hunting in tomb scenes is extremely accurate. Some behaviours – notably, suffocation and moving of prey by lions, orienting by cats, hovering by pied kingfishers, (possibly) the diminished vision of submerged crocodiles, and the use of camouflage by mantids - reveal a close study of subtle species-specific details.
- The postures of animals are well-observed – for example, the use of forepaws by attacking lions, stalking movements by leopards, the bodies of collapsed prey, the stretching of prey between members of dog packs, wing positions of hovering kingfishers, and the resting pose of mantids.
- Some images illustrate associations between different species that have only been noted in the scientific literature in recent years. These include the attraction of leopards to domestic dogs and their kills, the proximity of jackals to the kills of lions, and the predation of newborn calves by golden jackals.
- The depicted predators include both nocturnal and diurnal animals; these species often appear in the same scenes (e.g., nocturnal genets and diurnal mongooses in marsh scenes; nocturnal leopards and diurnal domestic dogs in a hunting scene). This temporal conflict is either due to lack of information by the artists or was deemed unimportant (i.e., the inclusion of particular predator types was considered to be more crucial than when they were active).
- Predators sometimes appear in pairs. This is an accurate observation of cooperative hunting by lions, but pairs of genets and Nile crocodiles, which are usually solitary, may instead indicate the breadth of the physical environment that is represented by marsh and fording scenes. Alternatively, pairs may be an artistic device intended to produce a symmetrical composition and to increase the perceived threat presented by the predators.
- The registers and surrounding framework of wall reliefs are utilised to represent features of some predation events – spilling the action across two registers to convey the tearing apart of prey by hunting dogs, and surmounting enclosure lines with the upper bodies of lions to represent the animals' movement across the barriers. This latter motif may also represent a sequence of spatial and temporal events within the one image.

CHAPTER 9

AGGRESSIVE BEHAVIOUR

DEFINITION

Aggression may be defined as actions that are directed by one animal towards another that express hostile intent, but are not for the purpose of predation.[1] The most extreme example of this behaviour is fighting, but most species try to avoid this in favour of asserting their dominance through ritualised displays - body postures, movements, and sounds - that can convey information about an individual's *potential* fighting ability. Some animals attack other species in order to defend themselves or family members. Aggressive interactions are often triggered when one species enters the territory of another. Animals either flee or resist vigorously when they are assaulted, and may often give vocalizations in response to an attack.

9.1 THREATENING

Threat displays allow animals to show off physical features that they might use in a fight. To do this, they perform distinctive actions that draw attention to the nature of their weapons. For example, ungulates may bend forward elaborately to reveal the size of their horns or animals with large teeth may open their mouths and pull back their lips.[2] Sometimes, however, the threat is only a bluff in which an animal attempts to trick an adversary into believing that they are bigger and more dangerous than they really are. All threat displays are intended to intimidate an opponent, to cause it to retreat and thereby avoid a fight.

9.1.1 Head display

Bulls

ḥmt-rꜥ (AB1: Giza, IV.4-V.1)
jj-mrjj (AB2: Giza, V.3L)
sšm-nfr III (AB3: Giza, V.8E)
ꜣḫtj-ḥtp (AB4-7: Saqqara, V.8L-9E)
ḫnwt (AB8: Saqqara, V.9)
Anon (AB9: Saqqara, V-VI)
nj-kꜣw-jzzj (AB10: Saqqara, VI.1M)
mrrw-kꜣ.j: mrj (AB11: Saqqara, VI.1M-L)
jnw-mnw (AB11a-11d: Saqqara, VI.2E)

Threat displays are an intrinsic part of the behavioural repertoire of bulls. When confronting an opponent, a male begins by standing across the path or parallel to its adversary, arching its neck, and lowering its head (see pl. 9A).[3] The hair on its back can be raised while in this position and it may also shake its head and protrude its eyeballs. Bulls often vocalize loudly when displaying in this manner.[4] To increase the threat, the animal will then turn its lowered head directly towards the source of its irritation.[5] Dropping its head enables the bull to display the size of its horns and shoulders, both of which are brought to bear when fighting. Bulls will also threaten people who approach too closely and invade their 'flight zone' – that is, within approximately six metres.[6]

---

1 Allaby, *Dictionary of Zoology*, 11.
2 Walther, F.R. (1977). "Artiodactyla" in Sebeok, *How Animals Communicate*, 675-687.
3 Fraser-Broom, *Farm Animal Behaviour*, 186; Albright-Arave, *Behaviour of Cattle*, 45, fig. 3.1.
4 Kiley, M. (1972). "The vocalizations of ungulates, their causation and function", *Z. Tierpsychol.* 31, 192-193.
5 Albright-Arave, *Behaviour of Cattle*, 45, fig. 3.2.
6 Kilgour-Dalton, *Livestock Behaviour*, 18.

Cattle depicted in wall scenes were found often to lower their heads below shoulder-level. Frequently, however, this position appears to be due to external manipulation by men who have either tied the bulls with ropes or grasp their horns. I have not considered these to be examples of threat behaviour (although see *Bulls* in section 9.4 below); only animals that depress their heads voluntarily are examined here.

Fifteen images from nine Memphite tombs depict bulls that may be adopting a low-level threat posture (AB1 to AB11d; see figs. 9-1 and 9-2). The animals, which appear in offering, pastoral, and agricultural scenes, are presented in groups and while alone. A few of them are hornless. The object of the animals' aggression is not always obvious, but in one example from the tomb of *ꜣḫtj-ḥtp* (AB7; fig. 9-1), a hornless bull at the head of a group appears to threaten a herdsman who wields a stick at them. The animal may be reacting to the close proximity of the man. It also protrudes its tongue, perhaps to indicate the roaring vocalization that bulls produce when threatening (see section 9.4 below). Protruding tongues are a feature of two other scenes from the same tomb. In one, a bull is again reacting to rough treatment by a herdsman (AB5). In the other, two members of a herd expose their tongues as all three animals lower their heads (AB4).[7] As noted above, bulls will threaten one another while standing parallel; it is thus tempting to see a threatening exchange between two or more individuals in this depiction.

Other examples are less straightforward. A bull in the tomb of *mrrw-kꜣ.j: mrj* is probably bending its head to reach a food trough on the ground, rather than threatening the herdsman standing before it (AB11). Three bulls also drop their heads seemingly for no reason while they are led or guided by their handlers (AB1 to AB3); the artists here may have chosen to present animals with lowered heads simply to avoid their horns conflicting with inscriptions positioned above them. So, too, in the tomb of *jnw-mnw*, it is likely that cattle with depressed heads (AB11a to AB11d; see fig. 9-2) have been inserted to fill gaps in the composition, as the scenes lack a clear source for an aggressive response from the animals. Thus, although the inclination of the heads and necks of these animals is suggestive of an aggressive attitude, it must be concluded that none of the bulls adopts an unambiguous threat posture – that is, to display its horns at an adversary.

Geese

ṯjj (AB12-13: Saqqara, V.8-9)
ptḥ-ḥtp II: ṯfj (AB14: Saqqara, V.9M-L)
wḏꜣ-ḥꜣ-ttj: ššj (AB15: Saqqara, VI.1L-2E)
mrjj-ttj: mrj (AB16-17: Saqqara, VI.2M-L)
ꜣḫtj-ḥtp (AB18: Saqqara, VI.5?)

Among flocks of geese depicted in scenes dating from the late 5th dynasty onward, a few individuals lower their heads and stretch their necks forward to hold them horizontally or at an incline. In some images, the geese appear to point their head directly at a nearby companion, while in others their attention is clearly focussed upon men who are endeavouring to force-feed them.

It is not possible to identify with certainty the species of goose depicted in poultry-yard and force-feeding scenes, but they are likely to be greylag geese, wild ancestors of the domesticated variety.[8] Greylag geese are highly social and live in large flocks. A strong hierarchy exists within groups, such that some individuals are more dominant or subordinate than others. The distance around each goose is maintained and defended; disputes often occur when one bird moves too close to another. Dominant greylag geese threaten one another by holding their head and neck stiffly in a linear manner (from an elevated angle to a horizontal position), while pointing it toward their opponent (cf. pl. 9B). Threats elicit a range of reactions, from escape to resistance. During fights, which are rare, rivals charge at one another to bite their opponent and beat them with their wings.[9]

In each of the depicted examples, the geese are packed closely together, a situation that is known to give rise to aggressive interactions. Their head positions closely resemble threat postures. This is particularly so for birds represented with horizontal necks (e.g., AB13, AB16, and AB17; see fig. 9-3),

7 Vandier (*Manuel v*, 80) suggests that these animals are feeding from a trough that has since disappeared.
8 Houlihan, *Birds*, 54-56.
9 Roselaar, C.S. (1977). "Greylag goose (*Anser anser*)" in Cramp-Simmons, *Handbook i*, 418.

but may equally account for the oddly angled heads of other geese (e.g., AB14, AB15, and AB18; see fig. 9-4). In the tomb of *ṯjj*, one bird leans forward to clearly threaten a nearby companion (AB12; see fig. 6-29 in Chapter 6), which flaps its wings in response. Threats are also directed towards humans in force-feeding scenes in the tomb of *mrjj-ttj: mrj*. Some geese even appear to peck their seated handlers upon the nose (AB17). Geese are known to charge and attack humans if they approach too closely, and so this behaviour is entirely feasible.

9.1.2 Leg display

Bulls

snḏm-jb: jntj (AB19: Giza, V.8M-L)
jnw-mnw (AB20: Saqqara, VI.2E)
mḥw (AB21: Saqqara, VI.2M-3?)

Bulls threaten one another by turning towards their opponent and lowering their head (see above). Sometimes this behaviour is also accompanied by 'pawing' in which the male signals his intention to charge by dragging one forefoot backwards. This motion causes dirt to fly up behind the bull and over its back (see pl. C).[10] If the recipient of the display responds with threatening movements, a fight will ensue. Pawing bulls often vocalize.[11]

Three wall reliefs were found to depict adults that appear to paw the ground.[12] In the latest image from the tomb of *mḥw* (AB21), a bull lowers its head to display its horns as it lifts one foreleg off the substrate. The animal also protrudes its tongue; this may indicate that it is vocalizing loudly. Altenmüller believed that this scene represented the roping of a bull prior to slaughter. This is a common motif in tomb scenes; frequently in such depictions, the animal is roped in such a way that one foreleg is pulled off the ground (see section 9.4 below). It is possible that the relief in question, which is unfinished, may ultimately have represented such an event. Indeed, Altenmüller, when describing the image, filled in the missing elements and stated that the animal's leg is lifted due to a rope.[13] Taken at face value, however, the bull clearly displays all the elements of pawing. Unfortunately, as the relief is incomplete and its intended context unknown, no firm conclusion can be drawn.

In the earliest scene, which appears in the tomb of *snḏm-jb: jntj* (AB21: see fig. 9-5), a bull raises one foreleg as it stands behind a mating pair. The animal seems to have a rope draped over its shoulders and attached to its lifted hoof. Schäfer compared this image with that of an oryx depicted in a desert scene in the sun-temple of Niuserre, which also lifts one foreleg.[14] The ungulate in the royal image is undoubtedly pawing the ground as a shower of red specks representing billowing soil is painted over the relief, arcing from the tip of its raised hoof and over its body. Schäfer thus concluded that the bull in the tomb of *snḏm-jb: jntj* is also pawing the ground and that the 'rope' is in fact a stylised representation of flying dirt.[15]

As further support for his interpretation, Schäfer noted the pastoral nature of the *snḏm-jb: jntj* scene, involving neither slaughter nor hunting, which indicated that the kicking animal was unlikely to have been lassoed with ropes. I would add that the context also provides an explanation for the

---

10 Hafez, E.S., Schein, M.W. and Ewbank, R. (1969). "The behaviour of cattle" in Hafez, *Behaviour of Domestic Animals*, 258-259, pl. 3; Fraser-Broom, *Farm Animal Behaviour*, 187; Albright-Arave, *Behaviour of Cattle*, 45, fig. 3.1.

11 Kiley, *Z. Tierpsychol.* 31, 192.

12 Six Saqqara tombs dating from the 6th dynasty were found to display calves, dorcas gazelle, scimitar oryx, Nubian ibex, and a barbary sheep that raise one foreleg as they process beside bearers in offering scenes. I have not considered these examples of pawing. In a small number of the images, it can be seen that their foreleg is lifted because it is attached to a short rope held by their human handler. This feature is missing in most of the remaining scenes, but may have been painted originally and since faded. The context in which these animals appear is also one in which pawing would not necessarily occur.

13 "Sein Kopf ist unter dem Druck eines für die Darstellung vorgesehenen Mannes gesenkt. Ein Vorderbein des Tieres wird vom Fussleitseil nach oben gezogen". See Altenmüller, *Mehu*, 219.

14 Schäfer, H. (1906). "Das Zeichen für *ṯwn*", *ZÄS* 43, 74-76.

15 Brovarski (*Senedjemib Complex Part I*, 51) has drawn the same conclusion.

male's threatening behaviour. Bulls become increasingly agitated when they detect a fertile female nearby. Studies have shown that they are stimulated by observing the sexual behaviour of others and that the presence of another male often spurs them into action.[16] The animal in question in the *snḏm-jb: jntj* scene is likely responding to the sight and smell of the mating bull and cow standing before it. This would also account for the behaviour of a bull in the tomb of *jnw-mnw* (AB20; see fig. 9-6), which similarly lifts one foreleg while facing a mating pair. Pawing is thus an indication of the males' heightened state of excitement. It is also notable that after mating has occurred, bulls often 'guard' the cow and will threaten other males by pawing the ground.

9.1.3 Ear display

Donkeys

Anon (AB22: Saqqara, V)
rˁ-ḫˁ.f-ˁnḫ (AB23: Giza, V.6?)
rˁ-m-kȝ.j (AB24: Saqqara, V.6? r/u - V.6-7)
ȝḫtj-ḥtp (AB25: Saqqara, V.6-8E)
kȝ.j-m-rḥw (AB26: Saqqara, V.8)
nfr-jrt-n.f (AB27: Saqqara, V.8-9)
ȝḫtj-mrw-nswt (AB28: Giza, V.9-VI.1)
ˁnḫ-m-ˁ-ḥr: zzj (AB29: Saqqara, VI.1L-2E)

Equids (e.g., horses, zebras, and donkeys) convey information about their motivational state via movements of their bodies, heads, necks, and tails.[17] Facial expressions communicate a range of emotions. The position of the ears, in particular, can indicate unambiguously whether an equid is in a benign or hostile condition: animals that hold their ears erect or forward indicate friendliness and interest, while those whose ears are pinned back convey a threatening or aggressive state (see pl. 9D).[18] When this latter posture is combined with an open mouth and exposed teeth, the animal may consequently bite.

Examination of donkeys depicted in a variety of wall scenes, such as upon threshing-floors, while transporting agricultural supplies, carrying riders, and being herded, revealed that most hold their ears backward. Ear positions range from slightly less than vertical through to flat against the neck. Some donkeys carry their ears fully upright or forward, but these images are rare. Most of the animals are working hard and so their drooping ears could simply be an expression of exhaustion. In eight threshing scenes, however, donkeys were found to hold their ears flat against their necks; this position is an unequivocal threat. Examination of these scenes gives the impression that the animals are being treated somewhat harshly – men wield sticks at them and wave their arms about to make them move (AB22, AB24, AB25, AB27 to AB29; see fig. 9-7), or alternatively, grab hold of them to keep them still (AB23). Often, the threatening animals lift their heads in the direction of these herdsmen as though directing their aggression towards them. Thus, the contexts in which they are placed supports the possibility that their posture is one of aggression. I suggest that their threatening demeanour is the result of their mistreatment.

Some animals are depicted with their mouths agape. In the tomb of *ˁnḫ-m-ˁ-ḥr: zzj* (AB29; fig. 9-7), a donkey opens its mouth to display its teeth at a herdsman wielding a stick. When combined with flattened ears, this would signal a preparedness to bite, but the position of its pinna is somewhat ambiguous, with one held vertically and the other somewhat inclined (not pressed against its neck). The donkey's raised foreleg, however, suggests that it is performing a 'strike', kicking out in an aggressive fashion.[19] The animal's posture, overall, indicates that it is indeed threatening its human handler.[20] In a fragmentary scene from the earlier tomb of *ȝḫtj-mrw-nswt* (AB28), two donkeys also open their mouths; unfortunately, the sketchy image does not reveal whether they bare their teeth.

---

16 Mader, D.R. and Price, E.O. (1984). "The effects of sexual stimulation on the sexual performance of Hereford bulls", *J. Anim. Sci.* 59(2), 294-300; Kilgour-Dalton, *Livestock Behaviour*, 20.

17 McDonnell, S.M. and Haviland, J.C.S. (1995). "Agonistic ethogram of the equid bachelor band", *App. Anim. Beh. Sci.* 43, 147-188.

18 Moehlman, P.D. (1998). "Behavioral patterns and communication in feral asses (*Equus africanus*)", *App. Anim. Beh. Sci.* 60(2-3), 162-165.

19 McDonnell-Haviland, *App. Anim. Beh. Sci.* 43, 163.

20 See also Power, R.K. (2004). "Deconstructing, deciphering and dating the donkey in Old Kingdom wall painting and relief", *BACE* 15, 134-135.

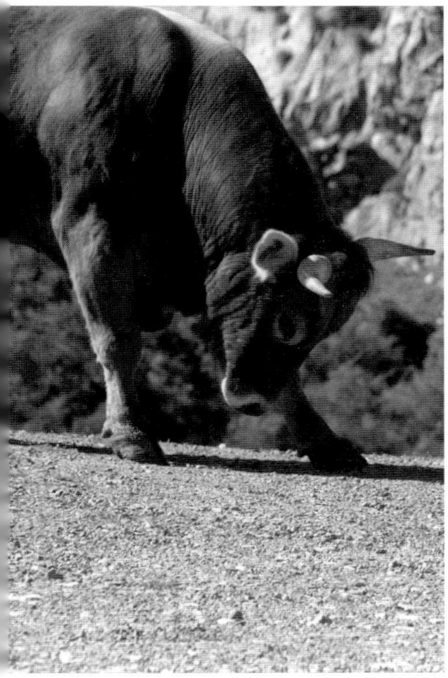

Pl. 9A Bull head display
© iStockphoto.com/Jesus Arias

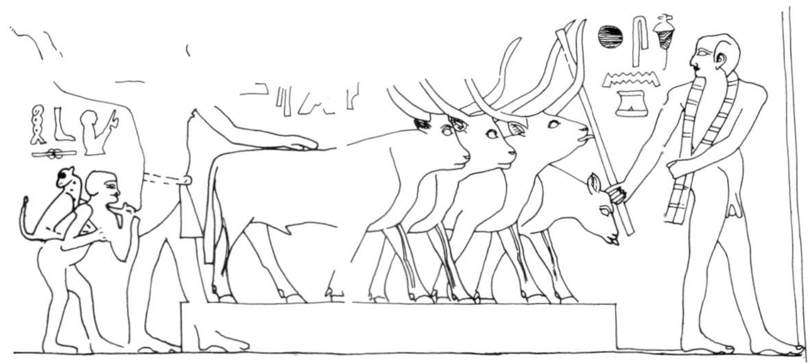

Fig. 9-1 (AB7)

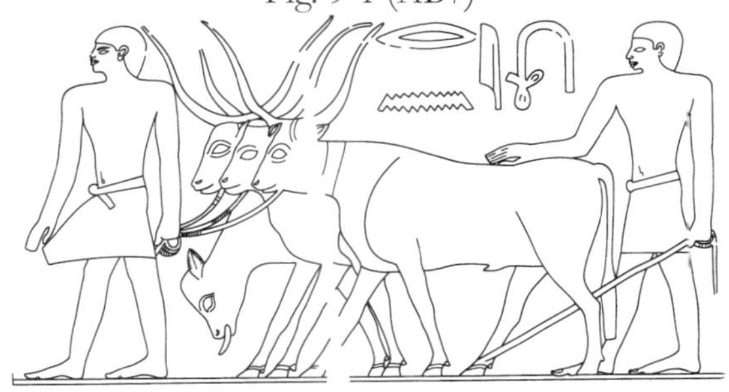

Fig. 9-2 (AB11c)

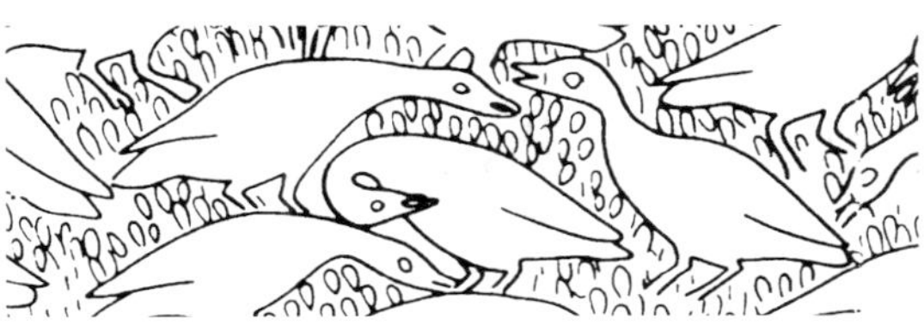

Fig. 9-3 (AB16)

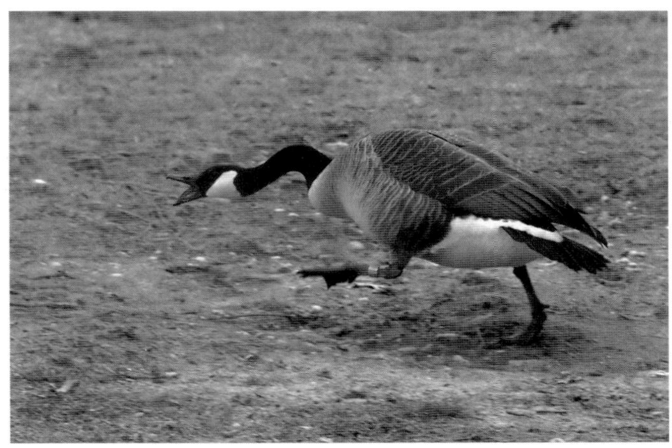

Pl. 9B Goose head display
© iStockphoto.com/Maxim Kazitov

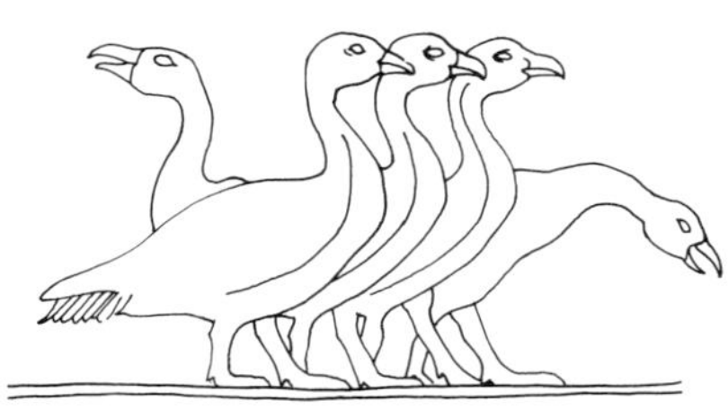

Fig. 9-4 (AB14)

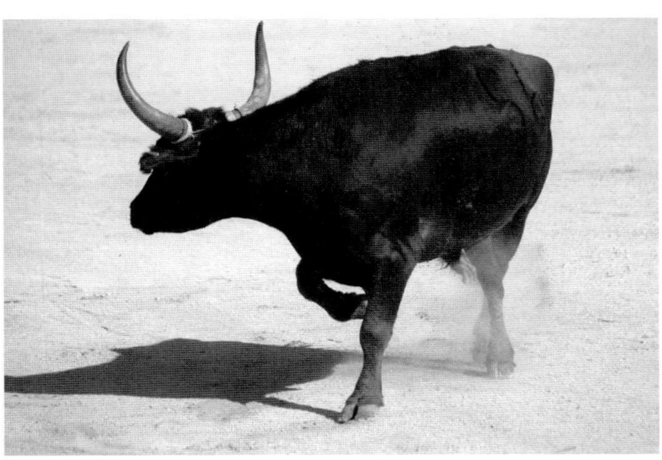

Pl. 9C Bull pawing
© iStockphoto.com/Diane White Rosier

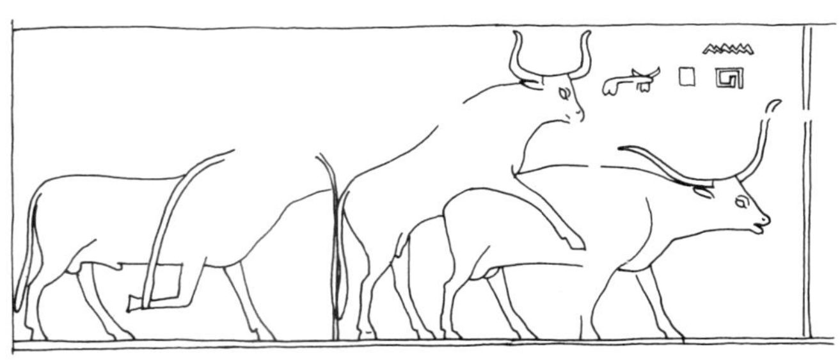

Fig. 9-5 (AB19)

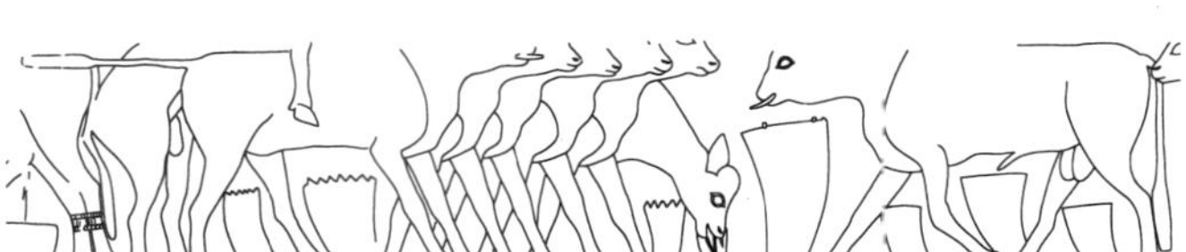

Fig. 9-6 (AB20)

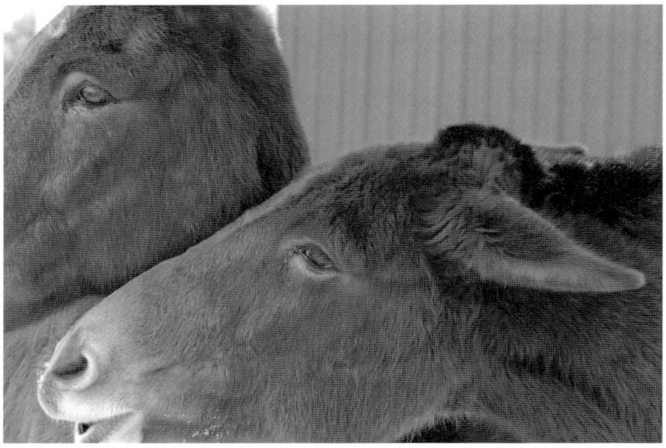

Pl. 9D Equid ear threat
© iStockphoto.com/Melissa Anne Galleries

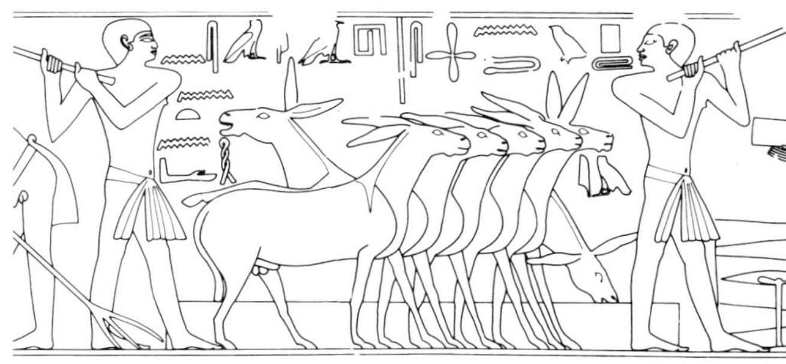

Fig. 9-7 (AB29)

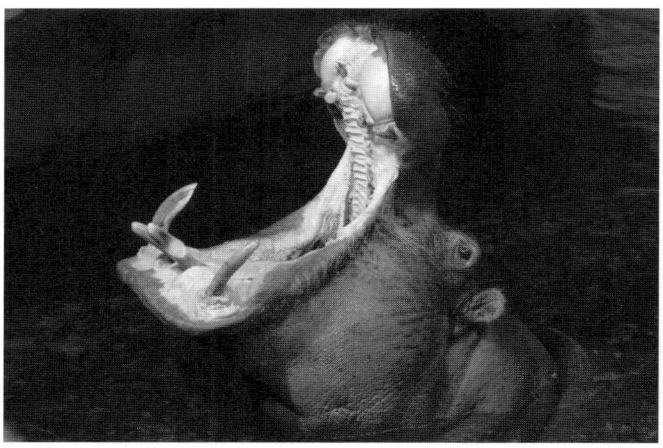

Pl. 9E Hippopotamus yawn display
© iStockphoto.com/Holger Ehlers

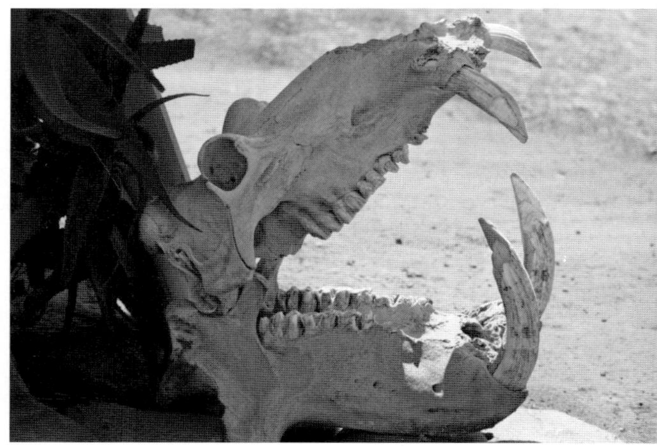

Pl. 9F Hippopotamus tusks
© iStockphoto.com/Liz Leyden

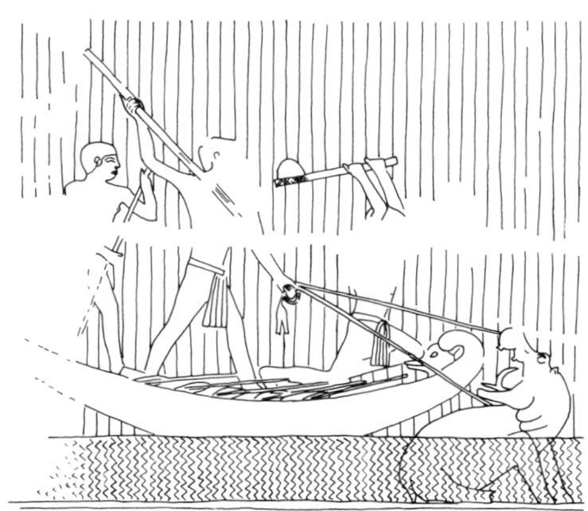

Fig. 9-8 (AB32)

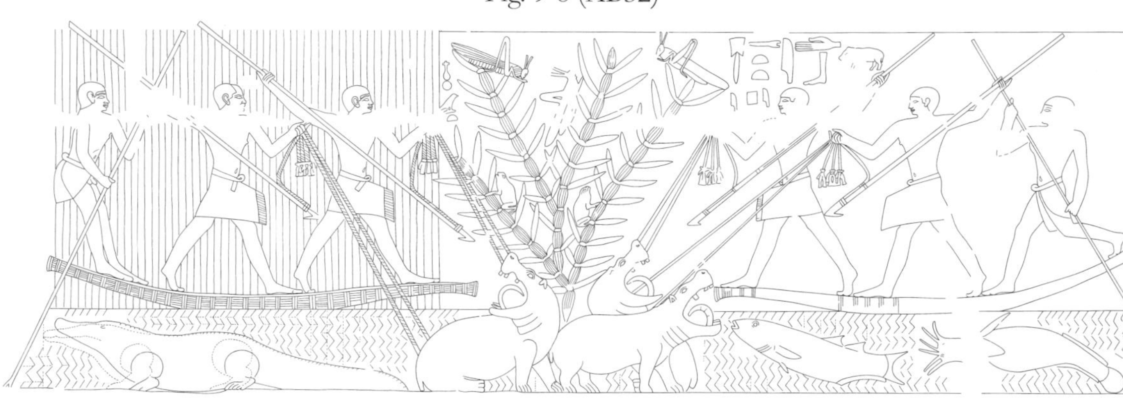

Fig. 9-9 (AB42)

Perhaps the clearest indication that the donkeys presented in wall reliefs are frequently in either a tired or aggressive state can be observed in scenes that feature foals. Ten images of foals were found in this study, ranging from the early 5th to the mid-6th dynasty.[21] The young animals accompany herds or single donkeys that may be their mothers. In every example, the ears of the foals are pricked vertically in an alert and friendly pose, the very antithesis of those of the adults that stand beside them (e.g., see fig. 4-1 in Chapter 4).

The lowered ears of some donkey images have undoubtedly been affected by the presence of inscriptions overhead. It is also possible that artists adjusted their placement of animal ears simply to inject some variety into the postures of herd members (particularly in threshing-floor scenes, in which the animals mill about). It cannot be stated with certainty whether they, or the Egyptian people generally, understood the significance of the donkeys' lowered ears, but the prevalence of this posture in wall reliefs suggests that they probably witnessed it constantly and so considered it an intrinsic part of the animals' physical appearance. Their behaviour in wall scenes thus hints at a somewhat unhappy relationship between donkeys and their human handlers.[22]

9.1.4 Mouth display

Hippopotamuses

dbḥnj (AB30: Giza, IV.5)
pr-sn (AB31: Saqqara, V.2-3)
ꜣḫtj-ḥtp (AB32: Saqqara, V.6-8E)
kꜣ.j-m-nfrt (AB33: Giza, V.6-8E?)
ẖnmw-ḥtp (AB34: Saqqara, V.6-8E?)
snḏm-jb: jntj (AB35: Giza, V.8M-L)
ṯjj (AB36: Saqqara, V.8-9)
snḏm-jb: mḥj (AB37: Giza, V.9)
nbt (AB38: Saqqara, V.9)
jj-nfrt: šꜣ.n.f (AB39: Saqqara, V.9)
Anon (AB40: Saqqara, V-VI)
kꜣ-gm-nj: mmj (AB41: Saqqara, VI.1M)
mrrw-kꜣ.j: mrj (AB42: Saqqara, VI.1M-L)
zšzšt: jdwt (AB43-46: Saqqara, V.9 r/u - VI.1-2E?)
sꜥnḫ-w(j)-ptḥ (AB47: Saqqara, VI.1L)
ṯfw (AB48: Saqqara, VI.1-2)
mḥw (AB49: Saqqara, VI.2M-3?)
jj-nfrt (AB50: Giza, VI.3-4E?)
nb-kꜣw-ḥr: jdw (AB51-52: Saqqara, V.9 r/u VI.7)

Marsh scenes in 19 tombs ranging from the late 4th until the late 6th dynasty were found to feature hippopotamuses that stand or sit with their jaws agape (e.g., AB32; see fig. 9-8).[23] When describing this behaviour by a hippo in the tomb of *mrrw-kꜣ.j: mrj* (AB42; fig. 9-9), Schäfer stated that it was "raising its head threateningly out of the water".[24] Schäfer's observation is correct.[25] Male hippopotamuses threaten one another - and other species - by opening their mouths to their fullest extent (see pl. 9E). This display is called 'gaping' or 'yawning'.[26]

The canines and incisors of hippopotamuses grow continuously throughout their lives and are consequently very large, particularly in adult males (see pl. 9F). The middle incisors on the hippo's lower jaw are long (c. 30-40 cm) and extend forward and slightly upward, while its upper incisors are short and project downward. Curved canine teeth ('tusks') on its lower jaw can measure up to 50 cm in length and are kept sharp by rubbing against its shorter upper canines. Hippos can open their jaws up to 150°, enabling them to fully reveal the state of their dentition.[27] When fighting, hippos lunge at one another to spar with their open jaws, and slash and bite at their opponent (see

21 Foals were depicted in the following tombs: *wr-jr.n-ptḥ* (ntn), *rꜥ-ḫꜥ.f-ꜥnḫ* (G7948=LG75), *nj-ꜥnḫ-hnmw/ẖnmw-ḥtp* (ntn), *pḥn-wj-kꜣ.j* (D70=LS15), *spd-ḥtp* (D15), *ptḥ-ḥtp I* (D62), *sḫm-ꜥnḫ-ptḥ* (D41), *ṯjj* (D22), *mḥw* (ntn), and *kꜣ-ḥj.f* (G2136).

22 This disagrees somewhat from the conclusions drawn by Power (2004). See *ibid.*, 134-138.

23 Hippos giving birth are sometimes depicted with open mouths, but these images are not considered here. Birth scenes are treated separately in Chapter 11.

24 Schäfer, *Principles*, 240.

25 Schäfer's translator, John Baines, dismissed this interpretation saying, "The hippopotamus is not in fact 'threatening', but is being pulled out of the water by lines attached to harpoons". See *ibid.*, n. 219.

26 Verheyen, *Hippopotame*, 60-61. This behaviour has also been noted in three-dimensional representations of hippopotamuses. See Hendrickx, S. and Depraeters, D. (2004). "A theriomorphic stone jar and hippopotamus symbolism" in Hendrickx et al., *Egypt at Its Origins*, 812-813.

27 Estes, *Behavior Guide*, 222.

pl. 9M and *Hippopotamuses vs crocodiles* below); 'yawning' thus allows males to assess the potential fighting ability of their opponents by viewing the size of their teeth. This display, which occurs on land and in the water, is often accompanied by a wheezing call and loud guttural honks. Hippos vocalize both when their head is above the water surface and when fully submerged.[28]

Male hippos are extremely territorial and threaten one another frequently by yawning. Females and sub-adults do not yawn; however, when threatened by a male, they may turn toward their aggressor, partially open their mouth without lifting their head, and remain immobile for a short period in an act of submission.[29] This is in contrast to the yawn of a dominant male in which the mouth is opened as widely as possible and the animal's head and jaws are held perpendicular to its back.[30]

All of the yawning hippos represented in wall scenes expose their curled lower canines. The curve of the canine has often been overemphasized (e.g., AB36, AB38, and AB42; see fig. 9-9 and also fig. 4-45 in Chapter 4). In the tomb of *zšzšt: jdwt*, the hippos' upper canines are also visible (AB44 to AB46; see fig. 9-10), and in some depictions, their peg-like upper incisors are apparent (e.g., AB32, AB35, AB36, AB38, AB40, AB42, and AB49; see figs. 9-8 and 9-11). In the tomb of *snḏm-jb: mḥj*, the upper canines of a pair of animals are curiously (and incorrectly) longer than their lower canines (AB37). The rolls of loose skin that envelop the hippos' necks when yawning are also depicted frequently.

The source of the animals' threat displays can often be seen nearby. In 11 images, men in canoes attack the yawning hippos with harpoons (AB32 to AB34, AB36, AB37, AB41, AB42, AB46, AB47, and AB50; figs. 9-8, 9-9, and 9-10).[31] In the tomb of *snḏm-jb: jntj* (AB35; fig. 9-11), the hunters float above a pair of displaying animals, but have not yet cast their harpoons, while in the tomb of *zšzšt: jdwt*, one hippo yawns in response to an attack upon its companions (AB45). Seven of the hunted animals have received harpoon lines directly into their open mouths. The skin of the hippopotamus is up to 6 cm thick[32] and consequently difficult to pierce; I suggest that hunters understood that these animals would respond to their presence by opening their mouths and that this then afforded them an easy and vulnerable target for their projectiles.

Two images may hint at knowledge of hippopotamus social behaviour. Although, as noted above, the animals depicted in the tomb of *snḏm-jb: jntj* (AB35; see fig. 9-11) may be threatening harpooners above them who have yet to begin their hunt,[33] I suggest that their arrangement indicates that they are actually signalling to each other. The two hippos stand opposite one another open-mouthed, separated by a small animal that likely represents a calf. The calf faces in the same direction as the adult on the left, suggesting that this may be its mother. The tusks of the animal on the right are clearly longer than those of the hippo on the left. The scene may thus represent a threatening male on the right and a female on the left that expresses her submission by facing him and opening her mouth. The similar tusk structure and placement of hippos in room B of the tomb of *zšzšt: jdwt* (AB46; see fig. 9-10), however, may indicate a sparring match between two males that has been interrupted by a harpooner above them. No other scenes were found in which yawning hippos face one another.

Almost all of the yawning animals are depicted in a seated position. I am unsure how frequently hippopotamuses rest on their haunches by choice, but, due to physical limitations, they must adopt this position briefly each time they lie down or rise (see Chapter 4). The seated hippos could thus represent their movement from a prone to a standing position in order to approach their human attackers. I suggest, however, that the posture is actually a deliberate attempt by artists to reduce or

---

28 Barklow, W.E. (2004). "Amphibious communication with sound in hippos, *Hippopotamus amphibius*", *Anim. Behav.* 68(5), 1125-1132.

29 Verheyen, *Hippopotame*, 60.

30 Karstad, E.L. and Hudson, R.J. (1986). "Social organization and communication of riverine hippopotami in southwestern Kenya", *Mammalia* 51, 160.

31 For a discussion of the secular and ritual killing of hippopotamuses, see Säve-Soderbergh, *Egyptian Representations of Hippopotamus Hunting.*

32 Estes, *Behavior Guide*, 225. The skin is particularly thick on the animal's back and hindquarters.

33 Brovarski (*Senedjemib Complex Part I*, 50) believes that they are roaring their "displeasure at the hunters".

neutralize the imminent threat presented by the yawning animals. It is likely that the Egyptians understood that yawning frequently precedes an attack. Many human fatalities and severe injuries are caused in Africa today by hippos overturning small boats and then mauling the passengers; these attacks are often unprovoked, but also occur in response to the animals being disturbed or hunted.[34] Hieroglyphs representing potentially dangerous creatures (e.g., snakes, insects, birds, etc.) in tomb inscriptions were often deliberately mutilated (particularly in burial chambers) to prevent the images from magically taking on a life of their own and either harming the body of the deceased or consuming food offerings.[35] In accordance with this expectation, I propose that the representation of gaping hippos in a seated posture may have been intended to undermine their ability to move quickly and potentially injure the human figures with which they share each scene. It is interesting to note that this treatment of yawning hippos is also expressed in three-dimensional art, as statuettes of standing hippos are usually closed-mouthed. Seated statues are less common, but frequently depict the animals with their jaws open.[36]

The yawning animals in 12 scenes twist their heads back fully to face their human aggressors behind them (e.g., see figs. 9-8 and 9-9).[37] The neck of the hippopotamus is remarkably mobile considering its bulk. When fighting, males often swing their heads from side to side to slash at their opponent's flanks with their tusks, and when yawning, their neck can be bent back at an extreme angle. Nevertheless, they are physically incapable of twisting their heads to look behind them, as represented. These animals must therefore be viewed as situated beside the boats of the harpooners, rather than in front of them, and turning their heads to one side to yawn at their attackers. This interpretation is supported by seated statuettes of hippos, which turn their heads to the left or right - not backward.

In many images, a plate-like object running parallel with their lower jaws projects from the animals' open mouths (e.g., AB37 to AB32, AB36, AB38 to AB40, AB42 (left), AB45, AB49, and AB50; see fig. 9-11). This horizontal structure possibly represents their lower incisors (see pl. 9E). Alternatively, it may depict the animals' tongues. Examination of three-dimensional representations of open-mouthed hippos, particularly images of the goddess Taweret, reveals that almost all protrude their tongues.[38] This is intriguing because the tongue of the hippopotamus lies flat against its lower jaw and is usually not visible when it opens its mouth (see pl. 9E).

As noted above, wheezing and honking accompany yawning; hippos also produce vocalizations when fighting, ranging from high-pitched squeals and grunting, to sounds similar to a lion's roar.[39] It is consequently tempting to see in both the object projecting from the mouths of the depicted hippos in wall scenes, and Taweret's exposed tongue, a motif by which the expulsion and experience of sound could be visualized (see other chapters for more examples).[40]

---

34 Eltringham, *The Hippos*, 120-121.

35 See Lacau, P. (1913). "Supressions et modifications de signes dans les textes funéraires", *ZÄS* 51, 1-64.

36 I am not proposing that this is a hard and fast rule. A brief examination of examples contained in Behrmann's (*Das Nilpferd i*) corpus revealed that a small number of hippo statuettes dating to the predynastic period are open-mouthed while standing. Also, Old Kingdom marsh scenes from provincial areas sometimes feature yawning hippos that stand. I do believe, however, that a general trend is suggested. It is also interesting to note that the legs of standing hippo statuettes have often been broken in antiquity to render them harmless. For one such example, see Fazzini et al., *Art for Eternity*, cat. 27 (Bost. Mus. no. 35.1276), 69.

37 Depictions of hunted animals often 'look back' at their attackers. For an interpretation of this motif in predynastic art, see Davis, *Masking the Blow*, 145-160.

38 This feature of the goddess is overlooked consistently in descriptions of her appearance. She also protrudes her tongue reliably in two-dimensional images, such as upon magic wands.

39 Karstad-Hudson, *Mammalia* 51, 161.

40 I note that Taweret is frequently associated with the god Bes (see Gundlach, R. (1986). "Thoeris" in *LÄ vi*, 495), and that both deities were responsible for protecting women during pregnancy and childbirth (see Pinch, *Magic in Ancient Egypt*). Bes repelled hostile forces via his hideous appearance and dancing, and by making noisy music. I propose that Taweret's exposed tongue indicates that the goddess was also capable of producing noise. The characteristic protruding tongue of Bes may thus have had a similar noise-making function, rather than simply to create an ugly visage.

9.1.5 Penile display

Baboons

nb.j-m-ꜣḫtj (AB53: Giza, IV.6-V.1)
tp-m-ꜥnḫ II (AB54: Saqqara, V.5-6?)
nfr/kꜣ-ḥꜣ.j (AB55-56: Saqqara, V.6)
nj-ꜥnḫ-ẖnmw/ẖnmw-ḥtp (AB57: Saqqara, V.6L-7)
jtj-sn (AB58: Giza, V.6-8?)
Anon (AB59: Saqqara, VI)

Baboons occur infrequently in tomb reliefs from Giza and Saqqara. The eight examples that were found appear in unusual contexts, including two market scenes, a boat-yard, and while using a sack press.[41] Both hamadryas and anubis baboons are depicted. One unambiguous depiction of a female anubis was found (see Chapter 11), but the remaining seven images represent males of both species. Male hamadryas baboons possess a distinctive mantle of fur across their shoulders, which artists have rendered clearly in six examples as a triangular-shaped cape (AB55, AB57 to AB61), however the clearest indication of their sex is their prominent genitalia. Although none of the baboons appears in a sexual context, six of the males display an extended penis as they walk quadrupedally (AB53 and AB58; see fig. 9-12) or stand on their hind legs (AB54 to AB57; see fig. 9-13).[42]

It must be emphasized that only baboons were found to be ithyphallic; the genitalia of the 49 representations of vervet monkeys that were examined in this study were never depicted. This suggests that the inclusion of this feature for baboons is significant and not simply to indicate their sex (indeed, the presence of a cape achieves this for images of hamadryas males).

Baboons live in large social groups known as 'troops'. Each troop is made up of smaller units consisting of a dominant adult male and his harem of females and their young. Each alpha male both protects and controls his group, asserting his dominance via threat displays and, occasionally, physical attack.[43] When their groups are foraging, dominant males usually sit in a conspicuous place nearby (e.g., on a mound or tree stump) and scan their surroundings. It has been observed that, while guarding their harem in this manner, anubis baboon males often sit with their thighs parted and their penis fully extended (see pl. 9G).[44] It was initially believed that the males' role was to alert their group to any signs of danger, but it has now been determined that they react only to the presence of unfamiliar troops. The penile display of sentinels thus acts as a signal and warning to other baboons that an adult male is present and in charge.[45]

The extended penis is an unambiguous feature of a dominant male baboon; it is also an indirect sign of their aggressive potential. The ithyphallic males in wall scenes may allude to this aspect of their behaviour. An aggressive meaning is certainly suggested by three of the depicted males that stand on their hind legs, as baboons adopt this position in agonistic situations, often just prior to an attack.[46] One male, for example, has grabbed the leg of a young man (AB54; see fig. 11-27 in Chapter 11). A rearing hamadryas in the tomb of *nfr/kꜣ-ḥꜣ.j* also waves its arms about in an excited manner while wielding an object (AB56; see fig. 9-14).

---

41 The depiction of this procedure in wall scenes is both rare and difficult to understand spatially. Only one image (AB55; located in the tomb of *nfr/kꜣ.ḥꜣ.j* at Saqqara) includes a baboon participating as one of the workers. For a discussion of this motif, see Klebs, L. (1914). "Die Tiefendimension in der Zeichnung des alten Reiches", *ZÄS* 32, 24-28.

42 The genitals of one hamadryas male are not visible (AB59) because the artist has drawn the leg closest to the viewer advancing first. This is one of only a small number of examples found in this study that break the rule of the limb furthermost advancing first (see Chapter 4).

43 Hall, K.R.L. and de Vore, I. (1965). "Baboon social behavior" in de Vore, *Primate Behavior*, 96-101.

44 Wickler, W. (1967). "Socio-sexual signals and their intra-specific imitation among primates" in Morris, *Primate Ethology*, 116-118; Estes, *Behavior Guide*, 514.

45 *ibid.*

46 Hall-de Vore in de Vore, *Primate Behavior*, 92.

Alternatively, the ithyphallic baboons may possess a magico-religious meaning.[47] Hamadryas males were viewed as a manifestation of the god Thoth;[48] they were also associated with the worship of the sun god Re. Religious representations of this animal frequently show it in a seated position, with its knees drawn up to its chest and legs parted to reveal its penis in precisely the same manner as baboon sentinels.[49] Hamadryas males that stand in adoration of the sun also display extended penises.[50] The religious iconography of these animals thus emphasizes this attribute and could point to a symbolic allusion for its depiction in wall scenes.

The ithyphallic baboons may also refer directly to a more obscure deity known from the Old Kingdom. In Egyptian mythology, the baboon god Babi was depicted with an erect penis and was associated with both virility and power. Babi represented an aggressive alpha male (*kꜣ jꜥn.w* 'bull of baboons') and as such was identified with the deceased king.[51] In particular, however, he was responsible for the sexual potency of the dead and allowed the deceased to enjoy sexual intercourse in the Afterlife. His penis is described as the latch to the doors of heaven (*Pyramid Texts*, Utt. 313); it was also said to form the mast of the ferryboat that transported souls to the Underworld.[52]

The depiction of virile baboons in wall scenes may thus allude specifically to Babi, to ensure that the souls of the tomb-owners would be carried to the next life as fully functioning and potent individuals. It is beyond the scope of this study to explore this possibility further. It is interesting to note, however, that one of the most enigmatic and detailed images of an ithyphallic baboon considered here stands upon a boat, thereby associating two of the features of Babi (AB56; see fig. 9-14). This scene is without parallel and has so far defied explanation.[53]

9.1.6 Wing display

Ostriches

nj-mꜣꜥt-rꜥ [*jzzj-mr-nṯr*] (AB60: Giza, V.9-VI.1)

The late 5th dynasty tomb of *nj-mꜣꜥt-rꜥ* [*jzzj-mr-nṯr*] at Giza yielded a unique image of an ostrich that adopts an aggressive posture (AB60; see fig. 9-15). The bird, which appears in a desert hunting scene, stands with its head and neck stretched vertically and both its wings and tail held erect. The ostrich is flanked by a pair of copulating canids and a stalking cat.

Ostriches threaten one another, and other species, by walking stiffly toward their adversary while alternately raising and lowering each of their wings (cf. pl. 9H). If the situation escalates, the birds will attack by running toward the recipient while flapping their wings or holding them vertically above their backs.[54] During both threat and attack, the birds' heads and tails are held erect.

---

47 For a discussion of the religious significance of baboon sexual behaviour, see Gordon-Schwabe, *Quick and the Dead*, 131-135.

48 Störk, L. (1986). "Pavian" in *LÄ iv*, 917-918.

49 The association between Egyptian imagery and baboon penile displays has also been pointed out by Gordon and Schwabe (*Quick and the Dead*, 131).

50 Herman te Velde suggested that their ithyphallic state was "an iconographic indication of the excitement or even religious extasy (sic) with which they sang and danced out their praises". See te Velde, H. (1988). "Some remarks on the mysterious language of the baboons," in Kamstra et al., *Funerary Symbols and Religion*, 131. For a discussion of Egyptian depictions of worshipping baboons and hamadryas sunning behaviour, see Thomas, E. (1979). "*Papio hamadryas* and the rising sun", *BES* 1, 91-94 and pl. 4.

51 An early dynastic statuette of a seated baboon (Berlin no. 22607) inscribed with the name of Narmer suggests a relationship between these animals and kingship. See Krauss, R. (1994). "Bemerkungen zum Narmer-Pavian (Berlin 22607) und seiner Inschrift", *MDAIK* 50, 223-230.

52 Wilkinson, *Complete Gods*, 196.

53 Moussa-Altenmüller, *Nefer and Ka-hay*, 27. The animal's raised arm, in association with its erect penis, may also be significant. For an examination of these features in relation to the god Min, see Ogdon, J. (1985/86). "Some notes on the iconography of the god Min", *BES* 7, 29-41 and Wilkinson, R.H. (1991/92). "Ancient near eastern raised-arm figures and the iconography of the Egyptian god Min", *BES* 11, 109-118.

54 Bolwig, N. (1973). "Agonistic and sexual behavior of the African ostrich (*Struthio camelus*)", *Condor* 75(1), 101; Cramp, S. (1977). "Ostrich (*Struthio camelus*)" in Cramp-Simmons, *Handbook i*, 39-40.

The posture of the *nj-m3ʿt-rʿ* [*jzzj-mr-nṯr*] ostrich is undoubtedly threatening, and may also signal its preparedness to attack. The bird's behaviour is thus in accord with the generally aggressive nature of the hunting scene in which it is placed. Although the source of its agitation cannot be identified, it is interesting to note the presence of a small, long-necked bird on a sub-register behind it, which bears a strong resemblance to an ostrich chick (see fig. 9-15).[55] Ostriches will defend their broods vigorously and often attempt to distract potential predators by running to and fro while flapping their wings; the *nj-m3ʿt-rʿ* [*jzzj-mr-nṯr*] image thus possibly represents this antipredator display.

Birds

ḥtp-ḥr-3ḫtj (AB61: Saqqara, V.6-8E)
jrj-n-k3-ptḥ (AB62: Saqqara, V.6-9)
rʿ-špss (AB63: Saqqara, V.8M)
snḏm-jb: jntj (AB64: Giza, V.8M-L)
ṯjj (AB65: Saqqara, V.8-9)
3ḫtj-ḥtp (AB66: Saqqara, V.8L-9E)
k3-gm-nj: mmj (AB67: Saqqara, VI.1M)
nj-k3w-jzzj (AB68: Saqqara, VI.1M)
mrrw-k3.j: mrj (AB69-70: Saqqara, VI.1M-L)
zšzšt: jdwt (AB71: Saqqara, V.9 r/u - VI.1-2E?)
sʿnḫ-w(j)-ptḥ (AB72: Saqqara, VI.1L)
ḥzj (AB73-74: Saqqara, VI.1L-2E)
ʿnḫ-m-ʿ-ḥr: zzj (AB75: Saqqara, VI.1L-2E)
mrjj-rʿ-mrj-ptḥ-ʿnḫ: nḫbw (AB76: Giza, VI.2)
mrrj (AB77: Saqqara, VI.2E?)
jnw-mnw (AB78: Saqqara, VI.2E)
ḫntj-k3.j: jḫḫj (AB79: Saqqara, VI.2M)
mḥw (AB80-82: Saqqara, VI.2M-3?)
3ḫtj-ḥtp (AB83: Saqqara, VI.5?)

Nineteen tombs, largely from Saqqara, were found to exhibit scenes in which birds adopt an unusual posture while incubating eggs.[56] Each animal is crouched upon a flattened dish-shaped nest that is either positioned on the ground (AB75 and AB82; see fig. 9-16), among papyrus umbels (e.g., AB61 to AB63, AB66, AB73, AB74, AB76, AB80, and AB81; see fig. 9-17), or within papyrus thickets (e.g., AB64 to AB74, AB77 to AB79, AB81, and AB83; see fig. 9-18). Their eggs are frequently visible and are usually deposited under their tails. The birds are represented with their wings stretched forward, either horizontally or at a slight incline, such that their primary feathers often project beyond the rim of the nest.

After laying a clutch, birds sit down and adjust their body until a featherless and highly vascularised section on their abdomen ('brood patch') makes contact with the eggs.[57] Warmth from their body enables the embryo within each egg to develop. These must be kept at an even temperature, so the eggs are incubated constantly until hatching. When birds sit upon their nests, their wings are folded and rest against the sides of their body to help shield the clutch from draughts (see pl. 9I). The outstretched wings exhibited by the birds in wall reliefs are thus incompatible with incubation and are incorrect.[58]

A brief examination of nesting images from other locations and time periods indicates that this unusual motif was widespread and lasting.[59] Nevertheless, the pose has drawn little comment from scholars. Some authors have referred to the birds' behaviour as "protecting their eggs",[60] but

---

55 Roth (*Cem. Pal. Att.*, 133) describes it as a "pheasant-like bird".

56 For a general discussion of nesting by birds, see section 11.1.3 in Chapter 11.

57 In most bird species, feathers over the brood patch fall out just prior to egg laying. Waterfowl pluck out the feathers and use them to line their nests. See Burton, *Bird Behaviour*, 173.

58 It should be noted that in very hot and arid environments, birds must work to keep their eggs cool, rather than warm. Some species stand above their nests to shade the eggs. Others adjust their body temperature by panting and holding their wings out from their bodies. The humid marsh habitat in which most of the nesting birds are depicted is not subject to the kind of extreme temperatures that result in this behaviour by birds.

59 The earliest example of which I am presently aware appears in a fragment from the mortuary complex of Userkaf. The image shows a goose brooding three eggs with its wings outstretched; another bird in a nest above the first may also be engaged in the same behaviour. See Labrousse-Lauer, *Complexes Funéraires d'Ouserkaf*, Doc 39 and fig. 108a-b.

60 For example, in describing the depiction of a nesting barn owl in a marsh scene from the 18th dynasty tomb of *nfr-ḥtp* (A5) at Thebes, Houlihan (*Birds*, 110) states that it is, "sheltering the clutch beneath its outstretched wings in an attempt to protect them from an Egyptian mongoose (*Herpestes ichneumon*), which

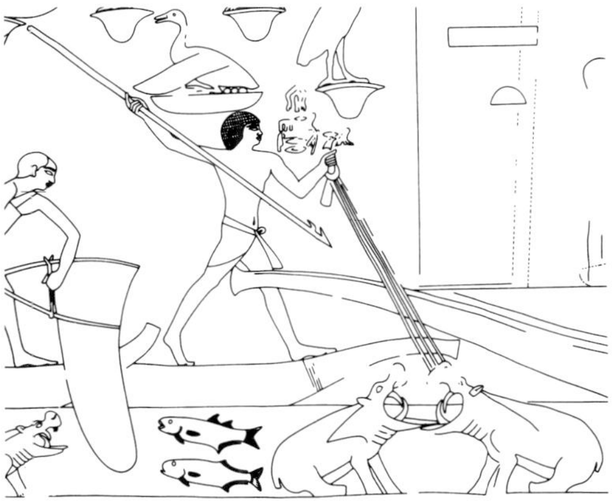

Fig. 9-10 (AB46)

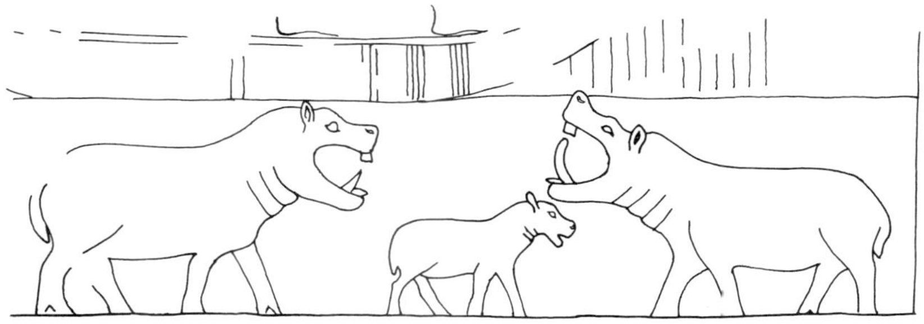

Fig. 9-11 (AB35)

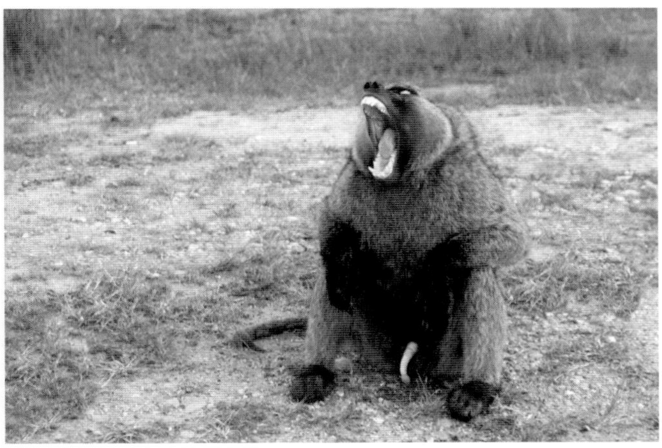

Pl. 9G Baboon penile display

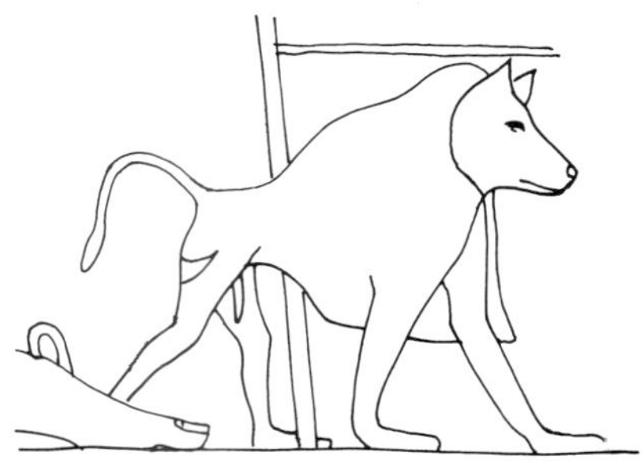

Fig. 9-12 (AB53)

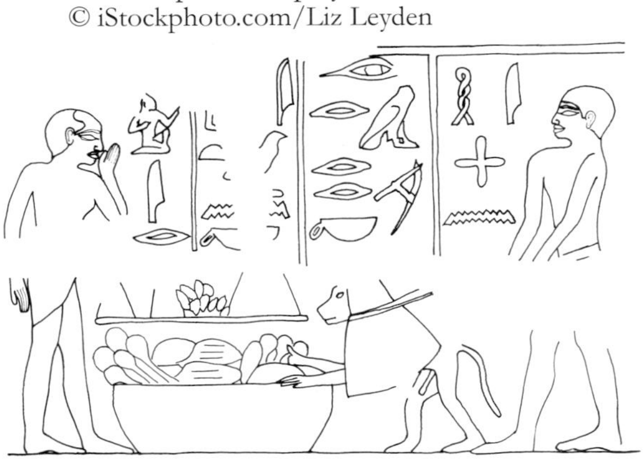

Fig. 9-13 (AB57)

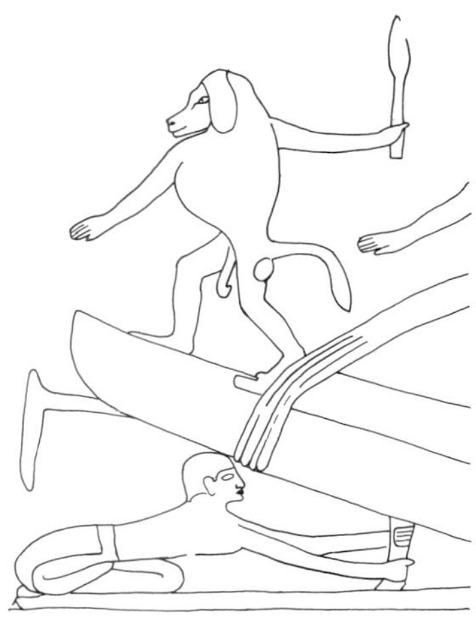

Fig. 9-14 (AB56)

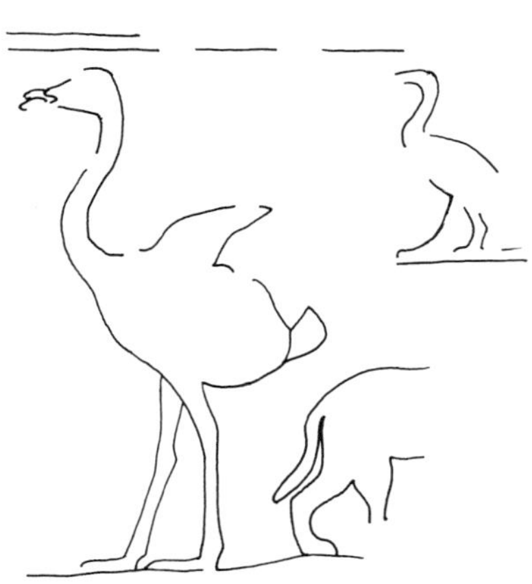

Fig. 9-15 (AB60)

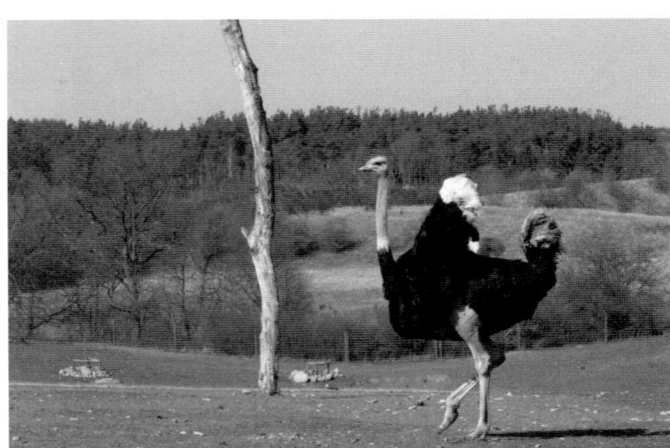

Pl. 9H Ostrich wing display

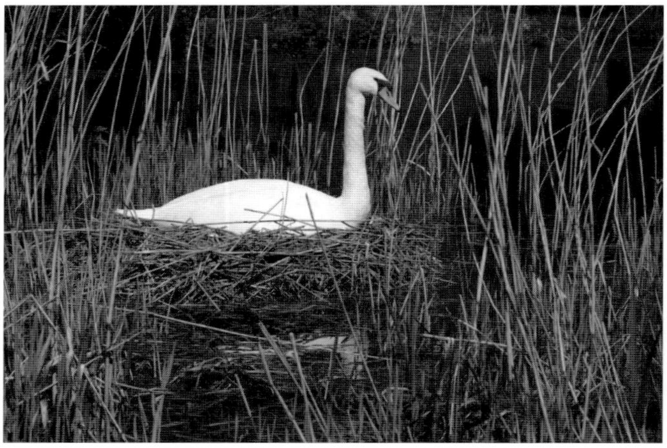

Pl. 9I Avian nesting posture
© iStockphoto.com/Katarzyna Fifielska

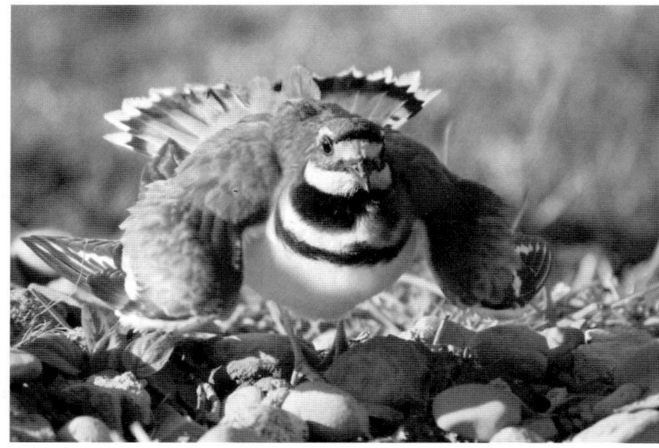

Pl. 9J Avian threat display
© iStockphoto.com/Cathleen Clapper

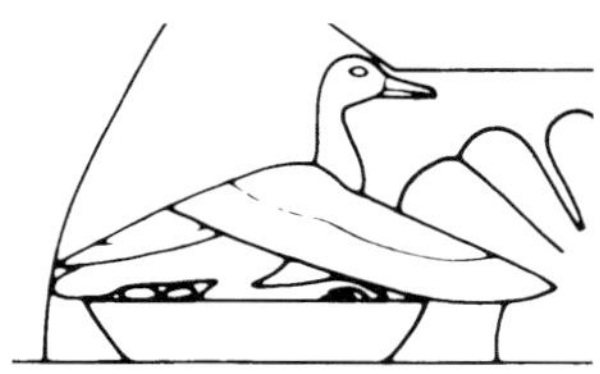

Fig. 9-16 (AB75)

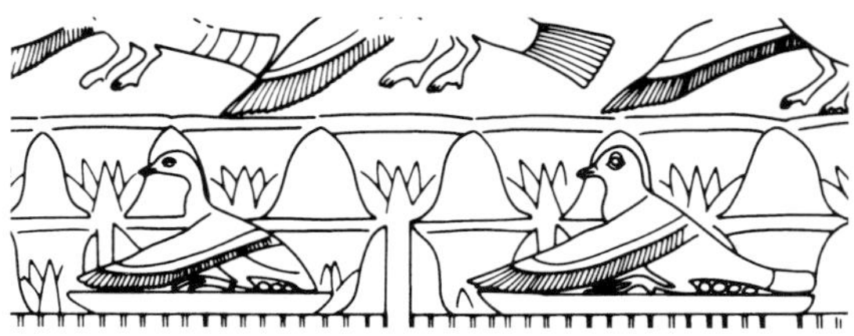

Fig. 9-17 (AB73)

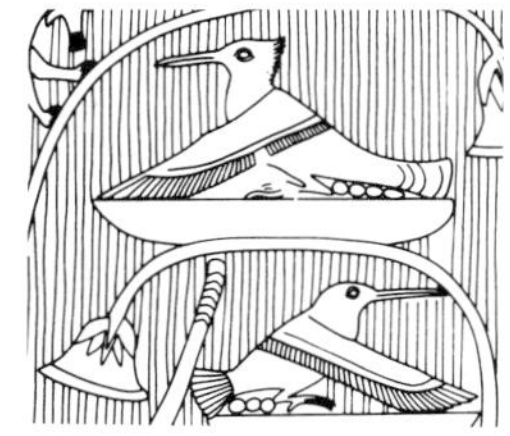

Fig. 9-18 (AB74)

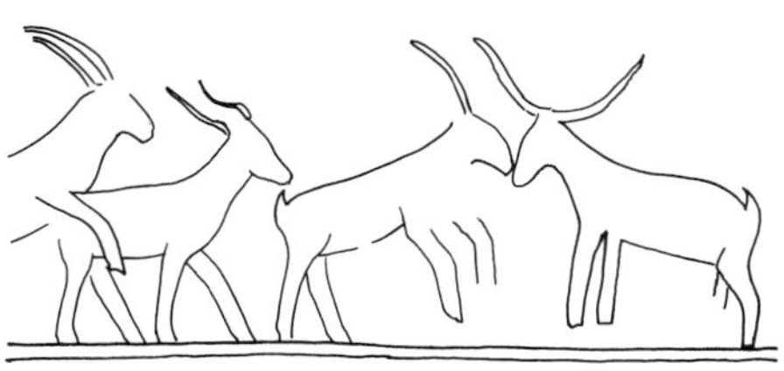

Fig. 9-19 (AB85)

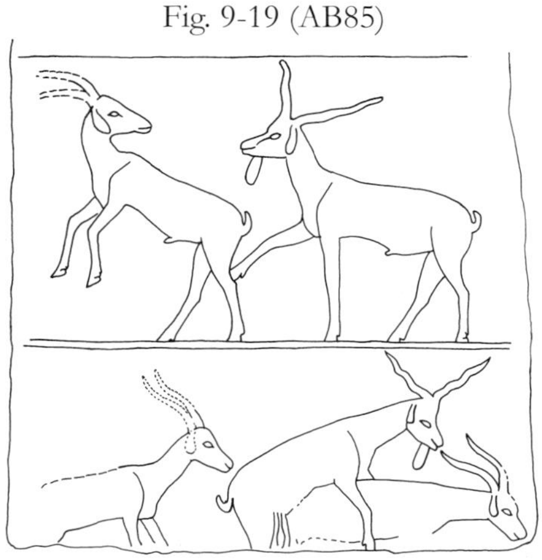

Fig. 9-20 (AB84)

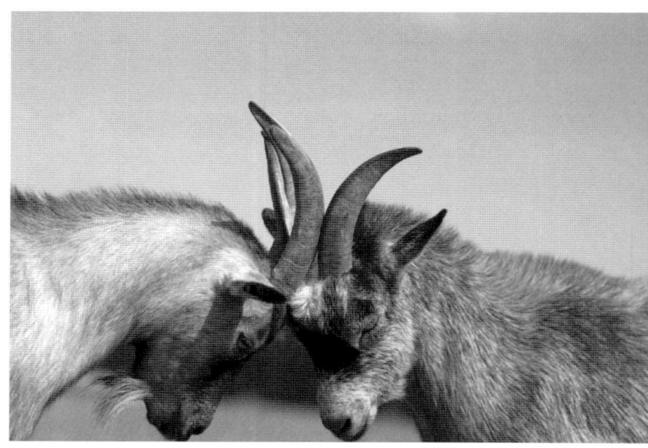

Pl. 9K Goats fighting
© iStockphoto.com/Paul Giehm

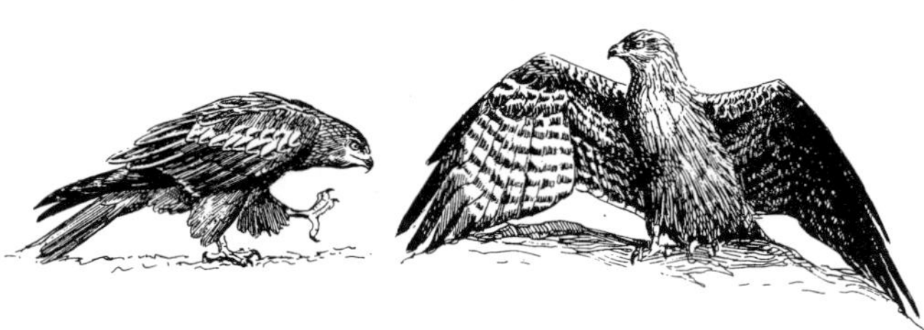

Pl. 9L Black kite threat display
Reprinted from: Cramp-Simmons, *Handbook ii*, 32
By permission of © Oxford University Press

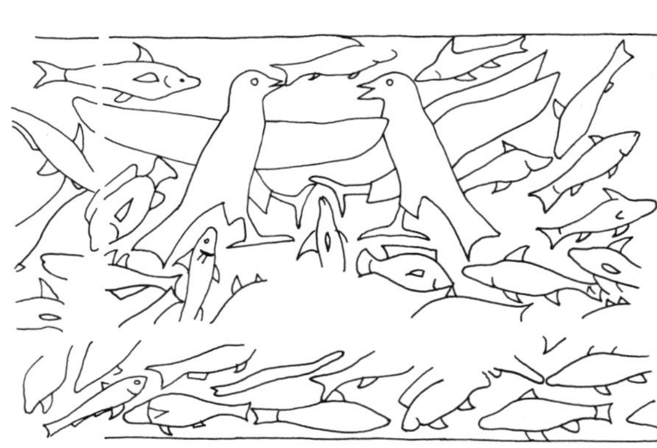

Fig. 9-21 (AB86)

otherwise their wing position has largely been ignored.[61] This is surprising, as the nesting habits of some of the depicted species, namely hoopoes and pied kingfishers, have been examined in detail.[62] As noted earlier in this volume, hoopoes place their nests in natural cavities, while pied kingfishers deposit their eggs at the end of tunnels that they excavate into riverbanks. The inclusion of these hole-nesting species among other incubating birds in papyrus thickets is thus incorrect and has been identified as such by several art historians, however the birds' strange wing postures have been accepted without question.

Why, then, did artists choose to represent these animals in such an odd posture, especially when it would have been more straightforward to depict them as they appear in nature, with their wings folded? The anti-predator behaviour of birds provides a solution. Birds are particularly vulnerable when incubating, especially those that nest upon the ground. Most try to avoid detection by hiding or camouflaging their nest. If a predator does approach, however, many species engage in a form of defensive behaviour. Some will attack an advancing predator (see *Birds vs predators* below), but most attempt to simply distract or intimidate it by exhibiting a threat display. The most common threat posture consists of erecting the feathers on the head and body (fluffing), fanning the tail, and spreading out the wings (see pl. 9J). This behaviour enables the bird to appear physically bigger and more imposing, and therefore potentially less attractive to a predator.[63]

I propose that the outstretched wings of nesting birds in wall reliefs represent a threat display that is directed towards the potential predators with which they share each scene. These are usually genets and Egyptian mongooses (both of which prey on birds and eggs; see Chapter 8), but in an agricultural scene, a quail also gestures to a nearby workman (AB82). The natural threat displays of some of the depicted species include: Egyptian goose – head and body erect with the wings lifted and fully opened;[64] sacred ibis – wings raised and opened to reveal distinctive red patches;[65] pied kingfishers – body erect with the wings half extended;[66] hoopoes – crest raised with the wings lifted and spread.[67] Wing spreading is thus a common response to danger. I believe that Old Kingdom artists adopted this unmistakable posture to convey the behaviour of threatened birds, but that due to the spatial constraints of their art, the animals hold their wings forward rather than out from the sides of their bodies. Each nesting bird adopts the same pose, regardless of the idiosyncrasies of the threat posture for their particular species. The images are thus intended to convey the concept of a threat response, rather than veridical representations of the behaviour of individual species.

The depicted birds reveal an understanding by artists of their natural response to threatening situations. This knowledge was probably acquired by the Egyptians from firsthand experience when collecting the eggs of both wild and domestic fowl for consumption, because humans - who are treated as potential predators by birds - also elicit threat displays. It is possible, in fact, that this posture was mistaken by the Egyptians for part of the normal incubation behaviour of birds. Thus, although derived from a threat response, this wing pose may have become associated with nesting generally and in so doing, provided an artistic convention by which to convey clearly to the viewer that a bird was incubating and protecting its eggs.

---

is approaching the nest . . .". Gaillard refers to the nesting behaviour of two birds in a marsh scene from the tomb of *tjj* as simply "couvent leurs oeufs". See Gaillard, C. (1931). "Quelques représentations du Martin-Pêcheur Pie (*Ceryle rudis*), sur les monuments de l'Égypte ancienne", *BIFAO* 30(1), 268.

61 Edel describes a number of images of nesting birds (e.g., pied kingfishers, common kingfishers, pelicans etc.) that appear in the Chamber of the Seasons in the mid-5th dynasty sun-temple of Niuserre at Abu Ghurab, but does not comment on the position of their wings. See Edel, E. (1961). "Zu den Inschriften auf den Jahreszeitenreliefs der 'Weltkammer' aus dem Sonnenheiligtum des Niuserre: Part I", *NAWG* 1961, 209-255 and Edel, E. (1963). Zu den Inschriften auf den Jahreszeitenreliefs der 'Weltkammer' aus dem Sonnenheiligtum des Niuserre: Part II", *NAWG* 1963, 89-217.

62 Gaillard, *BIFAO* 30(1), 269-270; Edel, *ibid.*, 137-140; Houlihan, P.F. (1996). "A guide to the wildlife represented in the great swampland scene in the offering-chapel of Ti (No. 60) at Saqqara", *GM* 155, 35.

63 Leahy, *Companion*, 182-183.

64 Roselaar, C.S. (1977). "Egyptian goose (*Alopochen aegyptiacus*)" in Cramp-Simmons, *Handbook i*, 449.

65 Roselaar, C.S. (1977). "Sacred ibis (*Threskiornis aethiopicus*)" in *ibid.*, 349.

66 Roselaar, C.S. (1985). "Pied kingfisher (*Ceryle rudis*)" in Cramp-Simmons, *Handbook iv*, 727.

67 Roselaar, C.S. (1985). "Hoopoe (*Upupa epops*)" in *ibid.*, 792.

9.2 FIGHTING

Animals often clash over resources such as food, territories, and access to mates. Fighting can also assist in establishing a dominance hierarchy within a social group. The female members of some species will occasionally spar, but it is a behaviour more commonly expressed by males. A fight consists of a physical interaction between two individuals with the intention of causing injury and/or forcing one of the combatants to submit to the strength and dominance of the other.

Goats

sšm-nfr IV (AB84: Giza, V.9-VI.1) *nb-k3w-ḥr: jdw* (AB85: Saqqara, V.9 r/u VI.7)

Dominance in goats is largely determined by age, such that older individuals tend to influence younger animals. If a young goat approaches an unrelated adult, the older animal may drop its head and 'butt' the subordinate[68] or simply threaten it by stretching its neck forward to display its horns. Submissive animals will be intimidated by these actions and move away, but equally matched goats may fight. Goats begin a fight by standing approximately two metres apart and rearing up simultaneously on their hind legs while facing away from one another. They then turn back and fall forward to strike their opponent with a lateral thrust of their head and horns (see pl. 9K).[69] Males are particularly aggressive during the breeding season when they compete with one another over access to fertile females.[70]

Two wall scenes dating from the late 5th to the early 6th dynasty depict pairs of fighting goats.[71] In the tomb of *nb-k3w-ḥr: jdw* (AB85; see fig. 9-19), two goats stand head to head, immediately in front of a male and female that are mating. The forequarters of the opposed goats are clearly higher than their hindquarters and the animal on the left also bends its forelegs slightly. The goats' orientation and proximity to one another, combined with the inclination of their backs, suggest that they have fallen forward to clash horns. I suggest that the presence of the mating pair nearby indicates that these are probably males that are fighting over fertile females.

An earlier image in the tomb of *sšm-nfr IV* (AB84; see fig. 9-20) is less straightforward. The animal on the right is a male that features spreading horns and a substantial beard. The goat on the left also appears to have male genitalia, but it lacks a beard[72] and its horns are rendered in a similar manner to those of females appearing in the same relief. The male on the right approaches the animal on the left and raises one leg. This action is part of the pre-copulatory behaviour of goats in which the male signals his interest in a female by kicking at her repeatedly with one foreleg (see pl. 10E in Chapter 10).[73] The goat on the left has reacted to this attention by rearing up on its hind legs to look back at its companion in an action akin to fighting. Thus two conflicting behaviours (sexual and aggressive) seem to be expressed in this interaction.[74]

An explanation for the apparent mismatch may be found in the complex dominance behaviour of sheep and goats.[75] Dominant males will occasionally treat subordinate males in the same manner as when courting fertile females – that is, by nudging, sniffing, and kicking out with a stiffened

---

68 Shackleton, D.M. and Shank, C.C. (1984). "A review of the social behavior of feral and wild sheep and goats", *J. Anim. Sci.* 58(2), 505; Hafez, E.S.E., Cairns, R.B., Hulet, C.V. and Scott, J. (1969). "The behaviour of sheep and goats" in Hafez, *Behaviour of Domestic Animals*, 335.

69 *ibid.*, 341.

70 Rutter, S.M. (2002). "Behaviour of sheep and goats" in Jensen, *Ethology*, 151.

71 Two pairs of goats in a damaged scene in the tomb of *mrrj* (Saqqara, ntn) rise up on their hind legs while facing one another in a manner similar to fighting (see IB79, middle and right, in Chapter 7), however the presence of a tree behind each pair suggests that they are more likely to be browsing.

72 Both male and female goats have beards, but those of males may be thicker and longer. Females lack beards in some modern breeds. Female goats in wall scenes are often depicted with beards.

73 Hart, *Behavior of Domestic Animals*, 113.

74 See *Goats* in section 10.1.4 in Chapter 10 for an alternative interpretation of this scene.

75 Although they differ morphologically, the aggressive behaviour of sheep and goats is very similar. See Shackleton-Shank, *J. Anim. Sci.* 58(2), 504-506.

foreleg.[76] In these circumstances, their behaviour is not sexual, but aggressive and intended to emphasize or assert their higher status. Like the image found in the tomb of *nb-k3w-ḥr: jdw*, a male and female are depicted mating in the register immediately below the pair in question. As noted earlier, this may indicate that the males are competing for access to females. Instead of presenting a clash between equals, however, the *sšm-nfr IV* scene may be a unique depiction of a dominant adult male attempting to intimidate a subordinate male, whose lower age has been indicated by its lack of a beard and 'feminine' horns. Rather than acting submissively, the younger animal has reared up to challenge the older male. This image may be the result of a chance observation of a male goat treating another male in an apparently sexual manner, which the witness found intriguing and possibly amusing. For further discussion of this scene, see *Goats* in section 10.1.4 in Chapter 10.

Black kites

nj-ʿnḫ-ẖnmw/ẖnmw-ḥtp (AB86: Saqqara, V.6L-7)

In a weir-fishing scene in the tomb of *nj-ʿnḫ-ẖnmw/ẖnmw-ḥtp* (AB86; see fig. 9-21), two birds have been depicted facing one another amidst a pile of fish on a riverbank. The bird on the left stands with its wings outstretched and beak closed while that on the right lifts both wings above its back and parts its mandibles. Both birds raise one foot. Each animal also displays a notched tail. This last feature is a diagnostic characteristic of black kites, large predatory raptors that are found frequently around human settlements in Egypt. This identification is confirmed by the birds' curious body postures, which also reveal that they are engaged in a fight.[77]

Black kites hunt for prey, but they will also scavenge whenever the opportunity arises. They are very aggressive animals and when confronted with a food source they will often attempt to intimidate competitors by engaging in threat displays and fighting. Black kites threaten one another by intermittently raising one foot and spreading their toes as they walk towards their opponent (pl. 9L, left), or by turning towards their adversary and quickly spreading both of their wings (pl. 9L, right). If the situation escalates to a fight, the aggressor may fly towards its competitor to attack it with its open claws. Subordinate birds will flee, but dominant animals usually hold their ground and kick out with one open clawed foot while leaning on their tail for support.[78]

The scene in the tomb of *nj-ʿnḫ-ẖnmw/ẖnmw-ḥtp* thus appears to depict a combination of threat displays and subsequent fighting by black kites.[79] Both birds show the distinctive open claws associated with threat and attack behaviour in this species. Furthermore, the posture of the kite on the right indicates that it is jumping forward to attack its opponent, which responds by opening its wings and holding up one open-clawed foot while supported by its tail. The image thus encompasses all the elements of black kite aggressive behaviour and reveals a remarkably accurate study of this unique avian interaction by the artist.[80]

---

76 *ibid.*, 505.

77 Moussa and Altenmüller (*Nianchchnum*, 100) were uncertain about the species depicted, but suggested that they might be seagulls attracted to the fish: "Da der Fischhaufen unbewacht ist, haben sich Raubvögel eingestellt. Zwei von verschiedenen Seiten herangeflogene Möwen(?) streiten sich um eine aus dem Fischhaufen herausgegriffene Meeräsche . . .".

78 Alvarez, F., Arias de Reyna, L. and Hiraldo, F. (1976). "Interactions among avian scavengers in southern Spain", *Ornis Scand.* 7, 221-222; Wattel, J. (1977). "Black kite (*Milvus migrans*)" in Cramp-Simmons, *Handbook ii*, 33.

79 For more details, see Evans, L. (2007). "Fighting kites: Behaviour as a key to species identity in wall scenes", *JEA* 93, 245-247.

80 A relief fragment from the mortuary complex of Userkaf depicting a bird that bites the beak of a companion while raising one foot against it may also depict a fight. See *Eg. Art in the Age of the Pyramids*, cat. 102 (Egypt. Mus. temp 6-9-32-1), 317; Houlihan (*Birds*, 130, fig. 184) suggests that they are "engaged in playful conflict".

9.3 ATTACKING

Animals will sometimes attack members of other species. Often this is an act of predation that results in the aggressor consuming the animal it has assaulted, however animals will also attack others as an act of self-defence, when they sense that their territory or personal space has been encroached upon, or if either they or their family members are threatened.

Hippopotamuses vs crocodiles

tjj (AB87: Saqqara, V.8-9)
3ḫtj-mrw-nswt (AB88: Giza, V.9-VI.1)
k3-gm-nj: mmj (AB89-91: Saqqara, VI.1M)
mrrw-k3.j: mrj (AB92: Saqqara, VI.1M-L)
jnw-mnw (AB93: Saqqara, VI.2E)
mḥw (AB94: Saqqara, VI.2M-3?)
nb-k3w-ḥr: jdw (AB94a: Saqqara, V.9 r/u VI.7)

Nine scenes in seven tombs were found to depict a hippopotamus as it bites a Nile crocodile.[81] The hippos frequently raise the reptiles' bodies so that their tails emerge above the water surface, suggesting a dynamic, shaking movement (see fig. 9-22). The crocodiles are usually grabbed around their trunk (between the fore- and hind legs) - ventrally in the tomb of *mrrw-k3.j: mrj* (AB92) and dorsally in the tombs of *tjj* (AB87) and *mḥw* (AB94). The crocodile in the tomb of *3ḫtj-mrw-nswt* (AB88) is bitten upon the tail, and by the foreleg in the tomb of *jnw-mnw* (AB93). In two examples, the hippos' tusks pierce their victims' bodies (AB87 and AB92; see fig. 9-22). The tomb of *k3-gm-nj: mmj* reveals three variations on this theme. In room I, a seated hippopotamus straddles the tail of a crocodile, mauling it on the lower back and lifting the reptile's hind legs off the substrate (AB89; see fig. 9-23). In room III, a standing hippopotamus bites the predator on the snout (AB90), while in room IV, yet another hippo grasps the upright body of a crocodile around its trunk, causing it to open its jaws and sweep its tail against its aggressor (AB91). Each of these interactions takes place while the animals are fully submerged.

Male hippos are fiercely territorial, highly aggressive, and unpredictable animals that often fight viciously with one another (see pl. 9M). Female hippos will also savagely attack any animal that strays too close to their young. The hippos' sharp tusks and incisors, combined with their powerful jaws and extremely wide gape, enable them to bite very large objects and to puncture the hides of most animals. Hippos fight by chasing one another, pushing their opponent with their open jaws, and slashing and biting with their tusks and incisors.[82] Fights can cause serious injuries and are occasionally fatal.

Nile crocodiles are carnivores and formidable predators (see Chapter 8), while hippopotamuses are vegetarians; nevertheless, in the wild, hippos tend to dominate Nile crocodiles.[83] The clearest evidence for an aggressive relationship between the two species comes from several reports indicating that the bodies of Nile crocodiles have been found bitten in half by hippos.[84] One animal killed in this manner originally measured over three metres in length. The images of hippopotamuses grasping the bodies of crocodiles are thus entirely feasible and either the result of direct observation of such attacks, or reconstructed from the physical remains of severed reptiles.[85] One modern report noted that the crocodile's body had been separated "just in front of the hind limbs", which matches the locus of attack in most wall scenes.[86]

---

81 Hassan (*Neb-Kaw-Her*, 8 and pl. v [c]) states that "a battle between a red hippopotamus and a yellow crocodile" is represented in a fragment from the tomb of *nb-k3w-ḥr: jdw*, but unfortunately this is impossible to discern in the photograph provided. A further example may be apparent in a scene from the north wall of room iv in the tomb of *ḫntj-k3.j: jḫḫj* at Saqqara. The image shows a hippopotamus facing an elongated structure that possibly depicts the sweeping tail of a crocodile, however the relief is too fragmentary to determine whether the two are engaged in a fight. See James, *Khentika*, pl. 15.

82 Estes, *Behavior Guide*, 225.

83 Cott, H.B. (1961). "Scientific results of an inquiry into the ecology and economic status of the Nile crocodile (*Crocodilus niloticus*) in Uganda and Northern Rhodesia", *Trans. Zool. Soc. Lond.* 29, 305.

84 *ibid.*, 306; Stevenson-Hamilton, *Wild Life*, 61.

85 The veracity of this motif has been doubted. For example, see Osborn-Osbornová, *Mammals*, 148. The scene has also been misinterpreted as an act of predation by the hippopotamuses. See Altenmüller, *Jagd im Alten Ägypten*, 22.

86 Cott, *Trans. Zool. Soc. Lond.* 29, 306.

An explanation for the hippos' aggression is found in the waters surrounding many of the images. In the tomb of *ṯjj* (AB87), a calf climbs upon the back of an adult near the attacking hippo. The aggressive interaction in the tomb of *mrrw-kꜣ.j: mrj* (AB92; see fig. 9-22) is flanked by a calf (right) and a hippopotamus giving birth (left). A hippopotamus appears to give birth before the attacking hippo in one scene from the tomb of *kꜣ-gm-nj: mmj* (AB91); in another image from the same tomb (AB89; fig. 9-23), the posture of a hippo directly before the crocodile's snout suggests that it may also be calving (although no foetus is visible). As noted above, female hippopotamuses protect their calves aggressively. The contexts in which crocodiles are grasped by hippos indicate that they have been attacked because they have ventured too close to females and their young. These scenes are thus the opposite (or the consequence) of those in which Nile crocodiles wait to prey upon newborn hippopotamus calves (see Chapter 8).[87]

Baboons vs humans

tp-m-ꜥnḫ II (AB95: Saqqara, V.5-6?) *nj-ꜥnḫ-ẖnmw/ẖnmw-ḥtp* (AB96: Saqqara, V.6L-7)

Two mid-5th dynasty market scenes feature baboons that attack young men.[88] Baboons are aggressive animals that use a range of threat displays to assert their dominance. Fights are infrequent in stable troops, but dominant males will occasionally chase and attack subordinates (see pl. 9N) by seizing and beating them with their hands and biting them savagely (usually on the neck).[89] Males possess long canine teeth that can inflict considerable injuries.

The earlier of the two scenes occurs in the tomb of *tp-m-ꜥnḫ II* (AB95; see fig. 11-27 in Chapter 11) and depicts two leashed baboons under the control of a male handler. On the left, a female walks quadrupedally while clutching an infant to her chest. The animal on the right is a male, however it lacks the shaggy fur cape characteristic of adult male hamadryas (a feature that is usually reproduced in Egyptian depictions of these animals), indicating that it is probably an anubis baboon.[90] The animal attacks a young man as it would a subordinate - rearing up on its hind legs, it strains against its leash to launch itself forward and grasp the boy by the thigh. The youth appears to have been caught while running, as his body is inclined to the right, his legs are wide apart, and his left knee is bent. In a later scene from the tomb of *nj-ꜥnḫ-ẖnmw/ẖnmw-ḥtp* (AB96; see fig. 9-24), a leashed baboon[91] also attacks a young man, but this time it has succeeded in biting the boy on his leg. This victim may also be moving quickly, as his right heel is lifted.[92]

Both boys try to push their attackers away, shoving the baboons on the arm or on the top of the head. It has been suggested elsewhere that these boys have been 'caught' by the baboons while trying to steal items in the marketplace and that, based on the accompanying inscriptions, the animals were directed to do so by their handlers. Since antiquity, primates have been kept as pets

87 The Egyptians noted the antagonistic relationship between Nile crocodiles and hippopotamuses as early as the predynastic period, as revealed by their frequent pairing as a decorative motif on white cross-lined pottery. For example, see Finkenstaedt, E. (1980). "Regional painting style in prehistoric Egypt", *ZÄS* 107, 116-120. The association between the two animals is also expressed visually in the form of the goddess Taweret.

88 Both of these images share many similarities with a market scene from the causeway of king Unas, which also features a pair of baboons. For details, see Smith, W.S. (1942). "The origin of some unidentified Old Kingdom reliefs", *AJA* 46, 515-518, fig. 8.

89 Hall-de Vore in de Vore, *Primate Behavior*, 96-99.

90 Depictions of anubis baboons are rare in Egyptian art. See Osborn-Osbornová, *Mammals*, 38-39.

91 Moussa and Altenmüller (*Nianchchnum*, 82) identified this animal as a long-tailed (Cercopithecine) monkey ("eine Meerkatze"), rather than a baboon. Indeed, it does not resemble a hamadryas male that appears further along on the same register (AB57). All Old World monkeys possess thick pads on their buttocks ('ischial callosities'); these are particularly large in baboons and often brightly coloured. I have observed that the rump area of depicted baboons is frequently defined in Old Kingdom scenes (cf. fig. 9-14), but that this feature never appears in images of vervet monkeys. The buttocks of the animal in question are quite apparent indicating that it is undoubtedly a baboon. I suggest that, as it lacks a fur cape or obvious genitalia, it is either a female hamadryas (and thus the mate of the male hamadryas on the same register), or, like those in the tomb of *tp-m-ꜥnḫ II*, an anubis baboon.

92 Schäfer, *Principles*, 16.

and also trained to provide services for humans,[93] however not all species are appropriate for these purposes. The Egyptians clearly kept baboons, even though they are known to be irritable and aggressive. They are also intelligent animals and so it may have been possible to train a baboon to chase and bite on command. I believe it is more likely, however, that the animals in these scenes have acted spontaneously in response to the sudden movements of the thieves, which have triggered their natural tendency to chase.[94]

Birds vs predators

nb.j-m-3ḫtj (AB97: Giza, IV.6-V.1)
ḥtp-ḥr-3ḫtj (AB98: Saqqara, V.6-8E)
jtj-sn (AB99: Giza, V.6-8?)
sndm-jb: jntj (AB100-101: Giza, V.8M-L)
ṯjj (AB102-104: Saqqara, V.8-9)
nfr-jrt-n.f (AB105: Saqqara, V.8-9)
k3.j-m-nfrt (AB106: Saqqara, V.8-9?)
3ḫtj-ḥtp (AB107-109: Saqqara, V.8L-9E)
ḫnwt (AB110: Saqqara, V.9)
nbt (AB111-112: Saqqara, V.9)
jj-nfrt: š3.n.f (AB113-114: Saqqara, V.9)
nj-k3w-jzzj (AB115-117: Saqqara, VI.1M)
mrrw-k3.j: mrj (AB118-120: Saqqara, VI.1M-L)
zšzšt: jdwt (AB121: Saqqara, V.9 r/u - VI.1-2E?)
sʿnḫ-w(j)-ptḥ (AB122-125: Saqqara, VI.1L)
ḥzj (AB126-129: Saqqara, VI.1L-2E)
mrjj-rʿ-mrj-ptḥ-ʿnḫ: nḫbw (AB130: Giza, VI.2)
mṯṯj (AB131: Saqqara, VI.2?)
mḥw (AB132-136: Saqqara, VI.2M-3?)
jj-nfrt (AB137-138: Giza, VI.3-4E?)
3ḫtj-ḥtp (AB139: Saqqara, VI.5?)

Twenty-one tombs dating from the mid-4th to the mid-6th dynasty exhibit scenes in which adult birds defend their nestlings against approaching predators. In each example, the chicks stand in nests that rest upon individual stalks or umbels situated in dense papyrus thickets. Many of the nestlings flap their wings, indicating that they are vigorous, but also alarmed. One or more adult birds are usually depicted flying above the nest or nearby; shared physical attributes with the nestlings (e.g., crests, beak shape) reveal that these are their parents. The adults' wing positions indicate that they either dive or hover above the genets or Egyptian mongooses that advance toward their nests; some adults also peck, bite, or scratch the predators.

Most bird species engage in some form of defensive behaviour to protect their nests, eggs, or hatched young from predators. As noted above, many adopt a threat posture by fluffing their feathers and spreading their wings (see *Wing display*, above). Some species, however, will attempt to distract an approaching predator by attacking it. Using speed and agility, attacking birds fly down and past an intruder repeatedly or flutter around it at close range, in order to confuse and intimidate it (see pl. 9O). Many species utter harsh cries or alarm calls as they swoop, which can increase the effectiveness of their display. Some birds defecate upon the predator. Others put themselves in great peril by hovering close enough to peck it with their beak or bill, scratch it with their feet, or strike it with their wing tips.[95] These actions place the parent bird at substantial risk of being caught, injured, or killed by the predator. Observational studies have determined that their defensive behaviour intensifies with the age of their young – that is, birds take some risks while incubating their eggs, but increase their antipredator activity dramatically after their young hatch, and up until they are old enough to leave the nest.[96]

Memphite wall scenes reveal a range of avian antipredator actions, but these are expressed by only a subset of species. The forty-five examples of parental defence include: Egyptian geese or ducks

93 For a discussion of the evidence for trained primates in ancient Egypt, see Vandier d'Abbadie, J. (1964). "Les singes familiers dans l'ancienne Égypte. I. L'Ancien Empire", *Rd'É* 16, 169ff; Houlihan, P.F. (1997). "Harvesters or monkey business?" *GM* 157, 31-47.

94 It is interesting to note that with the loss of habitat in Africa today, many baboon species are now forced to live in close proximity to humans. This has led to a marked increase in the number of aggressive interactions between people and baboons, which often enter settlements in search of food. Children, in particular, are frequently attacked.

95 Armstrong, E.A. (1964). "Parental care" in Thomson, *Dictionary of Birds*, 599; Brooke-Birkhead, *Ornithology*, 249-251

96 Montgomerie, R.D. and Weatherhead, P.J. (1988). "Risks and rewards of nest defence by parent birds", *Quart. Rev. Biol.* 63(2), 177-179.

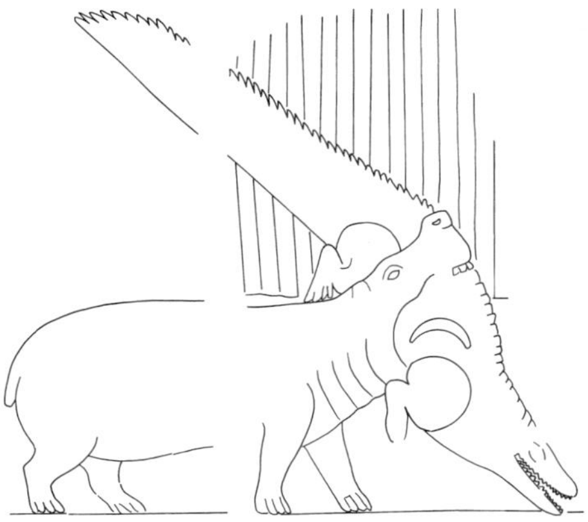

Fig. 9-22 (AB92)

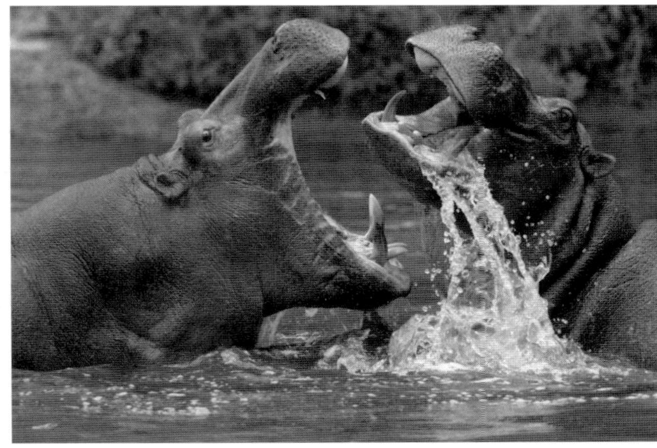

Pl. 9M Hippopotamuses fighting
© iStockphoto.com/Graeme Purdy

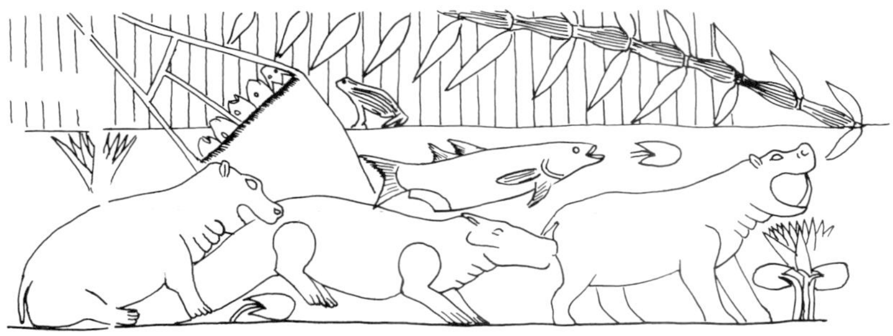

Fig. 9-23 (AB89)

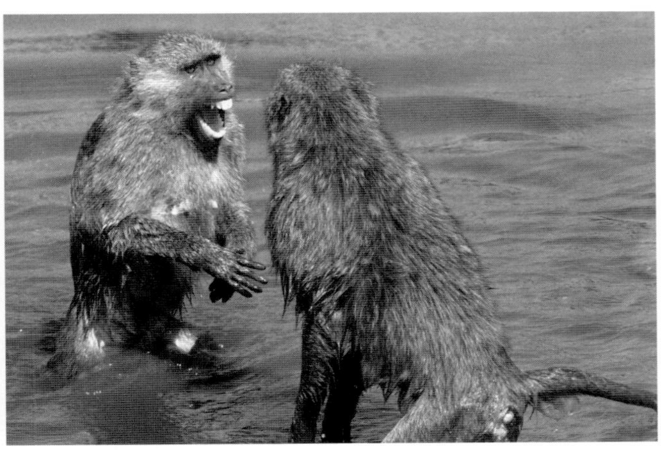

Pl. 9N Baboon aggression
© iStockphoto.com/DW Photos

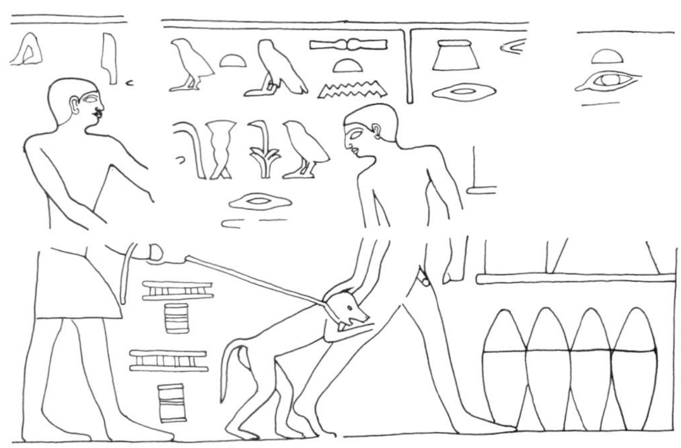

Fig. 9-24 (AB96)

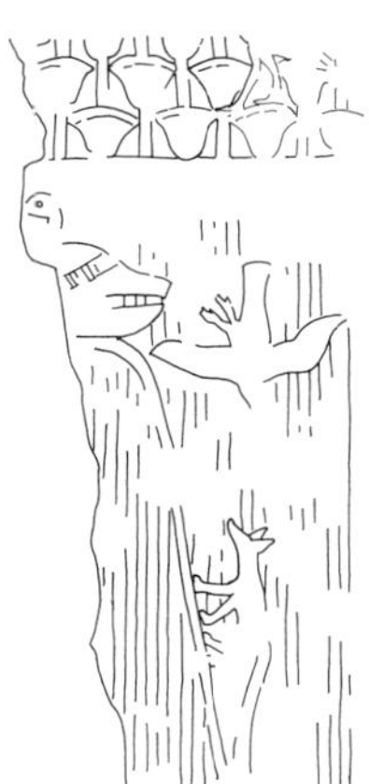

Fig. 9-25 (AB101)

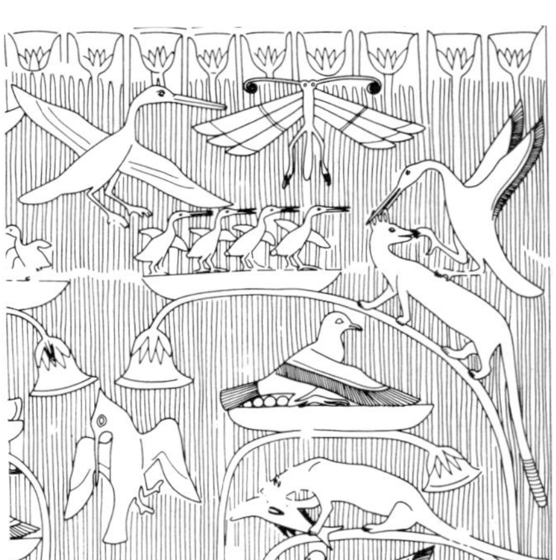

Fig. 9-26 (AB128)

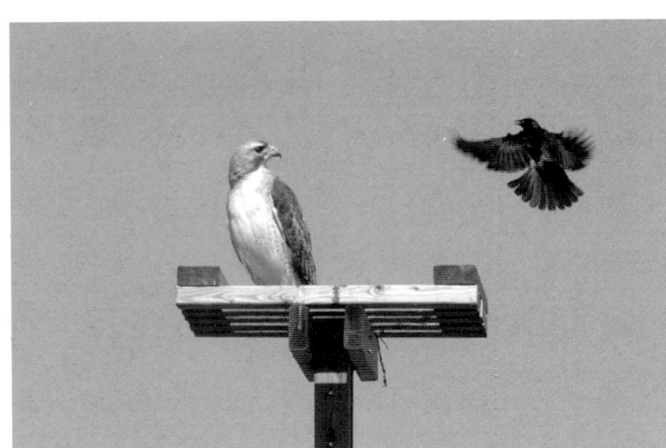

Pl. 9O Crow mobbing a hawk
© iStockphoto.com/Daniel Kourey

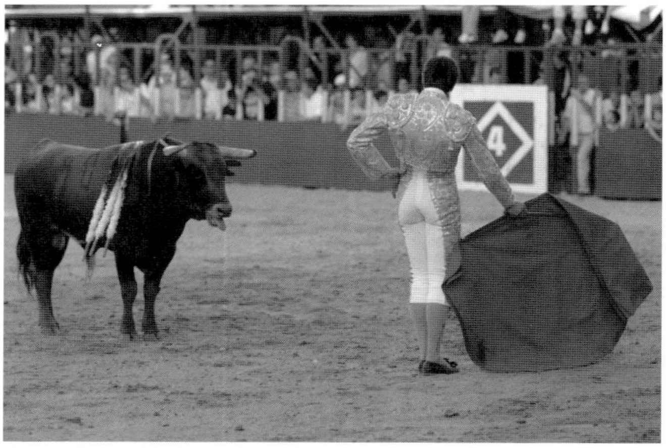

Pl. 9P Bull bellowing with protruded tongue
© iStockphoto.com/Ricardo Lopez

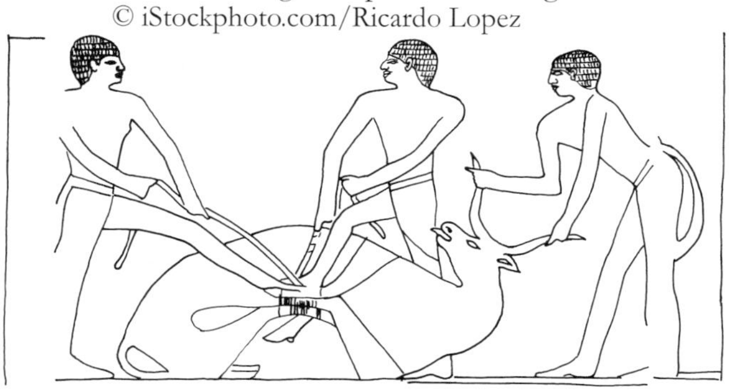

Fig. 9-29 (AB153)

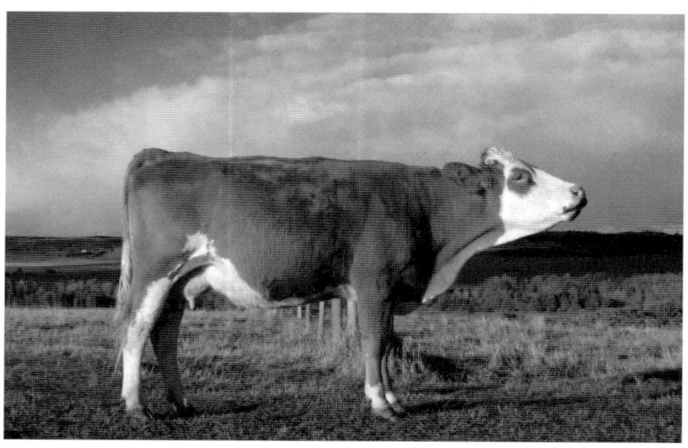

Pl. 9Q Cow vocalizing
© iStockphoto.com/Andrew Penner

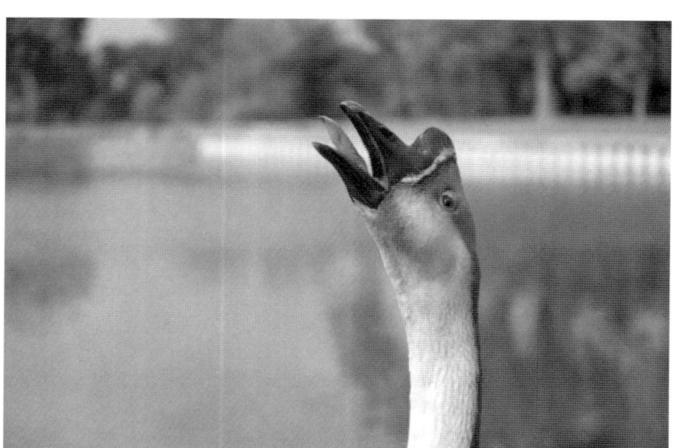

Pl. 9R Goose hissing
© iStockphoto.com/Tracy Tucker

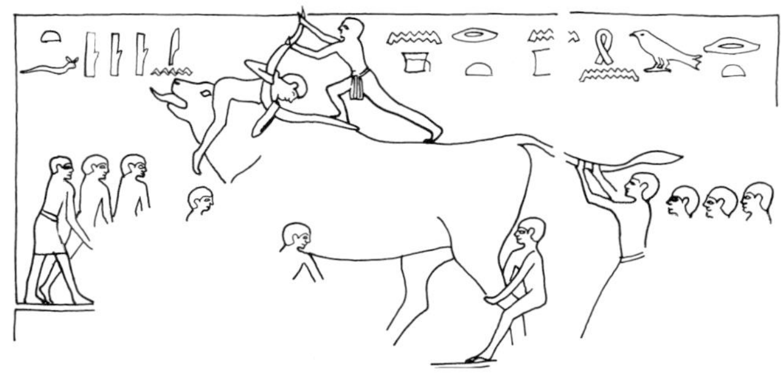

Fig. 9-27 (AB140)

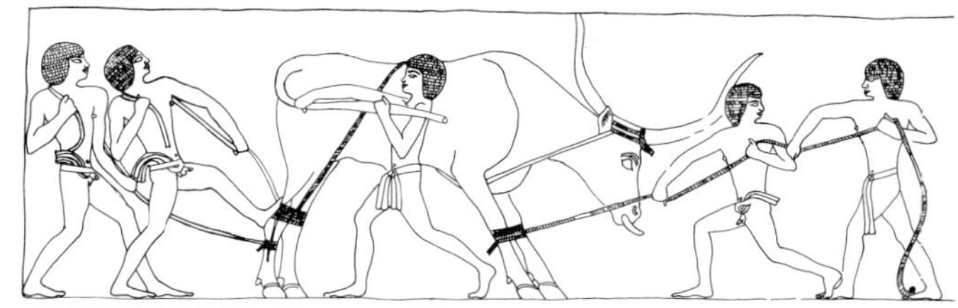

Fig. 9-28 (AB151)

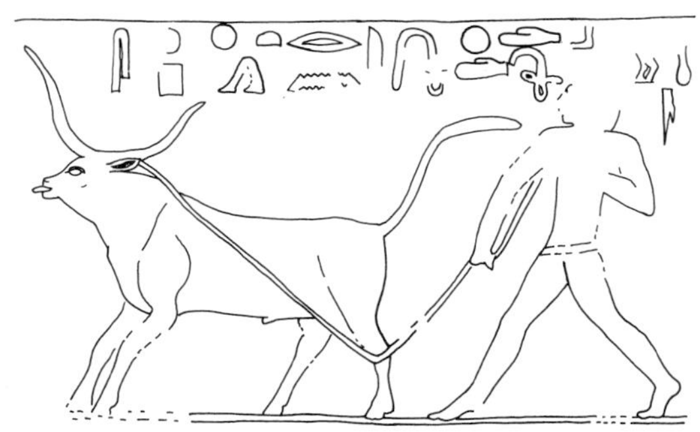

Fig. 9-30 (AB161)

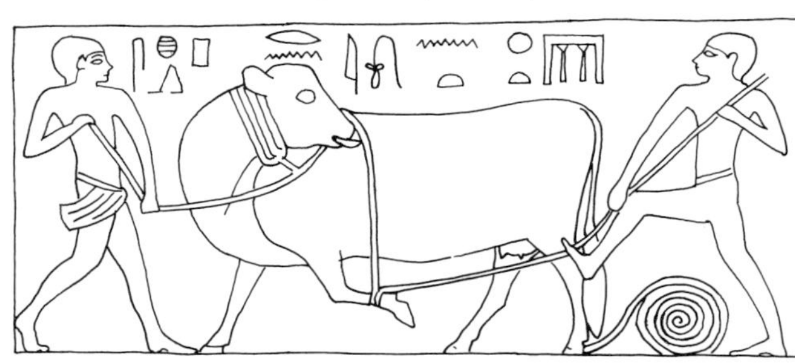

Fig. 9-31 (AB166)

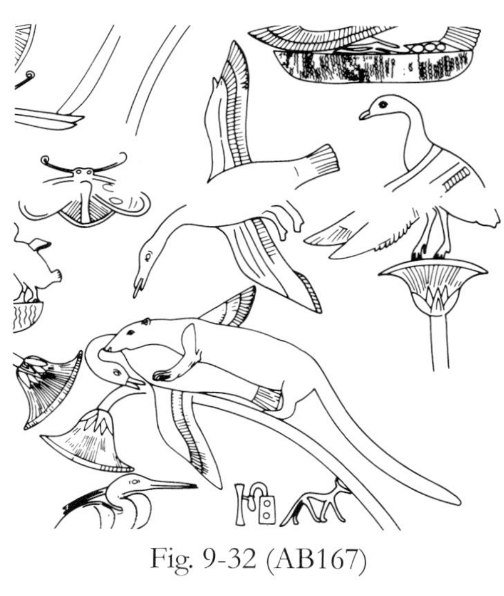

Fig. 9-32 (AB167)

(18), pied kingfishers (17), unidentified species (6), little egrets (3), and grey herons (1). There is thus a strong tendency for either geese or pied kingfishers to exhibit this behaviour in wall reliefs; indeed, in many tombs, these two species often appear side by side, defending their broods.

Egyptian geese and many duck species nest upon mounds of grass or reeds situated on the ground. Both ducks and geese usually form strong pair bonds when breeding. Nestlings are cared for by both parents, with whom they stay for many weeks after they have fledged. Waterfowl defend their young vigorously. If a potential predator approaches their nests, they will attempt to drive it away by flying towards it at speed to then bite it and beat it with their wings. Shelducks, of which the Egyptian goose is a member, possess bony sections on their wings that they also bring to bear when striking.[97]

As noted earlier in this study, pied kingfishers do not nest in the open as represented; however, the young stay close to their parents for some time after leaving the nesting burrow. They live in family groups in which juveniles from previous breeding seasons assist their parents to rear their younger siblings. When they detect a predator nearby, the adult birds fly in and circle it while flapping their wings in an exaggerated fashion and alarm calling. Pied kingfishers with eggs or young have been observed to swoop down in groups of two or more to strike snakes and mongooses on the tail, and a monitor lizard upon the head.[98]

Wall scenes occasionally show only a single bird engaged in parental defence; more commonly, however, two birds attack each intruder, suggesting that they are breeding pairs. The attacking birds either hover or dive at the predators. Hovering birds are usually depicted upright with both their wings above their backs; diving birds are often inverted and plummet downward with their wings outstretched (see fig. 9-25). Together, these two flying postures convey very effectively the fluttering swirl of activity that characterises nest defence. In the tomb of ***mḥw***, three geese dive, hover, and bite a mongoose upon its back as it attacks one of their nestlings (AB133). Elsewhere on the same wall, two pied kingfishers peck the eyes and tail of a genet as a third individual flies above their nest and adopts the distinctive front-facing hover usually associated with hunting in this species (AB132; see Chapter 8).

In early scenes, the parent birds do not make contact with the predators[99] (e.g., see figs. 8-10 and 8-14 in Chapter 8). In the late 5th dynasty, however, they begin to assault the intruders physically. This coincides closely with the first occurrence of predators biting nestlings (see Chapter 8). The earliest example appears in the tomb of ***nfr-jrt-n.f*** (see fig. 4-17 in Chapter 4) in which a pied kingfisher pecks an Egyptian mongoose on the rump, precisely as they have been observed to do in the wild. In later scenes, both genets and mongooses are pecked on the muzzle (AB106, AB122, AB130), ears (AB116 to AB118, AB128; see fig. 9-26), eyes (AB117, AB126, AB132, AB139), back (AB124, AB125, AB127, AB129, AB133; see figs. 8-12 and 8-16 in Chapter 8), and tail (AB132 and AB134). In the tomb of *ḥzj*, an egret also scratches the snout and back of a genet while it pecks its ears (AB128; see fig. 9-26). In another example from the same scene, a goose appears to use its feet to grab its nestling from the jaws of a genet (AB129; see fig. 4-16 in Chapter 4). Geese have small claws on the toes of their webbed feet, but it seems unlikely that they could use these to manipulate an object in this manner.

Most of the nests in these scenes contain three nestlings, yet the depicted species usually produce more young than this. Egyptian geese normally rear clutches of eight to nine goslings,[100] while pied kingfishers brood four to five nestlings.[101] A fourth chick is included in a few examples - often plucked from the nest and clasped in the jaws of a nearby predator (e.g., figs. AB115 to AB117, AB123, AB126, AB128, AB132 and AB133); the nests of Egyptian geese in the tombs of ***zšzšt: jdwt*** (AB121) and ***ḥzj*** (AB127) include five young.

---

97 Johnsgard, *Waterfowl*, 39.

98 Roselaar, "Pied kingfisher" in Cramp-Simmons, *Handbook iv*, 728.

99 It should be noted, however, that in a damaged relief from the early 5th dynasty mortuary complex of Sahure, a pied kingfisher(?) appears to peck the snout of a common genet as it ascends a papyrus stalk. See *Sahure ii*, pl. 74.

100 Roselaar, "Egyptian goose" in Cramp-Simmons, *Handbook i*, 450.

101 Roselaar, "Pied kingfisher" in Cramp-Simmons, *Handbook iv*, 729.

In an early example from the Giza tomb of *snḏm-jb: jntj* (AB101; see fig. 9-25), a bird dives toward an approaching genet to defend its mate incubating eggs upon a nest, rather than young.[102] The sitting bird holds its wings forward in a threatening pose, a display that is associated overwhelmingly with incubation (see *Wing display*, above). In all of the remaining wall scenes, adult birds only attack predators when they are protecting nestlings. As noted above, studies have shown that birds increase their defensive behaviour and become more aggressive towards predators after their eggs have hatched and as their nestlings develop. Remarkably, this is supported by the imagery in Old Kingdom wall reliefs for while birds incubating eggs simply engage in threat displays toward intruders, those depicted with young escalate their defensive behaviour to savagely attack the approaching predators.

9.4 PROTESTING

Many animals vocalize when they are threatening others. Calling may be part of the display itself, whereby the production of sounds helps to draw attention to the exhibiting animal. Vocalizations can also be a product of the animal's aggressive state, a spontaneous expression of their anger and frustration. Animals will vocalize when they are assaulted. This is often due to pain (see Chapter 8), but it can also be motivated by fear if their attempts to flee the situation are thwarted.

Bulls

nb.j-m-ꜣḫtj (AB140: Giza, IV.6-V.1)
pḥn-wj-kꜣ.j (AB141: Saqqara, V.6-8E)
nṯr-wsr (AB142: Saqqara, V.6L-7)
ṯjj (AB143-144: Saqqara, V.8-9)
ꜣḫtj-ḥtp (AB145-146: Saqqara, V.8L-9E)
ptḥ-ḥtp II: ṯfj (AB147-149: Saqqara, V.9M-L)
sšm-nfr IV (AB150: Giza, V.9-VI.1)
ḥtp-kꜣ.j: tp-kꜣ.j (AB151: Giza, V.9-VI.1)
sḫm-ꜥnḫ-ptḥ (AB152: Giza, V.9-VI.1?)
ṯtw: kꜣ.j-nswt (AB153: Giza, VI)
mrrw-kꜣ.j: mrj (AB154-157: Saqqara, VI.1M-L)
ꜥnḫ-m-ꜥ-ḥr: zzj (AB158: Saqqara, VI.1L-2E)
ḫntj-kꜣ.j: jḫḫj (AB159-160: Saqqara, VI.2M)
ꜣḫtj-ḥtp (AB161: Saqqara, VI.5?)
nb-kꜣw-ḥr: jdw (AB162: Saqqara, V.9 r/u VI.7)
zꜣt-jn-ttj (AB163: Saqqara, VI.7-FIP)

Among the many depictions of cattle in wall scenes are images in which individual bulls are grasped or roped and thrown to the ground prior to slaughter. The animals resist vigorously these physical assaults, moving angrily as they attempt to dislodge the shackles and men that tie them down. They also frequently protrude their tongues prominently.

These reactions are consistent with those shown by bulls when restrained, especially those that have had little experience with being handled; when approached, they will attempt to escape, and if caught, they will struggle.[103] In particular, in situations involving threat and conflict, bulls give a distinctive vocalization known as a 'roar' or screaming 'bellow'.[104] When uttering this particular sound, the mouth is held open and the tongue is frequently raised and protruded (see pl. 9P and Colour Plate A).[105] I suggest that the protesting bulls in wall scenes are thus depicted expressing this piercing sound of anger.

Images of bulls protruding their tongues while resisting were found in 16 tombs ranging from the mid-4th to the late 6th dynasty. In some scenes, the animals are set upon by groups of men who pull at their tails, horns, and legs to cause them to fall (AB140, AB150, AB154, and AB158; see fig. 9-27). This process is often assisted by the use of ropes (AB143, AB145, AB151, AB161 to AB163; see fig. 9-28). These may be used to bind the hind legs together to prevent escape or attached to a foreleg, which, upon lifting, causes the animal to topple. Ropes are also tied to their horns. In each of these examples, the bulls lift their heads away from their oppressors or lean well forward in a threatening manner (see *Head display*, above), while simultaneously protruding their tongues. In three late depictions, the animals

102 A bird flapping its wings in response to an approaching genet in the tomb of *jtj-sn* (AB99) may also be protecting eggs, rather than young. Alternatively, the bird itself may be a nestling. A firm conclusion cannot be drawn due to the poor state of the relief.

103 Ewbank, R. (1968). "The behavior of animals in restraint" in Fox, *Abnormal Behavior*, 166.

104 The '(m)enh' call. See Kiley, *Z. Tierpsychol.* 31, 192-193.

105 Hall, S.J.G., Vince, M.A., Walser, E.S. and Garson, P.J. (1988). "Vocalisations of the Chillingham cattle", *Behaviour* 104(1-2), 79.

elevate their tails, an action that is performed by bulls during fights and threatening, and also while running (AB161 to AB163; see fig. 9-30).[106]

Some bulls with protruding tongues have been brought to the ground (AB153, AB155 to AB157, AB159, and AB160; see fig. 9-29). In slaughter scenes, the tongues of butchered animals are frequently depicted as their bodies are severed and dismembered (see Chapter 4).[107] I suggest, however, that the fallen animals considered here are not yet dead, but instead fighting the attempts by men to rope their legs together or hold them down; their protruded tongues thus indicate that they are voicing their displeasure at this treatment. Unlike images of slaughtered animals, whose tongues invariably touch the substrate, the heads of these vocalizing animals are lifted up so that their tongues are free.

The remaining images show animals whose movements are also thwarted in some manner. Some are interrupted while turning their heads or grooming (AB142, AB144, and AB152; see fig. 6-5 in Chapter 6); others appear to be objecting to the behaviour of herdsmen who force them to move in a particular direction (AB141, AB146, AB148, and AB149). One animal vocalizes at a hunter who has lassoed it (AB147).

It is important to note that in each of these examples, the animal's tongue fills its entire mouth cavity. The images are thus not a veridical representation of the appearance of roaring bulls in which the organ is seen as an isolated shape within their gaping mouth, but instead match closely other depictions noted throughout this study in which calling animals effectively 'poke out' their tongues from between their lips. Consequently, although the scenes of protesting bulls may indeed reproduce their screaming bellow, I suggest that the artists responsible for them have nevertheless employed the same iconographic device - the protruding tongue - used elsewhere for conveying to the viewer the idea (and by extension, the sensation) of loud, abrasive animal sounds.

Cows

ḥtp-ḥr-ꜣḫtj (AB164: Saqqara, V.6-8E)
kꜣ.j-m-rḥw (AB165: Saqqara, V.8)
rꜥ-špss (AB166: Saqqara, V.8M)

In milking scenes, the hind legs of cows are frequently bound together (see Chapter 11), however only six images were found in which they are depicted while being roped. In three of these, the cows also protrude their tongues (AB164 to AB166; see fig. 9-31).[108] In each example, one of the animals' forelegs is lifted from the substrate via ropes that are pulled by men standing behind them.

I suggest that, like the images of restrained bulls, these animals should be perceived as vocalizing in response to their treatment. Cows give threatening 'bellows' less frequently than bulls,[109] but in frightening situations they are known to deliver a specific, high-pitched call.[110] Their tongues are not usually apparent when giving this or other vocalizations (see pl. 9Q; although see Chapter 11). The vocal behaviour of the depicted cows may thus either be an aggressive response to their restraint or an expression of their fear, a state that artists have conveyed by depicting them – like bellowing bulls - with protruded tongues.

Birds

nj-kꜣw-jzzj (AB167: Saqqara, VI.1M)

Among the many unique images found in the Saqqara tomb of *nj-kꜣw-jzzj*, is the depiction of an Egyptian mongoose that attacks an adult bird (AB167; see fig. 9-32). As noted above, mated pairs

106 Kiley-Worthington, M. (1976). "The tail movements of ungulates, canids and felids with particular reference to their causation and function as displays", *Behaviour* 56(1-2), 78.

107 See Eggebrecht, *Schlachtungsbräuche*, 70.

108 Note that the inscription in the scene from tomb of *rꜥ-špss* (AB166; fig. 9-31) refers to an ox, however the animal clearly has an udder!

109 Albright-Arave, *Behaviour of Cattle*, 47-48.

110 The 'men(h)' call. See Kiley, *Z. Tierpsychol.* 31, 191-192.

of birds are represented frequently while defending their nestlings from predators; here, however, it appears that a goose has flown too close to the intruder and has itself been captured. Above the unfortunate bird, another goose dives downward with its wings outstretched. Its behaviour indicates that this must be the mate of the seized goose and that it is trying desperately to deter the predator. As the defending bird plummets, it protrudes its long tongue directly at the biting mouth of the predator. Note that the bird does not open its beak to reveal its tongue; rather, it emerges from its closed bill. Geese are physically incapable of doing this, as their tongues do not exceed the length of their beaks (see pl. 9R and Colour Plate C).[111] The artist has thus not only drawn something that does not occur in nature, but has deliberately emphasized this feature.

Many bird species produce vocalizations in distressing situations. Egyptian geese give loud honks when alarmed,[112] and greylag geese produce a high-pitched vibrating call when under attack.[113] I suggest that the artist has represented the defending goose with a protruding tongue in order to convey its calling behaviour. The unusual depiction, in which the bird's tongue extends well beyond its closed beak,[114] both draws the viewer's attention to this feature and acts as a pointer to the source of its vocal agitation. The drama of this scene is consequently heightened by this small addition for not only do we observe the valiant efforts of the bird, we also effectively hear its plaintive cries.

SUMMARY

- Variations in the position of heads, legs, ears, mouths, and wings that occur in appropriate contexts indicate that many depicted animals in wall scenes exhibit threatening behaviour. The placement of the horned heads and ears of some animals, however, may have been dictated occasionally by overhead inscriptions. Consequently, it cannot always be determined whether the ancient Egyptians understood fully the significance of some subtle postural cues.
- The aggressive behaviour of birds is particularly well observed. The threatening wing posture adopted by all nesting birds indicates that incubation *per se* and individual differences between species were less important to artists than conveying the general concept of avian egg defence. Artists also observed a clear distinction in their depiction of protective behaviour: birds adopt threat postures to defend their eggs, but actually attack predators when they have young. This matches avian behaviour in the wild.
- Some postures, such as yawning by hippopotamuses while seated, may have been a deliberate, apotropaic attempt by artists to prevent the figures of the aggressive, and thus potentially dangerous, creatures from magically causing harm.
- Aggressive animals are frequently depicted with a protruded tongue. Loud vocalizations are a striking part of the threatening and defensive behaviour of many species. Some artists appear to have felt that this aspect of aggressive behaviour was as significant as its visual appearance and consequently found a method by which it could be conveyed clearly and unambiguously to the viewer.

---

[111] A few bird species are able to extend their tongues in this manner, notably woodpeckers and hummingbirds. See Leahy, *Companion*, 730-732.

[112] Roselaar, "Egyptian goose" in Cramp-Simmons, *Handbook i*, 449.

[113] Roselaar, "Greylag goose" in Cramp-Simmons, *Handbook i*, 419.

[114] Bird species are able to produce vocalizations while their beaks are closed or open. When their beak is open, their tongue is sometimes (but not always) visible between their parted mandibles.

CHAPTER 10

SEXUAL BEHAVIOUR

DEFINITION

Sexual behaviour includes all activities associated with acquiring a mate and the act of copulation. Some species experience a lengthy courtship period prior to mating. The postures adopted by animals during courtship displays are often quite distinctive and sometimes species-specific, as are those associated with copulation.

10.1 COURTSHIP

Prior to copulation, most animals engage in a series of behaviours to assess potential sexual partners. This information may be transmitted via visual, auditory, olfactory, or tactile signals. The actions and motivations of males and females differ. Females often advertise their reproductive state indirectly through olfactory cues, or passively by, for example, allowing a male to approach. The males of many species, however, produce vigorous displays to attract and impress a mate, and also carry out particular activities that allow them to determine the fertility level and receptiveness of females. It is possible that Egyptian pastoralists monitored these sexual behaviours in domesticated animals - as do modern farmers - to assist them with their breeding programmes.

10.1.1 Anovaginal sniffing

Pheromones are released in the skin and urine of female mammals as they become more fertile (a state known as 'oestrus'). The males of many species are often able to sense these olfactory signals and use them to track the reproductive status of potential partners. Pheromones may be detected by tasting the female's urine or by the male placing his snout under her tail and smelling her anovaginal area (see pl. 10A). If he determines that she is in heat, he may subsequently attempt to mount her.

Bulls

ṯtw: k3.j-nswt (SB1: Giza, VI) *zšzšt: jdwt* (SB2: Saqqara, V.9 r/u - VI.1-2E?)

Cattle following one another in tomb reliefs (particularly in offering scenes) are sometimes positioned so closely together that the muzzle of one animal touches the rump of the next in a manner similar to anovaginal sniffing. It is usually impossible to determine whether their placement is sexual or simply due to spatial constraints within the register, but in two scenes dating from the late 5th and early 6th dynasties this apparent behaviour by bulls is confirmed by both the accompanying inscriptions and the response of the cows.

In both examples, a bull stretches its head forward and places its nostrils directly beneath the tail of a cow standing before it.[1] The males thus appear to sniff the cows' external genitalia, a behaviour that bulls exhibit frequently as their female herd-mates approach oestrus (pl. 10A).[2] The cows in each scene react by lifting their tails - holding it directly out from her body in the tomb of *ṯtw: k3 j-nswt* (SB1; see fig. 10-1) and arching it away from her rump in the tomb of *zšzšt: jdwt* (SB2; see fig. 10-2).[3]

---

1 Neither of the cows possesses an obvious udder. It should be noted, however, that the udders of young heifers are considerably smaller than those of older cows that have produced a calf. The scenes may thus represent the first matings of immature cows.

2 Hafez, E.S.E., Schein, M.W. and Ewbank, R. (1969). "The behaviour of cattle" in Hafez, *Behaviour of Domestic Animals*, 258; Phillips, *Cattle Behaviour*, 137.

3 The cocked tail of the *zšzšt: jdwt* cow is biologically more accurate. The *ṯtw: k3.j-nswt* artist seems, in fact, to have positioned the female's tail in such a way as to playfully suggest one of the bull's horns, which is twisted against his flank.

Cows raise their tails in such a manner during oestrus;[4] the response of the depicted females is thus indicative of their receptive state. The cow in the tomb of *ṯtw: kꜣ.j-nswt* also looks back at the bull with interest, suggesting that she welcomes his attention.

Bulls will endeavour to copulate with cows as soon as these have entered full oestrus; vulval probing by a male therefore often signals an imminent mounting attempt.[5] The inscriptions associated with each sniffing scene indicate that the Egyptians understood the sexual nature of the interaction and the behaviour that could immediately follow it, as in both examples a herdsman is directed, with some urgency, to tie up the cow and separate her from the bull.[6] The texts imply, then, that the male is not to be trusted and that the cow must be kept from him.

Two conclusions follow from this. First, it appears that the ancient Egyptians were acutely aware of the reproductive behaviour of their cattle and that they could identify some of the signs of oestrus in cows.[7] Modern farmers also look for behavioural changes as an aid to breeding their cattle; however, the ability to detect oestrus reliably is notoriously difficult. The images and their labels also reveal that matings between some cows and bulls were not wanted. Egyptian herdsmen clearly exerted some control over the breeding of their animals and may consequently have chosen particular bulls to serve individual cows. It is likely that these practical concerns led to their detailed knowledge of the subtle signs and expressions of sexual behaviour in cattle.

10.1.2 Chin-resting

Attending males follow closely females that are approaching full oestrus and may often attempt to mount them. The females, however, will not stand still until they have come into heat. Some males test the female's readiness to mate by chin-resting - that is, by placing their muzzle on her back or upon the root of her tail. If she is receptive, she will accept this attention and hold her body rigid; mounting follows soon after. This behaviour is exhibited by many farm animals, but is particularly common in cattle (see pl. 10B).[8]

Bulls

jj-mrjj (SB3: Giza, V.3L)
ṯjj (SB4: Saqqara, V.8-9)
kꜣpj (SB5: Giza, V.9-VI)
ꜣḫtj-ḥtp (SB6: Saqqara, VI.5?)

Four tomb scenes may depict chin-resting by cattle, as in each case, a bull, while following another animal, lifts its head and stretches its neck over its companion's rump (SB3 to SB6). The sex of the

---

4 Kiley-Worthington, M. (1976). "The tail movements of ungulates, canids and felids with particular reference to their causation and function as displays", *Behaviour* 56(1-2), 78.

5 Vandier (*Manuel v*, 79-80) draws attention to the curious depiction of a *cow* approaching a bull in tomb No. 2 at Zawyet el-Amwat (Sauiet el Meitin). The scene (see *LD ii*, pl. 105) is virtually identical to those described here, as one animal places its muzzle directly under the lifted tail of another to lick the perianal region. Contrary to Vandier's description, however, the animal engaged in this probing behaviour has an erect penis, while the object of his attention appears to display testicles. With regard to this last animal, I suggest that either the original artist was confused about the sex of the animals involved or, more likely the modern copyist has mistaken the udder of the recipient for a scrotum. Indeed, the inscription above this scene confirms that a male and female are depicted as it reads, *mnjw pw m[k] rdjj nh[p] kꜣ pw ḥr sḳd jdt* 'This herdsman! Behold the one who causes this bull to copulate (by) travelling upon the cow'.

6 'Oh this herdsman, tether the cow quickly as you separate this bull'! Macramallah (*Idout*, 17) translated the text associated with the *zšzšt: jdwt* scene as *nhp* 'saillir' ('copulate'), however re-examination of this tomb by Kanawati and Abder-Raziq (*Iynefert*, pl. 54) has revealed more of the inscription (see fig. 10-2). Similarities between this fragmentary text and the second half of the fully preserved label that accompanies the scene in the tomb of *ṯtw: kꜣ.j-nswt* (fig. 10-2) indicate that it can instead be read as *m wp[=k] kꜣ [p]n* 'as you separate this bull'. It seems likely, then, that the first half of the text (*jj mnjw pw ḥnt jwꜣt wn* 'Oh this herdsman, tether the cow quickly!') was originally inscribed above the *zšzšt: jdwt* bull in the area that is now destroyed.

7 Precopulatory sniffing by goats is also clearly represented in Niuserre's Chamber of the Seasons at Abu Ghurab. See Edel, E. (1963). "Zu den Inschriften auf den Jahreszeitenreliefs der 'Weltkammer' aus dem Sonnenheiligtum des Niuserre: Part II", *NAWG* 1963, 180-181 and fig. 11.

8 Fraser-Broom, *Farm Animal Behaviour*, 188.

receiving animal is sometimes difficult to discern, but in the tomb of *k3pj* (SB5; see fig. 10-3), a bull clearly nuzzles the tail of a cow. In a late example from the tomb of *3ḫtj-ḥtp* (SB6), however, a male seems to rub the hindquarters of another male. Such same-sex interactions (including mounting) are common in single-sex herds, either due to the lack of appropriate sexual partners or as an expression of dominance.[9]

The artists responsible for each of these examples appear to have chosen deliberately to position the animals with their heads lifted instead of setting them away from, or overlapping, their companions. Nevertheless, a sexual meaning cannot be confirmed, as the juxtaposition of the animals may simply be the result of poor planning by the artists or spatial constraints within the register.

10.1.3 Nudging

A close association between males and females marks the courtship behaviour of most animal species. The males are generally more active in initiating and maintaining this contact, as a female's acquiescence at their approach is often a sign of her willingness to mate. Females usually signal their acceptance of a male in a passive manner (e.g., by standing still), but some may express their interest more overtly by pushing their body against his.

Cows

jnw-mnw (SB7: Saqqara, VI.2E)

Oestrus cows exhibit many signs that they are ready to mate. These can include increased vocalizing, general restlessness, and mounting of other cows. They will also actively seek out a bull, to which they react in an overtly physical manner - mounting (pl. 10C), sniffing and licking his genitalia, and nudging him - in order to attract his attention.[10]

A scene in the tomb of *jnw-mnw* (SB7; see fig. 10-4) may represent this attention-seeking behaviour. A cow[11] standing behind a bull presses her muzzle against the base of his tail. The male is in a highly agitated state - pawing the ground and vocalizing. This animal is probably reacting to the presence of a mating pair further along the register (see SB23; see fig. 7-4 in Chapter 7), as studies have shown that observing the sexual behaviour of others stimulates bulls. The aggressive demeanour of the *jnw-mnw* bull is thus an indication of its robust sex drive. This, in turn, would explain why a cow in heat might be drawn to such a male and endeavour to attract his attention by nosing at his hindquarters.

Goats

nj-ʿnḫ-ḫnmw/ḫnmw-ḥtp (SB8: Saqqara, V.6L-7) *3ḫtj-ḥtp* (SB9: Saqqara, V.6-8E)

Two mid-5th dynasty wall scenes were found in which female goats appear to nudge or rub themselves against nearby males. In one image, a doe pushes her snout against a male's hindquarters (SB8), while in the other the female rests her head heavily upon the back of a buck (SB9; see fig. 10-5). This behaviour matches that of oestrus does. Male goats produce a distinctive, musky odour that becomes more intense during the breeding season; when oestrus females are exposed to this pheromone they frequently respond by rubbing their body against that of the buck (cf. pl. 10D).[12]

Groups of goats in wall reliefs are often represented using 'lateral layering', in which the body of each individual partially overlaps that of the next (see Chapter 5). This composition often results in the head of each goat touching or resting upon the body of one of its companions. It is thus impossible to determine with certainty whether the placement of the females in the scenes in question was intended to represent a sexual interaction.

---

9 *ibid.*, 328-330.
10 Phillips, *Cattle Behaviour*, 123.
11 The animal's sex might be questioned, as it does not possess an udder, however a mating scene on the same register depicts a cow that also lacks this feature (see fig. 7-4 in Chapter 7). See also n. 1.
12 Jainudeen, M.R., Wahid, H. and Hafez, E.S.E. (2000). "Sheep and goats" in Hafez-Hafez, *Reproduction*, 177.

10.1.4 Foreleg kicking

A common courtship display among many species of hoofed animals is 'foreleg kicking', in which a male, while standing or walking behind a female, lifts one leg stiffly to touch her hind legs.[13]

Goats

sšm-nfr IV (SB10: Giza, V.9-VI.1)

A relief fragment from the Giza tomb of *sšm-nfr IV* depicts a male goat standing with one foreleg raised so that it touches the hind legs of the animal in front of it (SB10; see fig. 9-20 in Chapter 9).

Foreleg lifting is one of the many ways in which male goats express their sexual interest in a female (see pl. 10E). As he approaches a doe, a male may lift one forelimb and kick it towards her a number of times while he utters a low-pitched vocalization.[14] The posture of the buck on the *sšm-nfr IV* fragment clearly matches this distinctive behaviour; however, the animal towards whom this action is directed appears to be a male.

The second animal in this scene displays features in common with both sexes (see also Chapter 9): its horns resemble those of a mating female in the register below and, like her, it is beardless, but it seems also to display male genitals and has adopted a fighting stance. Inspection of a photograph of the fragment revealed that the area depicting the apparent genitalia is indistinct (either unfinished or damaged);[15] it may be possible, then, that the artist intended to represent a male courting a female, which has responded aggressively to his attentions. A mating pair depicted in the register below these animals supports the notion that their interaction may be sexual. Scientific studies have revealed that, like bulls, observation of the copulation behaviour of other males can increase the sexual performance of bucks.[16] Foreleg kicking by the depicted male is thus in accord with, and might be seen to have been motivated by, the sexual activities to which it is juxtaposed. Nevertheless, as the interaction between the buck and his companion can also be viewed as antagonistic (see Chapter 9), the true nature of this intriguing representation cannot be confirmed.

10.2 MATING

Mating or 'copulation' is the union of the sexual organs of male and female animals for the purpose of insemination and internal fertilization. The postures adopted during copulatory activity vary somewhat from species to species, but in many quadrupedal animals, the male rears onto his back legs and leans against the female's hindquarters in order to achieve intromission. Prior to actual mating, some males often test the responsiveness of females by 'mounting' - that is, rearing without genital contact. Many animals copulate frequently over a period of hours or days, while the female is receptive. There would thus be many opportunities for the ancient Egyptians to have witnessed this behaviour.

The earliest known examples of mating animals in wall scenes occur in the Chamber of the Seasons in the mid-5th dynasty sun-temple of Niuserre at Abu Ghurab.[17]

---

13 Walther, F.R. (1977). "Artiodactyla" in Sebeok, *How Animals Communicate*, 705-706.

14 Hart, *Behavior of Domestic Animals*, 113.

15 Junker, *Giza xi*, pl. 24 [b].

16 Price, E.O., Smith, V.M. and Katz, L.S. (1984/85). "Sexual stimulation of male dairy goats", *App. Anim. Beh. Sci.* 13, 83-92.

17 Salima Ikram has suggested that a pair of sheep in the valley temple of Sahure represents the earliest depiction of copulating animals (see Ikram, S. (1991). "Animal mating motifs in Egyptian funerary representations", *GM* 124, 54 and 57). As revealed in Chapter 5, however, I propose that apparent depictions of mating by herding sheep were instead intended to represent a leaping motion. Consequently, I believe that the earliest examples of mating behaviour are to be found in the sun-temple of Niuserre.

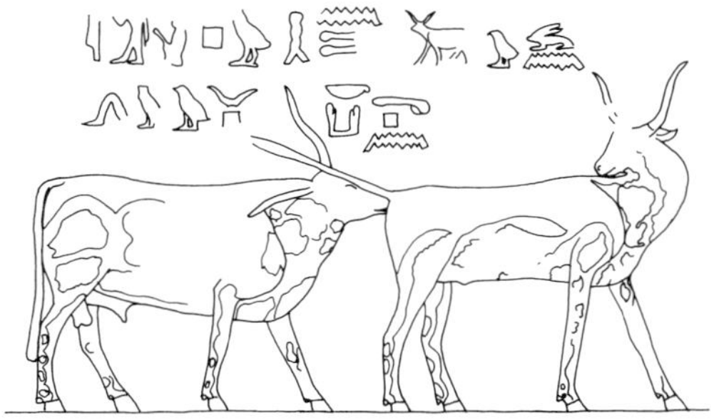

Fig. 10-1 (SB1)

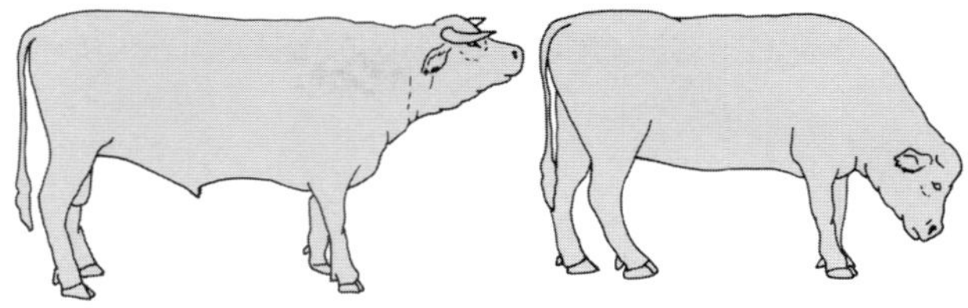

Pl. 10A Bull sniffing a cow
After: Hart, *Behavior of Domestic Animals*, fig. 3-13 A

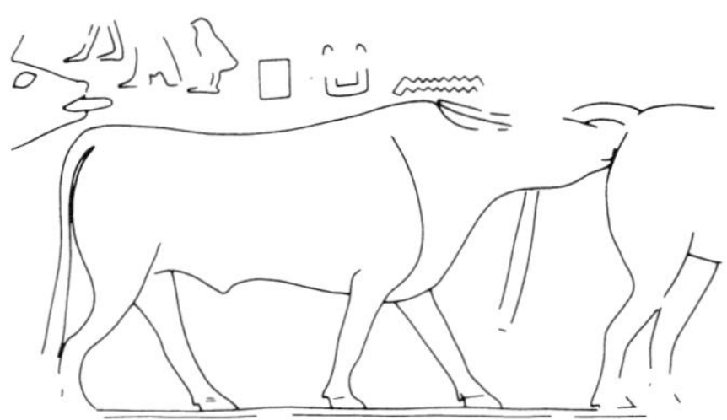

Fig. 10-2 (SB2)

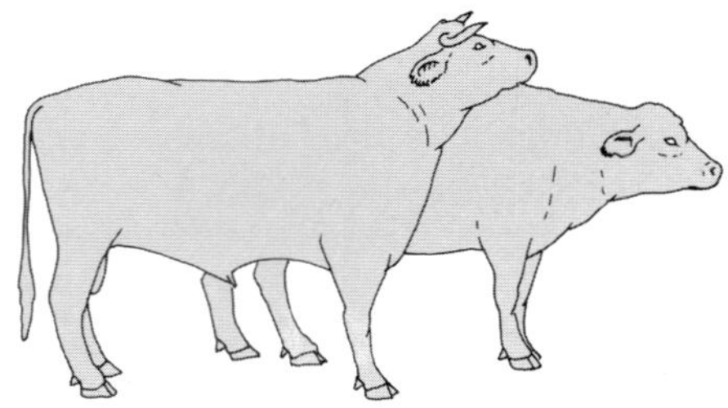

Pl. 10B Assessing readiness to mate
After: Hart, *Behavior of Domestic Animals*, fig. 3-13 B

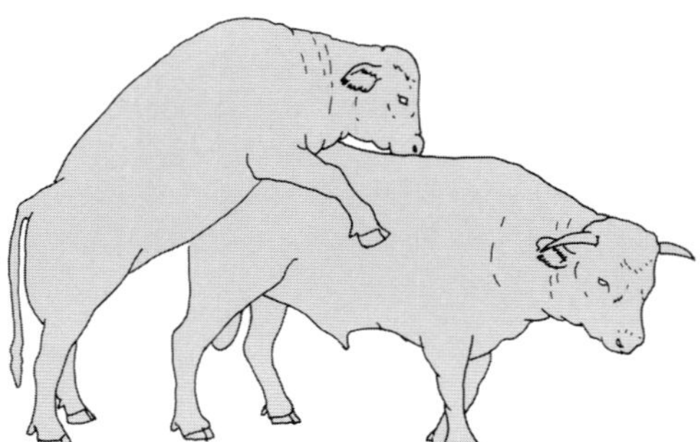

Pl. 10C Cow mounting a bull
After: Hart, *Behavior of Domestic Animals*, fig. 3-8

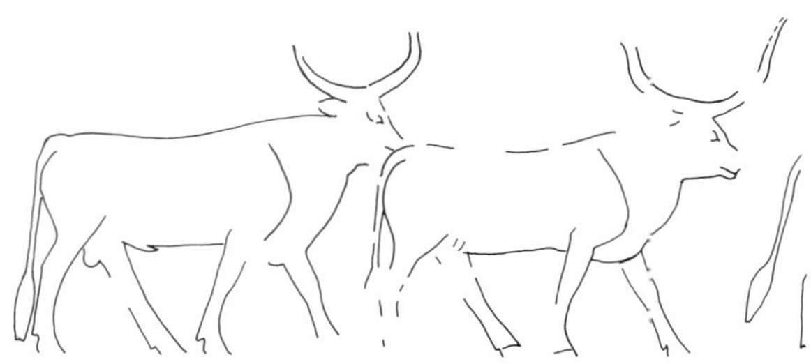

Fig. 10-3 (SB5)

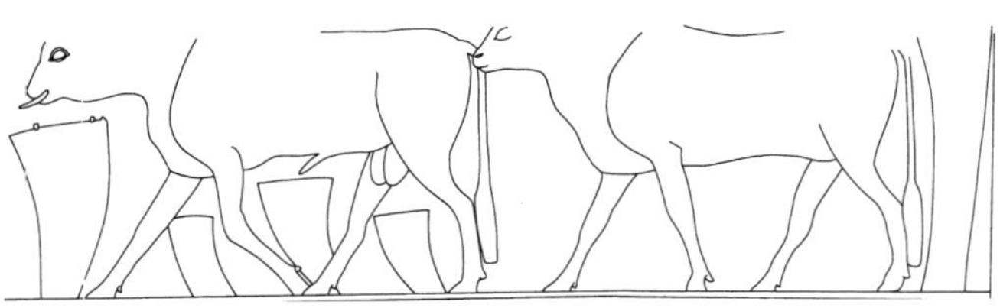

Fig. 10-4 (SB7)

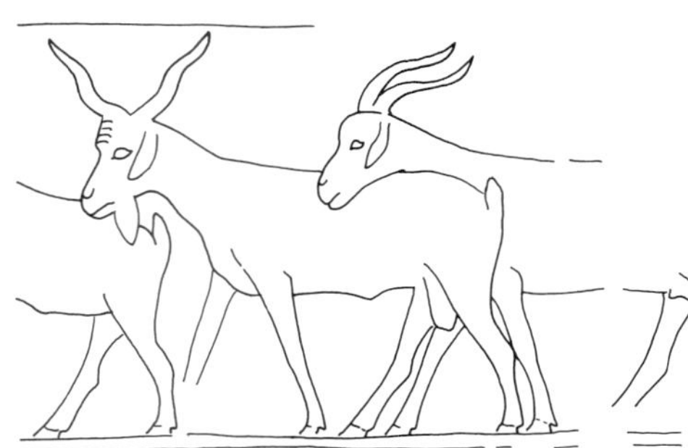

Fig. 10-5 (SB9)

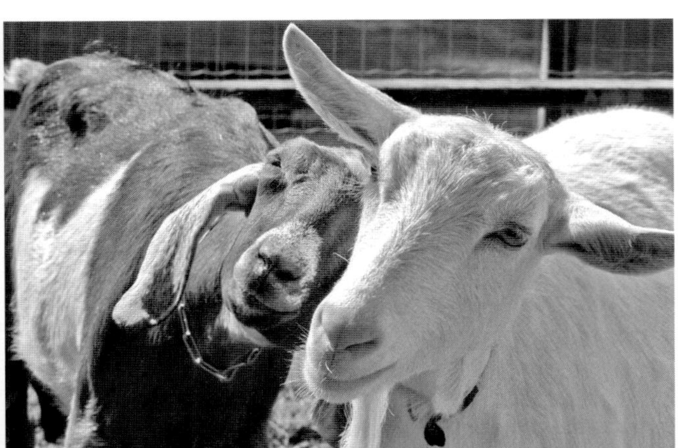

Pl. 10D Goat rubbing partner.

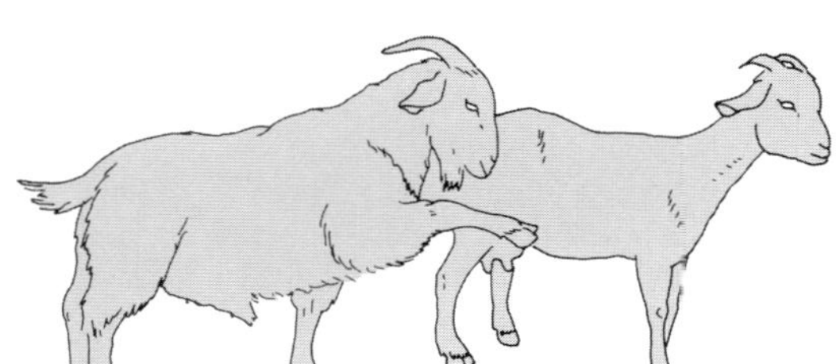

Pl. 10E Foreleg kicking
After: Hart, *Behavior of Domestic Animals*, fig. 3-15 B

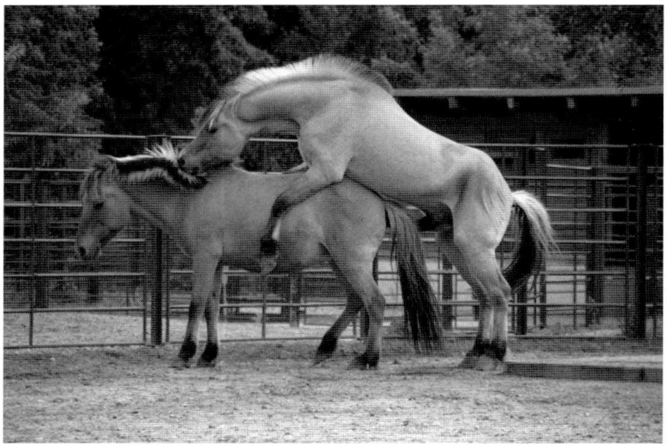

Pl. 10F Mating posture of equids
© iStockphoto.com/Andy Nowack

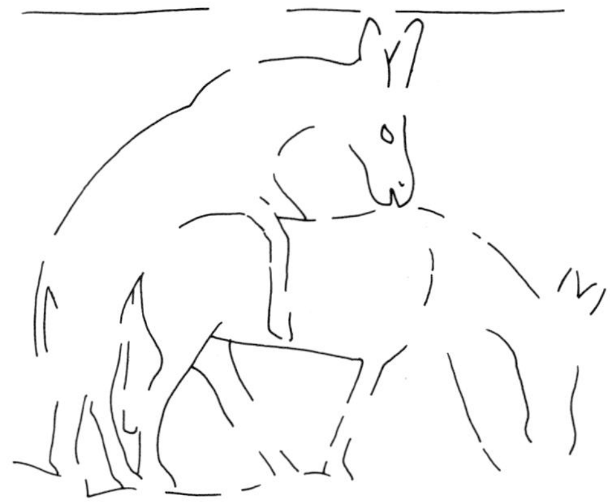

Fig. 10-6 (SB11)

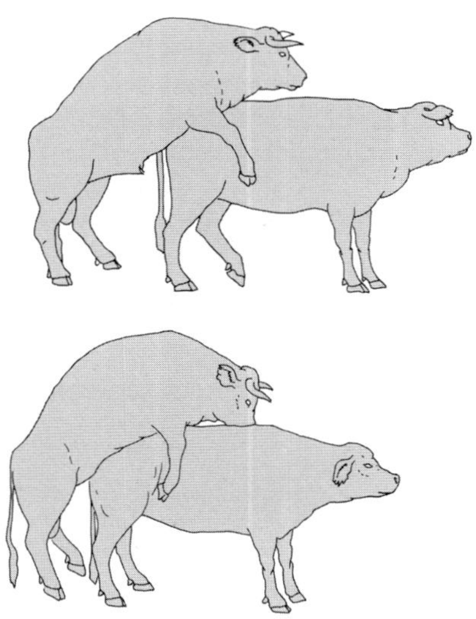

Pl. 10G Mounting and mating by cattle
After: Hart, *Behavior of Domestic Animals*, fig. 3-13 C, D

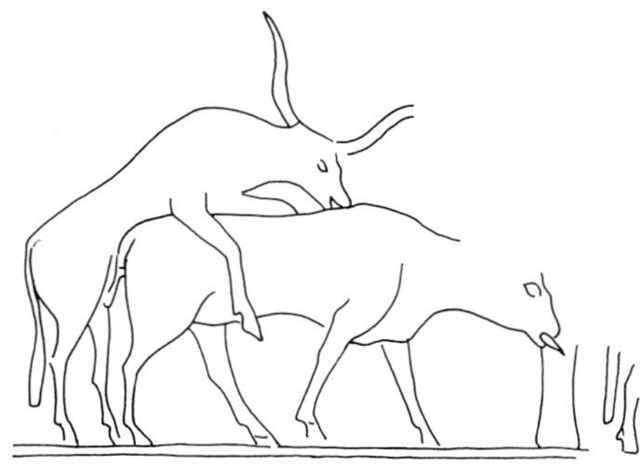

Fig. 10-7 (SB15)

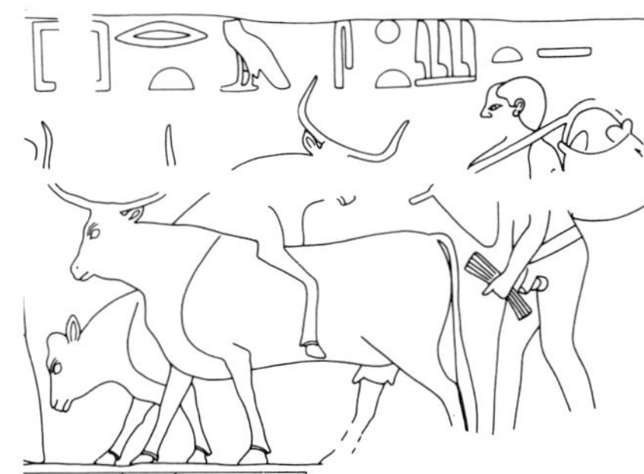

Fig. 10-8 (SB22)

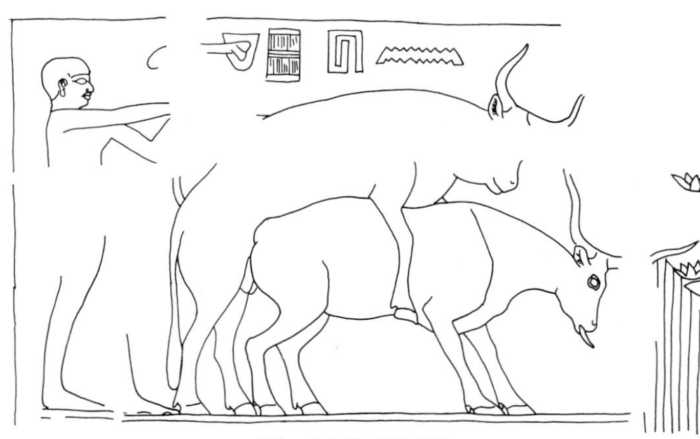

Fig. 10-9 (SB18))

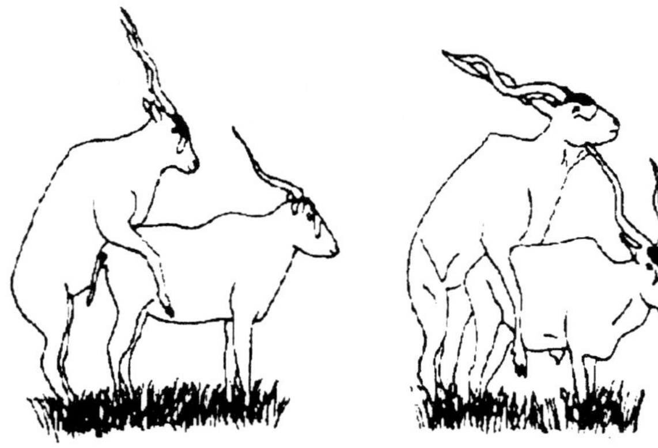

Pl. 10H Mounting and ejaculation postures of addax
Reprinted from: Manski, D. (1991). *App. Anim. Beh. Sci.* 29, fig. 3 E, F. By permission of © Elsevier

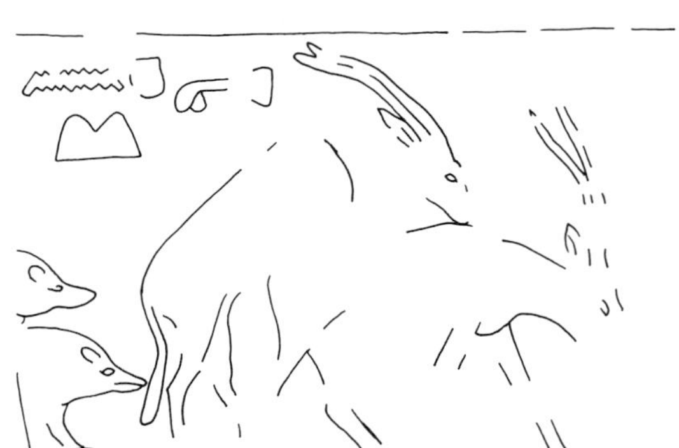

Fig. 10-10 (SB24)

Wild ass

nj-m3ˁt-rˁ [jzzj-mr-nṯr] (SB11: Giza, V.9-VI.1)

A pair of mating equids is depicted in a fragmentary desert scene located in the Giza tomb of *nj-m3ˁt-rˁ [jzzj-mr-nṯr]* (SB11; see fig. 10-6). The poorly preserved image provides few definitive morphological details, but the desert setting would suggest that the animals are wild asses, rather than domesticated donkeys.[18] The female stands with her head and neck lowered while the male rears up to grasp her with his forelegs. The male also bends his neck steeply so that his muzzle touches the female's shoulders. This pose is unusual and may represent neck biting, a common behaviour exhibited by equid males during copulation in which the female's withers or mane are grasped (cf. pl. 10F). Wild ass males, in particular, are reported to bite the female lightly upon the shoulders as they mate.[19] The *nj-m3ˁt-rˁ [jzzj-mr-nṯr]* image is thus a biologically accurate representation of the mating behaviour of this species.[20]

Cattle

rˁ-m-k3.j (SB12: Saqqara, V.6? r/u - V.6-7)
ḥtp-ḥr-3ḫtj (SB13: Saqqara, V.6-8E)
3ḫtj-ḥtp (SB14: Saqqara, V.6-8E)
jtj-sn (SB15: Giza, V.6-8?)
jrj-n-k3-ptḥ (SB16: Saqqara, V.6-9)
snḏm-jb: jntj (SB17: Giza, V.8M-L)
sḫm-ˁnḫ-ptḥ (SB18: Saqqara, V.8-9)
nfr-ḥr-nj-ptḥ (SB19: Saqqara, V.8-9)
k3.j-m-nfrt (SB20: Saqqara, V.8-9?)
nbt (SB21: Saqqara, V.9)
nj-k3w-jzzj (SB22: Saqqara, VI.1M)
jnw-mnw (SB23: Saqqara. VI.2E)

During this study, more examples were found of mating cattle than of any other species. The cows and bulls adopt characteristic, and biologically accurate, poses. In some images, the female's legs are braced and her neck is stretched forward, conveying well her response to the weight upon her back (e.g., SB14, SB17, and SB18; see fig. 9-5 in Chapter 9). The males largely hold their torsos upright, above those of the females. In the tomb of *sḫm-ˁnḫ-ptḥ* (SB18; see fig. 10-9), however, the bull leans well forward so that both his belly and brisket rest upon the cow's back, and in the earlier tomb of *jtj-sn* (SB15; see fig. 10-7), the male touches his muzzle upon the female's shoulders.

When a bull mounts a cow, he first shifts his weight onto his hind legs, lifts his forelegs off the ground, and then moves forward to straddle her back and clasp her flanks. As he manoeuvres into position, his upper body is largely elevated (pl. 10G, upper), however after intromission has occurred, the male leans forward as muscular contractions bring him to climax (see pl. 10G, lower). After ejaculation, he dismounts slowly and in the process, usually draws his chin across the female's back. I suggest, therefore, that most of the scenes considered here (in which the males are upright) depict mounting, rather than the act of copulation *per se*, but that the image in the tomb of *sḫm-ˁnḫ-ptḥ* (fig. 10-9) may represent the moment of ejaculation, while that in the tomb of *jtj-sn* (fig. 10-7) is post-coitus in which the bull drags its chin across its companion's back.

In these last two examples, the cows are protruding their tongues. As noted in Chapter 7, the *jtj-sn* cow extends her tongue so that it touches a food trough before her, suggesting that she may be eating (see fig. 10-7). It is unlikely, however, that the animal would continue to feed while being mated. Instead, I propose that the prominent tongues of both this and the *sḫm-ˁnḫ-ptḥ* cow represent their vocal protest at the extra weight inflicted upon them by the bulls during movements associated with copulation[21] and dismounting. The protruded tongue of the cow in a damaged

18 Keimer identified the pair as wild ass. See Keimer, L. (1948). "Les 'ânes sauvages' abattus par Aménophis II, près de Qadesh", *BIE* 30(1947-1948), 138.

19 Haltenorth-Diller, *Field Guide*, 110.

20 A pair of mating asses has also been found in the Chamber of the Seasons at Abu Ghurab, however the male in this fragmentary scene does not lean forward, but instead remains fully upright. See Edel, *NAWG* 1963, 191-192 and fig. 15.

21 The muscular contractions during coitus can be so strong that the bull's hind legs are lifted off the ground, thereby passing his entire weight onto the cow. See Hafez et al. in Hafez, *Behaviour of Domestic Animals*, 259.

scene in the tomb of *jnw-mnw* (SB23; see fig. 9-6 in Chapter 9) may also indicate that the bull originally rested heavily upon her back.

Only the distinctive leg positions of the animals in the tomb of *rʿ-m-kȝ.j* (SB12) indicate that a mating scene is depicted; that the image represents cattle is suggested strongly by the presence of a cow giving birth behind the two (see PB4 in Chapter 11). Mating pairs in the tombs of *snḏm-jb: jntj* and *jnw-mnw* (see figs. 9-5 and 9-6 in Chapter 9) are each adjacent to bulls that are reacting to their sexual activities by pawing the ground. Finally, the tomb of *nj-kȝw-jzzj* (SB22; see fig. 10-8) offers a curious depiction in which a mounting animal faces in the wrong direction. The artist here may have simply tried to fill the void between the cow in the foreground and the text inscribed above it by inserting the body of another animal, however this does not explain its unusual orientation. The image could instead depict 'coital disorientation', a form of sexual dysfunction in which cows are mounted from the wrong direction.[22] Males with this condition may approach a female at right angles to her body (i.e., at a 90° angle of deviation), or even face in the opposite direction (i.e., at a 135° angle of deviation). The disorder is found in all species of farm livestock, but is particularly common in young and inexperienced bulls. It seems more likely, however, that this depiction of apparent sexual activity has been inspired by scenes of seed treading goats (see Chapter 5) in which tight, milling groups are implied by facing some animals in the opposite direction and/or in a mounted position (cf. GB16). The postures of the *nj-kȝw-jzzj* animals may thus simply represent their close proximity to one another.

Addax

nj-mȝʿt-rʿ [*jzzj-mr-nṯr*] (SB24: Giza, V.9-VI.1)

After an elaborate courtship, male addax copulate by mounting females while holding their head and neck erect (see pl. 10H).[23] The fragmentary depiction of mating addax that appears in the tomb of *nj-mȝʿt-rʿ* [*jzzj-mr-nṯr*] (SB24; see fig. 10-10) is thus incorrect, as it shows the male leaning well forward so that his muzzle touches the female's shoulders. This head posture has been rendered correctly for wild ass (see above) and striped hyenas (see below) in the same tomb, and so it is curious that the artist has made a mistake here. One explanation for the animal's pose might be the length of its horns, for had the artist depicted the male correctly, these would have pierced the register above; however, the horns of a mating scimitar oryx on the same register have been presented this way (see *Scimitar oryx* below), so this was clearly not an impediment.

There is some evidence to suggest that Egyptian artists did not differentiate between addax and other antelope species. The *nj-mȝʿt-rʿ* [*jzzj-mr-nṯr*] animals are labelled *nw-ḏw*, a term that is most reliably associated with images of addax,[24] however this label has also been found above a depiction of a hartebeest in the tomb of *tjj*[25] and in conjunction with both a Sömmering's gazelle and an addax in the tomb of *nj-ʿnḫ-ẖnmw/ẖnmw-ḥtp*.[26] The term *šsȝw*, which is usually inscribed above images of hartebeest,[27] has also been found in conjunction with an addax in the Middle Kingdom tomb of *bȝkt* at Beni Hasan.[28]

Hartebeest copulate in an identical manner to the *nj-mȝʿt-rʿ* [*jzzj-mr-nṯr*] addax - that is, the males bend forward steeply to rest their muzzle upon the female's withers (see pl. 10J). Yet in the tomb of *pḥn-wj-kȝ.j*, a mating hartebeest male holds his body erect in a posture more akin to that of addax (see *Hartebeest* below). It may be possible therefore, that the Egyptians perceived some commonality between these two species that allowed them to render their mating postures interchangeably.

---

22 Fraser-Broom, *Farm Animal Behaviour*, 336-337.

23 Manski, D.A. (1991). "Reproductive behavior of addax antelope", *App. Anim. Beh. Sci.* 29(1-4), 41-45.

24 *Wb ii*, 226. See also Osborn-Osbornová, *Mammals*, 158-160.

25 See Epron et al., *Ti i*, pl. 24.

26 See Moussa-Altenmüller, *Nianchchnum*, pl. 85.

27 *Wb iv*, 543. See also Osborn-Osbornová, *Mammals*, 171-173.

28 See Champollion, *Monuments de l'Égypte iv*, pl. 428.

Scimitar oryx

nj-m3ˁt-rˁ [*jzzj-mr-nṯr*] (SB25: Giza, V.9-VI.1)

A unique depiction of mating scimitar oryx located in the Giza tomb of *nj-m3ˁt-rˁ* [*jzzj-mr-nṯr*] (SB25; see fig. 10-11) matches closely the copulatory behaviour of this species. Oryx males adopt a fully upright posture after mounting a female, grasping her hindquarters tightly with their forelegs (see pl. 10I).[29] The erect carriage of the depicted oryx has caused its horns to enter the register above, however its body is somewhat inclined, causing it to clasp the female around her flanks, rather than her loins.

Hartebeest

pḥn-wj-k3.j (SB26: Saqqara, V.6-8E)

As noted above (see *Addax*), the mating posture of a pair of hartebeest depicted in a desert scene in the Saqqara tomb of *pḥn-wj-k3.j* (SB26; see fig. 10-12) is incorrect, as males in this species do not hold their bodies erect, but instead lean forward when copulating (see pl. 10J).[30]

Dorcas gazelle

sšm-nfr IV (SB27: Giza, V.9-VI.1)

A single example of mating by dorcas gazelle was located in a desert hunting scene in the tomb of *sšm-nfr IV* (SB27; see fig. 10-13). The sex of neither animal is apparent, but one has mounted the other so that its forelegs straddle its companion's haunches. The pair is flanked by hunters who are each capturing desert ungulates.

After gazelle males have determined that a female is becoming fertile, they express their interest in her by approaching in an erect posture and walking stiffly. The males of some species also vocalize; dorcas gazelle males, for example, make growling sounds by inflating a pouch in their nose. The females initially try to escape these advances, but as they become more receptive, they will allow the male to draw near and begin preliminary mounting. Copulation by gazelles is particularly challenging for the males. During both mounting and mating, gazelle females hold their head and neck upright. If a female is moving when a male attempts to mount her, she will continue to walk forward, and if she is stationary, she will begin to move. Males must therefore achieve intromission while walking bipedally behind her; unlike other antelope males, they neither clasp the female with their forelegs nor lean against her (cf. pl. 10K).[31]

The image in the tomb of *sšm-nfr IV* appears to be an accurate representation of copulation by dorcas gazelle. The male's posture is slightly too inclined, but this may be a consequence of the register line above. Unlike mating female antelopes represented in other tombs, however, the gazelle doe stands with her head and neck fully erect; her demeanour may therefore indicate that she is moving. It is also interesting to note that her forelegs break the rule of the furthermost limb advancing first, possibly to draw attention to her movement. It is unlikely that the animals will have copulated in the hunting context shown and so the scene is probably not intended to be concurrent with those on either side of it. As dorcas gazelle are a desert-living species, however, the Egyptians will undoubtedly have encountered them during hunting expeditions and perhaps witnessed their somewhat comical mating style.

[29] Estes, *Behavior Guide*, 119 and fig. 7.16.
[30] *ibid.*, 141 and fig. 8.8.
[31] *ibid.*, 69.

Goats

sšm-nfr IV (SB28: Giza, V.9-VI.1)
ꜣḫtj-mrw-nswt (SB29: Giza, V.9-VI.1)
nb-kꜣw-ḥr: jdw (SB30: Saqqara, V.9 r/u VI.7)

Three examples of mating goats were found during this study, all of which date from the late 5th to the early 6th dynasty. During courtship, male goats mount females frequently without intromission (see pl. 10L, left). During copulation, the buck holds his body erect, arches his neck backward, and retracts his forelegs (see pl. 10L, right).[32] The doe remains standing throughout.[33]

While all of the Egyptian images are essentially accurate, the males depicted in the tombs of *ꜣḫtj-mrw-nswt* (SB29; see fig. 10-14) and *nb-kꜣw-ḥr: jdw* (SB30; see fig. 9-19, left, in Chapter 9) resemble more closely the upright posture adopted by bucks during copulation. The upper register line probably influenced the inclined body of the *sšm-nfr IV* male (SB28; see fig. 9-20, lower, in Chapter 9); nevertheless, his stance is not unlike that of a buck during mounting. The horns of the *ꜣḫtj-mrw-nswt* male have been allowed to pierce the upper register (see fig. 10-14), thereby achieving a more realistic representation of the mating behaviour of this species.

Unidentified ungulates

pḥn-wj-kꜣ.j (SB31: Saqqara, V.6-8E)
Anon (SB32: Saqqara, V-VI)

The number and distinctive positions of hoofed legs in two damaged relief fragments (figs. 10-15 and 10-16) indicate that the images may have originally displayed the mating behaviour of pairs of ungulates. In both cases, the four normally spaced legs of one animal are immediately followed by a single, closely aligned pair. In one scene (fig. 10-16), this latter set of hooves seems also to be tilted upward slightly. Each image would thus appear to depict the hind legs of a rearing ungulate mounted upon another.

The undulating substrate beneath the feet of the *pḥn-wj-kꜣ.j* animals (SB31; see fig. 10-15) suggests a desert environment. In the other example (SB32; see fig. 10-16), a pair of mating felines further along on the same register (see fig. 10-26), also points to a desert setting. It is likely, therefore, that both scenes depicted a species of wild ungulate, such as scimitar oryx, dorcas gazelle, or hartebeest.

Leopards

ptḥ-ḥtp II: ṯfj (SB33: Saqqara, V.9M-L)
nj-mꜣꜥt-rꜥ [*jzzj-mr-nṯr*] (SB34: Giza, V.9-VI.1)

The mating behaviour of leopards is displayed in two late 5th dynasty tombs. In the earlier relief, located in the Saqqara tomb of *ptḥ-ḥtp II: ṯfj* (SB33; see fig. 10-17), a male has mounted a female and, while elevating his torso, clasps her flanks with his forelimbs. The female stands alert, looking ahead. The tails of both animals arc away from their bodies, but that of the female is especially curved and erect. Only the tails remain in a fragmentary relief from the tomb of *nj-mꜣꜥt-rꜥ* [*jzzj-mr-nṯr*] (SB34; see fig. 10-18), however, as these are held in a similar fashion to those of the *ptḥ-ḥtp II: ṯfj* felines, it is very likely that the image also originally depicted either mating leopards or lions.[34]

The mating posture of these animals is incorrect. Female cats, including leopards, do not stand during copulation, but instead crouch upon the ground (see pl. 10M). Males do not grasp the female with their forelegs (as most carnivores do), but simply straddle her body while continuing to stand. Male cats also mouth frequently at the female's neck; smaller cat species maintain a neck grip

32 Hart, *Behavior of Domestic Animals*, 113.

33 Davies et al. (*Saqqara Tombs i*, 13 and pl. 7) suggest that an overlapping pair of rearing goats in the Saqqara tomb of *mrrj* is copulating, however females do not mate in an upright+ position. I suggest in Chapter 7, n. 25 that these animals are instead browsing on foliage.

34 Roth (*Cem. Pal. Att.*, 132) identified both these animals and the mating pair in the tomb of *ptḥ-ḥtp II: ṯfj* (fig. 10-17) as lions. Their slim build and the male's lack of a mane in the latter example is, however, more indicative of leopards.

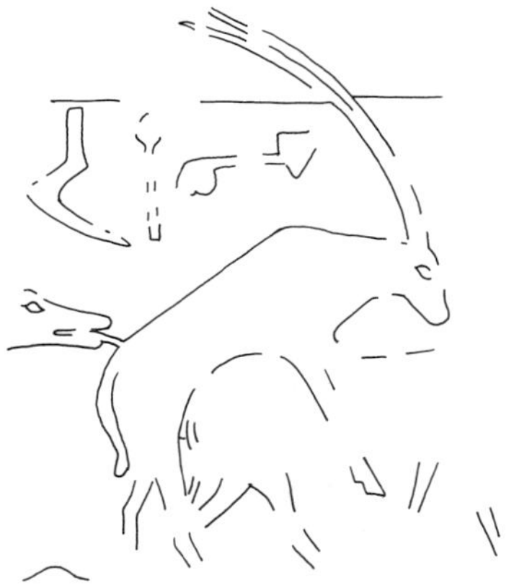

Fig. 10-11 (SB25)

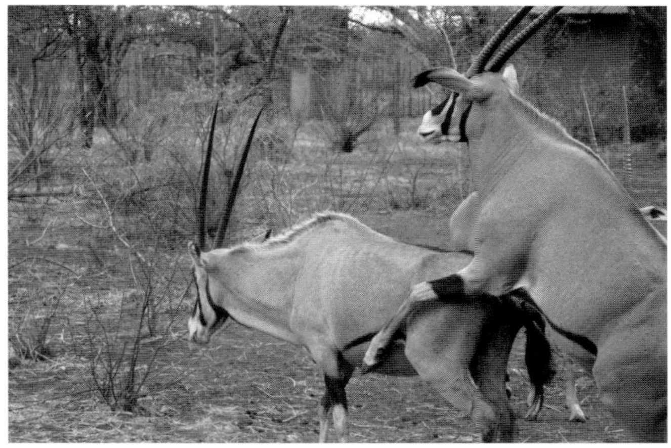

Pl. 10I Mating posture of oryx
© iStockphoto.com/Markus Divis

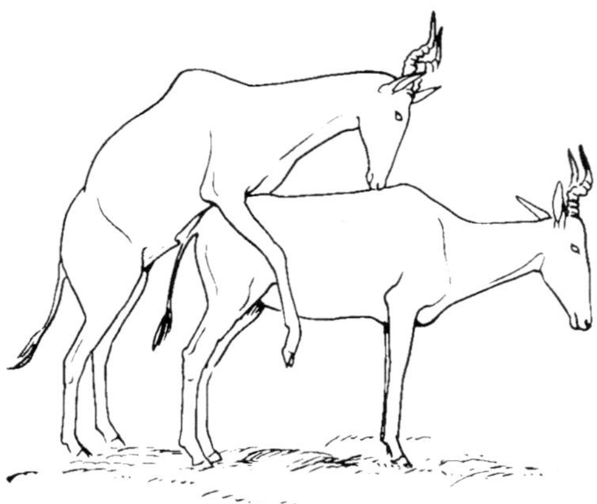

Pl. 10J Mating posture of hartebeest
Reprinted from: Estes, *Behavior Guide*, fig. 8.8
By permission of © University of California Press

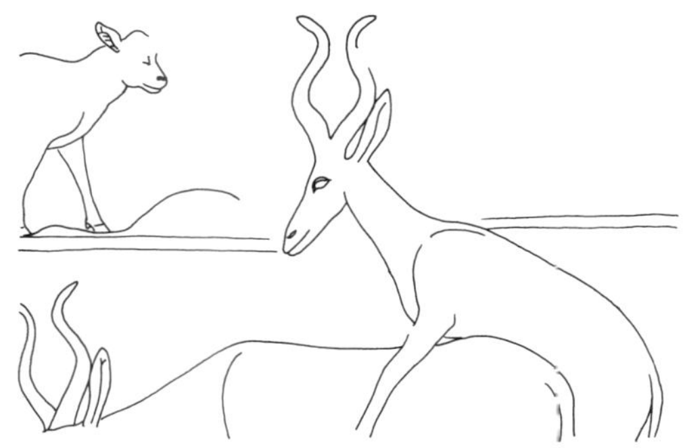

Fig. 10-12 (SB26)

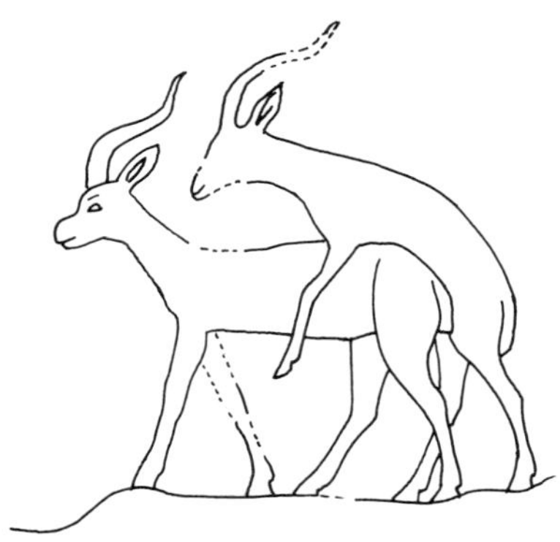

Fig. 10-13 (SB27)

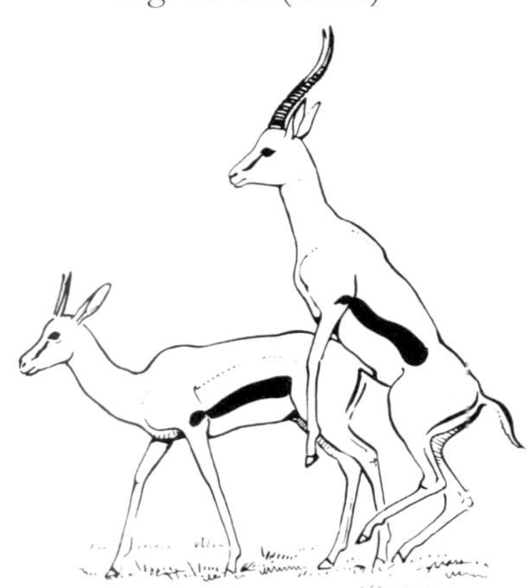

Pl. 10K Mating posture of gazelle
Reprinted from: Estes, *Behavior Guide*, fig. 5.5
By permission of © University of California Press

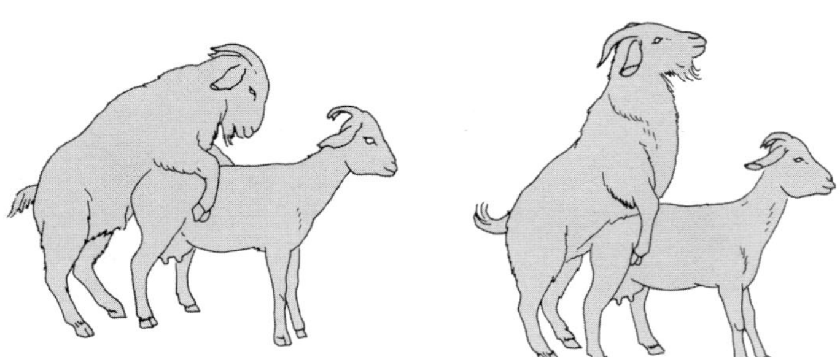

Pl. 10L Mating posture of goats
After: Hart, *Behavior of Domestic Animals*, fig. 3-15 C, D

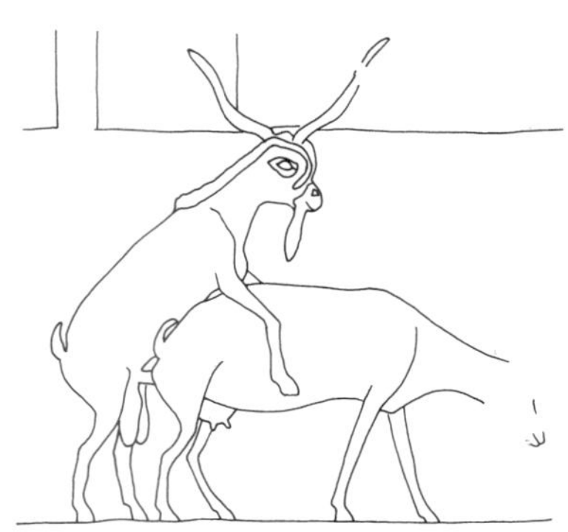

Fig. 10-14 (SB29)

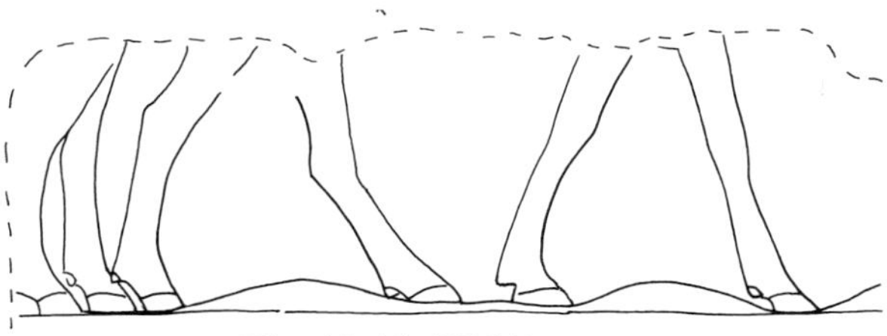

Fig. 10-15 (SB31)

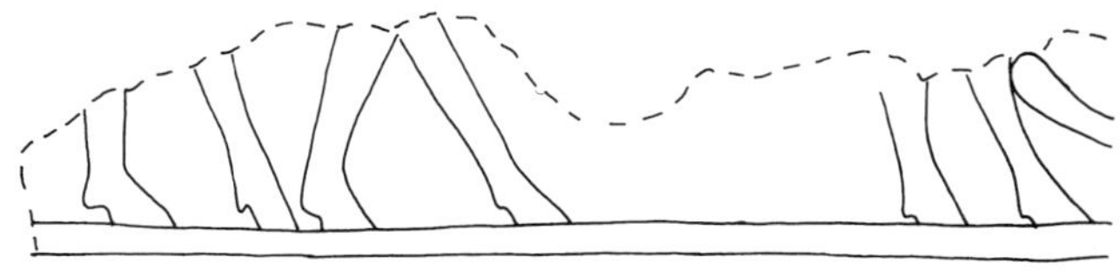

Fig. 10-16 (SB32)

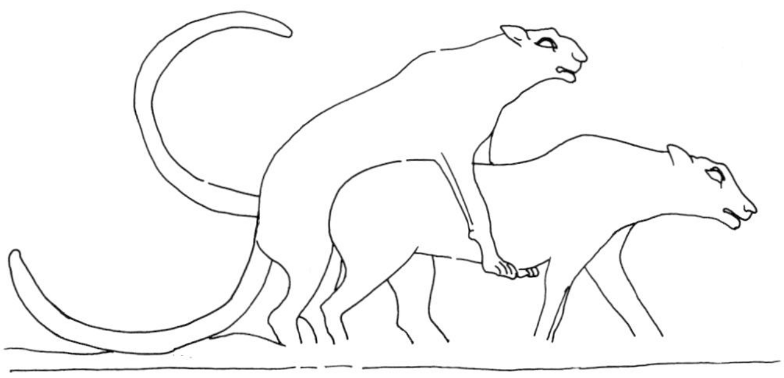

Fig. 10-17 (SB33)

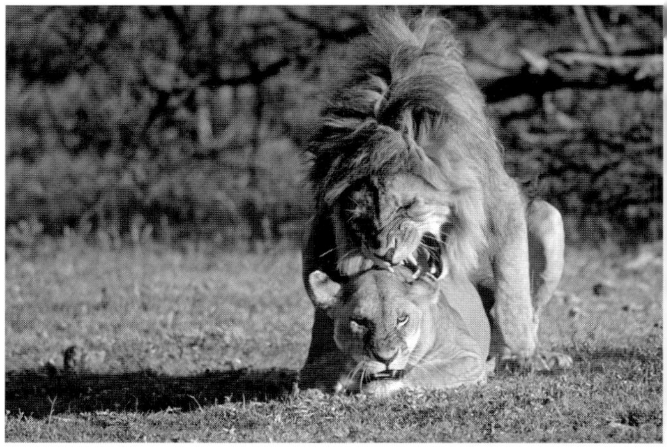

Pl. 10M Mating posture of lions
© iStockphoto.com/John Pitcher

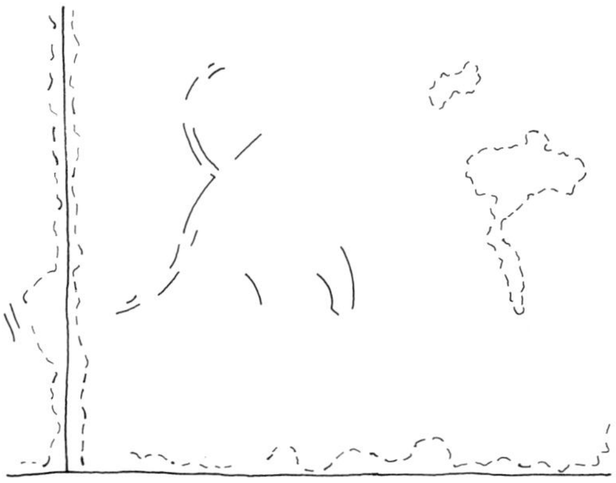

Fig. 10-18 (SB34)

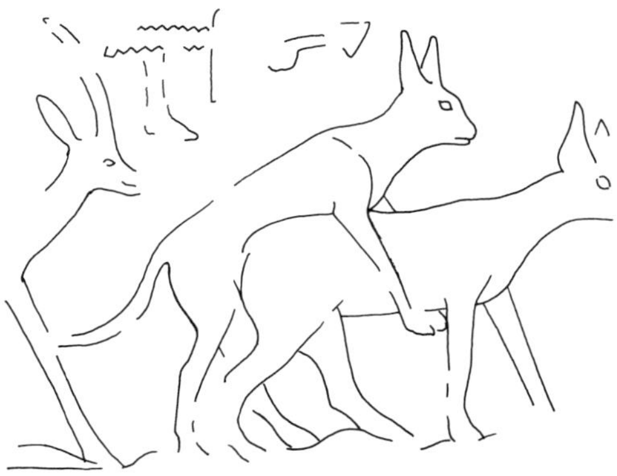

Fig. 10-19 (SB35)

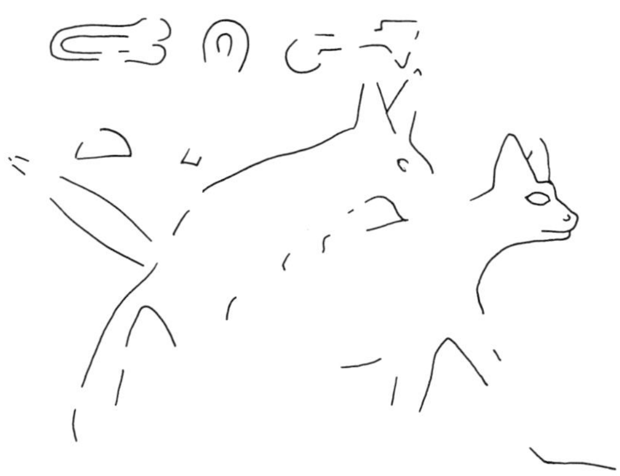

Fig. 10-20 (SB36)

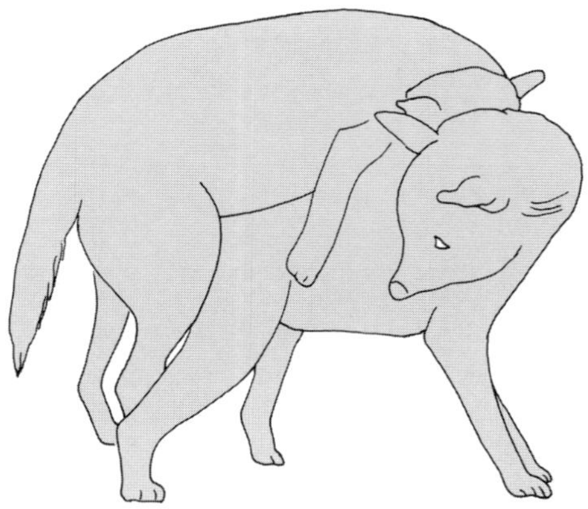

Pl. 10N Mating posture of striped hyenas
After: Rieger, I. (1979). *VNG Zürich* 124, fig. 2

throughout copulation, while larger animals, such as leopards and lions, usually bite the nape lightly or simply bare their teeth (see pl. 10M).[35]

The erroneous stance of the female leopard in the tomb of *ptḥ-ḥtp II: ṯfj* has been noted elsewhere.[36] Salima Ikram suggested that this mistake, which is also repeated in later depictions of lions, was due to lack of direct observation by the artists, who consequently modelled the mating posture of these animals upon that of domesticated species in which the females stand.[37] This is certainly feasible, however, the explanation is undermined by a depiction of mating lions located in the Middle Kingdom tomb of *wḫ-ḥtp* (B, no. 2), in which a mounted male leans forward to bite the neck of a female, who nevertheless remains upright.[38] The relief indicates that the artist had clearly witnessed the mating bite of male lions, yet chose to incorporate this correct feature in his image while simultaneously maintaining the incorrect stance for the female.

These scenes hint that the depiction of animal copulation was driven by representational/artistic considerations, rather than biological accuracy - that is, that images of mating in wall scenes followed a set pattern in which the male mounted while the female stood, regardless of species-specific differences. Although some observed details might be added - such as the male neck bite - artists were clearly reluctant to alter the fundamental composition.

The posture of the female leopards in the tombs of *ptḥ-ḥtp II: ṯfj* and *nj-m3ʿt-rʿ* [*jzzj-mr-nṯr*], although incorrect, thus cannot necessarily be viewed as evidence for the artists' lack of knowledge of, or familiarity with, the animals' mating habits.

Caracals

nj-m3ʿt-rʿ [*jzzj-mr-nṯr*] (SB35: Giza, V.9-VI.1)

A unique depiction of caracals has been noted in the Giza chapel of *nj-m3ʿt-rʿ* [*jzzj-mr-nṯr*] (SB35; see fig. 10-19).[39] The members of this feline species possess short tails and extremely long and pointed ears that are each surmounted by black tufts, features that have been reproduced in the *nj-m3ʿt-rʿ* [*jzzj-mr-nṯr*] image.

The scene depicts a male mounted upon a standing female. As noted above (see *Leopards*), the females of all cat species crouch during copulation (cf. pl. 10M). Consequently, the depiction of mating caracals is biologically incorrect; indeed, female caracals usually initiate mating by lying down upon the ground.[40]

Striped hyenas

nj-m3ʿt-rʿ [*jzzj-mr-nṯr*] (SB36: Giza, V.9-VI.1)

Among the many representations of mating animals found in the Giza tomb of *nj-m3ʿt-rʿ* [*jzzj-mr-nṯr*] is a clearly labelled example of striped hyenas (SB36; see fig. 10-20). Like other depictions of mating, the male has mounted the female from behind as she stands facing forward, however, in a manner similar to that reported for addax and wild ass (which appear on the same register), the male does not hold his body above that of the female, but instead leans over so that his muzzle touches her neck.

Striped hyenas are intensely secretive animals that are difficult to see in the wild. Only one

35 Estes, *Behavior Guide*, 356 and fig. 21.9.
36 Keimer, L. (1948). "*Inb*, Le caracal (*Caracal caracal* Schmitzi Matsch.) dans l'Égypte ancienne", *ASAE* 48, 377; Ikram, *GM* 124, 59.
37 *ibid.*
38 Blackman, *Rock Tombs of Meir ii*, pl. 7.
39 Keimer, *ASAE* 48, 374-379.
40 Bernard, R.T.F. and Stuart, C.T. (1987). "Reproduction of the caracal *Felis caracal* from the Cape Province of South Africa", *SAJZ* 22(3), 178.

observation has been made of their mating habits, which was recorded in a zoo in 1979. Remarkably, this description matches precisely the behaviour of the depicted animals as the male, while mounted, was observed to lick the neck of the female a number of times (see pl. 10N).[41]

Hyenas are featured frequently in tomb scenes, many of which suggest that the Egyptians not only kept these animals in captivity, but may also have tried to domesticate them. It is likely then, that they will have had many opportunities to witness the sexual behaviour of these animals and note the striped hyena male's curious tendency to mouth at the female's nape.[42]

Golden jackals

nj-m3ʿt-rʿ [*jzzj-mr-nṯr*] (SB37: Giza, V.9-VI.1)

The fragmentary label associated with a pair of mating canids depicted in the chapel of *nj-m3ʿt-rʿ* [*jzzj-mr-nṯr*] (SB37; see fig. 10-21) was interpreted by Ann Macy Roth as *nk wn*[*š*]. Although the spelling of *wnš* is unusual,[43] this caption probably indicates that the animals are jackals. The image shows the male reared and mounted upon a female who stands; the female also protrudes her tongue prominently. The short tail of the female is a characteristic of jackals.

The sexual behaviour of golden jackals is similar to that of other canids. Courtship commences after a male and female pair off and acquires a territory together. After scent marking and patrolling their patch, the two engage in a series of escalating physical interactions: approaching while whimpering, circling, sniffing, licking one another, and fighting briefly. The pair may also exhibit a 'T-formation', in which one animal stands at right angles to the other and presses its head upon the shoulders of its companion. As the female enters oestrus, these interactions increase and the male begins to make mounting attempts. Actual mating eventually occurs after many days of this precopulatory behaviour, and is repeated frequently for the following week. During copulation, the male rears and grasps the female's sides with his forelegs (cf. pl. 10O). After he has achieved intromission, the base of his penis swells, locking the pair in this position for approximately four minutes. Sometimes the male will dismount while connected and turn so that the two stand rear-to-rear.[44] This 'copulatory tie' is experienced by most canid species.

The image that appears in the tomb of *nj-m3ʿt-rʿ* [*jzzj-mr-nṯr*] is thus an accurate depiction of the mating posture of canids; however, the behaviour of the female is unusual. As noted in other chapters, Egyptian artists often portrayed vocalizing animals with protruded tongues. This may explain the female's demeanour, as a few canid species are known to vocalize during mating. Red foxes are particularly vocal during the breeding season[45] and Rüppell's sand fox, which is also found in Egypt, emits a series of barks while copulating.[46] To my knowledge, however, this behaviour has not been observed in golden jackals. Nevertheless, this species is generally quite vocal, and is particularly renowned for a howling wail that is often given in chorus by mated pairs.[47] The behaviour of the *nj-m3ʿt-rʿ* [*jzzj-mr-nṯr*] female possibly represents this bonding call.

---

41 Rieger, I. (1979). "Beobachtungen zur Aufzucht von Streifenhyänen, *Hyaena hyaena*", *VNG Zürich* 124(3), 171.

42 Ikram (*GM* 124, fig. 23) included in her study an ostracon from Deir el-Medina (Louvre E14311) upon which is depicted a pair of mating animals that she identified as either wild dogs or jackals. The mounted male in this sketch rests heavily upon the female's back, such that his muzzle is buried in her nape. The female protrudes her tongue prominently. Although the animals' muzzles are rather too elongated, I suggest, nevertheless, that the distinctive behaviour of the male indicates that the pair represents striped hyenas. The female's exposed tongue may possibly illustrate a cry given by this species when they are ready to copulate. See Rieger, I. (1981). "Hyaena hyaena", *Mam. Sp.* (150), 3.

43 The word is usually written with Gardiner E34, not M42 as appears in the inscription. See *Wb i*, 324.

44 Estes, *Behavior Guide*, 402-403.

45 Hoath, *Field Guide*, 75.

46 Larivière, S. and Seddon, P.J. (2001). "Vulpes rueppelli", *Mam. Sp.* (678), 3.

47 Estes, *Behavior Guide*, 401.

Red foxes

nj-ʿnḫ-ẖnmw/ẖnmw-ḥtp (SB38: Saqqara, V.6L-7)
ptḥ-ḥtp II: ṯfj (SB39: Saqqara, V.9M-L)
nj-mȝʿt-rʿ [jzzj-mr-nṯr] (SB40: Giza, V.9-VI.1)

Differentiating between depictions of unlabelled jackals and red foxes in wall scenes is extremely difficult as artists frequently rendered the animals in a very similar fashion: that is, with elongated snouts, large pricked ears, slender legs, and a tail of variable length. Indeed, the two species share many morphological features in common, although jackals are somewhat stockier animals and their tail is not as bushy or long as that of foxes.

Three pairs of mating canids examined during this study are believed to be red foxes (SB38 to SB40).[48] It must be admitted, however, that this conclusion is based entirely upon the length and shape of their tails (cf. *Golden jackals* above), as the images possess no other distinguishing cues. The animals have been identified differently elsewhere. For example, F.L. Griffith assumed that the *ptḥ-ḥtp II: ṯfj* pair (SB39; see fig. 10-22) depicts jackals.[49] Ikram, however, labelled these and the animals in the tomb of *nj-ʿnḫ-ẖnmw/ẖnmw-ḥtp* (SB38), "wild dogs".[50] Roth described the animals in the tomb of *nj-mȝʿt-rʿ [jzzj-mr-nṯr]* (SB40) as "dog-like", but could not offer a translation of the scene's label, *nk ẖn*.[51] The term *ẖn* is found in a clothing list in the Giza tomb of *snfrw-snb* (G4240) where it is believed to refer to a member of either the dog or cat family.[52] I suggest, however, that the similar morphology and pose of the *nj-mȝʿt-rʿ [jzzj-mr-nṯr]* animals to the other depicted pairs (SB38 and SB39) indicates that *ẖn* must instead describe a type of fox.

The sexual habits of red foxes resemble those of other canids (see *Golden jackals* above). Males mount females in a manner similar to that represented in the three reliefs (cf. pl. 10P). Like other canids, this species also forms a sexual lock during the final stage of copulation.[53]

Louis Keimer drew attention to small, wattle-like objects that appear on the lower jaws of each of the *ptḥ-ḥtp II: ṯfj* foxes (see fig. 10-22) and suggested that they represent saliva escaping from the mouths of the animals in response to their frenzied sexual state.[54] Although he presents a convincing case regarding the nature of these and similar projections depicted on the chins of forced-fed hyenas, the production of excessive saliva has not been noted as part of the copulatory behaviour of red foxes, or any canid species. It seems, therefore, that this detail - in this context at least - is a product of the artist's imagination.

Libyan striped weasels

nj-mȝʿt-rʿ [jzzj-mr-nṯr] (SB41: Giza, V.9-VI.1)

A pair of small, mating mammals located on a sub-register in the tomb of *nj-mȝʿt-rʿ [jzzj-mr-nṯr]* (SB41; see fig. 10-23) is labelled *[gs]fnw*.[55] Examination of similarly labelled animals in the Chamber of the Seasons situated in the sun-temple of Niuserre has led to the belief that they represent the Libyan striped weasel.[56] One example of this inscription (Berlin 20037) is modified by

48 See Osborn-Osbornová, *Mammals*, 68-72.
49 Griffith, F.L in Quibell, *Ptah-hetep*, 28.
50 Ikram, *GM* 124, 53. The Latin name, *Canis lupaster*, that Salima Ikram provides, however, indicates the Egyptian subspecies of golden jackal, *Canis aureus lupaster*.
51 Note that Roth only refers to these hieroglyphs in the text (*Cem. Pal. Att.*, 133). They are not apparent in the figure she provides (pl. 189).
52 See Wassell (*Ancient Egyptian Fauna i*, 98). The names of animals appear in many Old Kingdom clothing lists; these probably refer to their skins. See Brovarski, E. (1996). "An inventory list from 'Covington's tomb' and nomenclature for furniture in the Old Kingdom" in Der Manuelian-Freed, *William Kelly Simpson i*, 125, n. 47-50.
53 Bekoff, M. (1975). "Social behavior and ecology of the African canidae: A review" in Fox, *Wild Canids*, 123.
54 Keimer, L. (1955). "Notes de lecture", *BIFAO* 55, 10.
55 *Wb v*, 206, 8 translates this term as simply 'kleines Wüstentier'.
56 Keimer identified the animals as zorillas (*Ictonyx libyca*, now *Ictonyx striatus*), but following an extensive

the term *j3b*, which has been translated as 'evil-smelling' and interpreted as a reference to the repulsive odour emitted by members of this species.[57]

Weasels are very aggressive while mating. Courtship begins if a female answers the trilling calls of a male. As the pair frolic and sniff one another, the male will suddenly grasp the female with his forelegs and bite her nape. He then drags her about by the scruff of her neck until she stops resisting his attempts to mount. During copulation, the female lies on her belly with her tail raised while the male continues to maintain his hold on her neck (see pl. 10Q). This process is repeated many times an hour over a period of two or three days.[58]

The animals depicted in the *nj-m3ˁt-rˁ* [*jzzj-mr-nṯr*] scene are not mating in an appropriate manner for weasels.[59] The female is standing, rather than prostrate and, although the relief is damaged, the inclination of the male's back would seem to indicate that he is holding his head above the neck of his companion instead of touching it with his muzzle. The female does, however, lift her tail correctly.

Libyan striped weasels are intensely nocturnal animals that rarely leave their burrows during the day;[60] it is very likely, then, that the artist responsible for the *nj-m3ˁt-rˁ* [*jzzj-mr-nṯr*] image had not had the opportunity to witness the species' mating behaviour.

Hedgehogs

ṯjj (SB42: Saqqara, V.8-9) *nj-m3ˁt-rˁ* [*jzzj-mr-nṯr*] (SB43: Giza, V.9-VI.1)

The mating habits of hedgehogs are apparent in two images dating from the late 5th dynasty. In a scene located in the tomb *nj-m3ˁt-rˁ* [*jzzj-mr-nṯr*] (SB43; see fig. 10-24), one animal is perched upon the back of another. Unfortunately, the relief is damaged and so their postures cannot be determined. It is likely, however, that the mounting animal is standing on its hind legs. Indeed, in the tomb of *ṯjj* (SB42), a hedgehog appears to mount another by rearing up onto its companion's hindquarters.

Hedgehogs mate after an aggressive courtship in which the male circles about the female while she, in turn, butts him frequently with her head. Eventually, however, she will allow the male to approach and invites him to mount by pressing her belly to the ground and lifting her hindquarters. After rearing, the male often bites the spines around her shoulders in order to maintain his position. Long-eared hedgehog males usually adopt a more upright posture than those of other species, and may turn their head to one side against the female's rump (see pl. 10R).[61]

The male hedgehog in the *nj-m3ˁt-rˁ* [*jzzj-mr-nṯr*] relief does not mouth the female's spines, but instead elevates his head above her shoulders. This suggests that the artist was not familiar with the male's copulatory bite, as this behaviour has been included in depictions of other species that appear on the same register (see *Wild ass* and *Striped hyenas* above). The fragmentary state of the relief prevents a clear determination of the animals' postures, but the position of the female's foreleg indicates that she is probably standing, rather than prostrate. The *nj-m3ˁt-rˁ* [*jzzj-mr-nṯr*] image therefore bears little resemblance to the mating behaviour of hedgehogs.

---

study of extant images, Aufrère has suggested that they more closely resemble the Libyan striped weasel (*Poecilictis libyca* Lataste). See Keimer, L. (1942). "Quelques représentations rarissimes de mustélidés conservées sur des bas-reliefs de l'ancien empire", *Études* 4, 15-25; Aufrère, S. (1985). "Études de lexicologie et d'histoire naturelle VII: De nouvelles represéntations de zorilles de Libye, *Poecilitis libyca* Lataste", *BIFAO* 85, 23-32.

57 See Keimer, *Études* 4, 21-22; Edel, *NAWG* 1963, 182.

58 King, *Weasels and Stoats*, 127-128.

59 This has also been noted by Ikram (*GM* 124, 59).

60 Sitek, H. (1995). "Breeding of the Libyan striped weasel *Poecilictis libyca* at Poznan Zoo, Poland", *SCC* 13, 8; Hoath, *Field Guide*, 83.

61 Reeve, *Hedgehogs*, 189-191.

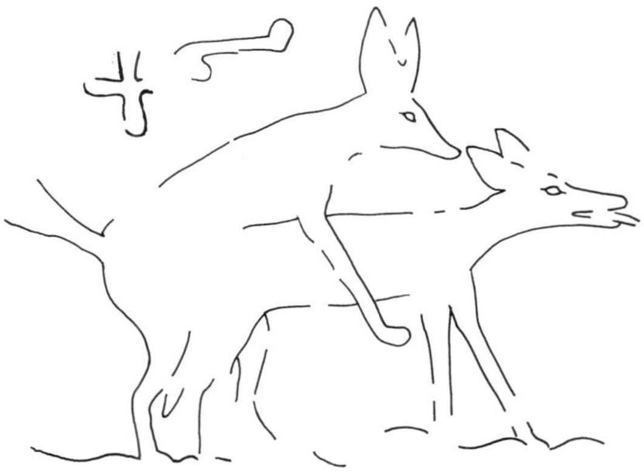

Fig. 10-21 (SB37)

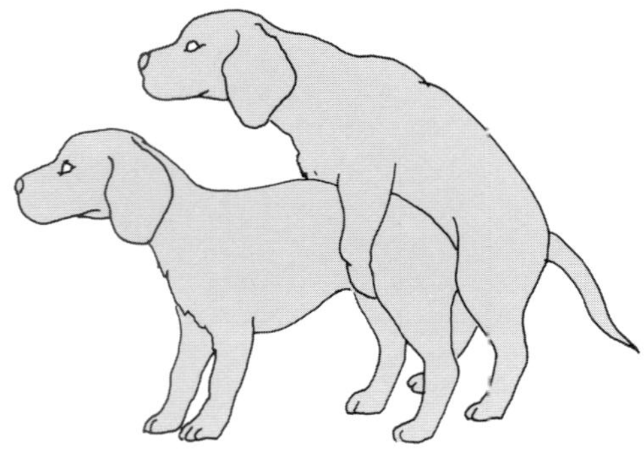

Pl. 10O Mating posture of canids
After: Hart, *Behavior of Domestic Animals*, fig. 3-10 B

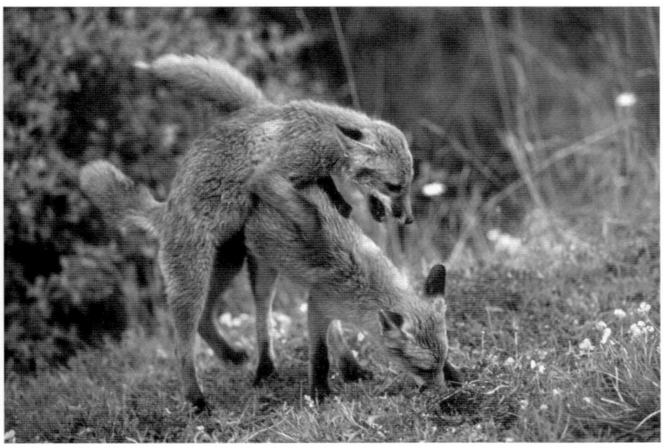

Pl. 10P Mating posture of red foxes
© ardea.com/Duncan Usher

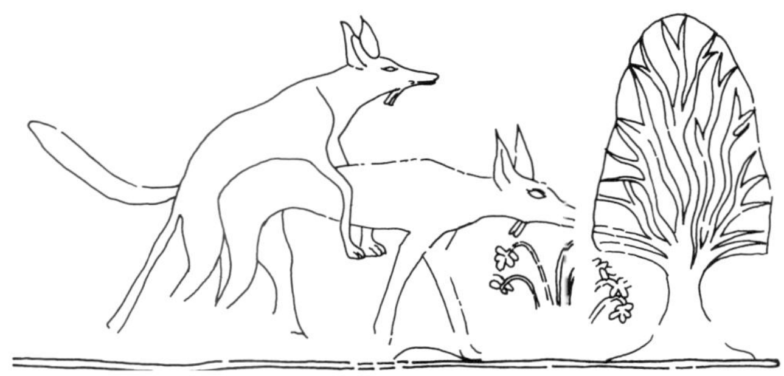

Fig. 10-22 (SB39)

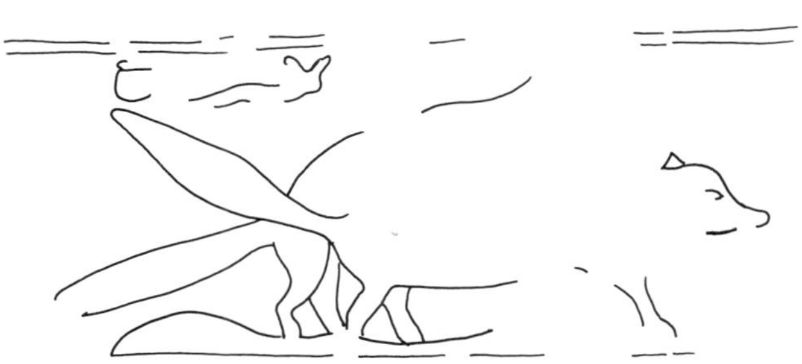

Fig. 10-23 (SB41)

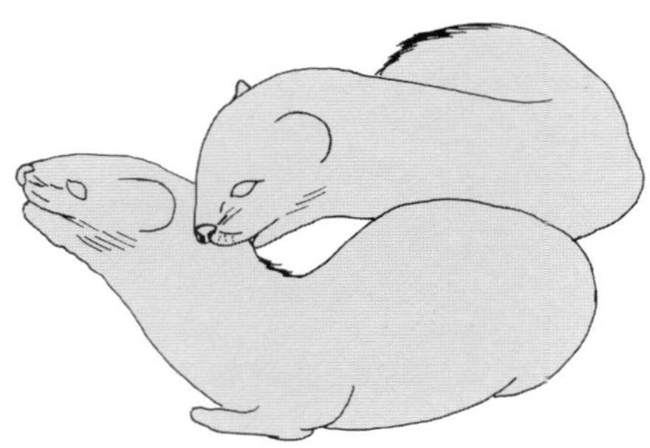

Pl. 10Q Mating posture of weasels
After: King, *Weasels and Stoats*, fig. 9.3

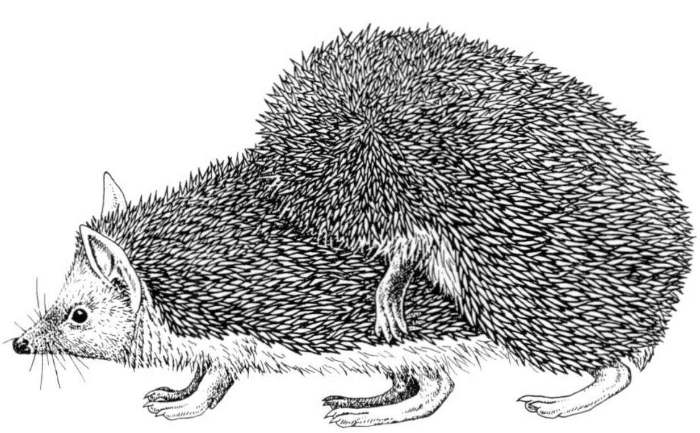

Pl. 10R Mating posture of long-eared hedgehogs
Reprinted from: Reeve, *Hedgehogs*, fig. 8.7
By permission of © A&C Black Publishers

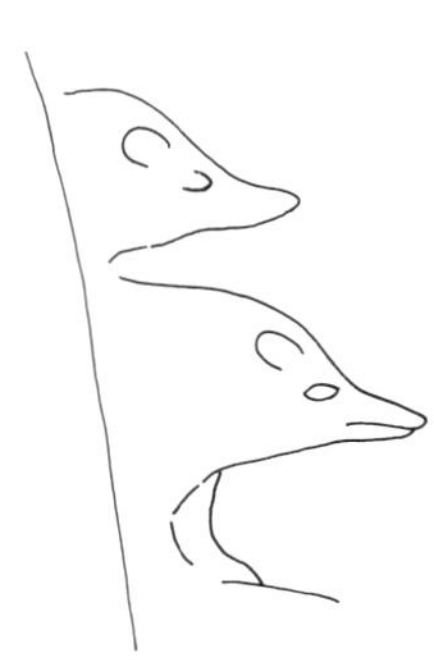

Fig. 10-24 (SB43)

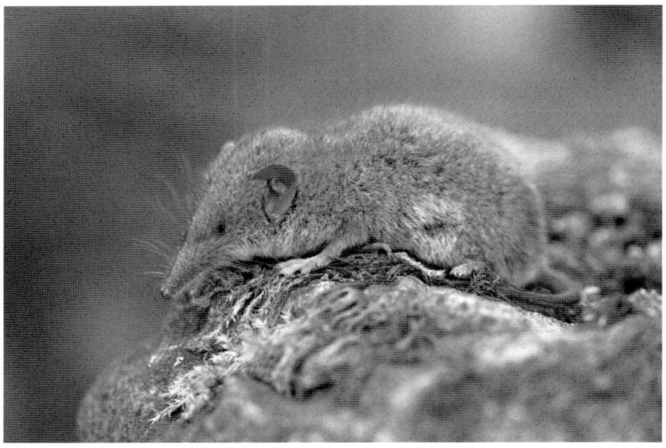

Pl. 10S Shrew

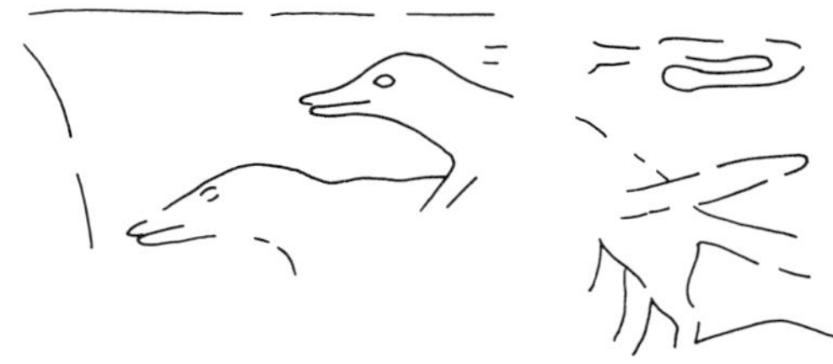

Fig. 10-25 (SB44)

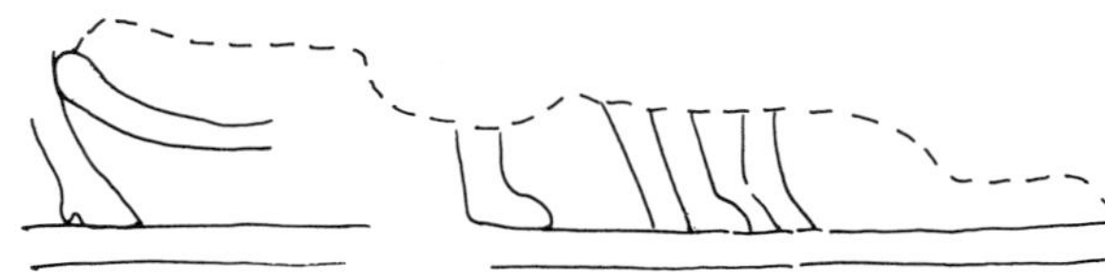

Fig. 10-26 (SB45)

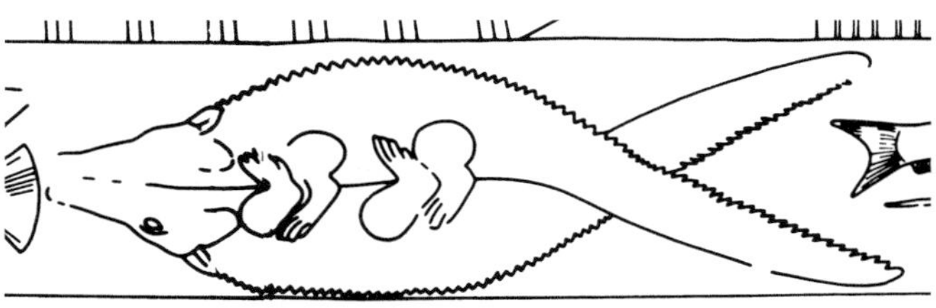

Fig. 10-27 (SB47)

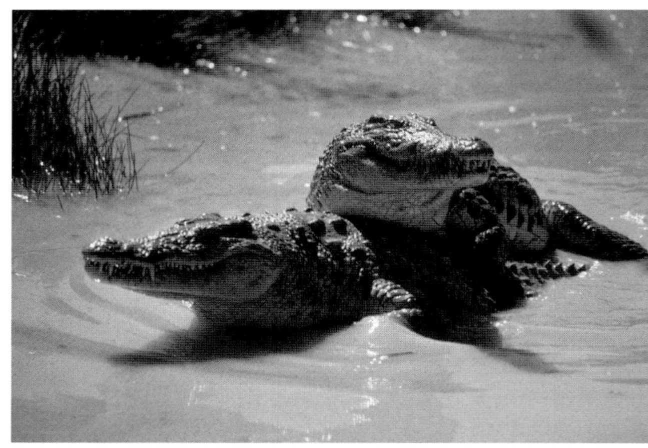

Pl. 10T Mating posture of Nile crocodiles

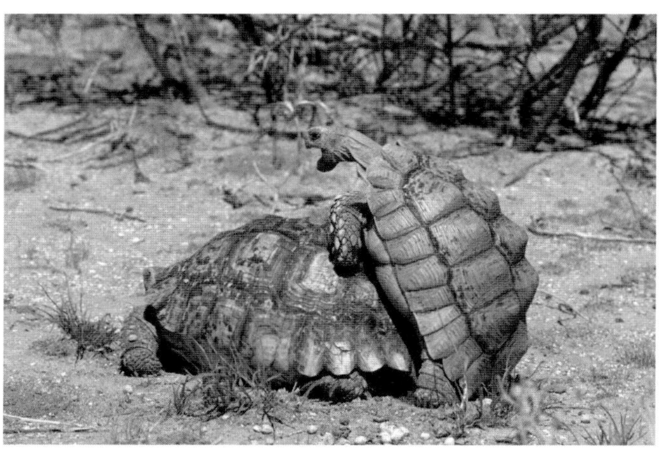

Pl. 10U Mating posture of tortoises

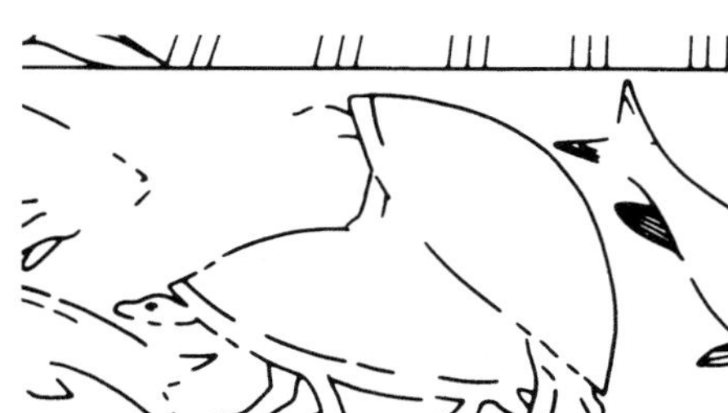

Fig. 10-28 (SB49)

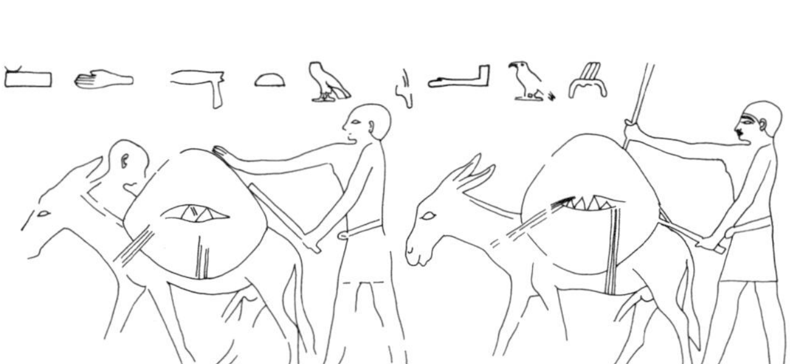

Fig. 10-29 (SB51)

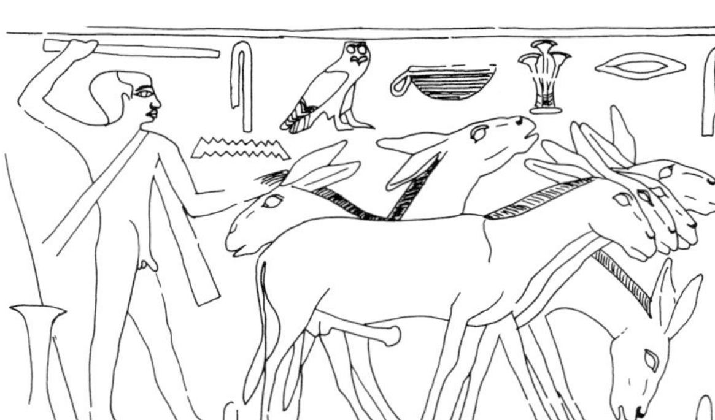

Fig. 10-30 (SB52)

Shrews

nj-mꜣꜥt-rꜥ [*jzzj-mr-nṯr*] (SB44: Giza, V.9-VI.1)

Two curious animals, depicted directly below a pair of mating Libyan striped weasels in the tomb of *nj-mꜣꜥt-rꜥ* [*jzzj-mr-nṯr*] (see above), are also copulating (SB44; see fig. 10-25). The pair, labelled *zṯ*, faces a small shrub on the left. Both possess beak-like snouts, small limbs, and short, narrow tails.

Edward Brovarski proposed that the animals might depict honey badgers (also known as ratels).[62] Ikram identified them tentatively as Egyptian mongooses,[63] however Roth described their appearance as "mole-like" and suggested that the term *zṯ* was possibly a "miswriting of *ḥṯs*, weasel".[64]

The most distinctive attribute of the *nj-mꜣꜥt-rꜥ* [*jzzj-mr-nṯr*] creatures is their snouts, which are extremely elongated. Combined with their apparent small size, I suggest that they closely resemble shrews.[65] Shrews are diminutive, mouse-like creatures that are active during both day and night and live in many different habitats. They are characterized by a long and tapered muzzle ending in a pair of nostrils that form a flexible protuberance similar to a proboscis (cf. pl. 10S). As they use vibrissae on their snout to locate prey, the tip is constantly in motion.[66] The muzzle shapes of the *nj-mꜣꜥt-rꜥ* [*jzzj-mr-nṯr*] pair clearly match those of shrews. Note also that the upper lip of each animal extends beyond its lower jaw; I believe that this projection was intended to represent the shrews' fleshy nostrils. The tails of the depicted animals are also shrew-like. Each is somewhat short relative to the length of their body. The tails of shrews are narrow, but robust, and in many species are often shorter than the animals' combined head and body length.[67]

Shrews are no longer common in Egypt, but mummified remains indicate that a number of species were present in the past.[68] These included the greater musk shrew, the most widespread species found in Egypt today. Like many other shrews, it emits an unpleasant, pungent odour from its skin and genital glands.[69] It is consequently tempting to see in the label associated with the *nj-mꜣꜥt-rꜥ* [*jzzj-mr-nṯr*] animals a possible mis-spelling of the word *sṯ(y)*, which is translated as 'odour' or 'smell'.[70] It is also interesting to note that greater musk shrews are often found in dense vegetation. Although the *nj-mꜣꜥt-rꜥ* [*jzzj-mr-nṯr*] pair occurs in a desert scene, they are pictured beside, and facing, a small bush, one of only two forms of vegetation that appear on the register.

If the *nj-mꜣꜥt-rꜥ* [*jzzj-mr-nṯr*] animals are accepted as shrews then the depiction of their mating habits is not entirely accurate. The sexual behaviour of the greater musk shrew has been studied in captivity. Males were observed to chase oestrus females and attempt to mount them many times before finally achieving intromission. During copulation, which lasted approximately seven

---

62 Brovarski took the label to read *šṯ*, and in so doing, compared the *nj-mꜣꜥt-rꜥ* [*jzzj-mr-nṯr*] pair with animal determinatives associated with this word that are inscribed in the 4th dynasty Saqqara tomb of *mṯn*, which others have variously identified as ichneumons, hyenas, dogs, and hunting dogs. Brovarski felt that the tail of the *nj-mꜣꜥt-rꜥ* [*jzzj-mr-nṯr*] pair was "too short for an ichneumon, the body too stout and thickly built". Like Roth, however, I believe that the inscription is to be read *zṯ*, not *šṯ*. See Brovarski, E. (1987). "Two Old Kingdom writing boards from Giza", *ASAE* 71, 33-35; Roth, *Cem. Pal. Att.*, 132.

63 Ikram, *GM* 124, 52.

64 Roth, *Cem. Pal. Att.*, 132.

65 For the significance of shrews in Egyptian religion, see Brunner-Traut, E. (1965). "Spitzmaus und Ichneumon als Tiere des Sonnengottes", *NAWG* 7, 123-163.

66 Fons, R. (1990). "Shews (Family Soricidae)" in *Grzimek's Encyclopedia i*, 480.

67 Hoath, *Field Guide,* pl. 2.

68 Osborn-Osbornová, *Mammals*, 24-25.

69 Fons in *Grzimek's Encyclopedia i*, 488 and 499.

70 The term that is usually associated with shrews is *ꜥmꜥmw*. See *Wb i*, 186, 10. The smell of the *ꜥmꜥmw* is referred to in the Kahun veterinary text: "the odour of its breath is like the amamu". See Griffith, *Hieratic Papyri*, 13, pl. 7. The word *sṯ(y)* occurs in the *Pyramid Texts* (PT 1790b, Utt. 632): "How harmful is your odour, how bad is your odour, how great is your odour"! Note that Libyan striped weasels, which appear in the panel immediately above these animals, are also known for their repulsive smell.

seconds, the female remained standing while the male leaned forward to bite her upon her back.[71] The female in the damaged *nj-m3ʿt-rʿ* [*jzzj-mr-nṯr*] image appears to be standing, but the male's body is elevated. The scene is thus largely incorrect, but might possibly represent the shrews' frequent precopulatory mounting behaviour.

Unidentified mammals

Anon (SB45: Saqqara, V-VI)

The distribution of legs on a damaged relief fragment found within the tomb of *ṯfw* revealed that a pair of ungulates is mating (SB32; see fig. 10-16 in *Unidentified ungulates* above). The legs and paws of mammals that appear next to this pair is less indicative of a mating pose, as some of the limbs are missing (SB45; see fig. 10-26). Nevertheless, the spacing of the remaining legs and the animals' close proximity to the mating ungulates hint at a depiction of copulation. The length, thickness, and elevated position of the animals' tails are similar to those of mating leopards (see above), indicating that this image may also depict a pair of felines.

Nile crocodiles

k3-gm-nj: mmj (SB46: Saqqara, VI.1M)
ḥzj (SB47: Saqqara, VI.1L-2E)
mḥw (SB48: Saqqara, VI.2M-3?)

Three 6th dynasty Saqqara tombs were found to display pairs of Nile crocodiles that embrace one another (SB46 to SB48; see fig. 10-27). The reptiles are situated underwater in marsh scenes and adopt a unique pose in which one animal, while fully inverted, is covered by the body of the other, such that their ventral surfaces (chin and belly) converge. Both animals reach out with their fore- and hind limbs to touch one another's torso. The tail of each crocodile crosses that of its partner.

The behaviour of these animals has been identified as mating and noted, as such, to be biologically incorrect.[72] Ikram suggested that the reptiles might instead be viewed as fighting one another, but ultimately concluded that their demeanour was too "peaceful" for this to be feasible.[73]

Nile crocodiles mate after an extensive and rather aggressive courtship. Females signal their receptiveness by adjusting the angle of their head at the water's surface. Males respond with a series of displays: expanding their neck, vocalizing, blowing bubbles from their mouth and nostrils, causing ripples to radiate outward from their head, slamming their head against the water, splashing with their tail, rolling over, and submerging and resurfacing repeatedly.[74] During the breeding season, males will often clash with one another over access to females; fights consist of the pair rolling together several times, thrashing their tails, and locking jaws.[75]

During copulation, which takes place in shallow water, the male usually moves over the female's dorsal surface and then curves his tail and vent under her to insert his curved penis into her cloaca (see pl. 10T). Nile crocodiles have been observed to mate in this manner many times.[76] Field studies have also revealed that after the male mounts a female, the pair sometimes roll in the water and frequently wrap their tails around one another. During one investigation, however, a single pair was seen to mate "venter to venter"[77] – that is, mating in a manner similar to the animals depicted in wall reliefs.

---

71 Baxter, R.M. and Meester, J. (1982). "The captive behaviour of the red musk shrew, *Crocidura f. flavescens* (I. Geoffroy, 1827) (Soricidae: Crocidurinae)", *Mammalia* 46, 18.

72 Ikram, *GM* 124, 59; Houlihan, *Animal World*, 115-116.

73 Ikram, *GM* 124, 59.

74 Kofron, C. (1991). "Courtship and mating of the Nile crocodile (*Crocodylus niloticus*)", *Amph. Rep.* 12, 45.

75 *ibid.*, 46-47.

76 Modha, M.L. (1967). "The ecology of the Nile crocodile (*Crocodylus niloticus* Laurenti) on Central Island, Lake Rudolf", *EAWJ* 5(August), 83; Kofron, *Amph. Rep.* 12, 45-46.

77 *ibid.*

There thus seems to be a number of possible explanations for the 'embracing' crocodiles: 1) the images are a product of artistic imagination; 2) all three images depict a rare observation of venter to venter mating; 3) the animals' rolling behaviour during mating was mistaken for their method of copulation; or 4) the images do not depict mating at all, but were intended to represent a fight between males.

I suggest that the confusion that these scenes create indicates that they do not represent mating. The assumption that the animals are thus engaged has, I believe, stemmed from the similarity of their pose to human intercourse. It is important to note, however, that only a few species mate in this manner. The evidence examined during this study has pointed consistently to a tendency by artists to show mating animals in a rigid style, regardless of species-specific differences - that is, with a male reared and mounted upon a standing female. This standard motif could thus communicate clearly to the viewer the intended message: these animals are copulating. Nothing prevented the artists responsible for the reliefs in question from showing the reptiles mating in a 'normal' manner – that is, with the male mounted upon the female (as they usually do so in nature; cf. pl. 10T) or by simply employing the mating motif. The extreme deviation from this pattern can only mean that the artists did not wish to represent mating, but something else. I propose therefore, that the images were intended to depict a fight between crocodiles in which the animals lock jaws and roll together. The Egyptians may have observed such behaviour quite often, particularly during the breeding season. It is also possible that they witnessed the rolling of copulating crocodiles and nevertheless interpreted their behaviour as an act of aggression.

Nile soft-shell turtles

ḥzj (SB49: Saqqara, VI.1L-2E)

Depictions of turtles are rare in wall scenes, but both the shape and the distinctive aquatic behaviour of those found in the tombs of *ḥzj* and *mḥw* indicate that they are probably Nile soft-shell turtles (see Chapter 4).[78] A further example, also found in the tomb of *ḥzj* (SB49; see fig. 10-28), appears to show these animals mating.[79]

The pair is located underwater in a marsh scene, standing upon a riverbed. The profile of their carapaces is somewhat high for soft-shell turtles, but their elongated and striated feet resemble the webbed and clawed toes of the species. The size of the two animals is closely matched. The male has reared up on his hind legs to place his forefeet upon the shell of his companion.

To my knowledge, the mating habits of Nile soft-shell turtles have not been recorded, but in related species, the male does not stand upon the substrate during mating, nor touches the female with his forelimbs, but instead achieves intromission while the pair swims together.[80] In some species, the male also hooks his hind-feet under the posterior rim of the female's shell while using his forelimbs to stay afloat.[81] Thus, both the posture and the location of the *ḥzj* soft-shell turtles are biologically inaccurate. It should also be noted that the size ratio of the depicted animals is also incorrect as female Nile soft-shell turtles are considerably larger than the males.

Nile soft-shell turtles are secretive creatures that spend most of their time hidden in the silt or sand of riverbeds. They are also nocturnal. Artists may consequently have been unable to witness the mating behaviour of these animals and instead modeled it upon either terrestrial mammals or, more likely, that of land tortoises, such as the Egyptian tortoise (cf. pl. 10U).

---

78 Fischer, *Turtles*, 9.

79 Although turtles were consumed during the predynastic and Old Kingdom periods, they are not mentioned in offering lists, indicating that they were probably considered impure. In later periods, the Egyptians viewed the turtle as the enemy of Re, possibly due to its secretive habits and nocturnal behaviour, however Fischer suggests that they may have deeply mistrusted the creature long before this. It is consequently surprising to find not only the image, but also the procreative behaviour of these animals depicted in a tomb. See *ibid.*, 6-12 and also van de Walle, B. (1953). "La tortue dans la religion et la magie égyptiennes", *NouvClio* 5, 173-189.

80 Carr, *Handbook of Turtles*, 8.

81 Plummer, M.V. (1977). "Notes on the courtship and mating behavior of the soft-shell turtle, *Trionyx muticus* (Reptilia, Testudines, Trionychidae)", *J. Herp.* 11(1), 92.

10.3 MASTURBATION

Male animals are occasionally depicted in tomb reliefs with an apparent erection. Sometimes, the length of the penis is the result of species-specific anatomical features of the sheath that covers the organ.[82] Other animals (e.g., baboons, bulls) display an erection during, and in response to, aggressive interactions (see Chapter 9). The males of some species, however, engage in masturbation. This behaviour is also found in wall scenes.

Donkeys

rꜥ-m-kꜣ.j (SB50-51: Saqqara, V.6? r/u - V.6-7)
nfr-jrt-n.f (SB52: Saqqara, V.8-9)
ḫww-wr (SB53: Giza, V.8-9M)
nj-kꜣw-ḥr (SB54: Saqqara, V.9)
nfr I (SB55: Giza, V.9-VI.1)
jꜣzn (SB56: Giza, VI.1M-2M)

Male donkeys were found to exhibit an erection in six tombs dating from the mid-5th until the early 6th dynasty.[83] In three examples, the animals are situated in close proximity to others; two of these herds occur on a threshing floor (SB50 and SB52; fig. 10-30), while the third group is returning to the field after delivering produce (SB54). In the remaining scenes, individual members of donkey trains display an erection as they transport large loads upon their backs (SB51, SB53, and SB56; fig. 10-29).

Both spontaneous erection and masturbation are common activities in male equids. Studies of this sexual behaviour by stallions have revealed that it occurs approximately once every one to three hours, in both pastured and stabled animals.[84] During erection, the penis is extended to its full length, while masturbation involves movements of the phallus (bouncing, pressing, or sliding) against the abdomen.[85] Ejaculation does not usually occur. The behaviour is socially facilitated such that groups of horses housed together often masturbate simultaneously.[86] Males usually engage in this activity when they are standing quietly, but it may also occur while they are feeding, drinking, grooming, or walking slowly. Erection is also sometimes observed in equids during aggressive interactions.[87]

None of the aroused males in wall scenes is adjacent to a female and so their physical state is not associated with, or an expression of, mating behaviour.[88] These images thus indicate that the Egyptians understood that male donkeys exhibit the behaviour spontaneously. The scenes also reveal an awareness of the social effect of the activity, as in many scenes, more than one animal is shown in this condition. In the tomb of *rꜥ-m-kꜣ.j*, for example, two donkeys at the head of a procession display an erect penis (SB51; see fig. 10-29); these animals are approaching a threshing floor in which two males in a crowded group are again shown with erections (SB50).[89] The phalluses of most of the animals hangs away from their bodies, but a male in a threshing scene in the tomb of *nfr-jrt-n.f* (SB52; see fig. 10-30) is clearly masturbating as he has elevated his penis so that it touches his belly.

All of the depicted donkeys are moving - transporting loads, marching over a threshing floor, or walking in a field. In each of these situations, men threaten them with sticks and wave their arms about. Some also tap the animals upon the back or grab hold of them. It is possible then, that the images do not display a spontaneous expression of pleasure by contented animals, but rather, an aggressive response to their mistreatment.

---

[82] For example, the apparently elongated penis of a dama deer depicted in the tomb of *ṯjj* is actually the result of a 15 cm tuft of hairs that extend from the sheath in this species. See Power, R.K. (2004). "Deconstructing, deciphering and dating the donkey in Old Kingdom wall paintings and relief", *BACE* 15, 133.

[83] For a recent study of the depiction of this behaviour by donkeys in tomb scenes, see *ibid.*, 131-143.

[84] McDonnell, S.M. (1992). "Normal and abnormal sexual behavior", *Vet. Clin. North Am. Equine Pract.* 8(1), 75.

[85] McDonnell, S.M. (1989). "Spontaneous erection and masturbation in equids", *Proc. AAEP* 35, 568.

[86] *ibid.*, 574-575.

[87] McDonnell, S.M. and Haviland, J.C.S. (1995). "Agonistic ethogram of the equid bachelor band", *App. Anim. Beh. Sci.* 43, 155.

[88] Curiously, no examples of mating by domesticated donkeys were found during this study (although see *Wild ass* above).

[89] Possibly these are intended to be the same animals. This would indicate that the donkeys that transported the sheaves were also employed to do the threshing.

It has been suggested that this impulsive and explicit behaviour will have offended the Egyptians' sense of decorum and that this, in turn, explains both their general dislike of the animals and the donkey's later association with the god of chaos, Seth.[90]

SUMMARY

- Comparatively few depictions of the sexual behaviour of animals were found in the wall reliefs examined. Many of the examples occurred in a small subset of tombs - notably those of *rʿ-m-kꜣ.j* and *nj-mꜣʿt-rʿ* [*jzzj-mr-nṯr*].
- Some images appear to depict precopulatory activity by domesticated animals, but apart from two labelled scenes of anovaginal sniffing, the sexual nature of most of these has been deemed questionable.
- The earliest example of mating was found in the Saqqara tomb of *rʿ-m-kꜣ.j*. This tomb dates to the reign of Niuserre in whose sun-temple the first known depictions of copulating animals are represented. The motif was thus likely copied from the royal context, but it is also possible that its inclusion in private tombs was the result of religious motivation, reflecting the rise of the worship of Re during the 5th dynasty.
- The tomb of *rʿ-m-kꜣ.j* is also the first to show masturbatory activity by donkeys. This suggests that either the artists responsible for the decoration, or possibly the tomb-owner, found the sexual behaviour of animals of particular interest.
- Often only one example of mating by a particular species was found; these unique examples are frequently located in the tomb chapel of *nj-mꜣʿt-rʿ* [*jzzj-mr-nṯr*].
- Many mating images appear to be biologically accurate. Some include idiosyncratic copulatory features - such as the neck bite of wild asses and striped hyenas - however species in which the female rests upon the substrate during copulation (leopards/lions, caracals, Libyan striped weasels) are reliably depicted incorrectly. The apparent mating behaviour of embracing Nile crocodiles was also found to be largely inaccurate and more likely to represent aggressive behaviour. The copulatory tie, one of the most distinctive features of canid mating behaviour, is never represented.
- The evidence examined here has indicated overwhelmingly that artists employed a standard design to represent mating. This 'mating motif' - which was based upon the most common pose adopted by animals in which the male rears while the female stands - could thus convey clearly and unambiguously to the viewer that a pair was copulating, irrespective of the species. It follows, therefore, that these reliefs do not necessarily depict observations of animal behaviour *per se*, but instead represent the procreative act generally. The ostensible accuracy of many of the images, then, may simply be the result of a chance correspondence between the motif and the mating style of individual species.

[90] Power, *BACE* 15, 134.

CHAPTER 11

PARENTAL BEHAVIOUR

DEFINITION

Parental behaviour involves the gestation and birth, or incubation and hatching, of offspring, and the care of young through to independence. Parental care occurs in all classes of vertebrates and in some invertebrates. It consists of a variety of activities, such as provision of nourishment for the young (by nursing or obtaining food), care of their physical state (e.g., by grooming), and protection from predators. Females most commonly perform parental care, but in many species both sexes play an active role.

11.1 REPRODUCTION

Animals produce progeny in one of two ways: either by giving birth to live young ('viviparous') or by laying eggs from which offspring eventually hatch ('oviparous'). Viviparous species include most mammals, but also some reptiles, amphibians, and fish. All birds and most reptiles, amphibians, fish, and invertebrates are oviparous.

11.1.1 Birth

The birth process for viviparous animals consists of three stages: a) dilation of the cervix, b) active labour in which the foetus enters the birth canal and is expelled from the mother's body, and c) ejection of the placenta ('afterbirth').[1] Images of birth in wall scenes from Giza and Saqqara depict the second stage.

Cows

nfr-msḏr-ḫwfw (PB1-2: Giza, V.2-8, possibly V.9)
jj-mrjj (PB3: Giza, V.3L)
rꜥ-m-kꜣ.j (PB4: Saqqara,V.6? r/u - V.6-7)
nj-ꜥnḫ-ẖnmw/ẖnmw-ḥtp (PB5: Saqqara, V.6L-7)
ꜣḫtj-ḥtp (PB6: Saqqara, V.6-8E)
jtj-sn (PB7: Giza, V.6-8?)
kꜣ.j-dwꜣ (PB8: Giza, V.6-9?)
sḫm-ꜥnḫ-ptḥ (PB9: Saqqara, V.8-9)
ṯjj (PB10: Saqqara, V.8-9)
kꜣ.j-m-nfrt (PB11: Saqqara, V.8-9?)
ꜣḫtj-ḥtp (PB12: Saqqara, V.8L-9E)
nbt (PB13: Saqqara, V.9)
ptḥ-ḥtp II: ṯfj (PB14: Saqqara, V.9M-L)
ꜣḫtj-mrw-nswt (PB15: Giza, V.9-VI.1)
kꜣ.j-m-ꜥnḫ (PB16: Giza, VI)
ṯtw: kꜣ.j-nswt (PB17: Giza, VI)
zšzšt: jdwt (PB18: Saqqara,V.9 r/u - VI.1-2E?)
jꜣzn (PB19: Giza, VI.1M-2M)
mṯtj (PB20: Saqqara, VI.2?)

Twenty scenes dating from the early 5th to the early 6th dynasty were found to depict cows as they give birth. In all but two of these images, a herdsman either stands or, more frequently, kneels behind the cow and grasps her emerging calf with both hands. The herdsmen's behaviour suggests that they are assisting the cows with their deliveries.

General restlessness is one of the first indications that a cow is beginning labour.[2] This is followed by signs of pain, such as treading of the hind legs or kicking at the abdomen. The cow then begins to strain, arching her back with each uterine contraction - an action that causes the first of two water bags (the 'allanto-chorion') to burst. This stage usually takes from four to ten hours, during which time the cow will lie down and rise repeatedly. Approximately one hour after

1 Fraser, *Reproductive Behaviour*, 113.
2 *ibid.*, 118.

the first bag has ruptured, the second sac (the 'amnion') emerges and breaks.[3] Abdominal contractions and straining then begin to increase in frequency and duration, until the foetus starts to emerge.[4] The cow is usually recumbent during this stage and often bellows in pain.[5] In a normal delivery, the calf's forelegs appear first and then its muzzle. After the head and shoulders have emerged, cows often stand up as the remainder of the body is expelled.

Normal delivery usually takes place within two hours after the second water bag has ruptured; if no progress occurs during this time, the cow may be experiencing difficulties. Calving problems ('dystocia') are especially common in cows that are delivering for the first time ('heifers'). The most frequent complications are: 1) the calf's head or legs are contorted, 2) the calf emerges backwards ('breech'), or 3) the calf is too big and its hips too broad to pass through its mother's pelvis in a normal orientation, a situation known as 'hip lock'.[6] All of these problems are potentially fatal and require human assistance to ensure the survival of the cow and/or calf. Cows often stand during difficult deliveries.[7]

The calf's abnormal posture can sometimes be corrected by reaching into the birth canal and gently manoeuvring its body by hand. If it is jammed in the uterus, however, obstetrical chains are attached to each foreleg and its body extracted by pulling first on one rope and then the other to slowly 'walk' its legs and shoulders forward and out (see pl. 11A). This procedure takes great effort (see pl. 11B), but must be done very slowly to avoid injuring the calf and cow. If hip lock has occurred, the calf's body must be pushed back a short distance and rotated a quarter turn so that its hips are positioned one above the other. Its forelegs are then pulled towards the cow's flank to slowly pass one hip and then the other through her pelvic opening.

Normal calving does not require human intervention. Consequently, the predominance of reliefs depicting assisted births must indicate that Egyptian wall scenes record cows that are experiencing difficulties. In the early tomb of *jj-mrjj* (PB3; see fig. 11-1), the attending herdsman stands behind the cow, but in the remaining images, the helper kneels upon the ground. Many of the men hold the calf's head with one hand and its forelegs with the other (see figs. 11-1 and 11-2), but in some scenes, only the legs are grasped (figs. PB1, PB6, PB9 to PB11, PB16, and PB19; see figs. 11-3 and 11-4). I suggest that all of the herdsmen are attempting to pull free calves that have become jammed in the birth canal.[8] When using obstetrical chains, the calf must be pulled in a manner that follows the curvature of the birth passage: first in a slightly upward direction until the head and shoulders have appeared and then downwards (cf. pl. 11B). By kneeling below the emerging foetus and pulling it down towards them, the men would thus be able to place each calf in the best position for extraction, while simultaneously using gravity to exert the maximum effort. Indeed, in the tomb of *nj-ꜥnḫ-ẖnmw/ẖnmw-ḥtp* (fig. 11-2), the herdsman has placed one foot against the cow's hind leg in order to achieve greater pulling strength with his upper body (cf. the use of obstetrical chains in pl. 11B).

The standing position of each of the cows while delivering may also indicate that they are experiencing difficulties. As noted in Chapter 4, cattle are represented in a recumbent posture while being fed, so it was certainly within the artists' capabilities to have depicted normal calving in which cows lie upon the ground until just prior to the full expulsion of their calves.

---

3 Montet (*Scènes*, 98) suggested that an elongated object that emerges from a calving cow in the tomb of *ṯjj* (PB10; see fig. 11-3) represents a flood of liquid falling to the ground; it may thus depict the bursting of the second water sac. This shape might also depict membranes from the ruptured bag, which often hang from the cow's vulva until the foetus begins to appear.

4 Fraser, *Reproductive Behaviour*, 119.

5 *ibid.*, 122.

6 See Cuneo et. al., *Injuries and Diseases*, 3.

7 Metz, J. and Metz, J.H.M. (1987). "Behavioural phenomena related to normal and difficult deliveries in dairy cows", *NJAS* 35(2), 98.

8 When describing the calving scene in the tomb of *ṯjj* (see fig. 11-3), Montet (*Scènes*, 98) stated that the herdsman holds the legs of the emerging calf in order to hasten its birth. Under normal circumstances, however, assistance is not necessary; indeed, human intervention during calving, when it is not required, is likely to harm both the cow and the calf.

The label *msjt*, 'giving birth'[9] sometimes accompanies calving scenes (PB3, PB5, PB9, PB16, and PB19); more frequently, though, these images are given the descriptor *sfḫ*, which refers to the act of 'loosening, releasing, or removing'[10] (PB3, PB5, PB6, PB7, PB9 to PB11, PB17, PB19, and PB20; see figs. 11-3 and 11-4). To the extent that the term *sfḫ* is associated with parturition in various texts, scholars have naturally interpreted it in the context of calving scenes to mean 'deliver' or 'give birth'.[11] In four scenes the term *sfḫ* (or *sfḫt*) appears alongside the label *msjt* (PB3, PB5, PB9, and PB19; see figs. 11-1 and 11-2). The draftsmen who wrote these inscriptions may have been attempting to emphasize the cows' behaviour by referring to it with two essentially identical labels.[12] It seems more likely, however, that the pairing of these terms indicates that *sfḫ* alludes to something other than giving birth *per se*. I propose that *sfḫ* instead describes the actions of the herdsmen who are endeavouring to 'loosen' and 'release' calves that are jammed inside their mothers by pulling on their heads and forelegs.[13] The unique label in the calving scene of *ꜣḫtj-ḥtp* (PB6) may refer directly to the calf's predicament with the statement *sfḫ wrt bḥz pw jm=f gꜣw*, 'Loosen/release greatly this calf in its constriction'.[14]

Other evidence in these scenes suggests that the cows are struggling to pass their calves. Some arch their backs abruptly (PB5 and PB15; see fig. 11-2). Many cows, particularly those depicted in reliefs from the late 5th dynasty onward, lock their forelegs and stretch their heads forward as though bracing in response to the herdsmen's tugging (PB10, PB12, PB13, PB17, PB18, and PB20; see fig. 11-3). In particular, however, almost all of the calving females protrude their tongues prominently (PB1, PB3, PB5, PB7, PB10 to PB13, PB16 to PB20; see figs. 11-1 to 11-4). As noted above, cows vocalize frequently during delivery. Although their tongues are not usually visible when calling (see pl. 9Q in Chapter 9), I propose that this motif in calving scenes was intended to convey to the viewer the notion that the cows are bellowing loudly in response to their prolonged and painful labours.[15] Other authors have also reached this conclusion. When describing the scene in the tomb of *zšzšt: jdwt*, Macramallah states that she is "poussant un meuglement de douleur",[16] Ziegler similarly says that the *mttj* cow "meugle de douleur",[17] and Simpson notes that in the tomb of *jꜣzn*, "The cow extends her tongue in pain".[18] It is interesting to note that in one of the two images of unassisted calving, the cow does not protrude her tongue (PB15; see fig. 8-20 in Chapter 8).

If my interpretation is correct and wall reliefs depict cows experiencing difficulties rather than calving normally, it must be asked why artists chose to depict such a dire and potentially fatal event? One possibility is that these images each record a successful assisted delivery made under extreme circumstances on the estates of individual tomb-owners - unique occasions of which they were particularly proud. Alternatively, the prevalence of these scenes may indicate that abnormal deliveries were actually a common occurrence for Egyptian herdsmen. Assistance during calving has increased considerably in recent years in modern farming practice. One of the major reasons for this rise in calving problems has been the tendency to mate cows of only average size with genetically large bulls in order to produce larger calves.[19] The pelvic area of cows is often not wide enough to allow them to deliver such heavy calves without assistance; this is particularly so for

---

9 *Wb ii*, 137-141.

10 *Wb iv*, 116-117. For a brief comment on this term, see Montet, *Scènes*, 98-99.

11 For example, see Simpson, *Sekhem-Ankh-Ptah*, 8; Simpson, *Kayemnofret*, 19; Simpson, *Mast. West. Cem. Part I*, 11 and 20; Moussa-Altenmüller, *Nianchchnum*, 154. Weeks (*Cemetery G6000*, 42) interprets the label as an instruction to the cow to 'Relax'!

12 Montet (*Scènes*, 98) believed that the two terms were interchangeable when referring to calving.

13 Some scholars have indeed translated *sfḫ* as 'delivre' 'free', but have not necessarily viewed this as an urgent command. For example, see Loret, V. (1896). "Les animaux reproducteurs dans l'Égypte ancienne" in *Deux Études*, 13, B. no. 1; Ziegler, *Catalogue*, 130; Ziegler, *Akhethetep*, 153.

14 Ziegler (*Akhethetep*, 153) translates the phrase, *jm=f gꜣw*, as a negative, 'that it does not choke', based upon a similar Late Period example.

15 Vandier (*Manuel v*, 64-67) describes this behaviour as licking, but the animals' tongues do not make contact with anything (except possibly the hand of a herdsman in PB16).

16 Macramallah, *Idout*, 17.

17 Ziegler, *Catalogue*, 130.

18 Simpson, *Mast. West. Cem. Part I*, 20.

19 Whittier-Thorne, *Assisting the Beef Cow*, 1.

immature heifers. The depiction of assisted calving in wall scenes may thus reflect a naïve or deliberate practice by Egyptian pastoralists to mate large bulls with small cows or heifers.

It is tempting to speculate whether these calving difficulties may be the result of breeding between domesticated cows and wild caught bulls. Wall scenes reveal that wild cattle were hunted and killed for sport throughout dynastic history,[20] but depictions of lassoing indicate that bulls were sometimes taken alive.[21] If these animals, which will have been genetically larger than domesticated stock, were added to existing herds and mated with cows, they would undoubtedly have contributed to an increase in calving difficulties.

Goats

sḫm-k3-rʿ (PB21: Giza, V.2)
nj-ʿnḫ-ẖnmw/ẖnmw-ḥtp (PB22: Saqqara, V.6L-7)
ḥtp-ḥr-3ḫtj (PB23: Saqqara, V.6-8E)
3ḫtj-ḥtp (PB24: Saqqara, V.6-8E)
3ḫtj-mrw-nswt (PB25: Giza, V.9-VI.1)

Goats give birth after a five-month gestation period, with both wild and domesticated breeds commonly producing twins (pl. 11C). During labour, the doe paws at the ground and lies down and rises repeatedly, straining until the amniotic sac emerges. During a normal delivery, the kid's nose and forelegs appear first, followed by its head (see pl. 11D); once its shoulders have passed, the rest of the body slides out quickly.[22] This is undoubtedly a painful process, as it has been observed that, "It is not uncommon for goats to scream as the kids are being born".[23]

Five depictions of goats giving birth were found during this study.[24] In four of the images, the females lean over and stretch their necks well forward, as the head of their newborn emerges beneath their tails (PB21, PB22, PB24, and PB25; see fig. 11-5). One doe arches her back (PB25; see fig. 8-8 in Chapter 8), while another protrudes her tongue prominently, indicating that she is vocalizing in response to the pain (PB22; see fig. 8-7 in Chapter 8). All of the females are thus labouring heavily to deliver their kids. In the remaining scene (PB23; see fig. 11-6), however, the goat stands with her head erect and seems to be more relaxed. Unlike the other images in which only the heads of the emerging foetuses are visible, both the head and shoulders of her kid have appeared, indicating that the worst of her labour has passed.

The resulting twins can be a problem for farmers, especially if the extra kids are male. Female goats are valued because they can provide both meat and milk. Males are more difficult to keep due to their greater size and strength, their aggressive behaviour, and also because of a distinctive, musky odour that they develop as they mature. Consequently, male kids are usually either castrated and reared for meat, or slaughtered shortly after birth.[25]

During this study, five scenes were found to depict the slaughter of a goat. These occur in four tombs: *nfr/k3-ḥ3.j*[26] *nj-ʿnḫ-ẖnmw/ẖnmw-ḥtp* (2 examples)[27] *nb-k3w-ḥr: jdw*,[28] and *3ḫtj-mrw-nswt*.[29] In each case, the animal's body is suspended from a tree branch while a herdsman dismembers it. Two of the five slaughter scenes occur on the same walls as goats giving birth (in the tombs of *nj-ʿnḫ-ẖnmw/ẖnmw-ḥtp* and *3ḫtj-mrw-nswt*). In a third slaughter scene (in the tomb of *nfr/k3-ḥ3.j*), the goat's body hangs from the same tree upon which a doe feeds while simultaneously nursing her

20 Brewer et al., *Domestic Plants and Animals*, 87.

21 In this study, scenes showing the lassoing of wild bulls in the desert were found in the Saqqara tombs of *nj-ʿnḫ-ẖnmw/ẖnmw-ḥtp* and *ptḥ-ḥtp II: tfj*.

22 Thear-Fraser, *Complete Book*, 90-91.

23 *ibid.*, 90.

24 The earliest of these images, in the tomb of *sḫm-k3-rʿ* (PB21), is in poor condition. Nevertheless, Hassan (*Giza iv*, 111) identifies the animal as a goat even though the accompanying inscription suggests that it could be a cow. The horns of the animal are also more akin to those of goats than cows.

25 Thear-Fraser, *Complete Book*, 92 and 94.

26 Chapel, east wall (see Moussa-Altenmüller, *Nefer and Ka-hay*, pl. 4).

27 Vestibule, north wall (see Moussa-Altenmüller, *Nianchchnum*, fig. 8) and outer hall, west wall (*ibid.*, pls. 76 and 78b).

28 Pillared hall, south wall (see Hassan, *Neb-Kaw-Her*, fig. 13).

29 Secondary offering-room, west wall (see *Giza Archives Project*, no. A779_NS).

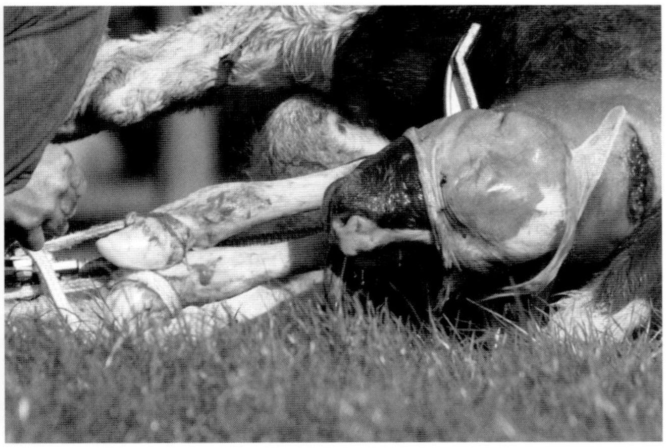

Pl. 11A Emergence of the head and forelegs
© iStockphoto.com/Rolf Nathaus

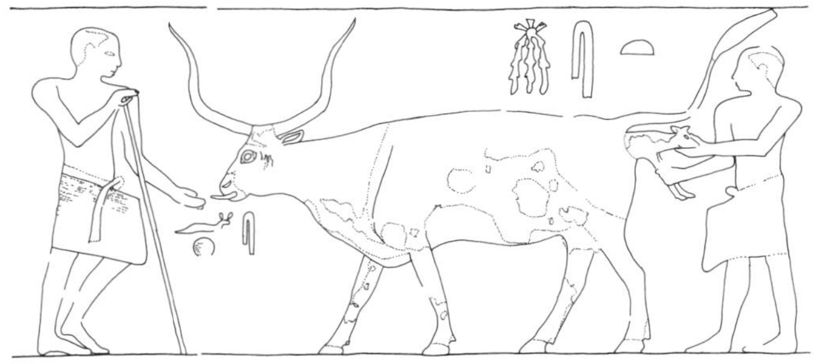

Fig. 11-1 (PB3)

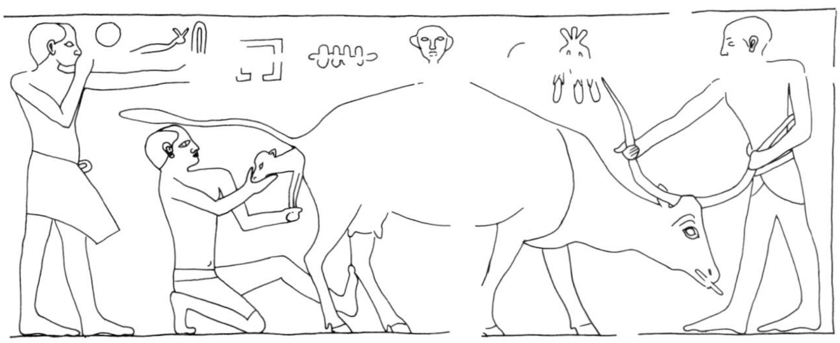

Fig. 11-2 (PB5)

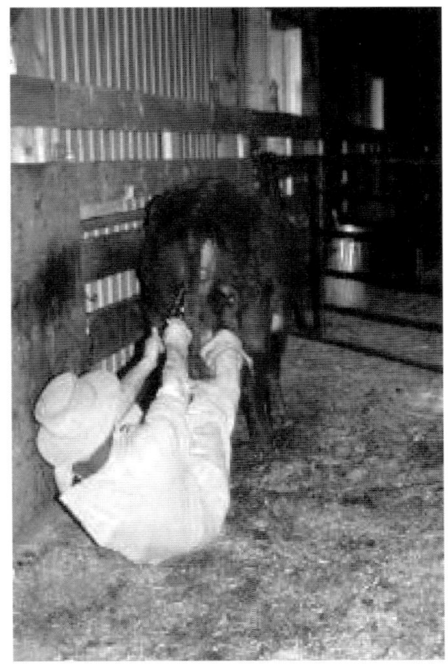

Pl. 11B Pulling a calf using chains
© Joe Stookey
Reproduced with permission

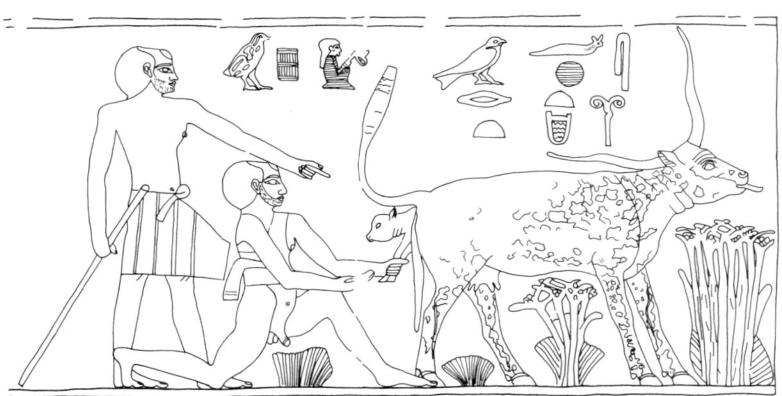

Fig. 11-3 (PB10)

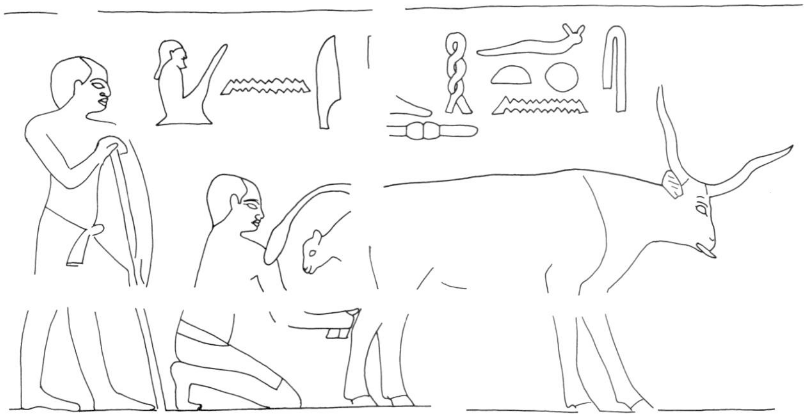

Fig. 11-4 (PB11)

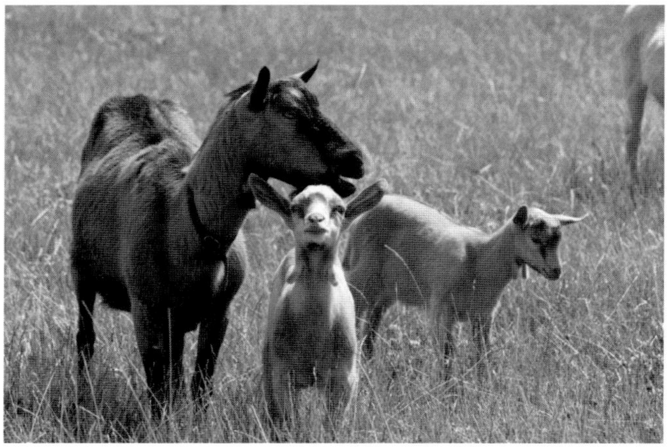

Pl. 11C Goat twins
© iStockphoto.com/Cornelia Pithart

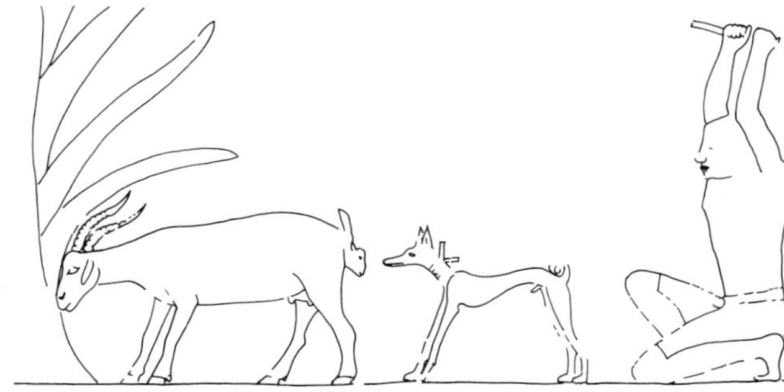

Fig. 11-5 (PB22)

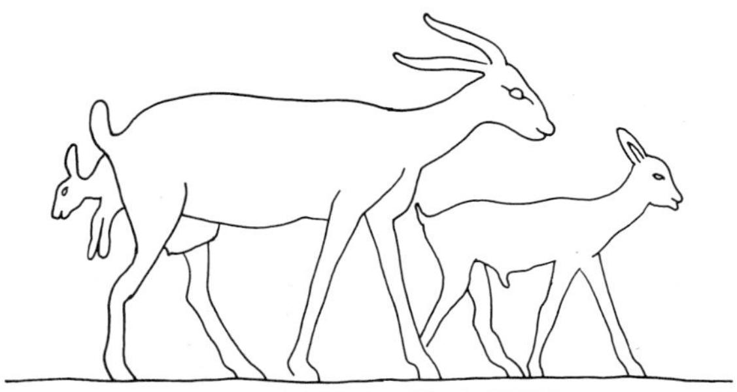

Fig. 11-6 (PB23)

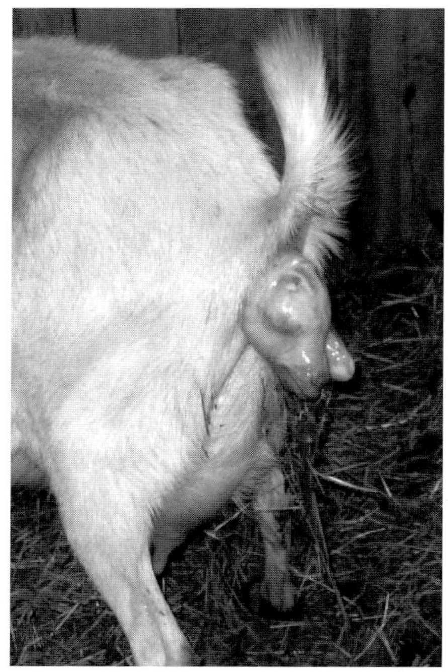

Pl. 11D Head and forelegs emerging
© Lynne Gelston, Dream Thyme Farm
Reproduced with permission

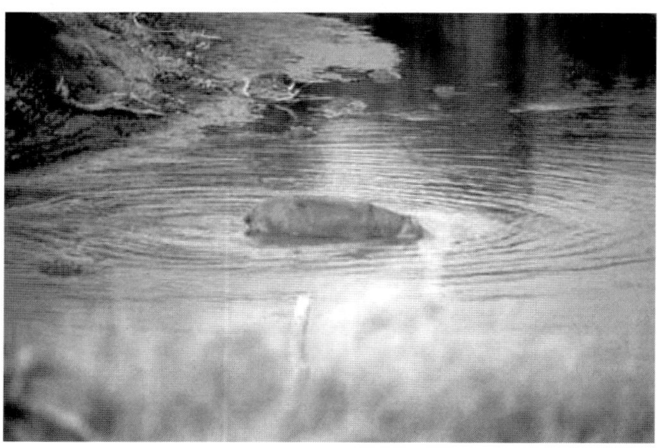

Pl. 11E Hippopotamus giving birth
© Dawn Dalton
Reproduced with permission

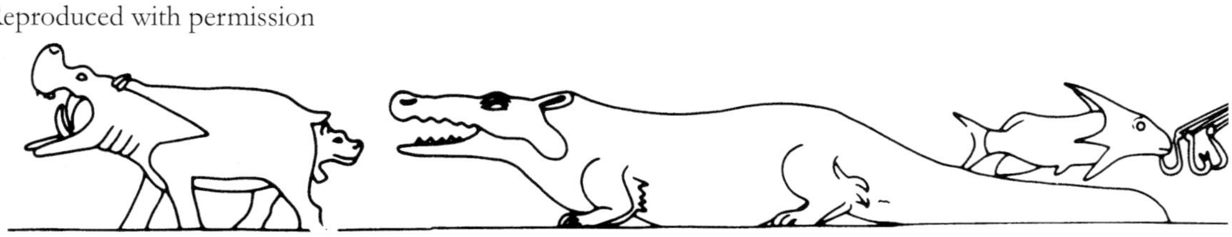

Fig. 11-7 (PB32)

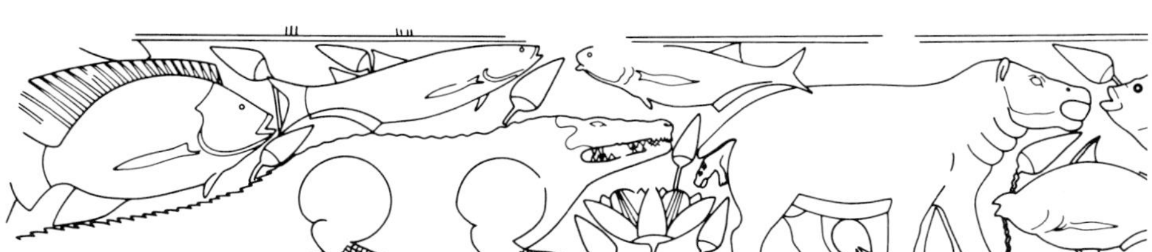

Fig.11-8 (PB34)

kid (see PB61). Finally, three of the slaughter scenes (in the tombs of *nfr/k3-ḥ3.j*, *nb-k3w-ḥr: jdw*, and *nj-ʿnḫ-ẖnmw/ẖnmw-ḥtp*) appear on the same walls upon which goat kids are depicted standing in small, dome-shaped cages (see fig. 11-28 and also fig. 7-22 in Chapter 7 and fig. 8-7 in Chapter 8). The tomb of *nj-ʿnḫ-ẖnmw/ẖnmw-ḥtp* depicts, on the same register, the birth, capture, and slaughter of a goat.

Ancient Egyptian goats, like wild and domesticated breeds today, probably produced twins frequently. I propose that the close compositional association of birth, slaughter, and trapping scenes involving goats in wall reliefs indicate that Egyptian herdsmen may have killed the extra (or male) kids for meat and other products. For further evidence regarding the removal of twins, see *Goats* in section 11.3.2 below.

Hippopotamuses

snḏm-jb: mḥj (PB26: Giza, V.9)
k3-gm-nj: mmj (PB27-29: Saqqara, VI.1M)
mrrw-k3.j: mrj (PB30: Saqqara, VI.1M-L)
zšzšt: jdwt (PB31-32: Saqqara, V.9 r/u - VI.1-2E?)
sʿnḫ-w(j)-ptḥ (PB33: Saqqara, VI.1L)
ʿnḫ-m-ʿ-ḥr: zzj (PB34: Saqqara, VI.1L-2E)
mrrj (PB35: Saqqara, VI.2E?)

Ten scenes were found to depict hippopotamuses giving birth. With one exception, these images occur exclusively in 6th dynasty tombs from Saqqara. Each mother stands underwater. In some examples, her head is held level or lowered with her mouth closed (PB28 to PB30, PB33, PB35; see fig. 8-35 in Chapter 8), while in others, her head is stretched forward with her jaws parted to reveal her tusks and/or tongue (PB26, PB27, PB32, and PB34; see fig. 11-7). The emerging newborn can be seen in seven images, but it is missing in three: that of *snḏm-jb: mḥj* (PB26) and in two of the three birth scenes that occur in the tomb of *k3-gm-nj: mmj* (PB27 and PB28). Nevertheless, the open mouths of two of the animals (PB26 and PB27), the arched back of one (PB26), and the presence of a crocodile behind each suggest that the artists intended to depict a birth.

Nile crocodiles lurk directly behind the mothers in eight examples (see Chapter 8). The crocodiles often part their jaws as though anticipating a strike on the emerging young; indeed, in the tomb of *mrrw-k3.j: mrj*, the reptile bites the newborn upon the head (PB30; see fig. 8-35 in Chapter 8). A crocodile menacing a straining hippo in the tomb of *k3-gm-nj: mmj* has itself been grasped by another hippopotamus (PB27; see fig. 9-23 in Chapter 9); perhaps the missing newborn in this scene indicates that the predatory reptile has been successful?

Hippopotamus calves are born after a gestation period of approximately eight months.[30] There is considerable disagreement in the literature regarding where hippos usually give birth, but it would appear that they can do so either on land or in shallow water (see pl. 11E).[31] Shortly before she delivers, the mother leaves her social group in search of a secluded area with dense cover.[32] Observations of births that have taken place in zoos indicate that labour lasts for approximately three hours. In one account, the female sat on her haunches or lay on her belly, and rose repeatedly as her labour progressed; most of this behaviour took place in a shallow pool in her enclosure. In the moments prior to the arrival of her calf, however, the female stood up, lunged forward, and strained heavily.[33] The calf was consequently expelled into the water. This matches closely the images of birth found in this study in which hippos stand underwater as the head and forelegs of their newborns emerge.

In another record of a zoo birth, the hippo was observed to suffer a great deal during labour, "opening and closing her jaws, and clattering her teeth".[34] In a small number of wall scenes, the females appear to protrude their tongues (PB26, PB27, PB32, and PB34?; see fig. 11-7 and possibly fig. 11-8). Hippos are very vocal animals and recent studies have shown that they also produce

30 Eltringham, *The Hippos*, 63.
31 Verheyen, *Hippopotame*, 45-47.
32 Apps, *Wild Ways*, 141.
33 Senior, M. and Tong, E.H. (1963). "Parturition in a hippopotamus", *Proc. Zool. Soc. Lond.* 140, 59.
34 Bartlett, A.D. (1871). "Notes on the birth of a hippopotamus in the Society's Gardens", *Proc. Zool. Soc. Lond.* 38, 255-257.

many sounds underwater.[35] Tongue presentation by cows and goats giving birth suggest that they are vocalizing in response to the pain of labour (see above); this feature may thus have a similar significance for depictions of the hippopotamus. As noted in Chapter 9, I suggest that the goddess Taweret, who is frequently depicted with a protruding tongue, may also have been perceived as vocalizing. This deity is represented as a pregnant hippopotamus and her primary role was the protection of women during childbirth. Her protruding tongue may thus be a visual expression of, and sympathetic allusion to, the noise produced by all females in response to the pain of birth.

Common genets

zšzšt: jdwt (PB35a: Saqqara,V.9 r/u - VI.1-2E?)

An isolated relief fragment found in the tomb of *zšzšt: jdwt* (PB35a; see fig. 11-9) depicts the birth of a common genet. The head and neck of an ibis and the remains of outstretched wings near this unique image reveal that the piece comes from a marsh scene. Although only a hind leg and part of the mother's long tail are visible, she is undoubtedly a genet, as her emerging young displays an elongated muzzle, small paws, and the characteristic pricked ears of this species.

Common genets usually give birth to a litter of one to three young in a hole or a nest hidden in vegetation; the papyrus thicket in which the image occurs is thus appropriate. Contrary to the alert appearance of the emerging newborn, the eyes of genet kittens remain closed until they are c. eight days old.[36] The large size of the animal (relative to its mother) is also incorrect, as genet kittens usually weigh less than 100 grams at birth.[37] Common genets are reported to breed in Egypt in March to April,[38] indicating that the depicted marsh scene may have been set during spring.

Unidentified mammal

pḥn-wj-k3.j (PB36: Saqqara, V.6-8E)

A fragmentary desert scene from the mid-5th dynasty tomb of *pḥn-wj-k3.j* was found to include a depiction of the birth by an unidentified animal (PB36; see fig. 11-10). Very little remains of the scene. Only a portion of the mother's lower back and cocked tail under which her newborn emerges; the young animal's shoulders and one of its ears are also visible. A small amount of red-brown paint is still adhered to the mother's hindquarters.

It is impossible to determine which species is depicted in this relief, but some observations can still be made. As the image occurs in a desert scene, it is unlikely to represent a domesticated animal, such as a cow. It has been suggested elsewhere that it may be an 'antelope', without reference to any particular species.[39] As it seems to have been of a similar size to a pair of mating hartebeest situated beside it (see fig. 10-12 in Chapter 10), which are also painted a reddish-brown, this might indicate that the animal in question is of the same species. The animal's tail appears to have extended beyond the break in the relief, suggesting that it was originally of considerable length. This allows dorcas gazelles to be eliminated from consideration as members of this species have short tails.

It is important to note that, during this study, no other depictions of birth were found to occur in a desert scene. Reliefs from the Chamber of the Seasons in the sun-temple of Niuserre, which are contemporary with the tomb of *pḥn-wj-k3.j*, include images of not only ungulates, but also cats that give birth while in a desert context.[40] The animal in question may thus represent a feline species.

35 Barklow, W.E. (2004). "Amphibious communication with sound in hippos, *Hippopotamus amphibius*", *Anim. Behav.* 68(5), 1125-1132.

36 MacDonald, *Encyclopaedia of Mammals*, 142; Schliemann, H. (1990). "Viverrids" in *Grzimek's Encyclopedia iii*, 519.

37 *ibid.*

38 Hoath, *Field Guide*, 88.

39 Arnold, D. (1999). "Cat. no. 146A,B. The hunt in the desert from the tomb of Pehen-wi-ka" in *Eg. Art in the Age of the Pyramids*, 398-400.

40 See von Bissing, F.W.F. (1956). "La chambre des trois saisons du sanctuaire solaire du roi Rathourès (Ve dynastie) à Abousir", *ASAE* 53, pl. 11.

11.1.2 Egg-laying

Oviparous animals produce young by laying fertilized eggs. In birds and reptiles, these pass through the female's oviduct and emerge via her cloaca. Some species lay one egg at a time; others produce many at one sitting. Eggs are frequently deposited in a concentrated group in a nest, depression, hole, or other location. The eggs of birds and reptiles are usually covered by a porous membrane and outer shell.

Nile crocodiles

k3.j-jrr (PB37: Saqqara, VI.2?)

In a unique image located in the Saqqara tomb of *k3.j-jrr*, a Nile crocodile lays four eggs (PB37; see fig. 11-11).[41] The animal arches its back and tilts its tail abruptly above the substrate in a manner not encountered in any other depictions of crocodiles examined during this study.[42] The line of eggs, which sit side by side upon the ground, begins directly behind the reptile's hind legs. This area is immediately below the cloaca, the orifice from which the eggs of crocodilians emerge. The animal is underwater as indicated by nearby water plants and fish.

Crocodiles lay their eggs in large holes that they excavate into sandy soil in riverbanks near water (see pl. 11F).[43] Heat from the surrounding soil causes the embryos to develop. After digging her nest, the female straddles the depression with her hind legs and deposits a clutch of c. 50 eggs. Once she has covered the hole with soil, the mother then guards her nest continuously during the next 12 weeks while the eggs incubate.[44]

The *k3.j-jrr* image, in which a crocodile, while fully submerged, elevates its tail to lay its eggs directly onto the substrate, is thus incorrect. When interpreting the animal's underwater activities, Khaled Daoud states that, "the scene displays a step ahead, before the eggs would be covered with a thick leathery shell and laid in rotting vegetation".[45] This would appear to imply that the scene shows the animal *prior* to egg-laying, however this does not explain the four spheres that are already situated outside the animal's body. Furthermore, when crocodile eggs are laid, they are covered by a smooth, brittle shell of calcium carbonate. Towards the end of the incubation period, however, the shells may occasionally crack to reveal the leathery inner membrane protecting each embryo.[46]

It is unlikely that the Egyptians did not know that these animals lay their eggs on land. When young crocodiles are ready to hatch, they emit vocalizations. The calls alert their mother who then unearths the nest-hole to release her buried hatchlings (see pl. 11F). This process leaves behind both the opened pit and many discarded shells, a clear marker for an observer of the crocodile's parental behaviour.[47] This species originally bred in large communal nesting sites in which the craters of many abandoned nests could be found in close proximity;[48] such sites will once have been located along the Nile. For example, a Late Period (?) crocodile mummy discovered in the

---

41 The full report for this tomb is not yet published; however, this image suggests that it may contain other unique scenes. See Daoud, K.A. (1995). "The tomb of Kairer: Preliminary report on the field work, season, 1993", *GM* 147, 35-52; Daoud, K.A. (1997). "Unusual scenes in the Saqqara tomb of Kairer", *EA* 10, 6-7; Daoud, K.A. (2000). "The mastaba of Kairer: Preliminary report on the field work seasons, 1995-1998", *ASAE* 75, 91-106.

42 A crocodile in the Giza tomb of *3ḫtj-mrw-nswt* lifts its tail in a similar manner (see fig. 4-22 in Chapter 4), but as it does not arch its back or display a clutch of eggs, the animal's posture clearly indicates a predatory activity, not reproductive behaviour.

43 Modha, M.L. (1967). "The ecology of the Nile crocodile (*Crocodylus niloticus*) on Central Island, Lake Rudolf", *EAWJ* 5(August), 84-88.

44 Cott, H.B. (1961). "Scientific results of an inquiry into the ecology and economic status of the Nile crocodile (*Crocodilus niloticus*) in Uganda and Northern Rhodesia", *Trans. Zool. Soc. Lond.* 29, 259-276; Modha, *EAWJ* 5(August), 90.

45 Daoud, *ASAE* 75, 97, n. 9.

46 Modha, *EAWJ* 5(August), 88.

47 Cott, *Trans. Zool. Soc. Lond.* 29, 276.

48 *ibid.*, 270-272.

Fayum, which was found to contain the bodies of 50 hatchlings and their shells,[49] reveal that an entire clutch had been exhumed.

It is important to note that every depicted Nile crocodile examined during this study was found to be situated in water. The bodies of some occasionally break the water surface (e.g., when grasped by an attacking hippopotamus), but none are depicted on land. The Egyptians were undoubtedly aware of the terrestrial activities of these creatures as they will have witnessed them walking and basking on riverbanks during the day. Within the tomb, however, their depiction was restricted to the aquatic environment. As noted in Chapter 8, I suggest that the Egyptians may have believed that the crocodile's vision was limited when submerged, thus rendering it less of a threat. Extending this idea further, it is possible that - within the context of the tomb - depictions of crocodiles were required to be submerged in order to protect the spirit of the deceased. It would follow, then, that even egg-laying - a terrestrial activity - had also to be depicted underwater. Heinrich Schäfer observed that depressions and pits never pierce the baseline in Egyptian art;[50] this constraint would explain why the *k3.j-jrr* crocodile deposits its eggs directly upon the substrate. This, in turn, required that the reptile's tail be lifted so that her laying behaviour could be viewed clearly.

11.1.3 Egg incubation

Eggs must be kept in a stable environment (e.g., appropriate temperature, humidity level, etc.) if the embryo within each is to develop. Warmth may be applied via the parent's body or by taking advantage of environmental sources of heat (e.g., the sun or generated by decomposing vegetation). Birds usually deposit and incubate their eggs in a nest. These can range from a simple scrape in the ground to elaborate structures made from plant and other materials. Nests are often elevated (in trees, on ledges, etc.), but they may also be situated upon the ground or even underground (in holes and burrows).

Birds

rʿ-špss (PB38: Saqqara, V.8M)
snḏm-jb: jntj (PB39: Giza, V.8M-L)
ꜣḫtj-ḥtp (PB40: Saqqara, V.8L-9E)
mḥw (PB41: Saqqara, VI.2M-3?)

Most of the images of birds incubating their eggs that were examined during this study were found to depict them in a threat posture with their wings spread (see Chapter 9). The normal wing position for birds when sitting upon their nests is folded and held close to the body (see pl. 11G), yet only four examples of this posture were detected in tombs dating from the late 5th to mid-6th dynasty.

The species of the seated birds can be determined in only two scenes: a hoopoe in the tomb of *snḏm-jb: jntj* (PB39; see fig. 11-12) and a sacred ibis in the tomb of *ꜣḫtj-ḥtp* (PB40). The birds clearly sit in bowl-shaped nests, two of which rest upon papyrus stalks (PB38 and PB41), and two upon umbels (PB39 and PB40; see fig. 11-12). As their eggs are not apparent in any of the scenes, however, it cannot be stated with confidence that they are incubating. In the earliest image (PB38), an approaching genet threatens the seated bird; another, somewhat smaller bird, which stands and flaps its wings in alarm, also shares the nest. This suggests that the seated bird is probably not incubating eggs, but is an adult with a hatched nestling. Its passive response to the predator is unusual.

The overwhelming tendency to depict birds in a threat posture, rather than sitting passively upon their nests, suggests that the act of incubation was of less interest to Egyptian artists than the active defence of eggs. The preoccupation with this avian behaviour explains why nesting 'mistakes' are found in wall reliefs. As many authors have pointed out, hoopoes and kingfishers are hole-nesting species, yet they are depicted on eggs out in the open. The location of their nests will have been unimportant to artists, however, if the primary purpose of these bird images was to represent their defensive, rather

---

49 de Gorostarzu, X. (1901). "Lettre sur deux tombeaux de crocodiles découverts au Fayoum", *ASAE* 2, 182-184.
50 Schäfer, *Principles*, 124.

than their reproductive, behaviour. This focus would also explain why it was acceptable to depict migrating species that do not breed in Egypt nesting alongside species that do.[51]

Another error that occurs in scenes of nesting birds is that many of the depicted species are colony nesters - birds that breed in close association with others of their own kind. Many waterbirds, such as pelicans, herons, ibises, egrets, and storks all nest in colonies.[52] These birds do not situate their nests among those of other species, as depicted in wall scenes, but in large single-species groups of up to 100 or more individuals. The Egyptian artists cannot have failed to notice these huge, noisy aggregations, yet they did not depict any species nesting in this manner.[53]

It is also important to note that the positioning of bird nests in marsh scenes is often incorrect. Depicted nests are reliably elevated above the substrate and situated upon either the curved stalk or umbel of a papyrus plant, however many of the birds shown nesting in this manner are ground-nesting species. These include: Egyptian geese, grey herons, little egrets, bitterns, sacred ibis, and cormorants. All of these birds place their nests on flattened vegetation close to the ground, in dense reed-beds or under bushes (cf. pl. 11G). Although these species will breed in papyrus thickets, their nests are never raised above the ground as they appear in marsh scenes. Furthermore, the structure of their nests precludes them from being placed upon a single stalk or on top of a delicate umbel.

This combined evidence leads to the conclusion that the reproductive habits of birds were largely irrelevant to the message that artists were attempting to convey in marsh scenes: namely, that all nesting birds, no matter what their species or where they place their nest, defend their eggs and young vigorously.

11.2 FEEDING YOUNG

One of the most important activities of parental behaviour is the provision of food for offspring. Some neonates are able to forage independently shortly after emerging. The young of many species, however, must rely on their parents to provide them with nourishment until they are old enough to obtain food for themselves.

11.2.1 Nursing

Female mammals produce milk with which to feed their newborn offspring. During nursing, females allow their young to access this milk ('suckle') via mammary glands located on their ventral surface.

Cows

ṯntj (PB42: Giza, V.1-5)
jj-mrjj (PB43: Giza, V.3L)
z3-jb (PB44: Giza, V.8)
snḏm-jb: jntj (PB45: Giza, V.8M-L)
3ḫtj-ḥtp (PB46: Saqqara, V.8L-9E)
k3pj (PB47: Giza, V.9-VI)
k3-gm-nj: mmj (PB48: Saqqara, VI.1M)
j3zn (PB49: Giza, VI.1M-2M)

Eight tombs dating from the early 5th until the mid-6th dynasty were found to depict scenes of nursing cows. In the earliest example, a small calf crouches between its mother's legs as it sucks from her udder (PB42). The cow stands passively, facing forward. In most of the remaining scenes, however, the cows adopt a curious posture in which they lift a hind leg closest to the viewer and hold it

51 For the Egyptians' awareness of the migratory habits of birds, see Edel, E. (1963). "Zu den Inschriften auf den Jahreszeitenreliefs der 'Weltkammer' aus dem Sonnenheiligtum des Niuserre: Part II", *NAWG* 1963, 105-115 and Goelet, O. (1983). "The migratory geese of Meidum and some Egyptian words for "migratory bird", *BES* 5, 41-60.

52 Leahy, *Companion*. 139.

53 I note, however, that in the Chamber of the Seasons of the sun-temple of Niuserre, three pelicans sitting upon nests with their wings held in a threat posture are situated on three separate registers, one above the other. It is tempting to see a depiction of colonial nesting; unfortunately, however, pelicans do not breed in Egypt. See Edel, *NAWG* 1961, 232-233 and fig. 9.

horizontally.[54] Scenes from the West Field Giza tombs of *jj-mrjj* (PB43; see fig. 11-13), *k3pj* (PB47), and *j3zn* (PB49; see fig. 11-14) are virtually identical to one another, suggesting that the later images may have been copied from the first.[55] In each of these examples, the cow turns its head back to face its lifted hoof as it protrudes its tongue. Each calf kneels upon the ground to reach its mother's udder, bending the foreleg closest to the viewer and steadying itself with the other as it stretches its head upward. An image in the tomb of *snḏm-jb: jntj* (PB45) is very similar; the hornless cow is depicted with only three legs and so it seems likely that the missing limb was also originally held horizontally.[56] Here, however, the calf does not kneel, and the cow, although looking back in the direction of her missing hoof, does not protrude her tongue. All of these scenes occur at Giza.

One of only two nursing images found at Saqqara may have been inspired by the earlier scene in the Giza tomb of *jj-mrj*, as the cow again looks back and protrudes her tongue (PB46; fig. 11-15). She does not lift her hind leg, however, and although her calf bends one foreleg slightly, it does not kneel. In the remaining scene, which appears in the tomb of *k3-gm-nj: mmj*, a calf attempts to nurse while its mother is being restrained (PB48; see fig. 11-21).

Cows do not lift their hind legs when nursing (see pl. 11H). Among ungulate species, only dik dik and impala are known to behave in this manner, neither of which is found in Egypt.[57] The behaviour has been observed occasionally among sheep, the ewe raising a hind leg to help her newborn lamb find her udder for the first time, however leg-lifting has not been documented in nursing cattle. When describing the scene in the tomb of *j3zn*, Simpson suggested that, "The device of the lifted leg provides the artist with the means of showing the entire calf",[58] however examination of images from the tombs of *ṯntj* (PB42) and *3ḫtj-ḥtp* (PB46; see fig. 11-15) reveals that the cows' legs do not need to be elevated for their young to be seen. The raised hind leg must therefore signify something else; clues are found in the behaviour of the suckling calves.

Calves begin to suckle from their mothers approximately five hours after birth, although cows must often help their newborns to find their udder by licking and nudging them in the correct direction. Once it begins to suck, the calf stands alongside its mother, lowering its shoulders and tilting its head up to reach her udder (see pl. 11H);[59] as they grow taller, however, they must bend their legs and eventually drop to their knees to reach her teats.[60] Newborn calves suckle five to ten times a day, for approximately ten minutes per session.[61] The number of suckling bouts per day decreases as they age, as does the cow's capacity to produce milk. Growing calves often respond to their mother's diminishing milk flow by butting her udder.[62] Cows may respond to this painful treatment by kicking.[63]

As noted above, three of the depicted calves adopt a kneeling posture (PB43, PB47, and PB49; see figs. 11-13 and 11-14), and a further two images show suckling animals bending their forelegs (PB45 and PB46; see fig. 11-15). This indicates that they are not newborns, but older calves; they would thus be of an age when they would be more inclined to butt the cow's udder, a behaviour that is known to elicit a kick. When cows kick, their hind leg is lifted forward and then out from their body. Images of nursing in which cows lift a back leg may thus depict their reflexive response to the insistent and painful demands of their calves. The Egyptians were clearly aware that cows

[54] It is interesting to note that this motif must break the rule of the limb furthermost from the viewer advancing first (although cf. fig. 6-4 in Chapter 6).

[55] Smith (*HESPOK*, 197) has also noted other similarities in decoration between the tomb of *j3zn* and that of *jj-mrjj*. Kent Weeks (*Cemetery G6000*, 7) dates the tomb of *jj-mrjj* to the reign of Niuserre, rather than Neferirkare, as proposed by Harpur (*Decoration*, 265); the tomb nevertheless predates those of *k3pj* and *j3zn*.

[56] It has been suggested that Lepsius simply failed to draw this detail. See Brovarski, *Senedjemib Complex Part I*, 51.

[57] Leuthold, *African Ungulates*, 172.

[58] Simpson, *Mast. West. Cem. Part I*, 20.

[59] Hafez, E.S.E., Schein, M.W. and Ewbank, R. (1969). "The behaviour of cattle" in Hafez, *Behaviour of Domestic Animals*, 253.

[60] Leuthold, *African Ungulates*, 172.

[61] Fraser-Broom, *Farm Animal Behaviour*, 232.

[62] Haley, D.B., Rushen, J., Duncan, I.J.H., Widowski, T.M. and de Passillé, A.M. (1998). "Butting by calves, *Bos taurus*, and rate of milk flow", *Anim. Behav.* 56, 1545-1551.

[63] Kilgour-Dalton, *Livestock Behaviour*, 11-12.

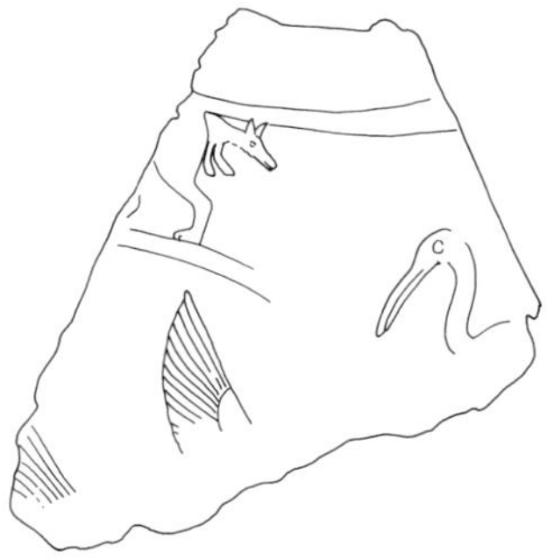

Fig. 11-9 (PB35a)

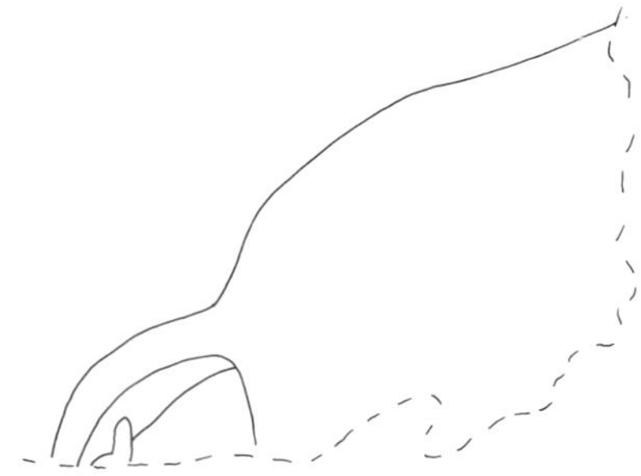

Fig. 11-10 (PB36)

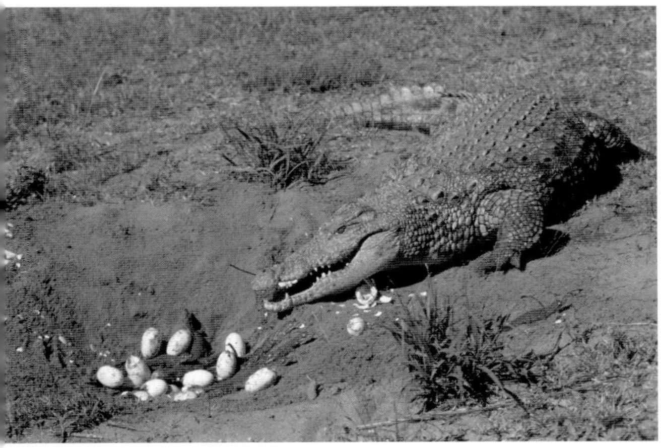

Pl. 11F Nile crocodiles lay eggs in holes
© ardea.com/M. Watson

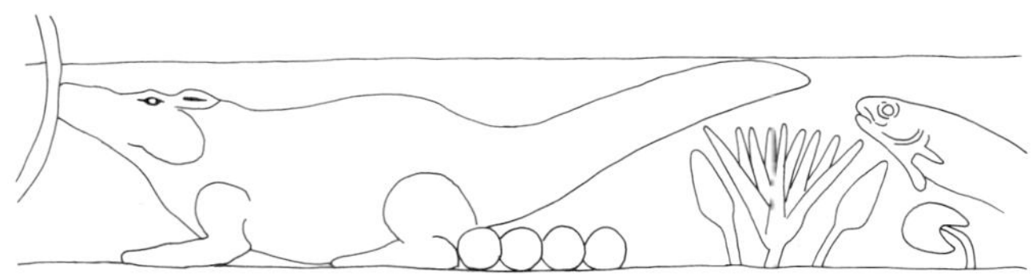

Fig. 11-11 (PB37)

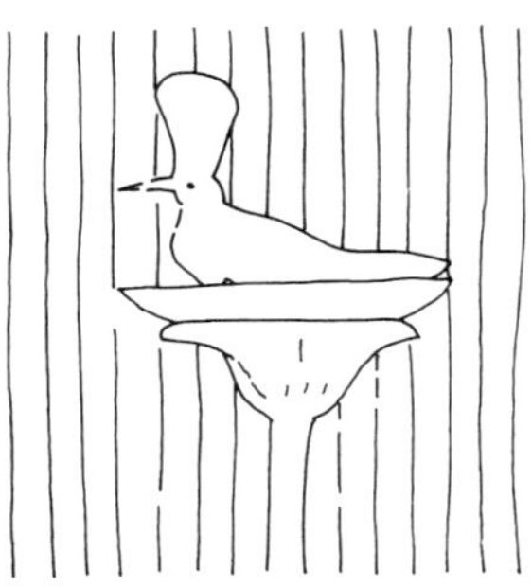

Fig. 11-12 (PB39)

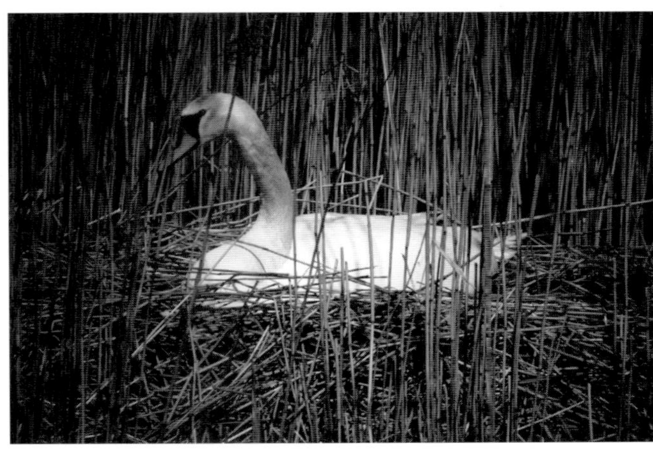

Pl. 11G Nesting in a reed bed
© iStockphoto.com/Greg Kingston

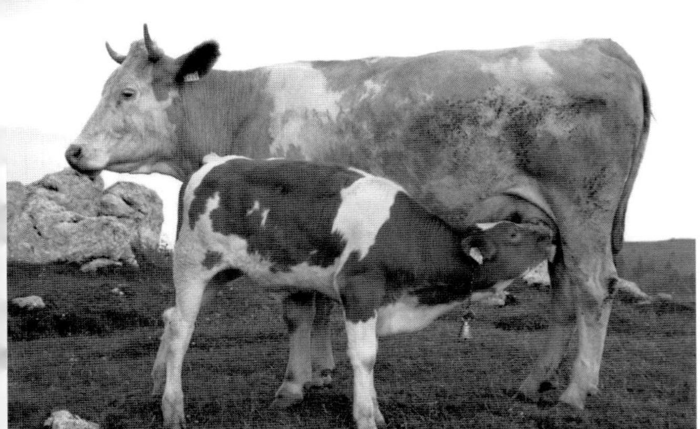

Pl. 11H Suckling posture of newborn calves
© iStockphoto.com/Bojan Tezak

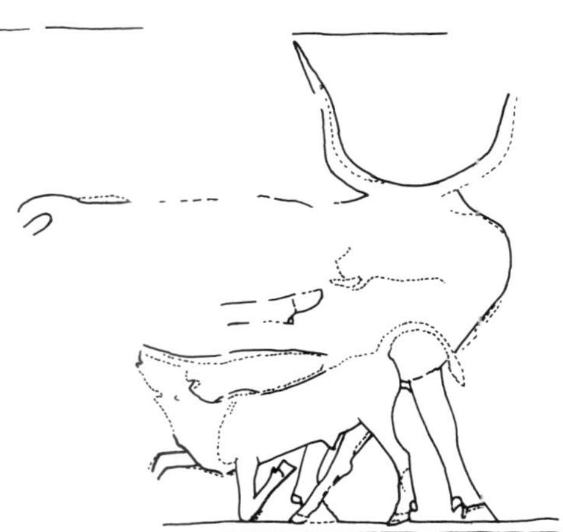

Fig. 11-13 (PB43)

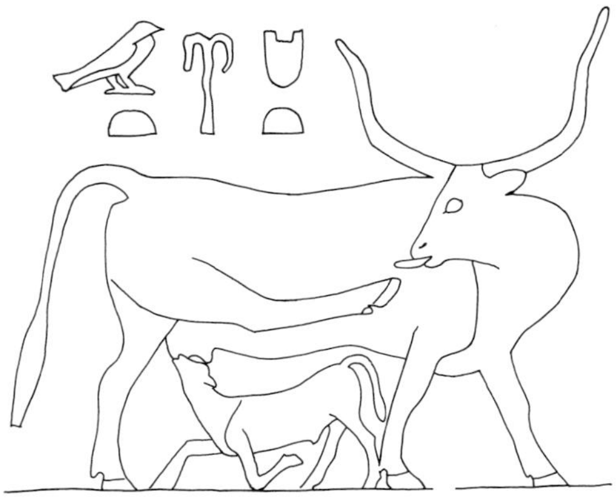

Fig. 11-14 (PB49)

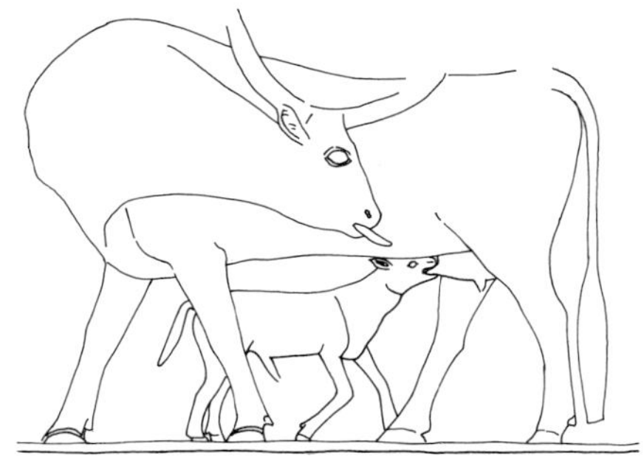

Fig. 11-15 (PB46)

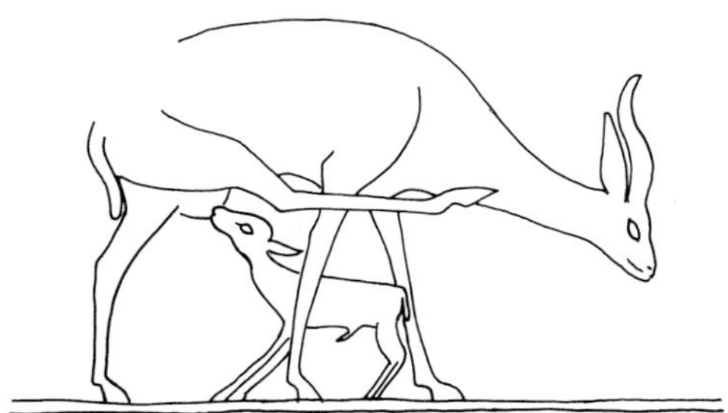

Fig. 11-16 (PB50)

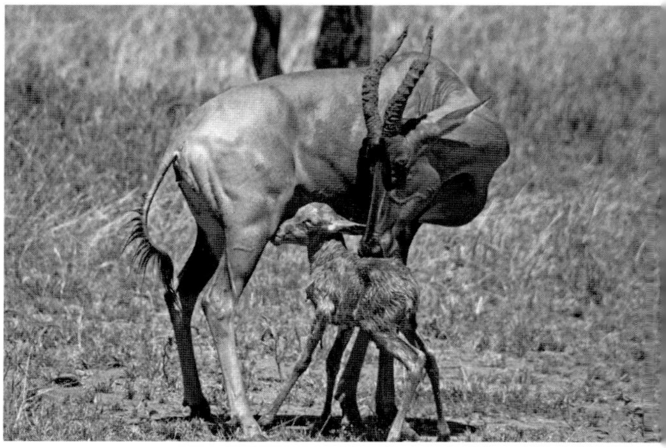

Pl. 11I Suckling posture of antelope species
© iStockphoto.com/Bruce Block

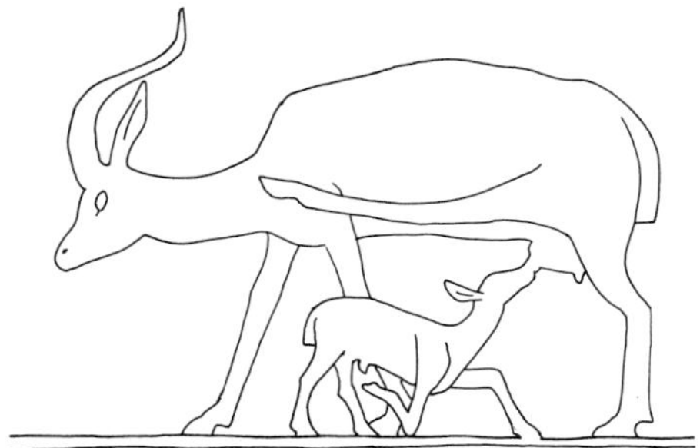

Fig. 11-17 (PB56)

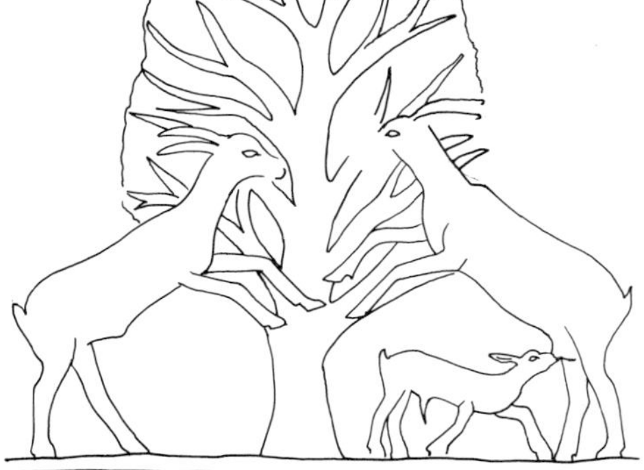

Fig. 11-18 (PB62)

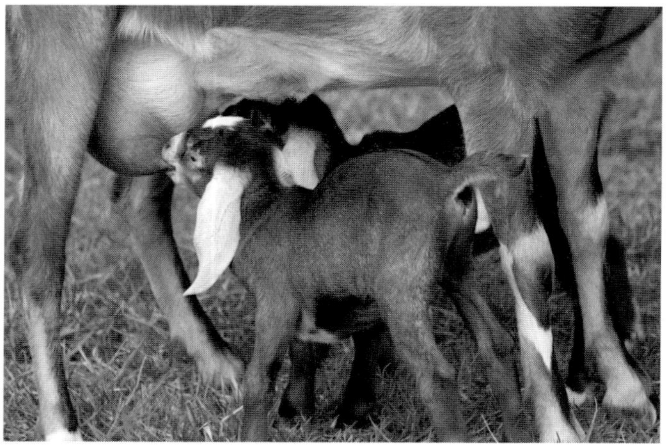

Pl. 11J Suckling posture of goat twins
© iStockphoto.com/Isaac Toy

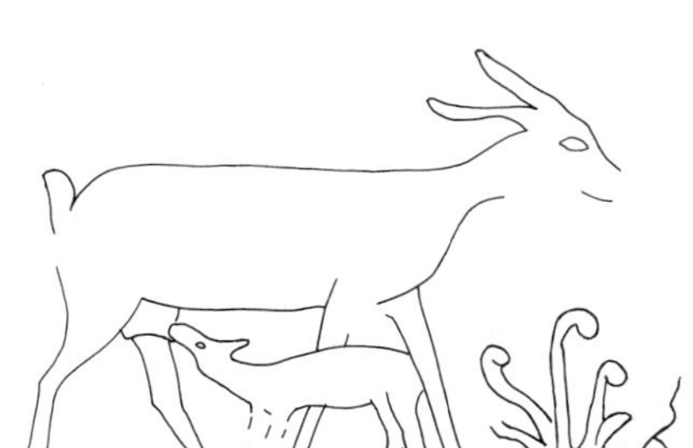

Fig. 11-19 (PB63)

could respond in this manner, because in milking scenes their hind legs are usually roped together tightly (see below).

Regarding the scene in the tomb of *jꜣzn* (fig. 11-14), Simpson also states that, "The second cow turns back to lick her rear right hoof while her calf sucks at her teats".[64] Note, however, that the tongue of this animal, and all other nursing cows, does not make contact with its hoof; thus, the behaviour cannot be described as licking.[65] Their protruded tongues may instead represent a vocal response to the pain and irritation that has caused them to kick (see Chapter 9 for a discussion of distress calling by cows). Both calling and kicking are common reactions to pain in cattle.

Dorcas gazelle

nb.j-m-ꜣḫtj (PB50: Giza, IV.6-V.1)
nj-ꜥnḫ-ḫnmw/ḫnmw-ḥtp (PB51: Saqqara, V.6L-7)
pḥn-wj-kꜣ.j (PB52: Saqqara, V.6-8E)
kꜣ.j-dwꜣ (PB53: Giza, V.6-9?)
rꜥ-wr II (PB54: Giza, V.8-9M)
ptḥ-ḥtp II: ṯfj (PB55: Saqqara, V.9M-L)
kꜣpj (PB56: Giza, V.9-VI)
Anon (PB57: Saqqara, VI)
kꜣ-gm-nj: mmj (PB58: Saqqara, VI.1M)
ꜥnḫ-m-ꜥ-ḥr: zzj (PB59-60: Saqqara, VI.1L-2E)

Dorcas gazelle nursing their calves was a popular motif in both desert and offering scenes from the late 4th until the early 6th dynasty.[66] The gazelles are occasionally depicted with one hind leg elevated (PB50, PB53, PB54[67] to PB57, and possibly PB59; see figs. 11-16 and 11-17). As noted above, the only antelope species that nurse their young in this fashion are dik dik and impala; the position is thus incorrect for dorcas gazelle. The normal nursing posture for this species, which consists of the mother and young standing in reverse parallel (cf. pl. 11I), occurs in the remaining reliefs (PB51, PB52, PB58, and PB60). Dorcas gazelle leave their newborn young hidden in the undergrowth for up to six weeks and only return for brief periods each day in order to suckle them (see Chapter 4).

The earliest tomb in which a gazelle lifts its hind leg is that of *nb.j-m-ꜣḫtj* (PB50; see fig. 11-16), which is dated to no later than the reign of Userkaf. The image thus precedes the first occurrence of the same motif in cows, that of *jj-mrjj* (cf. PB43; fig. 11-13). On this evidence it might be concluded that the images of cows adopting this posture were copied from, or inspired by, those of gazelle. From a behavioural perspective, however, I would argue that the motif will more likely have arisen first in depictions of cows and was then transferred to images of gazelle. As noted above, the raised leg probably represents a kicking motion exhibited by cows in response to the insistent and painful demands of older calves. This is a behaviour that the Egyptians will have witnessed in their domestic cattle. Opportunities to observe nursing by dorcas gazelle, on the other hand, will have been quite limited due to their intensely secretive behaviour following the birth of their young. Artists may consequently have been required to model their images of nursing gazelle on those of cows.

I propose, therefore, that the first occurrence of the raised hind leg motif must have represented a nursing cow[68] - possibly in a complementary scene in the tomb of *nb.j-m-ꜣḫtj*, which has since been destroyed, or in an earlier 4th dynasty tomb at Giza. Indeed, the influence of the cow motif on images of gazelle continued later in the 5th dynasty, as some of the latter share similarities with the earliest occurrence of this style of bovine nursing. For example, in the tomb of *kꜣ.j-dwꜣ* (PB53), the

64 Simpson, *Mast. West. Cem. Part I*, 20.

65 I note, however, that in a fresco found at Herculaneum depicting Telephus, the infant son of Heracles, being nursed by a deer, the female lifts her hind leg and turns back to lick the child. See Toynbee, *Animals in Roman Life*, 145 and pl. 66. In later wall reliefs, cows sometime lick their calves on the base of their tail (i.e., their protruded tongue touches their calf). Cows often sniff and lick the perianal region of their newborn calves when guiding them towards their udders. For example, see Blackman, *Rock Tombs of Meir ii*, pl. 7.

66 See Strandberg, *Gazelle*, 26-27 and passim.

67 Note that the raised leg is missing in the *rꜥ-wr II* image (PB54), but as the animal stands on only three legs, this feature was likely depicted originally.

68 Smith (*HESPOK*, 170) believes that the motif of a nursing gazelle may have originated in the tomb of *nb.j-m-ꜣḫtj*. I do not disagree with this, but believe that the posture must have been inspired by an earlier image(s) of a nursing cow.

gazelle looks back at her lifted hoof in the same manner as the *jj-mrjj* cow (see fig. 11-13),[69] and in the tomb of *k3pj* (PB56; see fig. 11-17), a young gazelle kneels under its mother in a similar manner to the *jj-mrjj* calf (cf. fig. 11-13).

Goats

nfr/k3-ḥ3.j (PB61-62: Saqqara, V.6) *ḥtp-ḥr-3ḫtj* (PB63: Saqqara, V.6-8E)

Goat kids begin to suckle shortly after birth. As noted above, twins are common and so, perhaps to avoid conflict, they quickly form a preference for a particular teat (see pl. 11J). The young animals stand or kneel in reverse parallel to their mother and tilt their heads up to reach her udder. They sweep their tails back and forth as they nurse, in a rhythm corresponding to their sucking movements.[70]

Only two tombs were found to depict goats nursing their young, both from the mid-late 5th dynasty. None of the females appear with twins. In the latest scene, found in the tomb of *ḥtp-ḥr-3ḫtj* (PB63; see fig. 11-19), a single kid stands alongside its mother, bracing its forelegs and lifting its head in a typical suckling posture. In the tomb of *nfr/k3-ḥ3.j*, however, two kids each suckle from their mothers while these browse from trees (PB61 and PB62), forcing the young animals to orient their bodies differently in order to reach the adults' teats. In one scene (PB61), the kid reaches from behind its kneeling mother to grasp hold of her udder; this position is probably a result of the inclination of the adult's torso, which precluded the kid from standing between her front and back legs lest it obscure her body. This conflict is avoided in the other relief in which the mother stands bipedally (PB62; see fig. 11-18). Although its access is unimpeded, the kid nevertheless bends its forelegs and tips back its head to latch onto her teats.

11.2.2 Milk ejection reflex

Cows produce milk for approximately six months after the birth of their calves. Prior to nursing, milk accumulates in specialized structures ('alveoli') in the cow's udder. The process by which her calf removes this milk is not a simple mechanical extraction, but depends upon the cow receiving particular signals from the environment. These include stimulation of the teats and vigorous butting of her udder by the suckling calf. Upon receiving these stimuli, the cow's brain sends a signal to her pituitary gland that prompts the release of the hormone, 'oxytocin', into her bloodstream. Once transported to the udder, this hormone causes muscles surrounding the alveoli to contract, squeezing the milk contained within them into ducts. The fluid then descends into each of the four glands and teats that form the udder, causing them to swell perceptibly. This automatic physiological response is known as the 'milk ejection reflex' (MER) or 'milk let-down' (see pl. 11K).[71]

Cows learn quickly to associate the sight and smell of their calf with the act of nursing so that eventually the presence of their young alone is enough to trigger the let-down reflex. When a cow is milked for human consumption, either by hand or by a modern milking machine, other stimuli can induce the MER. These include the touch of the milker's hands or machine cups upon the udder and teats, but even aspects associated with the milking procedure, such as the milk-shed, sight of the milker or the sounds of a milking machine, can be enough to cause a release of oxytocin.[72] To allow for more efficient milking in large-scale production, domesticated dairy cows have been bred so that they no longer require stimulation from their calves for the MER to occur, however in more primitive breeds, such as zebu cattle, cows are unable to let-down their milk unless their calf is present (see pl. 11L).[73]

---

69 Vandier (*Manuel v*, 67-68) suggests that the mother is scratching her muzzle.

70 Wolf, P.H. (1973). "Natural history of sucking patterns in infant goats: A comparative study", *J. Comp. Physiol. Psychol.* 84(2), 252-257; Fraser-Broom, *Farm Animal Behaviour*, 234.

71 *ibid.*, 224-225.

72 *ibid.*

73 Orihuela, A. (1990). "Effect of calf stimulus on the milk yield of Zebu-type cattle", *App. Anim. Beh. Sci.* 26, 187-190.

Ethnographic studies have revealed that pastoral peoples worldwide are aware of the importance of the calf's presence for effective milking. The most common milking practice allows the calf to suckle first for a short period of time (to induce the MER), followed by mechanical milking to extract the remaining fluid. This method has been found scientifically to result in a higher milk yield compared to milking without the calf present or if the calf suckles after milking.[74] After they have suckled, calves are often tied to their mother's foreleg so that she can still sense their presence while she is being milked (see pl. 11L).

It has been suggested that the people of the ancient Near East, including Egypt, were likely aware of the milk ejection reflex and that this is reflected in their art.[75]

Cows

mrs-ʿnḫ III (PB64: Giza, IV.6)
Anon (PB65: Saqqara, V)
nfr-msḏr-ḫwfw (PB66: Giza, V.2-8, possibly V.9)
jj-mrjj (PB67-68: Giza, V.3L)
nj-ʿnḫ-ḫnmw/ḫnmw-ḥtp (PB69: Saqqara, V.6L-7)
ḥtp-ḥr-ȝḫtj (PB70: Saqqara, V.6-8E)
kȝ.j-m-nfrt (PB71: Giza, V.6-8E?)
jrj-n-kȝ-ptḥ (PB72: Saqqara, V.6-9)
snḏm-jb: jntj (PB73: Giza, V.8M-L)
sḫm-ʿnḫ-ptḥ (PB74: Saqqara, V.8-9)
tjj (PB75-76: Saqqara, V.8-9)
nfr-ḥr-nj-ptḥ (PB77: Saqqara, V.8-9)
ȝḫtj-ḥtp (PB78: Saqqara, V.8L-9E)
nbt (PB79-80: Saqqara, V.9)
kȝpj (PB81: Giza, V.9-VI)
ȝḫtj-mrw-nswt (PB82: Giza, V.9-VI.1)
ptḥ-ḥtp: jj-n-ʿnḫ (PB83: Saqqara, V.9-VI.1)
mȝ-nfr (PB84: Saqqara, V.9-VI.1)
sḫm-ʿnḫ-ptḥ (PB85: Giza, V.9-VI.1?)
Anon (PB86: Saqqara, V-VI)
Anon (PB87: Saqqara, V-VI)
kȝ.j-m-ʿnḫ (PB88: Giza, VI)
sšm-nfr: jwfj (PB89: Saqqara, VI)
kȝ-gm-nj: mmj (PB90: Saqqara, VI.1M)
nj-kȝw-jzzj (PB91: Saqqara, VI.1M)
jȝzn (PB92: Giza, VI.1M-2M)
ḥzj (PB93: Saqqara, VI.1L-2E)
mttj (PB94: Saqqara, VI.2?)
jnw-mnw (PB95: Saqqara, VI.2E)
mḥw (PB96: Saqqara, VI.2M-3)

In contrast to the small number of scenes that feature cows nursing their calves, 40 images of milking were found in 30 tombs. In almost all of these examples, a calf stands directly beside the milked cow, either in front or behind her. The close relationship between the mothers and their young is expressed well in a scene from the 6th dynasty tomb of *mttj* (PB94), in which the cow licks the head of her gambolling calf. Most of the young animals are unrestrained, but in the tombs of *tjj* (PB75; see fig. 11-20) and *ȝḫtj-mrw-nswt* (PB82; see fig. 11-22), they are grasped by herdsmen, and in the tomb of *ȝḫtj-ḥtp* (PB78), a rope attached to its foreleg prevents a calf from reaching its mother. In each of these scenes, the calves protrude their tongues and appear to struggle against their restraints.[76]

The hind legs of almost all of the milked cows are bound together. Occasionally, a herdsman pulls vigorously on the restraining rope, as though labouring to keep the cow still (e.g., see fig. 11-23). Beginning in the 6th dynasty, the rope is also tied to the animal's horns before passing back to its hind legs (PB90, PB92, PB93, and PB96; see fig. 11-23). I suggest that milked cows were restrained in this manner to stop them from kicking (see below). This method is used by pastoral peoples today;[77] the use of 'cow hobbles' (a length of webbing tied between the hind legs), is also a common practice in modern farming to prevent kicking and pawing.

74 Combellas, J. and Tesorero, M. (2003). "Cow-calf relationship during milking and its effect on milk yield and calf live weight gain", *LRRD* 15(3), 1-10.

75 Amoroso, E.C. and Jewell, P.A. (1963). "The exploitation of the milk ejection reflex by primitive peoples" in Mourant-Zeuner, *Man and Cattle*, 132-134; Balasse, M. (2003). "Keeping the young alive to stimulate milk production? Differences between cattle and small stock", *Anthropozoologica* 37, 8; Evans, L. (2004). "Art and animal behavior" in Bekoff, *Enc. Anim. Behav. i*, 82.

76 This is also suggested by the label accompanying the scene in the tomb of *tjj* (PB75), which calls for the herdsman to 'Prevail over the calf'.

77 For example, see Evans-Pritchard, *The Nuer*, pl. 3.

In the remaining milking scenes, calves do not stand adjacent to the milked cow, but are instead situated further along in the same register (PB90), in a sub-register (PB70, PB71, and PB86), or in the panel immediately above (PB74, PB80, and PB84) or below (PB83 and PB95). The calves occur frequently in small groups and in many cases they are hobbled by ropes attached to their legs (see below).

I believe that the close proximity of calves to milked cows is not the result of arbitrary or sentimental placement by the artists, but that their presence is deliberate and reveals an awareness of the milk ejection reflex. Furthermore, I propose that these scenes document the practice of allowing calves to suckle first, prior to milking for human consumption.

The overwhelming tendency to depict calves beside milked cows is certainly suggestive that their occurrence is not random. Nevertheless, it could be argued that the calf of a lactating cow is likely to be found close to its dam simply because of their maternal bond. Images in which calves are adjacent to milked cows, but restrained, give the impression, however, that they are being denied access to the udder yet *also* held where their mothers can still see, smell, and hear them. In the tomb of *ꜣḫtj-ḥtp* (PB78), for example, a herdsman stands in front of the milked cow, patting her head with one hand while holding the leash of her roped calf in the other. Such scenes point to knowledge of the calf's crucial role in the milking process.

The most common label accompanying these images is *sšr jrṯt*, which is frequently translated as 'drawing milk' or 'milking milk',[78] but the term *sšr* is also associated with the concept of desiccation and 'drying up' of fluid.[79] If *sšr* in a milking context also had the underlying connotation of 'drying up', then it would imply that the act of milking was intended to remove *all* of the fluid from the cow's udder, leaving nothing with which to feed her calf.[80] It follows, therefore, that calves must have been allowed to suckle first, before milking began. As noted above, studies have shown that this method produces the greatest milk yield, particularly in more primitive breeds of cattle. It is also the most common practice amongst pastoralists today. The calm demeanour of most of the calves standing beside their mothers in milking scenes suggests that they are not hungry; examples in which they are physically restrained may consequently represent the moment shortly after they have been pulled from their mother's udder so that milking could commence.

A scene in the tomb of *kꜣ-gm-nj: mmj* (PB90; see fig. 11-21), in which a calf continues to suckle while its mother is being restrained, appears to capture this transition. Jacques Vandier has suggested that the cow is being prepared for milking.[81] Indeed, her hind legs are already encircled by ropes, which a herdsman is proceeding to pull tight (cf. the man's posture with that in fig. 11-23), while another handler prepares to lasso her horns (note also the roped horns in fig. 11-23). Her calf is thus endeavouring to suckle right up until the last moment, prior to being denied access to the udder.

The labelling and composition of milking scenes indicate that Egyptian pastoralists were probably aware of the importance of the calf for both the initiation and steady supply of milk. Two other factors support this notion. First, ancient Egyptian cattle were probably quite similar to less intensively-bred strains, such as zebu, for which the presence of the calf is essential for milk let-down. The milk yield of primitive cattle is considerably less than that of modern dairy cows, which are capable of producing approximately 20 litres of milk daily. After examining milk production by cattle in rural East Africa, Douglas Brewer et al. concluded that ancient Egyptian cows probably produced little more than one litre per day.[82] It seems likely, then, that Egyptian farmers will have tried various methods to increase milk yield; like pastoralists today, they will have determined,

[78] Montet, *Scènes*, 107-108. A similar term *sḫr jrṯt*, 'Milking milk' (Alenmüller, *Mehu*, 110) was found in a scene depicting a traveling herd of cows in the tomb of *mḥw* (see GB102), yet curiously this label does not accompany a milking scene that appears in the register directly below the first (see PB96). This may indicate that the draftsman has placed the inscription in the wrong panel.

[79] *Wb iv*, 295.

[80] Indeed, while later milking scenes are usually labelled, *sšr jrṯt* 'Drying up milk', earlier examples refer instead to, *sšr jdt* 'Drying up the cow'.

[81] Vandier, *Manuel v*, 70.

[82] Deshler, W.W. (1965). "Native cattle keeping in eastern Africa" in Leeds-Vayda, *Man, Culture, and Animals*, 161; Brewer et al., *Domestic Plants and Animals*, 85.

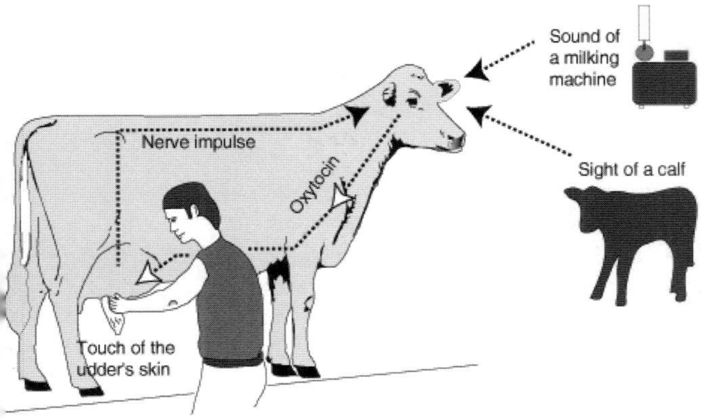

Pl. 11K The milk ejection reflex
© Babcock Institute for International Dairy Research and Development, University of Wisconsin
Reproduced with permission

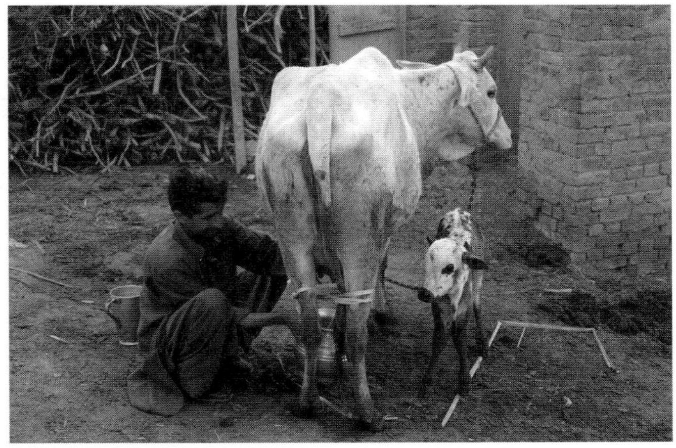

Pl. 11L Milking primitive cattle
© Brian Jones – www.thecompassedge.net
Reproduced with permission

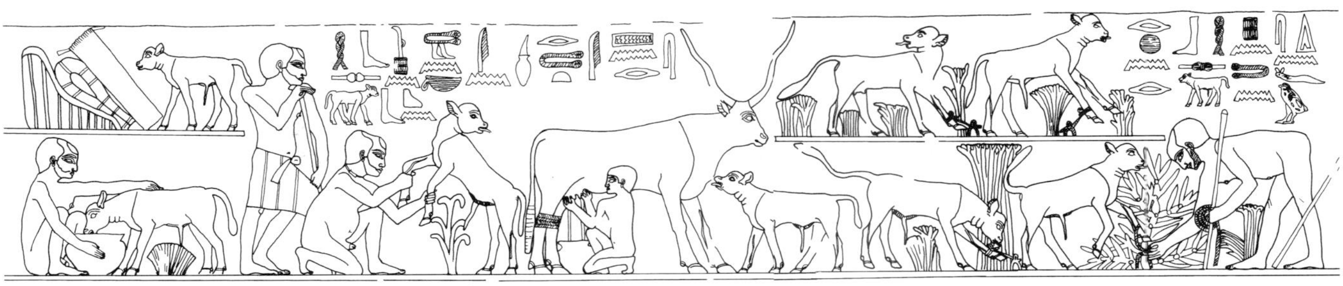

Fig. 11-20 (PB75)

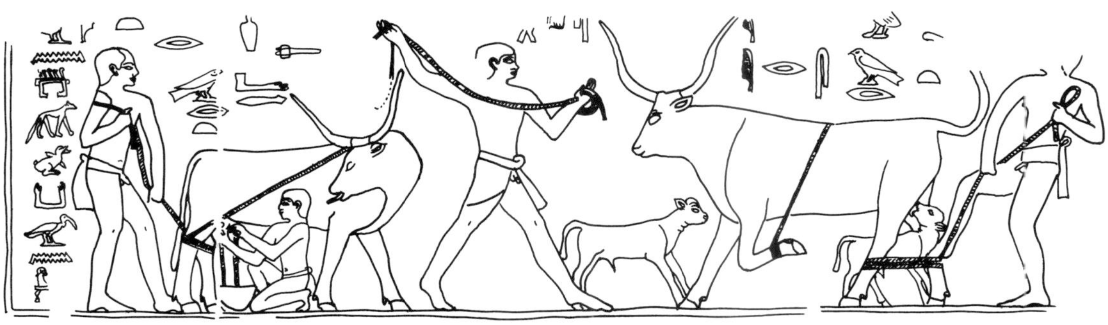

Fig. 11-21 (PB90)

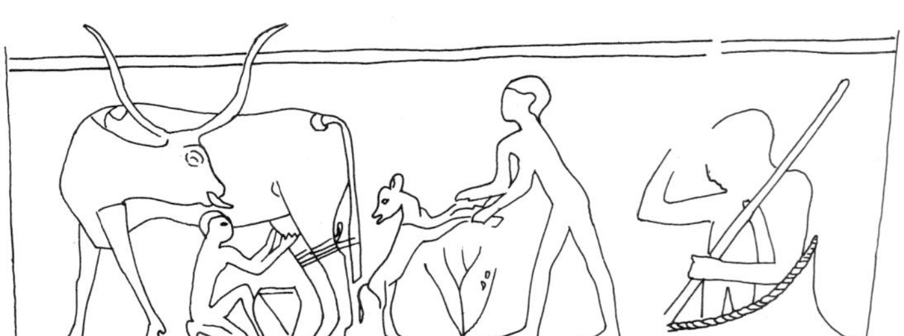

Fig. 11-22 (PB82)

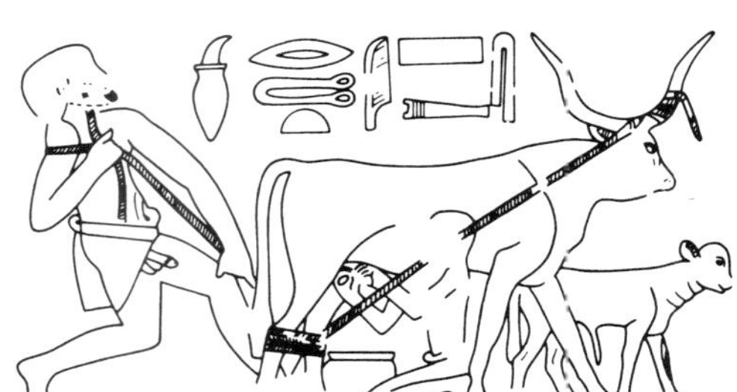

Fig. 11-23 (PB93)

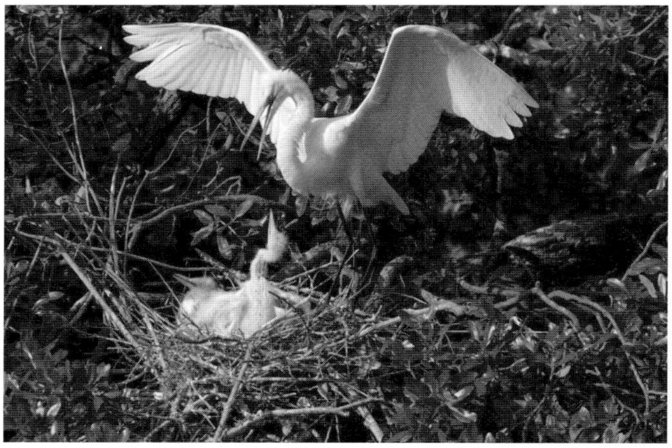

Pl. 11M Feeding begging chicks

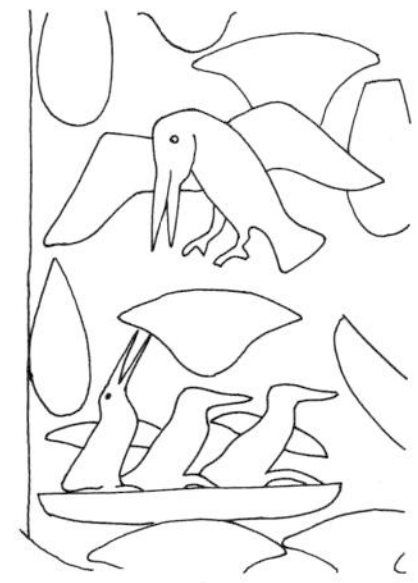

Fig. 11-24 (PB98)

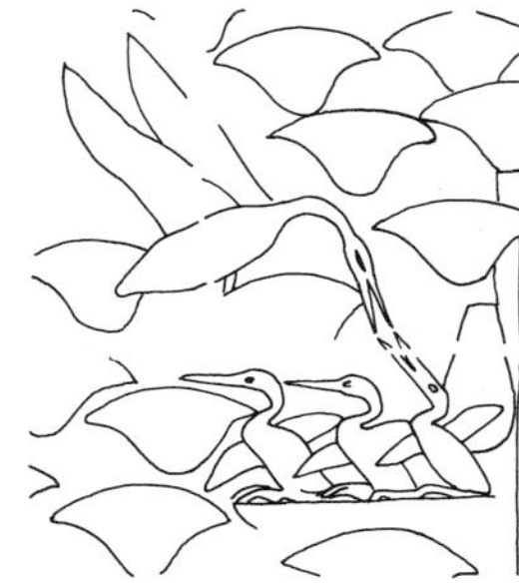

Fig. 11-25 (PB99)

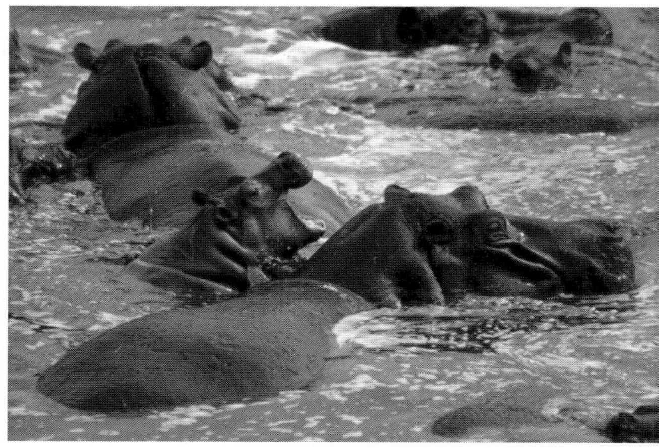

Pl. 11N Hippopotamus parental care

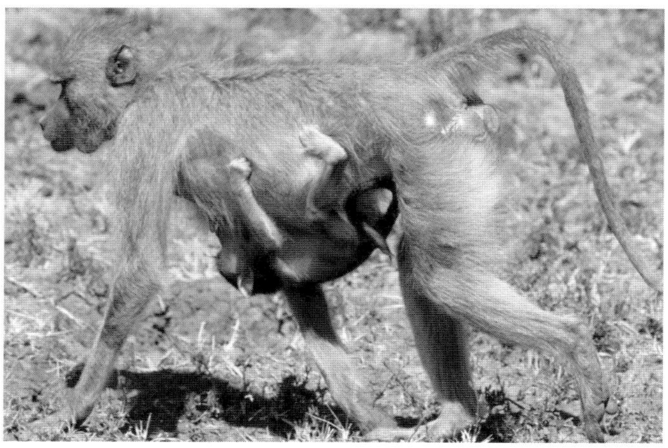

Pl. 11O Transportation of a baboon infant

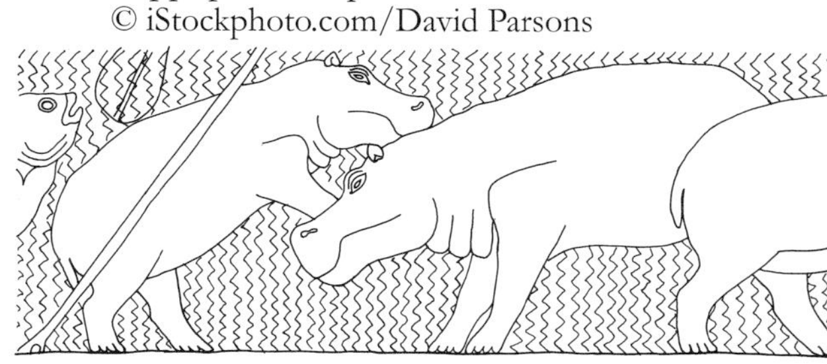

Fig. 11-26 (PB101)

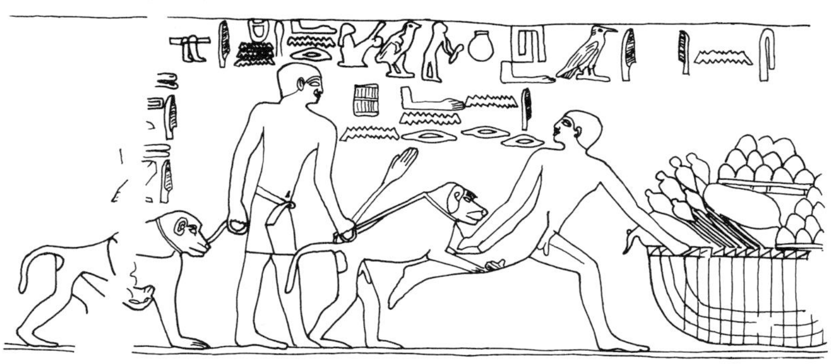

Fig. 11-27 (PB103)

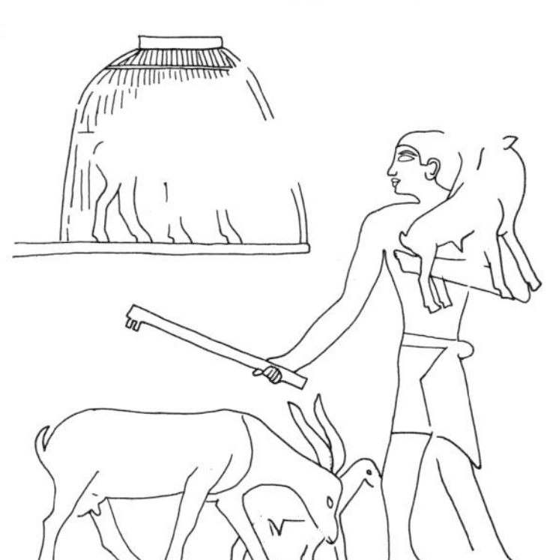

Fig. 11-28 (PB104)

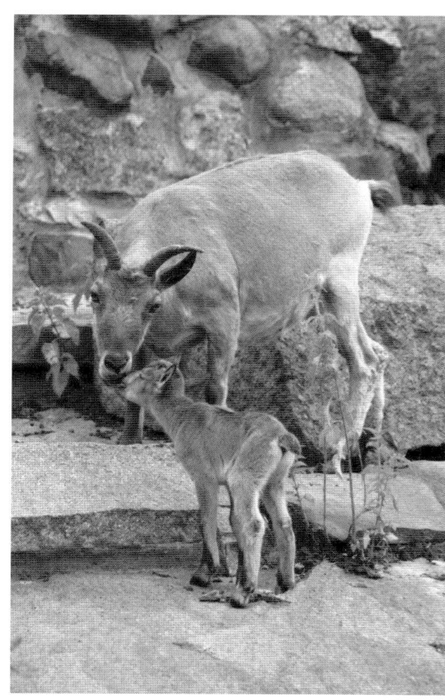

Pl. 11P Goat parental care

through trial and error, that both the presence of the calf during milking, and allowing it to suckle first, will have resulted in a greater supply of milk for their use.[83]

Images depicting small groups of hobbled calves in the vicinity of milked cows are intriguing. Under natural conditions, cows continue to nurse their calves until they are six months old. In modern dairy operations, however, newborns are usually separated from their mothers 24 hours after birth and then fed milk (or milk substitute) by bottle or bucket until they are four to 10 weeks old.[84] If the separation is delayed for days or weeks, the cow and calf become extremely distressed and vocalize excessively when parted;[85] indeed, young animals sometimes injure themselves in their frantic attempts to rejoin their mother. Thus, to help them deal with being isolated, recently weaned calves are sometimes housed together. When making the transition from suckling to bucket feeding, they must also be taught to drink from a container (see Chapter 7).

Egyptian farmers may have separated calves from their mothers permanently once the milk ejection reflex could be activated by stimuli associated with milking. I suggest, therefore, that the hobbled calves represent animals that have been removed from their mothers to be fed by container.[86] As noted in Chapter 7, ten scenes were found to depict calves drinking from bowls. Four of these images occur on the same walls as depictions of milking (in the tombs of *nj-ꜥnḫ-ẖnmw/ẖnmw-ḥtp* (PB69), *ṯjj* (PB75), *nbt* (PB79), and *jꜣzn* (PB92). In the tomb of *ṯjj*, a milked cow with a restrained calf beside her stands between four fettered calves and a single calf being fed from a bowl (PB75; see fig. 11-20). Thus, the entire weaning process is presented – removal of a calf from the udder, separation from the mother, and finally, feeding by container. It is interesting to note that a pair of roped calves in the tomb of *ḥtp-ḥr-ꜣḫtj* (PB70) is labelled *snḳtyw*, 'sucklers', informing the viewer that these isolated animals are still at this immature stage.

The trauma of the separation process is revealed by the demeanour of the fettered calves, as many struggle against their restraints, rear up, and protrude their tongues (e.g., PB70, PB71, and PB75; see fig. 11-20). Studies have shown that calves vocalize in response to separation from their mothers due to both social isolation and deprivation of a constant supply of milk.[87] The protruded tongues of the depicted young may thus represent their distress calling.

11.2.3 Provisioning

Non-mammalian animals feed dependent offspring by hunting or gathering appropriate food items and then transporting these to their young. Some species carry items in their mouths, while others ingest the food themselves and then regurgitate it once they have returned to their den or nest.

Birds

nj-ꜥnḫ-ẖnmw/ẖnmw-ḥtp (PB97-99: Saqqara, V.6L-7)

The tomb of *nj-ꜥnḫ-ẖnmw/ẖnmw-ḥtp* displays three vignettes that may illustrate the provisioning of nestlings by their parents. Two of the images depict families of pied kingfishers (PB97 and PB98; see fig. 11-24), while the third shows a little egret[88] with her brood (PB99; see fig. 11-25). In each

---

83 Brewer et al. (*ibid.*) state that milking scenes often reveal concern by the artists for the cows' feelings. Reference is made, in particular, to depictions showing the cow "shedding tears at the removal of her milk while the calf is denied its meal". It is important to note that while cows may signal their distress through vocalizations and other behaviours, they do not cry. The 11th dynasty relief that Brewer et al. use to illustrate crying by a milked cow (fig. 7.5) more likely depicts a facial marking below the animal's eye (cf. depictions of scimitar oryx).

84 Weary, D.M. (2002). "Four fallacies of dairy calf rearing" in *Proc. 37th Ann. Pac. Northwest Anim. Nutrit. Conf.*, 2.

85 Hart, *Behavior of Domestic Animals*, 152.

86 Vandier (*Manuel v*, 68-70) suggests that restrained calves are those that are being weaned.

87 Thomas, T.J., Weary, D.M. and Appleby, M.C. (2001). "Newborn and 5-week-old calves vocalize in response to milk deprivation", *App. Anim. Beh. Sci.* 74, 171.

88 Moussa and Altenmüller (*Nianchchnum*, 60) identify this bird as a glossy ibis, however the length and shape of its beak indicate that it is more likely to represent a member of the heron family. The lack of a crest indicates that it is probably a little egret. This species is known to nest in dense swamp vegetation and reeds. See

example, an adult bird hovers immediately above a nest containing chicks. The adult kingfishers spread their wings and lower their heads so that their beaks are held vertically; as noted in Chapter 8, this posture is adopted by this species when hunting for prey. The adult egret stretches both of its wings above its back while bending its neck at an abrupt angle. In each nest, a chick lifts its head to look directly at the adult fluttering above it. The beaks of the adults and their attentive nestlings are all open; those of the egret and her chick almost touch one another.

Adult birds in marsh scenes often fly near their nests in response to the approach of genets or Egyptian mongooses. No predators are in the vicinity of these birds, however, and their head positions indicate that their attention is devoted entirely to their chicks. A survey of the nestlings depicted in marsh scenes also reveals that they rarely look up at adults as these fly overhead. That the unusual composition of adults and nestlings considered here (which has not been encountered in any other tomb), is repeated three times at this site indicates that the artist was experimenting with a new motif.

The young of some bird species are able to leave the nest shortly after hatching ('precocial'); many nestlings, however, are under-developed when they emerge and are consequently fully dependent upon their parents for survival ('altricial'). Altricial chicks remain in the nest where they are fed and brooded by their parents until they have learned to fly. The adults leave many times a day to collect appropriate items with which to feed their young. The chicks of most altricial species 'beg' for food upon the arrival of their foraging parents by pointing their beaks upward and gaping their mouths wide open (see pl. 11M).[89] The chicks in the tomb of *nj-ꜥnḫ-ẖnmw/ẖnmw-ḥtp* adopt this posture.

Pied kingfishers rear their altricial nestlings in chambers that they excavate into riverbanks. The young are fed fish that their parents and juvenile helpers bring back to the nest. The chicks leave the burrow when they are approximately three weeks old, but stay near the entrance for the first few days, where their parents continue to feed them. As they grow stronger, they begin to fly towards the adults as these return with food. At this stage, the parents no longer feed them readily, but now encourage the chicks to pursue them. The food they carry is subsequently transferred mid-flight.[90]

Pied kingfisher adults also engage in a behaviour known as "courtship feeding" in which females, prior to and during egg-laying, are provided with food by their male partners. Upon the approach of their mate, the females beg for food with their bills raised in the same manner as chicks. The males either land beside or hover above their partner while transferring their food item.[91] This latter behaviour matches closely the depicted images; as courtship feeding takes place outside of the nesting burrow, it is thus possible that the Egyptians will have witnessed this intriguing exchange.

I have proposed in Chapter 8 that depictions of pied kingfishers defending their broods against predators may represent the time shortly after their young have left the burrow, but are still being fed by their parents; the nests upon which the chicks are depicted thus serve only to remind the viewer of their immaturity. I suggest that scenes in the tomb of *nj-ꜥnḫ-ẖnmw/ẖnmw-ḥtp* represent an amalgamation of activities associated with the provisioning of such newly fledged birds - the lifted and open beaks of the nestlings indicates their begging response, while the hovering, 'hunting' posture of the adults conveys both their foraging behaviour and subsequent feeding of their chicks. The motif as represented may also have been modelled on an observation of courtship feeding by adults.

Little egret chicks are semi-altricial after hatching and so are fed by their parents until they leave the nest (at around six weeks). Upon hearing the approach of an adult, egret nestlings begin to beg.[92] After the parent has settled on the nest rim, it regurgitates food (small fish, amphibians, and insects) from its stomach and oesophagus directly into the mouths of the chicks (see pl. 11M).[93] As the bill tips of the adult egret and nestling depicted in the tomb of *nj-ꜥnḫ-ẖnmw/ẖnmw-ḥtp* (fig. 11-

Cramp, S. (1977). "Little egret (*Egretta garzetta*)" in Cramp-Simmons, *Handbook i*, 290.

89 Burton, *Bird Behaviour*, 187-189.

90 Roselaar, C.S. (1985). "Pied kingfisher (*Ceryle rudis*)" in Cramp-Simmons, *Handbook iv*, 727-728.

91 *ibid.*, 727.

92 Cramp in Cramp-Simmons, *Handbook i*, 294.

93 *ibid.*, 295.

25) almost touch - an arrangement not found in any other depiction of birds - I suggest that this image may represent the transference of regurgitated food. It should be noted, however, that this behaviour is normally performed after adults have landed on the nest and so the hovering flight of the depicted egret is not accurate.[94]

11.3 PROTECTION

Some species abandon their young shortly after birth, leaving them to fend for themselves, however the parents of many others go to extraordinary lengths to protect their young, often sacrificing their own well-being to provide them with sufficient care and to defend them against potential predators and other environmental threats.

11.3.1 Care

Immature young are particularly vulnerable to disease, injury, and predation due to their reduced size, strength, and underdeveloped motor skills. Parents thus need to monitor the physical state of their offspring constantly and must often adapt their behaviour to accommodate their infant's needs. By acting as behavioural role models, adults also provide young with social and survival skills.

Hippopotamuses

k3.j-m-nfrt (PB100: Giza, V.6-8E?)
tjj (PB101: Saqqara, V.8-9)
k3-gm-nj: mmj (PB102: Saqqara, VI.1M)

Three reliefs were found to depict the caring behaviour of hippopotamuses. In the tomb of *tjj* (PB101; see fig. 11-26), a hippo calf situated underwater in a marsh scene, rears up to place its forelegs upon the snout of an adult; the larger animal appears to drop its head and welcome the calf's behaviour. Adult hippos are able to stay underwater for extended periods of time (approximately four to six minutes), but calves must come to the surface to breath more frequently (every two or three minutes). In deep water, they often rest upon their mother's back so that they have less distance to swim when resurfacing (pl. 11N).[95] I suggest, therefore, that the image in the tomb of *tjj* depicts a calf clambering onto its mother, a behaviour that the ancient Egyptians will have undoubtedly witnessed often.[96]

The remaining scenes are more ambiguous. In each case, a hippo calf prostrates itself upon the riverbed next to an adult. In the tomb of *k3.j-m-nfrt* (PB100), the young animal lies down beneath a canoe containing a harpooner as an adult hippo nearby lifts its head from the water to threaten the hunter (see Chapter 9). In the tomb of *k3.gm-nj:mmj* (PB102; see fig. 8-36 in Chapter 8), a calf stretches out below the gaping jaws of a swimming crocodile and at the feet of a standing hippopotamus that is probably its mother. The calves' recumbent postures are curious and have not been encountered in any other reliefs examined in this study.

Hippopotamus mothers are very protective of their young, both on land and in the water, however they will also punish their calf if it strays too far from them by head-swiping and biting it. Calves that are knocked over during this aggressive treatment will then prostrate themselves upon the ground as a sign of submission.[97] The calves in the two scenes appear to adopt this submissive posture. It should be noted, however, that aggressive interactions between mothers and calves usually take place on land, while both tomb scenes occur underwater. Possibly, the responsible artists witnessed a calf's fearful response to its mother's anger, but then chose to depict it reacting in a similar manner to the threatening situations presented in each image.

---

94 As nestlings age, their attempts to obtain regurgitated food from their returning parents becomes increasingly more aggressive. Adults in these situations often flap their wings to help keep their balance (see pl. 11M).

95 Eltringham, *The Hippos*, 4.

96 A clear depiction of a hippo calf riding upon its mother's back can be seen in the Middle Kingdom tomb of *snbj* at Meir (B1). See Blackman, *Rock Tombs of Meir i*, pl. 2.

97 Estes, *Behavior Guide*, 226; Verheyen, *Hippopotame*, 68.

Baboons

tp-m-ꜥnḫ II (PB103: Saqqara, V.5-6?)

The tomb of *tp-m-ꜥnḫ II* (PB103; see fig. 11-27) was found to display a unique depiction of maternal care by an anubis baboon. The female appears in a market scene in which a male anubis rears up to attack a young boy (see Chapter 9). The female is walking forward, as the heel of one of her feet is raised. She moves quadrupedally, but one of her forelimbs is missing. The head of a smaller baboon is depicted under her chest, indicating that she is clasping an infant to her body.[98]

Newborn baboons do not become entirely mobile until they are at least one month old. Until then, their mothers transport them by holding them against their ventral surface (pl. 11O) and supporting their bodies with one hand; the infants sometimes latch onto a nipple to achieve a better grip.[99]

The *tp-m-ꜥnḫ II* baboon is rendered with only three limbs. This may indicate that the artist intended to depict the female supporting the body of her infant as she walks. It cannot be determined with certainty, however, whether her missing forelimb was originally drawn beneath the infant, as both figures are damaged by a break in the relief. Note that the artist has ignored convention and shows the baboon with the forelimb closest to the viewer advancing first. If the image had been represented in the usual manner - with the furthermost limb advancing - this would have required that the arm closest to the viewer be bent back against the adult's flank and then over and under that of her infant to support it, thereby obscuring its body. I suggest that the artist chose the alternative arrangement to ensure that the infant remained visible while simultaneously implying the presence of the mother's supporting arm.

11.3.2 Defence

The adults of many species defend their young vigorously against potential predators, even when these are considerably larger or capable of inflicting substantial injuries. Mothers safeguard their offspring in many different ways; some may attack the perceived threat physically, but others engage in more indirect behaviours, such as distraction or aggressive displays.

Goats

nfr/kꜣ-ḥꜣ.j (PB104: Saqqara, V.6)

The composition of a scene located on the east wall of the niche in the tomb of *nfr/kꜣ-ḥꜣ.j* (PB104; see fig. 11-28) is suggestive of protective behaviour by a female goat. The doe, which is accompanied by a small kid, stands beside a herdsman who holds another kid in the crook of his left arm. Both he and the small animal he carries look back in the direction of the adult goat. The female drops her head in an aggressive manner; as noted in Chapter 9, goats threaten one another by lowering their head to display their horns (cf. pl. 11P). In response to this behaviour, the herdsman waves a stick above the animal as though trying to deter it. In a damaged sub-register above the female goat, a small animal stands inside a dome-shaped cage; similar depictions found during this study suggest that the trap contains a goat kid.

As noted above, the concurrence of scenes displaying the birth, caging, and slaughter of goats may indicate that Egyptian pastoralists killed male or extra kids born as twins. I propose that the *nfr/kꜣ-ḥꜣ.j* scene depicts the moment when a herdsman has removed one member of a female's set of twins, an action that has elicited her threatening behaviour. The separated kid, held in the arms of the herdsman, looks back at its mother, whose lowered head also draws attention to her one remaining infant. The trapped kid in the register above this poignant scene may be intended to convey clearly to the viewer the purpose of the enforced separation.[100]

---

[98] Hodjash and Berlev suggest that the adult is, instead, holding a piece of fruit, yet the infant's face is clearly visible. See Hodjash, S.I. and Berlev, O.D. (1980). "A market-scene in the mastaba of *dꜣdꜣ-m-ꜥnḫ* (*tp-m-ꜥnḫ*?)", *AOF* 7, 41.

[99] Estes, *Behavior Guide*, 518.

[100] The tomb of *nfr/kꜣ-ḥꜣ.j* includes a depiction of goat slaughtering on the east wall of the chapel.

SUMMARY

- The overwhelming tendency to depict human intervention during the birth of calves suggests that calving difficulties were a common occurrence during the Old Kingdom. This may reflect a husbandry practice in which cows of average size were mated with large bulls.
- Cows, goats, and hippopotamuses are often illustrated with protruded tongues while giving birth. This feature likely represents the vocal expression of pain made by these species during labour.
- The unique depiction of a crocodile laying its eggs underwater highlights the remarkable ingenuity of Egyptian artists in achieving a balance between representational constraints, magical requirements, and their desire to display observations from life.
- Nesting birds in marsh scenes are rarely shown with their wings folded in a normal roosting posture. The disregard for biological accuracy in the rendition of nesting - such as the representation of species that do not breed in Egypt, solitary depictions of colonial nesters, and the nests of ground-nesting birds balanced upon foliage - indicates that the artists' primary concern was the birds' protection and defence of their nestlings, rather than their reproductive habits.
- A curious nursing posture, in which one hind leg is elevated, may be derived from the kicking behaviour of cows in response to painful suckling by older calves. This posture was likely extended to depictions of dorcas gazelle due to lack of firsthand observations of the secretive nursing behaviour of this species.
- The composition and labelling of milking scenes suggest that the ancient Egyptians, like other pastoral peoples, were aware of, and used, the milk ejection reflex to effect a greater yield during milking.

CHAPTER 12

ANIMAL BEHAVIOUR IN OLD KINGDOM ART

12.1 THE OCCURRENCE OF ANIMAL BEHAVIOUR IN WALL SCENES

The primary aim of this study was to document the depiction of animal behaviour in wall scenes from private tombs at Giza and Saqqara. The demeanour of animal images derived from these two locations was consequently evaluated and, in reference to their surrounding contexts (scene themes), categorised accordingly. The apparent behaviour displayed by each depicted animal was then compared to that of contemporary fauna via consultation of the ethological literature, a process that allowed the identification, accuracy, and method of representation of species-specific activities and movements to be determined. This method ultimately yielded a wide range of behaviours - from general and immediately recognizable actions to extremely subtle and idiosyncratic expressions - which were exhibited by animals from each taxonomic class.

Approximately half of the animal images that were inspected revealed no distinctive activity beyond walking and/or standing still. This was a direct consequence of the contexts in which they were depicted, for many were found in offering scenes in which they were presented, controlled, and restrained by bearers, a situation that allowed limited scope for the expression of natural behaviour. The frequency with which this scene type occurred in the sample of wall reliefs examined also contributed to the large number of animals shown in this manner.[1]

In the remaining images, however, a rich variety of essential behaviours - i.e., those associated with survival and reproduction - were exhibited by various species. These included different forms of locomotion, feeding, drinking, grooming, hunting, threatening, fighting, courting, mating, and nursing. Indeed, the wealth of activities displayed suggests that artists took a special interest in the everyday lives of animals. These behaviours were easy to identify as they often involved gross changes in body posture that were distinctive and unmistakeable. For example, the limb and body positions associated with particular forms of locomotion enabled walking/standing to be differentiated from that of balking or running. The grooming actions of ungulates and birds were conveyed by characteristic placements of head, legs, and/or wings. The interaction of animals with elements in their immediate environment in combination with their pose also enabled the clear identification of such activities as feeding and drinking (head down, muzzle or tongue in contact with a food item or fluid), as well as mating by different species (male positioned above and behind a female).

The meaning of many of these basic behaviours could be determined at a glance, but closer examination revealed small details that showed how closely artists had observed their models. For example, swimming and (in one case) walking birds were found to purse their toes, a transitory, yet inherent, part of both methods of avian movement. The lowered heads and stretched necks of predatory cats reproduced their customary stalking pose, while the lifted tails of cows confirmed the pre-copulatory nature of the nosing behaviour of nearby bulls.

The range of species performing basic activities differed from one category to the next. Thus, many different animal types were seen to exhibit a form of locomotion or ingestion, but only a subset was represented while engaged in predation, aggressive displays, or nursing. The mating posture of some species was often illustrated by only a single example (e.g., hartebeest, caracals, turtles, etc.).

---

1 Of the 291 tombs and fragments examined during this study, 231 (79%) included processions of offered animals. The earliest example was found in the tomb of *nfr-mꜣꜥt* (G7060) at Giza, which dates from IV.2-4; the latest was located in the Saqqara tomb of *zꜣt-jn-ttj* (ntn), which Harpur dates to VI.7 or later. For the indispensable nature of some scenes in wall reliefs, see van Walsem, *Iconography*, 53.

Cattle were represented in almost all behavioural categories - an artistic bias that may reflect a particular interest in these domesticated animals[2] and/or their close availability to act as models.

In addition to frequently reproduced, general behaviours were found a number of other images that depicted the idiosyncratic activities of individual species. These motifs, which were often located in either a single tomb or just a handful, required consultation of the literature to determine their precise meaning. Images of this kind often highlighted specific behaviours that might strike a human observer as unusual or strange. Thus, a depiction of the fighting behaviour of black kites focused prominently upon this species' odd tendency to lift and show an open claw during aggressive disputes. The mating behaviour of striped hyenas and wild asses revealed the males' curious copulatory neck bite. And the suffocating grasp of lions on the mouths of their prey was captured with an astonishing degree of accuracy so that even the predators' steadying paws and their reclined postures were represented.

A remarkably diverse selection of unusual behaviours were deemed worthy of special treatment, such as the sexual proclivities of donkeys, the exaggerated leg movements of purple gallinules, the dramatic hover and dive of pied kingfishers, the Nile crocodile's headfirst method of ingesting fish, the secretive lurking of soft-shell turtles, the spontaneous leaping of mullet, and the astonishing camouflage of praying mantids. The limited occurrence of these striking species-specific activities might lead one to conclude that the recording of each was due to a memorable, but random, observation made by either the artist or the tomb-owner, which they wished to preserve.[3] It is important to note, however, that these images were not spread randomly throughout the entire sample, but were instead confined to a small number of tombs, mainly from Saqqara. In particular, those of *nj-ꜥnḫ-ẖnmw/ẖnmw-ḥtp*, *ꜣḫtj-mrw-nswt*, *nj-mꜣꜥt-rꜥ* [*jzzj-mr-nṯr*], *kꜣ-gm-nj: mmj*, *nj-kꜣw-jzzj*, *mrrw-kꜣ.j: mrj*, *sꜥnḫ-w(j)-ptḥ*, *mrjj-ttj: mrj*, *mḥw*, but especially *ḥzj*, yielded many unusual scenes. Indeed, often more than one unique image was located within the same tomb. The size and state of preservation of the structures involved varies considerably, so the restricted distribution of such unusual images cannot be due to these factors alone.[4] Instead, I suggest that, in each case, they reflect a special interest in the behaviour of different species by those responsible for the scene content. Their fascination with the natural world is indicated both by the variety of behaviours chosen and by the accuracy with which each is represented. Of course, some images were undoubtedly copied from others (i.e., those appearing in more than one tomb), but the decision to include such unusual motifs among the standard repertoire of scenes nevertheless reveals an underlying curiosity by the artist or tomb-owner.

A general interest in behaviour is also evident in a few representations that show the application of such knowledge in the handling of animals. For example, the consistent use of sheep for the purpose of treading in sown seed may indicate an awareness, and shrewd manipulation, of the fear-driven bunching of this particular species. So, too, herdsmen appear to have taken advantage of the social cohesion of cattle - especially the bond between cows and their calves - to control the movement of herds, while the physical response of ungulates to a nasal pinch, enabled individual animals to be similarly constrained and directed. Some images provided details of practices in which the natural behaviour of domestic animals was exploited for economic benefit. Thus, the reliable occurrence of calves beside their mothers during the process of milking, combined with the curious reference to the "drying up" of milk, may point to knowledge of the milk ejection reflex. The close

2 For the importance of cattle in Egyptian culture, see Frankfort, *Kingship and the Gods*, 162-180; Lobban, R.A. (1989). "Cattle and the rise of the Egyptian state", *Anthrozoös* 2(3), 194-201; Schwabe, C.W. (1994). "Animals in the ancient world" in Manning-Serpell, *Animals and Human Society*, 41-44.

3 It is difficult to decide whether it was the artist or the tomb-owner who acted as the original eyewitness to these behaviours. The details of some activities and movements - such as the open claws of fighting kites - could be communicated easily via verbal description; thus, some observations may have passed from the tomb-owner to the artist, who then attempted to reproduce the behaviour, sight unseen. I suggest, however, that some behaviours could not have been produced so accurately without firsthand experience by the artist - such as the particular placement of the praying mantis within its surroundings to achieve a camouflage effect.

4 For example, the tomb of *ḥzj* is tiny compared to that of *mrrw-kꜣ.j: mrj*, yet it contains an abundance of unique scenes.

association of lumberjacks with scenes of browsing goats may also indicate a mutually advantageous use of the animals' voracious appetites and climbing ability to strip trees.

Other scenes also hinted at an application of behavioural knowledge that affected how potentially dangerous animals were represented within the tomb environment. Thus, hippopotamuses were found to threaten with open mouths, but only while seated, possibly to ensure that any subsequent show of physical aggression might be slowed. And Nile crocodiles rarely emerge above the water surface (not even to lay their eggs), but are instead reliably depicted underwater where they were believed by the Egyptians to be 'blind-of-head'.

Although a broad range of everyday activities was represented within the sample of images, a few behaviours were nevertheless missing. For example, no animal was depicted while sleeping or playing. Often, too, only a subset of behaviours was shown. Thus, cattle were depicted eating at troughs and even ingesting water plants, but never while grazing, their most common feeding activity. Some species were also absent from certain behavioural categories. So, while mammals and reptiles were shown mating, the sexual behaviour of birds and insects was never depicted.

Despite the large number of animal images examined during this study, few were found to exhibit a posture for which a behavioural meaning could not be determined. A behavioural explanation was also able to account for some ostensible artistic mistakes. Indeed, an ethological approach occasionally presented a possible solution to some long-standing conundrums (e.g., the 'climbing' of genets and mongooses upon single papyrus stalks, the supposed fish-eating 'otters' in the tombs of *mrrw-k3.j: mrj* and *ḥzj*, the identity of the Anubis animal, etc.) and clarified other anomalous elements (e.g., the unusual wing position of nesting birds). Artists were not inclined to depict imaginary behaviours, however the constraints of Egyptian art sometimes created a mismatch between the actual appearance of an animal's behaviour in nature and how it was reproduced in wall scenes. These images could nevertheless be 'translated' with reference to the rules of Egyptian representation.

12.2 THE REPRESENTATION OF BEHAVIOUR IN TWO DIMENSIONS

The second major goal of this study was to examine how Egyptian artists of the Old Kingdom converted dynamic animal behaviour into static two-dimensional images. The approach undertaken enabled the artistic treatment of depicted postures and movements to be evaluated with an unprecedented level of confidence, because these could be compared directly to the known behaviour of living animals. This assessment not only verified that the depiction of animals in wall scenes was dependent upon the established principles of Egyptian representation, but that their conceptual method enabled artists to illustrate some aspects of behaviour that would otherwise have been impossible.

Contrary to general opinion, the representation of animals was not found to be less rule-bound than that of humans. In fact, despite the wide range of species with which artists had to contend, the variety of situations in which these were placed, and the activities in which they are engaged, animal images were nevertheless depicted in a consistent and orderly manner that stayed firmly within the boundaries of the Egyptian representational method.[5] Even isolated examples of species-specific behaviour, which may have been generated from direct observation, complied with established conventions. Thus, artists did not resort to free expression to show the behaviour of animals; however, the diverse array of postures that were used to illustrate various activities has undoubtedly contributed to the impression of a less rigid style.

The range of behaviour reproduced in wall scenes indicates that the Egyptians were acutely aware of the idiosyncratic postures and movements of animals. Nevertheless, representations of these were often stripped of any unnecessary details and reduced to their most essential attributes in an effort to communicate as clearly and unambiguously as possible their intended meaning. This resulted in a number of standardised motifs by which to represent particular behaviours

5 The physical attributes of individual species were not examined during this study; nevertheless, although a definitive statement cannot be made, I noted very little variation in body forms either.

consistently. Motifs were especially common when illustrating frequently occurring activities, such as types of locomotion. The use of repeated forms consequently enabled the straightforward and reliable differentiation of movements, such as the application of pursed or flat toes to distinguish swimming from standing birds, or fixed adjustments in the leg positions of ungulates to indicate different gaits.

Paradoxically, however, motifs sometimes caused artists to produce biologically inaccurate depictions. For example, a rearing male mounted behind a standing female invariably represented the act of mating. This motif was applied to all animals, regardless of species-specific differences. Consequently, depictions of cats, weasels, and shrews, in which the females recline during copulation, were found to be incorrect. The presence of additional details, such as the feline neck bite, indicated that these apparent mistakes were not due to lack of observation. By conforming to the standard motif, however, these unnatural images were nevertheless able to communicate the appropriate message (i.e., "These animals are mating") and be interpreted correctly by viewers.

The consistency with which behavioural motifs were applied helped occasionally to clarify the meaning of anomalous images. Thus, I suggest that the depiction of embracing crocodiles found in a small number of tombs must represent something other than copulation, because the animals' pose is unlike that of any other mating species. In fact, the very ambiguity of their depicted behaviour confirmed this.

Once established, behavioural motifs, like Heinrich Schäfer's "language of forms",[6] enabled the idea of an activity to be grasped immediately by a viewer; however, the effect of representational rules often resulted in motifs that actually deviated widely from the visual appearance of the behaviour they were intended to illustrate. For example, nesting birds were found to be depicted reliably with their wings stretched forward, a pose that bears no resemblance to their normal state when incubating eggs. The context in which these animals were represented revealed that they are exhibiting a common threat display in which the wings are held out from the sides of the body in order to intimidate an approaching predator. Due to the lack of foreshortening in Egyptian art, however, the spread wings of the depicted birds are instead held before them. The resulting motif, although distorted, nevertheless tells an informed viewer immediately what the animals are doing.

It might be argued that manipulated images such as these must have caused confusion among those confronted by them, thereby negating their ability to impart their message effectively. While modern observers, who are largely divorced from the natural world, are likely to struggle to ascertain the meaning of these motifs, I believe that ancient Egyptian viewers could decipher them at once, because they will have understood, based on their own firsthand experiences, how animals might be expected to behave in particular circumstances.

This also holds true for images that were found to display a contraction of time. The presence of this representational device was indicated by the occurrence of two antithetical behaviours within the same image, such as the simultaneous feeding and running of donkeys or the combination of feeding and recumbent rumination by cattle. So, too, some images appear to contain elements selected from a sequence of behaviour, such as the shallow-water capture and headfirst ingestion of fish by Nile crocodiles and the threat displays and subsequent fighting by black kites. I suggest that knowledgeable viewers would have realized at once that the pairing of these behaviours was impossible or incorrect and consequently intuit that they instead represent a series of events.

The behavioural capabilities and constraints of animals also suggested new interpretations for some well-known scenes. For example, hippopotamuses are physically unable to look behind them, indicating that this apparent behaviour in harpooning scenes instead represents a turn to the side. In particular, it was found that the inability of Egyptian mongooses to climb and the propensity by genets to hunt upon the ground meant that their representation in marsh scenes has been misunderstood. After the natural characteristics of papyrus thickets were considered, as well as the presence of other elements in the environment surrounding these predators, such as ground- and hole-nesting bird species and spreading pondweed, it was concluded that the marsh scene is best

[6] Schäfer, *Principles*, 150.

understood as folding away from the viewer. I have suggested, therefore, that it represents depth, rather than height, to reveal a hidden world beyond. This interpretation is in accord with the later well-established rule of distant objects appearing higher in the picture-field, providing evidence that - as Schäfer suspected, but could not confirm[7] - this convention was probably first applied prior to the Middle Kingdom.

Thus, like other aspects of Egyptian art, the representation of animal behaviour in wall scenes was undoubtedly affected by the particular ideas and information that artists wished to transmit to viewers, as images were adapted consistently in ways that ensured that their meaning was communicated unambiguously, but often at the expense of visual truth. The clearest indication of this was the intriguing discovery that Egyptian artists were compelled to represent an intangible and entirely non-visual behaviour - the vocalizations of animals.

12.3 A BEHAVIOURAL CASE STUDY: THE PROTRUDING TONGUE

During this study, 718 animals from 112 tombs were found to protrude their tongues. This occurs while they are engaged in a variety of activities and in many different contexts. The organ is usually not depicted within an open mouth, but instead, issues forth from between the lips. The animals thus effectively 'poke out' their tongues. The length of the exposed organ varies, but is often too large for the size of the animals' heads. The tip, which can be sharp or rounded, is either fully isolated or makes contact with a substance of some kind (see below).

Two hundred and sixty animals from 55 tombs were found to exhibit an isolated tongue, including: cattle (bulls, cows, and calves), addax, scimitar oryx, dorcas gazelle, goats, Nubian ibex, hippopotamuses, red foxes, golden jackals, striped hyenas, and birds. A consistent pattern was detected such that these animals appear exclusively in situations that might potentially cause them distress, fear, frustration, and/or pain. For example, calves protrude their tongues when separated from their mothers and when immersed in cold water. Cows react similarly to the forced removal of their young. Bulls expose their tongues when resisting angrily any attempts to restrain them with ropes. Animals of all kinds engage in this behaviour when giving birth or when under attack from predators. In nature, the normal response of the depicted species to such situations would be to vocalize loudly. As stated earlier in this volume, I suggest that the depiction of their isolated tongues was intended to impart this information.

It must be stressed that many species open their muzzle or beak when calling, which may occasionally reveal the tongue within the mouth cavity (Colour Plate A), but they do not normally vocalize by poking out their tongue from between their lips or mandibles (Colour Plate B). In fact, many species are physically incapable of extending their tongue in this manner (e.g., birds; see Colour Plate C). Consequently, although the animals in wall scenes appear in situations in which a vocal response would undoubtedly occur, artists have not produced a veridical representation of their calling behaviour, but have instead chosen to communicate this idea by exposing, and thus emphasizing, their tongues. I propose, therefore, that the isolated tongue was a graphic device by which to convey clearly and unambiguously to viewers the notion/experience of sound production - namely, strident animal noises.

A cursory examination of Egyptian art from other time periods has revealed that this motif was widespread and long lasting, yet scholars have paid little attention to it. Those who have described it often refer to the behaviour as 'licking'[8] when in fact the organ fails to make contact with a surface of any kind. To my knowledge, a specific vocal interpretation for this feature has hitherto not been suspected. The use of graphic symbols to represent non-visual information is known in the art of other ancient cultures, however. For example, Mayan and other Mesoamerican peoples depicted small scrolls or dotted lines immediately in front of the mouth, nose, or eyes of human figures in order to express speech, scent, and sight.[9] The meaning of these devices could also be

---

7 *ibid.*, 190-191.

8 For example, Vandier, *Manuel v*, 64-67.

9 See Houston, S. and Taube, K. (2000). "An archaeology of the senses: Perception and cultural expression in ancient Mesoamerica", *CAJ* 10, 261-294.

qualified by the addition of extra decorative elements; thus, in Classic Mayan art, song was communicated by attaching floral emblems to the speech scroll. The manner in which the speech scroll was drawn (from controlled to whip-like spirals) appears also to have indicated the apparent volume of the sound.[10]

Further examination of the animal images from Giza and Saqqara revealed that the tongue motif also received a similar codification. It became apparent that it ceased to have an association with sound production if the tip made contact with a substance. Four hundred and fifty-six animal images were found to exhibit this behaviour. The contexts in which these occurred differed from those in which animals with isolated tongues were depicted. Some were found in pastoral scenes in which they fed, drank, or groomed, while others were represented in slaughter and hunting scenes in which they appeared to be dead. In these situations, the behavioural meaning of the protruded tongue changed systematically depending upon the type of substance it touched. In short, contact with a food item appeared to indicate that an animal was feeding; contact with fluid implied drinking; if contact was with the animal's body, then a grooming motion was suggested; and if the tongue touched the ground, then the viewer was to understand that the animal was dead.

The coded use of the tongue motif was also suggested by its occasional biologically inaccurate use. For example, depicted cattle protrude their tongues into water even though they do not lap when drinking; sheep, which do not use their tongues when feeding, lick seeds from the hands of sowers; and a dorcas gazelle licks its leg instead of grooming with its teeth. Muscle relaxation ('primary flaccidity') shortly after death may cause an animal's tongue to loll (Colour Plate D), but this is not an invariable response.[11] Nevertheless, of the 405 images of slaughtered animals examined in which the head is still visible, 296 (73%) were found to extend their tongues so that they make contact with the floor; the tongues of the remaining ungulates were not depicted. When the tongue is shown in slaughter scenes, it often lies flat upon the substrate, but regardless of an animal's head position, artists always ensured that at least the tip made contact with the floor.

The contexts in which all of these animals were found indicated that none was to be perceived as calling - animals do not normally vocalize when eating or drinking, and the dismembered state of many of the slaughtered victims clearly confirmed that these, too, were silent. The placement of their protruded tongues thus served to link each one of them unambiguously to different aspects in their environment (food, water, self, ground), thereby clarifying the nature of their behaviour for the viewer.

Schäfer concluded that, "The great majority of the non-foreshortened views which dominate Egyptian drawing go back to data which the eye at some time provided".[12] Emma Brunner-Traut concurred, stating that, "although their rendering is not dependent on immediate visual impressions, the image was originally received by the organs of sight".[13] The tongue motif provides definitive proof, however, that Egyptian art was not simply derived from mental or memory images of visual data, but was instead based upon *ideas.* Although the biological use and physical movements of the tongue differ with regard to sound production, ingestion, grooming, and death, artists nevertheless produced images that used this anatomical feature in an identical fashion - i.e., protruded - to express a variety of behavioural states, even when the resulting images frequently do not match the actual appearance of the behaviour they represent. The systematic use of this graphic device, like other behavioural motifs examined during this study, confirms that the representational rules of Egyptian art were motivated above all by the need to communicate the *meaning* of images as clearly as possible to the viewer, no matter what type of information this entailed.

---

10 It is interesting to speculate whether Egyptian artists also modified protruded tongues in a similar manner to convey volume. For example, in the Saqqara tomb of *rꜥ-m-kꜣ.j*, the tongue of a fox issues forth prominently as it receives a fatal bite to the neck from a hunting dog (see fig 8-23), while, in the same scene, a dorcas gazelle that is also attacked, but only grasped by one foot, barely shows its tongue at all (see fig. 8-25).

11 Indeed, tongue protrusion is more likely to occur some time after death (once putrefaction has begun), due to gas build up in the corpse. See Iserson, K.V. (2002). "Rigor mortis and other postmortem changes" in Kastenbaum, *Encyclopedia of Death ii,* 723.

12 Schäfer, *Principles*, 90.

13 Brunner-Traut, E. (1986). "Aspective" in *ibid.*, 429.

12.4 THE EGYPTIAN CLASSIFICATION OF ANIMALS

What, then, did artists wish to communicate about the behaviour of animals? In short, animals do in wall scenes what humans do not. They move in ways that human figures rarely do or cannot. They mouth at food and slurp at fluid. They lick and scratch and defecate. They posture and threaten, and bite and maul one another. They mate openly, give birth loudly, and suckle their young freely. They open their mouths, bare their teeth, and stick out their tongues. They are unruly and so must be physically controlled and restrained. Animals thus break the rules of proper conduct that were valued so highly in Egyptian society,[14] and which human figures display consistently. Indeed, their uninhibited behaviour is the very antithesis of that of depicted Egyptians.[15]

This behavioural contrast suggests strongly that the Egyptian people perceived animals as being fundamentally different from humans. Such a view conflicts somewhat with the oft-repeated belief that they lived in equilibrium with the animal kingdom.[16] According to Philippe Germond, for example, the Egyptians held that, "Each element or being in creation occupied a clearly defined place and played its part within a perfectly balanced whole. There was no strict hierarchy, however. Man did not occupy the most important place . . . humans played their part alongside their fellow creatures and in harmony with them. No clear-cut distinctions or Cartesian divisions separated the animal, vegetable and mineral kingdoms".[17] Yet the images examined in this study indicate that the Egyptians undoubtedly categorised animals separately from humans. The lack of self-control exhibited by depicted species also hints that the Egyptians may have viewed themselves as at least morally superior to animals.

Further evidence for a perceived difference is again provided by the application of the protruding tongue motif. It has been noted that the figures of Egyptian people in wall scenes are rarely represented open-mouthed, even when they are engaged in conversation, singing, eating, drinking, or vomiting.[18] Neither their teeth nor their tongue is ever shown. The prohibition against using the tongue motif to represent human behaviour indicates that it expressed a uniquely *animal* attribute. I suggest that this reveals that the Egyptians did not experience a kinship with the animal kingdom, but instead perceived them as distinctly "other".

The difference between humans and animals has been explored by many other cultures,[19] none more so than the ancient Greeks. For these early philosophers, what set humans apart from animals was the ability to speak.[20] While animals had a voice and were capable of making noises that could convey various emotions, only humans possessed language by which they could, "communicate moral values, act politically, acquire knowledge, teach, persuade".[21] Indeed, the incomprehensible language of non-Greek speaking peoples ('barbarians'), who were generally thought by the Greeks to be less than human, was often equated with the noises of animals.[22] Intelligible speech was thus seen to be a defining attribute of being fully human.

It is consequently interesting to consider whether the ancient Egyptians recognised a similar distinction, which is expressed, in particular, in the exclusive use of the protruding tongue motif to express animal vocalizations. This graphic device might therefore not only have represented the roars and bellows of animals, but also signified, in a broader sense, their *inarticulate* state. The

14 The Wisdom texts make clear the importance that was placed upon modesty and decorum.

15 For a recent discussion of the open mouth in Mesopotamian art as an indicator of the uncivilized state of animals, see Cheng, J. (2009). "A review of Early Dynastic III music: Man's animal call", *JNES* 68(3), 170-171.

16 For an overview, see Malaise, M. (1987). "La perception du monde animal dans l'Égypte ancienne", *Anthropozoologica* 7, 28-48.

17 Germond, *Egyptian Bestiary*, 15.

18 Gilroy, T.D. (2002). "Outlandish outlanders: Foreigners and caricature in Egyptian art", *GM* 191, 36. Very occasionally, however, mourners and blind harpists are shown with their lips parted.

19 The question of human uniqueness continues to be one of the most hotly debated topics in science, religion, and philosophy. For a discussion of related issues, see Ingold, *What Is an Animal?*

20 Heath, *The Talking Greeks*, 1-35.

21 Buxton, *Persuasion in Greek Tragedy*, 61.

22 In particular, the speech of foreigners was often compared to the twittering of birds. See Heath, *The Talking Greeks*, 200-201.

inability to make oneself understood may thus have served as a general marker of "otherness", providing a fundamental criterion by which the Egyptians were able to define themselves as both human beings and a people.[23]

It is beyond the scope of the present study to speculate on this further, but a final piece of evidence offers some support for this notion. Unlike images of Egyptians, foreigners are occasionally depicted with open mouths (especially during the New Kingdom period). In particular, foreign prisoners are frequently represented in this manner in order to express their pain and fear in response to punishment meted out by their Egyptian captors. During this study, note was taken of any human figure that displayed a protruding tongue. To date, only four examples have been found and in every case these also depict foreigners. The images consist of: a fragment from the 6th dynasty mortuary temple of Pepi II, in which a fallen warrior protrudes his tongue as he is trampled upon by the king in the form of a sphinx;[24] a wall scene in TT58, dating from the reign of Amenhotep III, which displays a group of four Syrian prisoners lying prostate, three of whom open their mouths to bare their upper and lower teeth and show their tongues;[25] a fallen Nubian on the painted box of Tutankhamun, who appears face on and extends his tongue;[26] and an artist's test piece from the Late Period upon which is modelled the head of a foreign prince who reacts to the severing of his nose by opening his mouth, revealing his upper teeth, and protruding his tongue beyond his lips.[27]

Thus, foreigners are depicted doing something in Egyptian art that otherwise only animals do. The images were likely intended to show that the men are making noises like distressed animals in response to their physical pain. I suggest, however, that their protruding tongues also informed viewers that these foreign-speaking non-Egyptians - like animals - were unintelligible.[28]

12.5 A NEW APPROACH FOR THE ANALYSIS OF WALL SCENES

In his thought-provoking essay, "Art, word, and the Egyptian world view", Kent Weeks outlined three criteria that he believed were necessary for any investigation that endeavoured to understand Old Kingdom tomb scenes. These are: "an awareness of context when examining the content of relief; a recognition of the significant attributes of that content; and a willingness to study and correlate those features in terms other than those our own culture may try to impose upon them".[29] By this method, he suggested that clues might be found which would clarify, "how the Egyptians ordered their world". I suggest that a behavioural approach to tomb scenes satisfies these requirements and has, indeed, provided new insights about the manner in which animals were perceived and categorised in ancient Egyptian culture.

First, an analysis of behaviour considers both the depicted context and the definitive features of animal images, as it is the combination of these two factors that provides the key by which the artists' intended meaning may be ascertained. Context is crucial in this type of investigation as the consistent manner in which animals respond to particular circumstances in nature means that the scene in which depicted fauna are placed will increase the likelihood of some behaviours being expressed and rule out others.

---

23 Incomprehensibility may provide an explanation for the standard depiction of children in Egyptian art in which an index finger is held to the lower lip (see Feucht, E. (1986). "Kind" in *LÄ iii*, 424). They thus appear to point directly at their mouths. This gesture may have acted as a signal to an observer that, although their age prevented them from expressing themselves properly, they nevertheless possessed this inherent ability (i.e., because they were both human and Egyptian) and would, in time, eventually speak.

24 Jequier, *Monument Funéraire iii*, pl. 18.

25 Davies, *Ancient Egyptian Paintings ii*, pl. 60.

26 Davies-Gardiner, *Tutankhamun's Painted Box*, pl. 2. It is possible, however, that the man is instead showing his teeth, rather than his tongue.

27 Keimer, L. (1954). "Das Bildhauer-Modell eines Mannes mit abgeschnittener Nase", *ZÄS* 79, figs. 1a and 1b.

28 There is some evidence to suggest that the Egyptians, like the ancient Greeks, also viewed foreigners as animal-like. See Naguib, S.-A. (1994). "Interpreting abstract concepts: Towards an attempt to classify the ancient Egyptian notion of person", *DE* 29, 111.

29 Weeks, K.R. (1979). "Art, word, and the Egyptian world view" in Weeks, *Egyptology and the Social Sciences*, 65.

The "significant attributes" of an individual animal image are then evaluated in order to confirm the specific behaviour that is suggested by the context. As noted by Weeks, only a subset of the features of a given object or event is ever displayed in wall reliefs. Instead, artists selected those elements that they "considered essential for the correct identification of the culturally defined category represented by the figure".[30] Although behaviour inherently consists of a series of movements, this, too, was stripped down to its most important elements. To represent animation in two dimensions, artists were required to reduce it to a single pose made up of those features that could best relay the entire sequence. The postural features of a depicted animal effectively summarise its behaviour as a whole and so, consequently, each must be taken into account and also analysed in relation to one another. A focus upon depicted behaviour in wall reliefs thus ensures a particularly thorough and systematic assessment of animal imagery, because both the composition of individual figures and their relationship to their surroundings is valued and explored.

More importantly, however, a behavioural perspective offers a culturally neutral filter by which to evaluate animal images because interpretation is dependent upon, and guided by, objective criteria - the behaviour of contemporary fauna - rather than subjective impressions and reconstructions. We are all influenced by the culture in which we are immersed. The biases and values that this engenders in us unconsciously affect what we are drawn to in wall scenes, what we deem worthy of investigation, what we believe is correct and incorrect, and our interpretation of their meaning. The Egyptians were also a product of *their* culture, which affected what they chose to represent in their art. Weeks has argued convincingly that the subset of scenes chosen for depiction on tomb walls and the "significant attributes" of the objects placed within those contexts reveal how the Egyptians categorised their world, as the elements chosen for inclusion likely represented those aspects that they considered most relevant. Consequently, if we attempt to understand Egyptian representations by imposing our societal norms upon them (i.e., our categories, terminology, and rules of organization), then we are likely to misunderstand and distort their intended meaning.

No two cultures categorise the world in the same way; however, the behaviour of animals is a constant. Different cultures might disagree about what they regard as the most important features of an animal's behaviour,[31] but it is nevertheless a finite pool from which features can be plucked. Consequently, regardless of what the Egyptians selected to record in their art about the activities of different species, this can be determined objectively, because we can refer to the behaviour of modern animals as a guide. A behavioural analysis of wall scenes is thus comparative, rather than interpretative, and is relatively free of cultural bias.

This method has consequently yielded new information about the Egyptians' relationship with the animal world. The wealth of behaviours that were detected in their art have confirmed their profound interest in species of all kinds and thrown light upon their general and practical knowledge of biology and natural history. Indeed, their images suggest that the Egyptians were compelled to communicate as precisely as possible what they knew about animals, an endeavour that was undoubtedly aided by their conceptual method of representation. The freedom with which depicted animals behave stands in stark contrast to the restraint of human figures, a bias that has hinted at a perceived difference between the two. In fact, the Egyptians' curiosity about the behaviour of animals may conversely indicate an underlying concern with what it means to be human.

Perhaps, however, the most significant benefit of a behavioural approach to the analysis of art is that it offers an unparalleled opportunity to witness one aspect of life *exactly* as it was for the Egyptian people. When the meaning of their images finally becomes clear and we observe animals today engaging in identical behaviours, we can see what inspired the ancient spectators and experience what they felt. We share their wonder and awe about the animal kingdom and in that one magical moment, time effectively disappears.

---

30 *ibid.*

31 Cultures show a remarkable consistency in the way that they classify animals, however. Folk classification schemes also often correspond to established zoological taxonomies. See Diamond, J.M. (1966). "Zoological classification system of a primitive people", *Science* 151(3714), 1102-1104.

NAME	TOMB NO.	*P-M III*	LOCATION	DATE
ḥzjj-rʿ	A3=S2405	429-431	Saqqara	III
mtn	LS6	493-494	Saqqara	IV.1-2
kꜣ.j-wʿb	G7110+G7120	187-188	Giza	IV.2-3
nfr-mꜣʿt	G7060=LG57	183	Giza	IV.2-4
ḥmt-rʿ	ntn	243-244	Giza	IV.4-V.1
dbḥnj	LG90	235-236	Giza	IV.5
mrs-ʿnḫ III	G7530+G7540	197-199	Giza	IV.6
nb.j-m-ꜣḫtj	LG86	230-232	Giza	IV.6-V.1
bw-nfr	ntn	256	Giza	IV.6-V.1?
Anon	ntn	753-754	Saqqara	V
Anon	ntn	758	Saqqara	V
sꜣbw	C16	461	Saqqara	V?
kꜣ.j-nfr	G2150	77-78	Giza	V.1-3
ṯntj	G4920=LG47	141-142	Giza	V.1-5
kꜣ-pw-nswt: kꜣj	G4651	135	Giza	V.1-5
sḫm-kꜣ-rʿ	LG89	233-235	Giza	V.2
pr-sn	D45	577-578	Saqqara	V.2-3
nfr-msḏr-ḫwfw	G2240	nr	Giza	V.2-8 or V.9
jj-mrjj	G6020=LG16	170-174	Giza	V.3L
wr-jr.n-ptḥ	ntn	699-700	Saqqara	V.3-5?
tp-m-ʿnḫ II	D11	483-484	Saqqara	V.5-6?
nfr-bꜣw-ptḥ	G6010=LG15	169-170	Giza	V.6
nfr/kꜣ.ḥꜣ.j	ntn	639-641	Saqqara	V.6
rʿ-ḫʿ.f-ʿnḫ	G7948=LG75	207-208	Giza	V.6?
rʿ-m-kꜣ.j	D3=S903	487-488	Saqqara	V.6? r/u – V.6-7
nj-ʿnḫ-ḫnmw/ḫnmw-ḥtp	ntn	641-644	Saqqara	V.6L-7
nṯr-wsr	D1=S901	485	Saqqara	V.6L-7
pḥn-wj-kꜣ.j	D70=LS15	491-492	Saqqara	V.6-8E
ḥtp-ḥr-ꜣḫtj	D60	593-595	Saqqara	V.6-8E
ꜣḫtj-ḥtp	ntn	634-637	Saqqara	V.6-8E
kꜣ.j-m-nfrt	ntn	263-264	Giza	V.6-8E?
ḫnmw-ḥtp	D49	578-579	Saqqara	V.6-8E?
jtj-sn	ntn	252-523	Giza	V.6-8?
jrj-n-kꜣ-ptḥ	ntn	644	Saqqara	V.6-9
kꜣ.j-dwꜣ	ntn	244-245	Giza	V.6-9?
nḫt.kꜣ.j	ntn	240	Giza	V.6-9?
spd-ḥtp	D15	481	Saqqara	V.6L-9?
sšm-nfr III	G5170	153-154	Giza	V.8E
zꜣ-jb	G2092+G2093	nr	Giza	V.8
kꜣ.j-m-rḥw	D2=S905	485-487	Saqqara	V.8
rʿ-špss	LS16=S902	494-496	Saqqara	V.8M
snḏm-jb: jntj	G2370=LG27	85-87	Giza	V.8M-L
ptḥ-ḥtp I	D62	596-598	Saqqara	V.8M-L
sḫm-ʿnḫ-ptḥ	D41	454-455	Saqqara	V.8-9
ṯjj	D22	468-478	Saqqara	V.8-9
nfr-jrt-n.f	D55	583-584	Saqqara	V.8-9
nfr-ḥr-nj-ptḥ	ntn	637-638	Saqqara	V.8-9
kꜣ.j-m-nfrt	D23	467-468	Saqqara	V.8-9?
ꜣḫtj-ḥtp	D64	599-600	Saqqara	V.8L-9E
rʿ-wr II	G5470=LG32	162-163	Giza	V.8-9M
ḫww-wr	LG95	254-255	Giza	V.8-9M
snḏm-jb: mḥj	G2378=LG26	87-89	Giza	V.9
nj-mꜣʿt-rʿ	ntn	282-284	Giza	V.9
sḫntjw/nfr-sšm-ptḥ	ntn	645	Saqqara	V.9
nj-kꜣw-ḥr	S915	498	Saqqara	V.9
ḫnwt	ntn	623-624	Saqqara	V.9
nbt	ntn	624-625	Saqqara	V.9
jj-nfrt: šꜣ.n.f	ntn	616	Saqqara	V.9
ptḥ-ḥtp II: ṯfj	D64	600-604	Saqqara	V.9M-L
kꜣpj	G2091	69-70	Giza	V.9-VI
nfr I	G4761	137-138	Giza	V.9-VI.1
sšm-nfr IV	LG53	223-226	Giza	V.9-VI.1

ꜣḫtj-mrw-nswt	G2184	80-81	Giza	V.9-VI.1
nj-mꜣꜥt-rꜥ [*jzzj-mr-nṯr*]	G2097	70	Giza	V.9-VI.1
ḥtp-kꜣ.j: tp-kꜣ.j	S3509	447-448	Giza	V.9-VI.1
ptḥ-ḥtp: jj-n-ꜥnḫ	ntn	606-607	Saqqara	V.9-VI.1
mꜣ-nfr	H2=LS17	575-577	Saqqara	V.9-VI.1
sḫm-kꜣ.j	G1029	53	Giza	V.9-VI.1?
sḫm-ꜥnḫ-ptḥ	G7152	191	Giza	V.9-VI.1?
kꜣ.j-m-ḥzt	ntn	542-543	Saqqara	V.9-VI.1?
Anon	ntn	309/761	Giza/Saqqara	V/ V-VI
kꜣ.j-m-nfrt	LG63	208-209	Giza	V-VI
Anon	ntn	308-309	Giza	V-VI
Anon	ntn	605	Saqqara	V-VI
Anon	ntn	607-608	Saqqara	V-VI
Anon	ntn	654	Saqqara	V-VI
ḥtpt	ntn	298	Giza	V-VI
Anon	ntn	nr	Saqqara	late V-VI
kꜣ.j-m-ꜥnḫ	G4561	131-133	Giza	VI
ṯtw: kꜣ.j-nswt	G2001	66-67	Giza	VI
jrw-kꜣ-ptḥ: ḫnw	ntn	639	Saqqara	VI
Anon	ntn	613	Saqqara	VI
sšm-nfr: jwfj	ntn	614-615	Saqqara	VI
ẖnm-ntj	G2374	87	Giza	VI.1
nfr I: jdw	G5550	165	Giza	VI.1
kꜣ-gm-nj: mmj	LS10	521-525	Saqqara	VI.1M
nj-kꜣw-jzzj	ntn	nr	Saqqara	VI.1M
mrrw-kꜣ.j: mrj	ntn	525-534	Saqqara	VI.1M-L
nj-ꜥnḫ-nswt	ntn	694-696	Saqqara	VI.1-2E
zšzšt: jdwt	ntn	617-619	Saqqara	V.9 r/u - VI.1-2E?
jꜣzn	G2196	82	Giza	VI.1M-2M
sꜥnḫ-w(j)-ptḥ	ntn	nr	Saqqara	VI.1L
ḥzj	ntn	nr	Saqqara	VI.IL-2E
ꜥnḫ-m-ꜥ-ḥr: zzj	ntn	512-515	Saqqara	VI.1L-2E
wḏꜣ-ḥꜣ-ttj: ššj [*nfr-sšm-ptḥ*]	ntn	515-516	Saqqara	VI.1L-2E
ṯfw	ntn	605	Saqqara	VI.1-2
mrjj-rꜥ-mrj-ptḥ-ꜥnḫ: nḫbw	G2381	90-91	Giza	VI.2
mṯtj	ntn	646-648	Saqqara	VI.2?
kꜣ.j-jrr	ntn	631-632	Saqqara	VI.2?
ptḥ-ḥtp	LS31	653-654	Saqqara	VI.2? (or LV-VIE?)
mrrj	ntn	518-519	Saqqara	VI.2E?
jnw-mnw	ntn	nr	Saqqara	VI.2E
mrw: ttj-snb	ntn	520	Saqqara	VI.2E-M?
ḫntj-kꜣ.j: jḫḫj	ntn	508-511	Saqqara	VI.2M
mrjj-ttj: mrj	ntn	356-357	Saqqara	VI.2M-L
mḥw	ntn	619-622	Saqqara	VI.2M-3?
kꜣ-ḥj.f	G2136	76	Giza	VI.2M-3?
jj-nfrt	ntn	298-299	Giza	VI.3-4E?
ptḥ-špss II	E1,2+H3	460-461	Saqqara	VI.3-4
wr-nw	ntn	519	Saqqara	VI.3-4?
mrjj-rꜥ-nfr: ḳꜣr	G7101	184-185	Giza	VI.4
ḫꜥ-bꜣw-ḫnm	M.XIV	684	Saqqara	VI.4
ꜣḫtj-ḥtp	E17	633-634	Saqqara	VI.5?
nb-kꜣw-ḥr: jdw	ntn	627-629	Saqqara	V.9 r/u - VI.7
zꜣt-jn-ttj	ntn	539-540	Saqqara	VI.7-FIP

4.1 MOVEMENT

4.1.1 Walking/Standing

Quadrupeds

LB1 *nj-ꜥnḫ-ẖnmw/ẖnmw-ḥtp* - Saqqara, UPC (ntn), second vestibule, west: offering (dorcas gazelle herding)
Source: *P-M III*, 641-644; Moussa-Altenmüller, *Nianchchnum*, fig. 19
Label: *rn gḥs*
Young gazelle[1]
Date: Harpur, *Decoration*, [#431], 274: V.6L-7

LB2 *jtj-sn* - Giza, CF (ntn), chapel, east: marsh (common genet hunting)
Source: *P-M III*, 252-253; Hassan, *Giza v*, fig. 123
Label: none
Date: Harpur, *Decoration*, [#035], 265: V.6-8?

LB3 *nḫt.kꜣ.j* - Giza, CF (ntn), chapel, east: offering (striped hyena tethered)
Source: *P-M III*, 240; Hassan, *Giza vii*, fig. 22
Label: none
Date: Harpur, *Decoration*, [#143], 268: V.6-9?

LB4 *snḏm-jb: jntj* - Giza, WF (G2370=LG27) room II, west: marsh (opposed hippopotamuses)
Source: *P-M III*, 85-87; *LD ii*, pl. 77; Brovarski, *Senedjemib Complex Part I*, fig. 43
Label: none
Date: Brovarski, *Senedjemib Complex Part I*, 23: V.8M-L; Harpur, *Decoration*, [#219], 269: V.8M-L

LB5 *sḫm-ꜥnḫ-ptḥ* - Saqqara, NSP (D41), chapel, south: offering (scimitar oryx being led)
Source: *P-M III*, 454-455; Simpson, *Sekem-ankh-ptah*, pl. D
Label: *rn mꜣ-ḥḏ*
Young oryx
Date: Harpur, *Decoration*, [#504], 276: V.8-9

LB6 *ṯjj* - Saqqara, NSP (D22), room I/pillared hall, west: pastoral (goats herding)
Source: *P-M III*, 468-478; Epron et al., *Ti i*, pl. 26
Label: n/a
Date: Harpur, *Decoration,* [#543], 277: V.8-9

LB7 *ṯjj* - Saqqara, NSP (D22), room III, east: agricultural (donkeys transporting)
Source: *P-M III*, 468-478; Wild, *Ti iii*, pl. 154
Label: none
Date: Harpur, *Decoration*, [#543], 277: V.8-9

LB8 *nfr-jrt-n.f* - Saqqara, ESP (D55), chapel, east: marsh (Egyptian mongoose hunting)
Source: *P-M III*, 583-584; van de Walle, *Neferirtenef*, pl. 1
Label: none
Date: Harpur, *Decoration*, [#440], 274: V.8-9

LB9 *ꜣḫtj-ḥtp* - Saqqara, WSP (D64), relief fragment from corridor, east: desert (hedgehog entering burrow)
Source: *P-M III*, 599-600; Davies, *Five Theban*, pl. 40 [1]
Label: none
Date: Harpur, *Decoration*, [#338], 272: V.8L-9E

LB10 *ptḥ-ḥtp II: ṯfj* - Saqqara, WSP (D64), offering-room, east: post-hunt (lion/leopard trapped)
Source: *P-M III*, 600-604; Quibell, *Ptah-hetep*, pl. 33 [top]
Label: none
Date: Harpur, *Decoration*, [#400], 274: V.9M-L

---

1 I have followed general consensus and have translated *rn* as young (see Junker, *Giza vi*, 50), however others have proposed that the term instead refers to an animal's domesticated state (e.g., Montet, *Scènes*, 113). I suggest an alternative meaning of 'tamed'. The label occurs frequently in offering scenes and may therefore signify that *rn* animals are sufficiently tractable to allow them to be brought before the deceased.

LB11 *ptḥ-ḥtp II: ṯfj* - Saqqara, WSP (D64), offering-room, east: hunting (dogs leashed)
Source: *P-M III*, 600-604; Quibell, *Ptah-hetep*, pl. 32
Label: none
Date: Harpur, *Decoration*, [#400], 274: V.9M-L

LB12 Anon - Saqqara, WSP (ntn), relief fragment from shaft 6, 'Mastaba of Shaft no. 14', unknown: desert (hamadryas baboon being led)
Source: *P-M III*, 613; Hassan, *Hemet-Re*, pl. 14 [c]
Label: *s3w jʿn*
Guardian of the baboon
Date: *P-M III*, 613: VI; Harpur, *Decoration*, nr

LB13 *sḫm-k3.j* - Giza, WF (G1029), chapel, east: offering (Nubian ibex being led)
Source: *P-M III*, 53; Simpson, *Mast. West. Cem. Part I*, fig. 4
Label: *rn [n] j3(w)*
Young ibex
Date: Harpur, *Decoration*, [#225], 269: V.9-VI.1?

LB14 *sḫm-ʿnḫ-ptḥ* - Giza, EF (G7152), chapel, west: offering (hartebeest being led)
Source: *P-M III*, 191; Badawy, *Tombs of Iteti*, fig. 20
Label: none
Date: Harpur, *Decoration*, [#223], 269: V.9-VI.1?

LB15 *nfr I: jdw* - Giza, WF (G5550), chapel, south: domestic (vervet monkey leashed)
Source: *P-M III*, 165; Junker, *Giza viii*, fig. 35
Label: *mrrj*
Mereri
Date: Harpur, *Decoration*, [#129], 267: VI.1

LB16 *mrrw-k3.j: mrj* - Saqqara, TP (ntn) chamber A6, south: offering (addax being led)
Source: *P-M III*, 525-534; Duell, *Mereruka i*, pl. 51
Label: *rn nw-ḏw*
Young addax
Date: Harpur, *Decoration*, [#420], 274: VI.1M-L

LB17 *mrrw-k3.j: mrj* - Saqqara, TP (ntn), chamber A13, west: boating (vervet monkey on rigging)
Source: *P-M III*, 525-534; Duell, *Mereruka ii*, pl. 140
Label: none
Date: Harpur, *Decoration*, [#420], 274: VI.1M-L

LB18 *ḥzj* - Saqqara, TP (ntn), portico, south (west of entrance): marsh (vervet monkey being led)
Source: *P-M III*, nr; Kanawati-Abder-Raziq, *Hesi*, pl. 54
Label: none
Date: Kanawati-Abder-Raziq, *Hesi*, 16: VI.1L-2E; Harpur, *Decoration*, nr

LB19 *jnw-mnw* - Saqqara (ntn), room II, south: pastoral (cattle being led)
Source: *P-M III*, nr; Kanawati, *Inumin*, pl. 49 [a]
Label: *rn jw3*
Young oxen
Date: Kanawati, *Inumin*, 16-17: VI.2E; Harpur, *Decoration*, nr

LB20 *mrjj-ttj: mrj* - Saqqara, TP (ntn), room C1, west: offering (dama deer being led)
Source: *P-M III*, 356-357; Kanawati-Abder-Raziq, *Meryteti*, pl. 46
Label: none
Date: Kanawati-Abder-Raziq, *Meryteti*, 18: end VI.1-VI.2M; Harpur, *Decoration*, [#415], 274: VI.2M-L

LB21 *mḥw* - Saqqara, UPC (ntn), room II, south: boating (vervet monkey on rigging)
Source: *P-M III*, 619-622; Altenmüller, *Mehu*, pl. 19 [a]
Label: n/a
Date: Altenmüller, *Mehu*, 19: VI.2; Harpur, *Decoration*, [#424], 274: VI.2M-3?

Bipeds

LB22 *ṯjj* - Saqqara, NSP (D22), room III, south: offering (common and demoiselle cranes)
Source: *P-M III*, 468-478; Wild, *Ti iii*, pl. 169
Label: *ḏ3t*
Crane
*wḏ*ʿ
Crane
Date: Harpur, *Decoration,* [#543], 277: V.8-9

LB23 *ptḥ-ḥtp II: ṯfj* - Saqqara, WSP (D64), offering-room, east: offering (4 geese)
Source: *P-M III*, 600-604; Quibell, *Ptah-hetep*, pl. 31
Label: n/a
Date: Harpur, *Decoration*, [#400], 274: V.9M-L

LB24 *ʿnḫ-m-ʿ-ḥr: zzj* - Saqqara, TP (ntn), room II, west: clapnet (grey heron beside fowlers)
Source: *P-M III*, 512-515; Kanawati-Hassan, *Ankhmahor*, pl. 42
Label: none
Date: Kanawati-Hassan, *Ankhmahor*, 18: VI.1M-L; Harpur, *Decoration*, [#374], 273: VI.1L-2E

LB25 *ḥzj* - Saqqara, TP (ntn), portico, east: clapnet (sacred ibis below clapnet)
Source: *P-M III*, nr; Kanawati-Abder-Raziq, *Hesi*, pl. 55
Label: *dw sḫt jn jmj-r mḥw j mḥw pw ḥb r=k*
Setting the trap by the overseer of fowlers: Oh this fowler! There is a catch of birds for you!
Date: Kanawati-Abder-Raziq, *Hesi*, 16: VI.1L-2E; Harpur, *Decoration*, nr

LB26a *ḥzj* - Saqqara, TP (ntn), portico, south (west of entrance): marsh (purple gallinule walking left)
Source: *P-M III*, nr; Kanawati-Abder-Raziq, *Hesi*, pl. 54
Label: none
Date: Kanawati-Abder-Raziq, *Hesi*, 16: VI.1L-2E; Harpur, *Decoration*, nr

LB26b *ḥzj* - Saqqara, TP (ntn), portico, south (west of entrance): marsh (purple gallinule walking right)
Source: *P-M III*, nr; Kanawati-Abder-Raziq, *Hesi*, pl. 54
Label: none
Date: Kanawati-Abder-Raziq, *Hesi*, 16: VI.1L-2E; Harpur, *Decoration*, nr

---

4.1.2 Balking

Representative species

LB27 *nfr-m3ʿt* - Giza, EF (G7060=LG57), chapel, east: offering (scimitar oryx being pulled)
Source: *P-M III*, 183; *LD ii*, pl. 17
Label: *rn m3-ḥḏ*
Young oryx
Date: Harpur, *Decoration*, [#131], 267: IV.2-4

LB28 *mrs-ʿnḫ III* - Giza, EF (G7530 + G7540), door jamb to main chapel, north: offering (scimitar oryx being pulled)
Source: *P-M III*, 197-199; Dunham-Simpson, *Mersyankh III*, fig. 3 [a]
Label: none
Date: Harpur, *Decoration*, [#098], 267: IV.6

LB29 *k3-pw-nswt: k3j* - Giza, WF (G4651), chapel, north: offering (Nubian ibex being pulled)
Source: *P-M III*, 135; Junker, *Giza iii*, fig. 18
Label: *rn nj3(w)*
Young ibex
Date: Harpur, *Decoration*, [#277], 270: V.1-5

LB30 *jj-mrjj* - Giza, WF (G6020=LG16), first chamber, east: offering (Nubian ibex being pulled)
Source: *P-M III*, 170-174; Weeks, *Cemetery G6000*, fig. 29
Label: *rn nj3(w)*
Young ibex

Date: Weeks, *Cemetery G6000*, 7: V.6; Harpur, *Decoration*, [#014], 265: V.3L

LB31 *rʿ-špss* - Saqqara, NSP (LS16=S902), hall, south: offering (Nubian ibex being pulled)
Source: *P-M III*, 494-496; *LD ii*, pl. 61 [b]
Label: *njꜣ(w)*
Ibex
Date: Harpur, *Decoration*, [#456], 275: V.8M

LB32 *rʿ-špss* - Saqqara, NSP (LS16=S902), hall, east: offering (donkey being pulled?)
Source: *P-M III*, 494-496; *LD Erganz.*, pl. 39
Label: missing
Date: Harpur, *Decoration*, [#456], 275: V.8M

LB33 *tjj* - Saqqara, NSP (D22), room III, south: offering (Nubian ibex being pulled)
Source: *P-M III*, 468-478; Wild, *Ti iii*, pl. 165
Label: *njꜣw*
Ibex
sṯꜣ r=k
Drag/pull to you!
jrr(=j) ḥzjt=k
I am one who does (so that) it pleases you
Date: Harpur, *Decoration*, [#543], 277: V.8-9

LB34 *nbt* - Saqqara, UPC (ntn), room C, south: offering (scimitar oryx being pulled)
Source: *P-M III*, 624-625; Munro, *Unas-Friedhof*, pl. 17
Label: none
Date: Munro, *Unas-Friedhof*, 20: V.9; Harpur, *Decoration*, [#438], 274: V.9

LB35 *nbt* - Saqqara, UPC (ntn), room C, south: offering (Nubian ibex being pulled)
Source: *P-M III*, 624-625; Munro, *Unas-Friedhof*, pl. 17
Label: *rn [n]jꜣ(w)*
Young ibex
Date: Munro, *Unas-Friedhof*, 20: V.9; Harpur, *Decoration*, [#438], 274: V.9

LB36 *nbt* - Saqqara, UPC (ntn), room C, north: offering (scimitar oryx being pulled)
Source: *P-M III*, 624-625; Munro, *Unas-Friedhof*, pl. 19
Label: *rn mꜣ-ḥḏ*
Young oryx
Date: Munro, *Unas-Friedhof*, 20: V.9; Harpur, *Decoration*, [#438], 274: V.9

LB37 *nbt* - Saqqara, UPC (ntn), room C, north: offering (Nubian ibex being pulled)
Source: *P-M III*, 624-625; Munro, *Unas-Friedhof*, pl. 19
Label: missing
Date: Munro, *Unas-Friedhof*, 20: V.9; Harpur, *Decoration*, [#438], 274: V.9

LB38 *sšm-nfr IV* - Giza, GIS (LG53), doorway between chambers A and B, west: offering (ox being pulled)
Source: *P-M III*, 223-226; Junker, *Giza xi*, fig. 79
Label: *rn jwꜣ*
Young ox
Date: Harpur, *Decoration*, [#235], 270: V.9-VI.1

LB39 *jꜣzn* - Giza, WF (G2196), chapel, east: offering (dorcas gazelle being pulled)
Source: *P-M III*, 82; Simpson, *Mast. West. Cem. Part I*, fig. 31
Label: none
Date: Harpur, *Decoration*, [#013], 265: VI.1M-2M

LB40 *jꜣzn* - Giza, WF (G2196), chapel, east: offering (Nubian ibex being pulled)
Source: *P-M III*, 82; Simpson, *Mast. West. Cem. Part I*, fig. 31
Label: none
Date: Harpur, *Decoration*, [#013], 265: VI.1M-2M

LB41 *ptḥ-ḥtp* - Saqqara, AJS (LS31), door thickness, west: offering (Nubian ibex being pulled)
Source: *P-M III*, 653-654; *LD ii*, pl. 104 [d]

Label: *rn njꜣ(w)*
Young ibex
Date: Harpur, *Decoration*, [#397], 273: VI.2? or LV-VIE?

4.1.3 Galloping

Donkeys

LB42 *jj-mrjj* - Giza, WF (G6020=LG16), second chamber, north: agricultural
Source: *P-M III*, 170-174; Weeks, *Cemetery G6000*, fig. 39
Label: [ ]*ꜣ jm=s*[*n*][2]
Date: Weeks, *Cemetery G6000*, 7: V.6; Harpur, *Decoration*, [#014], 265: V.3L

LB43 *nj-ꜥnḫ-ẖnmw/ẖnmw-ḥtp* - Saqqara, UPC (ntn), outer hall, east: agricultural
Source: *P-M III*, 641-644; Moussa-Altenmüller, *Nianchchnum*, pl. 56
Label: none
Date: Harpur, *Decoration*, [#431], 274: V.6L-7

LB44 *pḥn-wj-kꜣ.j* - Saqqara, NSP (D70=LS15), antechamber, unknown: agricultural
Source: *P-M III*, 491-492; *LD ii*, pl. 47
Label: *hꜣ*[*j*] . . . [*s*]*ḥꜣt r* . . .
Descending . . . of the donkey pack to . . .
Date: Harpur, *Decoration*, [#393], 273: V.6-8E

LB45 *ḥtp-ḥr-ꜣḫtj* - Saqqara, WSP (D60), chapel, south: agricultural
Source: *P-M III*, 593-595; Mohr, *Hetep-her-akhti*, fig. 50
Label: *sꜥḳ(=sn) jm*
Drive (them) therein
jꜣt jr=k
A mound for you
Date: Harpur, *Decoration*, [#467], 275: V.6-8E

LB46 *jrj-n-kꜣ-ptḥ* - Saqqara, UPC (ntn), offering chamber, east: agricultural
Source: *P-M III*, 644; Moussa-Junge, *Two Tombs*, pl. 9
Label: *hꜣj*
Go down! Descend!
Date: Harpur, *Decoration*, [#362], 273: V.6-9

LB47 *zꜣ-jb* - Giza, WF (G2092 + G2093), chapel, east: agricultural
Source: *P-M III*, nr; Roth, *Cem. Pal. Att.*, pl. 183
Label: *sḥ*[*ꜣ*]*t*
Donkey pack
Date: Roth, *Cem. Pal. Att.*, 109: V.8; Harpur, *Decoration*, nr

LB48 *sḫm-ꜥnḫ-ptḥ* - Saqqara, NSP (D41), chapel, south: agricultural
Source: *P-M III*, 454-455; Simpson, *Sekem-ankh-ptah*, pl. D
Label: missing[3]
Date: Harpur, *Decoration*, [#504], 276: V.8-9

LB49 *tjj* - Saqqara, NSP (D22), room III, east: agricultural
Source: *P-M III*, 468-478; Wild, *Ti iii*, pl. 153
Label: n/a
Date: Harpur, *Decoration*, [#543], 277: V.8-9

LB50 *nfr-jrt-n.f* - Saqqara, ESP (D55), chapel, east: agricultural
Source: *P-M III*, 583-584; van de Walle, *Neferirtenef*, pl. 12
Label: *sḥꜣt nt* 1,200
Donkey pack of 1,200

---

2 Weeks (*Cemetery G6000*, 48) suggests [*jr ḥ*]*ꜣ-k jm=sn* 'Make the turning around it'.

3 The inscription is missing, but according to Simpson (*Sekem-ankh-ptah*, 14), Mariette repor:ed the following caption in this area [*d*]*m n pḥ nt*(*y*) *ḥ*[*nꜥ*] 'A prod to the rump, oh comrade!'

Date: Harpur, *Decoration*, [#440], 274: V.8-9

LB51 *k3.j-m-nfrt* - Saqqara, NSP (D23), chapel, east: agricultural
Source: *P-M III*, 467-468; Simpson, *Kayemnofret*, pl. F
Label: *sḥ3t nfr m33 wr*
Donkey pack
The sight is very beautiful
Date: Simpson, *Kayemnofret*, 1: V.8-9; Harpur, *Decoration*, [#522], 276: V.8-9?

LB52 *3ḫtj-ḥtp* - Saqqara, WSP (D64), corridor, west: agricultural
Source: *P-M III*, 599-600; Davies, *Ptahhetep ii*, pl. 8
Label: *sḥ3t* 2,500 *h3j*
Donkey pack of 2,500
Descend!
Date: Harpur, *Decoration*, [#338], 272: V.8L-9E

LB53 *nj-k3w-ḥr* - Saqqara, NSP (S915), offering-room, east: agricultural
Source: *P-M III*, 498; Quibell, *Saqqara (1907-1908)*, pl. 62 [2]
Label: . . . *sḥ*[*3t*] . . .
. . . donkey pack . . .
Date: Harpur, *Decoration*, [#435], 274: V.9

LB54 *ḫnwt* - Saqqara, UPC (ntn), relief fragment Kh/16 (from corridor and court areas), unknown: agricultural
Source: *P-M III*, 623-624; Munro, *Unas-Friedhof*, pl. 43
Label: [*s*]*ḥ3t ṯt*
Donkey pack gang
Date: Munro, *Unas-Friedhof*, 20: V.9; Harpur, *Decoration*, [#478], 275: V.9

LB55 *sḫm-k3.j* - Giza, WF (G1029), chapel, east: agricultural
Source: *P-M III*, 53; Simpson, *Mast. West. Cem. Part I*, fig. 4
Label: *sḥ3t*
Donkey pack
Date: Harpur, *Decoration*, [#225], 269: V.9-VI.1?

LB56 *k3.j-m-nfrt* - Giza, EF (LG63), chapel, east: agricultural
Source: *P-M III*, 208-209; *LD Erganz.*, pl. 32 [lower]
Label: none
Date: Harpur, *Decoration*, [#258], 270: V-VI

LB57 *mrrw-k3.j: mrj* - Saqqara, TP (ntn), chamber A13, east: agricultural
Source: *P-M III*, 525-534; Duell, *Mereruka ii*, pl. 170
Label: [*s*]*ḥ3t*
Donkey pack
Date: Harpur, *Decoration*, [#420], 274: VI.1M-L

LB58 *zšzšt: jdwt* - Saqqara, UPC (ntn), room IV, west: agricultural
Source: *P-M III*, 617-619; Macramallah, *Idout*, pl. 10 [a]; Kanawati-Abder-Raziq, *Iynefert*, pl. 59 [a]
Label: missing
Date: Kanawati-Abder-Raziq, *Iynefert*, 36-37: VI.1E; Harpur, *Decoration*, [#487], 275: V.9 r/u, VI.1-2E?

LB59 *mṯtj* - Saqqara, UPC (ntn), relief fragment (Roy. Ont. Mus. no. 953.116.2), unknown: agricultural
Source: *P-M III*, 646-648; Kaplony, *Studien Methethi*, no. 3
Label: *ḏwt n pḥt* 1,300
Balance of sheaves 1,300
sḥ[*3t*]
Donkey pack
Date: Ziegler, *Catalogue*, 123: probably VI; Harpur, *Decoration*, [#426], 274: VI.2?

LB60 *mḥw* - Saqqara, UPC (ntn), room II, south: agricultural
Source: *P-M III*, 619-622; Altenmüller, *Mehu*, pl. 24 [a]
Label: *ḥwjt sḥ3t r pḥt*
Driving in/drawing in of the donkey-pack to the pile of sheaves

h3 jzj jr=k nty ḥnʿ(=j)
Ho! Go my comrade!
dw3t ḫprty sḥ3t
The morning is here, oh herd![4]
Date: Altenmüller, *Mehu*, 19: VI.2; Harpur, *Decoration*, [#424], 274: VI.2M-3?

LB61 *k3-ḥj.f* - Giza, WF (G2136), chapel, east: agricultural
Source: *P-M III*, 76; Junker, *Giza vi*, fig. 45
Label: *h3j.t pḥrt sḥ3t*
Descending to the sheaves by the donkey-pack
Date: Harpur, *Decoration*, [#278], 271: VI.2M-3?

4.1.4 Jumping

Jerboas

LB62 *mṯn* - Saqqara, NSP (LS6), chapel, west: hunting (on hind legs, facing left)
Source: *P-M III*, 493-494; Smith, *HESPOK*, fig. 60
Label: none
Date: Harpur, *Decoration*, [#425], 274: IV.1-2

LB63 *nj-ʿnḫ-ẖnmw/ḫnmw-ḥtp* - Saqqara, UPC (ntn), vestibule, west: hunting (leaving burrow)
Source: *P-M III*, 641-644; Moussa-Altenmüller, *Nianchchnum*, pl. 40
Label: none
Date: Harpur, *Decoration*, [#431], 274: V.6L-7

LB64 *ptḥ-ḥtp II: ṯfj* - Saqqara, WSP (D64), offering-room, east: hunting (entering burrow)
Source: *P-M III*, 600-604; Quibell, *Ptah-hetep*, pl. 32
Label: none
Date: Harpur, *Decoration*, [#400], 274: V.9M-L

LB65 Anon - Saqqara, WSP (ntn), relief fragment from shaft 6, 'Mastaba of Shaft no. 14', unknown: desert (on hind legs, facing left)
Source: *P-M III*, 613; Hassan, *Hemet-Re*, pl. 14 [c]
Label: none
Date: *P-M III*, 613: VI; Harpur, *Decoration*, nr

LB66 *jnw-mnw* - Saqqara (ntn), room II, north: desert (leaping upward, facing left)
Source: *P-M III*, nr; Kanawati, *Inumin*, pl. 47
Label: none
Date: Kanawati, *Inumin*, 16-17: VI.2E; Harpur, *Decoration*, nr

LB67 *jnw-mnw* - Saqqara (ntn), room II, north: desert (on hind legs, facing right)
Source: *P-M III*, nr; Kanawati, *Inumin*, pl. 47
Label: none
Date: Kanawati, *Inumin*, 16-17: VI.2E; Harpur, *Decoration*, nr

Fish

LB68 *mrrw-k3.j: mrj* - Saqqara, TP (ntn), chamber A4, east: fishing (catfish leaping from hand-net)
Source: *P-M III*, 525-534; Duell, *Mereruka i*, pl. 43
Label: none
Date: Harpur, *Decoration*, [#420], 274: VI.1M-L

LB69 *mrrw-k3.j: mrj* - Saqqara, TP (ntn), chamber A4, east: fishing (catfish leaping from hand-net)
Source: *P-M III*, 525-534; Duell, *Mereruka i*, pl. 43
Label: none
Date: Harpur, *Decoration*, [#420], 274: VI.1M-L

4 Altenmüller (*Mehu*, 120) suggests that this is a reference to the crew of men.

LB70 *zšzšt: jdwt* - Saqqara, UPC (ntn), room I, north: fishing (mullet leaping from hand-net)
Source: *P-M III*, 617-619; Macramallah, *Idout*, pl. 5 [d]; Kanawati-Abder-Raziq, *Iynefert*, pl. 53 [f]
Label: none
Date: Kanawati-Abder-Raziq, *Iynefert*, 36-37: VI.1E; Harpur, *Decoration*, [#487], 275: V.9 r/u, VI.1-2E?

4.1.5 Climbing

Common genets

LB71 *nb.j-m-ȝḫtj* - Giza, CF (LG86), chapel I, west: marsh (ascending to right)
Source: *P-M III*, 230-232; *LD ii*, pl. 12 [b]
Label: none
Date: Harpur, *Decoration*, [#122], 267: IV.6-V.1

LB72 *nj-ʿnḫ-ẖnmw/ẖnmw-ḥtp* - Saqqara, UPC (ntn), forecourt, south: fishing (ascending to right)
Source: *P-M III*, 641-644; Moussa-Altenmüller, *Nianchchnum*, fig. 5
Label: none
Date: Harpur, *Decoration*, [#431], 274: V.6L-7

LB73 *nj-ʿnḫ-ẖnmw/ẖnmw-ḥtp* - Saqqara, UPC (ntn), outer hall, west: marsh (ascending to right)
Source: *P-M III*, 641-644; Moussa-Altenmüller, *Nianchchnum*, pl. 74[5]
Label: none
Date: Harpur, *Decoration*, [#431], 274: V.6L-7

LB74 *ḥtp-ḥr-ȝḫtj* - Saqqara, WSP (D60), chapel, north: marsh (ascending to left)
Source: *P-M III*, 593-595; Mohr, *Hetep-her-akhti*, fig. 34
Label: none
Date: Harpur, *Decoration*, [#467], 275: V.6-8E

LB75 *jtj-sn* - Giza, CF (ntn), chapel, east: marsh (ascending to right)
Source: *P-M III*, 252-253; Hassan, *Giza v*, fig. 123[6]
Label: none
Date: Harpur, *Decoration*, [#035], 265: V.6-8?

LB76 *jrj-n-kȝ-ptḥ* - Saqqara, UPC (ntn), offering chamber, south: marsh (ascending to left)
Source: *P-M III*, 644; Moussa-Junge, *Two Tombs*, pl. 12[7]
Label: none
Date: Harpur, *Decoration*, [#362], 273: V.6-9

LB77 *rʿ-špss* - Saqqara, NSP (LS16=S902), forecourt, west: marsh (ascending to right)
Source: *P-M III*, 494-496; Harpur, *Decoration*, fig. 187; *LD ii*, pl. 60 [left]
Label: none
Date: Harpur, *Decoration*, [#456], 275: V.8M

LB78 *snḏm-jb: jntj* - Giza, WF (G2370=LG27), portico, west: marsh (ascending to left)
Source: *P-M III*, 85-87; *LD Ergänz.*, pl. 18; Brovarski, *Senedjemib Complex Part I*, fig. 27
Label: none
Date: Brovarski, *Senedjemib Complex Part I*, 23: V.8M-L; Harpur, *Decoration*, [#219], 269: V.8M-L

LB79 *snḏm-jb: jntj* - Giza, WF (G2370=LG27), room II, west: marsh (ascending to left)
Source: *P-M III*, 85-87; *LD ii*, pl. 77[8]
Label: none
Date: Brovarski, *Senedjemib Complex Part I*, 23: V.8M-L; Harpur, *Decoration*, [#219], 269: V.8M-L

5 Moussa and Altenmüller (*Nianchchnum*, 152) identify this animal as a genet.

6 Hassan (*Giza v*, 267) states that this is an ichneumon, but the animal's pricked ears clearly indicate that it is a genet.

7 Moussa and Junge (*Two Tombs*, 42) suggest that this is a genet, but the form could equally depict a mongoose. It has been accepted elsewhere as a genet. See Aufrère, S.H. (1998). "La loutre, le chat, la genette et l'ichneumon, hôtes du fourré de papyrus: Présages, prédateurs des marécages et croyances funéraires", *DE* 41, 7-28, fig. 4.

8 According to Brovarski (*Senedjemib Complex Part I*, 49), this relief has now been destroyed.

LB80 *ṯjj* - Saqqara, NSP (D22), room III, north: marsh (ascending to left)
Source: *P-M III*, 468-478; Wild, *Ti ii*, pl. 119
Label: none
Date: Harpur, *Decoration*, [#543], 277: V.8-9

LB81 *ṯjj* - Saqqara, NSP (D22), room III, north: marsh (ascending to left)
Source: *P-M III*, 468-478; Wild, *Ti ii*, pl. 119
Label: none
Date: Harpur, *Decoration*, [#543], 277: V.8-9

LB82 *kꜣ.j-m-nfrt* - Saqqara, NSP (D23), chapel, north: marsh (ascending to right, head turned back)
Source: *P-M III*, 467-468; Simpson, *Kayemnofret*, pl. A
Label: none
Date: Simpson, *Kayemnofret*, 1: V.8-9; Harpur, *Decoration*, [#522], 276: V.8-9?

LB83 *ꜣḫtj-ḥtp* - Saqqara, WSP (D64), hall, east: marsh (ascending to right, with chick)
Source: *P-M III*, 599-600; Davies, *Ptahhetep ii*, pl. 13
Label: none
Date: Harpur, *Decoration*, [#338], 272: V.8L-9E

LB84 *ꜣḫtj-ḥtp* - Saqqara, WSP (D64), hall, east: marsh (ascending to right, head turned back)
Source: *P-M III*, 599-600; Davies, *Ptahhetep ii*, pl. 14
Label: none
Date: Harpur, *Decoration*, [#338], 272: V.8L-9E

LB85 *ḫnwt* - Saqqara, UPC (ntn), room A [I], east: marsh (ascending to left)
Source: *P-M III*, 623-624; Munro, *Unas-Friedhof*, pl. 33[9]
Label: none
Date: Munro, *Unas-Friedhof*, 20: V.9; Harpur, *Decoration*, [#478], 275: V.9

LB86 *jj-nfrt: šꜣ.n.f* - Saqqara, UPC (ntn), room I, north: marsh (ascending to right)
Source: *P-M III*, 616; Kanawati-Abder-Raziq, *Iynefert*, pl. 37
Label: none
Date: Kanawati-Abder-Raziq, *Iynefert*, 13: V.9E-M; Harpur, *Decoration*, [#345], 272: V.9

LB87 *nj-kꜣw-jzzj* - Saqqara, TP (ntn), room I, east: marsh (ascending to right, head turned back)
Source: *P-M III*, nr; Kanawati-Abder-Raziq, *Nikauisesi*, pl. 50
Label: none
Date: Kanawati-Abder-Raziq, *Nikauisesi*, 17-23: VI.1M; Harpur, *Decoration*, nr

LB88 *nj-kꜣw-jzzj* - Saqqara, TP (ntn), room I, east: marsh (ascending to right, with chick)
Source: *P-M III*, nr; Kanawati-Abder-Raziq, *Nikauisesi*, pl. 50
Label: none
Date: Kanawati-Abder-Raziq, *Nikauisesi*, 17-23: VI.1M; Harpur, *Decoration*, nr

LB89 *mrrw-kꜣ.j: mrj* - Saqqara, TP (ntn), chamber A1, north: marsh (ascending to right)
Source: *P-M III*, 525-534; Duell, *Mereruka i*, pl. 10
Label: none
Date: Harpur, *Decoration*, [#420], 274: VI.1M-L

LB90 *zšzšt: jdwt* - Saqqara, UPC (ntn), room III, west: marsh (ascending to right, with chick)
Source: *P-M III*, 617-619; Macramallah, *Idout*, pl. 7; Kanawati-Abder-Raziq, *Iynefert*, pl. 54
Label: none
Date: Kanawati-Abder-Raziq, *Iynefert*, 36-37: VI.1E; Harpur, *Decoration*, [#487], 275: V.9 r/u, VI.1-2E?

LB91 *sꜥnḫ-w(j)-ptḥ* - Saqqara, TP (ntn), room I, east: marsh (ascending to left, with chick)
Source: *P-M III*, nr; Kanawati-Abder-Raziq, *Neferseshemre*, pls. 69 and 76
Label: none
Date: Kanawati-Abder-Raziq, *Neferseshemre*, 40-41: VI.1L; Harpur, *Decoration*, nr

9 *P-M III*, 623 describe this animal as an ichneumon. Similarly, Munro (*Unas-Friedhof*, 52) states that the birds are "von einem Ichneumon belauert". The pricked ears indicate that it is actually a genet.

LB92 *sꜥnḫ-w(j)-ptḥ* - Saqqara, TP (ntn), room I, east: marsh (ascending to left, head turned back)
Source: *P-M III*, nr; Kanawati-Abder-Raziq, *Neferseshemre*, pls. 69 and 76
Label: none
Date: Kanawati-Abder-Raziq, *Neferseshemre*, 40-41: VI.1L; Harpur, *Decoration*, nr

LB93 *ḥzj* - Saqqara, TP (ntn), portico, south (west of entrance): marsh (ascending to left, head turned back)
Source: *P-M III*, nr; Kanawati-Abder-Raziq, *Hesi*, pl. 54
Label: none
Date: Kanawati-Abder-Raziq, *Hesi*, 16: VI.1L-2E; Harpur, *Decoration*, nr

LB94 *ḥzj* - Saqqara, TP (ntn), portico, south (west of entrance): marsh (descending to left, with chick)
Source: *P-M III*, nr; Kanawati-Abder-Raziq, *Hesi*, pl. 54
Label: none
Date: Kanawati-Abder-Raziq, *Hesi*, 16: VI.1L-2E; Harpur, *Decoration*, nr

LB95 *ḥzj* - Saqqara, TP (ntn), portico, south (east of entrance): marsh (ascending to right, with chick)
Source: *P-M III*, nr; Kanawati-Abder-Raziq, *Hesi*, pl. 53
Label: none
Date: Kanawati-Abder-Raziq, *Hesi*, 16: VI.1L-2E; Harpur, *Decoration*, nr

LB96 *mrjj-rꜥ-mrj-ptḥ-ꜥnḫ: nḫbw* - Giza, WF (G2381), relief fragment from forecourt (MFA no. 13.4332), unknown: marsh (ascending to left, head turned back)
Source: *P-M III*, 90-91; Smith, W.S. (1958). *Bost. Mus. Bull.* 56, fig. 2
Label: none
Date: Harpur, *Decoration*, [#089], 267: VI.2

LB97 *jnw-mnw* - Saqqara (ntn), room II, west: marsh (ascending to right, with chick)
Source: *P-M III*, nr; Kanawati, *Inumin*, pl. 46
Label: none
Date: Kanawati, *Inumin*, 16-17: VI.2E, Harpur, *Decoration*, nr

LB98 *mḥw* - Saqqara, UPC (ntn), room I, north: marsh (ascending to right, with chick)
Source: *P-M III*, 619-622; Altenmüller, *Mehu*, pl. 9
Label: none
Date: Altenmüller, *Mehu*, 19: VI.2; Harpur, *Decoration*, [#424], 274: VI.2M-3?

LB99 *mḥw* - Saqqara, UPC (ntn), room I, east: marsh (ascending to left, with chick)
Source: *P-M III*, 619-622; Altenmüller, *Mehu*, pl. 10
Label: none
Date: Altenmüller, *Mehu*, 19: VI.2; Harpur, *Decoration*, [#424], 274: VI.2M-3?

LB100 *mḥw* - Saqqara, UPC (ntn), room I, east: marsh (ascending to right, with adult bird)
Source: *P-M III*, 619-622; Altenmüller, *Mehu*, pl. 11
Label: none
Date: Altenmüller, *Mehu*, 19: VI.2; Harpur, *Decoration*, [#424], 274: VI.2M-3?

LB101 *mḥw* - Saqqara, UPC (ntn), room I, west: marsh (ascending to left, with chick)
Source: *P-M III*, 619-622; Altenmüller, *Mehu*, pl. 13
Label: none
Date: Altenmüller, *Mehu*, 19: VI.2; Harpur, *Decoration*, [#424], 274: VI.2M-3?

LB102 *jj-nfrt* - Giza (ntn), chapel, unknown: marsh (ascending to right)
Source: *P-M III*, 298-299; Schürmann, *Die Reliefs*, fig. 6 [b][10]
Label: none
Date: Harpur, *Decoration*, [#017], 265: VI.3-4E?

LB103 *jj-nfrt* - Giza (ntn), chapel, unknown: marsh (ascending to left)
Source: *P-M III*, 298-299; Schürmann, *Die Reliefs*, fig. 6 [b]
Label: none
Date: Harpur, *Decoration*, [#017], 265: VI.3-4E?

[10] Schürmann (*Die Reliefs*, 32) identifies the animal here as a common genet ("Ginsterkatze"), but as so little remains it is possible that it may equally represent an Egyptian mongoose.

LB104 *mrjj-rʿ-nfr: ḳ3r* - Giza, EF (G7101), lower stairway (fragment no. 25-5-16), north: fishing (ascending to left, with chick)
Source: *P-M III*, 184-185; Simpson, *Qar and Idu*, fig. 16[11]
Label: none
Date: Harpur, *Decoration*, [#090], 267: VI.4

LB105 *3ḫtj-ḥtp* - Saqqara, UPC (E17), offering-room, east: marsh (ascending to right, head turned back)
Source: *P-M III*, 633-634; Petrie-Murray, *Seven Memphite*, pl. 6 [lower]
Label: none
Date: Harpur, *Decoration*, [#339], 272: VI.5?

Egyptian mongooses

LB106 *nb.j-m-3ḫtj* - Giza, CF (LG86), chapel I, west: marsh (ascending to left)
Source: *P-M III*, 230-232; *LD ii*, pl. 12 [b]
Label: none
Date: Harpur, *Decoration*, [#122], 267: IV.6-V.1

LB107 *nj-ʿnḫ-ẖnmw/ẖnmw-ḥtp* - Saqqara, UPC (ntn), outer hall, west: marsh (ascending to left)
Source: *P-M III*, 641-644; Moussa-Altenmüller, *Nianchchnum*, pl. 74[12]
Label: none
Date: Harpur, *Decoration*, [#431], 274: V.6L-7

LB108 *3ḫtj-ḥtp* - Saqqara, UPC (ntn), chapel, east: marsh (ascending to right)
Source: *P-M III*, 634-637; Ziegler, *Akhethetep*, 130-131
Label: none
Date: Ziegler, *Akhethetep*, 9: V.6; Harpur, *Decoration*, [#340], 272: V.6-8E

LB109 *ẖnmw-ḥtp* - Saqqara, ESP (D49), relief fragment (Berlin (East) Mus. no. 14101), north: marsh (ascending to left)
Source: *P-M III*, 578-579; Decker-Herb, *Bildatlas*, pl. 192 [K.126]
Label: none
Date: Harpur, *Decoration*, [#481], 275: V.6-8E?

LB110 *k3.j-dw3* - Giza, CF (ntn), chapel, north: marsh (ascending to right)
Source: *P-M III*, 244-245; Hassan, *Giza vi*, fig. 80
Label: none
Date: Harpur, *Decoration*, [#272], 270: V.6-9?

LB111 *k3.j-dw3* - Giza, CF (ntn), chapel, north: marsh (ascending to left)
Source: *P-M III*, 244-245; Hassan, *Giza vi*, fig. 80
Label: none
Date: Harpur, *Decoration*, [#272], 270: V.6-9?

LB112 *rʿ-špss* - Saqqara, NSP (LS16=S902), forecourt, north: marsh (tail only, ascending to right)
Source: *P-M III*, 494-496; *LD ii*, pl. 60 [right]
Label: none
Date: Harpur, *Decoration*, [#456], 275: V.8M

LB113 *rʿ-špss* - Saqqara, NSP (LS16=S902), forecourt, north: marsh (ascending to left)
Source: *P-M III*, 494-496; *LD ii*, pl. 60 [right]
Label: none
Date: Harpur, *Decoration*, [#456], 275: V.8M

LB114 *snḏm-jb: jntj* - Giza, WF (G2370=LG27), portico, west: marsh (ascending to right)
Source: *P-M III*, 85-87; *LD Erganz.*, pl. 18[13]
Label: none

---

11 Simpson (*Qar and Idu*, 3) states that this is a cat. For a discussion regarding the identification of this animal, see Malek, *The Cat*, 38.

12 Moussa and Altenmüller (*Nianchchnum*, 152) identify this animal as a marsh mongoose (" Sumpf-ichneumon, *Atilax paludinosus*"). To my knowledge, this species does not occur in Egypt.

13 According to Brovarski (*Senedjemib Complex Part I*, 40), this section of the relief has now been destroyed.

Date: Brovarski, *Senedjemib Complex Part I*, 23: V.8M-L; Harpur, *Decoration*, [#219], 269: V.8M-L

LB115 *snḏm-jb: jntj* - Giza, WF (G2370=LG27), room II, west: marsh (ascending to right)
Source: *P-M III*, 85-87; *LD ii*, pl. 77[14]
Label: none
Date: Brovarski, *Senedjemib Complex Part I*, 23: V.8M-L; Harpur, *Decoration*, [#219], 269: V.8M-L

LB116 *tjj* - Saqqara, NSP (D22), room III, north: marsh (ascending to right)
Source: *P-M III*, 468-478; Wild, *Ti ii*, pl. 119
Label: none
Date: Harpur, *Decoration*, [#543], 277: V.8-9

LB117 *nfr-jrt-n.f* - Saqqara, ESP (D55), chapel, east: marsh (ascending to right)
Source: *P-M III*, 583-584; van de Walle, *Neferirtenef*, pl. 1
Label: none
Date: Harpur, *Decoration*, [#440], 274: V.8-9

LB118 *k3.j-m-nfrt* - Saqqara, NSP (D23), chapel, north: marsh (ascending to left)
Source: *P-M III*, 467-468; Simpson, *Kayemnofret*, pl. A
Label: none
Date: Simpson, *Kayemnofret*, 1: V.8-9; Harpur, *Decoration*, [#522], 276: V.8-9?

LB119 *3ḫtj-ḥtp* - Saqqara, WSP (D64), hall, east: marsh (ascending to right, head turned back)
Source: *P-M III*, 599-600; Davies, *Ptahhetep ii*, pl. 13
Label: none
Date: Harpur, *Decoration*, [#338], 272: V.8L-9E

LB120 *3ḫtj-ḥtp* - Saqqara, WSP (D64), hall, east: marsh (ascending to left)
Source: *P-M III*, 599-600; Davies, *Ptahhetep ii*, pl. 14
Label: none
Date: Harpur, *Decoration*, [#338], 272: V.8L-9E

LB121 *nbt* - Saqqara, UPC (ntn), room A [I], east: marsh (ascending to left)
Source: *P-M III*, 624-625; Munro, *Unas-Friedhof*, pl. 11
Label: missing
Date: Munro, *Unas-Friedhof*, 20: V.9; Harpur, *Decoration*, [#438], 274: V.9

LB122 *jj-nfrt: š3.n.f* - Saqqara, UPC (ntn), room I, north: marsh (ascending to right, head missing)
Source: *P-M III*, 616; Kanawati-Abder-Raziq, *Iynefert*, pl. 37[15]
Label: none
Date: Kanawati-Abder-Raziq, *Iynefert*, 13: V.9E-M; Harpur, *Decoration*, [#345], 272: V.9

LB123 *nj-k3w-jzzj* - Saqqara, TP (ntn), room I, east: marsh (ascending to right, with chick)
Source: *P-M III*, nr; Kanawati-Abder-Raziq, *Nikauisesi*, pl. 50
Label: none
Date: Kanawati-Abder-Raziq, *Nikauisesi*, 17-23: VI.1M; Harpur, *Decoration*, nr

LB124 *nj-k3w-jzzj* - Saqqara, TP (ntn), room I, east: marsh (ascending to to left, attacking adult bird)
Source: *P-M III*, nr; Kanawati-Abder-Raziq, *Nikauisesi*, pl. 50
Label: none
Date: Kanawati-Abder-Raziq, *Nikauisesi*, 17-23: VI.1M; Harpur, *Decoration*, nr

LB125 *mrrw-k3.j: mrj* - Saqqara, TP (ntn), chamber A1, north: marsh (ascending to left, head turned back)
Source: *P-M III*, 525-534; Duell, *Mereruka i*, pl. 10
Label: none
Date: Harpur, *Decoration*, [#420], 274: VI.1M-L

---

14 This relief is no longer extant. See *ibid.*, 49.

15 Kanawati and Abder-Raziq (*Iynefert*, 18) identify this animal as an Egyptian mongoose even though the head region is damaged. This identification is supported by the fact that it drags its tail on the substrate, whereas that of a genet higher up in the same scene (see LB86) is elevated.

LB126 *mrrw-kꜣ.j: mrj* - Saqqara, TP (ntn), chamber A1, south: marsh (descending to left, with chick)
Source: *P-M III*, 525-534; Duell, *Mereruka i*, pl. 19
Label: none
Date: Harpur, *Decoration*, [#420], 274: VI.1M-L

LB127 *mrrw-kꜣ.j: mrj* - Saqqara, TP (ntn), chamber A1, south: marsh (ascending to right, tail held)
Source: *P-M III*, 525-534; Duell, *Mereruka i*, pl. 19
Label: none
Date: Harpur, *Decoration*, [#420], 274: VI.1M-L

LB128 *zšzšt: jdwt* - Saqqara, UPC (ntn), room III, west: marsh (ascending to left)
Source: *P-M III*, 617-619; Macramallah, *Idout*, pl. 7; Kanawati-Abder-Raziq, *Iynefert*, pl. 54
Label: none
Date: Kanawati-Abder-Raziq, *Iynefert*, 36-37: VI.1E; Harpur, *Decoration*, [#487], 275: V.9 r/u, VI.1-2E?

LB129 *jꜣzn* - Giza, WF (G2196), chapel, north: fishing (ascending to right)
Source: *P-M III*, 82; Simpson, *Mast. West. Cem. Part I*, fig. 30
Label: none
Date: Harpur, *Decoration*, [#013], 265: VI.1M-2M

LB130 *sꜥnḫ-w(j)-ptḥ* - Saqqara, TP (ntn), room I, east: marsh (ascending to right)
Source: *P-M III*, nr; Kanawati-Abder-Raziq, *Neferseshemre*, pls. 69 and 76
Label: none
Date: Kanawati-Abder-Raziq, *Neferseshemre*, 40-41: VI.1L; Harpur, *Decoration*, nr

LB131 *sꜥnḫ-w(j)-ptḥ* - Saqqara, TP (ntn), room I, east: marsh (ascending to left, with chick)
Source: *P-M III*, nr; Kanawati-Abder-Raziq, *Neferseshemre*, pls. 69 and 76
Label: none
Date: Kanawati-Abder-Raziq, *Neferseshemre*, 40-41: VI.1L; Harpur, *Decoration*, nr

LB132 *ḥzj* - Saqqara, TP (ntn), portico, south (east of entrance): marsh (ascending to right, attacking adult bird)
Source: *P-M III*, nr; Kanawati-Abder-Raziq, *Hesi*, pl. 53
Label: none
Date: Kanawati-Abder-Raziq, *Hesi*, 16: VI.1L-2E; Harpur, *Decoration*, nr

LB133 *ḥzj* - Saqqara, TP (ntn), portico, south (west of entrance): marsh (ascending to right, with chick)
Source: *P-M III*, nr; Kanawati-Abder-Raziq, *Hesi*, pl. 54
Label: none
Date: Kanawati-Abder-Raziq, *Hesi*, 16: VI.1L-2E; Harpur, *Decoration*, nr

LB134 *mṯṯj* - Saqqara, UPC (ntn), relief fragment (Abegg-Stiftung, Riggisberg no. 12.39.67), unknown: marsh (ascending to left, head turned back)
Source: *P-M III*, 646-648; Kaplony, *Studien Methethi*, fig. on 10[16]
Label: none
Date: Ziegler, *Catalogue*, 123: probably VI; Harpur, *Decoration*, [#426], 274: VI.2?

LB135 *mrrj* - Saqqara, TP (ntn), room 1, south: marsh (ascending to right, head turned back)
Source: *P-M III*, 518-519; Davies et al., *Saqqara Tombs i*, pl. 5[17]
Label: none
Date: Harpur, *Decoration,* [#418], 274: VI.2E?

LB136 *mḥw* - Saqqara, UPC (ntn), room I, north: marsh (ascending to left, with chick)
Source: *P-M III*, 619-622; Altenmüller, *Mehu*, pl. 9
Label: none
Date: Altenmüller, *Mehu*, 19: VI.2; Harpur, *Decoration*, [#424], 274: VI.2M-3?

LB137 *mḥw* - Saqqara, UPC (ntn), room I, east: marsh (ascending to right)
Source: *P-M III*, 619-622; Altenmüller, *Mehu*, pl. 10
Label: none

16 Kaplony (*Studien Methethi*, 10-11) believes this to be a mongoose ("hält der Ichneumon ein Vögelchen im Maul"). The elevated tail, however, may indicate that it is a genet.

17 Davies et al. (*Saqqara Tombs i*, 11) identify this animal as an otter.

Date: Altenmüller, *Mehu*, 19: VI.2; Harpur, *Decoration*, [#424], 274: VI.2M-3?

LB138 *mḥw* - Saqqara, UPC (ntn), room I, east: marsh (ascending to right, tail held)
Source: *P-M III*, 619-622; Altenmüller, *Mehu*, pl. 11
Label: name and titles
Date: Altenmüller, *Mehu*, 19: VI.2; Harpur, *Decoration*, [#424], 274: VI.2M-3?

LB139 *mḥw* - Saqqara, UPC (ntn), room I, west: marsh (ascending to right, with chick)
Source: *P-M III*, 619-622; Altenmüller, *Mehu*, pl. 13
Label: name and titles
Date: Altenmüller, *Mehu*, 19: VI.2; Harpur, *Decoration*, [#424], 274: VI.2M-3?

LB140 *mrjj-rʿ-nfr: ḳ3r* - Giza, EF (G7101), lower stairway (fragment no. 25-5-8), north: fishing (ascending to right)
Source: *P-M III*, 184-185; Simpson, *Qar and Idu*, fig. 16[18]
Label: none
Date: Harpur, *Decoration*, [#090], 267: VI.4

LB141 *3ḫtj-ḥtp* - Saqqara, UPC (E17), offering-room, east: marsh (ascending to left)
Source: *P-M III*, 633-634; Petrie-Murray, *Seven Memphite*, pl. 6 [lower]
Label: none
Date: Harpur, *Decoration*, [#339], 272: VI.5?

4.1.6 Swimming

Birds

LB142 *ptḥ-ḥtp I* - Saqqara, WSP (D62), offering chamber, east: clapnet
Source: *P-M III*, 596-598; Murray, *Saqqara Mastabas*, pl. 11
Label: *rdjt sḫt 3pdw jn wḥʿ nw pr*
Making the catch of birds by the fowlers of the house
Date: Harpur, *Decoration*, [#398], 273: V.8M-L

LB143 *tjj* - Saqqara, NSP (D22), room I/pillared hall, west: pastoral/poultry-yard
Source: *P-M III*, 467-478; Epron et al., *Ti i*, pl. 28
Label: none
Date: Harpur, *Decoration,* [#543], 277: V.8-9

Nile crocodiles

LB144 *ḥtp-ḥr-3ḫtj* - Saqqara, WSP (D60), chapel, north: marsh
Source: *P-M III*, 593-595; Mohr, *Hetep-her-akhti*, fig. 34
Label: none
Date: Harpur, *Decoration*, [#467], 275: V.6-8E

LB145 *snḏm-jb: jntj* - Giza, WF (G2370=LG27), portico, west: marsh
Source: *P-M III*, 85-87; *LD Erganz.*, pl. 17; Brovarski, *Senedjemib Complex Part I*, fig. 25
Label: none
Date: Brovarski, *Senedjemib Complex Part I*, 23: V.8M-L; Harpur, *Decoration*, [#219], 269: V.8M-L

LB146 *snḏm-jb: jntj* - Giza, WF (G2370=LG27), room II, west: marsh
Source: *P-M III*, 85-87; *LD ii*, pl. 77; Brovarski, *Senedjemib Complex Part I*, fig. 43
Label: none
Date: Brovarski, *Senedjemib Complex Part I*, 23: V.8M-L; Harpur, *Decoration*, [#219], 269: V.8M-L

LB147 *tjj* - Saqqara, NSP (D22), room III, north: fording
Source: *P-M III*, 468-478; Wild, *Ti ii*, pl. 124
Label: *mnjw pw ʿnḫ ḥr=k r šjj pw ntj ḥr mw jw(=f) m šp tp*
This herdsman! May you be watchful against that aquatic which is in the water!
May he come as blind-of-head
Date: Harpur, *Decoration*, [#543], 277: V.8-9

---

18 Simpson (*Qar and Idu*, 3) states that this is a cat.

LB148 *sšm-nfr IV* - Giza, GIS (LG53), portico, north: marsh
Source: *P-M III*, 223-226; Junker, *Giza xi*, fig. 60
Label: none
Date: Harpur, *Decoration*, [#235], 270: V.9-VI.1

LB149 *sšm-nfr IV* - Giza, GIS (LG53), pillared hall, north: fording
Source: *P-M III*, 223-226; Junker, *Giza xi*, fig. 67
Label: missing
Date: Harpur, *Decoration*, [#235], 270: V.9-VI.1

LB150 *ꜣḫtj-mrw-nswt* - Giza, WF (G2184), secondary offering-room, west: fording
Source: *P-M III*, 80-81; Smith, *HESPOK*, fig. 229; *Giza Archives Project* (Photo no. C12742_NS)[19]
Label: *wꜣḥ mr j(n) mnjw kꜣw*
Entering the canal by the herdsmen and cattle
Date: Harpur, *Decoration*, [#008], 265: V.9-VI.1

LB151 *nj-mꜣꜥt-rꜥ [jzzj-mr-nṯr]* - Giza, WF (G2097), chapel, north: marsh
Source: *P-M III*, 70; Roth, *Cem. Pal. Att.*, pl. 185
Label: . . . *mḥt pḥw*
. . . the Delta and the marshes . . .
ꜥnḫ ḥ=k wrt r š[jj pw] ntj ḫr mw
May you be watchful greatly against that aquatic which is in the water!
Date: Roth, *Cem. Pal. Att.*, 129-130: V.9; Harpur, *Decoration*, [#033], 265: V.9-VI.1

LB152a *kꜣ-gm-nj: mmj* - Saqqara, TP (LS10), room III, north: marsh
Source: *P-M III*, 521-525; Firth-Gunn, *Teti Pyr. ii*, pl. 52; Harpur-Scremin, *Kagemni*, fig. 7
Label: *j mnjw pw dj jḥy n kꜣw nw sn mk wḏb(=sn) r=f*
Oh this herdsman! Show Ihy of the cattle to them
Behold! They turn against him![20]
jḥy n kꜣw
Ihy of the cattle
Date: Harpur, *Decoration*, [#534], 276: VI.1M

LB152b *kꜣ-gm-nj: mmj* - Saqqara, TP (LS10), room III, north: marsh
Source: *P-M III*, 521-525; Harpur-Scremin, *Kagemni*, fig. 9
Label: none
Date: Harpur, *Decoration*, [#534], 276: VI.1M

LB153 *ḥzj* - Saqqara, TP (ntn), portico, south (west of entrance): marsh
Source: *P-M III*, nr; Kanawati-Abder-Raziq, *Hesi*, pl. 54
Label: none
Date: Kanawati-Abder-Raziq, *Hesi*, 16: VI.1L-2E; Harpur, *Decoration*, nr

LB154 *ḥzj* - Saqqara, TP (ntn), portico, east: fording
Source: *P-M III*, nr; Kanawati-Abder-Raziq, *Hesi*, pl. 55
Label: *j mnjw pw ꜥnḫ ḥr=k r šjj pw ntj ḫr mw jw(=f) m šp tp tm jwjj n šjj pw*
Oh this herdsman! May you be watchful against that aquatic which is in the water!
May he come as blind-of-head, and may (the cattle) not come to that aquatic
ꜥ=k ḥr mw mnjw pw jw mzḥ pw gr(w) ḫr s . . . t nt wḏb wḏꜣ jb=k r=f
Your hand above the water, this herdsman!
This crocodile is still under the . . . of the riverbank.
May your heart prosper against it
jn r=k
Bring away
mk ḫr . . . tp n ḫt r . . .
Behold! . . . the end of the pole . . .
Date: Kanawati-Abder-Raziq, *Hesi*, 16: VI.1L-2E; Harpur, *Decoration*, nr

---

[19] Smith (*HESPOK*, fig. 229) omits the hieroglyphic label that appears in a watercolour sketch, produced c. 1912 by Frida Firth, which is reproduced here (see fig. 4-22).

[20] Firth and Gunn (*Teti Pyr. i*, 114) translate this inscription as 'Show Ihy to the cattle that I may keep them together'.

LB155 *jj-nfrt* - Giza (ntn), chapel, unknown: marsh
Source: *P-M III*, 298-299; Schürmann, *Die Reliefs*, fig. 6 [a,b]
Label: none
Date: Harpur, *Decoration*, [#017], 265: VI.3-4E?

Nile soft-shell turtles

LB156 *ḥzj* - Saqqara, TP (ntn), portico, south (east of entrance): marsh
Source: *P-M III*, nr; Kanawati-Abder-Raziq, *Hesi*, pl. 53
Label: none
Date: Kanawati-Abder-Raziq, *Hesi*, 16: VI.1L-2E, Harpur, *Decoration*, nr

LB157 *mḥw* - Saqqara, UPC (ntn), room II, south: boating
Source: *P-M III*, 619-622; Altenmüller, *Mehu*, pl. 21 [a]
Label: none
Date: Altenmüller, *Mehu*, 19: VI.2; Harpur, *Decoration,* [#424], 274: VI.2M-3?

LB158 *mḥw* - Saqqara, UPC (ntn), room II, north: boating
Source: *P-M III*, 619-622; Altenmüller, *Mehu*, pl. 32
Label: none
Date: Altenmüller, *Mehu*, 19: VI.2; Harpur, *Decoration,* [#424], 274: VI.2M-3?

Fish

LB159 *nj-mꜣꜥt-rꜥ* [*jzzj-mr-nṯr*] - Giza, WF (G2097), chapel, north: marsh
Source: *P-M III*, 70; Roth, *Cem. Pal. Att.*, pl. 185
Label: none
Date: Roth, *Cem. Pal. Att.*, 129-130: V.9; Harpur, *Decoration,* [#033], 265: V.9-VI.1

LB160 *ḥzj* - Saqqara, TP (ntn), portico, south (west of entrance): marsh
Source: *P-M III*, nr; Kanawati-Abder-Raziq, *Hesi*, pl. 54
Label: none
Date: Kanawati-Abder-Raziq, *Hesi*, 16: VI.1L-2E; Harpur, *Decoration,* nr

---

4.1.7 Flying

Birds

LB161 *jj-mrjj* - Giza, WF (G6020=LG16), second chamber, north: bird netting
Source: *P-M III*, 170-174; *LD ii*, pl. 53 [b]
Label: none
Date: Weeks, *Cemetery G6000*, 7: V.6; Harpur, *Decoration*, [#014], 265: V.3L

LB162 *nj-ꜥnḫ-ẖnmw/ẖnmw-ḥtp* - Saqqara, UPC (ntn), vestibule, north: bird netting
Source: *P-M III*, 641-644; Moussa-Altenmüller, *Nianchchnum*, fig. 9
Label: [*swḥ*] *gnw*
Capturing orioles
swḥ
Capturing
zṯz gnw
Stretching out orioles
Date: Harpur, *Decoration*, [#431], 274: V.6L-7

LB163 *ḥtp-ḥr-ꜣḫtj* - Saqqara, WSP (D60), chapel, east: bird netting
Source: *P-M III*, 593-595; Mohr, *Hetep-her-akhti*, fig. 20
Label: none
Date: Harpur, *Decoration*, [#467], 275: V.6-8E

LB164 *ꜣḫtj-ḥtp* - Saqqara, UPC (ntn), chapel, east: bird netting
Source: *P-M III*, 634-637; Ziegler, *Akhethetep*, 138-139
Label: none
Date: Ziegler, *Akhethetep*, 9: V.6; Harpur, *Decoration*, [#340], 272: V.6-8E

LB165 *rʿ-špss* - Saqqara, NSP (LS16=S902), hall, west: bird netting
Source: *P-M III*, 494-496; *LD Ergänz.*, pl. 39 [a]
Label: missing
Date: Harpur, *Decoration*, [#456], 275: V.8M

LB166 *ṯjj* - Saqqara, NSP (D22), room III, south: bird netting
Source: *P-M III*, 468-478; Wild, *Ti iii*, pl. 170
Label: missing
Date: Harpur, *Decoration*, [#543], 277: V.8-9

LB167 *nfr-jrt-n.f* - Saqqara, ESP (D55), chapel, north: bird netting
Source: *P-M III*, 583-584; van de Walle, *Neferirtenef*, pl. 9
Label: *wḥʿ*
Fowler
swḥ gnw
Capturing orioles
Date: Harpur, *Decoration*, [#440], 274: V.8-9

LB168 *mrrw-k3.j: mrj* - Saqqara, TP (ntn), chamber A13, north: bird netting
Source: *P-M III*, 546-547; Firth-Gunn, *Teti Pyr. i*, fig. on 6[21]
Label: [. . . *sẖt*] *tn*
. . . this trap
Date: Harpur, *Decoration*, [#420], 274: VI.1M-L

LB169 *nj-ʿnḫ-nswt* - Saqqara (ntn), relief fragment, unknown: bird netting
Source: *P-M III*, 694-696; Leahy, A. and Mathieson, I. (2001). *JEA* 87, pl. 5 [2]
Label: *swḥ gnw*
Capturing orioles
Date: Berman, *Catalogue*, 135: VI.1; Harpur, *Decoration*, [#429], 274: VI.1-2E

LB170 *ḥzj* - Saqqara, TP (ntn), portico, west: bird netting
Source: *P-M III*, nr; Kanawati-Abder-Raziq, *Hesi*, pl. 56
Label: *s(w)ḥ gnw ʿ3 n k3 n ḥrj-tp nswt jmj-r wp(w)t ḥtpt-nṯr* . . .
Capturing numerous orioles, for the ka of the royal chamberlain, the overseer of apportionments of the gods offerings . . .
Date: Kanawati-Abder-Raziq, *Hesi*, 16: VI.1L-2E; Harpur, *Decoration*, nr

LB171 *ʿnḫ-m-ʿ-ḥr: zzj* - Saqqara, TP (ntn), relief fragment (TNE95: 147), unknown: bird netting
Source: *P-M III*, 512-515; Kanawati-Hassan, *Ankhmahor*, pl. 72[22]
Label: missing
Date: Kanawati-Hassan, *Ankhmahor*, 18: VI.1M-L; Harpur, *Decoration*, [#374], 273: VI.1L-2E

Butterflies

LB172 *sḫm-k3-rʿ* - Giza, CF (LG89), partition, north: marsh
Source: *P-M III*, 233-235; Hassan, *Giza iv*, fig. 59
Label: none
Date: Harpur, *Decoration*, [#229], 269: V.2

LB173 *nfr/k3-ḥ3.j* - Saqqara, UPC (ntn), chapel, east: marsh
Source: *P-M III*, 639-641; Moussa-Altenmüller, *Nefer and Ka-Hay*, pl. 5
Label: none
Date: Harpur, *Decoration*, [#439], 274: V.6

LB174 *nj-ʿnḫ-ẖnmw/ẖnmw-ḥtp* - Saqqara, UPC (ntn), forecourt, south: fishing
Source: *P-M III*, 641-644; Moussa-Altenmüller, *Nianchchnum*, fig. 5
Label: none
Date: Harpur, *Decoration*, [#431], 274: V.6L-7

21 This block has been assigned to the tomb of *mrrw-k3.j: mrj* by Harpur (*Decoration*, 233, no. 3).

22 According to Kanawati and Hassan (*Ankhmahor*, 70-71), the fragment cannot be attributed with certainty to this tomb.

LB175 *ḥtp-ḥr-3ḫtj* - Saqqara, WSP (D60), chapel, north: marsh
Source: *P-M III*, 593-595; Mohr, *Hetep-her-akhti*, fig. 34
Label: none
Date: Harpur, *Decoration*, [#467], 275: V.6-8E

LB176 *3ḫtj-ḥtp* - Saqqara, UPC (ntn), chapel, east: marsh
Source: *P-M III*, 634-637; Ziegler, *Akhethetep*, 130-131
Label: none
Date: Ziegler, *Akhethetep*, 9: V.6; Harpur, *Decoration*, [#340], 272: V.6-8E

LB177 *jtj-sn* - Giza, CF (ntn), chapel, east: marsh
Source: *P-M III*, 252-253; Harpur, *Decoration*, fig. 186
Label: none
Date: Harpur, *Decoration*, [#035], 265: V.6-8?

LB178 *jrj-n-k3-ptḥ* - Saqqara, UPC (ntn), offering chamber, south: marsh
Source: *P-M III*, 644; Moussa-Junge, *Two Tombs*, pl. 12
Label: none
Date: Harpur, *Decoration*, [#362], 273: V.6-9

LB179 *snḏm-jb: jntj* - Giza, WF (G2370=LG27), room II, west: marsh
Source: *P-M III*, 85-87; *LD ii*, pl. 77; Brovarski, *Senedjemib Complex Part I*, fig. 42[23]
Label: none
Date: Brovarski, *Senedjemib Complex Part I*, 23: V.8M-L; Harpur, *Decoration*, [#219], 269: V.8M-L

LB180 *sḫm-ʿnḫ-ptḥ* - Saqqara, NSP (D41), chapel, east: marsh
Source: *P-M III*, 454-455; Simpson, *Sekem-ankh-ptah*, pl. C
Label: none
Date: Harpur, *Decoration*, [#504], 276: V.8-9

LB181 *tjj* - Saqqara, NSP (D22), room III, north: marsh
Source: *P-M III*, 468-478; Wild, *Ti ii*, pl. 115
Label: none
Date: Harpur, *Decoration*, [#543], 277: V.8-9

LB182 *k3.j-m-nfrt* - Saqqara, NSP (D23), chapel, north: marsh
Source: *P-M III*, 467-468; Simpson, *Kayemnofret*, pl. A
Label: none
Date: Simpson, *Kayemnofret*, 1: V.8-9; Harpur, *Decoration*, [#522], 276: V.8-9?

LB183 *3ḫtj-ḥtp* - Saqqara, WSP (D64), hall, east: marsh
Source: *P-M III*, 599-600; Davies, *Ptahhetep ii*, pl. 13
Label: none
Date: Harpur, *Decoration*, [#338], 272: V.8L-9E

LB184 *sḫntjw/nfr-sšm-ptḥ* - Saqqara, UPC (ntn), offering chamber, east: marsh
Source: *P-M III*, 645; Moussa-Junge, *Two Tombs*, pl. 6
Label: none
Date: Harpur, *Decoration*, [#507], 276: V.9

LB185 *ḫnwt* - Saqqara, UPC (ntn), room A [I], east: marsh
Source: *P-M III*, 623-624; Munro, *Unas-Friedhof*, pl. 33
Label: none
Date: Munro, *Unas-Friedhof*, 20: V.9; Harpur, *Decoration*, [#478], 275: V.9

LB186 *nbt* - Saqqara, UPC (ntn), room A [I], west: marsh
Source: *P-M III*, 624-625; Munro, *Unas-Friedhof*, pl. 10
Label: none
Date: Munro, *Unas-Friedhof*, 20: V.9; Harpur, *Decoration*, [#438], 274: V.9

23 Brovarski (*Senedjemib Complex Part I*, 49) reports that this relief has now been destroyed.

LB187 *jj-nfrt: šꜣ.n.f* - Saqqara, UPC (ntn), room I, north: marsh
Source: *P-M III*, nr; Kanawati-Abder-Raziq, *Iynefert*, pl. 37
Label: none
Date: Kanawati-Abder-Raziq, *Iynefert*, 13: V.9E-M; Harpur, *Decoration*, [#345], 272: V.9

LB188 *ptḥ-ḥtp II: ṯfj* - Saqqara, WSP (D64), offering-room, east: clapnet
Source: *P-M III*, 600-604; Quibell, *Ptah-hetep*, pl. 32
Label: *sḫt m ḥsbt*
Catching by signal
Date: Harpur, *Decoration*, [#400], 274: V.9M-L

LB189 *ptḥ-ḥtp II: ṯfj* - Saqqara, WSP (D64), offering-room, east: clapnet
Source: *P-M III*, 600-604; Quibell, *Ptah-hetep*, pl. 32
Label: *sḫt m ḥsbt*
Catching by signal
Date: Harpur, *Decoration*, [#400], 274: V.9M-L

LB190 *nj-mꜣꜥt-rꜥ [jzzj-mr-nṯr]* - Giza, WF (G2097), chapel, north: marsh
Source: *P-M III*, 70; Roth, *Cem. Pal. Att.*, pl. 185
Label: none
Date: Roth, *Cem. Pal. Att.*, 129-130: V.9; Harpur, *Decoration*, [#033], 265: V.9-VI.1

LB191 *ḥtpt* - Giza (ntn), relief fragment, unknown: marsh
Source: *P-M III*, 298; Schäfer-Andrae, *Kunst*, pl. 250 [1]
Label: none
Date: Harpur, *Decoration*, [#176], 268: V-VI

LB192 *kꜣ.j-m-ꜥnḫ* - Giza, WF (G4561), hall, north: marsh
Source: *P-M III*, 131-133; Junker, *Giza iv*, fig. 8; Kanawati, *Kaiemankh*, pl. 31
Label: none
Date: Kanawati, *Kaiemankh*, 18: late V, probably V.8; Harpur, *Decoration*, [#255], 270: VI[24]

LB193 *kꜣ.j-m-ꜥnḫ* - Giza, WF (G4561), hall, north: marsh
Source: *P-M III*, 131-133; Junker, *Giza iv*, fig. 8; Kanawati, *Kaiemankh*, pl. 31
Label: none
Date: Kanawati, *Kaiemankh*, 18: late V, probably V.8; Harpur, *Decoration*, [#255], 270: VI

LB194 *kꜣ.j-m-ꜥnḫ* - Giza, WF (G4561), burial chamber, west: clapnet
Source: *P-M III*, 131-133; Junker, *Giza iv*, pl. 11; Kanawati, *Kaiemankh*, pl. 36
Label: none
Date: Kanawati, *Kaiemankh*, 18: late V, probably V.8; Harpur, *Decoration*, [#255], 270: VI

LB195 *kꜣ-gm-nj: mmj* - Saqqara, TP (LS10), room III, north: marsh
Source: *P-M III*, 521-525; Firth-Gunn, *Teti Pyr. ii*, pl. 52; Harpur-Scremin, *Kagemni*, fig. 8
Label: none
Date: Harpur, *Decoration*, [#534], 276: VI.1M

LB196 *nj-kꜣw-jzzj* - Saqqara, TP (ntn), room I, east: marsh
Source: *P-M III*, nr; Kanawati-Abder-Raziq, *Nikauisesi*, pl. 50
Label: none
Date: Kanawati-Abder-Raziq, *Nikauisesi*, 17-23: VI.1M; Harpur, *Decoration*, nr

LB197 *mrrw-kꜣ.j: mrj* - Saqqara, TP (ntn), chamber A1, north: marsh
Source: *P-M III*, 525-534; Duell, *Mereruka i*, pl. 11
Label: none
Date: Harpur, *Decoration*, [#420], 274: VI.1M-L

LB198 *nj-ꜥnḫ-nswt* - Saqqara (ntn), relief fragment, unknown: marsh
Source: *P-M III*, 694-696; Leahy, A. and Mathieson, I. (2001). *JEA* 87, pl. 4 [3]
Label: none
Date: Berman, *Catalogue*, 135: VI.1; Harpur, *Decoration*, [#429], 274: VI.1-2E

[24] See Chapter 4, n. 98 for evidence supporting a late 5th dynasty date for this tomb.

LB199 *zšzšt: jdwt* - Saqqara, UPC (ntn), room III, west: marsh
Source: *P-M III*, 617-619; Macramallah, *Idout*, pl. 7; Kanawati-Abder-Raziq, *Iynefert*, pl. 54
Label: none
Date: Kanawati-Abder-Raziq, *Iynefert*, 36-37: VI.1E; Harpur, *Decoration*, [#487], 275: V.9 r/u, VI.1-2E?

LB200 *jꜣzn* - Giza, WF (G2196), chapel, north: fishing
Source: *P-M III*, 82; Simpson, *Mast. West. Cem. Part I*, fig. 30
Label: none
Date: Harpur, *Decoration*, [#013], 265: VI.1M-2M[25]

LB201 *sꜥnḫ-w(j)-ptḥ* - Saqqara, TP (ntn), room I, east: marsh
Source: *P-M III*, nr; Kanawati-Abder-Raziq, *Neferseshemre*, pls. 69 and 76
Label: none
Date: Kanawati-Abder-Raziq, *Neferseshemre*, 40-41: VI.1L; Harpur, *Decoration*, nr

LB202 *ḥzj* - Saqqara, TP (ntn), portico, south (east of entrance): marsh
Source: *P-M III*, nr; Kanawati-Abder-Raziq, *Hesi*, pl. 53
Label: none
Date: Kanawati-Abder-Raziq, *Hesi*, 16: VI.1L-2E; Harpur, *Decoration*, nr

LB203 *ḥzj* - Saqqara, TP (ntn), portico, south (west of entrance): marsh
Source: *P-M III*, nr; Kanawati-Abder-Raziq, *Hesi*, pl. 54
Label: none
Date: Kanawati-Abder-Raziq, *Hesi*, 16: VI.1L-2E; Harpur, *Decoration*, nr

LB204 *ḥzj* - Saqqara, TP (ntn), portico, south (west of entrance): marsh
Source: *P-M III*, nr; Kanawati-Abder-Raziq, *Hesi*, pl. 54
Label: none
Date: Kanawati-Abder-Raziq, *Hesi*, 16: VI.1L-2E; Harpur, *Decoration*, nr

LB205 *ḥzj* - Saqqara, TP (ntn), portico, east: clapnet
Source: *P-M III*, nr; Kanawati-Abder-Raziq, *Hesi*, pl. 55
Label: *dw sḫt jn jmj-r mḥw j mḥw pw ḥb r=k*
Setting the trap by the overseer of fowlers
Oh this fowler! There is a catch of birds for you!
Date: Kanawati-Abder-Raziq, *Hesi*, 16: VI.1L-2E; Harpur, *Decoration*, nr

LB206 *ꜥnḫ-m-ꜥ-ḥr: zzj* - Saqqara, TP (ntn), room II, west: clapnet
Source: *P-M III*, 512-515; Kanawati-Hassan, *Ankhmahor*, pl. 42
Label: none
Date: Kanawati-Hassan, *Ankhmahor*, 18: VI.1M-L; Harpur, *Decoration*, [#374], 273: VI.1L-2E

LB207 *mrjj-rꜥ-mrj-ptḥ-ꜥnḫ: nḫbw* - Giza, WF (G2381), relief fragment from forecourt (MFA no. 13.4332), unknown: marsh
Source: *P-M III*, 90-91; Smith, W.S. (1958). *Bost. Mus. Bull.* 56, fig. 2
Label: none
Date: Harpur, *Decoration*, [#089], 267: VI.2

LB208 *mḥw* - Saqqara, UPC (ntn), room I, north: marsh
Source: *P-M III*, 619-622; Altenmüller, *Mehu*, pl. 9
Label: none
Date: Altenmüller, *Mehu*, 19: VI.2; Harpur, *Decoration*, [#424], 274: VI.2M-3?

LB209 *mḥw* - Saqqara, UPC (ntn), room I, west: marsh
Source: *P-M III*, 619-622; Altenmüller, *Mehu*, pl. 12
Label: none
Date: Altenmüller, *Mehu*, 19: VI.2; Harpur, *Decoration*, [#424], 274: VI.2M-3?

LB210 *mḥw* - Saqqara, UPC (ntn), room II, north: clapnet
Source: *P-M III*, 619-622; Altenmüller, *Mehu*, pl. 34 [b]
Label: none

25 *ibid.*

Date: Altenmüller, *Mehu*, 19: VI.2; Harpur, *Decoration*, [#424], 274: VI.2M-3?

LB211 *wr-nw* - Saqqara, TP (ntn), hall, east: marsh
Source: *P-M III*, 519; Davies et al., *Saqqara Tombs i*, pl. 25
Label: none
Date: Harpur, *Decoration*, [#380], 273: VI.3-4?

LB212 *ꜣḫtj-ḥtp* - Saqqara, UPC (E17), offering-room, east: marsh
Source: *P-M III*, 633-634; Petrie-Murray, *Seven Memphite*, pl. 6 [lower]
Label: none
Date: Harpur, *Decoration*, [#339], 272: VI.5?

Dragonflies/Grasshoppers

LB213 *kꜣ.j-m-nfrt* - Saqqara, NSP (D23), chapel, north: marsh
Source: *P-M III*, 467-468; Simpson, *Kayemnofret*, pl. A
Label: none
Date: Simpson, *Kayemnofret*, 1: V.8-9; Harpur, *Decoration*, [#522], 276: V.8-9?

LB214 *ꜣḫtj-ḥtp* - Saqqara, WSP (D64), hall, east: marsh
Source: *P-M III*, 599-600; Davies, *Ptahhetep ii*, pl. 14
Label: none
Date: Harpur, *Decoration*, [#338], 272: V.8L-9E

LB215 *ꜣḫtj-ḥtp* - Saqqara, WSP (D64), hall, east: marsh
Source: *P-M III*, 599-600; Davies, *Ptahhetep ii*, pl. 14
Label: none
Date: Harpur, *Decoration*, [#338], 272: V.8L-9E

LB216 *jj-nfrt: šꜣ.n.f* - Saqqara, UPC (ntn), room I, north: marsh
Source: *P-M III*, 616; Kanawati-Abder-Raziq, *Iynefert*, pl. 37
Label: none
Date: Kanawati-Abder-Raziq, *Iynefert*, 13: V.9E-M; Harpur, *Decoration*, [#345], 272: V.9

LB217 *jj-nfrt: šꜣ.n.f* - Saqqara, UPC (ntn), room I, south: marsh
Source: *P-M III*, 616; Kanawati-Abder-Raziq, *Iynefert*, pl. 38
Label: none
Date: Kanawati-Abder-Raziq, *Iynefert*, 13: V.9E-M; Harpur, *Decoration*, [#345], 272: V.9

LB218 *ꜣḫtj-mrw-nswt* - Giza, WF (G2184), secondary offering-room, west (south of niche): marsh
Source: *P-M III*, 80-81; *Giza Archives Project* (Photo no. C12741_NS)
Label: none
Date: Harpur, *Decoration*, [#008], 265: V.9-VI.1

LB219 *ꜣḫtj-mrw-nswt* - Giza, WF (G2184), secondary offering-room, west (south of niche): marsh
Source: *P-M III*, 80-81; Smith, *HESPOK*, fig. 194
Label: none
Date: Harpur, *Decoration*, [#008], 265: V.9-VI.1

LB220 *kꜣ-gm-nj: mmj* - Saqqara, TP (LS10), room III, north: marsh
Source: *P-M III*, 521-525; von Bissing, *Gem-ni-kai i*, pl. 4; Harpur-Scremin, *Kagemni*, fig. 9
Label: none
Date: Harpur, *Decoration*, [#534], 276: VI.1M

LB221 *nj-kꜣw-jzzj* - Saqqara, TP (ntn), room I, east: marsh
Source: *P-M III*, nr; Kanawati-Abder-Raziq, *Nikauisesi*, pl. 50
Label: none
Date: Kanawati-Abder-Raziq, *Nikauisesi*, 17-23: VI.1M; Harpur, *Decoration*, nr

LB222 *nj-kꜣw-jzzj* - Saqqara, TP (ntn), room I, east: marsh
Source: *P-M III*, nr; Kanawati-Abder-Raziq, *Nikauisesi*, pl. 50
Label: none
Date: Kanawati-Abder-Raziq, *Nikauisesi*, 17-23: VI.1M; Harpur, *Decoration*, nr

LB223 *nj-k3w-jzzj* - Saqqara, TP (ntn), room I, east: marsh
Source: *P-M III*, nr; Kanawati-Abder-Raziq, *Nikauisesi*, pl. 50
Label: none
Date: Kanawati-Abder-Raziq, *Nikauisesi*, 17-23: VI.1M; Harpur, *Decoration*, nr

LB224 *nj-k3w-jzzj* - Saqqara, TP (ntn), room I, east: marsh
Source: *P-M III*, nr; Kanawati-Abder-Raziq, *Nikauisesi*, pl. 50
Label: none
Date: Kanawati-Abder-Raziq, *Nikauisesi*, 17-23: VI.1M; Harpur, *Decoration*, nr

LB225 *j3zn* - Giza, WF (G2196), chapel, north: fishing
Source: *P-M III*, 82; Simpson, *Mast. West. Cem. Part I*, fig. 30
Label: none
Date: Harpur, *Decoration,* [#013], 265: VI.1M-2M

LB226 *sˁnḫ-w(j)-ptḥ* - Saqqara, TP (ntn), room I, east: marsh
Source: *P-M III*, nr; Kanawati-Abder-Raziq, *Neferseshemre*, pls. 69 and 76
Label: none
Date: Kanawati-Abder-Raziq, *Neferseshemre*, 40-41: VI.1L; Harpur, *Decoration*, nr

LB227 *ḥzj* - Saqqara, TP (ntn), portico, south (west of entrance): marsh
Source: *P-M III*, nr; Kanawati-Abder-Raziq, *Hesi*, pl. 54
Label: none
Date: Kanawati-Abder-Raziq, *Hesi*, 16: VI.1L-2E; Harpur, *Decoration*, nr

LB228 *mrjj-rˁ-mrj-ptḥ-ˁnḫ: nḫbw* - Giza, WF (G2381), relief fragment from forecourt (MFA no. 13.4332), unknown: marsh
Source: *P-M III*, 90-91; Smith, W.S. (1958). *Bost. Mus. Bull.* 56, fig. 2
Label: none
Date: Harpur, *Decoration*, [#089], 267: VI.2

LB229 *mrrj* - Saqqara, TP (ntn), room 1, south: marsh
Source: *P-M III*, 518-519; Davies et al., *Saqqara Tombs i*, pl. 5
Label: none
Date: Harpur, *Decoration*, [#418], 274: VI.2E?

LB230 *mḥw* - Saqqara, UPC (ntn), room I, south: clapnet
Source: *P-M III*, 619-622; Altenmüller, *Mehu*, pl. 7
Label: none
Date: Altenmüller, *Mehu*, 19: VI.2; Harpur, *Decoration*, [#424], 274: VI.2M-3?

LB231 *mḥw* - Saqqara, UPC (ntn), room I, west: marsh
Source: *P-M III*, 619-622; Altenmüller, *Mehu*, pl. 12
Label: none
Date: Altenmüller, *Mehu*, 19: VI.2; Harpur, *Decoration*, [#424], 274: VI.2M-3?

LB232 *mḥw* - Saqqara, UPC (ntn), room I, west: marsh
Source: *P-M III*, 619-622; Altenmüller, *Mehu*, pl. 13
Label: none
Date: Altenmüller, *Mehu*, 19: VI.2; Harpur, *Decoration*, [#424], 274: VI.2M-3?

Wasps

LB233 *nj-ˁnḫ-ẖnmw/ẖnmw-ḥtp* - Saqqara, UPC (ntn), vestibule, south: fishing
Source: *P-M III*, 641-644; Moussa-Altenmüller, *Nianchchnum*, fig. 12 (bottom, left)
Label: none
Date: Harpur, *Decoration,* [#431], 274: V.6L-7

LB234 *nj-ˁnḫ-ẖnmw/ẖnmw-ḥtp* - Saqqara, UPC (ntn), vestibule, south: fishing
Source: *P-M III*, 641-644; Moussa-Altenmüller, *Nianchchnum*, fig. 12 (bottom, right)
Label: none
Date: Harpur, *Decoration*, [#431], 274: V.6L-7

4.2 REST

4.2.1 Sitting

Hippopotamuses

LB235 *dbḥnj* - Giza, CF (LG90), room II, east: marsh
Source: *P-M III*, 235-236; Hassan, *Giza iv*, fig. 121
Label: missing
Date: Harpur, *Decoration*, [#297], 271: IV.5

LB236 *pr-sn* - Saqqara, ESP (D45), relief fragment, unknown: marsh?
Source: *P-M III*, 577-578; Petrie-Murray, *Seven Memphite*, pl. 10 [upper]
Label: missing
Date: Harpur, *Decoration*, [#391], 273: V.2-3

LB237 *3ḫtj-ḥtp* - Saqqara, UPC (ntn), chapel, east: marsh
Source: *P-M III*, 634-637; Ziegler, *Akhethetep*, 132-133
Label: none
Date: Ziegler, *Akhethetep*, 9: V.6; Harpur, *Decoration*, [#340], 272: V.6-8E

LB238 *k3.j-m-nfrt* - Giza, CF (ntn), vestibule D, north: marsh
Source: *P-M III*, 263-264; Hassan, *Giza ii*, fig. 140
Label: none
Date: Harpur, *Decoration*, [#261], 270: V.6-8E?

LB239 *ṯjj* - Saqqara, NSP (D22), room III, north: marsh
Source: *P-M III*, 468-478; Wild, *Ti ii*, pl. 119
Label: none
Date: Harpur, *Decoration*, [#543], 277: V.8-9

LB240 *ḫnwt* - Saqqara, UPC (ntn), room A [I], east: marsh
Source: *P-M III*, 623-624; Munro, *Unas-Friedhof*, pl. 33
Label: none
Date: Munro, *Unas-Friedhof*, 20: V.9; Harpur, *Decoration*, [#478], 275: V.9

LB241 *nbt* - Saqqara, UPC (ntn), room A [I], west: marsh
Source: *P-M III*, 624-625; Munro, *Unas-Friedhof*, pl. 10
Label: none
Date: Munro, *Unas-Friedhof*, 20: V.9; Harpur, *Decoration*, [#438], 274: V.9

LB242 *sšm-nfr IV* - Giza, GIS (LG53), pillared hall, north: fording (facing left)
Source: *P-M III*, 223-226; Junker, *Giza xi*, fig. 67
Label: missing
Date: Harpur, *Decoration*, [#235], 270: V.9-VI.1

LB243 *sšm-nfr IV* - Giza, GIS (LG53), pillared hall, north: fording (facing right)
Source: *P-M III*, 223-226; Junker, *Giza xi*, fig. 67
Label: missing
Date: Harpur, *Decoration*, [#235], 270: V.9-VI.1

LB244 Anon - Saqqara, AJS (ntn), relief fragment (Cairo Mus. JE41572), unknown: marsh
Source: *P-M III*, 654; Quibell, *Saqqara (1907-1908)*, pl. 51 [3]
Label: none
Date: Harpur, *Decoration*, [#574], 277: V-VI

LB245 *k3-gm-nj: mmj* - Saqqara, TP (LS10), room I, west: marsh
Source: *P-M III*, 521-525; Firth-Gunn, *Teti Pyr. ii*, pl. 53 [2]; Harpur-Scremin, *Kagemni*, fig. 3
Label: none
Date: Harpur, *Decoration*, [#534], 276: VI.1M

LB246 *k3-gm-nj: mmj* - Saqqara, TP (LS10), room III, north: marsh
Source: *P-M III*, 521-525; Decker-Herb, *Bildatlas*, pl. 196 [K1.33]; Harpur-Scremin, *Kagemni*, fig. 9

Label: none
Date: Harpur, *Decoration,* [#534], 276: VI.1M

LB247 *k3-gm-nj: mmj* - Saqqara, TP (LS10), room IV, east: marsh
Source: *P-M III*, 521-525; von Bissing, *Gem-ni-kai i*, pl. 18; Harpur-Scremin, *Kagemni*, fig. 16
Label: none
Date: Harpur, *Decoration*, [#534], 276: VI.1M

LB248 *mrrw-k3.j: mrj* - Saqqara, TP (ntn), chamber A1, north: marsh
Source: *P-M III*, 525-534; Duell, *Mereruka i*, pls. 11 and 13
Label: none
Date: Harpur, *Decoration*, [#420], 274: VI.1M-L

LB249 *zšzšt: jdwt* - Saqqara, UPC (ntn), room I, north: fishing
Source: *P-M III*, 617-619; Macramallah, *Idout*, pl. 5 [d], Kanawati-Abder-Raziq, *Iynefert*, pl. 53 [f]
Label: none
Date: Kanawati-Abder-Raziq, *Iynefert*, 36-37: VI.1E; Harpur, *Decoration*, [#487], 275: V.9 r/u, VI.1-2E?

LB250 *zšzšt: jdwt* - Saqqara, UPC (ntn), room III, west: marsh
Source: *P-M III*, 617-619; Macramallah, *Idout*, pl. 7; Kanawati-Abder-Raziq, *Iynefert*, pl. 54
Label: none
Date: Kanawati-Abder-Raziq, *Iynefert*, 36-37: VI.1E; Harpur, *Decoration*, [#487], 275: V.9 r/u, VI.1-2E?

LB251 *sꜥnḫ-w(j)-ptḥ* - Saqqara, TP (ntn), room I, east: marsh
Source: *P-M III*, nr; Kanawati-Abder-Raziq, *Neferseshemre*, pls. 69 and 76
Label: none
Date: Kanawati-Abder-Raziq, *Neferseshemre*, 40-41: VI.1L; Harpur, *Decoration*, nr

LB252 *mḥw* - Saqqara, UPC (ntn), room II, north: boating
Source: *P-M III*, 619-622; Altenmüller, *Mehu*, pl. 31 [a]
Label: none
Date: Altenmüller, *Mehu*, 19: VI.2; Harpur, *Decoration*, [#424], 274: VI.2M-3?

LB253 *jj-nfrt* - Giza (ntn), chapel, unknown: marsh
Source: *P-M III*, 298-299; Schürmann, *Die Reliefs*, fig. 6 [a,b]
Label: none
Date: Harpur, *Decoration*, [#017], 265: VI.3-4E?

LB254 *nb-k3w-ḥr: jdw* - Saqqara, UPC (ntn), vestibule, north: marsh
Source: *P-M III*, 627-629; Hassan, *Neb-Kaw-Her*, pl. 5 [c]
Label: none
Date: Harpur, *Decoration*, [#437], 274: V.9 r/u VI.7

Frogs

LB255 *nj-ꜥnḫ-ẖnmw/ẖnmw-ḥtp* - Saqqara, UPC (ntn), outer hall, west
Source: *P-M III*, 641-644; Moussa-Altenmüller, *Nianchchnum*, pl. 74
Label: none
Date: Harpur, *Decoration*, [#431], 274: V.6L-7

LB256 *rꜥ-špss* - Saqqara, NSP (LS16=S902), forecourt, north: marsh
Source: *P-M III*, 494-496; *LD ii*, pl. 60 [right]
Label: none
Date: Harpur, *Decoration*, [#456], 275: V.8M

LB257 *nfr-jrt-n.f* - Saqqara, ESP (D55), chapel, east: marsh
Source: *P-M III*, 583-584; van de Walle, *Neferirtenef*, pl. 1
Label: none
Date: Harpur, *Decoration*, [#440], 274: V.8-9

LB258 *nbt* - Saqqara, UPC (ntn), room A [I], west: marsh
Source: *P-M III*, 624-625; Munro, *Unas-Friedhof*, pl. 10
Label: none

Date: Munro, *Unas-Friedhof*, 20: V.9; Harpur, *Decoration*, [#438], 274: V.9

LB259 *jj-nfrt: šꜣ.n.f* - Saqqara, UPC (ntn), room I, north: marsh
Source: *P-M III*, 616; Kanawati-Abder-Raziq, *Iynefert*, pl. 37
Label: none
Date: Kanawati-Abder-Raziq, *Iynefert*, 13: V.9E-M; Harpur, *Decoration*, [#345], 272: V.9

LB260 *ptḥ-ḥtp II: ṯfj* - Saqqara, WSP (D64), offering-room, east: boating
Source: *P-M III*, 600-604; Quibell, *Ptah-hetep*, pl. 32
Label: none
Date: Harpur, *Decoration*, [#400], 274: V.9M-L

LB261 *sšm-nfr IV* - Giza, GIS (LG53), portico, north: fishing
Source: *P-M III*, 223-226; Junker, *Giza xi*, fig. 60
Label: none
Date: Harpur, *Decoration*, [#235], 270: V.9-VI.1

LB262 *kꜣ.j-m-ꜥnḫ* - Giza, WF (G4561), chapel, north: marsh
Source: *P-M III*, 131-133; Junker, *Giza iv*, fig. 8, Kanawati, *Kaiemankh*, pl. 31
Label: none
Date: Kanawati, *Kaiemankh*, 18: late V; Harpur, *Decoration*, [#255], 270: VI

LB263 *kꜣ-gm-nj: mmj* - Saqqara, TP (LS10), room I, west: marsh (under bow)
Source: *P-M III*, 521-525; Firth-Gunn, *Teti Pyr. ii*, pl. 53 [2]; Harpur-Scremin, *Kagemni*, fig. 3
Label: none
Date: Harpur, *Decoration*, [#534], 276: VI.1M

LB264 *kꜣ-gm-nj: mmj* - Saqqara, TP (LS10), room I, west: marsh (on water surface, under bow)
Source: *P-M III*, 521-525; Firth-Gunn, *Teti Pyr. ii*, pl. 53 [2]; Harpur-Scremin, *Kagemni*, fig. 3
Label: none
Date: Harpur, *Decoration*, [#534], 276: VI.1M

LB265 *kꜣ-gm-nj: mmj* - Saqqara, TP (LS10), room I, west: marsh (under stern)
Source: *P-M III*, 521-525; Firth-Gunn, *Teti Pyr. ii*, pl. 53 [2]; Harpur-Scremin, *Kagemni*, fig. 3
Label: none
Date: Harpur, *Decoration*, [#534], 276: VI.1M

LB266 *kꜣ-gm-nj: mmj* - Saqqara, TP (LS10), room III, north: marsh (under stern)
Source: *P-M III*, 521-525; Firth-Gunn, *Teti Pyr. ii*, pl. 52; Harpur-Scremin, *Kagemni*, fig. 8
Label: none
Date: Harpur, *Decoration*, [#534], 276: VI.1M

LB267 *kꜣ-gm-nj: mmj* - Saqqara, TP (LS10), room III, north: marsh (under stern)
Source: *P-M III*, 521-525; von Bissing, *Gem-ni-kai i*, pl. 4; Harpur-Scremin, *Kagemni*, fig. 9
Label: none
Date: Harpur, *Decoration*, [#534], 276: VI.1M

LB268 *nj-kꜣw-jzzj* - Saqqara, TP (ntn), room I, east: marsh
Source: *P-M III*, nr; Kanawati-Abder-Raziq, *Nikauisesi*, pl. 50
Label: none
Date: Kanawati-Abder-Raziq, *Nikauisesi*, 17-23: VI.1M; Harpur, *Decoration*, nr

LB269 *mrrw-kꜣ.j: mrj* - Saqqara, TP (ntn), chamber A1, north: marsh
Source: *P-M III*, 525-534; Duell, *Mereruka i*, pl. 9
Label: none
Date: Harpur, *Decoration*, [#420], 274: VI.1M-L

LB270 *mrrw-kꜣ.j: mrj* - Saqqara, TP (ntn), chamber A1, north: marsh
Source: *P-M III*, 525-534; Duell, *Mereruka i*, pl. 13
Label: none
Date: Harpur, *Decoration*, [#420], 274: VI.1M-L

LB271 *mrrw-k3.j: mrj* - Saqqara, TP (ntn), chamber A1, south: marsh
Source: *P-M III*, 525-534; Duell, *Mereruka i*, pl. 18
Label: none
Date: Harpur, *Decoration,* [#420], 274: VI.1M-L

LB272 *mrrw-k3.j: mrj* - Saqqara, TP (ntn), chamber A13, south: marsh
Source: *P-M III*, 525-534; Duell, *Mereruka i*, pl. 128
Label: none
Date: Harpur, *Decoration*, [#420], 274: VI.1M-L

LB273 *ḥzj* - Saqqara, TP (ntn), portico, south (west of entrance): marsh
Source: *P-M III*, nr; Kanawati-Abder-Raziq, *Hesi*, pl. 54
Label: none
Date: Kanawati-Abder-Raziq, *Hesi*, 16: VI.1L-2E; Harpur, *Decoration*, nr

LB274 *ḥzj* - Saqqara, TP (ntn), portico, south (west of entrance): marsh
Source: *P-M III*, nr; Kanawati-Abder-Raziq, *Hesi*, pl. 54
Label: none
Date: Kanawati-Abder-Raziq, *Hesi*, 16: VI.1L-2E; Harpur, *Decoration*, nr

LB275 *mrrj* - Saqqara, TP (ntn), room 1, south: marsh
Source: *P-M III*, 518-519; Davies et al., *Saqqara Tombs i*, pl. 5
Label: none
Date: Harpur, *Decoration*, [#418], 274: VI.2E?

LB276 *mḥw* - Saqqara, UPC (ntn), room I, north: marsh
Source: *P-M III*, 619-622; Altenmüller, *Mehu*, pl. 9
Label: none
Date: Altenmüller, *Mehu*, 19: VI.2; Harpur, *Decoration*, [#424], 274: VI.2M-3?

LB277 *jj-nfrt* - Giza (ntn), chapel, unknown: boating
Source: *P-M III*, 298-299; Schürmann, *Die Reliefs*, fig. 7 [b]
Label: none
Date: Harpur, *Decoration*, [#017], 265: VI.3-4E?

LB278 *wr-nw* - Saqqara, TP (ntn), hall, east: boating
Source: *P-M III*, 519; Davies et al., *Saqqara Tombs i*, pl. 25
Label: none
Date: Harpur, *Decoration*, [#380], 273: VI.3-4?

4.2.2 Recumbent

Cattle

LB279 *bw-nfr* - Giza, CF (ntn), relief fragment found in main chapel, unknown: pastoral
Source: *P-M III*, 256; Hassan, *Giza iii*, fig. 154
Label: *wš3 jw3*
Fattening an ox
Date: Harpur, *Decoration*, [#065], 266: IV.6-V.1?

LB280 *jrj-n-k3-ptḥ* - Saqqara, UPC (ntn), offering chamber, south: pastoral
Source: *P-M III*, 644; Moussa-Junge, *Two Tombs*, pl. 13
Label: missing
Date: Harpur, *Decoration*, [#362], 273: V.6-9

LB281 *rʿ-špss* - Saqqara, NSP (LS16=S902), pillared hall, south: pastoral
Source: *P-M III*, 494-496; *LD ii*, pl. 62
Label: *rn jw3*
Young ox
Date: Harpur, *Decoration*, [#456], 275: V.8M

LB282 *sndm-jb: jntj* - Giza, WF (G2370=LG27), room III, east: pastoral
Source: *P-M III*, 85-87; *LD Erganz.*, pl. 23 [c][26]
Label: none
Date: Brovarski, *Senedjemib Complex Part I*, 23: V.8M-L; Harpur, *Decoration*, [#219], 269: V.8M-L

LB283 *tjj* - Saqqara, NSP (D22), room III, north: pastoral
Source: *P-M III*, 468-478; Wild, *Ti ii*, pl. 124
Label: *jwꜣ*
Ox
Date: Harpur, *Decoration*, [#543], 277: V.8-9

LB284 *nfr-jrt-n.f* - Saqqara, ESP (D55), chapel, east: pastoral
Source: *P-M III*, 583-584; van de Walle, *Neferirtenef*, pl. 12
Label: *rn jwꜣ*
Young ox
Date: Harpur, *Decoration*, [#440], 274: V.8-9

LB285 *kꜣ.j-m-nfrt* - Saqqara, NSP (D23), chapel, east: boating
Source: *P-M III*, 467-468; Simpson, *Kayemnofret*, pl. G
Label: *stꜣ* [. . .] *wšꜣ jwꜣ*
Hauling . . . and fattening an ox
Date: Simpson, *Kayemnofret*, 1: V.8-9; Harpur, *Decoration*, [#522], 276: V.8-9?

LB286 *ptḥ-ḥtp II*: *tfj* - Saqqara, WSP (D64), offering-room, east: pastoral
Source: *P-M III*, 600-604; Quibell, *Ptah-hetep*, pl. 31
Label: missing
Date: Harpur, *Decoration*, [#400], 274: V.9M-L

LB287 *nj-mꜣꜥt-rꜥ* [*jzzj-mr-nṯr*] - Giza, WF (G2097), chapel, north: pastoral
Source: *P-M III*, 70; Roth, *Cem. Pal. Att.*, pl. 185
Label: *wšꜣ*
Fattening
jwꜣt
Cow
Date: Roth, *Cem. Pal. Att.*, 129-130: V.9; Harpur, *Decoration*, [#033], 265: V.9-VI.1

LB288 *kꜣ.j-m-nfrt* - Giza, EF (LG63), chapel, east: pastoral
Source: *P-M III*, 208-209; *LD Erganz.*, pl. 32
Label: *wšꜣ jwꜣ*
Fattening an ox
Date: Harpur, *Decoration*, [#258], 270: V-VI

LB289 *kꜣ-gm-nj: mmj* - Saqqara, TP (LS10), room III, west: boating
Source: *P-M III*, 521-525; Firth-Gunn, *Teti Pyr. ii*, pl. 53 [1]; Harpur-Scremin, *Kagemni*, fig. 4
Label: none
Date: Harpur, *Decoration*, [#534], 276: VI.1M

LB290 *nj-kꜣw-jzzj* - Saqqara, TP (ntn), room I, south: pastoral
Source: *P-M III*, nr; Kanawati-Abder-Raziq, *Nikauisesi*, pl. 47
Label: *rn jwꜣ*
Young ox
Date: Kanawati-Abder-Raziq, *Nikauisesi*, 17-23: VI.1M; Harpur, *Decoration*, nr

LB291 *mrrw-kꜣ.j: mrj* - Saqqara, TP (ntn), chamber A13, north: pastoral
Source: *P-M III*, 525-534; Duell, *Mereruka ii*, pl. 153
Label: *rn jwꜣ*
Young ox
Date: Harpur, *Decoration*, [#420], 274: VI.1M-L

LB292 *ꜥnḫ-m-ꜥ-ḥr: zzj* - Saqqara, TP (ntn), room VI, north: pastoral
Source: *P-M III*, 512-515; Kanawati-Hassan, *Ankhmahor*, pl. 57 [b]

[26] According Brovarski (*Senedjemib Complex Part I*, 56), this relief has now been destroyed.

Label: missing
Date: Kanawati-Hassan, *Ankhmahor*, 18: VI.1M-L; Harpur, *Decoration*, [#374], 273: VI.1L-2E

LB293 *ptḥ-ḥtp* - Saqqara, AJS (LS31), pillared hall I, west: pastoral
Source: *P-M III*, 653-654; *LD ii*, pl. 102 [b]
Label: *wšꜣ rn jwꜣ*
Fattening a young ox
Date: Harpur, *Decoration*, [#397], 273: VI.2? or LV-VIE?

LB294 *mrrj* - Saqqara, TP (ntn), room 1, west: pastoral
Source: *P-M III*, 518-519; Davies et al., *Saqqara Tombs i*, pl. 6
Label: *rn jwꜣ mdꜣṯ*
Young ox of the stall
Date: Harpur, *Decoration*, [#418], 274: VI.2E?

LB295 *mrw: ttj-snb* - Saqqara, TP (ntn), chapel, south: pastoral
Source: *P-M III*, 520; Lloyd et al., *Saqqara Tombs ii*, pl. 8
Label: none
Date: Harpur, *Decoration*, [#417], 274: VI.2E-M?

LB296 *mḥw* - Saqqara, UPC (ntn), room II, south: boating
Source: *P-M III*, 619-622; Altenmüller, *Mehu*, pl. 22 [a]
Label: none
Date: Altenmüller, *Mehu*, 19: VI.2; Harpur, *Decoration*, [#424], 274: VI.2M-3?

Ungulate calves

LB297 *rʿ-m-kꜣ.j* - Saqqara, NSP (D3=S903), chapel, south: hunting (dorcas gazelle calf)
Source: *P-M III*, 487-488; Hayes, *Scepter i*, fig. 56
Label: *gḥs*
Gazelle
Date: Harpur, *Decoration*, [#453], 275: V.6? r/u - V.6-7

LB298 *nj-ʿnḫ-ẖnmw/ẖnmw-ḥtp* - Saqqara, UPC (ntn), vestibule, west: hunting (dorcas gazelle calf)
Source: *P-M III*, 641-644; Moussa-Altenmüller, *Nianchchnum*, pl. 40
Label: none
Date: Harpur, *Decoration*, [#431], 274: V.6L-7

LB299 *nj-ʿnḫ-ẖnmw/ẖnmw-ḥtp* - Saqqara, UPC (ntn), vestibule, west: hunting (scimitar oryx calf)
Source: *P-M III*, 641-644; Moussa-Altenmüller, *Nianchchnum*, pl. 40
Label: none
Date: Harpur, *Decoration*, [#431], 274: V.6L-7

LB300 *nj-ʿnḫ-ẖnmw/ẖnmw-ḥtp* - Saqqara, UPC (ntn), vestibule, west: hunting (dorcas gazelle calf)
Source: *P-M III*, 641-644; Moussa-Altenmüller, *Nianchchnum*, pl. 40
Label: none
Date: Harpur, *Decoration*, [#431], 274: V.6L-7

LB301 *nj-ʿnḫ-ẖnmw/ẖnmw-ḥtp* - Saqqara, UPC (ntn), vestibule, west: hunting (dorcas gazelle calf)
Source: *P-M III*, 641-644; Moussa-Altenmüller, *Nianchchnum*, pl. 40
Label: none
Date: Harpur, *Decoration*, [#431], 274: V.6L-7

LB302 *ꜣḫtj-ḥtp* - Saqqara, WSP (D64), relief fragment from corridor, east: desert (dorcas gazelle calf)
Source: *P-M III*, 599-600; Davies, *Five Theban*, pl. 40 [2]
Label: none
Date: Harpur, *Decoration*, [#338], 272: V.8L-9E

LB303 *ptḥ-ḥtp II: ṯfj* - Saqqara, WSP (D64), offering-room, east: hunting (dorcas gazelle calf)
Source: *P-M III*, 600-604; Quibell, *Ptah-hetep*, pl. 32
Label: none
Date: Harpur, *Decoration*, [#400], 274: V.9M-L

LB304 *sšm-nfr IV* - Giza, GIS (LG53), vestibule, west: hunting (dorcas gazelle calf)
Source: *P-M III*, 223-226; Junker, *Giza xi*, fig. 63
Label: none
Date: Harpur, *Decoration*, [#235], 270: V.9-VI.1

LB305 *sšm-nfr IV* - Giza, GIS (LG53), vestibule, west: hunting (dorcas gazelle calf)
Source: *P-M III*, 223-226; Junker, *Giza xi*, fig. 63
Label: none
Date: Harpur, *Decoration*, [#235], 270: V.9-VI.1

LB306 Anon - Saqqara, WSP (ntn), relief fragment from shaft 6, 'Mastaba of Shaft no. 14', unknown: desert (scimitar oryx calf)
Source: *P-M III*, 613; Hassan, *Hemet-Re*, pl. 14 [c]
Label: none
Date: *P-M III*, 613: VI; Harpur, *Decoration*, nr

LB307 *mrrw-k3.j: mrj* - Saqqara, TP (ntn), chamber A2, west: hunting (scimitar oryx calf)
Source: *P-M III*, 525-534; Duell, *Mereruka i*, pl. 25
Label: none
Date: Harpur, *Decoration*, [#420], 274: VI.1M-L

LB308 *sʿnḫ-w(j)-ptḥ* - Saqqara, TP (ntn), room I, west: hunting (dorcas gazelle calf)
Source: *P-M III*, nr; Kanawati-Abder-Raziq, *Neferseshemre*, pls. 67 and 71
Label: none
Date: Kanawati-Abder-Raziq, *Neferseshemre*, 40-41: VI.1L; Harpur, *Decoration*, nr

LB309 *sʿnḫ-w(j)-ptḥ* - Saqqara, TP (ntn), room I, west: hunting (dorcas gazelle calf)
Source: *P-M III*, nr; Kanawati-Abder-Raziq, *Neferseshemre*, pls. 67 and 71
Label: none
Date: Kanawati-Abder-Raziq, *Neferseshemre*, 40-41: VI.1L; Harpur, *Decoration*, nr

LB310 *sʿnḫ-w(j)-ptḥ* - Saqqara, TP (ntn), room I, west: hunting (scimitar oryx calf)
Source: *P-M III*, nr; Kanawati-Abder-Raziq, *Neferseshemre*, pls. 67 and 71
Label: none
Date: Kanawati-Abder-Raziq, *Neferseshemre*, 40-41: VI.1L; Harpur, *Decoration*, nr

LB311 *mrjj-ttj: mrj* - Saqqara, TP (ntn), room C1, west: desert (scimitar oryx calf)
Source: *P-M III*, 356-357; Kanawati-Abder-Raziq, *Meryteti*, pl. 46
Label: none
Date: Kanawati-Abder-Raziq, *Meryteti*, 18: end VI.1-VI.2M; Harpur, *Decoration*, [#415], 274: VI.2M-L

LB312 *mrjj-ttj: mrj* - Saqqara, TP (ntn), room C1, west: desert (dorcas gazelle calf)
Source: *P-M III*, 356-357; Kanawati-Abder-Raziq, *Meryteti*, pl. 46
Label: none
Date: Kanawati-Abder-Raziq, *Meryteti*, 18: end VI.1-VI.2M; Harpur, *Decoration*, [#415], 274: VI.2M-L

LB313 *mrjj-ttj: mrj* - Saqqara, TP (ntn), room C1, west: desert (dorcas gazelle calf)
Source: *P-M III*, 356-357; Kanawati-Abder-Raziq, *Meryteti*, pl. 46
Label: none
Date: Kanawati-Abder-Raziq, *Meryteti*, 18: end VI.1-VI.2M; Harpur, *Decoration*, [#415], 274: VI.2M-L

LB314 *mrjj-ttj: mrj* - Saqqara, TP (ntn), room C1, west: desert (scimitar oryx calf?)
Source: *P-M III*, 356-357; Kanawati-Abder-Raziq, *Meryteti*, pl. 46
Label: none
Date: Kanawati-Abder-Raziq, *Meryteti*, 18: end VI.1-VI.2M; Harpur, *Decoration*, [#415], 274: VI.2M-L

LB315 *mrjj-ttj: mrj* - Saqqara, TP (ntn), room C1, west: desert (dorcas gazelle calf)
Source: *P-M III*, 356-357; Kanawati-Abder-Raziq, *Meryteti*, pl. 46
Label: none
Date: Kanawati-Abder-Raziq, *Meryteti*, 18: end VI.1-VI.2M; Harpur, *Decoration*, [#415], 274: VI.2M-L

LB316 *mrjj-ttj: mrj* - Saqqara, TP (ntn), room C1, west: desert (dorcas gazelle calf?)
Source: *P-M III*, 356-357; Kanawati-Abder-Raziq, *Meryteti*, pl. 46

Label: none
Date: Kanawati-Abder-Raziq, *Meryteti*, 18: end VI.1-VI.2M; Harpur, *Decoration*, [#415], 274: VI.2M-L

LB317 *mrjj-ttj: mrj* - Saqqara, TP (ntn), room C1, west: desert (scimitar oryx calf)
Source: *P-M III*, 356-357; Kanawati-Abder-Raziq, *Meryteti*, pl. 46
Label: none
Date: Kanawati-Abder-Raziq, *Meryteti*, 18: end VI.1-VI.2M; Harpur, *Decoration*, [#415], 274: VI.2M-L

LB318 *mrjj-ttj: mrj* - Saqqara, TP (ntn), room C1, west: desert (dorcas gazelle calf)
Source: *P-M III*, 356-357; Kanawati-Abder-Raziq, *Meryteti*, pl. 46
Label: none
Date: Kanawati-Abder-Raziq, *Meryteti*, 18: end VI.1-VI.2M; Harpur, *Decoration*, [#415], 274: VI.2M-L

LB319 *mrjj-ttj: mrj* - Saqqara, TP (ntn), room C1, west: desert (Nubian ibex)
Source: *P-M III*, 356-357; Kanawati-Abder-Raziq, *Meryteti*, pl. 46
Label: none
Date: Kanawati-Abder-Raziq, *Meryteti*, 18: end VI.1-VI.2M; Harpur, *Decoration*, [#415], 274: VI.2M-L

Cape hares

LB320 *rꜥ-m-kꜣ.j* - Saqqara, NSP (D3=S903), chapel, south: hunting
Source: *P-M III*, 487-488; Hayes, *Scepter i*, fig. 56
Label: none
Date: Harpur, *Decoration*, [#453], 275: V.6? r/u - V.6-7

LB321 *nj-ꜥnḫ-ẖnmw/ẖnmw-ḥtp* - Saqqara, UPC (ntn), vestibule, west: hunting
Source: *P-M III*, 641-644; Moussa-Altenmüller, *Nianchchnum*, pl. 40
Label: none
Date: Harpur, *Decoration*, [#431], 274: V.6L-7

LB322 *nj-ꜥnḫ-ẖnmw/ẖnmw-ḥtp* - Saqqara, UPC (ntn), vestibule, west: hunting
Source: *P-M III*, 641-644; Moussa-Altenmüller, *Nianchchnum*, pl. 40
Label: none
Date: Harpur, *Decoration*, [#431], 274: V.6L-7

LB323 *pḥn-wj-kꜣ.j* - Saqqara, NSP (D70=LS15), antechamber, west: desert
Source: *P-M III*, 491-492; *LD ii*, pl. 46
Label: none
Date: Harpur, *Decoration*, [#393], 273: V.6-8E

LB324 *ṯjj* - Saqqara, NSP (D22), room III, east: desert
Source: *P-M III*, 468-478; Wild, *Ti ii*, pl. 127
Label: none
Date: Harpur, *Decoration*, [#543], 277: V.8-9

LB325 *mrrw-kꜣ.j: mrj* - Saqqara, TP (ntn), chamber A2, west: hunting
Source: *P-M III*, 525-534; Duell, *Mereruka i*, pl. 25
Label: none
Date: Harpur, *Decoration*, [#420], 274: VI.1M-L

4.2.3 Death

Cattle

LB326 *sḫm-ꜥnḫ-ptḥ* - Saqqara, NSP (D41), chapel, north: slaughter (cutting the foreleg)
Source: *P-M III*, 454-455; Simpson, *Sekem-ankh-ptah*, pl. A
Label: none
Date: Harpur, *Decoration*, [#504], 276: V.8-9

LB327 *ṯtw: kꜣ.j-nswt* - Giza, WF (G2001), recessed portico, west: slaughter (removing the foreleg)
Source: *P-M III*, 66-67; Simpson, *Mast. West. Cem. Part I*, fig. 19
Label: none

Date: Harpur, *Decoration*, [#292], 271: VI

LB328 *jrw-kꜣ-ptḥ: ḫnw* - Saqqara, UPC (ntn), chapel, east: slaughter (dismembering the body)
Source: *P-M III*, 639; McFarlane, *Irukaptah*, pl. 45
Label: none
Date: McFarlane, *Irukaptah*, 19: late V, V.7-8; Harpur, *Decoration*, [#364], 273: VI

LB329 *ḫntj-kꜣ.j: jḫḫj* - Saqqara, TP (ntn), main burial chamber, east: offering (left as an offering)
Source: *P-M III*, 508-511; James, *Khentika*, pl. 38
Label: *rn n jwꜣ jmj mḏt*
Young of an ox which is in the stall
Date: Harpur, *Decoration*, [#479], 275: VI.2M

5.1 HERDING

Sheep

GB1 *mrs-ꜥnḫ III* - Giza, EF (G7530 + G7540), main chapel (subterranean), east: agricultural
Source: *P-M III*, 197-199; Dunham-Simpson, *Mersyankh III*, fig. 4
Label: *skꜣ m sḥt*
Cultivating with a herd of sheep
stjt prt
Scattering seed
Date: Harpur, *Decoration*, [#098], 267: IV.6

GB2 *nb.j-m-ꜣḫtj* - Giza, CF (LG86), chapel I, south: agricultural
Source: *P-M III*, 230-232; *LD ii*, pl. 12 [a]
Label: missing
Date: Harpur, *Decoration*, [#122], 267: IV.6-V.1

GB3 *jj-mrjj* - Giza, WF (G6020=LG16), second chamber, north: agricultural
Source: *P-M III*, 170-174; Weeks, *Cemetery G6000*, fig. 39
Label: *skꜣ m sḥt jn jswt.f nt pr.f n ḏt*
Cultivating with a herd of sheep by his gang of his funerary estate
Date: Weeks, *Cemetery G6000*, 7: V.6; Harpur, *Decoration*, [#014], 265: V.3L

GB4 *wr-jr.n-ptḥ* - Saqqara (ntn), chapel, east: agricultural
Source: *P-M III*, 699-700; *Hiero. Texts vi*, pl. 11
Label: none
Date: Harpur, *Decoration*, [#379], 273: V.3-5?

GB5 *nfr-bꜣw-ptḥ* - Giza, WF (G6010=LG15), first chamber, east: agricultural
Source: *P-M III*, 169-170; Weeks, *Cemetery G6000*, fig. 11
Label: *skꜣ*
Cultivating
Date: Weeks, *Cemetery G6000*, 5-6: V.6-8; Harpur, *Decoration*, [#130], 267: V.6

GB6 *nfr-bꜣw-ptḥ* - Giza, WF (G6010=LG15), first chamber, east: agricultural
Source: *P-M III*, 169-170; Weeks, *Cemetery G6000*, fig. 11
Label: *skꜣ m sḥt jn jswt[=f] n[t] pr=f n ḏt*
Cultivating with a herd of sheep by his gang of his funerary estate
Date: Weeks, *Cemetery G6000*, 5-6: V.6-8; Harpur, *Decoration*, [#130], 267: V.6

GB7 *nfr/kꜣ-ḥꜣ.j* - Saqqara, UPC (ntn), chapel, east: agricultural
Source: *P-M III*, 639-641; Moussa-Altenmüller, *Nefer and Ka-Hay*, pl. 9
Label: none
Date: Harpur, *Decoration*, [#439], 274: V.6

GB8 *nj-ꜥnḫ-ḫnmw/ḫnmw-ḥtp* - Saqqara, UPC (ntn), outer hall, east: agricultural
Source: *P-M III*, 641-644; Moussa-Altenmüller, *Nianchchnum*, pls. 56 and 58
Label: *skꜣ m sḥt*
Cultivating with a herd of sheep
Date: Harpur, *Decoration*, [#431], 274: V.6L-7

GB9 *ḥtp-ḥr-ꜣḫtj* - Saqqara, WSP (D60), chapel, south: agricultural
Source: *P-M III*, 593-595; Mohr, *Hetep-her-akhti*, fig. 46
Label: *skꜣ m sḥt*
Cultivating with a herd of sheep
Date: Harpur, *Decoration*, [#467], 275: V.6-8E

GB10 *ꜣḫtj-ḥtp* - Saqqara, UPC (ntn), chapel, east: agricultural
Source: *P-M III*, 634-637; Ziegler, *Akhethetep*, 128-129
Label: *skꜣ jn jswt=f*
Cultivating by his gang
Date: Ziegler, *Akhethetep*, 9: V.6; Harpur, *Decoration*, [#340], 272: V.6-8E

GB11 *jrj-n-k3-ptḥ* - Saqqara, UPC (ntn), offering chamber, east: agricultural
Source: *P-M III*, 644; Moussa-Junge, *Two Tombs*, pl. 9
Label: missing
Date: Harpur, *Decoration*, [#362], 273: V.6-9

GB12 *z3-jb* - Giza, WF (G2092 + G2093), chapel, east: agricultural
Source: *P-M III*, nr; Roth, *Cem. Pal. Att.*, fig. 183
Label: missing
Date: Roth, *Cem. Pal. Att.*, 109: V.8; Harpur, *Decoration*, nr

GB13 *snḏm-jb: jntj* - Giza, WF (G2370=LG27), room III, west: agricultural
Source: *P-M III*, 85-87; *LD Ergänz.*, pl. 22 [b][27]
Label: none
Date: Brovarski, *Senedjemib Complex Part I*, 23: V.8M-L; Harpur, *Decoration*, [#219], 269: V.8M-L

GB14 *sḫm-ʿnḫ-ptḥ* - Saqqara, NSP (D41), chapel, south: agricultural
Source: *P-M III*, 454-455; Simpson, *Sekem-ankh-ptah*, pl. D
Label: n/a (Sower's Song)
Date: Harpur, *Decoration*, [#504], 276: V.8-9

GB15 *ṯjj* - Saqqara, NSP (D22), room III, north: agricultural
Source: *P-M III*, 468-478; Wild, *Ti ii*, pl. 113
Label: n/a (Sower's Song)
Date: Harpur, *Decoration*, [#543], 277: V.8-9

GB16 *nfr-jrt-n.f* - Saqqara, ESP (D55), chapel, east: agricultural
Source: *P-M III*, 583-584; van de Walle, *Neferirtenef*, pl. 11
Label: *sk3 m sḥt*
Cultivating with a herd of sheep
Date: Harpur, *Decoration*, [#440], 274: V.8-9

GB17 *k3.j-m-nfrt* - Saqqara, NSP (D23), chapel, east: agricultural
Source: *P-M III*, 467-468; Simpson, *Kayemnofret*, pl. G
Label: missing
Date: Simpson, *Kayemnofret*, 1: V.8-9; Harpur, *Decoration*, [#522], 276: V.8-9?

GB18 *nbt* - Saqqara, UPC (ntn), room B [II], east: agricultural
Source: *P-M III*, 624-625; Munro, *Unas-Friedhof*, pl. 13
Label: n/a (Sower's Song)
Date: Munro, *Unas-Friedhof*, 20: V.9; Harpur, *Decoration*, [#438], 274: V.9

GB19 *nfr I* - Giza, WF (G4761), chapel, east: agricultural
Source: *P-M III*, 137-138; Junker, *Giza vi*, fig. 14
Label: none
Date: Harpur, *Decoration*, [#126], 267: V.9-VI.1

GB20 *sḫm-k3.j* - Giza, WF (G1029), chapel, east: agricultural
Source: *P-M III*, 53; Simpson, *Mast. West. Cem. Part I*, fig. 4
Label: *sk3 m sḥt*
Cultivating with a herd of sheep
Date: Harpur, *Decoration*, [#225], 269: V.9-VI.1?

GB21 *k3.j-m-ḥzt* - Saqqara, TP (ntn), pillared hall, east: agricultural
Source: *P-M III*, 542-543; McFarlane, *Kaiemheset*, pl. 44
Label: *ḫsf*
Driving[28]

27 According to Brovarski (*Senedjemib Complex Part I*, 59), blocks recorded by Lepsius (including that featuring the sheep treading scene) were subsequently removed by tomb robbers.

28 McFarlane (*Kaiemheset*, 31) suggests that these are donkeys. However, the small size of the animals and the presence of a herdsman at the head of the line, whose posture is characteristic of sowers, indicate that this is a sheep treading scene.

Date: Cherpion, *Mastabas*, 112-115: no later than IV.3; McFarlane, *Kaiemheset*, 23: V.6-V.8; Harpur, *Decoration*, [#525], 276: V.9-VI.1?

GB22 *k3.j-m-nfrt* - Giza, EF (LG63), chapel, east: agricultural
Source: *P-M III*, 208-209; *LD Erganz.*, pl. 32 [lower]
Label: *sk3*
Cultivating
Date: Harpur, *Decoration*, [#258], 270: V-VI

GB23 *j3zn* - Giza, WF (G2196), chapel, north: agricultural
Source: *P-M III*, 82; Simpson, *Mast. West. Cem. Part I*, fig. 30
Label: none
Date: Harpur, *Decoration*, [#013], 265: VI.1M-2M

GB24 *mḥw* - Saqqara, UPC (ntn), room II, north: agricultural
Source: *P-M III*, 619-622; Altenmüller, *Mehu*, pl. 41 [b-a]
Label: *stjt jt bdt jn ḥk3 ḥwt*
Scattering barley and emmer by the farm foreman (plus part of the Sower's Song)
Date: Altenmüller, *Mehu*, 19: VI.2; Harpur, *Decoration*, [#424], 274: VI.2M-3?

GB25 *k3-ḥj.f* - Giza, WF (G2136), chapel, east: agricultural
Source: *P-M III*, 76; Junker, *Giza vi*, fig. 42
Label: missing
Date: Harpur, *Decoration*, [#278], 271: VI.2M-3?

GB26 *jj-nfrt* - Giza (ntn), chapel, unknown: agricultural
Source: *P-M III*, 298-299; Schürmann, *Die Reliefs*, fig. 11 [b]
Label: *sk3t m ḥst jn jst n pr-ḏt r-gs rḫ nswt wr 10 šmꜥw jj-nfrt*
Ploughing with (the herd of sheep) by the workers of the estate beside the acquaintance of the king, the greatest of the 10 of Upper Egypt, Iy-nefert (plus part of the Sower's Song)
Date: Harpur, *Decoration*, [#017], 265: VI.3-4E?

GB27 *nb-k3w-ḥr: jdw* - Saqqara, UPC (ntn), pillared hall, east: agricultural
Source: *P-M III*, 627-629; Hassan, *Neb-Kaw-Her*, pl. 23 [b]
Label: *sk3*
Cultivating
Date: Harpur, *Decoration*, [#437], 274: V.9 r/u VI.7

Mixed species

GB28 *ptḥ-ḥtp II: ṯfj* - Saqqara, WSP (D64), offering-room, east: hunting (dorcas gazelle, 2 bulls)
Source: *P-M III*, 600-604; Quibell, *Ptah-hetep*, pl. 32
Label: none
Date: Harpur, *Decoration*, [#400], 274: V.9M-L

GB29 *sšm-nfr IV* - Giza, GIS (LG53), vestibule, west: hunting (2 scimitar oryx, 2 hartebeest, Nubian ibex)
Source: *P-M III*, 223-226; Junker, *Giza xi*, fig. 63
Label: none
Date: Harpur, *Decoration*, [#235], 270: V.9-VI.1

GB30 *sšm-nfr IV* - Giza, GIS (LG53), vestibule, west: hunting (addax, scimitar oryx, 2 hartebeest, dorcas gazelle)
Source: *P-M III*, 223-226; Junker, *Giza xi*, fig. 63
Label: none
Date: Harpur, *Decoration*, [#235], 270: V.9-VI.1

GB31 *nj-m3ꜥt-rꜥ [jzzj-mr-nṯr]* - Giza, WF (G2097), chapel, east: hunting (2 dorcas gazelle, 2 Sömmerings gazelle)
Source: *P-M III*, 70; Roth, *Cem. Pal. Att.*, pl. 189
Label: none
Date: Roth, *Cem. Pal. Att.*, 129-130: V.9; Harpur, *Decoration*, [#033], 265: V.9-VI.1

GB32 *mrrw-k3.j: mrj* - Saqqara, TP (ntn), chamber A2, west: hunting (scimitar oryx, Nubian ibex, dorcas gazelle)
Source: *P-M III*, 525-534; Duell, *Mereruka i*, pl. 25
Label: none

Date: Harpur, *Decoration*, [#420], 274: VI.1M-L

GB33 *jnw-mnw* - Saqqara (ntn), room II, north: hunting (2 scimitar oryx, Nubian ibex, dorcas gazelle, jerboa)
Source: *P-M III*, nr; Kanawati, *Inumin*, pl. 47
Label: none
Date: Kanawati, *Inumin*, 16-17: VI.2E; Harpur, *Decoration*, nr

GB34 *mrjj-ttj: mrj* - Saqqara, TP (ntn), room C1, west: hunting (Sömmerings gazelle, dorcas gazelle, Nubian ibex)
Source: *P-M III*, 356-357; Kanawati-Abder-Raziq, *Meryteti*, pl. 46
Label: none
Date: Kanawati-Abder-Raziq, *Meryteti*, 18: end VI.1-VI.2M; Harpur, *Decoration*, [#415], 274: VI.2M-L

GB35 *mrjj-ttj: mrj* - Saqqara, TP (ntn), room C1, west: hunting (dama deer, Nubian ibex, scimitar oryx)
Source: *P-M III*, 356-357; Kanawati-Abder-Raziq, *Meryteti*, pl. 46
Label: none
Date: Kanawati-Abder-Raziq, *Meryteti*, 18: end VI.1-VI.2M; Harpur, *Decoration*, [#415], 274: VI.2M-L

GB36 *mrjj-ttj: mrj* - Saqqara, TP (ntn), room C1, west: hunting (dorcas gazelle, hartebeest, Sömmerings gazelle)
Source: *P-M III*, 356-357; Kanawati-Abder-Raziq, *Meryteti*, pl. 46
Label: none
Date: Kanawati-Abder-Raziq, *Meryteti*, 18: end VI.1-VI.2M; Harpur, *Decoration*, [#415], 274: VI.2M-L

GB37 *mrjj-ttj: mrj* - Saqqara, TP (ntn), room C1, west: hunting (dorcas gazelle, hartebeest, Nubian ibex)
Source: *P-M III*, 356-357; Kanawati-Abder-Raziq, *Meryteti*, pl. 46
Label: none
Date: Kanawati-Abder-Raziq, *Meryteti*, 18: end VI.1-VI.2M; Harpur, *Decoration*, [#415], 274: VI.2M-L

GB38 *wr-nw* - Saqqara, TP (ntn), hall, west: desert (hartebeest, dorcas gazelle)
Source: *P-M III*, 519; Davies et al., *Saqqara Tombs i*, pl. 31
Label: none
Date: Harpur, *Decoration*, [#380], 273: VI.3-4?

5.2 FLOCKING

Confidence decoys

GB39 *k3.j-wꜥb* - Giza, EF (G7110+7120), chamber B, east (no. 24-12-1001 = MFA acc. no. 34.59): fowling (grey heron upon crate)
Source: *P-M III*, 487-488; *HESPOK*, fig. 62; Simpson, *Kawab*, fig. 11 [g]
Label: *nw*
Hunter
Date: Harpur, *Decoration*, [#252], 270: IV.2-3

GB40 *rꜥ-m-k3.j* - Saqqara, NSP (D3=S903), chapel, west: clapnet
Source: *P-M III*, 487-488; Hayes, *Scepter i*, fig. 55
Label: none
Date: Harpur, *Decoration*, [#453], 275: V.6? r/u - V.6-7

GB41 *nj-ꜥnḫ-ẖnmw/ẖnmw-ḥtp* - Saqqara, UPC (ntn), outer hall, west: boating (grey heron upon crate)
Source: *P-M III*, 641-644; Moussa-Altenmüller, *Nianchchnum*, pl. 77
Label: n/a
Date: Harpur, *Decoration*, [#431], 274: V.6L-7

GB42 *ḥtp-ḥr-3ḫtj* - Saqqara, WSP (D60), chapel, north: fording
Source: *P-M III*, 593-595; Mohr, *Hetep-her-akhti*, fig. 33
Label: *mnjw pw ꜥnḫ ḥr(=k) r(=f) wrt*
This herdsman! May you be watchful against it greatly!
Date: Harpur, *Decoration*, [#467], 275: V.6-8E

GB43 *3ḫtj-ḥtp* - Saqqara, UPC (ntn), chapel, east: clapnet
Source: *P-M III*, 634-637; Ziegler, *Akhethetep*, 132-133
Label: *rdj sḫt*

Making the catch
Date: Ziegler, *Akhethetep*, 9: V.6; Harpur, *Decoration*, [#340], 272: V.6-8E

GB44 *jrj-n-k3-ptḥ* - Saqqara, UPC (ntn), offering chamber, south: clapnet
Source: *P-M III*, 644; Moussa-Junge, *Two Tombs*, pl. 13
Label: none
Date: Harpur, *Decoration*, [#362], 273: V.6-9

GB45 *rʿ-špss* - Saqqara, NSP (LS16=S902), forecourt, north: boating (grey heron and bittern upon crate)
Source: *P-M III*, 494-496; *LD ii*, pl. 60 [right]
Label: none
Date: Harpur, *Decoration*, [#456], 275: V.8M

GB46 *ptḥ-ḥtp I* - Saqqara, WSP (D62), offering chamber, east: clapnet
Source: *P-M III*, 596-598; Murray, *Saqqara Mastabas*, pl. 11
Label: *rdj sḫt 3pdw jn wḥʿ nw pr*
Making the catch of birds by the fowlers of the house
Date: Harpur, *Decoration*, [#398], 273: V.8M-L

GB47 *tjj* - Saqqara, NSP (D22), room III, north: clapnet
Source: *P-M III*, 468-478; Wild, *Ti ii*, pl. 120
Label: *šm r=k jm=f jw wn 3pd(w) ḥr-s3=f*
Go from it! There are birds upon it!
smr wʿtjj
Sole companion
Date: Harpur, *Decoration*, [#543], 277: V.8-9

GB48 *tjj* - Saqqara, NSP (D22), room III, north: clapnet (fettered grey heron)
Source: *P-M III*, 468-478; Wild, *Ti ii*, pl. 121
Label: none
Date: Harpur, *Decoration*, [#543], 277: V.8-9

GB49 *tjj* - Saqqara, NSP (D22), room III, north: fowling (grey heron upon crate)
Source: *P-M III*, 468-478; Wild, *Ti ii*, pl. 121
Label: none
Date: Harpur, *Decoration*, [#543], 277: V.8-9

GB50 *nfr-jrt-n.f* - Saqqara, ESP (D55), chapel, east: clapnet
Source: *P-M III*, 583-584; van de Walle, *Neferirtenef*, pl. 13
Label: *rdj sḫt*
Making the catch
Date: Harpur, *Decoration*, [#440], 274: V.8-9

GB51 *ḫww-wr* - Giza, CF (LG95), chapel, east: clapnet
Source: *P-M III*, 254-255; Hassan, *Giza v*, fig. 104
Label: *rdj sḫt*
Making the catch
Date: Harpur, *Decoration*, [#180], 268: V.8-9M

GB52 *ḫnwt* - Saqqara, UPC (ntn), room A [I], east: clapnet
Source: *P-M III*, 623-624; Munro, *Unas-Friedhof*, pl. 33
Label: *jmj-r wḥʿ(w)*
Overseer of fowlers
Date: Munro, *Unas-Friedhof*, 20: V.9; Harpur, *Decoration*, [#478], 275: V.9

GB53 *jj-nfrt: š3.n.f* - Saqqara, UPC (ntn), room I, south: clapnet
Source: *P-M III*, 623-624; Kanawati-Abder-Raziq, *Iynefert*, pl. 38
Label: none
Date: Kanawati-Abder-Raziq, *Iynefert*, 13: V.9E-M; Harpur, *Decoration*, [#345], 272: V.9

GB54 *ptḥ-ḥtp II: tfj* - Saqqara, WSP (D64), offering-room, east: clapnet
Source: *P-M III*, 600-604; Quibell, *Ptah-hetep*, pl. 32
Label: *sḫt m ḥsbt*

Catching in multitude
jtj r=k ntj ḥnꜥ jw ḥb n=k
Seize my companions! It is a catch for you!
Date: Harpur, *Decoration*, [#400], 274: V.9M-L

GB55 *sḫm-kꜣ.j* - Giza, WF (G1029), chapel, east: clapnet
Source: *P-M III*, 53; Simpson, *Mast. West. Cem. Part I*, fig. 4
Label: none
Date: Harpur, *Decoration*, [#225], 269: V.9-VI.1?

GB56 Anon - Giza (ntn), relief fragment (Munich Staatl. Sammlung #Gl. 115), unknown: clapnet
Source: *P-M III*, 308-309; *Munich. Äg. Sammlung*, fig. 22
Label: none
Date: Harpur, *Decoration*, [#325], 272: V-VI

GB57 *kꜣ-gm-nj: mmj* - Saqqara, TP (LS10), room IV, west: clapnet
Source: *P-M III*, 521-525; von Bissing, *Gem-ni-kai i*, pl. 9; Harpur-Scremin, *Kagemni*, fig. 14
Label: *wdjt sḫt*
Laying the trap
Date: Harpur, *Decoration*, [#534], 276: VI.1M

GB58 *sꜥnḫ-w(j)-ptḥ* - Saqqara, TP (ntn), room I, east: clapnet
Source: *P-M III*, nr; Kanawati-Abder-Raziq, *Neferseshemre*, pls. 69 and 75
Label: none
Date: Kanawati-Abder-Raziq, *Neferseshemre*, 40-41: VI.1L; Harpur, *Decoration*, nr

GB59 *ḥzj* - Saqqara, TP (ntn), portico, east: clapnet (grey heron upon crate and another beside clapnet)
Source: *P-M III*, nr; Kanawati-Abder-Raziq, *Hesi*, pl. 55
Label: *dw sḫt jn jmj-r wḥꜥ(w) j wḥꜥ pw ḥb r=k*
Setting the trap by the overseer of fowlers: Oh this fowler! There is a catch for you!
Date: Kanawati-Abder-Raziq, *Hesi*, 16: VI.1L-2E; Harpur, *Decoration*, nr

GB60 *ꜥnḫ-m-ꜥ-ḥr: zzj* - Saqqara, TP (ntn), room II, west: clapnet
Source: *P-M III*, 512-515; Kanawati-Hassan, *Ankhmahor*, pl. 42
Label: *jw ḥb n ꜥ=k wḥꜥ pw jgr tjwn*
There is a catch for your arm, this fowler
May you keep quiet
Date: Kanawati-Hassan, *Ankhmahor*, 18: VI.1M-L; Harpur, *Decoration*, [#374], 273: VI.1L-2E

GB61 *ꜥnḫ-m-ꜥ-ḥr: zzj* - Saqqara, TP (ntn), room II, west: clapnet (fettered grey heron beside crates)
Source: *P-M III*, 512-515; Kanawati-Hassan, *Ankhmahor*, pl. 42
Label: none
Date: Kanawati-Hassan, *Ankhmahor*, 18: VI.1M-L; Harpur, *Decoration*, [#374], 273: VI.1L-2E

GB62 *ꜥnḫ-m-ꜥ-ḥr: zzj* - Saqqara, TP (ntn), room II, west: clapnet (fettered grey heron beside fowlers)
Source: *P-M III*, 512-515; Kanawati-Hassan, *Ankhmahor*, pl. 42
Label: none
Date: Kanawati-Hassan, *Ankhmahor*, 18: VI.1M-L; Harpur, *Decoration*, [#374], 273: VI.1L-2E

GB63 *wḏꜣ-ḥꜣ-ttj: ššj [nfr-sšm-ptḥ]* - Saqqara, TP (ntn), room II, east: clapnet
Source: *P-M III*, 515-516; Capart, *Rue de Tombeaux*, pl. 85
Label: *j wḥꜥ pw ḥr=k nḏr nb jw ḥb jr=k*
Oh this fowler! May you hold fast everyone
It is a catch for you!
Date: Harpur, *Decoration*, [#384], 273: VI.1L-2E

GB64 *wḏꜣ-ḥꜣ-ttj: ššj [nfr-sšm-ptḥ]* - Saqqara, TP (ntn), room II, east: clapnet
Source: *P-M III*, 515-516; Capart, *Rue de Tombeaux*, pl. 85
Label: *js<t> wrt m ꜣpd(w)*
Lo! An excess of birds
jw ḥb m-ḫnw=s
It is a catch within it
Date: Harpur, *Decoration*, [#384], 273: VI.1L-2E

GB65 *mttj* - Saqqara, UPC (ntn), relief fragment (Louvre no. E. 25508), unknown: fowling (fettered grey heron)
Source: *P-M III*, 646-648; Ziegler, *Catalogue*, 143
Label: none
Date: Ziegler, *Catalogue*, 123: probably VI; Harpur, *Decoration*, [#426], 274: VI.2?

GB66 *mḥw* - Saqqara, UPC (ntn), room I, south: clapnet
Source: *P-M III*, 619-622; Altenmüller, *Mehu*, pl. 7
Label: *jnj jr=k ntj ḥnʿ(=j) ḥwt-dšr(t)*
Haul it, my comrades!
(It is) a Red House!
Date: Altenmüller, *Mehu*, 19: VI.2; Harpur, *Decoration*, [#424], 274: VI.2M-3?

GB67 *mḥw* - Saqqara, UPC (ntn), room I, south: clapnet
Source: *P-M III*, 619-622; Altenmüller, *Mehu*, pl. 7
Label: none
Date: Altenmüller, *Mehu*, 19: VI.2; Harpur, *Decoration*, [#424], 274: VI.2M-3?

GB68 *mḥw* - Saqqara, UPC (ntn), room I, south: pastoral (grey heron upon crate)
Source: *P-M III*, 619-622; Altenmüller, *Mehu*, pl. 6
Label: none
Date: Altenmüller, *Mehu*, 19: VI.2; Harpur, *Decoration*, [#424], 274: VI.2M-3?

GB69 *mḥw* - Saqqara, UPC (ntn), room II, north: clapnet
Source: *P-M III*, 619-622; Altenmüller, *Mehu*, pl. 31 [b]
Label: *jnj jr=k ntj ḥnʿ(=j) jw ḥb jr=n*
Haul it my comrades
It is a catch for us!
Date: Altenmüller, *Mehu*, 19: VI.2; Harpur, *Decoration*, [#424], 274: VI.2M-3?

5.3 SCHOOLING

Fish

GB70 *ḥzj* - Saqqara, TP (ntn), portico, south (east of entrance): marsh (3 tilapia)
Source: *P-M III*, nr; Kanawati-Abder-Raziq, *Hesi*, pl. 53
Label: none
Date: Kanawati-Abder-Raziq, *Hesi*, 16: VI.1L-2E; Harpur, *Decoration*, nr

GB71 *ḥzj* - Saqqara, TP (ntn), portico, south (west of entrance): marsh (4 tilapia)
Source: *P-M III*, nr; Kanawati-Abder-Raziq, *Hesi*, pl. 54
Label: none
Date: Kanawati-Abder-Raziq, *Hesi*, 16: VI.1L-2E; Harpur, *Decoration*, nr

GB72 *ḥzj* - Saqqara, TP (ntn), portico, east: fishing (4 tilapia (left) and 3 mullet (right) in net)
Source: *P-M III*, nr; Kanawati-Abder-Raziq, *Hesi*, pl. 55
Label: n/a
Date: Kanawati-Abder-Raziq, *Hesi*, 16: VI.1L-2E; Harpur, *Decoration*, nr

---

5.4 SEPARATION

Cattle

GB73 *s3bw* - Saqqara, NSP (C16), chapel, east: offering
Source: *P-M III*, 461; Mariette, *Voyage i*, pl. 5
Label: none
Date: Harpur, *Decoration*, [#490], 275: V?

GB74 *tntj* - Giza, WF (G4920=LG47), chapel, east: agricultural
Source: *P-M III*, 141-142; *LD ii*, pl. 31 [a]
Label: *mnjw jmy sḫt*
Herdsman belonging to the marshes

ḥwt wnḏt nt pr-ḏt
Driving the cattle of the funerary estate
Date: Harpur, *Decoration*, [#287], 271: V.1-5

GB75 *nfr-msḏr-ḫwfw* - Giza, WF (G2240), relief fragment, unknown: fording
Source: *P-M III*, nr; Roth, *Cem. Pal. Att.*, pl. 210
Label: none
Date: Roth, *Cem. Pal. Att.*, 163: V.2-8 or V.9; Harpur, *Decoration*, nr

GB76 *ḥtp-ḥr-ꜣḫtj* - Saqqara, WSP (D60), chapel, north: fording
Source: *P-M III*, 593-595; Mohr, *Hetep-her-akhti*, fig. 33
Label: *mnjw pw ꜥnḫ ḥr(=k) r(=f) wrt*
This herdsman! May you be watchful against him greatly!
Date: Harpur, *Decoration*, [#467], 275: V.6-8E

GB77 *ꜣḫtj-ḥtp* - Saqqara, UPC (ntn), chapel, east: pastoral
Source: *P-M III*, 634-637; Ziegler, *Akhethetep*, 130-131
Label: *mnjw*
Herdsman
Date: Ziegler, *Akhethetep*, 9: V.6; Harpur, *Decoration*, [#340], 272: V.6-8E

GB78 *kꜣ.j-m-nfrt* - Giza, CF (ntn), vestibule A, west: fording
Source: *P-M III*, 263-264; Hassan, *Giẓa ii*, fig. 124[29]
Label: missing
Date: Harpur, *Decoration*, [#261], 270: V.6-8E?

GB79 *zꜣ-jb* - Giza, WF (G2092 + G2093), chapel, east: fording
Source: *P-M III*, nr; Roth, *Cem. Pal. Att.*, pl. 179
Label: missing
Date: Roth, *Cem. Pal. Att.*, 109: V.8; Harpur, *Decoration*, nr

GB80 *kꜣ.j-m-rḥw* - Saqqara, NSP (D2=S905), chapel, north: fording
Source: *P-M III*, 485-487; Mogensen, *Ny Carlsberg*, fig. 4
Label: *j mnjw pw ḥr=k ꜥnḫ wrt r šjj pw ntj ḥr mw jw=f m šp tp*
Oh this herdsman! May you be watchful greatly against that aquatic which is in the water!
May he come as blind-of-head
Date: Harpur, *Decoration*, [#523], 276: V.8

GB81 *rꜥ-špss* - Saqqara, NSP (LS16=S902), forecourt, north: fording
Source: *P-M III*, 494-496; *LD ii*, pl. 60 [right]
Label: *mnjw ꜥnḫ ḥr(=k) r šjj pw ntj ḥr mw*
Herdsman! May you be watchful against that aquatic which is in the water!
Date: Harpur, *Decoration*, [#456], 275: V.8M

GB82 *ṯjj* - Saqqara, NSP (D22), room III, north: fording
Source: *P-M III*, 468-478; Wild, *Ti ii*, pl. 124
Label: *mnjw pw ꜥnḫ ḥr=k r šjj pw ntj ḥr mw jw(=f) m šp tp*
This herdsman! May you be watchful against that aquatic which is in the water!
May he come as blind-of-head
Date: Harpur, *Decoration*, [#543], 277: V.8-9

GB83 *ṯjj* - Saqqara, NSP (D22), room III, north: fording
Source: *P-M III*, 468-478; Wild, *Ti ii*, pl. 114
Label: *j mḥsḥs pw rdj=k šm kꜣw jptn prt m mḥt*
Oh this filthy one! Cause to go these cattle coming out from the North
jꜣꜥ bḥs pw (n) mnꜥt
Swing around this calf of the milch cows[30]

---

[29] Hassan (*Giẓa ii*, 116) suggests that both the roped animal, and the one that follows it, are calves; however, comparison with other fording scenes indicates that the second animal must be a cow.

[30] For an extensive discussion of the translation of this passage, see van Walsem, R. (1998). "The caption to a cattle-fording scene in a tomb at Saqqara and its implications for the seh/sinnbild discussion on Egyptian iconography," in Clarysse et al., *Egyptian Religion 2*, 1469-1485.

Date: Harpur, *Decoration*, [#543], 277: V.8-9

GB84 *ꜣḫtj-ḥtp* - Saqqara, WSP (D64), corridor, west: pastoral
Source: *P-M III*, 599-600; Davies, *Ptahhetep ii*, pl. 8
Label: *mnjw bḥs*
Herdsman and calf
Date: Harpur, *Decoration*, [#338], 272: V.8L-9E

GB85 *ꜣḫtj-ḥtp* - Saqqara, WSP (D64), hall, east: fording
Source: *P-M III*, 599-600; Davies, *Ptahhetep ii*, pl. 14
Label: n/a
Date: Harpur, *Decoration*, [#338], 272: V.8L-9E

GB86 *snḏm-jb: mḥj* - Giza, WF (G2378=LG26), forecourt, west: fording
Source: *P-M III*, 87-89; *LD Erganz.*, pl. 12 [lower]; Brovarski, *Senedjemib Complex Part I*, fig. 97
Label: [*ḫsf*] *mt*
Warding off death
ḫsf msḥ(*wt*)
Warding off crocodiles
Date: Brovarski, *Senedjemib Complex Part I*, 29-30: V.9; Harpur, *Decoration*, [#220], 269: V.9

GB87 *nj-mꜣꜥt-rꜥ* - Giza, CF (ntn), chapel, north: fording
Source: *P-M III*, 282-284; Hassan, *Giza ii*, fig. 236
Label: none
Date: Harpur, *Decoration*, [#110], 267: V.9

GB88 *ḫnwt* - Saqqara, UPC (ntn), room A, south: pastoral
Source: *P-M III*, 623-624; Munro, *Unas-Friedhof*, pl. 36 [b]
Label: *r*[*n*] *jwꜣ*
Young oxen
Date: Munro, *Unas-Friedhof*, 20: V.9; Harpur, *Decoration*, [#478], 275: V.9

GB89 *ḫnwt* - Saqqara, UPC (ntn), relief fragments from corridor and court areas (Kh/15 + Kh/14), unknown: agricultural
Source: *P-M III*, 623-624; Munro, *Unas-Friedhof*, pl. 42
Label: *ḫrṯ*(*j*) *bꜣkt*(*j*)
Go on you two workers![31]
Date: Munro, *Unas-Friedhof*, 20: V.9; Harpur, *Decoration*, [#478], 275: V.9

GB90 *ptḥ-ḥtp II: ṯfj* - Saqqara, WSP (D64), offering-room, east: fording
Source: *P-M III*, 600-604; Quibell, *Ptah-hetep*, pl. 33
Label: *j jḥs jw jb=k r šjj* [*pw*]
Oh filthy one! May your heart (prosper) against that aquatic!
Date: Harpur, *Decoration*, [#400], 274: V.9M-L

GB91 *nj-mꜣꜥt-rꜥ* [*jzzj-mr-nṯr*] - Giza, WF (G2097), chapel, north: fording
Source: *P-M III*, 70; Roth, *Cem. Pal. Att.*, pl. 185
Label: . . . *mḥt pḥw*
. . . the Delta and the marshes . . .
ꜥnḫ ḥ=k wrt r š[*jj pw*] *ntj ḫr mw*
May you be watchful greatly against that aquatic which is in the water
Date: Roth, *Cem. Pal. Att.*, 129-130: V.9; Harpur, *Decoration*, [#033], 265: V.9-VI.1

GB92 *kꜣ-gm-nj: mmj* - Saqqara, TP (LS10), room III, north: fording
Source: *P-M III*, 521-525; Firth-Gunn, *Teti Pyr. ii*, pl. 52; Harpur-Scremin, *Kagemni*, fig. 7
Label: none
Date: Harpur, *Decoration*, [#534], 276: VI.1M

GB93 *kꜣ-gm-nj: mmj* - Saqqara, TP (LS10), room III, north: fording
Source: *P-M III*, 521-525; Firth-Gunn, *Teti Pyr. ii*, pl. 52; Harpur-Scremin, *Kagemni*, fig. 7
Label: *j mnjw pw dj jḥy n kꜣw nw sn mk wḏb*(=*sn*) *r=f*

31 According to Berman (*Catalogue*, 137), "the exact translation of this phrase is unknown".

Oh this herdsman! Show Ihy of the cattle to them
Behold! They turn against him![32]
jḥy n k3w
Ihy of the cattle
Date: Harpur, *Decoration*, [#534], 276: VI.1M

GB94 *k3-gm-nj: mmj* - Saqqara, TP (LS10), room VII, south: offering (lower register)
Source: *P-M III*, 521-525; von Bissing, *Gem-ni-kai ii*, pl. 22; Harpur-Scremin, *Kagemni*, fig. 31
Label: none
Date: Harpur, *Decoration*, [#534], 276: VI.1M

GB95 *mrrw-k3.j: mrj* - Saqqara, TP (ntn), chamber A1, south: fording
Source: *P-M III*, 525-534; Duell, *Mereruka i*, pl. 21
Label: *mnjw pw ꜥnḫ ḥr=k r šjj pw ntj ḥr mw jw=f m šp tp jrj=ṯw r=f wrt*
This herdsman! May you be watchful against that aquatic which is in the water!
May he come as blind-of-head
Act against him, greatly!
ḫnj=t(w)? wnj ntj ḥnꜥ
Row quickly, my comrade!
nr pw ꜥ=k ḥr mw
This herdsman! Your arm above the water!
Date: Harpur, *Decoration*, [#420], 274: VI.1M-L

GB96 *nj-ꜥnḫ-nswt* - Saqqara (ntn), relief fragment (Detroit Inst. Arts no. 30.371), unknown: fording
Source: *P-M III*, 694-696; *Eg. Art in the Age of the Pyramids*, cat. no. 193
Label: missing
Date: Berman, *Catalogue*, 135: VI.1; Harpur, *Decoration*, [#429], 274: VI.1-2E

GB97 *zšzšt: jdwt* - Saqqara, UPC (ntn), room III, west: fording
Source: *P-M III*, 617-619; Macramallah, *Idout*, pl. 7; Kanawati-Abder-Raziq, *Iynefert*, pl. 54
Label: *mnjw p[w] ꜥnḫ ḥr=k r šjj pw ntj ḥr mw jw=f m šp tp jr=ṯw r=f wrt*
This herdsman! May you be watchful against that aquatic which is in the water!
May he come as blind-of-head
Act against him greatly!
j nr pw ꜥ=k ḥr mw
Oh this herdsman! Your arm above the water!
Date: Kanawati-Abder-Raziq, *Iynefert*, 36-37: VI.1E; Harpur, *Decoration*, [#487], 275: V.9 r/u, VI 1-2E?

GB98 *j3zn* - Giza, WF (G2196), chapel, north: agricultural
Source: *P-M III*, 82; Simpson, *Mast. West. Cem. Part I*, fig. 30
Label: none
Date: Harpur, *Decoration*, [#013], 265: VI.1M-2M

GB99 *ḥzj* - Saqqara, TP (ntn), portico, east: fording
Source: *P-M III*, nr; Kanawati-Abder-Raziq, *Hesi*, pl. 55
Label: *j mnjw pw ꜥnḫ ḥr=k r šjj pw ntj ḥr mw jw(=f) m šp tp tm jwjj n šjj pw*
Oh this herdsman! May you be watchful against that aquatic which is in the water!
May he come as blind-of-head, and may (the cattle) not come to that aquatic
ꜥ=k ḥr mw mnjw pw jw mzḥ pw gr(w) ḫr s . . . t nt wḏb wḏ3 jb=k r=f
Your arm above the water, this herdsman!
This crocodile is still under the . . . of the riverbank
May your heart prosper against it
jn r=k
Bring away
mk ḫr . . . tp n ḫt r . . .
Behold! . . . the end of the pole . . .
Date: Kanawati-Abder-Raziq, *Hesi*, 16: VI.1L-2E; Harpur, *Decoration*, nr

GB100 *ḥzj* - Saqqara, TP (ntn), portico, south: agricultural
Source: *P-M III*, nr; Kanawati-Abder-Raziq, *Hesi*, pl. 52

32 Firth and Gunn (*Teti Pyr. ii*, 114) translate this inscription as 'Show Ihy to the cattle that I may keep them together'.

Label: missing
Date: Kanawati-Abder-Raziq, *Hesi*, 16: VI.1L-2E; Harpur, *Decoration*, nr

GB101 *ꜥnḫ-m-ꜥ-ḥr: zzj* - Saqqara, TP (ntn), room I, east: fording
Source: *P-M III*, 512-515; Kanawati-Hassan, *Ankhmahor*, pl. 37 [a]
Label: *ẖsf msḥ*
Repelling the crocodile
j mnjw pw ꜥnḫ ḥr=k r šjj pw ntj ḥr mw tm jw nw n šjj pw jw=f m šp tp ꜥnḫ ḥr=k r=f wrt
Oh this herdsman! May you be watchful against that aquatic which is in the water!
May these not come to that aquatic and may he be as a blind-of-head
May you be watchful against him greatly!
Date: Kanawati-Hassan, *Ankhmahor*, 18: VI.1M-L; Harpur, *Decoration*, [#374], 273: VI.1L-2E

GB102 *mḥw* - Saqqara, UPC (ntn), door thickness (room I to II), north: pastoral
Source: *P-M III*, 619-622; Altenmüller, *Mehu*, pl. 17 [a]
Label: *pr(jt) mnꜥwt mḥt*
The coming out of the milch cows from the marshland
mk jr=k nbw(=j) ḥr sẖr jrṯt
Behold my lord, milking milk!
Date: Altenmüller, *Mehu*, 19: VI.2; Harpur, *Decoration*, [#424], 274: VI.2M-3?

GB103 *mḥw* - Saqqara, UPC (ntn), room II, north: fording
Source: *P-M III*, 619-622; Altenmüller, *Mehu*, pl. 40
Label: *mnjw pw ꜥnḫ ḥr=k r šjj pw ntj ḥr mw jw(=f) m šp tp jrj ṯw jr=f wrt mnjw pw ꜥ=k ḥr mw*
This herdsman! May you be watchful against that aquatic which is in the water!
May he come as blind-of-head
Pay attention to him very greatly this herdsman
Your arm above the water!
Date: Altenmüller, *Mehu*, 19: VI.2; Harpur, *Decoration*, [#424], 274: VI.2M-3?

GB104 *mḥw* - Saqqara, UPC (ntn), offering-room, south: offering
Source: *P-M III*, 619-622; Altenmüller, *Mehu*, pl. 62
Label: none
Date: Altenmüller, *Mehu*, 19: VI.2; Harpur, *Decoration*, [#424], 274: VI.2M-3?

6.1 GROOMING

6.1.1 Licking

Cattle

CB1 *rʿ-špss* - Saqqara, NSP (LS16=S902), hall, west: offering
Source: *P-M III*, 494-496; *LD Erganz.*, pl. 39 [a]
Label: *rn jwꜣ*
Young ox
Date: Harpur, *Decoration*, [#456], 275: V.8M

CB2 *snḏm-jb: jntj* - Giza, WF (G2370=LG27), room III, east: offering
Source: *P-M III*, 85-87; *LD Erganz.*, pl. 23 [c]; Brovarski, *Senedjemib Complex Part I*, fig. 51
Label: *rn jwꜣ*
Young ox
Date: Brovarski, *Senedjemib Complex Part I*, 23: V.8M-L; Harpur, *Decoration*, [#219], 269: V.8M-L

CB2a *ṯjj* - Saqqara, NSP (D22), room III, south: offering
Source: *P-M III*, 468-478; Wild, *Ti iii*, pl. 167
Label: *jwꜣ*
Ox
Date: Harpur, *Decoration*, [#543], 277: V.8-9

CB3 *nbt* - Saqqara, UPC (ntn), room C, north: offering
Source: *P-M III*, 624-625; Munro, *Unas-Friedhof*, pl. 19
Label: none
Date: Munro, *Unas-Friedhof*, 20: V.9; Harpur, *Decoration*, [#438], 274: V.9

CB4 *kꜣ-gm-nj: mmj* - Saqqara, TP (LS10), room VII, east: offering (top register)
Source: *P-M III*, 521-525; von Bissing, *Gem-ni-kai ii*, pl. 25; Harpur-Scremin, *Kagemni*, fig. 29
Label: none
Date: Harpur, *Decoration*, [#534], 276: VI.1M

CB5 *kꜣ-gm-nj: mmj* - Saqqara, TP (LS10), room VII, north: offering
Source: *P-M III*, 521-525; von Bissing, *Gem-ni-kai ii*, Suppl. I; Harpur-Scremin, *Kagemni*, fig. 34 (upper register, right)
Label: none
Date: Harpur, *Decoration*, [#534], 276: VI.1M

CB6 *wḏꜣ-ḥꜣ-ttj: ššj [nfr-sšm-ptḥ]* - Saqqara, TP (ntn), door thickness (room I to II), east: offering
Source: *P-M III*, 515-516; Capart, *Rue de Tombeaux*, pl. 81
Label: *rn jwꜣ [n] mḏt*
Young ox of the stall
Date: Harpur, *Decoration*, [#384], 273: VI.1L-2E

CB7 *wḏꜣ-ḥꜣ-ttj: ššj [nfr-sšm-ptḥ]* - Saqqara, TP (ntn), door thickness (room I to II), west: offering
Source: *P-M III*, 515-516; Capart, *Rue de Tombeaux*, pl. 80
Label: *rn jwꜣ mḏt n kꜣ ššj rʿ nb*
Young oxen of the stall for the ka of Sheshi every day
Date: Harpur, *Decoration*, [#384], 273: VI.1L-2E

CB8 *mṯtj* - Saqqara, UPC (ntn), relief fragment (Louvre no. E. 25 515), unknown: pastoral
Source: *P-M III*, 646-648; Ziegler, *Catalogue*, 143
Label: *sšr jrṯt*
Drying up milk[33]
Date: Ziegler, *Catalogue*, 123: probably VI; Harpur, *Decoration*, [#426], 274: VI.2?

CB9 *ḫntj-kꜣ.j: jḫḫj* - Saqqara, TP (ntn), door thickness (room VI to VII), west: offering
Source: *P-M III*, 508-511; James, *Khentika*, pl. 17
Label: none

[33] For a discussion of the term, *sšr jrṯt*, see section 11.2.2 in Chapter 11.

Date: Harpur, *Decoration*, [#479], 275: VI.2M

CB10 *ḫntj-k3.j: jḫḫj* - Saqqara, TP (ntn), room VII, south: offering
Source: *P-M III*, 508-511; James, *Khentika*, pl. 20 [middle]
Label: none
Date: Harpur, *Decoration*, [#479], 275: VI.2M

CB11 *ḫntj-k3.j: jḫḫj* - Saqqara, TP (ntn), room VII, north: offering
Source: *P-M III*, 508-511; James, *Khentika*, pl. 21
Label: none
Date: Harpur, *Decoration*, [#479], 275: VI.2M

Dorcas gazelle

CB12 *tjj* - Saqqara, NSP (D22), room III, south: offering
Source: *P-M III*, 468-478; Wild, *Ti iii*, pl. 165
Label: *jdt gḥs*
Female gazelle
Date: Harpur, *Decoration*, [#543], 277: V.8-9

6.1.2 Scratching

Cattle

CB13 *nṯr-wsr* - Saqqara, NSP (D1=S901), offering-room, east: offering
Source: *P-M III*, 485; Murray, *Saqqara Mastabas*, pl. 22
Label: *sḫpt rn jw3 nt ḫt-ḫ3wy*
Bringing a young ox of the night offerings
Date: Harpur, *Decoration*, [#451], 275: V.6L-7

CB14 *rʿ-špss* - Saqqara, NSP (LS16=S902), hall, west: offering
Source: *P-M III*, 494-496; *LD Erganz.*, pl. 39 [a]
Label: *jw3*
Ox
Date: Harpur, *Decoration*, [#456], 275: V.8M

CB15 *jj-nfrt: š3.n.f* - Saqqara, UPC (ntn), door thickness (room II to III), east: offering
Source: *P-M III*, 616; Kanawati-Abder-Raziq, *Iynefert*, pl. 44
Label: *rn jw3*
Young ox
Date: Kanawati-Abder-Raziq, *Iynefert*, 13: V.9E-M; Harpur, *Decoration*, [#345], 272: V.9

6.1.3 Rubbing

Cattle

CB16 *rʿ-ḫʿ.f-ʿnḫ* - Giza, EF (G7948=LG75), chapel, east: offering
Source: *P-M III*, 207-208; *LD ii*, pl. 9
Label: numerals 220
Date: Harpur, *Decoration*, [#156], 268: V.6?

6.2 PREENING

6.2.1 Preening movements

Birds

CB17 *nfr/k3-ḥ3.j* - Saqqara, UPC (ntn), chapel, east: fowling
Source: *P-M III*, 639-641; Moussa-Altenmüller, *Nefer and Ka-Hay*, pls. 5 and 13
Label: none

Date: Harpur, *Decoration*, [#439], 274: V.6

CB18 *ṯjj* - Saqqara, NSP (D22), portico, south: pastoral/force-feeding
Source: *P-M III*, 468-478; Epron et al., *Ti i*, pl. 7
Label: *pẖr ṯrp ḥḏ m-ḫt snmt*
Moving about (of) *ṯrp*-geese and *ḥḏ*-geese after feeding
Date: Harpur, *Decoration*, [#543], 277: V.8-9

CB19 *ṯjj* - Saqqara, NSP (D22), portico, south: pastoral/force-feeding
Source: *P-M III*, 468-478; Epron et al., *Ti i*, pls. 7-8
Label: *pẖr s(t) m-ḫt snmt*
Moving about (of) pintail ducks after feeding
Date: Harpur, *Decoration*, [#543], 277: V.8-9

CB20 *ṯjj* - Saqqara, NSP (D22), portico, south: pastoral/force-feeding
Source: *P-M III*, 468-478; Epron et al., *Ti i*, pl. 8
Label: *pẖr s mnwt m-ḫt snmt*
Moving about (of) *s*-geese and pigeons after feeding
Date: Harpur, *Decoration*, [#543], 277: V.8-9

CB21 *ṯjj* - Saqqara, NSP (D22), room I/pillared hall, west: pastoral/force-feeding
Source: *P-M III*, 468-478; Epron et al., *Ti i*, pl. 33
Label: *m33 snmt ḫt-ˁ3*
Watching the feeding (of) poultry
Date: Harpur, *Decoration*, [#543], 277: V.8-9

CB22 *3ḫtj-ḥtp* - Saqqara, WSP (D64), hall, east: marsh
Source: *P-M III*, 599-600; Davies, *Ptahhetep ii*, pl. 14
Label: none
Date: Harpur, *Decoration*, [#338], 272: V.8L-9E

CB23 *k3-gm-nj: mmj* - Saqqara, TP (LS10), room IV, west: pastoral/force-feeding (left)
Source: *P-M III*, 521-525; von Bissing, *Gem-ni-kai i*, pl. 11; Harpur-Scremin, *Kagemni*, fig. 13
Label: *pẖr ṯrp ḥḏ m-ḫt snm*
Moving about (of) *ṯrp*-geese and *ḥḏ*-geese after feeding
Date: Harpur, *Decoration*, [#534], 276: VI.1M

CB24 *k3-gm-nj: mmj* - Saqqara, TP (LS10), room IV, west: pastoral/force-feeding (right)
Source: *P-M III*, 521-525; von Bissing, *Gem-ni-kai i*, pl. 12; Harpur-Scremin, *Kagemni*, fig. 13
Label: *pẖr r m-ḫt snm*
Moving about (of) *r*-geese after feeding
Date: Harpur, *Decoration*, [#534], 276: VI.1M

CB25 *k3-gm-nj: mmj* - Saqqara, TP (LS10), room IV, west: pastoral/poultry-yard (above pool)
Source: *P-M III*, 521-525; von Bissing, *Gem-ni-kai i*, pl. 10; Harpur-Scremin, *Kagemni*, fig. 13
Label: *ṯrp*
ṯrp-goose
Date: Harpur, *Decoration*, [#534], 276: VI.1M

CB26 *k3-gm-nj: mmj* - Saqqara, TP (LS10), room IV, west: pastoral/poultry-yard
Source: *P-M III*, 521-525; von Bissing, *Gem-ni-kai i*, pls. 8 and 9; Harpur-Scremin, *Kagemni*, fig. 14
Label: *sṯjt n 3pdw*
Sowing for birds
Date: Harpur, *Decoration*, [#534], 276: VI.1M

CB27 *nj-k3w-jzzj* - Saqqara, TP (ntn), room I, north: pastoral
Source: *P-M III*, nr; Kanawati-Abder-Raziq, *Nikauisesi*, pl. 49
Label: *ṯtf jt n ḏ3t ẖr=sn*
Throwing down grain for cranes, beneath them
Date: Kanawati-Abder-Raziq, *Nikauisesi*, 17-23: VI.1M; Harpur, *Decoration*, nr

CB28 *mrrw-k3.j: mrj* - Saqqara, TP (ntn), chamber A6, west: pastoral/poultry-yard
Source: *P-M III*, 525-534; Duell, *Mereruka i*, pl. 53 [b]

Label: none
Date: Harpur, *Decoration*, [#420], 274: VI.1M-L

CB29 *mrrw-kꜣ.j: mrj* - Saqqara, TP (ntn), chamber A6, west: pastoral/force-feeding
Source: *P-M III*, 525-534; Duell, *Mereruka i*, pl. 52
Label: none
Date: Harpur, *Decoration*, [#420], 274: VI.1M-L

CB30 *mrrw-kꜣ.j: mrj* - Saqqara, TP (ntn), chamber A6, west: pastoral/force-feeding (cranes)
Source: *P-M III*, 525-534; Duell, *Mereruka i*, pl. 52
Label: none
Date: Harpur, *Decoration*, [#420], 274: VI.1M-L

CB31 *mrrw-kꜣ.j: mrj* - Saqqara, TP (ntn), chamber A6, west: pastoral/force-feeding (cranes)
Source: *P-M III*, 525-534; Duell, *Mereruka i*, pl. 52
Label: *snm ḏˁt*
Feeding cranes
Date: Harpur, *Decoration*, [#420], 274: VI.1M-L

CB32 *mrrw-kꜣ.j: mrj* - Saqqara, TP (ntn), chamber A6, west: pastoral/force-feeding (cranes)
Source: *P-M III*, 525-534; Duell, *Mereruka i*, pl. 52
Label: *snm ḏꜣt*
Feeding cranes
wšꜣ ḏꜣt
Fattening cranes
Date: Harpur, *Decoration*, [#420], 274: VI.1M-L

CB33 *mrrw-kꜣ.j: mrj* - Saqqara, TP (ntn), chamber A8, south: offering
Source: *P-M III*, 525-534; Duell, *Mereruka i*, pl. 57
Label: none
Date: Harpur, *Decoration*, [#420], 274: VI.1M-L

CB34 *ḥzj* - Saqqara, TP (ntn), portico, west: pastoral/poultry-yard
Source: *P-M III*, nr; Kanawati-Abder-Raziq, *Hesi*, pl. 56
Label: none
Date: Kanawati-Abder-Raziq, *Hesi*, 16: VI.1L-2E; Harpur, *Decoration*, nr

CB35 *mrrj* - Saqqara, TP (ntn), room 1, west: pastoral/poultry-yard
Source: *P-M III*, 518-519; Davies et al., *Saqqara Tombs i*, pl. 6
Label: none
Date: Harpur, *Decoration*, [#418], 274: VI.2E?

CB36 *mrjj-ttj: mrj* - Saqqara, TP (ntn), room C1, east: pastoral/poultry-yard
Source: *P-M III*, 356-357; Kanawati-Abder-Raziq, *Meryteti*, pl. 48
Label: none
Date: Kanawati-Abder-Raziq, *Meryteti*, 18: end VI.1-VI.2M; Harpur, *Decoration*, [#415], 274: VI.2M-L

CB37 *mḥw* - Saqqara, UPC (ntn), room I, south: pastoral
Source: *P-M III*, 619-622; Altenmüller, *Mehu*, pl. 6
Label: none
Date: Altenmüller, *Mehu*, 19: VI.2; Harpur, *Decoration*, [#424], 274: VI.2M-3?

6.2.2 Wing-flapping

Birds

CB38 *spd-ḥtp* - Saqqara, NSP (D15), chapel entrance (Äg. Mus. Berlin no. 14642), east: pastoral/force-feeding
Source: *P-M III*, 481; Klebs, *Reliefs*, fig. 53; Janssen-Janssen, *Egyptian Household Animals*, fig. 40
Label: *smn*
Goose
r
r-goose

Date: Harpur, *Decoration*, [#494], 276: V.6L-9?

CB39 *ṯjj* - Saqqara, NSP (D22), portico, south: pastoral/force-feeding
Source: *P-M III*, 468-478; Epron et al., *Ti i*, pls. 7-8
Label: *pẖr s(t) m-ḫt snmt*
Moving about (of) pintail ducks after feeding
Date: Harpur, *Decoration*, [#543], 277: V.8-9

CB40 *ṯjj* - Saqqara, NSP (D22), portico, south: pastoral/force-feeding
Source: *P-M III*, 468-478; Epron et al., *Ti i*, pl. 8
Label: *pẖr s mnwt m-ḫt snmt*
Moving about (of) *s*-geese and pigeons after feeding
Date: Harpur, *Decoration*, [#543], 277: V.8-9

CB41 *ṯjj* - Saqqara, NSP (D22), room I/pillared hall, west: pastoral/poultry-yard
Source: *P-M III*, 468-478; Epron et al., *Ti i*, pl. 38
Label: none
Date: Harpur, *Decoration*, [#543], 277: V.8-9

CB42 *ṯjj* - Saqqara, NSP (D22), room I/pillared hall, west: pastoral/force-feeding
Source: *P-M III*, 468-478; Epron et al., *Ti i*, pl. 38
Label: *m33 snmt ḫt-ꜥ3*
Watching the feeding (of) poultry
Date: Harpur, *Decoration*, [#543], 277: V.8-9

CB43 *k3-gm-nj: mmj* - Saqqara, TP (LS10), room IV, west: pastoral/force-feeding
Source: *P-M III*, 521-525; von Bissing, *Gem-ni-kai i*, pl. 11; Harpur-Scremin, *Kagemni*, fig. 13 (left)
Label: *pẖr ṯrp ḥd m-ḫt snm*
Turning around (of) *ṯrp*-geese and *ḥḏ*-geese after feeding
Date: Harpur, *Decoration*, [#534], 276: VI.1M

CB44 *k3-gm-nj: mmj* - Saqqara, TP (LS10), room IV, west: pastoral
Source: *P-M III*, 521-525; von Bissing, *Gem-ni-kai i*, pl. 12; Harpur-Scremin, *Kagemni*, fig. 13 (right)
Label: n/a
Date: Harpur, *Decoration*, [#534], 276: VI.1M

CB45 *k3-gm-nj: mmj* - Saqqara, TP (LS10), room IV, west: pastoral/poultry-yard
Source: *P-M III*, 521-525; von Bissing, *Gem-ni-kai i*, pl. 9; Harpur-Scremin, *Kagemni*, fig. 14
Label: *sṯjt n 3pdw*
Sowing for birds
Date: Harpur, *Decoration*, [#534], 276: VI.1M

CB46 *ḥzj* - Saqqara, TP (ntn), portico, west: pastoral/poultry-yard
Source: *P-M III*, nr; Kanawati-Abder-Raziq, *Hesi*, pl. 56
Label: none
Date: Kanawati-Abder-Raziq, *Hesi*, 16: VI.1L-2E; Harpur, *Decoration*, nr

CB47 *wḏ3-ḥ3-ttj: ššj [nfr-sšm-ptḥ]* - Saqqara, TP (ntn), room II, east: pastoral/poultry-yard
Source: *P-M III*, 515-516; Capart, *Rue de Tombeaux*, pl. 88
Label: none
Date: Harpur, *Decoration*, [#384], 273: VI.1L-2E

CB48 *mṯtj* - Saqqara, UPC (ntn), relief fragment (Louvre no. E. 25519), unknown: pastoral
Source: *P-M III*, 646-648; Ziegler, *Catalogue*, 145
Label: missing
Date: Ziegler, *Catalogue*, 123: probably VI; Harpur, *Decoration*, [#426], 274: VI.2?

CB49 *mrrj* - Saqqara, TP (ntn), room 1, west: pastoral/force-feeding
Source: *P-M III*, 518-519; Davies et al., *Saqqara Tombs i*, pl. 6
Label: none
Date: Harpur, *Decoration*, [#418], 274: VI.2E?

CB50 *mrrj* - Saqqara, TP (ntn), room 1, west: pastoral/poultry-yard
Source: *P-M III*, 518-519; Davies et al., *Saqqara Tombs i*, pl. 6
Label: none
Date: Harpur, *Decoration*, [#418], 274: VI.2E?

CB51 *mrjj-ttj: mrj* - Saqqara, TP (ntn), room C1, east: pastoral/poultry-yard
Source: *P-M III*, 356-357; Kanawati-Abder-Raziq, *Meryteti*, pl. 48
Label: none
Date: Kanawati-Abder-Raziq, *Meryteti*, 18: end VI.1-VI.2M; Harpur, *Decoration*, [#415], 274: VI.2M-L

6.3 ELIMINATION

Cattle

CB52 *ptḥ-ḥtp II: ṯfj* - Saqqara, WSP (D64), offering-room, east: hunting
Source: *P-M III*, 600-604; Quibell, *Ptah-hetep*, pl. 32
Label: none
Date: Harpur, *Decoration*, [#400], 274: V.9M-L

CB53 *ḥtp-kꜣ.j: tp-kꜣ.j* - Saqqara, NSP (S3509), door thickness, east: slaughter
Source: *P-M III*, 447-448; Martin, *Hetepka*, pl. 8 [3]
Label: *stp*
Cutting up
jṯj r=k
Pull!
Date: Harpur, *Decoration*, [#469], 275: V.9-VI.1

CB54 *ḥtp-kꜣ.j: tp-kꜣ.j* - Saqqara, NSP (S3509), chapel, south: slaughter
Source: *P-M III*, 447-448; Martin, *Hetepka*, pl. 13 [14]
Label: [*d*]*m* [*ds*]
Sharpening the knife
jṯj=k
Pull!
Date: Harpur, *Decoration*, [#469], 275: V.9-VI.1

CB55 *ḥtp-kꜣ.j: tp-kꜣ.j* - Saqqara, NSP (S3509), chapel, north: slaughter
Source: *P-M III*, 447-448; Martin, *Hetepka*, pl. 15 [16]
Label: *stp*
Cutting up
Date: Harpur, *Decoration*, [#469], 275: V.9-VI.1

CB56 *ḥtp-kꜣ.j: tp-kꜣ.j* - Saqqara, NSP (S3509), corridor, west: slaughter
Source: *P-M III*, 447-448; Martin, *Hetepka*, pl. 16 [18]
Label: n/a
Date: Harpur, *Decoration*, [#469], 275: V.9-VI.1

CB57 *nj-mꜣꜥt-rꜥ* [*jzzj-mr-nṯr*] - Giza, WF (G2097), chapel, east: hunting
Source: *P-M III*, 70; Roth, *Cem. Pal. Att.*, pl. 189
Label: none
Date: Roth, *Cem. Pal. Att.*, 129-130: V.9; Harpur, *Decoration*, [#033], 265: V.9-VI.1

CB58 *mrrw-kꜣ.j: mrj* - Saqqara, TP (ntn), chamber A2, west: hunting
Source: *P-M III*, 525-534; Duell, *Mereruka i*, pl. 25
Label: none
Date: Harpur, *Decoration*, [#420], 274: VI.1M-L

Dorcas gazelle

CB59 *ṯjj* - Saqqara, NSP (D22), room III, east: desert
Source: *P-M III*, 468-478; Wild, *Ti ii*, pl. 127
Label: none
Date: Harpur, *Decoration*, [#543], 277: V.8-9

7.1 FEEDING

Donkeys

IB1 *rꜥ-ḫꜥ.f-ꜥnḫ* - Giza, EF (G7948=LG75), chapel, east: agricultural (individual donkey)
Source: *P-M III*, 207-208; *LD ii*, pl. 9
Label: none
Date: Harpur, *Decoration*, [#156], 268: V.6?

IB2 *rꜥ-ḫꜥ.f-ꜥnḫ* - Giza, EF (G7948=LG75), chapel, east: agricultural (group on threshing-floor)
Source: *P-M III*, 207-208; *LD ii*, pl. 9
Label: none
Date: Harpur, *Decoration*, [#156], 268: V.6?

IB3 *rꜥ-m-kꜣ.j* - Saqqara, NSP (D3=S903), chapel, east: agricultural
Source: *P-M III*, 487-488; Hayes, *Scepter i*, fig. 57
Label: *ḥwj zpt nt jt*
Threshing on the threshing floor of barley
Date: Harpur, *Decoration*, [#453], 275: V.6? r/u - V.6-7

IB4 *nj-ꜥnḫ-ẖnmw/ẖnmw-ḥtp* - Saqqara, UPC (ntn), second court, south: riding donkeys
Source: *P-M III*, 641-644; Moussa-Altenmüller, *Nianchchnum*, pl. 42
Label: n/a
Date: Harpur, *Decoration*, [#431], 274: V.6L-7

IB5 *nj-ꜥnḫ-ẖnmw/ẖnmw-ḥtp* - Saqqara, UPC (ntn), second court, north: riding donkeys
Source: *P-M III*, 641-644; Moussa-Altenmüller, *Nianchchnum*, pl. 43
Label: n/a
Date: Harpur, *Decoration*, [#431], 274: V.6L-7

IB6 *pḥn-wj-kꜣ.j* - Saqqara, NSP (D70=LS15), antechamber, unknown: agricultural
Source: *P-M III*, 491-492; *LD ii*, pl. 47
Label: *jr ḥꜣ=k jm=sn*
Get back among them![34]
ḥꜣ(=k) mꜣꜣ=k jrt=k
Would that you watch what you are doing!
Date: Harpur, *Decoration*, [#393], 273: V.6-8E

IB7 *ḥtp-ḥr-ꜣḫtj* - Saqqara, WSP (D60), chapel, south: agricultural
Source: *P-M III*, 593-595; Mohr, *Hetep-her-akhti*, fig. 51
Label: *h[ꜣ] js r=k*
Oh hurry up!
Date: Harpur, *Decoration*, [#467], 275: V.6-8E

IB8 *ꜣḫtj-ḥtp* - Saqqara, UPC (ntn), chapel, east: agricultural
Source: *P-M III*, 634-637; Ziegler, *Akhethetep*, 134-135
Label: *jr ḥꜣ(=k) jm=sn*
Get back among them!
Date: Ziegler, *Akhethetep*, 9: V.6; Harpur, *Decoration*, [#340], 272: V.6-8E

IB9 *jrj-n-kꜣ-ptḥ* - Saqqara, UPC (ntn), offering chamber, east: agricultural
Source: *P-M III*, 644; Moussa-Junge, *Two Tombs*, pl. 9
Label: none
Date: Harpur, *Decoration*, [#362], 273: V.6-9

IB10 *kꜣ.j-m-rḥw* - Saqqara, NSP (D2=S905), chapel, east: agricultural
Source: *P-M III*, 485-487; Mogensen, *Ny Carlsberg*, fig. 18
Label: missing

[34] This phrase, which occurs frequently in donkey threshing scenes, has been translated in many different ways. I have interpreted it as a direction to one of the animals to move closer to its companions. For other translations see Erman, *Reden*, 26-28; Montet, *Scènes*, 218-219; Junker, *Giza vi*, 148; Vandier, *Manuel vi*, 174-175; Simpson, *Sekhem-ankh-ptah*, 16 and Simpson, *Kayemnofret*, 18.

Date: Harpur, *Decoration*, [#523], 276: V.8

IB11 *snḏm-jb: jntj* - Giza, WF (G2370=LG27), room III, west wall: agricultural
Source: *P-M III*, 85-87; Brovarski, *Senedjemib Complex Part I*, fig. 53
Label: *jr ḥ3(=k) jm=sn*
Get back among them!
ḥ3=k m33=k [jrt=k]
Would that you watch (what you are doing)!
Date: Brovarski, *Senedjemib Complex Part I*, 23: V.8M-L; Harpur, *Decoration*, [#219], 269: V.8M-L

IB12 *ptḥ-ḥtp I* - Saqqara, WSP (D62), offering chamber, east: fowling
Source: *P-M III*, 596-598; Murray, *Saqqara Mastabas*, pl. 11
Label: *jr(=k) mj ḥ3=k jm=sn*
Get back among them!
j3 ḥ3 m33=k jrt=k
Oh! Would that you watch what you are doing!
Date: Harpur, *Decoration*, [#398], 273: V.8M-L

IB13 *sḫm-ʿnḫ-ptḥ* - Saqqara, NSP (D41), chapel, south: agricultural
Source: *P-M III*, 454-455; Simpson, *Sekem-ankh-ptah*, pl. D
Label: *m33=k [jrt=k]*
Watch (what you are doing)!
jr(=k) mj ḥ3=k jm=sn
Get back among them!
Date: Harpur, *Decoration*, [#504], 276: V.8-9

IB14 *tjj* - Saqqara, NSP (D22), room III, east: agricultural
Source: *P-M III*, 468-478; Wild, *Ti iii*, pl. 155
Label: *ḥ3=k m33=k jrt=k*
Would that you watch what you are doing!
jr ḥ3=k jm=sn
Get back among them!
Date: Harpur, *Decoration*, [#543], 277: V.8-9

IB15 *nfr-jrt-n.f* - Saqqara, ESP (D55), chapel, east: agricultural
Source: *P-M III*, 583-584; van de Walle, *Neferirtenef*, pl. 12
Label: *jr ḥ3=k jm=sn*
Get back among them!
Date: Harpur, *Decoration*, [#440], 274: V.8-9

IB16 *k3.j-m-nfrt* - Saqqara, NSP (D23), chapel, east: agricultural
Source: *P-M III*, 467-468; Simpson, *Kayemnofret*, pl. F
Label: *jr ḥ3=k jm=sn*
Get back among them!
Date: Simpson, *Kayemnofret*, 1: V.8-9; Harpur, *Decoration*, [#522], 276: V.8-9?

IB17 *ḫww-wr* - Giza, CF (LG95), chapel, east: agricultural
Source: *P-M III*, 254-255; Hassan, *Giza v*, fig. 104
Label: missing
Date: Harpur, *Decoration*, [#180], 268: V.8-9M

IB18 *ḫww-wr* - Giza, CF (LG95), chapel, east: riding donkeys
Source: *P-M III*, 254-255; Hassan, *Giza v*, fig. 104
Label: n/a
Date: Harpur, *Decoration*, [#180], 268: V.8-9M

IB19 *snḏm-jb: mḥj* - Giza, WF (G2378= LG26), room II, west: agricultural
Source: *P-M III*, 87-89; *LD ii*, pl. 73; Brovarski, *Senedjemib Complex Part I*, fig. 113
Label: missing
Date: Brovarski, *Senedjemib Complex Part I*, 29-30: V.9; Harpur, *Decoration*, [#220], 269: V.9

IB20 *nj-k3w-ḥr* - Saqqara, NSP (S915), offering-room, east: agricultural
Source: *P-M III*, 498; Quibell, *Saqqara (1907-1908)*, pl. 62 [2]

Label: missing
Date: Harpur, *Decoration*, [#435], 274: V.9

IB21 *ȝḫtj-mrw-nswt* - Giza, WF (G2184), secondary offering-room, west (north end): agricultural
Source: *P-M III*, 80-81; *Giza Archives Project* (Photo no. A779_NS)
Label: missing
Date: Harpur, *Decoration*, [#008], 265: V.9-VI.1

IB22 *ḥtp-kȝ.j: tp-kȝ.j* - Saqqara, NSP (S3509), corridor, east: agricultural
Source: *P-M III*, 447-448; Martin, *Hetepka*, pl. 10 [7]
Label: *[jr] ḥȝ=k jm[=sn]*
Get back among them!
Date: Harpur, *Decoration*, [#469], 275: V.9-VI.1

IB23 *sḫm-kȝ.j* - Giza, WF (G1029), chapel, east: agricultural
Source: *P-M III*, 53; Simpson, *Mast. West. Cem. Part I*, fig. 4
Label: *sḥȝt*
Donkey pack
Date: Harpur, *Decoration*, [#225], 269: V.9-VI.1?

IB24 *mrrw-kȝ.j: mrj* - Saqqara, TP (ntn), chamber A13, east: agricultural
Source: *P-M III*, 525-534; Duell, *Mereruka ii*, pl. 169
Label: *jr ḥȝ=k jm=sn*
Get back among them!
Date: Harpur, *Decoration*, [#420], 274: VI.1M-L

IB25 *ʿnḫ-m-ʿ-ḥr: zzj* - Saqqara, TP (ntn), room I, east: agricultural
Source: *P-M III*, 512-515; Kanawati-Hassan, *Ankhmahor*, pl. 37 [a]
Label: *hȝ jm=sn ntj ḥnʿ(=j)*
Get down with them comrade!
swn(=j) ṯw pḫr jm
I will persuade you to turn therein!
Date: Kanawati-Hassan, *Ankhmahor*, 18: VI.1M-L; Harpur, *Decoration*, [#374], 273: VI.1L-2E

IB26 *kȝ-ḥj.f* - Giza, WF (G2136), chapel, east: agricultural
Source: *P-M III*, 76; Junker, *Giza vi*, fig. 46
Label: *jr ḥȝ=k jm=sn*
Get back among them!
Date: Harpur, *Decoration*, [#278], 271: VI.2M-3?

Cattle

IB27 *nfr-msḏr-ḫwfw* - Giza, WF (G2240), chapel, north: agricultural
Source: *P-M III*, nr; Roth, *Cem. Pal. Att.*, pl. 204
Label: none
Date: Roth, *Cem. Pal. Att.*, 163: V.2-8 or V.9; Harpur, *Decoration*, nr

IB28 *nj-ʿnḫ-ẖnmw/ẖnmw-ḥtp* - Saqqara, UPC (ntn), outer hall, east: agricultural
Source: *P-M III*, 641-644; Moussa-Altenmüller, *Nianchchnum*, pls. 56 and 58
Label: *ḥrṯ(j) bȝkt(j)*
Go on you two cows!
Date: Harpur, *Decoration*, [#431], 274: V.6L-7

IB29 *nj-ʿnḫ-ẖnmw/ẖnmw-ḥtp* - Saqqara, UPC (ntn), outer hall, east: agricultural
Source: *P-M III*, 641-644; Moussa-Altenmüller, *Nianchchnum*, pls. 57 and 59
Label: *ḥmj(t) m zpt*
Threshing on the threshing-floor
Date: Harpur, *Decoration*, [#431], 274: V.6L-7

IB30 *nj-ʿnḫ-ẖnmw/ẖnmw-ḥtp* - Saqqara, UPC (ntn), outer hall, west: pastoral
Source: *P-M III*, 641-644; Moussa-Altenmüller, *Nianchchnum*, pl. 77
Label: none
Date: Harpur, *Decoration*, [#431], 274: V.6L-7

IB31 *nj-ꜥnḫ-ẖnmw/ẖnmw-ḥtp* - Saqqara, UPC (ntn), outer hall, west: fording
Source: *P-M III*, 641-644; Moussa-Altenmüller, *Nianchchnum*, pl. 76
Label: *prt m mḥt r ḥrj-tp jn nrw ḥnꜥ kꜣ(w)t=sn kꜣw wrw*
Coming from the marshes to the high land by the herdsmen with their produce and great cattle
Date: Harpur, *Decoration*, [#431], 274: V.6L-7

IB32 *ꜣḫtj-ḥtp* - Saqqara, UPC (ntn), chapel, east: fording
Source: *P-M III*, 634-637; Ziegler, *Akhethetep*, 132-133
Label: *prt m mḥt*
Coming from the marshes
Date: Ziegler, *Akhethetep*, 9: V.6; Harpur, *Decoration*, [#340], 272: V.6-8E

IB33 *jtj-sn* - Giza, CF (ntn), chapel, east: pastoral (individual at a trough)
Source: *P-M III*, 252-253; Hassan, *Giza v*, fig. 123
Label: none
Date: Harpur, *Decoration*, [#035], 265: V.6-8?

IB34 *jtj-sn* - Giza, CF (ntn), chapel, east: pastoral (pair at a trough)
Source: *P-M III*, 252-253; Hassan, *Giza v*, fig. 123
Label: none
Date: Harpur, *Decoration*, [#035], 265: V.6-8?

IB35 *jtj-sn* - Giza, CF (ntn), chapel, east: pastoral (mating pair)
Source: *P-M III*, 252-253; Hassan, *Giza v*, fig. 123
Label: none
Date: Harpur, *Decoration*, [#035], 265: V.6-8?

IB36 *zꜣ-jb* - Giza, WF (G2092 + G2093), chapel, east: marsh
Source: *P-M III*, nr; Roth, *Cem. Pal. Att.*, pl. 179
Label: none
Date: Roth, *Cem. Pal. Att.*, 109: V.8; Harpur, *Decoration*, nr

IB37 *ṯjj* - Saqqara, NSP (D22), room III, north: pastoral
Source: *P-M III*, 468-478; Wild, *Ti ii*, pl. 124
Label: *jwꜣ*
Ox
Date: Harpur, *Decoration*, [#543], 277: V.8-9

IB38 *ṯjj* - Saqqara, NSP (D22), room III, east: agricultural
Source: *P-M III*, 468-478; Wild, *Ti iii*, pl. 155
Label: n/a
Date: Harpur, *Decoration*, [#543], 277: V.8-9

IB39 *kꜣ.j-m-nfrt* - Saqqara, NSP (D23), chapel, east: boating
Source: *P-M III*, 467-468; Simpson, *Kayemnofret*, pl. G
Label: *sṯꜣ* [. . .] *wš jwꜣ*
Hauling . . . and feeding an ox
Date: Simpson, *Kayemnofret*, 1: V.8-9; Harpur, *Decoration*, [#522], 276: V.8-9?

IB40 *kꜣ-gm-nj: mmj* - Saqqara, TP (LS10), room VII, south: offering (left section, upper register)
Source: *P-M III*, 521-525; von Bissing, *Gem-ni-kai ii*, pl. 24; Harpur-Scremin, *Kagemni*, fig. 31
Label: none
Date: Harpur, *Decoration*, [#534], 276: VI.1M

IB41 *kꜣ-gm-nj: mmj* - Saqqara, TP (LS10), room VII, north: offering (right section, upper register)
Source: *P-M III*, 521-525; von Bissing, *Gem-ni-kai ii*, pl. 31; Harpur-Scremin, *Kagemni*, fig. 34
Label: none
Date: Harpur, *Decoration*, [#534], 276: VI.1M

IB42 *kꜣ-gm-nj: mmj* - Saqqara, TP (LS10), room VII, north: offering (right section, bottom register)
Source: *P-M III*, 521-525; von Bissing, *Gem-ni-kai ii*, pl. 33; Harpur-Scremin, *Kagemni*, fig. 34
Label: none
Date: Harpur, *Decoration*, [#534], 276: VI.1M

IB43 *ꜥnḫ-m-ꜥ-ḥr: zzj* - Saqqara, TP (ntn), room III, west: offering
Source: *P-M III*, 512-515; Kanawati-Hassan, *Ankhmahor*, pl. 46
Label: none
Date: Kanawati-Hassan, *Ankhmahor*, 18: VI.1M-L; Harpur, *Decoration*, [#374], 273: VI.1L-2E

IB44 *ꜥnḫ-m-ꜥ-ḥr: zzj* - Saqqara, TP (ntn), door thickness (room III to IV), west: offering
Source: *P-M III*, 512-515; Kanawati-Hassan, *Ankhmahor*, pl. 48 [b]
Label: n/a
Date: Kanawati-Hassan, *Ankhmahor*, 18: VI.1M-L; Harpur, *Decoration*, [#374], 273: VI.1L-2E

IB44a *jnw-mnw* - Saqqara (ntn), room II, south: pastoral
Source: *P-M III*, nr; Kanawati, *Inumin*, pl. 49 [a]
Label: missing
Date: Kanawati, *Inumin*, 16-17: VI.2E; Harpur, *Decoration*, nr

IB45 *ḫntj-kꜣ.j: jḫḫj* - Saqqara, TP (ntn), room VII, north: offering
Source: *P-M III*, 508-511; James, *Khentika*, pl. 21
Label: none
Date: Harpur, *Decoration*, [#479], 275: VI.2M

IB46 *mḥw* - Saqqara, UPC (ntn), offering-room of Ankhmerire, north: offering
Source: *P-M III*, 619-622; Altenmüller, *Mehu*, pl. 89
Label: none
Date: Altenmüller, *Mehu*, 19: VI.2; Harpur, *Decoration*, [#424], 274: VI.2M-3?

IB47 *mḥw* - Saqqara, UPC (ntn), offering-room of Ankhmerire, north: offering
Source: *P-M III*, 619-622; Altenmüller, *Mehu*, pl. 89
Label: none
Date: Altenmüller, *Mehu*, 19: VI.2; Harpur, *Decoration*, [#424], 274: VI.2M-3?

Addax

IB48 *mrrw-kꜣ.j: mrj* - Saqqara, TP (ntn), chamber A13, north: pastoral
Source: *P-M III*, 525-534; Duell, *Mereruka ii*, pl. 153
Label: *rn nw-ḏw*
Young addax
Date: Harpur, *Decoration*, [#420], 274: VI.1M-L

Scimitar oryx

IB49 *ṯjj* - Saqqara, NSP (D22), portico, south: pastoral
Source: *P-M III*, 468-478; Epron et al., *Ti i*, pl. 6
Label: *mꜣ-ḥḏ*
Oryx
Date: Harpur, *Decoration*, [#543], 277: V.8-9

IB50 *ṯjj* - Saqqara, NSP (D22), portico, south: pastoral
Source: *P-M III*, 468-478; Epron et al., *Ti i*, pl. 6
Label: *rn mꜣ-ḥḏ*
Young oryx
Date: Harpur, *Decoration*, [#543], 277: V.8-9

IB51 *ṯjj* - Saqqara, NSP (D22), portico, south: pastoral
Source: *P-M III*, 468-478; Epron et al., *Ti i*, pl. 6
Label: *rn mꜣ-ḥḏ*
Young oryx
Date: Harpur, *Decoration*, [#543], 277: V.8-9

IB52 *ṯjj* - Saqqara, NSP (D22), portico, south: pastoral
Source: *P-M III*, 468-478; Epron et al., *Ti i*, pl. 6
Label: *rn mꜣ-ḥḏ*
Young oryx
Date: Harpur, *Decoration*, [#543], 277: V.8-9

IB53 *mrrw-kȝ.j: mrj* - Saqqara, TP (ntn), chamber A13, north: pastoral
Source: *P-M III*, 525-534; Duell, *Mereruka ii*, pl. 153
Label: *rn mȝ-ḥḏ*
Young oryx
Date: Harpur, *Decoration*, [#420], 274: VI.1M-L

IB54 *ptḥ-špss II* - Saqqara, NSP (E1, 2 + H3), chapel, south: offering
Source: *P-M III*, 460-461; Murray, *Saqqara Mastabas*, pl. 30
Label: none
Date: Harpur, *Decoration*, [#404], 274: VI.3-4

Dorcas gazelle

IB55 *mrrw-kȝ.j: mrj* - Saqqara, TP (ntn), chamber A13, north: pastoral
Source: *P-M III*, 525-534; Duell, *Mereruka ii*, pl. 153
Label: *rn gḥs*
Young gazelle
Date: Harpur, *Decoration*, [#420], 274: VI.1M-L

IB56 *ʿnḫ-m-ʿ-ḥr: zzj* - Saqqara, TP (ntn), room III, west: offering
Source: *P-M III*, 512-515; Kanawati-Hassan, *Ankhmahor*, pl. 46
Label: n/a
Date: Kanawati-Hassan, *Ankhmahor*, 18: VI.1M-L; Harpur, *Decoration*, [#374], 273: VI.1L-2E

IB57 *ʿnḫ-m-ʿ-ḥr: zzj* - Saqqara, TP (ntn), room III, west: offering
Source: *P-M III*, 512-515; Kanawati-Hassan, *Ankhmahor*, pl. 46
Label: n/a
Date: Kanawati-Hassan, *Ankhmahor*, 18: VI.1M-L; Harpur, *Decoration*, [#374], 273: VI.1L-2E

IB58 *ʿnḫ-m-ʿ-ḥr: zzj* - Saqqara, TP (ntn), room III, south: offering
Source: *P-M III*, 512-515; Kanawati-Hassan, *Ankhmahor*, pl. 47
Label: none
Date: Kanawati-Hassan, *Ankhmahor*, 18: VI.1M-L; Harpur, *Decoration*, [#374], 273: VI.1L-2E

IB59 *ʿnḫ-m-ʿ-ḥr: zzj* - Saqqara, TP (ntn), room III, west: offering
Source: *P-M III*, 512-515; Kanawati-Hassan, *Ankhmahor*, pl. 48 [b]
Label: none
Date: Kanawati-Hassan, *Ankhmahor*, 18: VI.1M-L; Harpur, *Decoration*, [#374], 273: VI.1L-2E

IB60 *mḥw* - Saqqara, UPC (ntn), offering-room, south: offering
Source: *P-M III*, 619-622; Altenmüller, *Mehu*, pl. 60
Label: none
Date: Altenmüller, *Mehu*, 19: VI.2; Harpur, *Decoration*, [#424], 274: VI.2M-3?

Nubian ibex

IB61 *ṯjj* - Saqqara, NSP (D22), portico, south: pastoral
Source: *P-M III*, 468-478; Epron et al., *Ti i*, pl. 6
Label: missing
Date: Harpur, *Decoration*, [#543], 277: V.8-9

IB62 *mrrw-kȝ.j: mrj* - Saqqara, TP (ntn), chamber A13, north: pastoral
Source: *P-M III*, 525-534; Duell, *Mereruka ii*, pl. 153
Label: *rn njȝ(w)*
Young ibex
Date: Harpur, *Decoration*, [#420], 274: VI.1M-L

Goats

IB63 *nb.j-m-ȝḫtj* - Giza, CF (LG86), chapel I, east: pastoral
Source: *P-M III*, 230-232; Hassan, *Giza iv*, fig. 75
Label: *jdt*
Cow/Female

Date: Harpur, *Decoration*, [#122], 267: IV.6-V.1

IB63a *nfr/kꜣ-ḥꜣ.j* - Saqqara, UPC (ntn), chapel, east: pastoral (goat on sub-register)
Source: *P-M III*, 639-641; Moussa-Altenmüller, *Nefer and Ka-hay*, pl. 4
Label: none
Date: Harpur, *Decoration*, [#439], 274: V.6

IB64 *nfr/kꜣ-ḥꜣ.j* - Saqqara, UPC (ntn), chapel, east: pastoral (pair on hind legs)
Source: *P-M III*, 639-641; Moussa-Altenmüller, *Nefer and Ka-hay*, pl. 4
Label: none
Date: Harpur, *Decoration*, [#439], 274: V.6

IB65 *nfr/kꜣ-ḥꜣ.j* - Saqqara, UPC (ntn), niche, east: pastoral (group left)
Source: *P-M III*, 639-641; Moussa-Altenmüller, *Nefer and Ka-hay*, pl. 18
Label: none
Date: Harpur, *Decoration*, [#439], 274: V.6

IB66 *nfr/kꜣ-ḥꜣ.j* - Saqqara, UPC (ntn), niche, east: pastoral (group right, goat on sub-register)
Source: *P-M III*, 639-641; Moussa-Altenmüller, *Nefer and Ka-hay*, pl. 18
Label: none
Date: Harpur, *Decoration*, [#439], 274: V.6

IB67 *nfr/kꜣ-ḥꜣ.j* - Saqqara, UPC (ntn), niche, east: pastoral
Source: *P-M III*, 639-641; Moussa-Altenmüller, *Nefer and Ka-hay*, pl. 19
Label: none
Date: Harpur, *Decoration*, [#439], 274: V.6

IB68 *nj-ꜥnḫ-ẖnmw/ẖnmw-ḥtp* - Saqqara, UPC (ntn), vestibule, north: pastoral
Source: *P-M III*, 641-644; Moussa-Altenmüller, *Nianchchnum*, fig. 8
Label: none
Date: Harpur, *Decoration*, [#431], 274: V.6L-7

IB69 *ḥtp-ḥr-ꜣḫtj* - Saqqara, WSP (D60), chapel, east: pastoral (goat left)
Source: *P-M III*, 593-595; Mohr, *Hetep-her-akhti*, fig. 21
Label: none
Date: Harpur, *Decoration*, [#467], 275: V.6-8E

IB70 *ḥtp-ḥr-ꜣḫtj* - Saqqara, WSP (D60), chapel, east: pastoral (goat centre)
Source: *P-M III*, 593-595; Mohr, *Hetep-her-akhti*, fig. 21
Label: none
Date: Harpur, *Decoration*, [#467], 275: V.6-8E

IB71 *ḥtp-ḥr-ꜣḫtj* - Saqqara, WSP (D60), chapel, east: pastoral (goat right)
Source: *P-M III*, 593-595; Mohr, *Hetep-her-akhti*, fig. 21
Label: none
Date: Harpur, *Decoration*, [#467], 275: V.6-8E

IB72 *ḥtp-ḥr-ꜣḫtj* - Saqqara, WSP (D60), chapel, east: pastoral (goats flanking tree on sub-registers)
Source: *P-M III*, 593-595; Mohr, *Hetep-her-akhti*, fig. 21
Label: none
Date: Harpur, *Decoration*, [#467], 275: V.6-8E

IB73 *ꜣḫtj-ḥtp* - Saqqara, UPC (ntn), chapel, east: pastoral
Source: *P-M III*, 634-637; Ziegler, *Akhethetep*, 138-139
Label: none
Date: Ziegler, *Akhethetep*, 9: V.6; Harpur, *Decoration*, [#340], 272: V.6-8E

IB74 *ꜣḫtj-mrw-nswt* - Giza, WF (G2184), secondary offering-room, west: pastoral
Source: *P-M III*, 80-81; Smith, *HESPOK*, fig. 239
Label: none
Date: Harpur, *Decoration*, [#008], 265: V.9-VI.1

IB75 *mrrw-k3.j: mrj* - Saqqara, TP (ntn), chamber A13, north: offering (recumbent, left)
Source: *P-M III*, 525-534; Duell, *Mereruka ii*, pl. 152
Label: *rn wnḏw*
Young goat
Date: Harpur, *Decoration*, [#420], 274: VI.1M-L

IB76 *mrrw-k3.j: mrj* - Saqqara, TP (ntn), chamber A13, north: offering (recumbent, centre left)
Source: *P-M III*, 525-534; Duell, *Mereruka ii*, pl. 152
Label: *rn wnḏw*
Young goat
Date: Harpur, *Decoration*, [#420], 274: VI.1M-L

IB77 *mrrw-k3.j: mrj* - Saqqara, TP (ntn), chamber A13, north: offering (recumbent, centre right)
Source: *P-M III*, 525-534; Duell, *Mereruka ii*, pl. 152
Label: *rn wnḏw*
Young goat
Date: Harpur, *Decoration*, [#420], 274: VI.1M-L

IB78 *mrrw-k3.j: mrj* - Saqqara, TP (ntn), chamber A13, north: offering (recumbent, right)
Source: *P-M III*, 525-534; Duell, *Mereruka ii*, pl. 152
Label: *rn wnḏw*
Young goat
Date: Harpur, *Decoration*, [#420], 274: VI.1M-L

IB79 *mrrj* - Saqqara, TP (ntn), room 1, north: offering
Source: *P-M III*, 518-519; Davies et al., *Saqqara Tombs i*, pl. 7
Label: missing
Date: Harpur, *Decoration*, [#418], 274: VI.2E?

IB80 *nb-k3w-ḥr: jdw* - Saqqara, UPC (ntn), pillared hall, south: pastoral
Source: *P-M III*, 627-629; Hassan, *Neb-Kaw-Her*, fig. 12
Label: n/a
Date: Harpur, *Decoration*, [#437], 274: V.9 r/u VI.7

IB81 *nb-k3w-ḥr: jdw* - Saqqara, UPC (ntn), pillared hall, south: pastoral (goat on hind legs)
Source: *P-M III*, 627-629; Hassan, *Neb-Kaw-Her*, fig. 13
Label: none
Date: Harpur, *Decoration*, [#437], 274: V.9 r/u VI.7

IB82 *nb-k3w-ḥr: jdw* - Saqqara, UPC (ntn), pillared hall, south: pastoral (goats on sub-registers)
Source: *P-M III*, 627-629; Hassan, *Neb-Kaw-Her*, fig. 13
Label: none
Date: Harpur, *Decoration*, [#437], 274: V.9 r/u VI.7

IB83 *nb-k3w-ḥr: jdw* - Saqqara, UPC (ntn), pillared hall, south: pastoral (goat within tree canopy)
Source: *P-M III*, 627-629; Hassan, *Neb-Kaw-Her*, pl. 19 [c]
Label: n/a
Date: Harpur, *Decoration*, [#437], 274: V.9 r/u VI.7

Sheep

IB84 *nb.j-m-3ḫtj* - Giza, CF (LG86), chapel I, south: agricultural
Source: *P-M III*, 230-232; *LD ii*, pl. 12 [a]
Label: missing
Date: Harpur, *Decoration*, [#122], 267: IV.6-V.1

IB84a *wr-jr.n-ptḥ* - Saqqara (ntn), chapel, east: agricultural
Source: *P-M III*, 699-700; *Hiero. Texts vi*, pl. 11
Label: none
Date: Harpur, *Decoration*, [#379], 273: V.3-5?

IB85 *nfr/k3-ḥ3.j* - Saqqara, UPC (ntn), chapel, east: agricultural
Source: *P-M III*, 639-641; Moussa-Altenmüller, *Nefer and Ka-Hay*, pl. 9

Label: none
Date: Harpur, *Decoration*, [#439], 274: V.6

IB86 *ḥtp-ḥr-3ḫtj* - Saqqara, WSP (D60), chapel, south: agricultural
Source: *P-M III*, 593-595; Mohr, *Hetep-her-akhti*, fig. 46
Label: *sk3 m sḥt*
Cultivating with a herd of sheep
Date: Harpur, *Decoration*, [#467], 275: V.6-8E

IB87 *3ḫtj-ḥtp* - Saqqara, UPC (ntn), chapel, east: agricultural
Source: *P-M III*, 634-637; Ziegler, *Akhethetep*, 128-129
Label: *sk3 jn jswt=f*
Cultivating by his gang
Date: Ziegler, *Akhethetep*, 9: V.6; Harpur, *Decoration*, [#340], 272: V.6-8E

IB88 *jrj-n-k3-ptḥ* - Saqqara, UPC (ntn), offering chamber, east: agricultural (upper register)
Source: *P-M III*, 644; Moussa-Junge, *Two Tombs*, pl. 9
Label: missing
Date: Harpur, *Decoration*, [#362], 273: V.6-9

IB89 *jrj-n-k3-ptḥ* - Saqqara, UPC (ntn), offering chamber, east: agricultural (lower register)
Source: *P-M III*, 644; Moussa-Junge, *Two Tombs*, pl. 9
Label: . . . *n ḥwj zp*
. . . threshing time?
Date: Harpur, *Decoration*, [#362], 273: V.6-9

IB90 *sḫm-ʿnḫ-ptḥ* - Saqqara, NSP (D41), chapel, south: agricultural
Source: *P-M III*, 454-455; Simpson, *Sekem-ankh-ptah*, pl. D
Label: n/a (Sower's Song)
Date: Harpur, *Decoration*, [#504], 276: V.8-9

IB91 *ṯjj* - Saqqara, NSP (D22), room III, north: agricultural
Source: *P-M III*, 468-478; Wild, *Ti ii*, pl. 113
Label: n/a (Sower's Song)
Date: Harpur, *Decoration*, [#543], 277: V.8-9

IB92 *nfr-jrt-n.f* - Saqqara, ESP (D55), chapel, east: agricultural
Source: *P-M III*, 583-584; van de Walle, *Neferirtenef*, pl. 11
Label: *sk3 m zḥt*
Cultivating with a herd of sheep
Date: Harpur, *Decoration*, [#440], 274: V.8-9

IB93 *k3.j-m-nfrt* - Saqqara, NSP (D23), chapel, east: agricultural
Source: *P-M III*, 467-468; Simpson, *Kayemnofret*, pl. G
Label: missing
Date: Simpson, *Kayemnofret*, 1: V.8-9; Harpur, *Decoration*, [#522], 276: V.8-9?

IB94 *nfr I* - Giza, WF (G4761), chapel, east: agricultural
Source: *P-M III*, 137-138; Junker, *Giza vi*, fig. 14
Label: none
Date: Harpur, *Decoration*, [#126], 267: V.9-VI.1

IB95 *sḫm-k3.j* - Giza, WF (G1029), chapel, east: agricultural
Source: *P-M III*, 53; Simpson, *Mast. West. Cem. Part I*, fig. 4
Label: *sk3 m sḥt*
Cultivating with a herd of sheep
Date: Harpur, *Decoration*, [#225], 269: V.9-VI.1?

IB96 *k3.j-m-ḥzt* - Saqqara, TP (ntn), pillared hall, east: agricultural
Source: *P-M III*, 542-543; McFarlane, *Kaiemheset*, pl. 44
Label: *ḫsf*

Driving[35]
Date: McFarlane, *Kaiemheset*, 23: V.6-V.8; Harpur, *Decoration*, [#525], 276: V.9-VI.1?

IB96a *mḥw* - Saqqara, UPC (ntn), room II, north: agricultural
Source: *P-M III*, 619-622; Altenmüller, *Mehu*, pl. 41 [b-a]
Label: n/a (Sower's Song)
Date: Altenmüller, *Mehu*, 19: VI.2; Harpur, *Decoration*, [#424], 274: VI.2M-3?

Egyptian mongooses

IB97 *mrrw-k3.j: mrj* - Saqqara, TP (ntn), chamber A13, south: marsh
Source: *P-M III*, 525-534; Duell, *Mereruka ii*, pl. 128
Label: none
Date: Harpur, *Decoration*, [#420], 274: VI.1M-L

IB98 *ḥzj* - Saqqara, TP (ntn), portico, south (west of entrance): marsh
Source: *P-M III*, nr; Kanawati-Abder-Raziq, *Hesi*, pl. 54
Label: none
Date: Kanawati-Abder-Raziq, *Hesi*, 16: VI.1L-2E; Harpur, *Decoration*, nr

Vervet monkeys

IB99 *k3.j-nfr* - Giza, WF (G2150), facade, east: domestic
Source: *P-M III*, 77-78; Reisner, *Giza Necropolis i*, fig. 263
Label: *gf*
Monkey
Date: Harpur, *Decoration*, [#268], 270: V.1-3

IB100 *jj-nfrt: š3.n.f* - Saqqara, UPC (ntn), room II, north: domestic
Source: *P-M III*, 616; Kanawati-Abder-Raziq, *Iynefert*, pl. 43
Label: none
Date: Kanawati-Abder-Raziq, *Iynefert*, 13: V.9E-M; Harpur, *Decoration*, [#345], 272: V.9

IB101 *nj-k3w-jzzj* - Saqqara, TP (ntn), room I, west: domestic
Source: *P-M III*, nr; Kanawati-Abder-Raziq, *Nikauisesi*, pl. 48
Label: n/a
Date: Kanawati-Abder-Raziq, *Nikauisesi*, 17-23: VI.1M; Harpur, *Decoration*, nr

IB102 *nj-k3w-jzzj* - Saqqara, TP (ntn), room II, north: domestic
Source: *P-M III*, nr; Kanawati-Abder-Raziq, *Nikauisesi*, pl. 55
Label: n/a
Date: Kanawati-Abder-Raziq, *Nikauisesi*, 17-23: VI.1M; Harpur, *Decoration*, nr

IB103 *ꜥnḫ-m-ꜥ-ḥr: zzj* - Saqqara, TP (ntn), room II, west: fowling
Source: *P-M III*, 512-515; Kanawati-Hassan, *Ankhmahor*, pl. 42
Label: missing
Date: Kanawati-Hassan, *Ankhmahor*, 18: VI.1M-L; Harpur, *Decoration*, [#374], 273: VI.1L-2E

IB104 *ꜥnḫ-m-ꜥ-ḥr: zzj* - Saqqara, TP (ntn), fragment (TNE95: F146), unknown: domestic
Source: *P-M III*, 512-515; Kanawati-Hassan, *Ankhmahor*, pl. 71
Label: missing
Date: Kanawati-Hassan, *Ankhmahor*, 18: VI.1M-L; Harpur, *Decoration*, [#374], 273: VI.1L-2E

IB105 *mḥw* - Saqqara, UPC (ntn), room II, south: boating
Source: *P-M III*, 619-622; Altenmüller, *Mehu*, pl. 19 [b]
Label: none
Date: Altenmüller, *Mehu*, 19: VI.2; Harpur, *Decoration*, [#424], 274: VI.2M-3?

[35] McFarlane (*Kaiemheset*, 31) suggests that these are donkeys. However, the small size of the animals and the presence of a herdsman at the head of the line, whose posture is characteristic of sowers, indicate that this is a sheep treading scene.

Cormorants

IB106 *mrrw-k3.j: mrj* - Saqqara, TP (ntn), chamber A4, east: fishing
Source: *P-M III*, 525-534; Duell, *Mereruka i*, pl. 43
Label: none
Date: Harpur, *Decoration*, [#420], 274: VI.1M-L

Herons

IB107 *ptḥ-ḥtp I* - Saqqara, WSP (D62), offering chamber, east: fowling (below clapnet, left)
Source: *P-M III*, 596-598; Murray, *Saqqara Mastabas*, pl. 11
Label: *rdjt sẖt 3pdw jn wḥ*ʿ *nw pr*
Making the catch of birds by the fowlers of the house
Date: Harpur, *Decoration*, [#398], 273: V.8M-L

IB108 *ptḥ-ḥtp I* - Saqqara, WSP (D62), offering chamber, east: fowling (below clapnet, centre)
Source: *P-M III*, 596-598; Murray, *Saqqara Mastabas*, pl. 11
Label: *rdjt sẖt 3pdw jn wḥ*ʿ *nw pr*
Making the catch of birds by the fowlers of the house
Date: Harpur, *Decoration*, [#398], 273: V.8M-L

IB109 *k3-gm-nj: mmj* - Saqqara, TP (LS10), room IV, west: fowling (below righthand clapnet)
Source: *P-M III*, 521-525; von Bissing, *Gem-ni-kai i*, pl. 9, Harpur-Scremin, *Kagemni*, fig. 14
Label: none
Date: Harpur, *Decoration*, [#534], 276: VI.1M

IB110 *ḥzj* - Saqqara, TP (ntn), portico, east: fowling (below clapnet, right)
Source: *P-M III*, nr; Kanawati-Abder-Raziq, *Hesi*, pl. 55
Label: *dw sẖt jn jmj-r wḥ*ʿ*(w) j wḥ*ʿ *pw ḥb r=k*
Setting the trap by the overseer of fowlers: Oh this fowler! There is a catch for you!
Date: Kanawati-Abder-Raziq, *Hesi*, 16: VI.1L-2E; Harpur, *Decoration*, nr

Spoonbills

IB111 *nfr-jrt-n.f* - Saqqara, ESP (D55), chapel, east: fowling (below clapnet)
Source: *P-M III*, 583-584; van de Walle, *Neferirtenef*, pl. 13
Label: none
Date: Harpur, *Decoration*, [#440], 274: V.8-9

IB112 *ẖww-wr* - Giza, CF (LG95), chapel, east: fowling (below clapnet)
Source: *P-M III*, 254-255; Hassan, *Giza v*, fig. 104
Label: *rdj sẖt*
Making the catch
Date: Harpur, *Decoration*, [#180], 268: V.8-9M

IB113 *wḏ3-ḥ3-ttj: ššj [nfr-sšm-ptḥ]* - Saqqara, TP (ntn), room II, east: fowling (below clapnet, right)
Source: *P-M III*, 515-516; Capart, *Rue de Tombeaux*, pl. 85
Label: none
Date: Harpur, *Decoration*, [#384], 273: VI.1L-2E

Sacred ibis

IB114 *ḥzj* - Saqqara, TP (ntn), portico, east: fowling (below clapnet, left)
Source: *P-M III*, nr; Kanawati-Abder-Raziq, *Hesi*, pl. 55
Label: *dw sẖt jn jmj-r wḥ*ʿ*(w) j wḥ*ʿ *pw ḥb r=k*
Setting the trap by the overseer of fowlers: Oh this fowler! There is a catch for you!
Date: Kanawati-Abder-Raziq, *Hesi*, 16: VI.1L-2E; Harpur, *Decoration*, nr

IB115 *ʿnḫ-m-ʿ-ḥr: zzj* - Saqqara, TP (ntn), room II, west: fowling (below clapnet, centre)
Source: *P-M III*, 512-515; Kanawati-Hassan, *Ankhmahor*, pl. 42
Label: none
Date: Kanawati-Hassan, *Ankhmahor*, 18: VI.1M-L; Harpur, *Decoration*, [#374], 273: VI.1L-2E

Geese

IB116 *spd-ḥtp* - Saqqara, NSP (D15), chapel entrance (Äg. Mus. Berlin no. 14642), east: pastoral/force-feeding
Source: *P-M III*, 481; Klebs, *Reliefs*, fig. 53; Janssen-Janssen, *Egyptian Household Animals*, fig. 40
Label: *r*
r-goose
ṯrp
ṯrp-goose
Date: Harpur, *Decoration*, [#494], 276: V.6L-9?

IB117 *ṯjj* - Saqqara, NSP (D22), portico, south: pastoral/poultry-yard
Source: *P-M III*, 467-478; Epron et al., *Ti i*, pl. 7
Label: *szwr ẖtꜥ3*
Causing *ẖtꜥ3*-geese to drink
Date: Harpur, *Decoration*, [#543], 277: V.8-9

IB118 *ṯjj* - Saqqara, NSP (D22), portico, south: pastoral/poultry-yard
Source: *P-M III*, 467-478; Epron et al., *Ti i*, pl. 8
Label: *szwr ḥrt mw*
Causing water birds to drink water
Date: Harpur, *Decoration*, [#543], 277: V.8-9

IB119 *ṯjj* - Saqqara, NSP (D22), room I/pillared hall, west: pastoral/poultry-yard
Source: *P-M III*, 467-478; Epron et al., *Ti i*, pl. 28
Label: missing
Date: Harpur, *Decoration*, [#543], 277: V.8-9

IB120 *ṯjj* - Saqqara, NSP (D22), room I/pillared hall, west: pastoral/poultry-yard
Source: *P-M III*, 467-478; Epron et al., *Ti i*, pl. 28
Label: missing
Date: Harpur, *Decoration*, [#543], 277: V.8-9

IB121 *k3-gm-nj: mmj* - Saqqara, TP (LS10), room IV, west: pastoral/poultry-yard (left enclosure, fragment)
Source: *P-M III*, 521-525; von Bissing, *Gem-ni-kai i*, pl. 8; Harpur-Scremin, *Kagemni*, fig. 14
Label: none
Date: Harpur, *Decoration*, [#534], 276: VI.1M

IB122 *k3-gm-nj: mmj* - Saqqara, TP (LS10), room IV, west: pastoral/poultry-yard (middle enclosure)
Source: *P-M III*, 521-525; von Bissing, *Gem-ni-kai i*, pl. 9, Harpur-Scremin, *Kagemni*, fig. 14
Label: none
Date: Harpur, *Decoration*, [#534], 276: VI.1M

IB123 *k3-gm-nj: mmj* - Saqqara, TP (LS10), room IV, west: pastoral/poultry-yard (right enclosure)
Source: *P-M III*, 521-525; von Bissing, *Gem-ni-kai i*, pl. 9, Harpur-Scremin, *Kagemni*, fig. 14
Label: none
Date: Harpur, *Decoration*, [#534], 276: VI.1M

IB124 *k3-gm-nj: mmj* - Saqqara, TP (LS10), room IV, west: pastoral (surrounding pool)
Source: *P-M III*, 521-525; von Bissing, *Gem-ni-kai i*, pl. 10, Harpur-Scremin, *Kagemni*, fig. 13
Label: none
Date: Harpur, *Decoration*, [#534], 276: VI.1M

IB125 *nj-k3w-jzzj* - Saqqara, TP (ntn), room I, north: pastoral
Source: *P-M III*, nr; Kanawati-Abder-Raziq, *Nikauisesi*, pl. 49
Label: *r*
r-goose
ṯrp
ṯrp-goose
Date: Kanawati-Abder-Raziq, *Nikauisesi*, 17-23: VI.1M; Harpur, *Decoration*, nr

IB126 *nj-k3w-jzzj* - Saqqara, TP (ntn), room I, north: pastoral
Source: *P-M III*, nr; Kanawati-Abder-Raziq, *Nikauisesi*, pl. 49
Label: *wnm jt jn r ṯrp*

Feeding on grain by the *r*-geese and the *ṯrp*-geese
Date: Kanawati-Abder-Raziq, *Nikauisesi*, 17-23: VI.1M; Harpur, *Decoration*, nr

IB127 *mrrw-kꜣ.j: mrj* - Saqqara, TP (ntn), chamber A6, west: pastoral/poultry yard
Source: *P-M III*, 525-534; Duell, *Mereruka i*, pl. 53 [b]
Label: none
Date: Harpur, *Decoration*, [#420], 274: VI.1M-L

IB128 *mrrw-kꜣ.j: mrj* - Saqqara, TP (ntn), chamber A6, west: pastoral/force-feeding (left)
Source: *P-M III*, 525-534; Duell, *Mereruka i*, pl. 52
Label: none
Date: Harpur, *Decoration*, [#420], 274: VI.1M-L

IB129 *mrrw-kꜣ.j: mrj* - Saqqara, TP (ntn), chamber A6, west: pastoral/force-feeding (right)
Source: *P-M III*, 525-534; Duell, *Mereruka i*, pl. 52
Label: none
Date: Harpur, *Decoration*, [#420], 274: VI.1M-L

IB130 *ḥzj* - Saqqara, TP (ntn), portico, west: pastoral/poultry-yard
Source: *P-M III*, nr; Kanawati-Abder-Raziq, *Hesi*, pl. 56
Label: none
Date: Kanawati-Abder-Raziq, *Hesi*, 16: VI.1L-2E; Harpur, *Decoration*, nr

IB131 *wḏꜣ-ḥꜣ-ttj: ššj* [*nfr-sšm-ptḥ*] - Saqqara, TP (ntn), room II, east: pastoral/poultry-yard
Source: *P-M III*, 515-516; Capart, *Rue de Tombeaux*, pl. 88
Label: none
Date: Harpur, *Decoration*, [#384], 273: VI.1L-2E

IB132 *kꜣ.j-jrr* - Saqqara, UPC (ntn), room II, west: pastoral: poultry-yard
Source: *P-M III*, 631-632; Daoud, K. (2000). *ASAE* 75, pl. 3
Label: missing
Date: Daoud, *ASAE* 75 (2000), 106: mid-VI, possibly VI.5; Harpur, *Decoration*, [#520], 276: VI.2?

IB133 *mrrj* - Saqqara, TP (ntn), room 1, west: pastoral
Source: *P-M III*, 517-519; Davies et al., *Saqqara Tombs i*, pl. 6
Label: none
Date: Harpur, *Decoration*, [#418], 274: VI.2E?

IB134 *mrrj* - Saqqara, TP (ntn), room 1, west: pastoral
Source: *P-M III*, 517-519; Davies et al., *Saqqara Tombs i*, pl. 6
Label: none
Date: Harpur, *Decoration*, [#418], 274: VI.2E?

IB135 *mrjj-ttj: mrj* - Saqqara, TP (ntn), room C1, east: pastoral/poultry-yard (surrounding pool)
Source: *P-M III*, 356-357; Kanawati-Abder-Raziq, *Meryteti*, pl. 48
Label: none
Date: Kanawati-Abder-Raziq, *Meryteti*, 18: end VI.1-VI.2M; Harpur, *Decoration*, [#415], 274: VI.2M-L

IB136 *mrjj-ttj: mrj* - Saqqara, TP (ntn), room C1, east: pastoral/force-feeding
Source: *P-M III*, 356-357; Kanawati-Abder-Raziq, *Meryteti*, pl. 48
Label: none
Date: Kanawati-Abder-Raziq, *Meryteti*, 18: end VI.1-VI.2M; Harpur, *Decoration*, [#415], 274: VI.2M-L

Ducks

IB137 *nb.j-m-ꜣḫtj* - Giza, CF (LG86), chapel I, south: clapnet
Source: *P-M III*, 230-232; *LD ii*, pl. 12 [a]
Label: none
Date: Harpur, *Decoration*, [#122], 267: IV.6-V.1

IB138 *wr-jr.n-ptḥ* - Saqqara (ntn), chapel, east: clapnet
Source: *P-M III*, 699-700; *Hiero. Texts vi*, pl. 11
Label: none

Date: Harpur, *Decoration*, [#379], 273: V.3-5?

IB139 *nfr/k3-h3.j* - Saqqara, UPC (ntn), chapel, east: clapnet
Source: *P-M III*, 639-641; Moussa-Altenmüller, *Nefer and Ka-hay*, pls. 5-6
Label: none
Date: Harpur, *Decoration*, [#439], 274: V.6

IB140 *rʿ-ḫʿ.f-ʿnḫ* - Giza, EF (G7948=LG75), chapel, east: clapnet
Source: *P-M III*, 207-208; *LD ii*, pl. 9
Label: none
Date: Harpur, *Decoration*, [#156], 268: V.6?

IB141 *rʿ-m-k3.j* - Saqqara, NSP (D3=S903), chapel, west: clapnet
Source: *P-M III*, 487-488; Hayes, *Scepter i*, fig. 55
Label: none
Date: Harpur, *Decoration*, [#453], 275: V.6? r/u - V.6-7

IB142 *nj-ʿnḫ-ẖnmw/ẖnmw-ḥtp* - Saqqara, UPC (ntn), vestibule, south: clapnet
Source: *P-M III*, 641-644; Moussa-Altenmüller, *Nianchchnum*, fig. 12
Label: *mj pḥr r zp sẖn 3pd(w) mj š3s ḥn=k m ʿnḫ*
Come! Run at once. Birds alight! Come! Travel. Go with haste!
Date: Harpur, *Decoration*, [#431], 274: V.6L-7

IB143 *ḥtp-ḥr-3ḫtj* - Saqqara, WSP (D60), chapel, north: clapnet
Source: *P-M III*, 593-595; Mohr, *Hetep-her-akhti*, fig. 25
Label: *rdjt sẖt*
Making the catch
Date: Harpur, *Decoration*, [#467], 275: V.6-8E

IB144 *jrj-n-k3-ptḥ* - Saqqara, UPC (ntn), offering chamber, south: fowling
Source: *P-M III*, 644; Moussa-Junge, *Two Tombs*, pl. 13
Label: none
Date: Harpur, *Decoration*, [#362], 273: V.6-9

IB145 *spd-ḥtp* - Saqqara, NSP (D15), hall, north: clapnet
Source: *P-M III*, 481; Borchardt, *Denkmäler ii*, pl. 60 (CG 1671)
Label: *jr=ṯw wrt wḥʿ pw*
Act greatly! This fowler . . .
Date: Harpur, *Decoration*, [#494], 276: V.6L-9?

IB146 *ptḥ-ḥtp I* - Saqqara, WSP (D62), offering chamber, east: clapnet
Source: *P-M III*, 596-598; Murray, *Saqqara Mastabas*, pl. 11
Label: *rdjt sẖt 3pdw jn wḥʿ nw pr*
Making the catch of birds by the fowlers of the house
Date: Harpur, *Decoration*, [#398], 273: V.8M-L

IB147 *tjj* - Saqqara, NSP (D22), room III, north: clapnet
Source: *P-M III*, 468-478; Wild, *Ti ii*, pl. 120
Label: *šm r=k jm=f jw wn 3pd(w) ḥr-s3=f*
Go from it! There are birds upon it!
smr wʿtjj
Sole companion
Date: Harpur, *Decoration*, [#543], 277: V.8-9

IB148 *nfr-jrt-n.f* - Saqqara, ESP (D55), chapel, east: clapnet
Source: *P-M III*, 583-584; van de Walle, *Neferirtenef*, pl. 13
Label: *rdj sẖt*
Making the catch
Date: Harpur, *Decoration*, [#440], 274: V.8-9

IB149 *k3.j-m-nfrt* - Saqqara, NSP (D23), chapel, east: clapnet
Source: *P-M III*, 467-468; Simpson, *Kayemnofret*, pl. G
Label: *rdj sẖt nḏr nw r mnḫ sẖt 3pdw jn wḥʿw*

Making the catch and grasping these well, trapping birds by the fowlers
Date: Simpson, *Kayemnofret*, 1: V.8-9; Harpur, *Decoration*, [#522], 276: V.8-9?

IB150 *ꜣḫtj-ḥtp* - Saqqara, WSP (D64), corridor, west: clapnet
Source: *P-M III*, 599-600; Davies, *Ptahhetep ii*, pl. 8
Label: missing
Date: Harpur, *Decoration*, [#338], 272: V.8L-9E

IB151 *ptḥ-ḥtp II: ṯfj* - Saqqara, WSP (D64), offering-room, east: clapnet
Source: *P-M III*, 600-604; Quibell, *Ptah-hetep* pl. 32
Label: *sẖt m ḥsbt*
Catching in multitude
jṯj r=k ntj ḥnꜥ jw ḥb n=k
Seize my companions! It is a catch for you!
Date: Harpur, *Decoration*, [#400], 274: V.9M-L

IB152 *sḫm-kꜣ.j* - Giza, WF (G1029), chapel, east: clapnet
Source: *P-M III*, 53; Simpson, *Mast. West. Cem. Part I*, fig. 4
Label: none
Date: Harpur, *Decoration*, [#225], 269: V.9-VI.1?

IB153 Anon - Giza (ntn), relief fragment (Munich Staatl. Sammlung #Gl. 115), unknown: clapnet
Source: *P-M III*, 308-309; *Munich. Äg. Sammlung*, fig. 22
Label: none
Date: Harpur, *Decoration*, [#325], 272: V-VI

IB154 *kꜣ-gm-nj: mmj* - Saqqara, TP (LS10), room IV, west: clapnet
Source: *P-M III*, 521-525; von Bissing, *Gem-ni-kai i*, pl. 8; Harpur-Scremin, *Kagemni*, fig. 14
Label: *sgrt jn jmj-r wḥꜥ(w)*
Silencing by the overseer of the fowlers
Date: Harpur, *Decoration*, [#534], 276: VI.1M

IB155 *sꜥnḫ-w(j)-ptḥ* - Saqqara, TP (ntn), room I, east: clapnet
Source: *P-M III*, nr; Kanawati-Abder-Raziq, *Neferseshemre*, pls. 69 and 75
Label: none
Date: Kanawati-Abder-Raziq, *Neferseshemre*, 40-41: VI.1L; Harpur, *Decoration*, nr

IB156 *ḥzj* - Saqqara, TP (ntn), portico, east: clapnet
Source: *P-M III*, nr; Kanawati-Abder-Raziq, *Hesi*, pl. 55
Label: *dw sẖt jn jmj-r wḥꜥ(w) j wḥꜥ pw ḥb r=k*
Setting the trap by the overseer of fowlers: Oh this fowler! There is a catch for you!
Date: Kanawati-Abder-Raziq, *Hesi*, 16: VI.1L-2E; Harpur, *Decoration*, nr

IB157 *ꜥnḫ-m-ꜥ-ḥr: zzj* - Saqqara, TP (ntn), room II, west: clapnet
Source: *P-M III*, 512-515; Kanawati-Hassan, *Ankhmahor*, pl. 42
Label: *jw ḥb n ꜥ=k wḥꜥ pw jgr tjwn*
There is a catch for your arm, this fowler. May you keep quiet
Date: Kanawati-Hassan, *Ankhmahor*, 18: VI.1M-L; Harpur, *Decoration*, [#374], 273: VI.1L-2E

IB158 *mḥw* - Saqqara, UPC (ntn), room I, south: clapnet
Source: *P-M III*, 619-622; Altenmüller, *Mehu*, pl. 7
Label: *nḏr jr=k wnj jw ḥb jr=n*
Grab hold quickly! It gives a good catch for us!
jṯj jr=k nty ḥnꜥ(=j)
Pull my comrades!
Date: Altenmüller, *Mehu*, 19: VI.2; Harpur, *Decoration*, [#424], 274: VI.2M-3?

IB159 *mḥw* - Saqqara, UPC (ntn), room I, south: clapnet
Source: *P-M III*, 619-622; Altenmüller, *Mehu*, pl. 7
Label: *jnj jr=k nty ḥnꜥ(=j) ḥwt-dšr(t)*
Haul it, my comrades! (It is) a Red House!
Date: Altenmüller, *Mehu*, 19: VI.2; Harpur, *Decoration*, [#424], 274: VI.2M-3?

Quail

IB160 *sḫm-kꜣ.j* - Giza, WF (G1029), chapel, east: agricultural
Source: *P-M III*, 53; Smith, *HESPOK*, fig. 73
Label: none
Date: Harpur, *Decoration*, [#225], 269: V.9-VI.1?

IB161 *mrrw-kꜣ.j: mrj* - Saqqara, TP (ntn), chamber A13, east: agricultural (bird, left)
Source: *P-M III*, 525-534; Duell, *Mereruka ii*, pl. 168
Label: n/a
Date: Harpur, *Decoration*, [#420], 274: VI.1M-L

IB162 *mrrw-kꜣ.j: mrj* - Saqqara, TP (ntn), chamber A13, east: agricultural (group in net)
Source: *P-M III*, 525-534; Duell, *Mereruka ii*, pls. 168-169
Label: n/a
Date: Harpur, *Decoration*, [#420], 274: VI.1M-L

IB163 *mrrw-kꜣ.j: mrj* - Saqqara, TP (ntn), chamber A13, east: agricultural (bird, right)
Source: *P-M III*, 525-534; Duell, *Mereruka ii*, pl. 168
Label: n/a
Date: Harpur, *Decoration*, [#420], 274: VI.1M-L

IB164 *ḥzj* - Saqqara, TP (ntn), portico, south: agricultural (bird, left)
Source: *P-M III*, nr; Kanawati-Abder-Raziq, *Hesi*, pl. 52
Label: none
Date: Kanawati-Abder-Raziq, *Hesi*, 16: VI.1L-2E; Harpur, *Decoration*, nr

IB165 *ḥzj* - Saqqara, TP (ntn), portico, south: agricultural (bird, middle)
Source: *P-M III*, nr; Kanawati-Abder-Raziq, *Hesi*, pl. 52
Label: *j ṯꜣj wbꜣ m h(w)t.f*
Oh fellow who is steady in his place!
Date: Kanawati-Abder-Raziq, *Hesi*, 16: VI.1L-2E; Harpur, *Decoration*, nr

IB166 *ḥzj* - Saqqara, TP (ntn), portico, south: agricultural (group in net)
Source: *P-M III*, nr; Kanawati-Abder-Raziq, *Hesi*, pl. 52
Label: *ꜥ[jw] r ḏr*
Birds to the limit
Date: Kanawati-Abder-Raziq, *Hesi*, 16: VI.1L-2E; Harpur, *Decoration*, nr

IB167 *mḥw* - Saqqara, UPC (ntn), room II, south: agricultural
Source: *P-M III*, 619-622; Altenmüller, *Mehu*, pl. 23 [a]
Label: n/a
Date: Altenmüller, *Mehu*, 19: VI.2; Harpur, *Decoration*, [#424], 274: VI.2M-3?

Cranes

IB168 *ṯjj* - Saqqara, NSP (D22), room I/pillared hall, west: offering
Source: *P-M III*, 467-478; Epron et al., *Ti i*, pl. 22
Label: n/a
Date: Harpur, *Decoration*, [#543], 277: V.8-9

IB169 *ṯjj* - Saqqara, NSP (D22), room I/pillared hall, west: pastoral/force-feeding
Source: *P-M III*, 467-478; Epron et al., *Ti i*, pl. 33
Label: *snmt ḏꜣt*
Feeding cranes
Date: Harpur, *Decoration*, [#543], 277: V.8-9

IB170 *nj-kꜣw-jzzj* - Saqqara, TP (ntn), room I, north: pastoral
Source: *P-M III*, nr; Kanawati, *Nikauisesi*, pl. 49
Label: *ṯṯf jt n ḏꜣt ẖr=sn*
Throwing down grain for cranes, beneath them
Date: Kanawati-Abder-Raziq, *Nikauisesi*, 17-23: VI.1M; Harpur, *Decoration*, nr

IB171 *mrrw-k3.j: mrj* - Saqqara, TP (ntn), chamber A6, west: pastoral/force-feeding
Source: *P-M III*, 525-534; Duell, *Mereruka i*, pl. 52
Label: none
Date: Harpur, *Decoration*, [#420], 274: VI.1M-L

IB172 *mrrw-k3.j: mrj* - Saqqara, TP (ntn), chamber A6, west: pastoral/force-feeding
Source: *P-M III*, 525-534; Duell, *Mereruka i*, pl. 52
Label: *snm d̠3t*
Feeding cranes
Date: Harpur, *Decoration*, [#420], 274: VI.1M-L

IB173 *mrrw-k3.j: mrj* - Saqqara, TP (ntn), chamber A6, west: pastoral/force-feeding
Source: *P-M III*, 525-534; Duell, *Mereruka i*, pl. 52
Label: *snm d̠3t*
Feeding cranes
wš3 d̠3t
Fattening cranes
Date: Harpur, *Decoration*, [#420], 274: VI.1M-L

IB174 *wd̠3-ḥ3-ttj: ššj [nfr-ssm-ptḥ]* - Saqqara, TP (ntn), room II, east: pastoral/force-feeding
Source: *P-M III*, 515-516; Capart, *Rue de Tombeaux*, pl. 85
Label: none
Date: Harpur, *Decoration*, [#384], 273: VI.1L-2E

IB175 *mrjj-ttj: mrj* - Saqqara, TP (ntn), room C1, east: pastoral/force-feeding
Source: *P-M III*, 356-357; Kanawati-Abder-Raziq, *Meryteti*, pl. 48
Label: *snm d̠3t*
Feeding cranes
Date: Kanawati-Abder-Raziq, *Meryteti*, 18: end VI.1-VI.2M; Harpur, *Decoration*, [#415], 274: VI.2M-L

IB176 *mrjj-ttj: mrj* - Saqqara, TP (ntn), room C1, east: pastoral/force-feeding (left)
Source: *P-M III*, 356-357; Kanawati-Abder-Raziq, *Meryteti*, pl. 48
Label: none
Date: Kanawati-Abder-Raziq, *Meryteti*, 18: end VI.1-VI.2M; Harpur, *Decoration*, [#415], 274: VI.2M-L

IB177 *mrjj-ttj: mrj* - Saqqara, TP (ntn), room C1, east: pastoral/force-feeding (right)
Source: *P-M III*, 356-357; Kanawati-Abder-Raziq, *Meryteti*, pl. 48
Label: none
Date: Kanawati-Abder-Raziq, *Meryteti*, 18: end VI.1-VI.2M; Harpur, *Decoration*, [#415], 274: VI.2M-L

Golden orioles

IB178 *rꜥ-špss* - Saqqara, NSP (LS16=S902), hall, west: bird netting
Source: *P-M III*, 494-496; *LD Erganz.*, pl. 39 [a]
Label: missing
Date: Harpur, *Decoration*, [#456], 275: V.8M

Nile crocodiles

IB179 *h̠nm-ntj* - Giza, WF (G2374), room I, west: marsh
Source: *P-M III*, 87; Brovarski, *Senedjemib Complex Part I*, fig. 84 [b]
Label: none
Date: Brovarski, *Senedjemib Complex Part I*, 30-31: V.9-VI.1; Harpur, *Decoration*, [#205], 269: VI.1

IB179a *k3-gm-nj: mmj* - Saqqara, TP (LS10), room I, west: marsh
Source: *P-M III*, 521-525; Firth-Gunn, *Teti Pyr. ii*, pl. 53 [2]; Harpur-Scremin, *Kagemni*, fig. 3
Label: none
Date: Harpur, *Decoration*, [#534], 276: VI.1M

IB180 *k3-gm-nj: mmj* - Saqqara, TP (LS10), room IV, east: fishing
Source: *P-M III*, 521-525; von Bissing, *Gem-ni-kai i*, pl. 17; Harpur-Scremin, *Kagemni*, fig. 16
Label: none
Date: Harpur, *Decoration*, [#534], 276: VI.1M

IB181 *mrrj* - Saqqara, TP (ntn), room 1, south: marsh
Source: *P-M III*, 518-519; Davies et al., *Saqqara Tombs i*, pl. 5
Label: none
Date: Harpur, *Decoration*, [#418], 274: VI.2E?

IB182 *mḥw* - Saqqara, UPC (ntn), room I, east: marsh
Source: *P-M III*, 619-622; Altenmüller, *Mehu*, pl. 11
Label: none
Date: Altenmüller, *Mehu*, 19: VI.2; Harpur, *Decoration*, [#424], 274: VI.2M-3?

IB183 *mḥw* - Saqqara, UPC (ntn), room I, west: marsh
Source: *P-M III*, 619-622; Altenmüller, *Mehu*, pl. 13
Label: none
Date: Altenmüller, *Mehu*, 19: VI.2; Harpur, *Decoration*, [#424], 274: VI.2M-3?

Catfish

IB184 *snḏm-jb: jntj* - Giza, WF (G2370=LG27), portico, west: marsh
Source: *P-M III*, 85-87; *LD Ergänz.*, pl. 17; Brovarski, *Senedjemib Complex Part I*, fig. 25
Label: none
Date: Brovarski, *Senedjemib Complex Part I*, 23: V.8M-L; Harpur, *Decoration*, [#219], 269: V.8M-L

IB185 *ṯjj* - Saqqara, NSP (D22), room III, north: marsh
Source: *P-M III*, 468-47; Wild, *Ti ii*, pl. 119
Label: none
Date: Harpur, *Decoration*, [#543], 277: V.8-9

IB186 *ḫnwt* - Saqqara, UPC (ntn), room A[I], east: marsh
Source: *P-M III*, 623-624; Munro, *Unas-Friedhof*, pl. 33
Label: none
Date: Munro, *Unas-Friedhof*, 20: V.9; Harpur, *Decoration*, [#478], 275: V.9

IB187 Anon - Saqqara (ntn), block from debris of Temple Terrace, unknown: marsh
Source: *P-M III*, nr; Martin, *Hetepka*, pl. 23 [24]
Label: none
Date: Harpur, *Decoration*, [#601], 279: late V-VI

IB188 *k3.j-m-ʿnḫ* - Giza, WF (G4561), chapel, north: marsh
Source: *P-M III*, 131-133; Kanawati, *Kaiemankh*, pl. 31
Label: none
Date: Kanawati, *Kaiemankh*, 18: late V; Harpur, *Decoration*, [#255], 270: VI

IB189 *k3-gm-nj: mmj* - Saqqara, TP (LS10), room I, west: marsh
Source: *P-M III*, 521-525; Firth-Gunn, *Teti Pyr. ii*, pl. 53 [2]; Harpur-Scremin, *Kagemni*, fig. 3
Label: none
Date: Harpur, *Decoration*, [#534], 276: VI.1M

IB190 *k3-gm-nj: mmj* - Saqqara, TP (LS10), room III, north: marsh
Source: *P-M III*, 521-525; von Bissing, *Gem-ni-kai i*, pl. 4; Harpur-Scremin, *Kagemni*, fig. 9
Label: none
Date: Harpur, *Decoration*, [#534], 276: VI.1M

IB191 *mrrw-k3.j: mrj* - Saqqara, TP (ntn), chamber A1, north: marsh
Source: *P-M III*, 525-534; Duell, *Mereruka i*, pl. 11
Label: none
Date: Harpur, *Decoration*, [#420], 274: VI.1M-L

IB192 *mrrw-k3.j: mrj* - Saqqara, TP (ntn), chamber A1, south: marsh
Source: *P-M III*, 525-534; Duell, *Mereruka i*, pl. 15
Label: none
Date: Harpur, *Decoration*, [#420], 274: VI.1M-L

IB193 *sʿnḫ-w(j)-ptḥ* - Saqqara, TP (ntn), room I, east: marsh
Source: *P-M III*, nr; Kanawati, *Neferseshemre*, pls. 69 and 76
Label: none
Date: Kanawati-Abder-Raziq, *Neferseshemre*, 40-41: VI.1L; Harpur, *Decoration*, nr

IB194 *ḥzj* - Saqqara, TP (ntn), portico, south (east of entrance): marsh
Source: *P-M III*, nr; Kanawati-Abder-Raziq, *Hesi*, pl. 53
Label: none
Date: Kanawati-Abder-Raziq, *Hesi*, 16: VI.1L-2E; Harpur, *Decoration*, nr

IB195 *ḥzj* - Saqqara, TP (ntn), portico, south (west of entrance): marsh (left)
Source: *P-M III*, nr; Kanawati-Abder-Raziq, *Hesi*, pl. 54
Label: none
Date: Kanawati-Abder-Raziq, *Hesi*, 16: VI.1L-2E; Harpur, *Decoration*, nr

IB196 *ḥzj* - Saqqara, TP (ntn), portico, south (west of entrance): marsh (right)
Source: *P-M III*, nr; Kanawati-Abder-Raziq, *Hesi*, pl. 54
Label: none
Date: Kanawati-Abder-Raziq, *Hesi*, 16: VI.1L-2E; Harpur, *Decoration*, nr

IB196a *jnw-mnw* - Saqqara (ntn), room I, west: marsh
Source: *P-M III*, nr; Kanawati, *Inumin*, pl. 44
Label: none
Date: Kanawati, *Inumin*, 16-17: VI.2E; Harpur, *Decoration*, nr

IB197 *mḥw* - Saqqara, UPC (ntn), room I, east: marsh
Source: *P-M III*, 619-622; Altenmüller, *Mehu*, pl. 11
Label: none
Date: Altenmüller, *Mehu*, 19: VI.2; Harpur, *Decoration*, [#424], 274: VI.2M-3?

IB198 *mḥw* - Saqqara, UPC (ntn), room I, west: marsh
Source: *P-M III*, 619-622; Altenmüller, *Mehu*, pl. 13
Label: none
Date: Altenmüller, *Mehu*, 19: VI.2; Harpur, *Decoration*, [#424], 274: VI.2M-3?

IB199 *mḥw* - Saqqara, UPC (ntn), room II, south: boating
Source: *P-M III*, 619-622; Altenmüller, *Mehu*, pl. 21 [b]
Label: none
Date: Altenmüller, *Mehu*, 19: VI.2; Harpur, *Decoration*, [#424], 274: VI.2M-3?

IB200 *mḥw* - Saqqara, UPC (ntn), room II, north: boating
Source: *P-M III*, 619-622; Altenmüller, *Mehu*, pl. 32
Label: none
Date: Altenmüller, *Mehu*, 19: VI.2; Harpur, *Decoration*, [#424], 274: VI.2M-3?

IB201 *mḥw* - Saqqara, UPC (ntn), room II, north: fishing
Image source: *P-M III*, 619-622; Altenmüller, *Mehu*, pl. 38
Label: n/a
Date: Altenmüller, *Mehu*, 19: VI.2; Harpur, *Decoration*, [#424], 274: VI.2M-3?

IB202 *wr-nw* - Saqqara, TP (ntn), hall, east: marsh
Source: *P-M III*, 519; Davies et al., *Saqqara Tombs i*, pl. 25
Label: none
Date: Harpur, *Decoration*, [#380], 273: VI.3-4?

IB203 *ꜣḫtj-ḥtp* - Saqqara, UPC (E17), offering-room, east: marsh
Source: *P-M III*, 633-634; Petrie-Murray, *Seven Memphite*, pl. 6 [lower]
Label: none
Date: Harpur, *Decoration*, [#339], 272: VI.5?

7.2 DRINKING

Cattle

IB204 *nb.j-m-ꜣḫtj* - Giza, CF (LG86), chapel I, west: fording
Source: *P-M III*, 230-232; *LD ii*, pl. 12 [b]
Label: none
Date: Harpur, *Decoration*, [#122], 267: IV.6-V.1

IB205 Anon - Saqqara (Giza[36]) (ntn), relief fragment (Albertinum, Dresden ZV. 1748-50), unknown: fording
Source: *P-M III*, 309/761; Breasted, *Geschichte Aegyptens*, fig. 194; Harpur, *Decoration*, fig. 122
Label: *jsṯ smḥ*
Behold the flood
mnjw
Herdsman
ẖnj [ ] *mḥt*
Rowing? . . . the Delta marshes
Date: Harpur, *Decoration*, [#324], 272: V

IB206 *nfr/kꜣ-ḥꜣ.j* - Saqqara, UPC (ntn), chapel, east: pastoral
Source: *P-M III*, 639-641; Moussa-Altenmüller, *Nefer and Ka-hay*, pl. 5
Label: none
Date: Harpur, *Decoration*, [#439], 274: V.6

IB207 *nfr/kꜣ-ḥꜣ.j* - Saqqara, UPC (ntn), niche, east: pastoral
Source: *P-M III*, 639-641; Moussa-Altenmüller, *Nefer and Ka-hay*, pl. 20
Label: none
Date: Harpur, *Decoration*, [#439], 274: V.6

IB208 *nfr/kꜣ-ḥꜣ.j* - Saqqara, UPC (ntn), niche, east: pastoral
Source: *P-M III*, 639-641; Moussa-Altenmüller, *Nefer and Ka-hay*, pl. 20
Label: none
Date: Harpur, *Decoration*, [#439], 274: V.6

IB209 *nj-ꜥnḫ-ẖnmw/ẖnmw-ḥtp* - Saqqara, UPC (ntn), outer hall, west: pastoral
Source: *P-M III*, 641-644; Moussa-Altenmüller, *Nianchchnum*, pl. 77
Lable: none
Date: Harpur, *Decoration*, [#431], 274: V.6L-7

IB210 *nj-ꜥnḫ-ẖnmw/ẖnmw-ḥtp* - Saqqara, UPC (ntn), outer hall, west: fording
Source: *P-M III*, 641-644; Moussa-Altenmüller, *Nianchchnum*, pl. 76
Lable: *prjt m mḥjt r ḥrj-tp jn mnjw ḥnꜥ kꜣ(w)t=sn kꜣw wrw*
Coming out of the marshland to the high land by the herdsmen with their produce and great cattle
Date: Harpur, *Decoration*, [#431], 274: V.6L-7

IB211 *kꜣ.j-dwꜣ* - Giza, CF (ntn), chapel, north: pastoral
Source: *P-M III*, 244-245; Hassan, *Giza vi*, fig. 80
Label: none
Date: Harpur, *Decoration*, [#272], 270: V.6-9?

IB212 *rꜥ-špss* - Saqqara, NSP (LS16=S902), pillared hall, south: pastoral
Source: *P-M III*, 494-496; *LD ii*, pl. 62
Label: *ḥry-ḏbꜥ*
Hornless cattle
Date: Harpur, *Decoration*, [#456], 275: V.8M

IB213 *rꜥ-špss* - Saqqara, NSP (LS16=S902), forecourt, west: fording
Source: *P-M III*, 494-496; *LD ii*, pl. 60 [left]
Label: *prt ngꜣw r ḥrj-tp*
Coming out of the long-horned bulls to the high land

---

[36] Harpur (*Decoration*, 234) has determined that this fragment joins another (Barracco Collection, Rome, no. 2; *P-M III*, 761) for which the provenance is known to be Saqqara.

Date: Harpur, *Decoration*, [#456], 275: V.8M

IB214 *sḫm-ꜥnḫ-ptḥ* - Saqqara, NSP (D41), chapel, east: pastoral
Source: *P-M III*, 454-455; Simpson, *Sekem-ankh-ptah*, pl. C
Label: *zwr r(n) jwꜣ*
Young ox drinking
Date: Harpur, *Decoration*, [#504], 276: V.8-9

IB215 *tjj* - Saqqara, NSP (D22), room III, north: pastoral
Source: *P-M III*, 467-478; Wild, *Ti ii*, pl. 124
Label: none
Date: Harpur, *Decoration*, [#543], 277: V.8-9

IB216 *tjj* - Saqqara, NSP (D22), room III, north: fording
Source: *P-M III*, 467-478; Wild, *Ti ii*, pl. 114
Label: *j mḥsḥs pw rdj=k šm kꜣw jptn prt m mḥt*
Oh this filthy one! Cause to go these cattle coming out from the North
jꜣꜥ bḥs pw (n) mnꜥt
Swing around this calf of the milch cows[37]
Date: Harpur, *Decoration*, [#543], 277: V.8-9

IB217 *nfr-jrt-n.f* - Saqqara, ESP (D55), chapel, east: agricultural
Source: *P-M III*, 583-584; van de Walle, *Neferirtenef*, pl. 12
Label: none
Date: Harpur, *Decoration*, [#440], 274: V.8-9

IB218 *ꜣḫtj-ḥtp* - Saqqara, WSP (D64), hall, east: fording
Source: *P-M III*, 599-600; Davies, *Ptahhetep ii*, pl. 14
Label: n/a
Date: Harpur, *Decoration*, [#338], 272: V.8L-9E

IB219 *nbt* - Saqqara, UPC (ntn), room C, east: pastoral
Source: *P-M III*, 624-625; Munro, *Unas-Friedhof*, pl. 21
Label: missing
Date: Munro, *Unas-Friedhof*, 20: V.9; Harpur, *Decoration*, [#438], 274: V.9

IB220 *sḫm-ꜥnḫ-ptḥ* - Giza, EF (G7152), chapel, north: pastoral
Source: *P-M III*, 191; Badawy, *Tombs of Iteti*, fig. 23
Label: none
Date: Harpur, *Decoration*, [#223], 269: V.9-VI.1?

IB221 *mrrw-kꜣ.j: mrj* - Saqqara, TP (ntn), chamber A13, north: pastoral
Source: *P-M III*, 525-534; Duell, *Mereruka ii*, pl. 153
Label: *rn jwꜣ*
Young ox
Date: Harpur, *Decoration*, [#420], 274: VI.1M-L

IB222 *nj-ꜥnḫ-nswt* - Saqqara (ntn), relief fragment (Detroit Inst. Arts no. 30.371), unknown: fording
Source: *P-M III*, 694-696; *Eg. Art in the Age of the Pyramids*, cat. no. 193
Label: none
Date: Berman, *Catalogue*, 135: VI.1; Harpur, *Decoration*, [#429], 274: VI.1-2E

IB223 *jꜣzn* - Giza, WF (G2196), chapel?, north: agricultural
Source: *P-M III*, 82; Simpson, *Mast. West. Cem. Part I*, fig. 30
Label: none
Date: Harpur, *Decoration*, [#013], 265: VI.1M-2M

IB224 *mrjj-rꜥ-nfr: ḳꜣr* - Giza, EF (G7101), lower stairway (relief fragment no. 25-5-48), south: offering
Source: *P-M III*, 184-185; Simpson, *Qar and Idu*, fig. 19 [a]

---

37 For an extensive discussion of the translation of this passage see van Walsem, R. (1998). "The caption to a cattle-fording scene in a tomb at Saqqara and its implications for the seh/sinnbild discussion on Egyptian iconography" in Clarysse et al., *Egyptian Religion 2*, 1469-1485.

Label: missing
Date: Harpur, *Decoration*, [#090], 267: VI.4

Dogs

IB225 *k3-gm-nj: mmj* - Saqqara, TP (LS10), room III, north: pastoral
Source: *P-M III*, 521-525; Firth-Gunn, *Teti Pyr. ii*, pl. 52; Harpur-Scremin, *Kagemni*, fig. 7
Label: *s[wr]*
Drinking
Date: Harpur, *Decoration*, [#534], 276: VI.1M

Birds

IB226 *spd-ḥtp* - Saqqara, NSP (D15), chapel (Äg. Mus. Berlin no. 14642), east: pastoral/force-feeding
Source: *P-M III*, 481; Klebs, *Reliefs*, fig. 53; Janssen-Janssen, *Egyptian Household Animals*, fig. 40
Label: *bḏ3*
Duck
Date: Harpur, *Decoration*, [#494], 276: V.6L-9?

IB227 *ṯjj* - Saqqara, NSP (D22), portico, south: pastoral/force-feeding
Source: *P-M III*, 467-478; Epron et al., *Ti i*, pl. 7
Label: *pẖr m-ḫt snm*
Moving about after feeding
Date: Harpur, *Decoration*, [#543], 277: V.8-9

IB228 *ṯjj* - Saqqara, NSP (D22), portico, south: pastoral/force-feeding
Source: *P-M III*, 467-478; Epron et al., *Ti i*, pl. 7
Label: *pẖr ṯrp ḥḏ m-ḫt snmt*
Moving about (of) *ṯrp*-geese and *ḥḏ*-geese after feeding
Date: Harpur, *Decoration*, [#543], 277: V.8-9

IB229 *ṯjj* - Saqqara, NSP (D22), portico, south: pastoral/force-feeding
Source: *P-M III*, 467-478; Epron et al., *Ti i*, pls. 7-8
Label: *pẖr s(t) m-ḫt snmt*
Moving about (of) pintail ducks after feeding
Date: Harpur, *Decoration*, [#543], 277: V.8-9

IB230 *ṯjj* - Saqqara, NSP (D22), portico, south: pastoral/force-feeding
Source: *P-M III*, 467-478; Epron et al., *Ti i*, pl. 8
Label: *pẖr s mnwt m-ḫt snmt*
Moving about (of) *s*-geese and pigeons after feeding
Date: Harpur, *Decoration*, [#543], 277: V.8-9

IB231 *k3-gm-nj: mmj* - Saqqara, TP (LS10), room IV, west (right): pastoral/force-feeding
Source: *P-M III*, 521-525; von Bissing, *Gem-ni-kai i*, pl. 11; Harpur-Scremin, *Kagemni*, fig. 13
Label: *pẖr ṯrp ḥḏ m-ḫt snm*
Moving about (of) *ṯrp*-geese and *ḥḏ*-geese after feeding
Date: Harpur, *Decoration*, [#534], 276: VI.1M

IB232 *k3-gm-nj: mmj* - Saqqara, TP (LS10), room IV, west (right): pastoral/force-feeding
Source: *P-M III*, 521-525; von Bissing, *Gem-ni-kai i*, pl. 12; Harpur-Scremin, *Kagemni*, fig. 13
Label: *pẖr r m-ḫt snm*
Moving about (of) *r*-geese after feeding
Date: Harpur, *Decoration*, [#534], 276: VI.1M

IB233 *mrrw-k3.j: mrj* - Saqqara, TP (ntn), chamber A6, west: pastoral/force-feeding
Source: *P-M III*, 525-534; Duell, *Mereruka i*, pl. 52
Label: *snm ḏ3t*
Feeding cranes
wš ḏ3t
Fattening cranes
Date: Harpur, *Decoration*, [#420], 274: VI.1M-L

IB234 *mrjj-ttj: mrj* - Saqqara, TP (ntn), room C1, east: pastoral/poultry-yard
Source: *P-M III*, 356-357; Kanawati-Abder-Raziq, *Meryteti*, pl. 48
Label: none
Date: Kanawati-Abder-Raziq, *Meryteti*, 18: end VI.1-VI.2M; Harpur, *Decoration*, [#415], 274: VI.2M-L

8.1 PREDATION

Lions

HB1 *nj-ʿnḫ-ẖnmw/ẖnmw-ḥtp* - Saqqara, UPC (ntn), vestibule, west: hunting
Source: *P-M III*, 641-644; Moussa-Altenmüller, *Nianchchnum*, pl. 40
Label: none
Date: Harpur, *Decoration*, [#431], 274: V.6L-7

HB2 *nj-ʿnḫ-ẖnmw/ẖnmw-ḥtp* - Saqqara, UPC (ntn), vestibule, west: hunting
Source: *P-M III*, 641-644; Moussa-Altenmüller, *Nianchchnum*, pl. 40
Label: none
Date: Harpur, *Decoration*, [#431], 274: V.6L-7

HB3 *ṯjj* - Saqqara, NSP (D22), room III, east: desert
Source: *P-M III*, 468-478; Wild, *Ti ii*, pl. 127
Label: none
Date: Harpur, *Decoration*, [#543], 277: V.8-9

HB4 *ꜣḫtj-ḥtp* - Saqqara, WSP (D64), relief fragment from corridor, east: desert
Source: *P-M III*, 599-600; Davies, *Five Theban*, pl. 40 [1]
Label: none
Date: Harpur, *Decoration*, [#338], 272: V.8L-9E

HB5 *ptḥ-ḥtp II: ṯfj* - Saqqara, WSP (D64), offering-room, east: hunting
Source: *P-M III*, 600-604; Quibell, *Ptah-hetep*, pl. 32
Label: none
Date: Harpur, *Decoration*, [#400], 274: V.9M-L

HB6 *sšm-nfr IV* - Giza, GIS (LG53), vestibule, west: hunting
Source: *P-M III*, 223-226; Junker, *Giza xi*, fig. 63
Label: none
Date: Harpur, *Decoration*, [#235], 270: V.9-VI.1

HB6a *nj-mꜣʿt-rʿ [jzzj-mr-nṯr]* - Giza, WF (G2097), chapel, east: hunting
Source: *P-M III*, 70; Roth, *Cem. Pal. Att.*, pl. 189
Label: none
Date: Roth, *Cem. Pal. Att.*, 129-130: V.9; Harpur, *Decoration*, [#033], 265: V.9-VI.1

HB7 Anon - Saqqara, WSP (ntn), location: relief fragment found in tomb of *ṯfw*, unknown: desert
Source: *P-M III*, 605; Hassan, *Ny-Ankh-Pepy*, fig. 59
Label: none
Date: *P-M III*, 605: V-VI; Harpur, *Decoration*, nr

HB8 *mrrw-kꜣ.j: mrj* - Saqqara, TP (ntn), chamber A2, west: hunting
Source: *P-M III*, 525-534; Duell, *Mereruka i*, pl. 25
Label: none
Date: Harpur, *Decoration*, [#420], 274: VI.1M-L

HB9 *mrjj-ttj: mrj* - Saqqara, TP (ntn), room C1, west: desert
Source: *P-M III*, 356-357; Kanawati-Abder-Raziq, *Meryteti*, pl. 46
Label: none
Date: Kanawati-Abder-Raziq, *Meryteti*, 18: end VI.1-VI.2M; Harpur, *Decoration*, [#415], 274: VI.2M-L

HB10 *mrjj-ttj: mrj* - Saqqara, TP (ntn), room C1, west: desert
Source: *P-M III*, 356-357; Kanawati-Abder-Raziq, *Meryteti*, pl. 46
Label: none
Date: Kanawati-Abder-Raziq, *Meryteti*, 18: end VI.1-VI.2M; Harpur, *Decoration*, [#415], 274: VI.2M-L

HB11 *mrjj-ttj: mrj* - Saqqara, TP (ntn), room C1, west: desert
Source: *P-M III*, 356-357; Kanawati-Abder-Raziq, *Meryteti*, pl. 46
Label: none
Date: Kanawati-Abder-Raziq, *Meryteti*, 18: end VI.1-VI.2M; Harpur, *Decoration*, [#415], 274: VI.2M-L

HB12 *wr-nw* - Saqqara, TP (ntn), hall, west: desert
Source: *P-M III*, 519; Davies et al., *Saqqara Tombs i*, pl. 31
Label: none
Date: Harpur, *Decoration*, [#380], 273: VI.3-4?

HB13 *ḫꜥ-bꜣw-ḫnm* - Saqqara, EPII M.XIV, cylinder seal: desert
Source: *P-M III*, 684; Jequier, *Monument Funéraire iii*, fig. 68
Label: none
Date: *P-M III*, 684: VI.4; Harpur, *Decoration*, nr

Leopards

HB14 *nj-mꜣꜥt-rꜥ [jzzj-mr-nṯr]* - Giza, WF (G2097), chapel, east: hunting
Source: *P-M III*, 70; Roth, *Cem. Pal. Att.*, pl. 189
Label: none
Date: Roth, *Cem. Pal. Att.*, 129-130: V.9; Harpur, *Decoration*, [#033], 265: V.9-VI.1

Wild cats

HB15 *nj-ꜥnḫ-ḫnmw/ḫnmw-ḥtp* - Saqqara, UPC (ntn), vestibule, north: pastoral
Source: *P-M III*, 641-644; Moussa-Altenmüller, *Nianchchnum*, fig. 8
Label: none
Date: Harpur, *Decoration*, [#431], 274: V.6L-7

HB16 *ꜣḫtj-mrw-nswt* - Giza, WF (G2184), chapel, west: pastoral
Source: *P-M III*, 80-81; *Giza Archives Project* (Photo no. A779_NS)
Label: none
Date: Harpur, *Decoration*, [#008], 265: V.9-VI.1

Striped hyenas

HB17 *nj-ꜥnḫ-ḫnmw/ḫnmw-ḥtp* - Saqqara, UPC (ntn), vestibule, west: hunting
Source: *P-M III*, 641-644; Moussa-Altenmüller, *Nianchchnum*, pl. 40
Label: none
Date: Harpur, *Decoration*, [#431], 274: V.6L-7

Common genets

HB18 *nb.j-m-ꜣḫtj* - Giza, CF (LG86), chapel I, west: marsh (ascending to right)
Source: *P-M III*, 230-232; *LD ii*, pl. 12 [b]
Label: none
Date: Harpur, *Decoration*, [#122], 267: IV.6-V.1

HB19 *nj-ꜥnḫ-ḫnmw/ḫnmw-ḥtp* - Saqqara, UPC (ntn), forecourt, south: fishing (ascending to right)
Source: *P-M III*, 641-644; Moussa-Altenmüller, *Nianchchnum*, fig. 5
Label: none
Date: Harpur, *Decoration*, [#431], 274: V.6L-7

HB20 *nj-ꜥnḫ-ḫnmw/ḫnmw-ḥtp* - Saqqara, UPC (ntn), outer hall, west: marsh (ascending to right)
Source: *P-M III*, 641-644; Moussa-Altenmüller, *Nianchchnum*, pl. 74[38]
Label: none
Date: Harpur, *Decoration*, [#431], 274: V.6L-7

HB21 *ḥtp-ḥr-ꜣḫtj* - Saqqara, WSP (D60), chapel, north: marsh (ascending to left)
Source: *P-M III*, 593-595; Mohr, *Hetep-her-akhti*, fig. 34
Label: none
Date: Harpur, *Decoration*, [#467], 275: V.6-8E

HB22 *jtj-sn* - Giza, CF (ntn), chapel, east: marsh (ascending to right)
Source: *P-M III*, 252-253; Hassan, *Giza v*, fig. 123[39]

[38] Moussa and Altenmüller (*Nianchchnum*, 152) identify this animal as a genet.

[39] Hassan (*Giza v*, 267) states that this is an ichneumon, but the animal's pricked ears clearly indicate that it is a genet.

Label: none
Date: Harpur, *Decoration*, [#035], 265: V.6-8?

HB23 *jrj-n-k3-ptḥ* - Saqqara, UPC (ntn), offering chamber, south: marsh (ascending to left)
Source: *P-M III*, 644; Moussa-Junge, *Two Tombs*, pl. 12[40]
Label: none
Date: Harpur, *Decoration*, [#362], 273: V.6-9

HB24 *rʿ-špss* - Saqqara, NSP (LS16=S902), forecourt, west: marsh (ascending to right)
Source: *P-M III*, 494-496; Harpur, *Decoration*, fig. 187; *LD ii*, pl. 60 [left]
Label: none
Date: Harpur, *Decoration*, [#456], 275: V.8M

HB25 *snḏm-jb: jntj* - Giza, WF (G2370=LG27), portico, west: marsh (ascending to left)
Source: *P-M III*, 85-87; *LD Erganz.*, pl. 18; Brovarski, *Senedjemib Complex Part I*, fig. 27
Label: none
Date: Brovarski, *Senedjemib Complex Part I*, 23: V.8M-L; Harpur, *Decoration*, [#219], 269: V.8M-L

HB26 *snḏm-jb: jntj* - Giza, WF (G2370=LG27), room II, west: marsh (ascending to left)
Source: *P-M III*, 85-87; *LD ii*, pl. 77[41]
Label: none
Date: Brovarski, *Senedjemib Complex Part I*, 23: V.8M-L; Harpur, *Decoration*, [#219], 269: V.8M-L

HB27 *ṯjj* - Saqqara, NSP (D22), room III, north: marsh (ascending to left)
Source: *P-M III*, 468-478; Wild, *Ti ii*, pl. 119
Label: none
Date: Harpur, *Decoration*, [#543], 277: V.8-9

HB28 *ṯjj* - Saqqara, NSP (D22), room III, north: marsh (ascending to left)
Source: *P-M III*, 468-478; Wild, *Ti ii*, pl. 119
Label: none
Date: Harpur, *Decoration*, [#543], 277: V.8-9

HB29 *k3.j-m-nfrt* - Saqqara, NSP (D23), chapel, north: marsh (ascending to right, head turned back)
Source: *P-M III*, 467-468; Simpson, *Kayemnofret*, pl. A
Label: none
Date: Simpson, *Kayemnofret*, 1: V.8-9; Harpur, *Decoration*, [#522], 276: V.8-9?

HB30 *3ḫtj-ḥtp* - Saqqara, WSP (D64), hall, east: marsh (ascending to right, with chick)
Source: *P-M III*, 599-600; Davies, *Ptahhetep ii*, pl. 13
Label: none
Date: Harpur, *Decoration*, [#338], 272: V.8L-9E

HB31 *3ḫtj-ḥtp* - Saqqara, WSP (D64), hall, east: marsh (ascending to right, head turned back)
Source: *P-M III*, 599-600; Davies, *Ptahhetep ii*, pl. 14
Label: none
Date: Harpur, *Decoration*, [#338], 272: V.8L-9E

HB32 *ḫnwt* - Saqqara, UPC (ntn), room A [I], east: marsh (ascending to left)
Source: *P-M III*, 623-624; Munro, *Unas-Friedhof*, pl. 33[42]
Label: none
Date: Munro, *Unas-Friedhof*, 20: V.9; Harpur, *Decoration*, [#478], 275: V.9

HB33 *jj-nfrt: š3.n.f* - Saqqara, UPC (ntn), room I, north: marsh (ascending to right)
Source: *P-M III*, 616; Kanawati-Abder-Raziq, *Iynefert*, pl. 37

40 Moussa and Junge (*Two Tombs*, 42) suggest this is a genet, but the form could equally depict a mongoose. It has been accepted elsewhere as a genet. See Aufrère, S.H. (1998). "La loutre, le chat, la genette et l'ichneumon, hôtes du fourré de papyrus: Présages, prédateurs des marécages et croyances funéraires", *DE* 41, 7-28, fig. 4, 22.

41 According to Brovarski (*Senedjemib Complex Part I*, 49) this relief has now been destroyed.

42 *P-M III* (623) describe this animal as an ichneumon. Similarly, Munro (*Unas-Friedhof*, 52) states that the birds are "von einem Ichneumon belauert". The pricked ears indicate that it is actually a genet.

Label: none
Date: Kanawati-Abder-Raziq, *Iynefert*, 13: V.9E-M; Harpur, *Decoration*, [#345], 272: V.9

HB34 *nj-kꜣw-jzzj* - Saqqara, TP (ntn), room I, east: marsh (ascending to right, head turned back)
Source: *P-M III*, nr; Kanawati-Abder-Raziq, *Nikauisesi*, pl. 50
Label: none
Date: Kanawati-Abder-Raziq, *Nikauisesi*, 17-23: VI.1M; Harpur, *Decoration*, nr

HB35 *nj-kꜣw-jzzj* - Saqqara, TP (ntn), room I, east: marsh (ascending to right, with chick)
Source: *P-M III*, nr; Kanawati-Abder-Raziq, *Nikauisesi*, pl. 50
Label: none
Date: Kanawati-Abder-Raziq, *Nikauisesi*, 17-23: VI.1M; Harpur, *Decoration*, nr

HB36 *mrrw-kꜣ.j: mrj* - Saqqara, TP (ntn), chamber A1, north: marsh (ascending to right)
Source: *P-M III*, 525-534; Duell, *Mereruka i*, pl. 10
Label: none
Date: Harpur, *Decoration*, [#420], 274: VI.1M-L

HB37 *zšzšt: jdwt* - Saqqara, UPC (ntn), room III, west: marsh (ascending to right, with chick)
Source: *P-M III*, 617-619; Macramallah, *Idout*, pl. 7; Kanawati-Abder-Raziq, *Iynefert*, pl. 54
Label: none
Date: Kanawati-Abder-Raziq, *Iynefert*, 36-37: VI.1E; Harpur, *Decoration*, [#487], 275: V.9 r/u, VI.1-2E?

HB38 *sꜥnḫ-w(j)-ptḥ* - Saqqara, TP (ntn), room I, east: marsh (ascending to left, with chick)
Source: *P-M III*, nr; Kanawati-Abder-Raziq, *Neferseshemre*, pls. 69 and 76
Label: none
Date: Kanawati-Abder-Raziq, *Neferseshemre*, 40-41: VI.1L; Harpur, *Decoration*, nr

HB39 *sꜥnḫ-w(j)-ptḥ* - Saqqara, TP (ntn), room I, east: marsh (ascending to left, head turned back)
Source: *P-M III*, nr; Kanawati-Abder-Raziq, *Neferseshemre*, pls. 69 and 76
Label: none
Date: Kanawati-Abder-Raziq, *Neferseshemre*, 40-41: VI.1L; Harpur, *Decoration*, nr

HB40 *ḥzj* - Saqqara, TP (ntn), portico, south (west of entrance): marsh (ascending to left, head turned back)
Source: *P-M III*, nr; Kanawati-Abder-Raziq, *Hesi*, pl. 54
Label: none
Date: Kanawati-Abder-Raziq, *Hesi*, 16: VI.1L-2E; Harpur, *Decoration*, nr

HB41 *ḥzj* - Saqqara, TP (ntn), portico, south (west of entrance): marsh (descending to left, with chick)
Source: *P-M III*, nr; Kanawati-Abder-Raziq, *Hesi*, pl. 54
Label: none
Date: Kanawati-Abder-Raziq, *Hesi*, 16: VI.1L-2E; Harpur, *Decoration*, nr

HB42 *ḥzj* - Saqqara, TP (ntn), portico, south (east of entrance): marsh (ascending to right, with chick)
Source: *P-M III*, nr; Kanawati-Abder-Raziq, *Hesi*, pl. 53
Label: none
Date: Kanawati-Abder-Raziq, *Hesi*, 16: VI.1L-2E; Harpur, *Decoration*, nr

HB43 *mrjj-rꜥ-mrj-ptḥ-ꜥnḫ: nḫbw* - Giza, WF (G2381), relief fragment from forecourt (MFA no. 13.4332), unknown: marsh (ascending to left, head turned back)
Source: *P-M III*, 90-91; Smith, W.S. (1958). *Bost. Mus. Bull.* 56, fig. 2
Label: none
Date: Harpur, *Decoration*, [#089], 267: VI.2

HB43a *jnw-mnw* - Saqqara (ntn), room II, west: marsh (ascending to right, with chick)
Source: *P-M III*, nr; Kanawati, *Inumin*, pl. 46
Label: none
Date: Kanawati, *Inumin*, 16-17: VI.2E; Harpur, *Decoration*, nr

HB44 *mḥw* - Saqqara, UPC (ntn), room I, north: marsh (ascending to right, with chick)
Source: *P-M III*, 619-622; Altenmüller, *Mehu*, pl. 9
Label: none
Date: Altenmüller, *Mehu*, 19: VI.2; Harpur, *Decoration*, [#424], 274: VI.2M-3?

HB45 *mḥw* - Saqqara, UPC (ntn), room I, east: marsh (ascending to left, with chick)
Source: *P-M III*, 619-622; Altenmüller, *Mehu*, pl. 10
Label: none
Date: Altenmüller, *Mehu*, 19: VI.2; Harpur, *Decoration*, [#424], 274: VI.2M-3?

HB46 *mḥw* - Saqqara, UPC (ntn), room I, east: marsh (ascending to right, with adult bird)
Source: *P-M III*, 619-622; Altenmüller, *Mehu*, pl. 11
Label: none
Date: Altenmüller, *Mehu*, 19: VI.2; Harpur, *Decoration*, [#424], 274: VI.2M-3?

HB47 *mḥw* - Saqqara, UPC (ntn), room I, west: marsh (ascending to left, with chick)
Source: *P-M III*, 619-622; Altenmüller, *Mehu*, pl. 13
Label: none
Date: Altenmüller, *Mehu*, 19: VI.2; Harpur, *Decoration*, [#424], 274: VI.2M-3?

HB48 *jj-nfrt* - Giza (ntn), chapel, unknown: marsh (ascending to right)
Source: *P-M III*, 298-299; Schürmann, *Die Reliefs*, fig. 6 [b][43]
Label: none
Date: Harpur, *Decoration*, [#017], 265: VI.3-4E?

HB49 *jj-nfrt* - Giza (ntn), chapel, unknown: marsh (ascending to left)
Source: *P-M III*, 298-299; Schürmann, *Die Reliefs*, fig. 6 [b]
Label: none
Date: Harpur, *Decoration*, [#017], 265: VI.3-4E?

HB50 *mrjj-rˁ-nfr: ḳ3r* - Giza, EF (G7101), lower stairway (fragment no. 25-5-16), north: fishing (ascending to left, with chick)
Source: *P-M III*, 184-185; Simpson, *Qar and Idu*, fig. 16[44]
Label: none
Date: Harpur, *Decoration*, [#090], 267: VI.4

HB51 *3ḫtj-ḥtp* - Saqqara, UPC (E17), offering-room, east: marsh (ascending to right, head turned back)
Source: *P-M III*, 633-634; Petrie-Murray, *Seven Memphite*, pl. 6 [lower]
Label: none
Date: Harpur, *Decoration*, [#339], 272: VI.5?

Egyptian mongooses

HB52 *nb.j-m-3ḫtj* - Giza, CF (LG86), chapel I, west: marsh (ascending to left)
Source: *P-M III*, 230-232; *LD ii*, pl. 12 [b]
Label: none
Date: Harpur, *Decoration*, [#122], 267: IV.6-V.1

HB52a *nj-ˁnḫ-ẖnmw/ẖnmw-ḥtp* - Saqqara, UPC (ntn), outer hall, west: marsh (ascending to left)
Source: *P-M III*, 641-644; Moussa-Altenmüller, *Nianchchnum*, pl. 74[45]
Label: none
Date: Harpur, *Decoration*, [#431], 274: V.6L-7

HB53 *3ḫtj-ḥtp* - Saqqara, UPC (ntn), chapel, east: marsh (ascending to right)
Source: *P-M III*, 634-637; Ziegler, *Akhethetep*, 130-131
Label: none
Date: Ziegler, *Akhethetep*, 9: V.6; Harpur, *Decoration*, [#340], 272: V.6-8E

HB54 *ẖnmw-ḥtp* - Saqqara, ESP (D49), relief fragment (Berlin (East) Mus. no. 14101), north: marsh (ascending to left)
Source: *P-M III*, 578-579; Decker-Herb, *Bildatlas*, pl. 192 [K.126]

43 Schürmann (*Die Reliefs*, 32) identifies the animal here as a common genet ("Ginsterkatze"), but so little remains, I suggest that it is impossible to determine with certainty the species depicted.

44 Simpson (*Qar and Idu*, 3) states that this is a cat. For a discussion regarding the identification of this animal, see Malek, *The Cat*, 38.

45 Moussa and Altenmüller (*Nianchchnum*, 152) identify this animal as a marsh mongoose ("Sumpf-ichneumon, *Atilax paludinosus*"). To my knowledge, this species is not found in Egypt.

Label: none
Date: Harpur, *Decoration*, [#481], 275: V.6-8E?

HB55 *k3.j-dw3* - Giza, CF (ntn), chapel, north: marsh (ascending to right)
Source: *P-M III*, 244-245; Hassan, *Giza vi*, fig. 80
Label: none
Date: Harpur, *Decoration*, [#272], 270: V.6-9?

HB56 *k3.j-dw3* - Giza, CF (ntn), chapel, north: marsh (ascending to left)
Source: *P-M III*, 244-245; Hassan, *Giza vi*, fig. 80
Label: none
Date: Harpur, *Decoration*, [#272], 270: V.6-9?

HB57 *rʿ-špss* - Saqqara, NSP (LS16=S902), forecourt, north: marsh (ascending to left)
Source: *P-M III*, 494-496; *LD ii*, pl. 60 [right]; Harpur, *Decoration*, fig. 120
Label: none
Date: Harpur, *Decoration*, [#456], 275: V.8

HB58 *snḏm-jb: jntj* - Giza, WF (G2370=LG27), portico, west: marsh (ascending to right)
Source: *P-M III*, 85-87; *LD Erganz.*, pl. 18[46]
Label: none
Date: Brovarski, *Senedjemib Complex Part I*, 23: V.8M-L; Harpur, *Decoration*, [#219], 269: V.8M-L

HB59 *snḏm-jb: jntj* - Giza, WF (G2370=LG27), room II, west: marsh (ascending to right)
Source: *P-M III*, 85-87; *LD ii*, pl. 77[47]
Label: none
Date: Brovarski, *Senedjemib Complex Part I*, 23: V.8M-L; Harpur, *Decoration*, [#219], 269: V.8M-L

HB60 *ṯjj* - Saqqara, NSP (D22), room III, north: marsh (ascending to right)
Source: *P-M III*, 468-478; Wild, *Ti ii*, pl. 119
Label: none
Date: Harpur, *Decoration*, [#543], 277: V.8-9

HB61 *nfr-jrt-n.f* - Saqqara, ESP (D55), chapel, east: marsh (ascending to right)
Source: *P-M III*, 583-584; van de Walle, *Neferirtenef*, pl. 1
Label: none
Date: Harpur, *Decoration*, [#440], 274: V.8-9

HB62 *k3.j-m-nfrt* - Saqqara, NSP (D23), chapel, north: marsh (ascending to left)
Source: *P-M III*, 467-468; Simpson, *Kayemnofret*, pl. A
Label: none
Date: Simpson, *Kayemnofret*, 1: V.8-9; Harpur, *Decoration*, [#522], 276: V.8-9?

HB63 *3ḫtj-ḥtp* - Saqqara, WSP (D64), hall, east: marsh (ascending to right, head turned back)
Source: *P-M III*, 599-600; Davies, *Ptahhetep ii*, pl. 13
Label: none
Date: Harpur, *Decoration*, [#338], 272: V.8L-9E

HB64 *3ḫtj-ḥtp* - Saqqara, WSP (D64), hall, east: marsh (ascending to left)
Source: *P-M III*, 599-600; Davies, *Ptahhetep ii*, pl. 14
Label: none
Date: Harpur, *Decoration*, [#338], 272: V.8L-9E

HB65 *nbt* - Saqqara, UPC (ntn), room A [I], east: marsh (ascending to left)
Source: *P-M III*, 624-625; Munro, *Unas-Friedhof*, pl. 11
Label: missing
Date: Munro, *Unas-Friedhof*, 20: V.9; Harpur, *Decoration*, [#438], 274: V.9

HB66 *jj-nfrt: š3.n.f* - Saqqara, UPC (ntn), room I, north: marsh (ascending to right, head missing)
Source: *P-M III*, 616; Kanawati-Abder-Raziq, *Iynefert*, pl. 37[48]

---

46 According to Brovarski (*Senedjemib Complex Part I*, 40), this section of the relief has now been destroyed.
47 This relief is no longer extant. See *ibid.*, 49.

Label: none
Date: Kanawati-Abder-Raziq, *Iynefert*, 13: V.9E-M; Harpur, *Decoration*, [#345], 272: V.9

HB67 *nj-k3w-jzzj* - Saqqara, TP (ntn), room I, east: marsh (ascending to right, with chick)
Source: *P-M III*, nr; Kanawati-Abder-Raziq, *Nikauisesi*, pl. 50
Label: none
Date: Kanawati-Abder-Raziq, *Nikauisesi*, 17-23: VI.1M; Harpur, *Decoration*, nr

HB68 *nj-k3w-jzzj* - Saqqara, TP (ntn), room I, east: marsh (ascending to left, attacking adult bird)
Source: *P-M III*, nr; Kanawati-Abder-Raziq, *Nikauisesi*, pl. 50
Label: none
Date: Kanawati-Abder-Raziq, *Nikauisesi*, 17-23: VI.1M; Harpur, *Decoration*, nr

HB69 *mrrw-k3.j: mrj* - Saqqara, TP (ntn), chamber A1, north: marsh (ascending to left, head turned back)
Source: *P-M III*, 525-534; Duell, *Mereruka i*, pl. 10
Label: none
Date: Harpur, *Decoration*, [#420], 274: VI.1M-L

HB70 *mrrw-k3.j: mrj* - Saqqara, TP (ntn), chamber A1, south: marsh (descending to left, with chick)
Source: *P-M III*, 525-534; Duell, *Mereruka i*, pl. 19
Label: none
Date: Harpur, *Decoration*, [#420], 274: VI.1M-L

HB71 *mrrw-k3.j: mrj* - Saqqara, TP (ntn), chamber A1, south: marsh (ascending to right, tail held)
Source: *P-M III*, 525-534; Duell, *Mereruka i*, pl. 19
Label: none
Date: Harpur, *Decoration*, [#420], 274: VI.1M-L

HB72 *zšzšt: jdwt* - Saqqara, UPC (ntn), room III, west: marsh (ascending to left)
Source: *P-M III*, 617-619; Macramallah, *Idout*, pl. 7; Kanawati-Abder-Raziq, *Iynefert*, pl. 54
Label: none
Date: Kanawati-Abder-Raziq, *Iynefert*, 36-37: VI.1E; Harpur, *Decoration*, [#487], 275: V.9 r/u, VI.1-2E?

HB73 *j3zn* - Giza, WF (G2196), chapel, north: fishing (ascending to right)
Source: *P-M III*, 82; Simpson, *Mast. West. Cem. Part I*, fig. 30
Label: none
Date: Harpur, *Decoration*, [#013], 265: VI.1M-2M

HB74 *sꜥnḫ-w(j)-ptḥ* - Saqqara, TP (ntn), room I, east: marsh (ascending to right)
Source: *P-M III*, nr; Kanawati-Abder-Raziq, *Neferseshemre*, pls. 69 and 76
Label: none
Date: Kanawati-Abder-Raziq, *Neferseshemre*, 40-41: VI.1L; Harpur, *Decoration*, nr

HB75 *sꜥnḫ-w(j)-ptḥ* - Saqqara, TP (ntn), room I, east: marsh (ascending to left, with chick)
Source: *P-M III*, nr; Kanawati-Abder-Raziq, *Neferseshemre*, pls. 69 and 76
Label: none
Date: Kanawati-Abder-Raziq, *Neferseshemre*, 40-41: VI.1L; Harpur, *Decoration*, nr

HB76 *ḥzj* - Saqqara, TP (ntn), portico, south (east of entrance): marsh (ascending to right, attacking adult bird)
Source: *P-M III*, nr; Kanawati-Abder-Raziq, *Hesi*, pl. 53
Label: none
Date: Kanawati-Abder-Raziq, *Hesi*, 16: VI.1L-2E; Harpur, *Decoration*, nr

HB77 *ḥzj* - Saqqara, TP (ntn), portico, south (west of entrance): marsh (ascending to right, with chick)
Source: *P-M III*, nr; Kanawati-Abder-Raziq, *Hesi*, pl. 54
Label: none
Date: Kanawati-Abder-Raziq, *Hesi*, 16: VI.1L-2E; Harpur, *Decoration*, nr

---

[48] Kanawati and Abder-Raziq (*Iynefert*, 18) identify this animal as an Egyptian mongoose even though the head region is damaged. This identification is supported by the fact that it drags its tail on the substrate, whereas that of a genet higher up in the same scene (see LB86 in Chapter 4) is elevated.

HB78 *mṯtj* - Saqqara, UPC (ntn), relief fragment (Abegg-Stiftung, Riggisberg no. 12.39.67), unknown: marsh (ascending to left, head turned back)
Source: *P-M III*, 646-648; Kaplony, *Studien Methethi*, fig. on 10[49]
Label: none
Date: Ziegler, *Catalogue*, 123: probably VI; Harpur, *Decoration*, [#426], 274: VI.2?

HB79 *mrrj* - Saqqara, TP (ntn), room 1, south: marsh (ascending to right, head turned back)
Source: *P-M III*, 518-519; Davies et al., *Saqqara Tombs i*, pl. 5[50]
Label: none
Date: Harpur, *Decoration*, [#418], 274: VI.2E?

HB80 *mḥw* - Saqqara, UPC (ntn), room I, north: marsh (ascending to left, with chick)
Source: *P-M III*, 619-622; Altenmüller, *Mehu*, pl. 9
Label: none
Date: Altenmüller, *Mehu*, 19: VI.2; Harpur, *Decoration*, [#424], 274: VI.2M-3?

HB81 *mḥw* - Saqqara, UPC (ntn), room I, east: marsh (ascending to right)
Source: *P-M III*, 619-622; Altenmüller, *Mehu*, pl. 10
Label: none
Date: Altenmüller, *Mehu*, 19: VI.2; Harpur, *Decoration*, [#424], 274: VI.2M-3?

HB82 *mḥw* - Saqqara, UPC (ntn), room I, east: marsh (ascending to right, tail held)
Source: *P-M III*, 619-622; Altenmüller, *Mehu*, pl. 11
Label: name and titles
Date: Altenmüller, *Mehu*, 19: VI.2; Harpur, *Decoration*, [#424], 274: VI.2M-3?

HB83 *mḥw* - Saqqara, UPC (ntn), room I, west: marsh (ascending to right, with chick)
Source: *P-M III*, 619-622; Altenmüller, *Mehu*, pl. 13
Label: none
Date: Altenmüller, *Mehu*, 19: VI.2; Harpur, *Decoration*, [#424], 274: VI.2M-3?

HB84 *mrjj-rˁ-nfr: ḳ3r* - Giza, EF (G7101), lower stairway (fragment no. 25-5-8), north: fishing (ascending to right)
Source: *P-M III*, 184-185; Simpson, *Qar and Idu*, fig. 16[51]
Label: none
Date: Harpur, *Decoration*, [#090], 267: VI.4

HB85 *3ḫtj-ḥtp* - Saqqara, UPC (E17), offering-room, east: marsh (ascending to left)
Source: *P-M III*, 633-634; Petrie-Murray, *Seven Memphite*, pl. 6 [lower]
Label: none
Date: Harpur, *Decoration*, [#339], 272: VI.5?

Golden jackals

HB86 *3ḫtj-mrw-nswt* - Giza, WF (G2184), secondary offering-room, west: agricultural
Source: *P-M III*, 80-81; Smith, *HESPOK*, fig. 226 [c]
Label: none
Date: Harpur, *Decoration*, [#008], 265: V.9-VI.1

HB87 *mrjj-ttj: mrj* - Saqqara, TP (ntn), room C1, west: desert (attacking Nubian ibex, left)
Source: *P-M III*, 356-357; Kanawati-Abder-Raziq, *Meryteti*, pl. 46
Label: none
Date: Kanawati-Abder-Raziq, *Meryteti*, 18: end VI.1-VI.2M; Harpur, *Decoration*, [#415], 274: VI.2M-L

HB88 *mrjj-ttj: mrj* - Saqqara, TP (ntn), room C1, west: desert (group on a sub-register)
Source: *P-M III*, 356-357; Kanawati-Abder-Raziq, *Meryteti*, pl. 46
Label: none
Date: Kanawati-Abder-Raziq, *Meryteti*, 18: end VI.1-VI.2M; Harpur, *Decoration*, [#415], 274: VI.2M-L

---

49 Kaplony (*Studien Methethi*, 10-11) believes this to be a mongoose ("hält der Ichneumon ein Vögelchen im Maul"). The elevated tail, however, may indicate that it is a genet.

50 Davies et al. (*Saqqara Tombs i*, 11) identify this animal as an otter.

51 Simpson (*Qar and Idu*, 3) states that this is a cat.

Dogs

HB89 *mṯn* - Saqqara, NSP (LS6), chapel, south: hunting (prey: Nubian ibex)
Source: *P-M III*, 493-494; Smith, *HESPOK*, fig. 60
Label: none
Date: Harpur, *Decoration*, [#425], 274: IV.1-2

HB90 *mṯn* - Saqqara, NSP (LS6), chapel, south: hunting (prey: Sömmerings gazelle)
Source: *P-M III*, 493-494; Smith, *HESPOK*, fig. 60
Label: none
Date: Harpur, *Decoration*, [#425], 274: IV.1-2

HB91 *mṯn* - Saqqara, NSP (LS6), chapel, south: hunting (prey: hartebeest)
Source: *P-M III*, 493-494; Smith, *HESPOK*, fig. 60
Label: none
Date: Harpur, *Decoration*, [#425], 274: IV.1-2

HB92 *rʿ-m-k3.j* - Saqqara, NSP (D3=S903), chapel, south: hunting (prey: red fox)
Source: *P-M III*, 487-488; Hayes, *Scepter i*, fig. 56
Label: *ṯsm*
Dog
Date: Harpur, *Decoration*, [#453], 275: V.6? r/u - V.6-7

HB93 *rʿ-m-k3.j* - Saqqara, NSP (D3=S903), chapel, south: hunting (prey: dorcas gazelle)
Source: *P-M III*, 487-488; Hayes, *Scepter i*, fig. 56
Label: *ṯsm*
Dog
Date: Harpur, *Decoration*, [#453], 275: V.6? r/u - V.6-7

HB94 *nj-ʿnḫ-ẖnmw/ḫnmw-ḥtp* - Saqqara, UPC (ntn), vestibule, west: hunting (prey: hare)
Source: *P-M III*, 641-644; Moussa-Altenmüller, *Nianchchnum*, pl. 40
Label: none
Date: Harpur, *Decoration*, [#431], 274: V.6L-7

HB95 *pḥn-wj-k3.j* - Saqqara, NSP (D70=LS15), relief fragment (Staat. Mus. Berlin no. 1132), probably from antechamber (room I), west: hunting (prey: red fox?)
Source: *P-M III*, 491-492; *LD ii*, pl. 46
Label: none
Date: Harpur, *Decoration*, [#393], 273: V.6-8E

HB96 *pḥn-wj-k3.j* - Saqqara, NSP (D70=LS15), antechamber (room I), west: hunting (prey: unidentified ungulate; only the legs of the dog are visible)
Source: *P-M III*, 491-492; *LD ii*, pl. 46
Label: missing
Date: Harpur, *Decoration*, [#393], 273: V.6-8E

HB97 *3ḫtj-ḥtp* - Saqqara, UPC (ntn), chapel, east: pastoral (watching a goat giving birth)
Source: *P-M III*, 634-637; Ziegler, *Akhethetep*, 138-139
Label: none
Date: Ziegler, *Akhethetep*, 9: V.6; Harpur, *Decoration*, [#340], 272: V.6-8E

HB98 *3ḫtj-ḥtp* - Saqqara, WSP (D64), relief fragment from corridor, east: hunting (prey: not visible)
Source: *P-M III*, 599-600; Davies, *Five Theban*, pl. 40 [3]
Label: none
Date: Harpur, *Decoration*, [#338], 272: V.8L-9E

HB99 *ptḥ-ḥtp II: ṯfj* - Saqqara, WSP (D64), offering-room, east: hunting (prey: striped hyenas)
Source: *P-M III*, 600-604; Quibell, *Ptah-hetep*, pl. 32
Label: none
Date: Harpur, *Decoration*, [#400], 274: V.9M-L

HB100 *ptḥ-ḥtp II: ṯfj* - Saqqara, WSP (D64), offering-room, east: hunting (prey: scimitar oryx, left)
Source: *P-M III*, 600-604; Quibell, *Ptah-hetep*, pl. 32

Label: none
Date: Harpur, *Decoration*, [#400], 274: V.9M-L

HB101 *ptḥ-ḥtp II: ṯfj* - Saqqara, WSP (D64), offering-room, east: hunting (prey: Nubian ibex)
Source: *P-M III*, 600-604; Quibell, *Ptah-hetep*, pl. 32
Label: none
Date: Harpur, *Decoration*, [#400], 274: V.9M-L

HB102 *ptḥ-ḥtp II: ṯfj* - Saqqara, WSP (D64), offering-room, east: hunting (prey: dorcas gazelle)
Source: *P-M III*, 600-604; Quibell, *Ptah-hetep*, pl. 32
Label: none
Date: Harpur, *Decoration*, [#400], 274: V.9M-L

HB103 *ptḥ-ḥtp II: ṯfj* - Saqqara, WSP (D64), offering-room, east: hunting (prey: scimitar oryx, right)
Source: *P-M III*, 600-604; Quibell, *Ptah-hetep*, pl. 32
Label: none
Date: Harpur, *Decoration*, [#400], 274: V.9M-L

HB104 *sšm-nfr IV* - Giza, GIS (LG53), vestibule, west: hunting (prey: unidentified ungulate)
Source: *P-M III*, 223-226; Junker, *Giza xi*, fig. 63
Label: none
Date: Harpur, *Decoration*, [#235], 270: V.9-VI.1

HB105 *ȝḫtj-mrw-nswt* - Giza, WF (G2184), courtyard, unknown: hunting (2 dogs; prey: scimitar oryx)
Source: *P-M III*, 80-81; *Giza Archives Project* (Photo no. C12743_NS)[52]
Label: none
Date: Harpur, *Decoration*, [#008], 265: V.9-VI.1

HB106 *nj-mȝꜥt-rꜥ* [*jzzj-mr-nṯr*] - Giza, WF (G2097), chapel, east: hunting (prey: dorcas gazelle)
Source: *P-M III*, 70; Roth, *Cem. Pal. Att.*, pl. 189
Label: none
Date: Roth, *Cem. Pal. Att.*, 129-130: V.9; Harpur, *Decoration*, [#033], 265: V.9-VI.1

HB107 *nj-mȝꜥt-rꜥ* [*jzzj-mr-nṯr*] - Giza, WF (G2097), chapel, east: hunting (2 dogs; prey: scimitar oryx)
Source: *P-M III*, 70; Roth, *Cem. Pal. Att.*, pl. 189
Label: none
Date: Roth, *Cem. Pal. Att.*, 129-130: V.9; Harpur, *Decoration*, [#033], 265: V.9-VI.1

HB108 *nj-mȝꜥt-rꜥ* [*jzzj-mr-nṯr*] - Giza, WF (G2097), chapel, east: hunting (prey: golden jackal?)
Source: *P-M III*, 70; Roth, *Cem. Pal. Att.*, pl. 189
Label: none
Date: Roth, *Cem. Pal. Att.*, 129-130: V.9; Harpur, *Decoration*, [#033], 265: V.9-VI.1

HB109 *nj-mȝꜥt-rꜥ* [*jzzj-mr-nṯr*] - Giza, WF (G2097), chapel, east: hunting (prey: unidentified ungulate)
Source: *P-M III*, 70; Roth, *Cem. Pal. Att.*, pl. 189
Label: none
Date: Roth, *Cem. Pal. Att.*, 129-130: V.9; Harpur, *Decoration*, [#033], 265: V.9-VI.1

HB110 *nj-mȝꜥt-rꜥ* [*jzzj-mr-nṯr*] - Giza, WF (G2097), chapel, east: hunting (2 dogs; prey: unidentified ungulate)
Source: *P-M III*, 70; Roth, *Cem. Pal. Att.*, pl. 189
Label: none
Date: Roth, *Cem. Pal. Att.*, 129-130: V.9; Harpur, *Decoration*, [#033], 265: V.9-VI.1

HB111 *nj-mȝꜥt-rꜥ* [*jzzj-mr-nṯr*] - Giza, WF (G2097), chapel, east: hunting (prey: red fox or golden jackal)
Source: *P-M III*, 70; Roth, *Cem. Pal. Att.*, pl. 189
Label: none
Date: Roth, *Cem. Pal. Att.*, 129-130: V.9; Harpur, *Decoration*, [#033], 265: V.9-VI.1

52 Watercolour sketches made c. 1912 by Frida Firth are the only remaining records of many of the scenes from the tomb of *ȝḫtj-mrw-nswt*. HB105 is unpublished, but was recently made available on the Giza Archives Project website. A note accompanying the image questions its association with tomb G1284 and states that, "It is not part of (a detail of) any other scenes copied by Firth". However, D'Auria et al. (*Mummies and Magic*, 86) report that "the vignette has been copied from the (courtyard) area" of the tomb.

HB112 Anon - Saqqara, WSP (ntn), block found in the tomb of *ṯfw*, unknown: hunting (prey: Nubian ibex)
Source: *P-M III*, 605; Hassan, *Ny-Ankh-Pepy*, fig. 59
Label: none
Date: *P-M III*, 605: V-VI; Harpur, *Decoration*, nr

HB113 Anon - Saqqara, WSP (ntn), relief fragment from shaft 6, 'Mastaba of Shaft no. 14', unknown: hunting (prey: unidentified ungulate)
Source: *P-M III*, 613; Hassan, *Hemet-Re*, pl. 14 [c]
Label: none
Date: *P-M III*, 613: VI; Harpur, *Decoration*, nr

HB114 *mrrw-kꜣ.j: mrj* - Saqqara, TP (ntn), chamber A2, west: hunting (8 dogs; prey: Nubian ibex)
Source: *P-M III*, 525-534; Duell, *Mereruka i*, pl. 25
Label: none
Date: Harpur, *Decoration*, [#420], 274: VI.1M-L

HB115 *mrrw-kꜣ.j: mrj* - Saqqara, TP (ntn), chamber A2, west: hunting (prey: scimitar oryx)
Source: *P-M III*, 525-534; Duell, *Mereruka i*, pl. 25
Label: none
Date: Harpur, *Decoration*, [#420], 274: VI.1M-L

HB116 *mrrw-kꜣ.j: mrj* - Saqqara, TP (ntn), chamber A2, west: hunting (prey: Nubian ibex)
Source: *P-M III*, 525-534; Duell, *Mereruka i*, pl. 25
Label: none
Date: Harpur, *Decoration*, [#420], 274: VI.1M-L

HB117 *sꜥnḫ-w(j)-ptḥ* - Saqqara, TP (ntn), room I, west: hunting (prey: Nubian ibex)
Source: *P-M III*, nr; Kanawati-Abder-Raziq, *Neferseshemre*, pls. 67 and 71
Label: none
Date: Kanawati-Abder-Raziq, *Neferseshemre*, 40-41: VI.1L; Harpur, *Decoration*, nr

HB118 *jnw-mnw* - Saqqara (ntn), room II, north: hunting (10 dogs; prey: Nubian ibex)
Source: *P-M III*, nr; Kanawati, *Inumin*, pl. 47
Label: none
Date: Kanawati, *Inumin*, 16-17: VI.2E; Harpur, *Decoration*, nr

HB118a *jnw-mnw* - Saqqara (ntn), room II, north: hunting (prey: Nubian ibex)
Source: *P-M III*, nr; Kanawati, *Inumin*, pl. 47
Label: none
Date: Kanawati, *Inumin*, 16-17: VI.2E; Harpur, *Decoration*, nr

HB119 *mrjj-ttj: mrj* - Saqqara, TP (ntn), room C1, west: hunting (2 dogs; prey: dorcas gazelle)
Source: *P-M III*, 356-357; Kanawati-Abder-Raziq, *Meryteti*, pl. 46
Label: none
Date: Kanawati-Abder-Raziq, *Meryteti*, 18: end VI.1-VI.2M; Harpur, *Decoration*, [#415], 274: VI.2M-L

HB120 *mrjj-ttj: mrj* - Saqqara, TP (ntn), room C1, west: hunting (9 dogs; prey: Nubian ibex)
Source: *P-M III*, 356-357; Kanawati-Abder-Raziq, *Meryteti*, pl. 46
Label: none
Date: Kanawati-Abder-Raziq, *Meryteti*, 18: end VI.1-VI.2M; Harpur, *Decoration*, [#415], 274: VI.2M-L

HB121 *mrjj-ttj: mrj* - Saqqara, TP (ntn), room C1, west: hunting (prey: dorcas gazelle)
Source: *P-M III*, 356-357; Kanawati-Abder-Raziq, *Meryteti*, pl. 46
Label: none
Date: Kanawati-Abder-Raziq, *Meryteti*, 18: end VI.1-VI.2M; Harpur, *Decoration*, [#415], 274: VI.2M-L

HB122 *mrjj-ttj: mrj* - Saqqara, TP (ntn), room C1, west: hunting (prey: scimitar oryx)
Source: *P-M III*, 356-357; Kanawati-Abder-Raziq, *Meryteti*, pl. 46
Label: none
Date: Kanawati-Abder-Raziq, *Meryteti*, 18: end VI.1-VI.2M; Harpur, *Decoration*, [#415], 274: VI.2M-L

Hedgehogs

HB123 *pḥn-wj-k3.j* - Saqqara, NSP (D70=LS15), antechamber (room I), west: desert
Source: P-M III, 491-492; *LD ii*, pl. 46
Label: none
Date: Harpur, *Decoration*, [#393], 273: V.6-8E

HB124 *ptḥ-ḥtp II*: *ṯfj* - Saqqara, WSP (D64), offering-room, east: hunting
Source: *P-M III*, 600-604; Quibell, *Ptah-hetep*, pl. 32
Label: none
Date: Harpur, *Decoration*, [#400], 274: V.9M-L

HB125 Anon - Saqqara, WSP (ntn), relief fragment from shaft 6, 'Mastaba of Shaft no. 14', unknown: desert
Source: *P-M III*, 613; Hassan, *Hemet-Re*, pl. 14 [c]
Label: none
Date: *P-M III*, 613: VI; Harpur, *Decoration*, nr

Sacred ibis

HB126 *mḥw* - Saqqara, UPC (ntn), room I, east: marsh
Source: *P-M III*, 619-622; Altenmüller, *Mehu*, pl. 10
Label: none
Date: Altenmüller, *Mehu*, 19: VI.2; Harpur, *Decoration*, [#424], 274: VI.2M-3?

Pied kingfishers

HB127 *nfr/k3-ḥ3.j* - Saqqara, UPC (ntn), chapel, east: marsh
Source: *P-M III*, 639-641; Moussa-Altenmüller, *Nefer and Ka-hay*, pl. 5
Label: none
Date: Harpur, *Decoration*, [#439], 274: V.6

HB128 *nj-ʿnḫ-ẖnmw/ẖnmw-ḥtp* - Saqqara, UPC (ntn), forecourt, south: fishing
Source: *P-M III*, 641-644; Moussa-Altenmüller, *Nianchchnum*, fig. 5
Label: none
Date: Harpur, *Decoration*, [#431], 274: V.6L-7)

HB129 *ṯjj* - Saqqara, NSP (D22), room III, north: marsh
Source: *P-M III*, 468-478; Wild, *Ti ii*, pl. 115
Label: none
Date: Harpur, *Decoration*, [#543], 277: V.8-9

HB130 *3ḫtj-ḥtp* - Saqqara, WSP (D64), hall, east: marsh
Source: *P-M III*, 599-600; Davies, *Ptahhetep ii*, pl. 13
Label: none
Date: Harpur, *Decoration*, [#338], 272: V.8L-9E

HB131 *3ḫtj-ḥtp* - Saqqara, Saqqara, WSP (D64), hall, east: marsh
Source: *P-M III*, 599-600; Davies, *Ptahhetep ii*, pl. 14
Label: none
Date: Harpur, *Decoration*, [#338], 272: V.8L-9E

HB132 *k3.j-m-ʿnḫ* - Giza, WF (G4561), hall, north: marsh
Source: *P-M III*, 131-133; Junker, *Giza iv*, fig. 8; Kanawati, *Kaiemankh*, pl. 31
Label: none
Date: Kanawati, *Kaiemankh*, 18: late V; Harpur, *Decoration*, [#255], 270: VI

HB133 *ḥzj* - Saqqara, TP (ntn), portico, south (west of entrance): marsh
Source: *P-M III*, nr; Kanawati-Abder-Raziq, *Hesi*, pl. 54
Label: none
Date: Kanawati-Abder-Raziq, *Hesi*, 16: VI.1L-2E; Harpur, *Decoration*, nr

Nile crocodiles

HB134 *ḥzjj-rʿ* - Saqqara, NSP (A3=S2405), corridor, west: fording
Source: *P-M III*, 429-431; Quibell, *Saqqara (1911-1912)*, pl. 7 [2]
Label: none
Date: Harpur, *Decoration*, [#462], 275: III

HB135 *k3.j-m-nfrt* - Giza, CF (ntn), vestibule D, north: fording
Source: *P-M III*, 263-264; Hassan, *Giza ii*, fig. 140
Label: missing
Date: Harpur, *Decoration*, [#261], 270: V.6-8E?

HB136 *k3.j-m-rḥw* - Saqqara, NSP (D2=S905), chapel, north: fording
Source: *P-M III*, 485-487; Mogensen, *Ny Carlsberg*, fig. 4
Label: *j mnjw pw ḥr=k ʿnḫ wrt r šjj pw ntj ḥr mw jw=f m šp tp*
Oh this herdsman! May you be watchful greatly against that aquatic which is in the water! May he come as blind-of-head
Date: Harpur, *Decoration*, [#523], 276: V.8

HB137 *tjj* - Saqqara, NSP (D22), room III, north: fording
Source: *P-M III*, 468-478; Wild, *Ti ii*, pl. 124
Label: *mnjw pw ʿnḫ ḥr=k r šjj pw ntj ḥr mw jw(=f) m šp tp*
This herdsman! May you be watchful against that aquatic which is in the water! May he come as blind-of-head
Date: Harpur, *Decoration*, [#543], 277: V.8-9

HB138 *snḏm-jb: mḥj* - Giza, WF (G2378=LG26), forecourt, north: marsh
Source: *P-M III*, 87-89; *LD Erganz.*, pl. 12 [upper]; Brovarski, *Senedjemib Complex Part I*, fig. 101[53]
Label: none
Date: Brovarski, *Senedjemib Complex Part I*, 29-30: V.9; Harpur, *Decoration*, [#220], 269: V.9

HB139 *ptḥ-ḥtp II: ṯfj* - Saqqara, WSP (D64), offering-room, east: fording
Source: *P-M III*, 600-604; Quibell, *Ptah-hetep*, pl. 33
Label: missing
Date: Harpur, *Decoration*, [#400], 274: V.9M-L

HB140 *sšm-nfr IV* - Giza, GIS (LG53), portico, north: marsh
Source: *P-M III*, 223-226; Junker, *Giza xi*, fig. 60
Label: none
Date: Harpur, *Decoration*, [#235], 270: V.9-VI.1

HB141 *sšm-nfr IV* - Giza, GIS (LG53), pillared hall, north: fording
Source: *P-M III*, 223-226; Junker, *Giza xi*, fig. 67
Label: none
Date: Harpur, *Decoration*, [#235], 270: V.9-VI.1

HB142 *3ḫtj-mrw-nswt* - Giza, WF (G2184), secondary offering-room, west: fording
Source: *P-M III*, 80-81; Smith, *HESPOK*, fig. 229; *Giza Archives Project* (Photo no. C12742_NS)[54]
Label: *w3ḥ mr j(n) mnjw k3w*
Entering the canal by the herdsmen and cattle
Date: Harpur, *Decoration*, [#008], 265: V.9-VI.1

HB143 *nj-m3ʿt-rʿ [jzzj-mr-nṯr]* - Giza, WF (G2097), chapel, north: fording
Source: *P-M III*, 70; Roth, *Cem. Pal. Att.*, pl. 185
Label: . . . *mḥt pḥw*
. . . the Delta and the marshes . . .
ʿnḫ ḥ=k wrt r š[jj pw] ntj ḥr mw
May you be watchful greatly against that aquatic which is in the water

---

[53] During a recent examination of this tomb by Brovarski (*Senedjemib Complex Part I*, fig. 101), this scene was found to be in a fragmentary condition.

[54] Smith (*HESPOK*, fig. 229) omits the hieroglyphic label, reproduced here, which appears in a watercolour sketch by Frida Firth (c. 1912).

Date: Roth, *Cem. Pal. Att.*, 129-130: V.9; Harpur, *Decoration*, [#033], 265: V.9-VI.1

HB144 *jrw-kꜣ-ptḥ: ḫnw* - Saqqara, UPC (ntn), chapel, east: fording
Source: *P-M III*, 639; McFarlane, *Irukaptah*, pl. 46
Label: none
Date: McFarlane, *Irukaptah*, 19: late V, V.7-8; Harpur, *Decoration*, [#364], 273: VI

HB145 *kꜣ-gm-nj: mmj* - Saqqara, TP (LS10), room I, west: marsh (behind standing hippopotamus)
Source: *P-M III*, 521-525; Firth-Gunn, *Teti Pyr. ii*, pl. 53 [2]; Harpur-Scremin, *Kagemni,* fig. 3
Label: none
Date: Harpur, *Decoration*, [#534], 276: VI.1M

HB146 *kꜣ-gm-nj: mmj* - Saqqara, TP (LS10), room III, north: fording
Source: *P-M III*, 521-525; Firth-Gunn, *Teti Pyr. ii*, pl. 52; Harpur-Scremin, *Kagemni,* fig. 7
Label: *j nrw pw dj jḥy n kꜣw nw sn mk wḏb(=sn) r=f*
Oh this herdsman! Show Ihy of the cattle to them
Behold! They turn against him![55]
jḥy n kꜣw
Ihy of the cattle
Date: Harpur, *Decoration*, [#534], 276: VI.1M

HB147 *kꜣ-gm-nj: mmj* - Saqqara, TP (LS10), room III, north: marsh (facing hippopotamus with calf)
Source: *P-M III*, 521-525; Harpur-Scremin, *Kagemni,* fig. 9
Label: missing
Date: Harpur, *Decoration*, [#534], 276: VI.1M

HB148 *kꜣ-gm-nj: mmj* - Saqqara, TP (LS10), room IV, east: fishing (behind standing hippopotamus)
Source: *P-M III*, 521-525; von Bissing, *Gem-ni-kai*, pl. 18; Harpur-Scremin, *Kagemni,* fig. 16
Label: none
Date: Harpur, *Decoration*, [#534], 276: VI.1M

HB149 *mrrw-kꜣ.j: mrj* - Saqqara, TP (ntn), chamber A1, south: fording
Source: *P-M III*, 525-534; Duell, *Mereruka i*, pl. 21
Label: *mnjw pw ꜥnḫ ḥr=k r šjj pw ntj ḥr mw jw=f m šp tp jrj=ṯw r=f wrt*
This herdsman! May you be watchful against that aquatic which is in the water!
May he come as blind-of-head
Act against him, greatly!
ḫnj=t(w)? wnj ntj ḥnꜥ
Row quickly, my comrade!
mnjw pw ꜥ=k ḥr mw
This herdsman! Your arm above the water!
Date: Harpur, *Decoration*, [#420], 274: VI.1M-L

HB150 *mrrw-kꜣ.j: mrj* - Saqqara, TP (ntn), chamber A1, south: marsh
Source: *P-M III*, 525-534; Duell, *Mereruka i*, pl. 19
Label: none
Date: Harpur, *Decoration*, [#420], 274: VI.1M-L

HB151 *zšzšt: jdwt* - Saqqara, UPC (ntn), room I, north: marsh
Source: *P-M III*, 617-619; Macramallah, *Idout*, pl. 5 [d]; Kanawati-Abder-Raziq, *Iynefert*, pl. 53 [f]
Label: none
Date: Kanawati-Abder-Raziq, *Iynefert*, 36-37: VI.1E; Harpur, *Decoration*, [#487], 275: V.9 r/u, VI.1-2E?

HB152 *zšzšt: jdwt* - Saqqara, UPC (ntn), room B, west: marsh
Source: *P-M III*, 617-619; Macramallah, *Idout*, pl. 7; Kanawati-Abder-Raziq, *Iynefert*, pl. 54
Label: none
Date: Kanawati-Abder-Raziq, *Iynefert*, 36-37: VI.1E; Harpur, *Decoration*, [#487], 275: V.9 r/u, VI.1-2E?

HB153 *zšzšt: jdwt* - Saqqara, UPC (ntn), room III, west: fording
Source: *P-M III*, 617-619; Macramallah, *Idout*, pl. 7; Kanawati-Abder-Raziq, *Iynefert*, pl. 54

[55] Firth and Gunn (*Teti Pyr. ii*, 114) translate this inscription as 'Show Ihy to the cattle that I may keep them together'.

Label: *mnjw p[w] ꜥnḫ ḥr=k r šjj pw ntj ḥr mw jw=f m šp tp jr=ṯw r=f wrt*
This herdsman! May you be watchful against that aquatic which is in the water!
May he come as blind-of-head
Act against him greatly!
j nr pw ꜥ=k ḥr mw
Oh this herdsman! Your arm above the water!
Date: Kanawati-Abder-Raziq, *Iynefert*, 36-37: VI.1E; Harpur, *Decoration*, [#487], 275: V.9 r/u, VI.1-2E?

HB154 *ḥzj* - Saqqara, TP (ntn), portico, east: fording
Source: *P-M III*, nr; Kanawati-Abder-Raziq, *Hesi*, pl. 55
Label: *j mnjw pw ꜥnḫ ḥr=k r šjj pw ntj ḥr mw jw(=f) m šp tp tm jwjj n šjj pw*
Oh this herdsman! May you be watchful against that aquatic which is in the water!
May he come as blind-of-head, and may (the cattle) not come to that aquatic
ꜥ=k ḥr mw mnjw pw jw mzḥ pw gr(w) ẖr s . . . t nt wḏb wḏꜣ jb=k r=f
Your arm above the water, this herdsman!
This crocodile is still under the . . . of the riverbank
May your heart prosper against it
jn r=k
Bring away
mk ḫr . . . tp n ḫt r . . .
Behold! . . . the end of the pole . . .
Date: Kanawati-Abder-Raziq, *Hesi*, 16: VI.1L-2E; Harpur, *Decoration*, nr

HB155 *ꜥnḫ-m-ꜥ-ḥr: zzj* - Saqqara, TP (ntn), room I, east: fording
Source: *P-M III*, 512-515; Kanawati-Hassan, *Ankhmahor*, pl. 37 [a]
Label: *ḫsf msḥ*
Repelling the crocodile
j mnjw pw ꜥnḫ ḥr=k r šjj pw ntj ḥr mw tm jw nw n šjj pw jw=f m šp tp ꜥnḫ ḥr=k r=f wrt
Oh this herdsman! May you be watchful against that aquatic which is in the water!
May these not come to that aquatic and may he be as a blind-of-head
May you be watchful against him, greatly!
Date: Kanawati-Hassan, *Ankhmahor*, 18: VI.1M-L; Harpur, *Decoration*, [#374], 273: VI.1L-2E

HB156 *ꜥnḫ-m-ꜥ-ḥr: zzj* - Saqqara, TP (ntn), room I, west: marsh
Source: *P-M III*, 512-515; Kanawati-Hassan, *Ankhmahor*, pl. 37 [c]
Label: none
Date: Kanawati-Hassan, *Ankhmahor*, 18: VI.1M-L; Harpur, *Decoration*, [#374], 273: VI.1L-2E

HB157 *mṯtj* - Saqqara, UPC (ntn), relief fragment (Louvre no. E. 25532), unknown: marsh
Source: *P-M III*, 5646-648; Ziegler, *Catalogue*, 144
Label: none
Date: Ziegler, *Catalogue*, 123: probably VI; Harpur, *Decoration*, [#426], 274: VI.2?

HB158 *mrrj* - Saqqara, TP (ntn), room 1, south: marsh
Source: *P-M III*, 518-519; Davies et al., *Saqqara Tombs i*, pl. 5
Label: none
Date: Harpur, *Decoration*, [#418], 274: VI.2E?

HB159 *mḥw* - Saqqara, UPC (ntn), room II, north: fording
Source: *P-M III*, 619-622; Altenmüller, *Mehu*, pl. 40 [a] and [b]
Label: *mnjw pw ꜥnḫ ḥr=k r šjj pw ntj ḥr mw jw(=f) m šp tp jrj ṯw jr=f wrt mnjw pwꜥ=k ḥr mw*
This herdsman! May you be watchful against that aquatic which is in the water!
May he come as blind-of-head
Pay attention to him very greatly this herdsman
Your arm above the water!
Date: Altenmüller, *Mehu*, 19: VI.2; Harpur, *Decoration*, [#424], 274: VI.2M-3?

Praying mantids

HB160 *ḥzj* - Saqqara, TP (ntn), portico, south (west of entrance): marsh
Source: *P-M III*, nr; Kanawati-Abder-Raziq, *Hesi*, pl. 54
Label: none
Date: Kanawati-Abder-Raziq, *Hesi*, 16: VI.1L-2E; Harpur, *Decoration*, nr

9.1 THREATENING

9.1.1 Head display

Bulls

AB1 *ḥmt-rˁ* - Giza, CF (ntn), entrance thicknesses, north: offering
Source: *P-M III*, 243-244; Hassan, *Giza vi*, fig. 37
Label: *jnj.t jw3 prt-ḫrw*
Bringing oxen as an invocation offering
Date: Harpur, *Decoration*, [#168], 268: IV.4-V.1

AB2 *jj-mrjj* - Giza, WF (G6020=LG16), third chamber, east: offering
Source: *P-M III*, 170-174; Weeks, *Cemetery G6000*, fig. 42
Label: *ng*(*3w*)
Long-horned bull
Date: Weeks, *Cemetery G6000*, 7: V.6; Harpur, *Decoration*, [#014], 265: V.3L

AB3 *sšm-nfr III* - Giza, WF (G5170), interior offering-room, east offering
Source: *P-M III*, 153-154; Brunner-Traut, *Seschemnofers III*, Suppl. 1
Label: *jnj.t jw3 m* [*nḏ*]*t-ḥr prt-ḫrw*
Bringing cattle as gifts for invocation offerings
Date: Harpur, *Decoration*, [#234], 270: V.8E

AB4 *3ḫtj-ḥtp* - Saqqara, WSP (D64), hall, east: pastoral
Source: *P-M III*, 599-600; Davies, *Ptahhetep ii*, pl. 13
Label: missing
Date: Harpur, *Decoration*, [#338], 272: V.8L-9E

AB5 *3ḫtj-ḥtp* - Saqqara, WSP (D64), hall, south: slaughter
Source: *P-M III*, 599-600; Davies, *Ptahhetep ii*, pl. 22
Label: *ḳ3s ng*(*3w*) *wr*
Binding a long-horned bull greatly!
Date: Harpur, *Decoration*, [#338], 272: V.8L-9E

AB6 *3ḫtj-ḥtp* - Saqqara, WSP (D64), hall, west: offering
Source: *P-M III*, 599-600; Davies, *Ptahhetep ii*, pl. 21
Label: *ng*(*3w*)
Long-horned cattle
Date: Harpur, *Decoration*, [#338], 272: V.8L-9E

AB7 *3ḫtj-ḥtp* - Saqqara, WSP (D64), corridor, west: agricultural
Source: *P-M III*, 599-600; Davies, *Ptahhetep ii*, pl. 8
Label: *ḥsf ng*(*3w*)
Driving the long-horned bulls
Date: Harpur, *Decoration*, [#338], 272: V.8L-9E

AB8 *ḫnwt* - Saqqara, UPC (ntn), room A [I], south: pastoral?
Source: *P-M III*, 623-624; Munro, *Unas-Friedhof*, pl. 36 [b]
Label: missing
Date: Munro, *Unas-Friedhof*, 20: V.9; Harpur, *Decoration*, [#478], 275: V.9

AB9 Anon - Saqqara, WSP (ntn), relief fragment found in the tomb of *tfw*, unknown: desert
Source: *P-M III*, 605; Hassan, *Ny-ankh-Pepy*, fig. 59
Label: none
Date: *P-M III*, 605: V-VI; Harpur, *Decoration*, nr

AB10 *nj-k3w-jzzj* - Saqqara, TP (ntn), room I, north: pastoral
Source: *P-M III*, nr; Kanawati, *Nikauisesi*, pl. 49
Label: *prt m sḫt r t3*
Coming out from the marshes to the land
Date: Kanawati-Abder-Raziq, *Nikauisesi*, 17-23: VI.1M; Harpur, *Decoration*, nr

AB11 *mrrw-k3.j: mrj* - Saqqara, TP (ntn), chamber A13, north: pastoral
Source: *P-M III*, 525-534; Duell, *Mereruka ii*, pl. 152
Label: *rn jw3*
Young oxen
Date: Harpur, *Decoration*, [#420], 274: VI.1M-L

AB11a *jnw-mnw* - Saqqara (ntn), room II, south: offering (upper register; in-curved horns)
Source: *P-M III*, nr; Kanawati, *Inumin*, pl. 49 [a]
Label: *rn jw3*
Young oxen
Date: Kanawati, *Inumin*, 16-17: VI.2E; Harpur, *Decoration*, nr

AB11b *jnw-mnw* - Saqqara (ntn), room II, south: offering (middle register; hornless)
Source: *P-M III*, nr; Kanawati, *Inumin*, pl. 49 [a]
Label: *rn jw3*
Young oxen
Date: Kanawati, *Inumin*, 16-17: VI.2E; Harpur, *Decoration*, nr

AB11c *jnw-mnw* - Saqqara (ntn), room II, south: offering (middle register; deformed horns)
Source: *P-M III*, nr; Kanawati, *Inumin*, pl. 49 [a]
Label: *rn jw3*
Young oxen
Date: Kanawati, *Inumin*, 16-17: VI.2E; Harpur, *Decoration*, nr

AB11d *jnw-mnw* - Saqqara (ntn), room II, south: offering (bottom register; hornless)
Source: *P-M III*, nr; Kanawati, *Inumin*, pl. 49 [a]
Label: *rn jw3*
Young oxen
Date: Kanawati, *Inumin*, 16-17: VI.2E; Harpur, *Decoration*, nr

Geese

AB12 *ṯjj* - Saqqara, NSP (D22), portico, south: pastoral/force-feeding
Source: *P-M III*, 468-478; Epron et al., *Ti i*, pl. 8
Label: *pẖr s mnwt m-ḫt snmt*
Moving about (of) *s*-geese and pigeons after feeding
Date: Harpur, *Decoration*, [#543], 277: V.8-9

AB13 *ṯjj* - Saqqara, NSP (D22), room I/pillared hall, west: pastoral
Source: *P-M III*, 468-478; Epron et al., *Ti i*, pl. 28
Label: none
Date: Harpur, *Decoration*, [#543], 277: V.8-9

AB14 *ptḥ-ḥtp II: ṯfj* - Saqqara, WSP (D64), offering-room, east: offering
Source: *P-M III*, 600-604; Quibell, *Ptah-hetep*, pl. 21
Label: n/a
Date: Harpur, *Decoration*, [#400], 274: V.9M-L

AB15 *wḏ3-ḥ3-ttj: ššj [nfr-sšm-ptḥ]* - Saqqara, TP (ntn), room II, east: pastoral/force-feeding
Source: *P-M III*, 515-516; Capart, *Rue de Tombeaux*, pl. 85
Label: none
Date: Harpur, *Decoration*, [#384], 273: VI.1L-2E

AB16 *mrjj-ttj: mrj* - Saqqara, TP (ntn), room C1, east: pastoral/poultry-yard
Source: *P-M III*, 356-357; Kanawati-Abder-Raziq, *Meryteti*, pl. 48
Label: none
Date: Kanawati-Abder-Raziq, *Meryteti*, 18: end VI.1-VI.2M; Harpur, *Decoration*, [#415], 274: VI.2M-L

AB17 *mrjj-ttj: mrj* - Saqqara, TP (ntn), room C1, east: pastoral/force-feeding
Source: *P-M III*, 356-357; Kanawati-Abder-Raziq, *Meryteti*, pl. 48
Label: none
Date: Kanawati-Abder-Raziq, *Meryteti*, 18: end VI.1-VI.2M; Harpur, *Decoration*, [#415], 274: VI.2M-L

AB18 *ꜣḫtj-ḥtp* - Saqqara, UPC (E17), offering-room, east: offering
Source: *P-M III*, 633-634; Petrie-Murray, *Seven Memphite*, pl. 7 [lower]
Label: *sḫpt r* [ ] *nt ḫ*[ ]
Conducting *r*-geese . . .
Date: Harpur, *Decoration*, [#339], 272: VI.5?

9.1.2 Leg display

Bulls

AB19 *snḏm-jb: jntj* - Giza, WF (G2370=LG27), room II, west: agricultural
Source: *P-M III*, 85-87; *LD ii*, pl. 77[56]
Label: *nḥp*
Mating
Date: Brovarski, *Senedjemib Complex Part I*, 23: V.8M-L; Harpur, *Decoration*, [#219], 269: V.8M-L

AB20 *jnw-mnw* - Saqqara (ntn), room II, south: pastoral
Source: *P-M III*, nr; Kanawati, *Inumin*, pl. 49 [a]
Label: missing
Date: Kanawati, *Inumin*, 16-17: VI.2E; Harpur, *Decoration*, nr

AB21 *mḥw* - Saqqara, UPC (ntn), doorway (room III to magazine), north: slaughter?
Source: *P-M III*, 619-622; Altenmüller, *Mehu*, pl. 54 [b]
Label: none
Date: Altenmüller, *Mehu*, 19: VI.2; Harpur, *Decoration*, [#424], 274: VI.2M-3?

9.1.3 Ear display

Donkeys

AB22 Anon - Saqqara (ntn), relief fragment (Rijksmuseum, Leiden no. AM.102), unknown: agricultural
Source: *P-M III*, 758; *Eg. Art in the Age of the Pyramids*, cat. no. 149
Label: missing
Date: *Eg. Art in the Age of the Pyramids*, 402-403: late V; Harpur, *Decoration*, [#584], 278: V

AB23 *rꜥ-ḫꜥ.f-ꜥnḫ* - Giza, EF (G7948=LG75), chapel, east: agricultural
Source: *P-M III*, 207-208; *LD ii*, pl. 9
Label: none
Date: Harpur, *Decoration*, [#156], 268: V.6?

AB24 *rꜥ-m-kꜣ.j* - Saqqara, NSP (D3=S903), chapel, east: agricultural
Source: *P-M III*, 487-488; Hayes, *Scepter i*, fig. 57
Label: *ḥwj zpt nt jt*
Threshing on the threshing floor of barley
Date: Harpur, *Decoration*, [#453], 275: V.6? r/u - V.6-7

AB25 *ꜣḫtj-ḥtp* - Saqqara, UPC (ntn), chapel, east: agricultural
Source: *P-M III*, 634-637; Ziegler, *Akhethetep*, 134-135
Label: *jr ḥꜣ(=k) jm=sn*
Get back among them![57]
Date: Ziegler, *Akhethetep*, 9: V.6; Harpur, *Decoration*, [#340], 272: V.6-8E

AB26 *kꜣ.j-m-rḥw* - Saqqara, NSP (D2=S905), chapel, east: agricultural
Source: *P-M III*, 485-487; Mogensen, *Ny Carlsberg*, fig. 18

56 According to Brovarski (*Senedjemib Complex Part I*, 49), this scene has now been destroyed.

57 This phrase, which occurs frequently in donkey threshing scenes, has been translated in many different ways. I have interpreted it as a direction to one of the animals to move closer to its companions. For other translations see Erman, *Reden*, 26-28; Montet, *Scènes*, 218-219; Junker, *Giza vi*, 148; Vandier, *Manuel vi*, 174-175; Simpson, *Sekhem-ankh-ptah*, 16 and Simpson, *Kayemnofret*, 18.

Label: missing
Date: Harpur, *Decoration*, [#523], 276: V.8

AB27 *nfr-jrt-n.f* - Saqqara, ESP (D55), chapel, east: agricultural
Source: *P-M III*, 583-584; van de Walle, *Neferirtenef*, pl. 12
Label: *jr ḥ3=k jm=sn*
Get back among them!
Date: Harpur, *Decoration*, [#440], 274: V.8-9

AB28 *3ḫtj-mrw-nswt* - Giza, WF (G2184), secondary offering-room, west: agricultural
Source: *P-M III*, 80-81; *Giza Archives Project* (Photo no. A779_NS)
Label: missing
Date: Harpur, *Decoration*, [#008], 265: V.9-VI.1

AB29 *ʿnḫ-m-ʿ-ḥr: zzj* - Saqqara, TP (ntn), room I, east: agricultural
Source: *P-M III*, 512-515; Kanawati-Hassan, *Ankhmahor*, pl. 37 [a]
Label: *h3 jm=sn ntj ḥnʿ(=j)*
Get down with them comrade!
swn(=j) ṯw pḫr jm
I will persuade you to turn therein!
Date: Kanawati-Hassan, *Ankhmahor*, 18: VI.1M-L; Harpur, *Decoration*, [#374], 273: VI.1L-2E

9.1.4 Mouth display

Hippopotamuses

AB30 *dbḥnj* - Giza, CF (LG90), room II, east: marsh
Source: *P-M III*, 235-236; Hassan, *Giza iv*, fig. 121
Label: none
Date: Harpur, *Decoration*, [#297], 271: IV.5

AB31 *pr-sn* - Saqqara, ESP (D45), relief fragment, unknown: marsh?
Source: *P-M III*, 577-578; Petrie-Murray, *Seven Memphite*, pl. 10 [upper]
Label: missing
Date: Harpur, *Decoration*, [#391], 273: V.2-3

AB32 *3ḫtj-ḥtp* - Saqqara, UPC (ntn), chapel, east: marsh
Source: *P-M III*, 634-637; Ziegler, *Akhethetep*, 132-133
Label: none
Date: Ziegler, *Akhethetep*, 9: V.6; Harpur, *Decoration*, [#340], 272: V.6-8E

AB33 *k3.j-m-nfrt* - Giza, CF (ntn), vestibule D, north: marsh
Source: *P-M III*, 263-264; Hassan, *Giza ii*, fig. 140
Label: none
Date: Harpur, *Decoration*, [#261], 270: V.6-8E?

AB34 *ẖnmw-ḥtp* - Saqqara, ESP (D49), relief fragment (Berlin (East) Museum, 14101), north: marsh
Source: *P-M III*, 578-579; Decker-Herb, *Bildatlas*, pl. 192 [K.126]
Label: none
Date: Harpur, *Decoration*, [#481], 275: V.6-8E?

AB35 *snḏm-jb: jntj* - Giza, WF (G2370=LG27), room II, west: marsh
Source: *P-M III*, 85-87; *LD ii*, pl. 77; Brovarski, *Senedjemib Complex Part I*, fig. 43
Label: none
Date: Brovarski, *Senedjemib Complex Part I*, 23: V.8M-L; Harpur, *Decoration*, [#219], 269: V.8M-L

AB36 *tjj* - Saqqara, NSP (D22), room III, north: marsh
Source: *P-M III*, 468-478; Wild, *Ti ii*, pl. 119
Label: none
Date: Harpur, *Decoration*, [#543], 277: V.8-9

AB37 *snḏm-jb: mḥj* - Giza, WF (G2378= LG26), portico, north: marsh
Source: *P-M III*, 624-625; *LD Erganz.*, pl. 11 [upper]; Brovarski, *Senedjemib Complex Part I*, fig. 103
Label: none
Date: Brovarski, *Senedjemib Complex Part I*, 29-30: V.9; Harpur, *Decoration*, [#220], 269: V.9

AB38 *nbt* - Saqqara, UPC (ntn), room A [I], west: marsh
Source: *P-M III*, 624-625; Munro, *Unas-Friedhof*, pl. 10
Label: none
Date: Munro, *Unas-Friedhof*, 20: V.9; Harpur, *Decoration*, [#438], 274: V.9

AB39 *jj-nfrt: šꜣ.n.f* - Saqqara, UPC (ntn), room I, north: marsh
Source: *P-M III*, 616; Kanawati-Abder-Raziq, *Iynefert*, pl. 37
Label: none
Date: Kanawati-Abder-Raziq, *Iynefert*, 13: V.9E-M; Harpur, *Decoration*, [#345], 272: V.9

AB40 Anon - Saqqara, AJS (ntn), relief fragment (JE41572), unknown: marsh
Source: *P-M III*, 654; Quibell, *Saqqara (1907-1908)*, pl. 51 [3]
Label: none
Date: Harpur, *Decoration*, [#574], 277: V-VI

AB41 *kꜣ-gm-nj: mmj* - Saqqara, TP (LS10), pillared hall, north: marsh
Source: *P-M III*, 521-525; Decker-Herb, *Bildatlas*, pl. 196 [K1.33]; Harpur-Scremin, *Kagemni*, fig. 9
Label: none
Date: Harpur, *Decoration*, [#534], 276: VI.1M

AB42 *mrrw-kꜣ.j: mrj* - Saqqara, TP (ntn), chamber A1, north: marsh
Source: *P-M III*, 525-534; Duell, *Mereruka i*, pls. 11 and pl. 13
Label: none
Date: Harpur, *Decoration*, [#420], 274: VI.1M-L

AB43 *zšzšt: jdwt* - Saqqara, UPC (ntn), room I, south: fishing
Source: *P-M III*, 617-619; Macramallah, *Idout*, pl. 5 [c]; Kanawati-Abder-Raziq, *Iynefert*, pl. 53 [d]
Label: none
Date: Kanawati-Abder-Raziq, *Iynefert*, 36-37: VI.1E; Harpur, *Decoration*, [#487], 275: V.9 r/u, VI.1-2E?

AB44 *zšzšt: jdwt* - Saqqara, UPC (ntn), room I, north: fishing
Source: *P-M III*, 617-619; Macramallah, *Idout*, pl. 5 [d]; Kanawati-Abder-Raziq, *Iynefert*, pl. 53 [f]
Label: none
Date: Kanawati-Abder-Raziq, *Iynefert*, 36-37: VI.1E; Harpur, *Decoration*, [#487], 275: V.9 r/u, VI.1-2E?

AB45 *zšzšt: jdwt* - Saqqara, UPC (ntn), room III, west: marsh (standing, left)
Source: *P-M III*, 617-619; Macramallah, *Idout*, pl. 7; Kanawati-Abder-Raziq, *Iynefert*, pl. 54
Label: none
Date: Kanawati-Abder-Raziq, *Iynefert*, 36-37: VI.1E; Harpur, *Decoration*, [#487], 275: V.9 r/u, VI.1-2E?

AB46 *zšzšt: jdwt* - Saqqara, UPC (ntn), room III, west: marsh (pair sitting, right)
Source: *P-M III*, 617-619; Macramallah, *Idout*, pl. 7; Kanawati-Abder-Raziq, *Iynefert*, pl. 54
Label: none
Date: Kanawati-Abder-Raziq, *Iynefert*, 36-37: VI.1E; Harpur, *Decoration*, [#487], 275: V.9 r/u, VI.1-2E?

AB47 *sʿnḫ-w(j)-ptḥ* - Saqqara, TP (ntn), room I, east: marsh
Source: *P-M III*, nr; Kanawati-Abder-Raziq, *Neferseshemre*, pls. 69 and 76
Label: none
Date: Kanawati-Abder-Raziq, *Neferseshemre*, 40-41: VI.1L; Harpur, *Decoration*, nr

AB48 *ṯfw* - Saqqara, WSP (ntn), pillared hall, south: fishing
Source: *P-M III*, 605; Hassan, *Ny-Ankh-Pepy*, fig. 57
Label: n/a
Date: Harpur, *Decoration*, [#544], 277: VI.1-2

AB49 *mḥw* - Saqqara, UPC (ntn), room II, north: boating
Source: *P-M III*, 619-622; Altenmüller, *Mehu*, pl. 31 [a]
Label: n/a

Date: Altenmüller, *Mehu*, 19: VI.2; Harpur, *Decoration*, [#424], 274: VI.2M-3?

AB50 *jj-nfrt* - Giza (ntn), chapel, unknown: marsh
Source: *P-M III*, 298-299; Schürmann, *Die Reliefs*, fig. 6 [b]
Label: none
Date: Harpur, *Decoration*, [#017], 265: VI.3-4E?

AB51 *nb-k3w-ḥr: jdw* - Saqqara, UPC (ntn), vestibule, north: marsh (standing, left)
Source: *P-M III*, 627-629; Hassan, *Neb-Kaw-Her*, pl. 5 [c]
Label: none
Date: Harpur, *Decoration*, [#437], 274: V.9 r/u VI.7

AB52 *nb-k3w-ḥr: jdw* - Saqqara, UPC (ntn), vestibule, north: marsh (sitting, right)
Source: *P-M III*, 627-629; Hassan, *Neb-Kaw-Her*, pl. 5 [c]
Label: none
Date: Harpur, *Decoration*, [#437], 274: V.9 r/u VI.7

9.1.5 Penile display

Baboons

AB53 *nb.j-m-3ḫtj* - Giza, CF (LG86), chapel II, east: domestic
Source: *P-M III*, 230-232; *LD ii*, pl. 13
Label: none
Date: Harpur, *Decoration*, [#122], 267: IV.6-V.1

AB54 *tp-m-ʿnḫ II* - Saqqara, NSP (D11), corridor, east: market
Source: *P-M III*, 483-484; Borchardt, *Denkmäler ii*, pl. 61 (CG 1556); *Eg. Art in the Age of the Pyramids*, fig. 126
Label: *j h3j nḏ ḥwj jry jʿn ḥmw js jr nrj ʿnr pn*
Oh charge down to protect and strike!
Keeper of baboons, Hemu: Go! as to make fearful this baboon
Date: Harpur, *Decoration*, [#541], 277: V.5-6?

AB55 *nfr/k3-ḥ3.j* - Saqqara, UPC (ntn), chapel, east: sack press
Source: *P-M III*, 639-641; Moussa-Altenmüller, *Nefer and Ka-hay*, pl. 12
Label: none
Date: Harpur, *Decoration*, [#439], 274: V.6

AB56 *nfr/k3-ḥ3.j* - Saqqara, UPC (ntn), niche, east: boat building
Source: *P-M III*, 639-641; Moussa-Altenmüller, *Nefer and Ka-hay*, pl. 19
Label: none
Date: Harpur, *Decoration*, [#439], 274: V.6

AB57 *nj-ʿnḫ-ẖnmw/ẖnmw-ḥtp* - Saqqara, UPC (ntn), vestibule, north: market
Source: *P-M III*, 641-644; Moussa-Altenmüller, *Nianchchnum*, fig. 10
Label: *j ḥwn jrjw jmj-r3 mrr=k jn.tw n=k nb jrj*
Oh young man who plays the leader. Do you want that one fetches for you his master?
Date: Harpur, *Decoration*, [#431], 274: V.6L-7

AB58 *jtj-sn* - Giza, CF (ntn), chapel, east: domestic
Source: *P-M III*, 252-253; Hassan, *Giza v*, fig. 123
Label: none
Date: Harpur, *Decoration*, [#035], 265: V.6-8?

AB59 Anon - Saqqara, WSP (ntn), block from shaft 6, 'Mastaba of Shaft no. 14', unknown: offering?
Source: *P-M III*, 613; Hassan, *Hemet-Re*, pl. 14 [c]
Label: *s3w jʿn*
Guardian of the baboon
Date: *P-M III*, 613: VI; Harpur, *Decoration*, nr

9.1.6 Wing display

Ostriches

AB60 *nj-mꜣꜥt-rꜥ* [*jzzj-mr-nṯr*] - Giza, WF (G2097), chapel, east: hunting
Source: *P-M III*, 70; Roth, *Cem. Pal. Att.*, pl. 189
Label: none
Date: Roth, *Cem. Pal. Att.*, 129-130: V.9; Harpur, *Decoration*, [#033], 265: V.9-VI.1

Birds

AB61 *ḥtp-ḥr-ꜣḫtj* - Saqqara, WSP (D60), chapel, north: marsh (1 bird among umbels)
Source: *P-M III*, 593-595; Mohr, *Hetep-her-akhti*, fig. 34
Label: none
Date: Harpur, *Decoration*, [#467], 275: V.6-8E

AB62 *jrj-n-kꜣ-ptḥ* - Saqqara, UPC (ntn), offering chamber, south: marsh (2 birds among umbels)
Source: *P-M III*, 644; Moussa-Junge, *Two Tombs*, pl. 12
Label: none
Date: Harpur, *Decoration*, [#362], 273: V.6-9

AB63 *rꜥ-špss* - Saqqara, NSP (LS16=S902), forecourt, west: marsh (1 bird among umbels)
Source: *P-M III*, 494-496; *LD ii*, pl. 60 [left]
Label: none
Date: Harpur, *Decoration*, [#456], 275: V.8M

AB64 *snḏm-jb: jntj* - Giza, WF (G2370=LG27), portico, west: marsh (2 birds within thicket)
Source: *P-M III*, 85-87; *LD Erganz.*, pl. 18[58]
Label: none
Date: Brovarski, *Senedjemib Complex Part I*, 23: V.8M-L Harpur, *Decoration*, [#219], 269: V.8M-L

AB65 *ṯjj* - Saqqara, NSP (D22), room III, north: marsh (13 birds within thicket, 8 with eggs)
Source: *P-M III*, 468-478; Wild, *Ti ii*, pl. 119
Label: none
Date: Harpur, *Decoration*, [#543], 277: V.8-9

AB66 *ꜣḫtj-ḥtp* - Saqqara, WSP (D64), hall, east: marsh (3 birds among umbels, 2 birds within thicket)
Source: *P-M III*, 599-600; Davies, *Ptahhetep ii*, pl. 13
Label: none
Date: Harpur, *Decoration*, [#338], 272: V.8L-9E

AB67 *kꜣ-gm-nj: mmj* - Saqqara, TP (LS10), room III, north: marsh (1 bird in thicket)
Source: *P-M III*, 521-525; Decker-Herb, *Bildatlas*, pl. 219 [K 2.39]; Harpur-Scremin, *Kagemni*, fig. 9
Label: none
Date: Harpur, *Decoration*, [#534], 276: VI.1M

AB68 *nj-kꜣw-jzzj* - Saqqara, TP (ntn), room I, east: marsh (5 birds within thicket)
Source: *P-M III*, nr; Kanawati-Abder-Raziq, *Nikauisesi*, pl. 50
Label: none
Date: Kanawati-Abder-Raziq, *Nikauisesi*, 17-23: VI.1M; Harpur, *Decoration*, nr

AB69 *mrrw-kꜣ.j: mrj* - Saqqara, TP (ntn), chamber A1, north: marsh (10 birds within thicket)
Source: *P-M III*, 525-534; Duell, *Mereruka i*, pls. 10 and 11
Label: none
Date: Harpur, *Decoration*, [#420], 274: VI.1M-L

AB70 *mrrw-kꜣ.j: mrj* - Saqqara, TP (ntn), chamber A1, south: marsh (6 birds within thicket)
Source: *P-M III*, 525-534; Duell, *Mereruka i*, pls. 19 and 128
Label: none
Date: Harpur, *Decoration*, [#420], 274: VM-L

[58] According to Brovarski (*Senedjemib Complex Part I*, 40), parts of this scene are now either destroyed or poorly preserved.

AB71 *zšzšt: jdwt* - Saqqara, UPC (ntn), room III, west: marsh (4 birds within thicket)
Source: *P-M III*, 617-619; Macramallah, *Idout*, pl. 7; Kanawati-Abder-Raziq, *Iynefert*, pl. 54
Label: none
Date: Kanawati-Abder-Raziq, *Iynefert*, 36-37: VI.1E; Harpur, *Decoration*, [#487], 275: V.9 r/u, VI.1-2E?

AB72 *sʿnḫ-w(j)-ptḥ* - Saqqara, TP (ntn), room I, east: marsh (4 birds within thicket)
Source: *P-M III*, nr; Kanawati-Abder-Raziq, *Neferseshemre*, pls. 69 and 76
Label: none
Date: Kanawati-Abder-Raziq, *Neferseshemre*, 40-41: VI.1L; Harpur, *Decoration*, nr

AB73 *ḥzj* - Saqqara, TP (ntn), portico, south (east of entrance): marsh (3 birds among umbels, 2 birds within thicket)
Source: *P-M III*, nr; Kanawati-Abder-Raziq, *Hesi*, pl. 53
Label: none
Date: Kanawati-Abder-Raziq, *Hesi*, 16: VI.1L-2E; Harpur, *Decoration*, nr

AB74 *ḥzj* - Saqqara, TP (ntn), portico, south (west of entrance): marsh (3 birds among umbels, 5 birds within thicket)
Source: *P-M III*, nr; Kanawati-Abder-Raziq, *Hesi*, pl. 54
Label: none
Date: Kanawati-Abder-Raziq, *Hesi*, 16: VI.1L-2E; Harpur, *Decoration*, nr

AB75 *ʿnḫ-m-ʿ-ḥr: zzj* - Saqqara, TP (ntn), room II, west: fowling (1 bird, below clapnet, on ground)
Source: *P-M III*, 512-515; Kanawati-Hassan, *Ankhmahor*, pl. 42
Label: none
Date: Kanawati-Hassan, *Ankhmahor*, 18: VI.1M-L; Harpur, *Decoration*, [#374], 273: VI.1L-2E

AB76 *mrjj-rʿ-mrj-ptḥ-ʿnḫ: nḫbw* - Giza, WF (G2381), relief fragment (MFA no. 13.4332), unknown: marsh (1 bird among umbels)
Source: *P-M III*, 90-91; Smith, W.S. (1958). *Bost. Mus. Bull.* 56, fig. 2
Label: none
Date: Harpur, *Decoration*, [#089], 267: VI.2

AB77 *mrrj* - Saqqara, TP (ntn), room 1, south: marsh (1 bird within thicket)
Source: *P-M III*, 518-519; Davies et al., *Saqqara Tombs i*, pl. 5
Label: none
Date: Harpur, *Decoration*, [#418], 274: VI.2E?

AB78 *jnw-mnw* - Saqqara (ntn), room II, west: marsh (2 birds within thicket)
Source: *P-M III*, nr; Kanawati, *Inumin*, pl. 46
Label: none
Date: Kanawati, *Inumin*, 16-17: VI.2E; Harpur, *Decoration*, nr

AB79 *ḫntj-kꜣ.j: jḫḫj* - Saqqara, TP (ntn), relief fragments, unknown: marsh (2 birds within thicket)
Source: *P-M III*, 508-511; James, *Khentika*, pl. 42 (XVI, iii and XVI, iv)
Label: none
Date: Harpur, *Decoration*, [#479], 275: VI.2M

AB80 *mḥw* - Saqqara, UPC (ntn), room I, north: marsh (4 birds among umbels)
Source: *P-M III*, 619-622; Altenmüller, *Mehu*, pl. 9
Label: none
Date: Altenmüller, *Mehu*, 19: VI.2; Harpur, *Decoration*, [#424], 274: VI.2M-3?

AB81 *mḥw* - Saqqara, UPC (ntn), room I, west: marsh (4 birds among umbels, 3 birds within thicket)
Source: *P-M III*, 619-622; Altenmüller, *Mehu*, pls. 12 and 13
Label: none
Date: Altenmüller, *Mehu*, 19: VI.2; Harpur, *Decoration*, [#424], 274: VI.2M-3?

AB82 *mḥw* - Saqqara, UPC (ntn), room II, south: agricultural (1 bird, in field on ground)
Source: *P-M III*, 619-622; Altenmüller, *Mehu*, pl. 23 [a]
Label: n/a
Date: Altenmüller, *Mehu*, 19: VI.2; Harpur, *Decoration*, [#424], 274: VI.2M-3?

AB83 *ꜣḫtj-ḥtp* - Saqqara, UPC (E17), offering-room, east: marsh (2 birds within thicket)
Source: *P-M III*, 633-634; Petrie-Murray, *Seven Memphite*, pl. 6 [lower]
Label: none
Date: Harpur, *Decoration*, [#339], 272: VI.5?

9.2 FIGHTING

Goats

AB84 *sšm-nfr IV* - Giza, GIS (LG53), relief fragment, unknown: pastoral
Source: *P-M III*, 223-226; Junker, *Giza xi*, fig. 93
Label: none
Date: Harpur, *Decoration*, [#235], 270: V.9-VI.1

AB85 *nb-kꜣw-ḥr: jdw* - Saqqara, UPC (ntn), pillared hall, south: pastoral
Source: *P-M III*, 627-629; Hassan, *Neb-Kaw-Her*, pl. 19 [c]
Label: none
Date: Harpur, *Decoration*, [#437], 274: V.9 r/u VI.7

Black kites

AB86 *nj-ꜥnḫ-ẖnmw/ẖnmw-ḥtp* - Saqqara, UPC (ntn), vestibule, south: fishing
Source: *P-M III*, 641-644; Moussa-Altenmüller, *Nianchchnum*, fig. 12
Label: none
Date: Harpur, *Decoration*, [#431], 274: V.6L-7

9.3 ATTACKING

Hippopotamuses vs Nile crocodiles

AB87 *ṯjj* - Saqqara, NSP (D22), room III, north: marsh
Source: *P-M III*, 468-478; Wild, *Ti ii*, pl. 119
Label: none
Date: Harpur, *Decoration*, [#543], 277: V.8-9

AB88 *ꜣḫtj-mrw-nswt* - Giza, WF (G2184), secondary offering-room (south of niche), west: marsh
Source: *P-M III*, 80-81; *Giza Archives Project* (Photo no. C12741_NS)
Label: none
Date: Harpur, *Decoration*, [#008], 265: V.9-VI.1

AB89 *kꜣ-gm-nj: mmj* - Saqqara, TP (LS10), room I, west: marsh
Source: *P-M III*, 521-525; Firth-Gunn, *Teti Pyr. ii*, pl. 53 [2]; Harpur-Scremin, *Kagemni*, fig. 3
Label: none
Date: Harpur, *Decoration*, [#534], 276: VI.1M

AB90 *kꜣ-gm-nj: mmj* - Saqqara, TP (LS10), room III, north: marsh
Source: *P-M III*, 521-525; Decker-Herb, *Bildatlas*, pl. 219 [K 2.39]; Harpur-Scremin, *Kagemni*, fig. 9
Label: none
Date: Harpur, *Decoration*, [#534], 276: VI.1M

AB91 *kꜣ-gm-nj: mmj* - Saqqara, TP (LS10), room IV, east: fishing
Source: *P-M III*, 521-525; von Bissing, *Gem-ni-kai i*, pl. 18; Harpur-Scremin, *Kagemni*, fig. 16
Label: missing
Date: Harpur, *Decoration*, [#534], 276: VI.1M

AB92 *mrrw-kꜣ.j: mrj* - Saqqara, TP (ntn), chamber A1, south: marsh
Source: *P-M III*, 525-534; Duell, *Mereruka i*, pl. 19
Label: none
Date: Harpur, *Decoration*, [#420], 274: VI.1M-L

AB93 *jnw-mnw* - Saqqara (ntn), room I, west: marsh
Source: *P-M III*, nr; Kanawati, *Inumin*, pl. 44
Label: none
Date: Kanawati, *Inumin*, 16-17: VI.2E; Harpur, *Decoration*, nr

AB94 *mḥw* - Saqqara, UPC (ntn), room I, west: marsh
Source: *P-M III*, 619-622; Altenmüller, *Mehu*, pl. 13
Label: none
Date: Altenmüller, *Mehu*, 19: VI.2; Harpur, *Decoration*, [#424], 274: VI.2M-3?

AB94a *nb-k3w-ḥr: jdw* - Saqqara, UPC (ntn), court, north: march
Source: *P-M III*, 627-629; Hassan, *Neb-Kaw-Her*, pl. 5 [c]
Label: missing
Date: Harpur, *Decoration*, [#437], 274: V.9 r/u VI.7

Baboons vs humans

AB95 *tp-m-ʿnḫ II* - Saqqara, NSP (D11), corridor, east: market
Source: *P-M III*, 483-484; Borchardt, *Denkmäler ii*, pl. 61 (CG 1556); *Eg. Art in the Age of the Pyramids*, fig. 126
Label: *j h3j nḏ ḥwj jry jʿn ḥmw js jr nrj ʿnr pn*
Oh charge down to protect and strike!
Keeper of baboons, Hemu: Go! as to make fearful this baboon
Date: Harpur, *Decoration*, [#541], 277: V.5-6?

AB96 *nj-ʿnḫ-ẖnmw/ẖnmw-ḥtp* - Saqqara, UPC (ntn), vestibule, north: market
Source: *P-M III*, 641-644; Moussa-Altenmüller, *Nianchchnum*, fig. 10
Label: *šzp šzp*
Catch him! Catch him!
j [3]wtj (?) m sṯ3 (?) n=j w3ḥ sw r t3 ḏr r . . . . r jr . . .
Oh official person! Do not lead the (monkey) to me!
Let her go to the ground
Keep her away from (my) . . .
Date: Harpur, *Decoration*, [#431], 274: V.6L-7

Birds vs predators

AB97 *nb.j-m-3ḫtj* - Giza, CF (LG86), chapel I, west: marsh (goose vs genet)
Source: *P-M III*, 230-232; *LD ii*, pl. 12 [b]
Label: none
Date: Harpur, *Decoration*, [#122], 267: IV.6-V.1

AB98 *ḥtp-ḥr-3ḫtj* - Saqqara, WSP (D60), chapel, north: marsh (2 pied kingfishers vs genet)
Source: *P-M III*, 593-595; Mohr, *Hetep-her-akhti*, fig. 34
Label: none
Date: Harpur, *Decoration*, [#467], 275: V.6-8E

AB99 *jtj-sn* - Giza, CF (ntn), chapel, east: marsh (bird vs genet)
Source: *P-M III*, 252-253; Hassan, *Giza v*, fig. 123
Label: none
Date: Harpur, *Decoration*, [#035], 265: V.6-8?

AB100 *snḏm-jb: jntj* - Giza, WF (G2370=LG27), portico, west: marsh (goose vs mongoose)
Source: *P-M III*, 85-87; *LD Erganz.*, pl. 18[59]
Label: none
Date: Brovarski, *Senedjemib Complex Part I*, 23: V.8M-L; Harpur, *Decoration*, [#219], 269: V.8M-L

AB101 *snḏm-jb: jntj* - Giza, WF (G2370=LG27), portico, west: marsh (goose vs genet)
Source: *P-M III*, 85-87; *LD Erganz.*, pl. 18; Brovarski, *Senedjemib Complex Part I*, fig. 27
Label: none
Date: Brovarski, *Senedjemib Complex Part I*, 23: V.8M-L; Harpur, *Decoration*, [#219], 269: V.8M-L

[59] According to Brovarski (*Senedjemib Complex Part I*, 40), this image has now been destroyed.

AB102 *ṯjj* - Saqqara, NSP (D22), room III, north: marsh (2 egrets vs mongoose)
Source: *P-M III*, 468-478; Wild, *Ti ii*, pl. 119
Label: none
Date: Harpur, *Decoration*, [#543], 277: V.8-9

AB103 *ṯjj* - Saqqara, NSP (D22), room III, north: marsh (goose vs genet)
Source: *P-M III*, 468-478; Wild, *Ti ii*, pl. 119
Label: none
Date: Harpur, *Decoration*, [#543], 277: V.8-9

AB104 *ṯjj* - Saqqara, NSP (D22), room III, north: marsh (2 pied kingfishers vs genet)
Source: *P-M III*, 468-478; Wild, *Ti ii*, pl. 119
Label: none
Date: Harpur, *Decoration*, [#543], 277: V.8-9

AB105 *nfr-jrt-n.f* - Saqqara, ESP (D55), chapel, east: marsh (goose pecks mongoose; rump)
Source: *P-M III*, 583-584; van de Walle, *Neferirtenef*, pl. 1
Label: none
Date: Harpur, *Decoration*, [#440], 274: V.8-9

AB106 *kȝ.j-m-nfrt* - Saqqara, NSP (D23), chapel, north: marsh (goose pecks genet; nose)
Source: *P-M III*, 467-468; Simpson, *Kayemnofret*, pl. A
Label: none
Date: Simpson, *Kayemnofret*, 1: V.8-9; Harpur, *Decoration*, [#522], 276: V.8-9?

AB107 *ȝḫtj-ḥtp* - Saqqara, WSP (D64), hall, east: marsh (goose vs mongoose)
Source: *P-M III*, 599-600; Davies, *Ptahhetep ii*, pl. 13
Label: none
Date: Harpur, *Decoration*, [#338], 272: V.8L-9E

AB108 *ȝḫtj-ḥtp* - Saqqara, WSP (D64), hall, east: marsh (2 pied kingfishers vs genet)
Source: *P-M III*, 599-600; Davies, *Ptahhetep ii*, pl. 13
Label: none
Date: Harpur, *Decoration*, [#338], 272: V.8L-9E

AB109 *ȝḫtj-ḥtp* - Saqqara, WSP (D64), hall, east: marsh (2 geese vs mongoose)
Source: *P-M III*, 599-600; Davies, *Ptahhetep ii*, pl. 14
Label: none
Date: Harpur, *Decoration*, [#338], 272: V.8L-9E

AB110 *ḫnwt* - Saqqara, UPC (ntn), room A [I], east: marsh (2 geese vs genet)
Source: *P-M III*, 623-624; Munro, *Unas-Friedhof*, pl. 33
Label: none
Date: Munro, *Unas-Friedhof*, 20: V.9; Harpur, *Decoration*, [#478], 275: V.9

AB111 *nbt* - Saqqara, UPC (ntn), room A [I], west: marsh (2 swooping geese?)
Source: *P-M III*, 624-625; Munro, *Unas-Friedhof*, pl. 10
Label: none
Date: Munro, *Unas-Friedhof*, 20: V.9; Harpur, *Decoration*, [#438], 274: V.9

AB112 *nbt* - Saqqara, UPC (ntn), room A [I], east: marsh (pied kingfisher vs mongoose)
Source: *P-M III*, 624-625; Munro, *Unas-Friedhof*, pl. 11
Label: missing
Date: Munro, *Unas-Friedhof*, 20: V.9; Harpur, *Decoration*, [#438], 274: V.9

AB113 *jj-nfrt: šȝ.n.f* - Saqqara, UPC (ntn), room I, north: marsh (goose vs genet)
Source: *P-M III*, 616; Kanawati-Abder-Raziq, *Iynefert*, pl. 37
Label: none
Date: Kanawati-Abder-Raziq, *Iynefert*, 13: V.9E-M; Harpur, *Decoration*, [#345], 272: V.9

AB114 *jj-nfrt: šȝ.n.f* - Saqqara, UPC (ntn), room I, north: marsh (pied kingfisher vs mongoose)
Source: *P-M III*, 616; Kanawati-Abder-Raziq, *Iynefert*, pl. 37
Label: none

Date: Kanawati-Abder-Raziq, *Iynefert*, 13: V.9E-M; Harpur, *Decoration*, [#345], 272: V.9

AB115 *nj-k3w-jzzj* - Saqqara, TP (ntn), room I, east: marsh (2 pied kingfishers vs mongoose)
Source: *P-M III*, nr; Kanawati-Abder-Raziq, *Nikauisesi*, pl. 50
Label: none
Date: Kanawati-Abder-Raziq, *Nikauisesi*, 17-23: VI.1M; Harpur, *Decoration*, nr

AB116 *nj-k3w-jzzj* - Saqqara, TP (ntn), room I, east: marsh (goose pecks genet; ears)
Source: *P-M III*, nr; Kanawati-Abder-Raziq, *Nikauisesi*, pl. 50
Label: none
Date: Kanawati-Abder-Raziq, *Nikauisesi*, 17-23: VI.1M; Harpur, *Decoration*, nr

AB117 *nj-k3w-jzzj* - Saqqara, TP (ntn), room I, east: marsh (2 grey herons peck genet; ears and eyes)
Source: *P-M III*, nr; Kanawati-Abder-Raziq, *Nikauisesi*, pl. 50[60]
Label: none
Date: Kanawati-Abder-Raziq, *Nikauisesi*, 17-23: VI.1M; Harpur, *Decoration*, nr

AB118 *mrrw-k3.j: mrj* - Saqqara, TP (ntn), chamber A1, north: marsh (2 pied kingfishers peck mongoose; ears)
Source: *P-M III*, 525-534; Duell, *Mereruka i*, pl. 10
Label: none
Date: Harpur, *Decoration*, [#420], 274: VI.1M-L

AB119 *mrrw-k3.j: mrj* - Saqqara, TP (ntn), chamber A1, north: marsh (2 geese? vs genet)
Source: *P-M III*, 525-534; Duell, *Mereruka i*, pl. 10
Label: none
Date: Harpur, *Decoration*, [#420], 274: VI.1M-L

AB120 *mrrw-k3.j: mrj* - Saqqara, TP (ntn), chamber AI, south: marsh (2 pied kingfishers vs mongoose)
Source: *P-M III*, 525-534; Duell, *Mereruka i*, pl. 19
Label: none
Date: Harpur, *Decoration*, [#420], 274: VI.1M-L

AB121 *zšzšt: jdwt* - Saqqara, UPC (ntn), room III, west: marsh (2 geese vs genet)
Source: *P-M III*, 617-619; Macramallah, *Idout*, pl. 7; Kanawati-Abder-Raziq, *Iynefert*, pl. 54
Label: none
Date: Kanawati-Abder-Raziq, *Iynefert*, 36-37: VI.1E; Harpur, *Decoration*, [#487], 275: V.9 r/u, VI.1-2E?

AB122 *sʿnḫ-w(j)-ptḥ* - Saqqara, TP (ntn), room I, east: marsh (2 geese peck mongoose; nose)
Source: *P-M III*, nr; Kanawati-Abder-Raziq, *Neferseshemre*, pls. 69 and 76
Label: none
Date: Kanawati-Abder-Raziq, *Neferseshemre*, 40-41: VI.1L; Harpur, *Decoration*, nr

AB123 *sʿnḫ-w(j)-ptḥ* - Saqqara, TP (ntn), room I, east: marsh (2 pied kingfishers vs genet)
Source: *P-M III*, nr; Kanawati-Abder-Raziq, *Neferseshemre*, pls. 69 and 76
Label: none
Date: Kanawati-Abder-Raziq, *Neferseshemre*, 40-41: VI.1L; Harpur, *Decoration*, nr

AB124 *sʿnḫ-w(j)-ptḥ* - Saqqara, TP (ntn), room I, east: marsh (2 egrets peck mongoose; back)
Source: *P-M III*, nr; Kanawati-Abder-Raziq, *Neferseshemre*, pls. 69 and 76
Label: none
Date: Kanawati-Abder-Raziq, *Neferseshemre*, 40-41: VI.1L; Harpur, *Decoration*, nr

AB125 *sʿnḫ-w(j)-ptḥ* - Saqqara, TP (ntn), room I, east: marsh (2 geese peck genet; back)
Source: *P-M III*, nr; Kanawati-Abder-Raziq, *Neferseshemre*, pls. 69 and 76
Label: none
Date: Kanawati-Abder-Raziq, *Neferseshemre*, 40-41: VI.1L; Harpur, *Decoration*, nr

60 Kanawati (*Nikauisesi*, 40) identifies these birds as herons (*Ardea* sp.), probably because of the distinctive plumes upon the head of the animal on the left. Although this is usually a diagnostic feature for grey herons, it should be noted that little egrets develop long occipital feathers during the breeding season. As the birds are clearly rearing young, it is possible that these may be breeding egrets, rather than herons.

AB126 *ḥzj* - Saqqara, TP (ntn), portico, south (east of entrance): marsh (2 pied kingfishers peck genet; eyes)
Source: *P-M III*, nr; Kanawati-Abder-Raziq, *Hesi*, pl. 53
Label: none
Date: Kanawati-Abder-Raziq, *Hesi*, 16: VI.1L-2E; Harpur, *Decoration*, nr

AB127 *ḥzj* - Saqqara, TP (ntn), portico, south (west of entrance): marsh (2 geese peck mongoose; back)
Source: *P-M III*, nr; Kanawati-Abder-Raziq, *Hesi*, pl. 54
Label: none
Date: Kanawati-Abder-Raziq, *Hesi*, 16: VI.1L-2E; Harpur, *Decoration*, nr

AB128 *ḥzj* - Saqqara, TP (ntn), portico, south (west of entrance): marsh (2 egrets peck and scratch genet; ears, nose, and back)
Source: *P-M III*, nr; Kanawati-Abder-Raziq, *Hesi*, pl. 54
Label: none
Date: Kanawati-Abder-Raziq, *Hesi*, 16: VI.1L-2E; Harpur, *Decoration*, nr

AB129 *ḥzj* - Saqqara, TP (ntn), portico, south (west of entrance): marsh (goose pecks genet; back)
Source: *P-M III*, nr; Kanawati-Abder-Raziq, *Hesi*, pl. 54
Label: none
Date: Kanawati-Abder-Raziq, *Hesi*, 16: VI.1L-2E; Harpur, *Decoration*, nr

AB130 *mrjj-rˁ-mrj-ptḥ-ˁnḫ: nḫbw* - Giza, WF (G2381), relief fragment (MFA no. 13.4332) from forecourt, unknown: marsh (2 pied kingfishers peck genet; nose)
Source: *P-M III*, 90-91; Smith, W.S. (1958). *Bost. Mus. Bull.* 56, fig. 2
Label: none
Date: Harpur, *Decoration*, [#089], 267: VI.2

AB131 *mṯtj* - Saqqara, UPC (ntn), relief fragment (Abegg-Stiftung, Riggisberg no. 12.39.67), unknown: marsh (bird? vs mongoose?)
Source: *P-M III*, 646-648; Kaplony, *Studien Methethi*, fig. on 10
Label: none
Date: Ziegler, *Catalogue*, 123: probably VI; Harpur, *Decoration*, [#426], 274: VI.2?

AB132 *mḥw* - Saqqara, UPC (ntn), room I, north: marsh (3 pied kingfishers peck genet; tail and eyes)
Source: *P-M III*, 619-622; Altenmüller, *Mehu*, pl. 9
Label: none
Date: Altenmüller, *Mehu*, 19: VI.2; Harpur, *Decoration*, [#424], 274: VI.2M-3?

AB133 *mḥw* - Saqqara, UPC (ntn), room I, north: marsh (3 geese peck and scratch mongoose; back and tail)
Source: *P-M III*, 619-622; Altenmüller, *Mehu*, pl. 9
Label: none
Date: Altenmüller, *Mehu*, 19: VI.2; Harpur, *Decoration*, [#424], 274: VI.2M-3?

AB134 *mḥw* - Saqqara, UPC (ntn), room I, east: marsh (goose pecks genet; tail)
Source: *P-M III*, 619-622; Altenmüller, *Mehu*, pl. 10
Label: none
Date: Altenmüller, *Mehu*, 19: VI.2; Harpur, *Decoration*, [#424], 274: VI.2M-3?

AB135 *mḥw* - Saqqara, UPC (ntn), room I, west: marsh (2 geese vs genet)
Source: *P-M III*, 619-622; Altenmüller, *Mehu*, pl. 13
Label: none
Date: Altenmüller, *Mehu*, 19: VI.2; Harpur, *Decoration*, [#424], 274: VI.2M-3?

AB136 *mḥw* - Saqqara, UPC (ntn), room I, west: marsh (2 pied kingfishers vs mongoose)
Source: *P-M III*, 619-622; Altenmüller, *Mehu*, pl. 13
Label: none
Date: Altenmüller, *Mehu*, 19: VI.2; Harpur, *Decoration*, [#424], 274: VI.2M-3?

AB137 *jj-nfrt* - Giza (ntn), chapel, unknown: marsh (2 geese? vs genet?)
Source: *P-M III*, 298-299; Schürmann, *Die Reliefs*, fig. 6 [b]
Label: none
Date: Harpur, *Decoration*, [#017], 265: VI.3-4E?

AB138 *jj-nfrt* - Giza (ntn), chapel, unknown: marsh (2 pied kingfishers? vs genet)
Source: *P-M III*, 298-299; Schürmann, *Die Reliefs*, fig. 6 [b]
Label: none
Date: Harpur, *Decoration*, [#017], 265: VI.3-4E?

AB139 *ꜣḫtj-ḥtp* - Saqqara, UPC (E17), offering-room, east: marsh (pied kingfisher pecks genet; eyes)
Source: *P-M III*, 633-634; Petrie-Murray, *Seven Memphite*, pl. 6 [lower]
Label: none
Date: Harpur, *Decoration*, [#339], 272: VI.5?

9.4 PROTESTING

Bulls

AB140 *nb.j-m-ꜣḫtj* - Giza, CF (LG86), entrance to chapel II, north: slaughter
Source: *P-M III*, 230-232; Hassan, *Giza iv*, fig. 79
Label: *nḏrt nḏ sḥ wrt*
Holding (the bull) and turning steadily
Date: Harpur, *Decoration*, [#122], 267: IV.6-V.1

AB141 *pḥn-wj-kꜣ.j* - Saqqara, NSP (D70=LS15), antechamber, unknown: agricultural
Source: *P-M III*, 491-492; *LD ii*, pl. 47
Label: *jr ḥꜣ=k jm=sn*
Get back among them!
jꜣ ḥwj(=k) pẖr=k
Oh! Thresh and move about!
Date: Harpur, *Decoration*, [#393], 273: V.6-8E

AB142 *nṯr-wsr* - Saqqara, NSP (D1=S901), offering-room, east: offering
Source: *P-M III*, 485; Murray, *Saqqara Mastabas*, pl. 22
Label: *sḫpt rn jwꜣ nt ḫt-ḫꜣwy*
Bringing a young ox of the night offerings
Date: Harpur, *Decoration*, [#451], 275: V.6L-7

AB143 *tjj* - Saqqara, NSP (D22), room I/pillared hall, north: slaughter
Source: *P-M III*, 468-478; Epron et al., *Ti i*, pl. 14
Label: *nḏrt rn jwꜣ ḏwt 50 rḫs*
Holding fast the young ox of the mountain(?)
50 slaughtered
Date: Harpur, *Decoration*, [#543], 277: V.8-9

AB144 *tjj* - Saqqara, NSP (D22), room III, south: offering
Source: *P-M III*, 468-478; Wild, *Ti iii*, pl. 167
Label: *jwꜣ*
Ox
Date: Harpur, *Decoration*, [#543], 277: V.8-9

AB145 *ꜣḫtj-ḥtp* - Saqqara, WSP (D64), hall, south: slaughter
Source: *P-M III*, 599-600; Davies, *Ptahhetep ii*, pl. 22
Label: *ḳꜣs ng(ꜣw) wrt*
Binding a long-horned bull greatly!
Date: Harpur, *Decoration*, [#338], 272: V.8L-9E

AB146 *ꜣḫtj-ḥtp* - Saqqara, WSP (D64), corridor, west: agricultural
Source: *P-M III*, 599-600; Davies, *Ptahhetep ii*, pl. 8
Label: *ḫsf ng(ꜣw)*
Driving the long-horned bulls
Date: Harpur, *Decoration*, [#338], 272: V.8L-9E

AB147 *ptḥ-ḥtp II: tfj* - Saqqara, WSP (D64), offering-room, east: hunting
Source: *P-M III*, 600-604; Quibell, *Ptah-hetep*, pl. 32
Label: n/a

Date: Harpur, *Decoration*, [#400], 274: V.9M-L

AB148 *ptḥ-ḥtp II: ṯfj* - Saqqara, WSP (D64), offering-room, east: offering
Source: *P-M III*, 600-604; Quibell, *Ptah-hetep*, pl. 31
Label: n/a
Date: Harpur, *Decoration*, [#400], 274: V.9M-L

AB149 *ptḥ-ḥtp II: ṯfj* - Saqqara, WSP (D64), offering-room, east: offering
Source: *P-M III*, 600-604; Quibell, *Ptah-hetep*, pl. 31
Label: n/a
Date: Harpur, *Decoration*, [#400], 274: V.9M-L

AB150 *sšm-nfr IV* - Giza, GIS (LG53), doorway between chambers A and B, west: offering
Source: *P-M III*, 223-226; Junker, *Giza xi*, fig. 79
Label: *rn jwꜣ*
Young ox
Date: Harpur, *Decoration*, [#235], 270: V.9-VI.1

AB151 *ḥtp-kꜣ.j: tp-kꜣ.j* - Saqqara, NSP (S3509), chapel, south: slaughter
Source: *P-M III*, 447-448; Martin, *Hetepka*, pl. 13 [14]
Label: *spḥ gw jn jswt pr ḏt=f*
Lassoing the *gw*-bull by the gangs of the funerary estate
jṯ r=k wrt
Pull well then!
sp[ḥ]=j g[w]
I am lassoing the bull
Date: Harpur, *Decoration*, [#469], 275: V.9-VI.1

AB152 *sḫm-ꜥnḫ-ptḥ* - Giza, EF (G7152), chapel, north: pastoral
Source: *P-M III*, 191; Badawy, *Tombs of Iteti*, fig. 23
Label: missing
Date: Harpur, *Decoration*, [#223], 269: V.9-VI.1?

AB153 *ṯtw: kꜣ.j-nswt* - Giza, WF (G2001), recessed portico, west: slaughter
Source: *P-M III*, 66-67; Simpson, *Mast. West. Cem. Part I*, fig. 19
Label: none
Date: Harpur, *Decoration*, [#292], 271: VI

AB154 *mrrw-kꜣ.j: mrj* - Saqqara, TP (ntn), chamber A1, south: slaughter
Source: *P-M III*, 525-534; Duell, *Mereruka i*, pl. 21
Label: none
Date: Harpur, *Decoration*, [#420], 274: VI.1M-L

AB155 *mrrw-kꜣ.j: mrj* - Saqqara, TP (ntn), chamber A1, south: slaughter
Source: *P-M III*, 525-534; Duell, *Mereruka i*, pl. 21
Label: none
Date: Harpur, *Decoration*, [#420], 274: VI.1M-L

AB156 *mrrw-kꜣ.j: mrj* - Saqqara, TP (ntn), chamber A6, north: slaughter
Source: *P-M III*, 525-534; Duell, *Mereruka i*, pl. 54
Label: *pns=f ntj ḥnꜥ*
He turns my comrade!
jrj=j ḥzt=k
I will do (it) so that you praise it!
Date: Harpur, *Decoration*, [#420], 274: VI.1M-L

AB157 *mrrw-kꜣ.j: mrj* - Saqqara, TP (ntn), chamber A11, east: slaughter
Source: *P-M III*, 525-534; Duell, *Mereruka ii*, pl. 109 [a]
Label: none
Date: Harpur, *Decoration*, [#420], 274: VI.1M-L

AB158 *ꜥnḫ-m-ꜥ-ḥr: zzj* - Saqqara, TP (ntn), door thickness between rooms II and III, north: offer.ng
Source: *P-M III*, 512-515; Kanawati-Hassan, *Ankhmahor*, pl. 43 [b]

Label: *rn jwꜣ*
Young ox
Date: Kanawati-Hassan, *Ankhmahor*, 18: VI.1M-L; Harpur, *Decoration*, [#374], 273: VI.1L-2E

AB159 *ḫntj-kꜣ.j: jḫḫj* - Saqqara, TP (ntn), room VII, east: slaughter
Source: *P-M III*, 508-511; James, *Khentika*, pl. 22
Label: *mk ḫrj-ḥbt ḥr jrt jḫt*
Behold! The lector priest is doing things!
dj jwt ḫpš wn
Let a haunch come away; hurry up!
Date: Harpur, *Decoration*, [#479], 275: VI.2M

AB160 *ḫntj-kꜣ.j: jḫḫj* - Saqqara, TP (ntn), room VII, east: slaughter
Source: *P-M III*, 508-511; James, *Khentika*, pl. 22
Label: *r nn wn ntj ḥnꜥ(=j)*
Do this; hurry up, my companion!
Date: Harpur, *Decoration*, [#479], 275: VI.2M

AB161 *ꜣḫtj-ḥtp* - Saqqara, UPC (E17), offering-room, east: slaughter
Source: *P-M III*, 633-634; Petrie-Murray, *Seven Memphite*, pl. 7 [lower]
Label: *sḫpt rn jwꜣ ḫ[t]-dwꜣjt*
Bringing a young ox for the morning offerings
Date: Harpur, *Decoration*, [#339], 272: VI.5?

AB162 *nb-kꜣw-ḥr: jdw* - Saqqara, UPC (ntn), doorway into mortuary chapel, east: slaughter?
Source: *P-M III*, 627-629; Hassan, *Neb-Kaw-Her*, pl. 33 [a]
Label: *sḫpt rn jwꜣ ḫt-dwꜣjt*
Bringing a young ox for the morning offerings
Date: Harpur, *Decoration*, [#437], 274: V.9 r/u VI.7

AB163 *zꜣt-jn-ttj* - Saqqara, TP (ntn), niche, unknown: offering
Source: *P-M III*, 539-540; Firth-Gunn, *Teti Pyr. ii*, pl. 21 [b]
Label: *rsw wšb pw*
This southern bull
Date: Harpur, *Decoration*, [#484], 275: VI.7-FIP

Cows

AB164 *ḥtp-ḥr-ꜣḫtj* - Saqqara, WSP (D60), corridor, north: agricultural
Source: *P-M III*, 593-595; Mohr, *Hetep-her-akhti*, fig. 4
Label: *ḥnṯ jwꜣt*
Binding the cow
jṯ jr=k ntj ḥnꜥ(=j)
Pull towards yourself, my companion!
Date: Harpur, *Decoration*, [#467], 275: V.6-8E

AB165 *kꜣ.j-m-rḥw* - Saqqara, NSP (D2=S905), door thickness (entrance to chapel), unknown: offering
Source: *P-M III*, 485-487; Borchardt, *Denkmäler ii*, pl. 62 (CG 1560)
Label: unable to determine
Date: Harpur, *Decoration*, [#523], 276: V.8

AB166 *rꜥ-špss* - Saqqara, NSP (LS16=S902), door thickness (between forecourt and hall), south: offering
Source: *P-M III*, 494-496; *LD Erganz.*, pl. 41
Label: *sḫpt rn jwꜣ ntj ḫꜣt*
Bringing a young ox of the office
Date: Harpur, *Decoration*, [#456], 275: V.8M

Birds

AB167 *nj-kꜣw-jzzj* - Saqqara, TP (ntn), room I, east: marsh
Source: *P-M III*, nr; Kanawati-Abder-Raziq, *Nikauisesi*, pl. 50
Label: none
Date: Kanawati-Abder-Raziq, *Nikauisesi*, 17-23: VI.1M; Harpur, *Decoration*, nr

10.1 COURTSHIP

10.1.1 Anovaginal sniffing

Bulls

SB1 *ṯtw: kꜣ.j-nswt* - Giza, WF (G2001), recessed portico, west: agricultural
Source: *P-M III*, 66-67; Simpson, *Mast. West. Cem. Part I*, fig. 22
Label: *jj mnjw pw ḥnṯ jwꜣt wn m wp=k kꜣ pn*
Oh this herdsman, tether the cow quickly as you separate this bull!
Date: Harpur, *Decoration,* [#292], 271: VI

SB2 *zšzšt: jdwt* - Saqqara, UPC (ntn), room III, west: agricultural
Source: *P-M III*, 617-619; Macramallah, *Idout*, pl. 7; Kanawati-Abder-Raziq, *Iynefert*, pl. 54
Label: *[ḥnṯ jwꜣt wn] m wp[=k] kꜣ [p]n*[61]
[Tether the cow quickly] as you separate this bull!
Date: Kanawati-Abder-Raziq, *Iynefert*, 36-37: VI.1E; Harpur, *Decoration,* [#487], 275: V.9 r/u, VI.1-2E?

10.1.2 Chin-resting

Bulls

SB3 *jj-mrjj* - Giza, WF (G6020=LG16), second chamber, south: agricultural
Source: *P-M III*, 170-174; Weeks, *Cemetery G6000*, fig. 34
Label: *kꜣ*
Bull
Date: Weeks, *Cemetery G6000*, 7: V.6; Harpur, *Decoration,* [#014], 265: V.3L

SB4 *ṯjj* - Saqqara, NSP (D22), room I/pillared hall, west: pastoral
Source: *P-M III*, 468-478; Epron et al., *Ti i*, pl. 26
Label: missing
Date: Harpur, *Decoration,* [#543], 277: V.8-9

SB5 *kꜣpj* - Giza, WF (G2091), corridor, west: agricultural
Source: *P-M III*, 69-70; Roth, *Pal. Cem. Att.*, pl. 156
Label: none
Date: Harpur, *Decoration,* [#275], 270: V.9-VI

SB6 *ꜣḫtj-ḥtp* - Saqqara, UPC (E17), passage, south: offering
Source: *P-M III*, 633-634; Petrie-Murray, *Seven Memphite*, pl. 7 [upper]
Label: . . . *r jnjt*
.. in order to bring . . .
Date: Harpur, *Decoration,* [#339], 272: VI.5?

10.1.3 Nudging

Cows

SB7 *jnw-mnw* - Saqqara (ntn), room II, south: pastoral
Source: *P-M III*, nr; Kanawati, *Inumin*, pl. 49
Label: missing
Date: Kanawati, *Inumin*, 16-17: VI.2E; Harpur, *Decoration,* nr

---

61 Macramallah (*Idout*, 17) translated this inscription as *nḥp* 'saillir'. However, the bull is not mating, but instead, sniffing under the tail of the animal in front of it. The translation that I suggest is based upon similarities between the remaining text and the second half of the fully preserved inscription that accompanies an identical scene in the tomb of *ṯtw: kꜣ.j-nswt* (see text for details).

Goats

SB8 *nj-ʿnḫ-ẖnmw/ẖnmw-ḥtp* - Saqqara, UPC (ntn), vestibule, north: pastoral
Source: *P-M III*, 641-644; Moussa-Altenmüller, *Nianchchnum*, fig. 8
Label: none
Date: Harpur, *Decoration,* [#431], 274: V.6L-7

SB9 *3ḫtj-ḥtp* - Saqqara, UPC (ntn), chapel, east: pastoral
Source: *P-M III*, 634-637; Ziegler, *Akhethetep*, 138-139
Label: none
Date: Ziegler, *Akhethetep*, 9: V.6; Harpur, *Decoration,* [#340], 272: V.6-8E

10.1.4 Foreleg kicking

Goats

SB10 *sšm-nfr IV* - Giza, GIS (LG53), relief fragment, unknown: pastoral
Source: *P-M III*, 223-226; Junker, *Giza xi*, fig. 93
Label: none
Date: Harpur, *Decoration*, [#235], 270: V.9-VI.1

10.2 MATING

Wild ass

SB11 *nj-m3ʿt-rʿ* [*jzzj-mr-nṯr*] - Giza, WF (G2097), chapel, east: hunting
Source: *P-M III*, 70; Roth, *Cem. Pal. Att.*, pl. 189
Label: none
Date: Roth, *Cem. Pal. Att.*, 129-130: V.9; Harpur, *Decoration,* [#033], 265: V.9-VI.1

Cattle

SB12 *rʿ-m-k3.j* - Saqqara, NSP (D3=S903), chapel, north: pastoral
Source: *P-M III*, 487-488; Hayes, *Scepter i*, fig. 54
Label: missing
Date: Harpur, *Decoration,* [#453], 275: V.6? r/u - V.6-7

SB13 *ḥtp-ḥr-3ḫtj* - Saqqara, WSP (D60), chapel, north: pastoral
Source: *P-M III*, 593-595; Mohr, *Hetep-her-akhti*, fig. 32
Label: *nhp*
Copulating
Date: Harpur, *Decoration,* [#467], 275: V.6-8E

SB14 *3ḫtj-ḥtp* - Saqqara, UPC (ntn), chapel, east: pastoral
Source: *P-M III*, 634-637; Ziegler, *Akhethetep*, 130-131
Label: *k3 nhp*
Bull copulating
jdt
Cow
Date: Ziegler, *Akhethetep*, 9: V.6; Harpur, *Decoration,* [#340], 272: V.6-8E

SB15 *jtj-sn* - Giza, CF (ntn), chapel, east: pastoral
Source: *P-M III*, 252-253; Hassan, *Giza v*, fig. 123
Label: none
Date: Harpur, *Decoration,* [#035], 265: V.6-8?

SB16 *jrj-n-k3-ptḥ* - Saqqara, UPC (ntn), offering chamber, south: pastoral
Source: *P-M III*, 644; Moussa-Junge, *Two Tombs*, pl. 13
Label: *nhp*
Copulating

jdt
Cow
Date: *P-M III*, 644: mid-late V; Harpur, *Decoration*, [#362], 273: V.6-9

SB17 *snḏm-jb: jntj* - Giza, WF (G2370=LG27), room II, west: agricultural
Source: *P-M III*, 85-87; *LD ii*, pl. 77[62]
Label: *nhp*
Copulating
Date: Brovarski, *Senedjemib Complex Part I*, 23: V.8M-L; Harpur, *Decoration*, [#219], 269: V.8M-L

SB18 *sḫm-ʿnḫ-ptḥ* - Saqqara, NSP (D41), chapel, east: pastoral
Source: *P-M III*, 454-455; Simpson, *Sekem-ankh-ptah*, pl. C
Label: *nhp*
Copulating
Date: Harpur, *Decoration*, [#504], 276: V.8-9

SB19 *nfr-ḥr-nj-ptḥ* - Saqqara, UPC (ntn), chapel, north: pastoral
Source: *P-M III*, 637-638; © Su Barfield (*www.egyptsites.co.uk*)
Label: *nh[p]*
Copulating
Date: Harpur, *Decoration*, [#441], 274: V.8-9

SB20 *kꜣ.j-m-nfrt* - Saqqara, NSP (D23), chapel, east: pastoral
Source: *P-M III*, 467-468; Simpson, *Kayemnofret*, pl. G
Label: *nhp*
Copulating
Date: Simpson, *Kayemnofret*, 1: V.8-9; Harpur, *Decoration*, [#522], 276: V.8-9?

SB21 *nbt* - Saqqara, UPC (ntn), room B [II], east: pastoral
Source: *P-M III*, 624-625; Munro, *Unas-Friedhof*, pl. 13
Label: none
Date: Munro, *Unas-Friedhof*, 20: V.9; Harpur, *Decoration*, [#438], 274: V.9

SB22 *nj-kꜣw-jzzj* - Saqqara, TP (ntn), room I, north: pastoral
Source: *P-M III*, nr; Kanawati-Abder-Raziq, *Nikauisesi*, pl. 49
Label: *prt m sḫt r tꜣ*
Coming out from the marshes to the land
Date: Kanawati-Abder-Raziq, *Nikauisesi*, 17-23: VI.1M; Harpur, *Decoration*, nr

SB23 *jnw-mnw* - Saqqara (ntn), room II, south: pastoral
Source: *P-M III*, nr; Kanawati, *Inumin*, pl. 49 [a]
Label: missing
Date: Kanawati, *Inumin*, 16-17: VI.2E; Harpur, *Decoration*, nr

Addax

SB24 *nj-mꜣʿt-rʿ [jzzj-mr-nṯr]* - Giza, WF (G2097), chapel, east: hunting
Source: *P-M III*, 70; Roth, *Cem. Pal. Att.*, pl. 189
Label: *nk nwḏw*
Copulating addax
Date: Roth, *Cem. Pal. Att.*, 129-130: V.9; Harpur, *Decoration*, [#033], 265: V.9-VI.1

Scimitar oryx

SB25 *nj-mꜣʿt-rʿ [jzzj-mr-nṯr]* - Giza, WF (G2097), chapel, east: hunting
Source: *P-M III*, 70; Roth, *Cem. Pal. Att.*, pl. 189
Label: *nk mꜣ-ḥḏ*
Copulating oryx
Date: Roth, *Cem. Pal. Att.*, 129-130: V.9; Harpur, *Decoration*, [#033], 265: V.9-VI.1

---

62 According to Brovarski (*Senedjemib Complex Part I*, 49), this scene has now been destroyed.

Hartebeest

SB26 *pḥn-wj-k3.j* - Saqqara, NSP (D70=LS15), relief fragment (Brooklyn Mus. no. 64.147), unknown (probably from the west wall of the antechamber): desert
Source: *P-M III*, 491-492; *Eg. Art in the Age of the Pyramids*, cat. no. 146 [b]
Label: none
Date: Harpur, *Decoration*, [#393], 273: V.6-8E

Dorcas gazelle

SB27 *sšm-nfr IV* - Giza, GIS (LG53), vestibule, west: hunting
Source: *P-M III*, 223-226; Junker, *Giza xi*, fig. 63
Label: none
Date: Harpur, *Decoration*, [#235], 270: V.9-VI.1

Goats

SB28 *sšm-nfr IV* - Giza, GIS (LG53), relief fragment, unknown: pastoral
Source: *P-M III*, 223-226; Junker, *Giza xi*, fig. 93
Label: none
Date: Harpur, *Decoration*, [#235], 270: V.9-VI.1

SB29 *3ḫtj-mrw-nswt* - Giza, WF (G2184), secondary offering-room, west (north end): pastoral
Source: *P-M III*, 80-81; *Giza Archives Project* (Photo no. A779_NS)
Label: none
Date: Harpur, *Decoration*, [#008], 265: V.9-VI.1

SB30 *nb-k3w-ḥr: jdw* - Saqqara, UPC (ntn), pillared hall, south: pastoral
Source: *P-M III*, 627-629; Hassan, *Neb-Kaw-Her*, pl. 19 [c]
Label: none
Date: Harpur, *Decoration*, [#437], 274: V.9 r/u VI.7

Unidentified ungulates

SB31 *pḥn-wj-k3.j* - Saqqara, NSP (D70=LS15), relief fragment (Brooklyn Mus. no. 64.147), unknown (probably from the west wall of the antechamber): desert
Source: *P-M III*, 491-492; *Eg. Art in the Age of the Pyramids*, cat. no. 146 [b]
Label: missing
Date: Harpur, *Decoration*, [#393], 273: V.6-8E

SB32 Anon - Saqqara, WSP (ntn), relief fragment found in the tomb of *ṯfw*, unknown
Source: *P-M III*, 605; Hassan, *Ny-Ankh-Pepy*, fig. 59
Label: missing
Date: *P-M III*, 605: V-VI; Harpur, *Decoration*, nr

Leopards

SB33 *ptḥ-ḥtp II: ṯfj* - Saqqara, WSP (D64), offering-room, east: hunting
Source: *P-M III*, 600-604; Quibell, *Ptah-hetep*, pl. 32
Label: none
Date: Harpur, *Decoration*, [#400], 274: V.9M-L

SB34 *nj-m3ʿt-rʿ [jzzj-mr-nṯr]* - Giza, WF (G2097), chapel, east: hunting
Source: *P-M III*, 70; Roth, *Cem. Pal. Att.*, pl. 189
Label: missing
Date: Roth, *Cem. Pal. Att.*, 129-130: V.9; Harpur, *Decoration*, [#033], 265: V.9-VI.1

Caracals

SB35 *nj-m3ʿt-rʿ [jzzj-mr-nṯr]* - Giza, WF (G2097), chapel, east: hunting
Source: *P-M III*, 70; Roth, *Cem. Pal. Att.*, pl. 189
Label: *nk jnb*
Copulating caracals

Date: Roth, *Cem. Pal. Att.*, 129-130: V.9; Harpur, *Decoration,* [#033], 265: V.9-VI.1

Striped hyenas

SB36 *nj-mꜣʿt-rʿ* [*jzzj-mr-nṯr*] - Giza, WF (G2097), chapel, east: hunting
Source: *P-M III*, 70; Roth, *Cem. Pal. Att.*, pl. 189
Label: *nk ḥtt*
Copulating hyenas
Date: Roth, *Cem. Pal. Att.*, 129-130: V.9; Harpur, *Decoration,* [#033], 265: V.9-VI.1

Golden jackals

SB37 *nj-mꜣʿt-rʿ* [*jzzj-mr-nṯr*] - Giza, WF (G2097), chapel, east: hunting
Source: *P-M III*, 70; Roth, *Cem. Pal. Att.*, pl. 189
Label: *nk wn*[*š*]
Copulating jackals
Date: Roth, *Cem. Pal. Att.*, 129-130: V.9; Harpur, *Decoration,* [#033], 265: V.9-VI.1

Red foxes

SB38 *nj-ʿnḫ-ẖnmw/ẖnmw-ḥtp* - Saqqara, UPC (ntn), vestibule, west: hunting
Source: *P-M III*, 641-644; Moussa-Altenmüller, *Nianchchnum*, pl. 40
Label: none
Date: Harpur, *Decoration,* [#431], 274: V.6L-7

SB39 *ptḥ-ḥtp II: ṯfj* - Saqqara, WSP (D64), offering-room, east: hunting
Source: *P-M III*, 600-604; Quibell, *Ptah-hetep*, pl. 32
Label: none
Date: Harpur, *Decoration,* [#400], 274: V.9M-L

SB40 *nj-mꜣʿt-rʿ* [*jzzj-mr-nṯr*] - Giza, WF (G2097), chapel, east: hunting
Source: *P-M III*, p.70; Roth, *Cem. Pal. Att.*, pl. 189
Label: *n*[*k*]
Copulating
Date: Roth, *Cem. Pal. Att.*, 129-130: V.9; Harpur, *Decoration,* [#033], 265: V.9-VI.1

Libyan striped weasels

SB41 *nj-mꜣʿt-rʿ* [*jzzj-mr-nṯr*] - Giza, WF (G2097), chapel, east: hunting
Source: *P-M III*, 70; Roth, *Cem. Pal. Att.*, pl. 189
Label: [*gs*]*fnw*
Striped weasels[63]
Date: Roth, *Cem. Pal. Att.*, 129-130: V.9; Harpur, *Decoration,* [#033], 265: V.9-VI.1

Hedgehogs

SB42 *ṯjj* - Saqqara, NSP (D22), portico, south: pastoral
Source: *P-M III*, 468-478; Epron et al., *Ti i*, pl. 6
Label: missing
Date: Harpur, *Decoration,* [#543], 277: V.8-9

SB43 *nj-mꜣʿt-rʿ* [*jzzj-mr-nṯr*] - Giza, WF (G2097), chapel, east: hunting
Source: *P-M III*, 70; Roth, *Cem. Pal. Att.*, pl. 189
Label: none
Date: Roth, *Cem. Pal. Att.*, 129-130: V.9; Harpur, *Decoration,* [#033], 265: V.9-VI.1

Shrews

SB44 *nj-mꜣʿt-rʿ* [*jzzj-mr-nṯr*] - Giza, WF (G2097), chapel, east: hunting
Source: *P-M III*, 70; Roth, *Cem. Pal. Att.*, pl. 189

[63] Roth (*Cem. Pal. Att.*, 132) suggested a translation of 'small foxes'. However, examination of other examples has indicated that it may instead refer to the Libyan striped weasel. See text for details.

Label: *zṯ*[64]
Shrew?
Date: Roth, *Cem. Pal. Att.*, 129-130: V.9; Harpur, *Decoration,* [#033], 265: V.9-VI.1

Unidentified mammals

SB45 Anon - Saqqara, WSP (ntn), relief fragment found in the tomb of *ṯfw*, unknown: desert
Source: *P-M III*, 605; Hassan, *Ny-Ankh-Pepy*, fig. 59
Label: missing
Date: *P-M III*, 605: V-VI; Harpur, *Decoration,* nr

Nile crocodiles

SB46 *kꜣ-gm-nj: mmj* - Saqqara, TP (LS10), room III, north: marsh
Source: *P-M III*, 521-525; von Bissing, *Gem-ni-kai i*, pl. 4; Harpur-Scremin, *Kagemni*, fig. 9
Label: none
Date: Harpur, *Decoration,* [#534], 276: VI.1M

SB47 *ḥzj* - Saqqara, TP (ntn), portico, south: marsh
Source: *P-M III*, nr; Kanawati-Abder-Raziq, *Hesi*, pl. 53
Label: none
Date: Kanawati-Abder-Raziq, *Hesi*, 16: VI.1L-2E; Harpur, *Decoration,* nr

SB48 *mḥw* - Saqqara, UPC (ntn), room II, north: boating
Source: *P-M III*, 619-622; Altenmüller, *Mehu*, pl. 32
Label: none
Date: Altenmüller, *Mehu*, 19: VI.2; Harpur, *Decoration,* [#424], 274: VI.2M-3?

Nile soft-shelled turtles

SB49 *ḥzj* - Saqqara, TP (ntn), portico, south: marsh
Source: *P-M III*, nr; Kanawati-Abder-Raziq, *Hesi*, pl. 53
Label: none
Date: Kanawati-Abder-Raziq, *Hesi*, 16: VI.1L-2E; Harpur, *Decoration,* nr

---

10.3 MASTURBATION

Donkeys

SB50 *rꜥ-m-kꜣ.j* - Saqqara, NSP (D3=S903), chapel, east: agricultural
Source: *P-M III*, 487-488; Hayes, *Scepter i*, fig. 57
Label: *ḥwj zpt nt jt*
Threshing on the threshing floor of barley
Date: Harpur, *Decoration,* [#453], 275: V.6? r/u - V.6-7

SB51 *rꜥ-m-kꜣ.j* - Saqqara, NSP (D3=S903), chapel, east: agricultural
Source: *P-M III*, 487-488; Hayes, *Scepter i*, fig. 57
Label: *šdjt m jꜥꜣw*
Taking away with nets
Date: Harpur, *Decoration,* [#453], 275: V.6? r/u - V.6-7

SB52 *nfr-jrt-n.f* - Saqqara, ESP (D55), chapel, east: agricultural
Source: *P-M III*, 583-584; van de Walle, *Neferirtenef*, pl. 12
Label: *jr ḥꜣ=k jm=sn*
Get back among them!
Date: Harpur, *Decoration,* [#440], 274: V.8-9

SB53 *ḫww-wr* - Giza, CF (LG95), chapel, east: agricultural
Source: *P-M III*, 254-255; Hassan, *Giza v*, fig. 104
Label: none

[64] Roth (*ibid.*) believes that this is a mis-writing of the word *ḥts*, 'weasel'. See text for details.

Date: Harpur, *Decoration,* [#180], 268: V.8-9M

SB54 *nj-kȝw-ḥr* - Saqqara, NSP (S915), offering-room, east: agricultural
Source: *P-M III*, 498; Quibell, *Saqqara (1907-1908)*, pl. 62 [2]
Label: . . *sḥ*[*ȝt*] . . .
. . donkey pack . . .
Date: Harpur, *Decoration,* [#435], 274: V.9

SB55 *nfr I* - Giza, WF (G4761), chapel, east: agricultural
Source: *P-M III*, 137-138; Junker, *Giza vi*, fig. 14
Label: none
Date: Harpur, *Decoration,* [#126], 267: V.9-VI.1

SB56 *jȝzn* - Giza, WF (G2196), chapel, north: agricultural
Source: *P-M III*, 82; Simpson, *Mast. West. Cem. Part I*, fig. 30
Label: none
Date: Harpur, *Decoration,* [#013], 265: VI.1M-2M

11.1 REPRODUCTION

11.1.1 Birth

Cattle

PB1 *nfr-msḏr-ḫwfw* - Giza, WF (G2240), chapel, north: agricultural (assisted calving, left)
Source: *P-M III*, nr; Roth, *Cem. Pal. Att.*, pl. 204
Label: none
Date: Roth, *Cem. Pal. Att.*, 163: V.2-8 or V.9; Harpur, *Decoration*, nr

PB2 *nfr-msḏr-ḫwfw* - Giza, WF (G2240), chapel, north: agricultural (natural calving, right)
Source: *P-M III*, nr; Roth, *Cem. Pal. Att.*, pl. 204
Label: none
Date: Roth, *Cem. Pal. Att.*, 163: V.2-8 or V.9; Harpur, *Decoration*, nr

PB3 *jj-mrjj* - Giza, WF (G6020=LG16), second chamber, south: agricultural
Source: *P-M III*, 170-174; Weeks, *Cemetery G6000*, fig. 34
Label: *sfḫ*
Loosen/release!
msjt
Giving birth
Date: Weeks, *Cemetery G6000*, 7: V.6; Harpur, *Decoration*, [#014], 265: V.3L

PB4 *rʿ-m-kꜣ.j* - Saqqara, NSP (D3=S903), chapel, north: pastoral
Source: *P-M III*, 487-488[65]; Hayes, *Scepter i*, fig. 54
Label: missing
Date: Harpur, *Decoration*, [#453], 275: V.6? r/u - V.6-7

PB5 *nj-ʿnḫ-ẖnmw/ẖnmw-ḥtp* - Saqqara, UPC (ntn), outer hall, west: pastoral
Source: *P-M III*, 641-644; Moussa-Altenmüller, *Nianchchnum*, pl. 79
Label: *msjt ḥr zꜣ-pr sfḫt*
Giving birth in the field/byre and loosening/releasing
Date: Harpur, *Decoration*, [#431], 274: V.6L-7

PB6 *ꜣḫtj-ḥtp* - Saqqara, UPC (ntn), chapel, east: pastoral
Source: *P-M III*, 634-637; Ziegler, *Akhethetep*, 130-131
Label: *sfḫ wrt bḥz pw jmj=f gꜣw*
Loosen/release greatly this calf in its constriction[66]
jdt
Cow
Date: Ziegler, *Akhethetep*, 9: V.6; Harpur, *Decoration*, [#340], 272: V.6-8E

PB7 *jtj-sn* - Giza, CF (ntn), chapel, east: pastoral
Source: *P-M III*, 252-253; Hassan, *Giza v*, fig. 123
Label: *sfḫ wrt [j]r=t*
Loosen/release greatly you![67]
Date: Harpur, *Decoration*, [#035], 265: V.6-8?

PB8 *kꜣ.j-dwꜣ* - Giza, CF (ntn), chapel, north: pastoral
Source: *P-M III*, 244-245; Hassan, *Giza vi*, fig. 80
Label: missing
Date: Harpur, *Decoration*, [#272], 270: V.6-9?

---

65 *P-M III*, 488 states that this is a milking scene; note, however, that the herdsman is positioned behind the cow, rather than beside her udder and that her hind legs are not roped together. See *Milk ejection reflex* in Chapter 11 for details.

66 Ziegler (*Akhethetep*, 153) translates this as a negative, 'that it does not choke', based upon a similar Late Period example.

67 Hassan (*Giza v*, 268) translates this label as 'Untie the cow from the rope'.

PB9 *sḫm-ꜥnḫ-ptḥ* - Saqqara, NSP (D41), chapel, east: pastoral
Source: *P-M III*, 454-455; Simpson, *Sekem-ankh-ptah*, pl. C
Label: *smsj jdt sfḫ wr(t)*
Causing the cow to give birth and loosen/release greatly
Date: Harpur, *Decoration*, [#504], 276: V.8-9

PB10 *tjj* - Saqqara, NSP (D22), room III, north: pastoral
Source: *P-M III*, 468-478; Wild, *Ti ii*, pl. 124
Label: *sfḫ jdt wrt mnjw pw*
Loosen/release the cow greatly, this herdsman
Date: Harpur, *Decoration*, [#543], 277: V.8-9

PB11 *kꜣ.j-m-nfrt* - Saqqara, NSP (D23), chapel, east: pastoral
Source: *P-M III*, 467-468; Simpson, *Kayemnofret*, pl. G
Label: *sfḫt n bḥs jn mnjw*
Loosening/releasing of a calf by the herdsman
Date: Simpson, *Kayemnofret*, 1: V.8-9; Harpur, *Decoration*, [#522], 276: V.8-9?

PB12 *ꜣḫtj-ḥtp* - Saqqara, WSP (D64), hall, east: pastoral
Source: *P-M III*, 599-600; Davies, *Ptahhetep ii*, pl. 32
Label: unable to determine
Date: Harpur, *Decoration*, [#338], 272: V.8L-9E

PB13 *nbt* - Saqqara, UPC (ntn), room B [II], east: pastoral
Source: *P-M III*, 624-625; Munro, *Unas-Friedhof*, pl. 13
Label: missing
Date: Munro, *Unas-Friedhof*, 20: V.9; Harpur, *Decoration*, [#438], 274: V.9

PB14 *ptḥ-ḥtp II: ṯfj* - Saqqara, WSP (D64), offering-room, east: pastoral
Source: *P-M III*, 600-604; Quibell, *Ptah-hetep*, pl. 31
Label: missing
Date: Harpur, *Decoration*, [#400], 274: V.9M-L

PB15 *ꜣḫtj-mrw-nswt* - Giza, WF (G2184), secondary offering-room, west: agricultural
Source: *P-M III*, 80-81; Smith, *HESPOK*, fig. 226 [c]
Label: none
Date: Harpur, *Decoration*, [#008], 265: V.9-VI.1

PB16 *kꜣ.j-m-ꜥnḫ* - Giza, WF (G4561), chapel, west: agricultural
Source: *P-M III*, 131-133; Junker, *Giza iv*, pl. 12; Kanawati, *Kaiemankh*, pl. 36
Label: *jdt msj*
Cow giving birth
Date: Kanawati, *Kaiemankh*, 18: late V; Harpur, *Decoration*, [#255], 270: VI

PB17 *ṯtw: kꜣ.j-nswt* - Giza, WF (G2001), recessed portico, west: agricultural
Source: *P-M III*, 66-67; Simpson, *Mast. West. Cem. Part I*, fig. 22
Label: *[s]fḫ jdt wrt mnjw pw*
Loosen/release the cow greatly, this herdsman
Date: Harpur, *Decoration*, [#292], 271: VI

PB18 *zšzšt: jdwt* - Saqqara, UPC (ntn), room III, west: pastoral
Source: *P-M III*, 617-619; Macramallah, *Idout*, pl. 7; Kanawati-Abder-Raziq, *Iynefert*, pl. 54
Label: missing
Date: Kanawati-Abder-Raziq, *Iynefert*, 36-37: VI.1E; Harpur, *Decoration*, [#487], 275: V.9 r/u, VI.1-2E?

PB19 *jꜣzn* - Giza, WF (G2196), chapel, north: agricultural
Source: *P-M III*, 82; Simpson, *Mast. West. Cem. Part I*, fig. 30
Label: *msjt*
Giving birth
sfḫt
Loosening/releasing
Date: Harpur, *Decoration*, [#013], 265: VI.1M-2M

PB20 *mttj* - Saqqara, UPC (ntn), relief fragment (Louvre no. E. 25515), unknown: pastoral
Source: *P-M III*, 646-648; Ziegler, *Catalogue*, 143
Label: *sfḫ wrt mnjw [pw]*
Loosen/release greatly, [this] herdsman!
Date: Ziegler, *Catalogue*, 123: probably VI; Harpur, *Decoration*, [#426], 274: VI.2?

Goats

PB21 *sḫm-kꜣ-rˁ* - Giza, CF (LG89), chapel, south: pastoral
Source: *P-M III*, 233-235; Hassan, *Giza iv*, fig. 57
Label: *jdt msj ṯs . . .*
Female giving birth . . .
Date: Harpur, *Decoration*, [#229], 269: V.2

PB22 *nj-ˁnḫ-ḫnmw/ḫnmw-ḥtp* - Saqqara, UPC (ntn), vestibule, north: pastoral
Source: *P-M III*, 641-644; Moussa-Altenmüller, *Nianchchnum*, fig. 8
Label: none
Date: Harpur, *Decoration*, [#431], 274: V.6L-7

PB23 *ḥtp-ḥr-ꜣḫtj* - Saqqara, WSP (D60), chapel, east: pastoral
Source: *P-M III*, 593-595; Mohr, *Hetep-her-akhti*, fig. 21
Label: none
Date: Harpur, *Decoration*, [#467], 275: V.6-8E

PB24 *ꜣḫtj-ḥtp* - Saqqara, UPC (ntn), chapel, east: pastoral
Source: *P-M III*, 634-637; Ziegler, *Akhethetep*, 138-139
Label: none
Date: Ziegler, *Akhethetep*, 9: V.6; Harpur, *Decoration*, [#340], 272: V.6-8E

PB25 *ꜣḫtj-mrw-nswt* - Giza, WF (G2184), secondary offering-room, west (north end): pastoral
Source: *P-M III*, 80-81; *Giza Archives Project* (Photo no. A779_NS)
Label: none
Date: Harpur, *Decoration*, [#008], 265: V.9-VI.1

Hippopotamuses

PB26 *snḏm-jb: mḥj* - Giza, WF (G2378=LG26), forecourt, north: fishing
Source: *P-M III*, 87-89; *LD Erganz.*, pl. 12 [upper]; Brovarski, *Senedjemib Complex Part I*, fig. 101
Label: none
Date: Brovarski, *Senedjemib Complex Part I*, 29-30: V.9; Harpur, *Decoration*, [#220], 269: V.9

PB27 *kꜣ-gm-nj: mmj* - Saqqara, TP (LS10), room I, west: marsh
Source: *P-M III*, 521-525; Firth-Gunn, *Teti Pyr. ii*, pl. 53 [2], Harpur-Scremin, *Kagemni*, fig. 3
Label: none
Date: Harpur, *Decoration*, [#534], 276: VI.1M

PB28 *kꜣ-gm-nj: mmj* - Saqqara, TP (LS10), room III, north: marsh (beneath harpooners)
Source: *P-M III*, 521-525; Decker-Herb, *Bildatlas*, pl. 196 [K1.33]; Harpur-Scremin, *Kagmeni*, fig. 9
Label: none
Date: Harpur, *Decoration*, [#534], 276: VI.1M

PB29 *kꜣ-gm-nj: mmj* - Saqqara, TP (LS10), room IV, east: fishing
Source: *P-M III*, 521-525; von Bissing, *Gem-ni-kai i*, pl. 18; Harpur-Scremin, *Kagemni*, fig. 16
Label: missing
Date: Harpur, *Decoration*, [#534], 276: VI.1M

PB30 *mrrw-kꜣ.j: mrj* - Saqqara, TP (ntn), chamber A1, south: marsh
Source: *P-M III*, 525-534; Duell, *Mereruka i*, pl. 19
Label: none
Date: Harpur, *Decoration*, [#420], 274: VI.1M-L

PB31 *zšzšt: jdwt* - Saqqara, UPC (ntn), room I, north: fishing
Source: *P-M III*, 617-619; Macramallah, *Idout*, pl. 5 [d]; Kanawati-Abder-Raziq, *Iynefert*, pl. 53 [f]

Label: none
Date: Kanawati-Abder-Raziq, *Iynefert*, 36-37: VI.1E; Harpur, *Decoration*, [#487], 275: V.9 r/u, VI.1-2E?

PB32 *zšzšt: jdwt* - Saqqara, UPC (ntn), room III, west: marsh
Source: *P-M III*, 617-619; Macramallah, *Idout*, pl. 7; Kanawati-Abder-Raziq, *Iynefert*, pl. 54
Label: none
Date: Kanawati-Abder-Raziq, *Iynefert*, 36-37: VI.1E; Harpur, *Decoration*, [#487], 275: V.9 r/u, VI.1-2E?

PB33 *sꜥnḫ-w(j)-ptḥ* - Saqqara, TP (ntn), room I, east: marsh
Source: *P-M III*, nr; Kanawati-Abder-Raziq, *Neferseshemre*, pls. 69 and 76
Label: none
Date: Kanawati-Abder-Raziq, *Neferseshemre*, 40-41: VI.1L; Harpur, *Decoration*, nr

PB34 *ꜥnḫ-m-ꜥ-ḥr: zzj* - Saqqara, TP (ntn), room I, west: marsh
Source: *P-M III*, 512-515; Kanawati-Hassan, *Ankhmahor*, pl. 37 [c]
Label: none
Date: Kanawati-Hassan, *Ankhmahor*, 18: VI.1M-L; Harpur, *Decoration*, [#374], 273: VI.1L-2E

PB35 *mrrj* - Saqqara, TP (ntn), room 1, south: marsh
Source: *P-M III*, 518-519; Davies et al., *Saqqara Tombs i*, pl. 5
Label: none
Date: Harpur, *Decoration*, [#418], 274: VI.2E?

Common genets

PB35a *zšzšt: jdwt* – Saqqara, UPC (ntn), relief fragment among rubble in the burial shaft, unknown: marsh
Source: *P-M III*, 617-619; Macramallah, *Idout*, pl. 26
Label: none
Date: Kanawati-Abder-Raziq, *Iynefert*, 36-37: VI.1E; Harpur, *Decoration*, [#487], 275: V.9 r/u, VI.1-2E?

Unidentified

PB36 *pḥn-wj-kꜣ.j* - Saqqara, NSP (D70=LS15), relief fragment (Brooklyn Mus. no. 64.147), unknown: desert
Source: *P-M III*, 233-235; *Eg. Art in the Age of the Pyramids*, cat. no. 146 [b]
Label: missing
Date: Harpur, *Decoration*, [#393], 273: V.6-8E

11.1.2 Egg laying

Nile crocodiles

PB37 *kꜣ.j-jrr* - Saqqara, UPC (ntn), room I, north: marsh
Source: *P-M III*, 631-632; Lauer, *Saqqara*, pl. 67
Label: none
Date: Daoud, K. (2000). *ASAE* 75, 106: mid-VI, possibly VI.5; Harpur, *Decoration*, [#520], 276: VI.2?

11.1.3 Egg incubation

PB38 *rꜥ-špss* - Saqqara, NSP (LS16=S902), forecourt, west: marsh
Source: *P-M III*, 494-496; *LD ii*, pl. 60 [left]
Label: none
Date: Harpur, *Decoration*, [#456], 275: V.8M

PB39 *snḏm-jb: jntj* - Giza, WF (G2370=LG27), portico, west: marsh
Source: *P-M III*, 85-87; *LD Erganz.*, pl. 18[68]
Label: none
Date: Brovarski, *Senedjemib Complex Part I*, 23: V.8M-L; Harpur, *Decoration*, [#219], 269: V.8M-L

---

68 According to Brovarski (*Senedjemib Complex Part I*, 40 and fig. 27), this scene has been destroyed.

PB40 *ꜣẖtj-ḥtp* - Saqqara, WSP (D64), hall, east: marsh
Source: *P-M III*, 599-600; Davies, *Ptahhetep ii*, pl. 14
Label: none
Date: Harpur, *Decoration*, [#338], 272: V.8L-9E

PB41 *mḥw* - Saqqara, UPC (ntn), room I, west: marsh
Source: *P-M III*, 619-622; Altenmüller, *Mehu*, 19
Label: none
Date: Altenmüller, *Mehu*, 19: VI.2; Harpur, *Decoration*, [#424], 274: VI.2M-3?

11.2 FEEDING YOUNG

11.2.1 Nursing

Cattle

PB42 *ṯntj* - Giza, WF (G4920=LG47), chapel, east: pastoral
Source: *P-M III*, 141-142; *LD ii*, pl. 31 [a]
Label: n/a
Date: Harpur, *Decoration*, [#287], 271: V.1-5

PB43 *jj-mrjj* - Giza, WF (G6020=LG16), second chamber, south: pastoral
Source: *P-M III*, 170-174; Weeks, *Cemetery G6000*, fig. 34
Label: none
Date: Weeks, *Cemetery G6000*, 7: V.6; Harpur, *Decoration*, [#014], 265: V.3L

PB44 *zꜣ-jb* - Giza, WF (G2092 + G2093), chapel, east: pastoral
Source: *P-M III*, nr; Roth, *Cem. Pal. Att.*, pl. 179
Label: missing
Date: Roth, *Cem. Pal. Att.*, 109: V.8; Harpur, *Decoration*, nr

PB45 *snḏm-jb: jntj* - Giza, WF (G2370=LG27), room II, west: pastoral
Source: *P-M III*, 85-87; *LD ii*, pl. 77[69]
Label: none
Date: Brovarski, *Senedjemib Complex Part I*, 23: V.8M-L; Harpur, *Decoration*, [#219], 269: V.8M-L

PB46 *ꜣẖtj-ḥtp* - Saqqara, WSP (D64), hall, north: pastoral
Source: *P-M III*, 599-600; Davies, *Ptahhetep ii*, pl. 32
Label: none
Date: Harpur, *Decoration*, [#338], 272: V.8L-9E

PB47 *kꜣpj* - Giza, WF (G2091), corridor, west: pastoral
Source: *P-M III*, 69-70; Roth, *Pal. Cem. Att.*, pl. 156
Label: *jdt*
Cow
Date: Roth, *Pal. Cem. Att.*, 100: V.6-8; Harpur, *Decoration*, [#275], 270: V.9-VI

PB48 *kꜣ-gm-nj: mmj* - Saqqara, TP (LS10), room III, north: pastoral
Source: *P-M III*, 521-525; Firth-Gunn, *Teti Pyr. ii*, pl. 52; Harpur-Scremin, *Kagmeni*, fig. 7
Label: *jr=ṯw r=s wrt*
Pay great heed to her
Date: Harpur, *Decoration*, [#534], 276: VI.1M

PB49 *jꜣzn* - Giza, WF (G2196), chapel, north: pastoral
Source: *P-M III*, 82; Simpson, *Mast. West. Cem. Part I*, fig. 30
Label: *jdt wrt*
Large cow
Date: Harpur, *Decoration*, [#013], 265: VI.1M-2M

---

[69] This scene is no longer extant (*ibid.*, 51).

Dorcas gazelle

PB50 *nb.j-m-ꜣḫtj* - Giza, CF (LG86), chapel I, west: offering
Source: *P-M III*, 230-232; *LD ii*, pl. 12 [b]
Label: none
Date: Harpur, *Decoration*, [#122], 267: IV.6-V.1

PB51 *nj-ꜥnḫ-ẖnmw/ẖnmw-ḥtp* - Saqqara, UPC (ntn), vestibule, west: desert
Source: *P-M III*, 641-644; Moussa-Altenmüller, *Nianchchnum*, pl. 40
Label: none
Date: Harpur, *Decoration*, [#431], 274: V.6L-7

PB52 *pḥn-wj-kꜣ.j* - Saqqara, NSP (D70=LS15), antechamber (room I); west: desert
Source: *P-M III*, 491-492; *LD ii*, pl. 46
Label: none
Date: Harpur, *Decoration*, [#393], 273: V.6-8E

PB53 *kꜣ.j-dwꜣ* - Giza, CF (ntn), chapel, south: offering
Source: *P-M III*, 244-245; Hassan, *Giza vi*, fig. 82
Label: *r*[*n*] *gḥs*
Young gazelle
Date: Harpur, *Decoration*, [#272], 270: V.6-9?

PB54 *rꜥ-wr II* - Giza, WF (G5470=LG32), chapel, east: offering
Source: *P-M III*, 162-163; Junker, *Giza iii*, fig. 48
Label: *rn g*[*ḥs*] 1,000
Young gazelle 1,000
Date: Harpur, *Decoration*, [#153], 268: V.8-9M

PB55 *ptḥ-ḥtp II: ṯfj* - Saqqara, WSP (D64), offering-room, east: desert
Source: *P-M III*, 600-604; Quibell, *Ptah-hetep*, pl.32
Label: none
Date: Harpur, *Decoration*, [#400], 274: V.9M-L

PB56 *kꜣpj* - Giza, WF (G2091), northern pilaster, north: offering
Source: *P-M III*, 69-70; Roth, *Pal. Cem. Att.*, pl. 168
Label: none
Date: Roth, *Pal. Cem. Att.*, 100: V.6-8; Harpur, *Decoration*, [#275], 270: V.9-VI

PB57 Anon - Saqqara, WSP (ntn), relief fragment from shaft 6, 'Mastaba of Shaft no. 14', unknown: desert
Source: *P-M III*, 613; Hassan, *Hemet-Re*, pl. 14 [c]
Label: none
Date: *P-M III*, 613: VI; Harpur, *Decoration*, nr

PB58 *kꜣ-gm-nj: mmj* - Saqqara, TP (LS10), room IV, south: offering
Source: *P-M III*, 521-525; von Bissing, *Gem-ni-kai i*, pl. 7 [2]; Harpur, *Kagmeni*, fig. 12
Label: none
Date: Harpur, *Decoration*, [#534], 276: VI.1M

PB59 *ꜥnḫ-m-ꜥ-ḥr: zzj* - Saqqara, TP (ntn), room I, south
Source: *P-M III*, 512-515; Kanawati-Hassan, *Ankhmahor*, pl. 37 [b]
Label: n/a
Date: Kanawati-Hassan, *Ankhmahor*, 18: VI.1M-L; Harpur, *Decoration*, [#374], 273: VI.1L-2E

PB60 *ꜥnḫ-m-ꜥ-ḥr: zzj* - Saqqara, TP (ntn), door thickness (rooms II to III), north: offering
Source: *P-M III*, 512-515; Kanawati-Hassan, *Ankhmahor*, pl. 43 [b]
Label: *rn n gḥs*
Young (of a) gazelle
nḏr sw r mnḫ
Hold it well
Date: Kanawati-Hassan, *Ankhmahor*, 18: VI.1M-L; Harpur, *Decoration*, [#374], 273: VI.1L-2E

Goats

PB61 *nfr/k3-ḥ3.j* - Saqqara, UPC (ntn), chapel, east: pastoral
Source: *P-M III*, 639-641; Moussa-Altenmüller, *Nefer and Ka-hay*, pl. 4
Label: none
Date: Harpur, *Decoration*, [#439], 274: V.6

PB62 *nfr/k3-ḥ3.j* - Saqqara, UPC (ntn), niche, east: pastoral
Source: *P-M III*, 639-64; Moussa-Altenmüller, *Nefer and Ka-hay*, pl. 18
Label: none
Date: Harpur, *Decoration*, [#439], 274: V.6

PB63 *ḥtp-ḥr-3ḫtj* - Saqqara, WSP (D60), chapel, east: pastoral
Source: *P-M III*, 593-595; Mohr, *Hetep-her-akhti*, fig. 21
Label: none
Date: Harpur, *Decoration*, [#467], 275: V.6-8E

11.2.2 Milk ejection reflex

Cattle

PB64 *mrs-ʿnḫ III* - Giza, EF (G7530 + G7540), relief fragment from interior chapel (no. 28-5-36), unknown: pastoral
Source: *P-M III*, 197-199; Dunham-Simpson, *Mersyankh III*, pl. 13 [e]
Label: missing
Date: Harpur, *Decoration*, [#098], 267: IV.6

PB65 Anon - Saqqara (ntn), relief fragment (CG II.1562), unknown: pastoral
Source: *P-M III*, 753-754; Borchardt, *Denkmäler ii*, pl. 63
Label: missing
Date: Harpur, *Decoration*, [#569], 277: V

PB66 *nfr-msḏr-ḫwfw* - Giza, WF (G2240), relief fragment, unknown: agricultural
Source: *P-M III*, nr; Roth, *Cem. Pal. Att.*, pl. 210
Label: missing
Date: Roth, *Cem. Pal. Att.*, 163: V.2-8 or V.9; Harpur, *Decoration*, nr

PB67 *jj-mrjj* - Giza, WF (G6020=LG16), second chamber, south: agricultural (cow, far left)
Source: *P-M III*, 170-174; Weeks, *Cemetery G6000*, fig. 34
Label: missing
Date: Weeks, *Cemetery G6000*, 7: V.6; Harpur, *Decoration*, [#014], 265: V.3L

PB68 *jj-mrjj* - Giza, WF (G6020=LG16), second chamber, south: agricultural (cow and calf, far right)
Source: *P-M III*, 170-174; Weeks, *Cemetery G6000*, fig. 34
Label: *sšr jdt*
Drying up the cow
Date: Weeks, *Cemetery G6000*, 7: V.6; Harpur, *Decoration*, [#014], 265: V.3L

PB69 *nj-ʿnḫ-ḫnmw/ḫnmw-ḥtp* - Saqqara, UPC (ntn), outer hall, west: pastoral
Source: *P-M III*, 641-644; Moussa-Altenmüller, *Nianchchnum*, pl. 76
Label: *h(r) jrṯt*
Drawing milk
Date: Harpur, *Decoration*, [#431], 274: V.6L-7

PB70 *ḥtp-ḥr-3ḫtj* - Saqqara, WSP (D60), chapel, north: pastoral
Source: *P-M III*, 593-595; Mohr, *Hetep-her-akhti*, fig. 32
Label: *sšr jdt [j]n mnḥ*
Drying up the cow by the herdsman
Date: Harpur, *Decoration*, [#467], 275: V.6-8E

PB71 *k3.j-m-nfrt* - Giza, CF (ntn), vestibule A, west: pastoral
Source: *P-M III*, 263-264; Hassan, *Giza ii*, fig. 123

Label: *[sšr] jdt*
[Drying up] the cow
Date: Harpur, *Decoration*, [#261], 270: V.6-8E?

PB72 *jrj-n-kꜣ-ptḥ* - Saqqara, UPC (ntn), offering chamber, south: pastoral
Source: *P-M III*, 644; Moussa-Junge, *Two Tombs*, pl. 13
Label: *[s]š[r] j[r]ṯt*
Drying up milk
Date: Harpur, *Decoration*, [#362], 273: V.6-9

PB73 *snḏm-jb: jntj* - Giza, WF (G2370=LG27), room II, west: agricultural
Source: *P-M III*, 85-87; *LD ii*, pl. 77[70]
Label: *sšr jrṯt*
Drying up milk
Date: Brovarski, *Senedjemib Complex Part I*, 23: V.8M-L; Harpur, *Decoration*, [#219], 269: V.8M-L

PB74 *sḫm-ꜥnḫ-ptḥ* - Saqqara, NSP (D41), chapel, east: pastoral
Source: *P-M III*, 454-455; Simpson, *Sekem-ankh-ptah*, pl. C
Label: *sšr jrṯt*
Drying up milk
Date: Harpur, *Decoration*, [#504], 276: V.8-9

PB75 *ṯjj* - Saqqara, NSP (D22), room III, north: pastoral
Source: *P-M III*, 468-478; Wild, *Ti ii*, pl. 124
Label: *sšr n jrṯt jn jtj n=k rnpj bḥs*
Drying up of milk
Indeed! Seize to you the young calf
ḳnj b[ḥs]
Prevail over the calf
Date: Harpur, *Decoration*, [#543], 277: V.8-9

PB76 *ṯjj* - Saqqara, NSP (D22), room III, north: pastoral
Source: *P-M III*, 468-478; Wild, *Ti ii*, pl. 112
Label: *sšr wn ṯw(t) ḏr jj.t rf ḥḳꜣ pw*
Dry up quickly before indeed comes this administrator!
Date: Harpur, *Decoration*, [#543], 277: V.8-9

PB77 *nfr-ḥr-nj-ptḥ* - Saqqara, UPC (ntn), chapel, north: pastoral
Source: *P-M III*, 637-638; © Su Barfield (*www.egyptsites.co.uk*)
Label: *sšr jrṯt*
Drying up milk
Date: Harpur, *Decoration*, [#441], 274: V.8-9

PB78 *ꜣḫtj-ḥtp* - Saqqara, WSP (D64), hall, north: pastoral
Source: *P-M III*, 599-600; Davies, *Ptahhetep ii*, pl. 17
Label: *sšr* [ ]
Drying up [ ]
Date: Harpur, *Decoration*, [#338], 272: V.8L-9E

PB79 *nbt* - Saqqara, UPC (ntn), room B [II], east: pastoral
Source: *P-M III*, 624-625; Munro, *Unas-Friedhof*, pl. 13
Label: missing
Date: Munro, *Unas-Friedhof*, 20: V.9; Harpur, *Decoration*, [#438], 274: V.9

PB80 *nbt* - Saqqara, UPC (ntn), room C, east: pastoral
Source: *P-M III*, 624-625; Munro, *Unas-Friedhof*, pl. 21
Label: *rdj sdꜣ jrṯt*
Causing the milk to flow
jmj-r sšr
Overseer of drying up
Date: Munro, *Unas-Friedhof*, 20: V.9; Harpur, *Decoration*, [#438], 274: V.9

---

[70] According to Brovarski (*ibid.*, 49), this scene has now been destroyed.

PB81 *k3pj* - Giza, WF (G2091), corridor, west: agricultural
Source: *P-M III*, 69-70; Roth, *Pal. Cem. Att.*, pl. 156
Label: *sšr[t jrṯ]t*
Drying up milk
Date: Harpur, *Decoration*, [#275], 270: V.9-VI

PB82 *3ḫtj-mrw-nswt* - Giza, WF (G2184), secondary offering-room, west: agricultural
Source: *P-M III*, 80-81; Smith, *HESPOK*, fig. 79
Label: none
Date: Harpur, *Decoration*, [#008], 265: V.9-VI.1

PB83 *ptḥ-ḥtp: jj-n-ʿnḫ* - Saqqara, WSP (ntn), offering-room, east: pastoral
Source: *P-M III*, 606-607; Hassan, *Ny-Ankh-Pepy*, fig. 39
Label: *h(r)y jrṯt*
The one who draws milk
Date: Harpur, *Decoration*, [#399], 274: V.9-VI.1

PB84 *m3-nfr* - Saqqara, ESP (H2=LS17), hall, east: pastoral
Source: *P-M III*, 575-577; *LD ii*, pl. 66 [left]
Label: *šsr jrṯt*
Drying up milk
Date: Harpur, *Decoration*, [#406], 274: V.9-VI.1

PB85 *sḫm-ʿnḫ-ptḥ* - Giza, EF (G7152), chapel, north: pastoral
Source: *P-M III*, 191; Badawy, *Tombs of Iteti*, fig. 23
Label: missing
Date: Harpur, *Decoration*, [#223], 269: V.9-VI.1?

PB86 Anon - Saqqara (Giza[71]) (ntn), relief fragment (Museo Barracco, Rome), unknown
Source: *P-M III*, 309/761; Ranke, *Art of ancient Egypt*, no. 200; Harpur, *Decoration*, fig. 122.
Label: *jdt sšr*
Cow being dried up
Date: Harpur, *Decoration*, [#592], 278: V-VI

PB87 Anon - Saqqara, WSP (ntn), relief fragment found in tomb of *mrrj*, unknown: pastoral
Source: *P-M III*, 607-608; Hassan, *Hemet-Re*, fig. 23
Label: *sšr jrṯt*
Drying up milk
Date: *P-M III*, 607: V-VI; Harpur, *Decoration*, nr

PB88 *k3.j-m-ʿnḫ* - Giza, WF (G4561), burial chamber, west: agricultural
Source: *P-M III*, 131-133; Junker, *Giza iv*, pl. 12; Kanawati, *Kaiemankh*, pl. 36
Label: *jdt s[šr]*
Cow being dried up
Date: Kanawati, *Kaiemankh*, 18: late V; Harpur, *Decoration*, [#255], 270: VI

PB89 *sšm-nfr: jwfj* - Saqqara, UPC (ntn), offering-room, east: pastoral
Source: *P-M III*, 614-615; Barsanti, A. (1900). *ASAE* 1, fig. 9
Label: none
Date: Harpur, *Decoration*, [#509], 276: VI

PB90 *k3-gm-nj: mmj* - Saqqara, TP (LS10), pillared hall, north: agricultural
Source: *P-M III*, 521-525; Firth-Gunn, *Teti Pyr. ii*, pl. 52; Harpur-Scremin, *Kagemni*, fig. 7
Label: *sšr jrṯt ʿ3 wrt*
Drying up milk very greatly
Date: Harpur, *Decoration*, [#534], 276: VI.1M

PB91 *nj-k3w-jzzj* - Saqqara, TP (ntn), room I, north: pastoral
Source: *P-M III*, nr; Kanawati-Abder-Raziq, *Nikauisesi*, pl. 49

---

71 Harpur (*Decoration*, 234) has found that this fragment (for which a Saqqara provenance is well-established), adjoins another that had been originally thought to derive from Giza (Albertinum, Dresden no. ZV. 1748-50). See IB205 in Chapter 7.

Label: *sšr jrṯt jdt jmꜣt*
Drying up the milk of the gracious cow
Date: Kanawati-Abder-Raziq, *Nikauisesi*, 17-23: VI.1M; Harpur, *Decoration*, nr

PB92 *jꜣzn* - Giza, WF (G2196), chapel, north: agricultural
Source: *P-M III*, 82; Simpson, *Mast. West. Cem. Part I*, fig. 30
Label: *jdt*
Cow
Date: Harpur, *Decoration*, [#013], 265: VI.1M-2M

PB93 *ḥzj* - Saqqara, TP (ntn), portico, east: pastoral
Source: *P-M III*, nr; Kanawati-Abder-Raziq, *Hesi*, pl. 55
Label: *sšr jrṯt*
Drying up milk
Date: Kanawati-Abder-Raziq, *Hesi*, 16: VI.1L-2E; Harpur, *Decoration*, nr

PB94 *mṯtj* - Saqqara, UPC (ntn), relief fragment (Louvre no. E. 25515), unknown: pastoral
Source: *P-M III*, 646-648; Ziegler, *Catalogue*, 143
Label: *sšr jrṯt*
Drying up milk
Date: Ziegler, *Catalogue*, 123: probably VI; Harpur, *Decoration*, [#426], 274: VI.2?

PB95 *jnw-mnw* - Saqqara (ntn), room II, south: pastoral
Source: *P-M III*, nr; Kanawati, *Inumin*, pl. 49 [a]
Label: missing
Date: Kanawati, *Inumin*, 16-17: VI.2E; Harpur, *Decoration*, nr

PB96 *mḥw* - Saqqara, UPC (ntn), door thickness (room I to II), north: pastoral
Source: *P-M III*, 619-622; Altenmüller, *Mehu*, pl. 17 [b]
Label: n/a[72]
Date: Altenmüller, *Mehu*, 19: VI.2; Harpur, *Decoration*, [#424], 274: VI.2M-3?

11.2.3 Provisioning

Birds

PB97 *nj-ꜥnḫ-ẖnmw/ẖnmw-ḥtp* - Saqqara, UPC (ntn), outer hall, west: marsh
Source: *P-M III*, 641-644; Moussa-Altenmüller, *Nianchchnum*, pl. 74
Label: none
Date: Harpur, *Decoration*, [#431], 274: V.6L-7

PB98 *nj-ꜥnḫ-ẖnmw/ẖnmw-ḥtp* - Saqqara, UPC (ntn), forecourt, south: fowling (hovering pied kingfisher)
Source: *P-M III*, 641-644; Moussa-Altenmüller, *Nianchchnum*, fig. 6
Label: none
Date: Harpur, *Decoration*, [#431], 274: V.6L-7

PB99 *nj-ꜥnḫ-ẖnmw/ẖnmw-ḥtp* - Saqqara, UPC (ntn), forecourt, south: fowling (hovering egret)
Source: *P-M III*, 641-644; Moussa-Altenmüller, *Nianchchnum*, fig. 6
Label: none
Date: Harpur, *Decoration*, [#431], 274: V.6L-7

---

72 Curiously, the inscription accompanying this scene does not refer to the milking procedure (although see Altenmüller, *Mehu*, 111), while a scene depicting a traveling herd of cows in the register directly above this (see GB102) is labeled *mk jr=k nbw(=j) ḥr sẖr jrṯt* 'Behold my lord! Milking milk'! Possibly the draftsman has placed this label in the wrong panel.

11.3 PROTECTION

11.3.1 Care

Hippopotamuses

PB100 *k3.j-m-nfrt* - Giza, CF (ntn), vestibule D, north: marsh
Source: *P-M III*, 263-264; Hassan, *Giza ii*, fig. 140
Label: none
Date: Harpur, *Decoration*, [#261], 270: V.6-8E?

PB101 *tjj* - Saqqara, NSP (D22), room III, north: marsh
Source: *P-M III*, 468-478; Wild, *Ti ii*, pl. 119
Label: none
Date: Harpur, *Decoration*, [#543], 277: V.8-9

PB102 *k3-gm-nj: mmj* - Saqqara, TP (LS10), room III, north: marsh
Source: *P-M III*, 521-525; Harpur-Scremin, *Kagemni*, fig. 9
Label: none
Date: Harpur, *Decoration*, [#534], 276: VI.1M

Baboons

PB103 *tp-m-ꜥnḫ II* - Saqqara, NSP (D11), corridor, east: market
Source: *P-M III*, 483-484; Borchardt, *Denkmäler ii*, pl. 61 (CG 1556); *Eg. Art in the Age of the Pyramids*, fig. 126
Label: *ṯ(wt) jšt=ṯ jm*
You! Your belongings are there!
Date: Harpur, *Decoration*, [#541], 277: V.5-6?

Goats

PB104 *nfr/k3-ḥ3.j* - Saqqara, UPC (ntn), niche, east: pastoral
Source: *P-M III*, 639-641; Moussa-Altenmüller, *Nefer and Ka-hay*, pl. 19
Label: none
Date: Harpur, *Decoration*, [#439], 274: V.6

Note: All figures were re-drawn by Mary Hartley, unless otherwise specified. See the Bibliography for full citation details.

CHAPTER 4

4-1 Wild, *Ti iii*, pl. 154
4-2 Hassan, *Giza vii*, fig. 22
4-3 Duell, *Mereruka ii*, pl. 140
4-4 Quibell, *Ptah-hetep*, pl. 31
4-5 Reprinted with permission from: Kanawati-Hassan, *Ankhmahor*, pl. 42
4-6 Reprinted with permission from: Kanawati-Abder-Raziq, *Hesi*, pl. 54
4-7 *LD ii*, pl. 17
4-8 Junker, *Giza iii*, fig. 18
4-9 Weeks, *Cemetery G6000*, fig. 29
4-10 Simpson, *Kayemnofret*, pl. F
4-11 Simpson, *Mast. West. Cem. Part I*, fig. 4
4-12 Reprinted with permission from: Kanawati, *Inumin*, pl. 47
4-13 Quibell, *Ptah-hetep*, pl. 32
4-14 Reprinted with permission from: Kanawati-Abder-Raziq, *Iynefert*, pl. 53 [f]
4-15 Wild, *Ti ii*, pl. 119
4-16 Reprinted with permission from: Kanawati-Abder-Raziq, *Hesi*, pl. 54
4-17 van de Walle, *Neferirtenef*, pl. 1
4-18 Epron et al.,*Ti i*, pl. 28
4-19 Murray, *Saqqara Mastabas*, pl. 11
4-20 Brovarski, *Senedjemib Complex Part I*, fig. 43
4-21 Mohr, *Hetep-her-akhti*, fig. 34
4-22 *Giza Archives Project* (Photo no. C12742_NS)
4-23 Reprinted with permission from: Kanawati-Abder-Raziq, *Hesi*, pl. 53
4-24 Altenmüller, *Mehu*, pl. 32
4-25 Altenmüller, *Mehu*, pl. 21 [a]
4-26 Roth, *Cem. Pal. Att.*, pl. 185
4-27 Reprinted with permission from: Kanawati-Abder-Raziq, *Hesi*, pl. 54
4-28 van de Walle, *Neferirtenef*, pl. 9
4-29 Ziegler, *Akhethetep*, 138-139
4-30 Moussa-Altenmüller, *Nianchchnum*, fig. 5
4-31 Wild, *Ti ii*, pl. 115
4-32 Simpson, *Kayemnofret*, pl. A
4-33 Reprinted with permission from: Kanawati-Abder-Raziq, *Iynefert*, pl. 37
4-34 Reprinted with permission from: Kanawati-Abder-Raziq, *Nikauisesi*, pl. 50
4-35 Reprinted with permission from: Kanawati-Abder-Raziq, *Iynefert*, pl. 54
4-36 Munro, *Unas-Friedhof*, pl. 10
4-37 Reprinted with permission from: Kanawati-Abder-Raziq, *Hesi*, pl. 54
4-38 Altenmüller, *Mehu*, pl. 12
4-39 Reprinted with permission from: Kanawati-Abder-Raziq, *Hesi*, pl. 53
4-40 Simpson, *Kayemnofret*, pl. A
4-41 Harpur-Scremin, *Kagemni*, pl. 133
4-42 Reprinted with permission from: Kanawati-Abder-Raziq, *Hesi*, pl. 54
4-43 Moussa-Altenmüller, *Nianchchnum*, fig. 12
4-44 Moussa-Altenmüller, *Nianchchnum*, fig. 12
4-45 Munro, *Unas-Friedhof*, pl. 10
4-46 Reprinted with permission from: Kanawati-Abder-Raziq, *Hesi*, pl. 54
4-47 Duell, *Mereruka i*, pl. 9
4-48 Firth-Gunn, *Teti Pyr. ii*, pl. 52
4-49 Wild, *Ti ii*, pl. 124
4-50 *LD Erganz.*, pl. 32
4-51 Duell, *Mereruka ii*, pl. 153
4-52 Junker, *Giza xi*, fig. 63

4-53 Duell, *Mereruka i*, pl. 25
4-54 Hayes, *Scepter i*, fig. 56
4-55 Moussa-Altenmüller, *Nianchchnum*, pl. 40
4-56 Duell, *Mereruka i*, pl. 25
4-57 Simpson, *Sekem-ankh-ptah*, pl. A
4-58 Simpson, *Mast. West. Cem. Part I*, fig. 19
4-59 McFarlane, *Irukaptah*, pl. 45
4-60 James, *Khentika*, pl. 38

CHAPTER 5
5-1 Dunham-Simpson, *Mersyankh III*, fig. 4
5-2 Simpson, *Kayemnofret*, pl. G
5-3 Simpson, *Sekem-ankh-ptah*, pl. D
5-4 Reprinted with permission from: Kanawati, *Inumin*, pl. 47
5-5 Junker, *Giza xi*, fig. 63
5-6 Simpson, *Kawab*, fig. 11 [g]
5-7 Ziegler, *Akhethetep*, 132-133
5-8 Wild, *Ti ii*, pl. 120
5-9 Reprinted with permission from: Kanawati-Abder-Raziq, *Neferseshemre*, pl. 75
5-10 Reprinted with permission from: Kanawati-Abder-Raziq, *Hesi*, pl. 53
5-11 Reprinted with permission from: Kanawati-Abder-Raziq, *Hesi*, pl. 54
5-12 Reprinted with permission from: Kanawati-Abder-Raziq, *Hesi*, pl. 55
5-13 *LD ii*, pl. 31 [a]
5-14 Hassan, *Giza ii*, fig. 124
5-15 Wild, *Ti ii*, pl. 114
5-16 Harpur-Scremin, *Kagemni*, fig. 7
5-17 Reprinted with permission from: Kanawati-Abder-Raziq, *Iynefert*, pl. 54
5-18 Reprinted with permission from: Kanawati-Hassan, *Ankhmahor*, pl. 37 [a]

CHAPTER 6
6-1 Munro, *Unas-Friedhof*, pl. 19
6-2 *LD Erganz.*, pl. 39 [a]
6-3 James, *Khentika*, pl. 21
6-4 Wild, *Ti iii*, pl. 165
6-5 Murray, *Saqqara Mastabas*, pl. 22
6-6 *LD Erganz.*, pl. 39 [a]
6-7 Reprinted with permission from: Kanawati-Abder-Raziq, *Iynefert*, pl. 44
6-8 *LD ii*, pl. 9
6-9 Moussa-Altenmüller, *Nefer and Ka-Hay*, pl. 5
6-10 Epron et al.,*Ti i*, pls. 7-8
6-11 Epron et al.,*Ti i*, pl. 33
6-12 Davies, *Ptahhetep ii*, pl. 14
6-13 Harpur-Scremin, *Kagemni*, fig. 13
6-14 Harpur-Scremin, *Kagemni*, pl. 200
6-15 Harpur-Scremin, *Kagemni*, pl. 209
6-16 Harpur-Scremin, *Kagemni*, fig. 13
6-17 Harpur-Scremin, *Kagemni*, fig. 13
6-18 Harpur-Scremin, *Kagemni*, pl. 201
6-19 Harpur-Scremin, *Kagemni*, pl. 202
6-20 Harpur-Scremin, *Kagemni*, fig. 13
6-21 Reprinted with permission from: Kanawati-Abder-Raziq, *Nikauisesi*, pl. 49
6-22 Duell, *Mereruka i*, pl. 53 [b]
6-23 Duell, *Mereruka i*, pl. 52
6-24 Reprinted with permission from: Kanawati-Abder-Raziq, *Hesi*, pl. 56
6-25 Duell, *Mereruka i*, pl. 52
6-26 Reprinted with permission from: Kanawati-Abder-Raziq, *Meryteti*, pl. 48

6-27 Reprinted with permission from: Kanawati-Abder-Raziq, *Meryteti*, pl. 48
6-28 Altenmüller, *Mehu*, pl. 6
6-29 Epron et al.,*Ti i*, pl. 8
6-30 Epron et al.,*Ti i*, pl. 38
6-31 Janssen-Janssen, *Egyptian Household Animals*, fig. 40
6-32 Quibell, *Ptah-hetep*, pl. 32
6-33 Wild, *Ti ii*, pl. 127
6-34 Martin, *Hetepka*, pl. 13 [14]

CHAPTER 7
7-1 Martin, *Hetepka*, pl. 10 [7]
7-2 *LD ii*, pl. 47
7-3 Hassan, *Giza v*, fig. 104
7-4 Reprinted with permission from: Kanawati, *Inumin*, pl. 49 [a]
7-5 Roth, *Cem. Pal. Att.*, pl. 179
7-6 Moussa-Altenmüller, *Nianchchnum*, pl. 76
7-7 Ziegler, *Akhethetep*, 132-133
7-8 Hassan, *Giza v*, fig. 123
7-9 Roth, *Cem. Pal. Att.*, pl. 204
7-10 Kanawati-Hassan, *Ankhmahor*, pl. 48 [b]
7-11 Duell, *Mereruka ii*, pl. 153
7-12 Epron et al.,*Ti i*, pl. 6
7-13 Duell, *Mereruka ii*, pl. 153
7-14 Murray, *Saqqara Mastabas*, pl. 30
7-15 Duell, *Mereruka ii*, pl. 153
7-16 Reprinted with permission from: Kanawati-Hassan, *Ankhmahor*, pl. 46
7-17 Epron et al.,*Ti i*, pl. 6
7-18 Duell, *Mereruka ii*, pl. 153
7-19 Mohr, *Hetep-her-akhti*, fig. 21
7-20 Moussa-Altenmüller, *Nefer and Ka-hay*, pl. 18
7-21 Smith, *HESPOK*, fig. 239
7-22 Hassan, *Neb-Kaw-Her*, fig. 12
7-23 *LD ii*, pl. 12 [a]
7-24 *Hiero. Texts vi*, pl. 11
7-25 Moussa-Altenmüller, *Nefer and Ka-Hay*, pl. 9
7-26 Duell, *Mereruka ii*, pl. 128
7-27 Reprinted with permission from: Kanawati-Abder-Raziq, *Hesi*, pl. 54
7-28 Reprinted with permission from: Kanawati-Abder-Raziq, *Nikauisesi*, pl. 48
7-29 Reprinted with permission from: Kanawati-Hassan, *Ankhmahor*, pl. 42
7-30 Duell, *Mereruka i*, pl. 43
7-31 Murray, *Saqqara Mastabas*, pl. 11
7-32 Murray, *Saqqara Mastabas*, pl. 11
7-33 Harpur-Scremin, *Kagemni*, pl. 236
7-34 Reprinted with permission from: Kanawati-Abder-Raziq, *Hesi*, pl. 55
7-35 van de Walle, *Neferirtenef*, pl. 13
7-36 Hassan, *Giza v*, fig. 104
7-37 *Pers. comm.* Alexandra Woods, Macquarie University
7-38 Reprinted with permission from: Kanawati-Abder-Raziq, *Hesi*, pl. 55
7-39 Reprinted with permission from: Kanawati-Hassan, *Ankhmahor*, pl. 42
7-40 Reprinted with permission from: Kanawati-Abder-Raziq, *Nikauisesi*, pl. 49
7-41 Reprinted with permission from: Kanawati-Abder-Raziq, *Nikauisesi*, pl. 49
7-42 Moussa-Junge, *Two Tombs*, pl. 13
7-43 Duell, *Mereruka ii*, pl. 168
7-44 Duell, *Mereruka ii*, pl. 168-169
7-45 Epron et al.,*Ti i*, pl. 22
7-46 Reprinted with permission from: Kanawati-Abder-Raziq, *Meryteti*, pl. 48

7-47 *LD Ergänz.*, pl. 39 [a]
7-48 Harpur-Scremin, *Kagemni*, pl. 28
7-49 Harpur-Scremin, *Kagemni*, pl. 255
7-50 Firth-Gunn, *Teti Pyr. ii*, pl. 53 [2]
7-51 Reprinted with permission from: Kanawati-Abder-Raziq, *Hesi*, pl. 54
7-52 Davies et al., *Saqqara Tombs i*, pl. 25
7-53 Reprinted with permission from: Kanawati-Abder-Raziq, *Hesi*, pl. 53
7-54 *LD ii*, pl. 12 [b]
7-55 *Eg. Art in the Age of the Pyramids*, cat. no. 193
7-56 Harpur, *Decoration*, fig. 122
7-57 Moussa-Altenmüller, *Nefer and Ka-hay*, pl. 20
7-58 *LD ii*, pl. 62
7-59 Harpur-Scremin, *Kagemni*, pls. 98, 99
7-60 Harpur-Scremin, *Kagemni*, pl. 203

CHAPTER 8
8-1 Junker, *Giza xi*, fig. 63
8-2 Davies et al., *Saqqara Tombs i*, pl. 31
8-3 Duell, *Mereruka i*, pl. 25
8-4 Moussa-Altenmüller, *Nianchchnum*, pl. 40
8-5 Hassan, *Ny-Ankh-Pepy*, fig. 59
8-6 Roth, *Cem. Pal. Att.*, pl. 189
8-7 Moussa-Altenmüller, *Nianchchnum*, fig. 8
8-8 *Giza Archives Project* (Photo no. A779_NS)
8-9 Moussa-Altenmüller, *Nianchchnum*, pl. 40
8-10 Davies, *Ptahhetep ii*, pl. 13
8-11 Davies, *Ptahhetep ii*, pl. 14
8-12 Reprinted with permission from: Kanawati-Abder-Raziq, *Neferseshemre*, pl. 76
8-13 Altenmüller, *Mehu*, pl. 11
8-14 Davies, *Ptahhetep ii*, pl. 13
8-15 *Pers. comm.* Naguib Kanawati, Macquarie University, Sydney
8-16 Reprinted with permission from: Kanawati-Abder-Raziq, *Neferseshemre*, pl. 76
8-17 *Pers. comm.* Naguib Kanawati, Macquarie University, Sydney
8-18 Reprinted with permission from: Kanawati-Abder-Raziq, *Hesi*, pl. 53
8-19 Altenmüller, *Mehu*, pl. 11
8-20 Smith, *HESPOK*, fig. 226 [c]
8-21 Reprinted with permission from: Kanawati-Abder-Raziq, *Meryteti*, pl. 46
8-22 Reprinted with permission from: Kanawati-Abder-Raziq, *Meryteti*, pl. 46
8-23 Hayes, *Scepter i*, fig. 56
8-24 Smith, *HESPOK*, fig. 60
8-25 Hayes, *Scepter i*, fig. 56
8-26 Quibell, *Ptah-hetep*, pl. 32
8-27 Quibell, *Ptah-hetep*, pl. 32
8-28 Reprinted with permission from: Kanawati, *Inumin*, pl. 47
8-29 *LD ii*, pl. 46
8-30 Quibell, *Ptah-hetep*, pl. 32
8-31 Altenmüller, *Mehu*, pl. 10
8-32 Moussa-Altenmüller, *Nefer and Ka-hay*, pl. 5
8-33 Junker, *Giza iv*, fig. 8
8-34 Reprinted with permission from: Kanawati-Abder-Raziq, *Hesi*, pl. 54
8-35 *Pers. comm.* Naguib Kanawati, Macquarie University, Sydney
8-36 Harpur-Scremin, *Kagemni*, fig. 9
8-37 Reprinted with permission from: Kanawati-Abder-Raziq, *Hesi*, pl. 55
8-38 Reprinted with permission from: Kanawati-Abder-Raziq, *Hesi*, pl. 54

CHAPTER 9

9-1 Davies, *Ptahhetep ii*, pl. 8
9-2 Reprinted with permission from: Kanawati, *Inumin*, pl. 49 [a]
9-3 Reprinted with permission from: Kanawati-Abder-Raziq, *Meryteti*, pl. 48
9-4 Quibell, *Ptah-hetep*, pl. 21
9-5 *LD ii*, pl. 77
9-6 Reprinted with permission from: Kanawati, *Inumin*, pl. 49 [a]
9-7 Reprinted with permission from: Kanawati-Hassan, *Ankhmahor*, pl. 37 [a]
9-8 Ziegler, *Akhethetep*, 132-133
9-9 *Pers. comm.* Naguib Kanawati, Macquarie University, Sydney
9-10 Reprinted with permission from: Kanawati-Abder-Raziq, *Iynefert*, pl. 54
9-11 Brovarski, *Senedjemib Complex Part I*, fig. 43
9-12 *LD ii*, pl. 13
9-13 Moussa-Altenmüller, *Nianchchnum*, fig. 10
9-14 Moussa-Altenmüller, *Nefer and Ka-hay*, pl. 19
9-15 Roth, *Cem. Pal. Att.*, pl. 189
9-16 Reprinted with permission from: Kanawati-Hassan, *Ankhmahor*, pl. 42
9-17 Reprinted with permission from: Kanawati-Abder-Raziq, *Hesi*, pl. 53
9-18 Reprinted with permission from: Kanawati-Abder-Raziq, *Hesi*, pl. 54
9-19 Hassan, *Neb-Kaw-Her*, pl. 19 [c]
9-20 Junker, *Giza xi*, fig. 93
9-21 Moussa-Altenmüller, *Nianchchnum*, fig. 12
9-22 *Pers. comm.* Naguib Kanawati, Macquarie University, Sydney
9-23 Harpur-Scremin, *Kagemni*, pl. 27
9-24 Moussa-Altenmüller, *Nianchchnum*, fig. 10
9-25 Brovarski, *Senedjemib Complex Part I*, fig. 27
9-26 Reprinted with permission from: Kanawati-Abder-Raziq, *Hesi*, pl. 54
9-27 Hassan, *Giza iv*, fig. 79
9-28 Martin, *Hetepka*, pl. 13 [14]
9-29 Simpson, *Mast. West. Cem. Part I*, fig. 19
9-30 Petrie-Murray, *Seven Memphite*, pl. 7 [lower]
9-31 *LD Erganz.*, pl. 41
9-32 Reprinted with permission from: Kanawati-Abder-Raziq, *Nikauisesi*, pl. 50

CHAPTER 10

10-1 Simpson, *Mast. West. Cem. Part I*, fig. 22
10-2 Reprinted with permission from: Kanawati-Abder-Raziq, *Iynefert*, pl. 54
10-3 Roth, *Pal. Cem. Att.*, pl. 156
10-4 Reprinted with permission from: Kanawati, *Inumin*, pl. 49
10-5 Ziegler, *Akhethetep*, 138-139
10-6 Roth, *Cem. Pal. Att.*, pl. 189
10-7 Hassan, *Giza v*, fig. 123
10-8 Reprinted with permission from: Kanawati-Abder-Raziq, *Nikauisesi*, pl. 49
10-9 Simpson, *Sekem-ankh-ptah*, pl. C
10-10 Roth, *Cem. Pal. Att.*, pl. 189
10-11 Roth, *Cem. Pal. Att.*, pl. 189
10-12 *Eg. Art in the Age of the Pyramids*, cat. no 146 [b]
10-13 Junker, *Giza xi*, fig. 63
10-14 *Giza Archives Project* (Photo no. A779_NS)
10-15 *Eg. Art in the Age of the Pyramids*, cat. no. 146 [b]
10-16 Hassan, *Ny-Ankh-Pepy*, fig. 59
10-17 Quibell, *Ptah-hetep*, pl. 32
10-18 Roth, *Cem. Pal. Att.*, pl. 189
10-19 Roth, *Cem. Pal. Att.*, pl. 189
10-20 Roth, *Cem. Pal. Att.*, pl. 189
10-21 Roth, *Cem. Pal. Att.*, pl. 189

10-22 Quibell, *Ptah-hetep*, pl. 32
10-23 Roth, *Cem. Pal. Att.*, pl. 189
10-24 Roth, *Cem. Pal. Att.*, pl. 189
10-25 Roth, *Cem. Pal. Att.*, pl. 189
10-26 Hassan, *Ny-Ankh-Pepy*, fig. 59
10-27 Reprinted with permission from: Kanawati-Abder-Raziq, *Hesi*, pl. 53
10-28 Reprinted with permission from: Kanawati-Abder-Raziq, *Hesi*, pl. 53
10-29 Hayes, *Scepter i*, fig. 57
10-30 van de Walle, *Neferirtenef*, pl. 12

CHAPTER 11
11-1 Weeks, *Cemetery G6000*, fig. 34
11-2 Moussa-Altenmüller, *Nianchchnum*, pl. 79
11-3 Wild, *Ti ii*, pl. 124
11-4 Simpson, *Kayemnofret*, pl. G
11-5 Ziegler, *Akhethetep*, 138-139
11-6 Mohr, *Hetep-her-akhti*, fig. 21
11-7 Reprinted with permission from: Kanawati-Abder-Raziq, *Iynefert*, pl. 54
11-8 Reprinted with permission from: Kanawati-Hassan, *Ankhmahor*, pl. 37 [c]
11-9 Macramallah, *Idout*, pl. 26
11-10 *Eg. Art in the Age of the Pyramids*, cat. no. 146 [b]
11-11 Lauer, *Saqqara*, pl. 67
11-12 *LD Erganz.*, pl. 18
11-13 Weeks, *Cemetery G6000*, fig. 34
11-14 Simpson, *Mast. West. Cem. Part I*, fig. 30
11-15 Davies, *Ptahhetep ii*, pl. 32
11-16 *LD ii*, pl. 12 [b]
11-17 Roth, *Pal. Cem. Att.*, pl. 168
11-18 Moussa-Altenmüller, *Nefer and Ka-hay*, pl. 18
11-19 Mohr, *Hetep-her-akhti*, fig. 21
11-20 Wild, *Ti ii*, pl. 124
11-21 Harpur-Scremin, *Kagemni*, fig. 7
11-22 Smith, *HESPOK*, fig. 79
11-23 Reprinted with permission from: Kanawati-Abder-Raziq, *Hesi*, pl. 55
11-24 Moussa-Altenmüller, *Nianchchnum*, fig. 6
11-25 Moussa-Altenmüller, *Nianchchnum*, fig. 6
11-26 Wild, *Ti ii*, pl. 119
11-27 *Eg. Art in the Age of the Pyramids*, fig. 126
11-28 Moussa-Altenmüller, *Nefer and Ka-hay*, pl. 19

Note: Bracketed page numbers refer to material in the footnotes.

TOMB-OWNERS

KINGS

DEITIES

Note: Bracketed page numbers refer to material in the footnotes.

MAMMALS

BIRDS

REPTLIES

AMPHIBIANS

FISH

INVERTEBRATES

Abegg-Stiftung	Abegg-Stiftung Museum, Riggisberg
Äg. Mus.	Ägyptisches Museum, Berlin
AJS	Between the Monastery of Apa Jeremias and Enclosure of Sekhemkhet, Saqqara
Albertinum	Albertinum Museum of Art, Dresden
Brooklyn Mus.	Brooklyn Museum, New York
CF	Central Field, Giza
CG	*Catalogue General*, Cairo Museum
Detroit Inst. Arts	The Detroit Institute of Arts
E	Early
EF	East Field, Giza
EPII	East of Pepi II Pyramid, Saqqara
ESP	East of Step Pyramid, Saqqara
FIP	First Intermediate Period
GIS	Cemetery GIS, Saqqara
JE	*Journal d'entré*, Cairo Museum
L	Late
M	Middle
MFA	Museum of Fine Arts, Boston
n/a	Not applicable
nr	Not recorded
ntn	No tomb number
NSP	North of Step Pyramid, Saqqara
P-M III	Porter-Moss, *Topographical Bibliography of Ancient Egyptian Hieroglyphic Texts, Reliefs and Paintings. Vol. iii. Memphis*
Rijksmuseum	Rijksmuseum van Oudheden, Leiden
Roy. Ont. Mus.	Royal Ontario Museum, Toronto
r/u	Tomb re-used by another
Staat. Mus.	Staatliche Museen zu Berlin
TP	Around Teti Pyramid, Saqqara
UPC	Around Unas Pyramid Complex, Saqqara
WF	West Field, Giza
WSP	West of Step Pyramid, Saqqara
IV.1	Snefru
IV.2	Khufu
IV.3	Djedefre
IV.4	Khafre
IV.5	Menkaure
IV.6	Shepseskaf
V.1	Userkaf
V.2	Sahure
V.3	Neferirkare
V.4	Shepseskare
V.5	Raneferef
V.6	Niuserre
V.7	Menkauhor
V.8	Djedkare/Isesi
V.9	Unas
VI.1	Teti
VI.2	Pepi I
VI.3	Merenre
VI.4	Pepi II (years 1-34)
VI.5	Pepi II (years 35-54)
VI.6	Pepi II (years 55-85)
VI.7	VI late-VIII

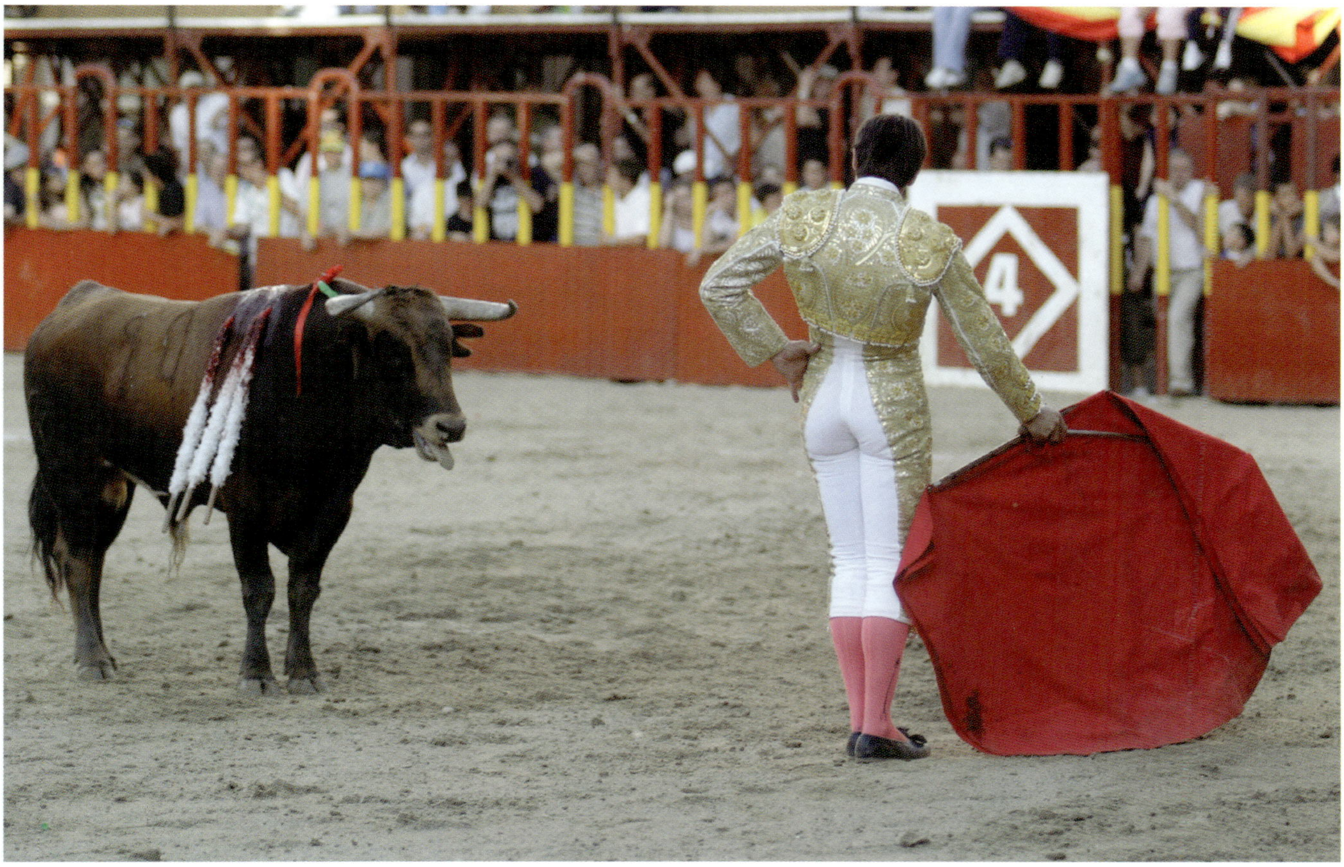

Colour Plate A The tongue of a vocalizing animal can sometimes be seen within its open mouth cavity or may be apparent during rare calls, such as the screaming bellow of bulls. © iStockphoto.com/Ricardo Lopez

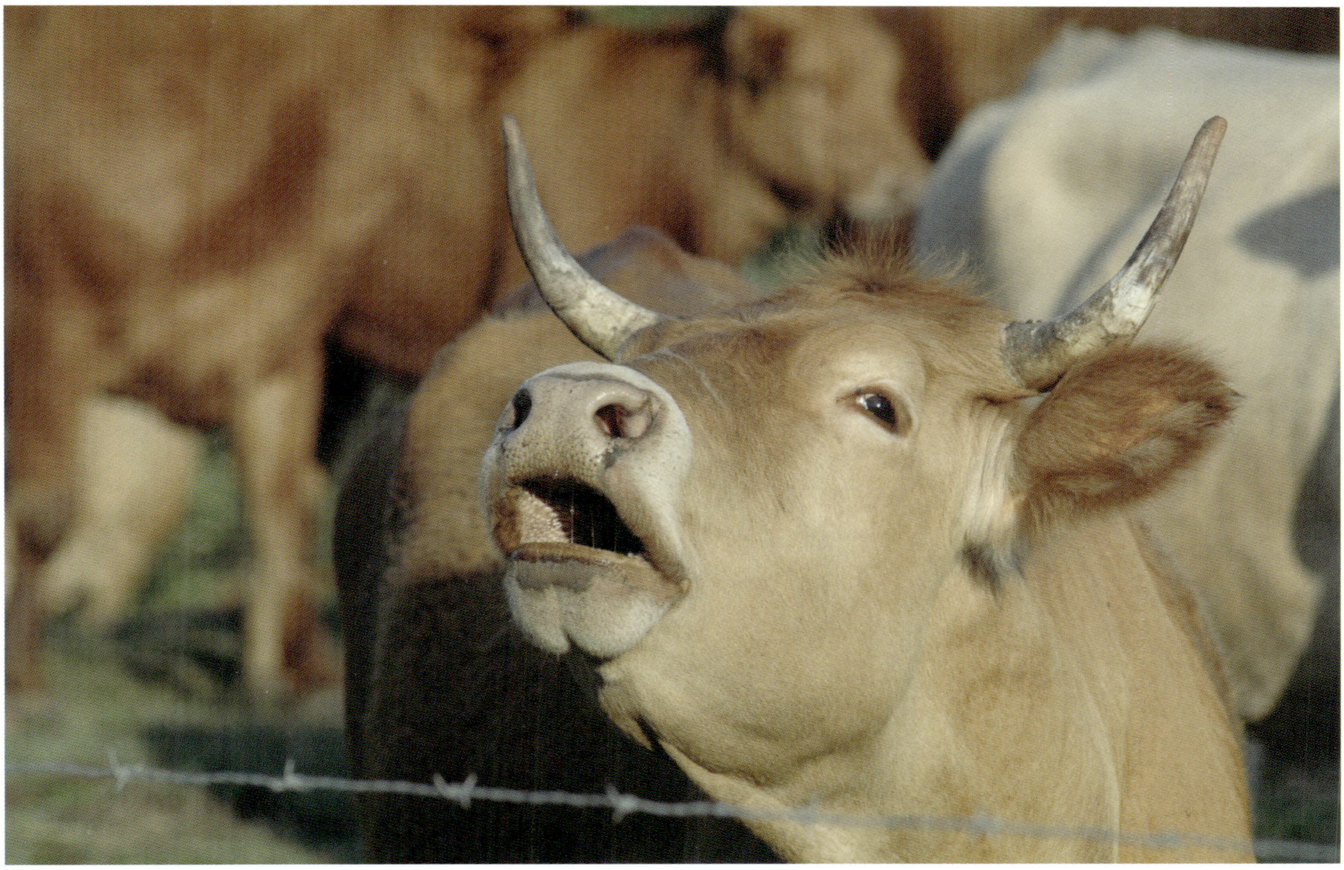

Colour Plate B Cow mooing. Animals do not usually protrude their tongues from between their lips or mandibles when vocalizing. © iStockphoto.com/Steve Simzer

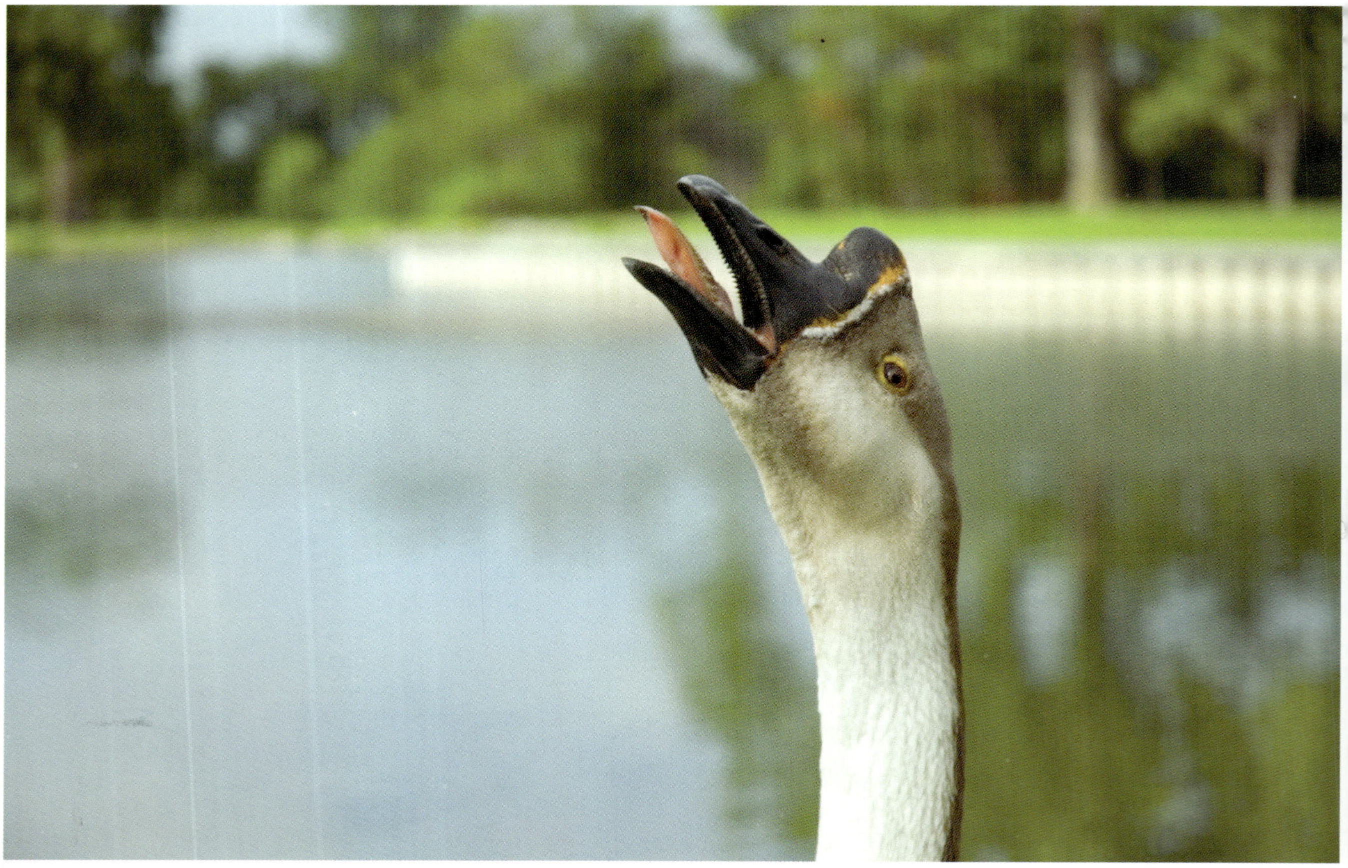

Colour Plate C Goose hissing. Most bird species are physically incapable of extending their tongues beyond the limits of their beaks. © iStockphoto.com/Tracy Tucker

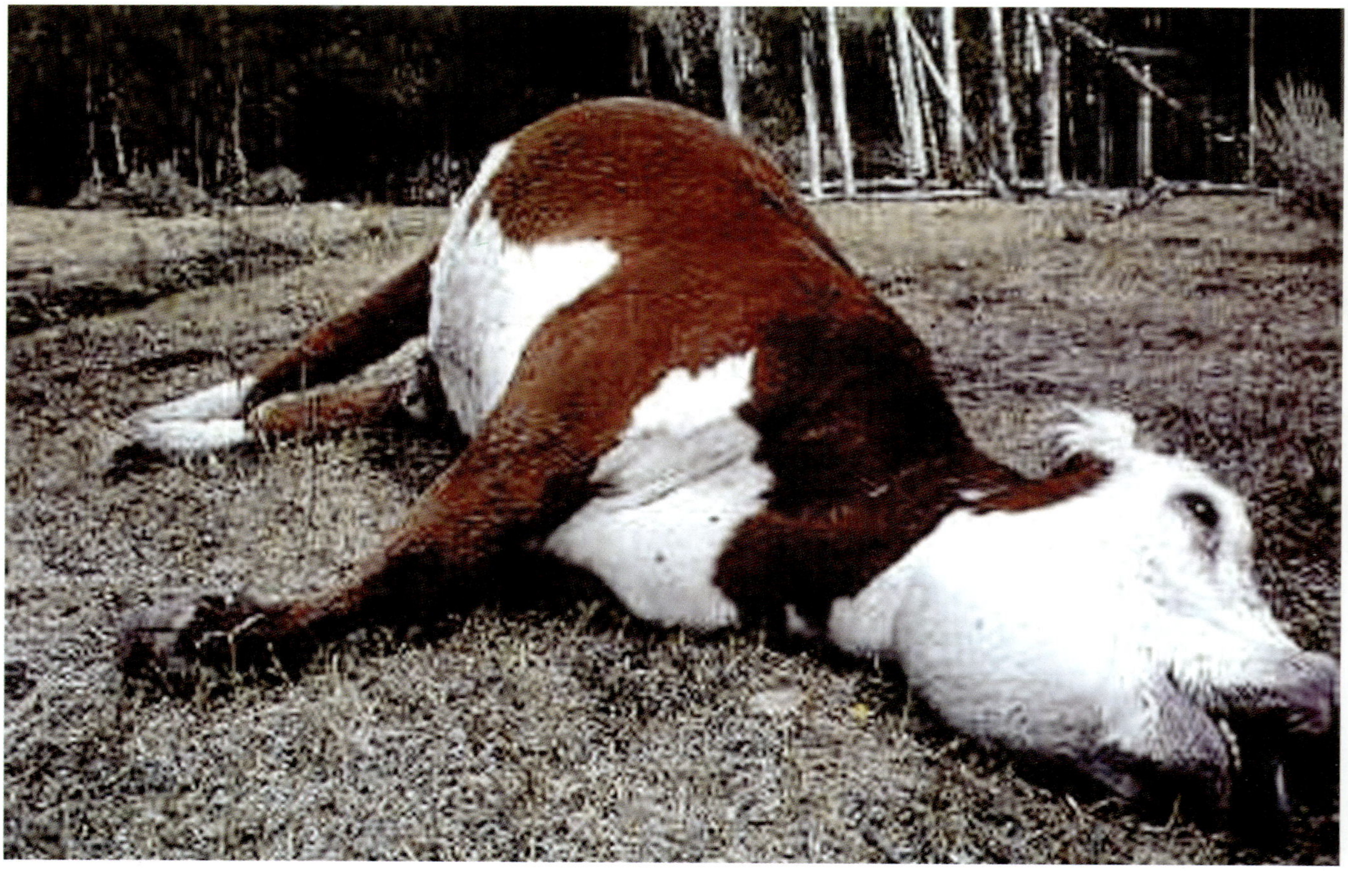

Colour Plate D Muscle relaxation will occasionally cause an animal's tongue to loll immediately after death, but this reponse is more likely to occur later, once putrefaction has begun. © Western Watersheds Project